Lecture Notes in Computer Science 14447

Founding Editors

Gerhard Goos
Juris Hartmanis

The series Lecture Notes in Computer Science (LNCS), including its subseries Lecture Notes in Artificial Intelligence (LNAI) and Lecture Notes in Bioinformatics (LNBI), has established itself as a medium for the publication of new developments in computer science and information technology research, teaching, and education.

LNCS enjoys close cooperation with the computer science R & D community, the series counts many renowned academics among its volume editors and paper authors, and collaborates with prestigious societies. Its mission is to serve this international community by providing an invaluable service, mainly focused on the publication of conference and workshop proceedings and postproceedings. LNCS commenced publication in 1973.

Biao Luo · Long Cheng · Zheng-Guang Wu ·
Hongyi Li · Chaojie Li
Editors

Neural Information Processing

30th International Conference, ICONIP 2023
Changsha, China, November 20–23, 2023
Proceedings, Part I

Editors
Biao Luo
Central South University
Changsha, China

Long Cheng
Chinese Academy of Sciences
Beijing, China

Zheng-Guang Wu
Zhejiang University
Hangzhou, China

Hongyi Li
Guangdong University of Technology
Guangzhou, China

Chaojie Li
UNSW Sydney
Sydney, NSW, Australia

ISSN 0302-9743 ISSN 1611-3349 (electronic)
Lecture Notes in Computer Science
ISBN 978-981-99-8078-9 ISBN 978-981-99-8079-6 (eBook)
https://doi.org/10.1007/978-981-99-8079-6

This Springer imprint is published by the registered company Springer Nature Singapore Pte Ltd.
The registered company address is: 152 Beach Road, #21-01/04 Gateway East, Singapore 189721, Singapore

Preface

Welcome to the 30th International Conference on Neural Information Processing (ICONIP2023) of the Asia-Pacific Neural Network Society (APNNS), held in Changsha, China, November 20–23, 2023.

The mission of the Asia-Pacific Neural Network Society is to promote active interactions among researchers, scientists, and industry professionals who are working in neural networks and related fields in the Asia-Pacific region. APNNS has Governing Board Members from 13 countries/regions – Australia, China, Hong Kong, India, Japan, Malaysia, New Zealand, Singapore, South Korea, Qatar, Taiwan, Thailand, and Turkey. The society's flagship annual conference is the International Conference of Neural Information Processing (ICONIP). The ICONIP conference aims to provide a leading international forum for researchers, scientists, and industry professionals who are working in neuroscience, neural networks, deep learning, and related fields to share their new ideas, progress, and achievements.

ICONIP2023 received 1274 papers, of which 256 papers were accepted for publication in Lecture Notes in Computer Science (LNCS), representing an acceptance rate of 20.09% and reflecting the increasingly high quality of research in neural networks and related areas. The conference focused on four main areas, i.e., "Theory and Algorithms", "Cognitive Neurosciences", "Human-Centered Computing", and "Applications". All the submissions were rigorously reviewed by the conference Program Committee (PC), comprising 258 PC members, and they ensured that every paper had at least two high-quality single-blind reviews. In fact, 5270 reviews were provided by 2145 reviewers. On average, each paper received 4.14 reviews.

We would like to take this opportunity to thank all the authors for submitting their papers to our conference, and our great appreciation goes to the Program Committee members and the reviewers who devoted their time and effort to our rigorous peer-review process; their insightful reviews and timely feedback ensured the high quality of the papers accepted for publication. We hope you enjoyed the research program at the conference.

October 2023

Biao Luo
Long Cheng
Zheng-Guang Wu
Hongyi Li
Chaojie Li

Organization

Honorary Chair

Weihua Gui	Central South University, China

Advisory Chairs

Jonathan Chan	King Mongkut's University of Technology Thonburi, Thailand
Zeng-Guang Hou	Chinese Academy of Sciences, China
Nikola Kasabov	Auckland University of Technology, New Zealand
Derong Liu	Southern University of Science and Technology, China
Seiichi Ozawa	Kobe University, Japan
Kevin Wong	Murdoch University, Australia

General Chairs

Tingwen Huang	Texas A&M University at Qatar, Qatar
Chunhua Yang	Central South University, China

Program Chairs

Biao Luo	Central South University, China
Long Cheng	Chinese Academy of Sciences, China
Zheng-Guang Wu	Zhejiang University, China
Hongyi Li	Guangdong University of Technology, China
Chaojie Li	University of New South Wales, Australia

Technical Chairs

Xing He	Southwest University, China
Keke Huang	Central South University, China
Huaqing Li	Southwest University, China
Qi Zhou	Guangdong University of Technology, China

Local Arrangement Chairs

Wenfeng Hu	Central South University, China
Bei Sun	Central South University, China

Finance Chairs

Fanbiao Li	Central South University, China
Hayaru Shouno	University of Electro-Communications, Japan
Xiaojun Zhou	Central South University, China

Special Session Chairs

Hongjing Liang	University of Electronic Science and Technology, China
Paul S. Pang	Federation University, Australia
Qiankun Song	Chongqing Jiaotong University, China
Lin Xiao	Hunan Normal University, China

Tutorial Chairs

Min Liu	Hunan University, China
M. Tanveer	Indian Institute of Technology Indore, India
Guanghui Wen	Southeast University, China

Publicity Chairs

Sabri Arik	Istanbul University-Cerrahpaşa, Turkey
Sung-Bae Cho	Yonsei University, South Korea
Maryam Doborjeh	Auckland University of Technology, New Zealand
El-Sayed M. El-Alfy	King Fahd University of Petroleum and Minerals, Saudi Arabia
Ashish Ghosh	Indian Statistical Institute, India
Chuandong Li	Southwest University, China
Weng Kin Lai	Tunku Abdul Rahman University of Management & Technology, Malaysia
Chu Kiong Loo	University of Malaya, Malaysia

Qinmin Yang	Zhejiang University, China
Zhigang Zeng	Huazhong University of Science and Technology, China

Publication Chairs

Zhiwen Chen	Central South University, China
Andrew Chi-Sing Leung	City University of Hong Kong, China
Xin Wang	Southwest University, China
Xiaofeng Yuan	Central South University, China

Secretaries

Yun Feng	Hunan University, China
Bingchuan Wang	Central South University, China

Webmasters

Tianmeng Hu	Central South University, China
Xianzhe Liu	Xiangtan University, China

Program Committee

Rohit Agarwal	UiT The Arctic University of Norway, Norway
Hasin Ahmed	Gauhati University, India
Harith Al-Sahaf	Victoria University of Wellington, New Zealand
Brad Alexander	University of Adelaide, Australia
Mashaan Alshammari	Independent Researcher, Saudi Arabia
Sabri Arik	Istanbul University, Turkey
Ravneet Singh Arora	Block Inc., USA
Zeyar Aung	Khalifa University of Science and Technology, UAE
Monowar Bhuyan	Umeå University, Sweden
Jingguo Bi	Beijing University of Posts and Telecommunications, China
Xu Bin	Northwestern Polytechnical University, China
Marcin Blachnik	Silesian University of Technology, Poland
Paul Black	Federation University, Australia

Xing He	Southwest University, China
Akira Hirose	University of Tokyo, Japan
Yin Hongwei	Huzhou Normal University, China
Md Zakir Hossain	Curtin University, Australia
Zengguang Hou	Chinese Academy of Sciences, China
Lu Hu	Jiangsu University, China
Zeke Zexi Hu	University of Sydney, Australia
He Huang	Soochow University, China
Junjian Huang	Chongqing University of Education, China
Kaizhu Huang	Duke Kunshan University, China
David Iclanzan	Sapientia University, Romania
Radu Tudor Ionescu	University of Bucharest, Romania
Asim Iqbal	Cornell University, USA
Syed Islam	Edith Cowan University, Australia
Kazunori Iwata	Hiroshima City University, Japan
Junkai Ji	Shenzhen University, China
Yi Ji	Soochow University, China
Canghong Jin	Zhejiang University, China
Xiaoyang Kang	Fudan University, China
Mutsumi Kimura	Ryukoku University, Japan
Masahiro Kohjima	NTT, Japan
Damian Kordos	Rzeszow University of Technology, Poland
Marek Kraft	Poznań University of Technology, Poland
Lov Kumar	NIT Kurukshetra, India
Weng Kin Lai	Tunku Abdul Rahman University of Management & Technology, Malaysia
Xinyi Le	Shanghai Jiao Tong University, China
Bin Li	University of Science and Technology of China, China
Hongfei Li	Xinjiang University, China
Houcheng Li	Chinese Academy of Sciences, China
Huaqing Li	Southwest University, China
Jianfeng Li	Southwest University, China
Jun Li	Nanjing Normal University, China
Kan Li	Beijing Institute of Technology, China
Peifeng Li	Soochow University, China
Wenye Li	Chinese University of Hong Kong, China
Xiangyu Li	Beijing Jiaotong University, China
Yantao Li	Chongqing University, China
Yaoman Li	Chinese University of Hong Kong, China
Yinlin Li	Chinese Academy of Sciences, China
Yuan Li	Academy of Military Science, China

Piotr Milczarski	Lodz University of Technology, Poland
Malek Mouhoub	University of Regina, Canada
Nankun Mu	Chongqing University, China
Wenlong Ni	Jiangxi Normal University, China
Anupiya Nugaliyadde	Murdoch University, Australia
Toshiaki Omori	Kobe University, Japan
Babatunde Onasanya	University of Ibadan, Nigeria
Manisha Padala	Indian Institute of Science, India
Sarbani Palit	Indian Statistical Institute, India
Paul Pang	Federation University, Australia
Rasmita Panigrahi	Giet University, India
Kitsuchart Pasupa	King Mongkut's Institute of Technology Ladkrabang, Thailand
Dipanjyoti Paul	Ohio State University, USA
Hu Peng	Jiujiang University, China
Kebin Peng	University of Texas at San Antonio, USA
Dawid Połap	Silesian University of Technology, Poland
Zhong Qian	Soochow University, China
Sitian Qin	Harbin Institute of Technology at Weihai, China
Toshimichi Saito	Hosei University, Japan
Fumiaki Saitoh	Chiba Institute of Technology, Japan
Naoyuki Sato	Future University Hakodate, Japan
Chandni Saxena	Chinese University of Hong Kong, China
Jiaxing Shang	Chongqing University, China
Lin Shang	Nanjing University, China
Jie Shao	University of Science and Technology of China, China
Yin Sheng	Huazhong University of Science and Technology, China
Liu Sheng-Lan	Dalian University of Technology, China
Hayaru Shouno	University of Electro-Communications, Japan
Gautam Srivastava	Brandon University, Canada
Jianbo Su	Shanghai Jiao Tong University, China
Jianhua Su	Institute of Automation, China
Xiangdong Su	Inner Mongolia University, China
Daiki Suehiro	Kyushu University, Japan
Basem Suleiman	University of New South Wales, Australia
Ning Sun	Shandong Normal University, China
Shiliang Sun	East China Normal University, China
Chunyu Tan	Anhui University, China
Gouhei Tanaka	University of Tokyo, Japan
Maolin Tang	Queensland University of Technology, Australia

Shu Tian	University of Science and Technology Beijing, China
Shikui Tu	Shanghai Jiao Tong University, China
Nancy Victor	Vellore Institute of Technology, India
Petra Vidnerová	Institute of Computer Science, Czech Republic
Shanchuan Wan	University of Tokyo, Japan
Tao Wan	Beihang University, China
Ying Wan	Southeast University, China
Bangjun Wang	Soochow University, China
Hao Wang	Shanghai University, China
Huamin Wang	Southwest University, China
Hui Wang	Nanchang Institute of Technology, China
Huiwei Wang	Southwest University, China
Jianzong Wang	Ping An Technology, China
Lei Wang	National University of Defense Technology, China
Lin Wang	University of Jinan, China
Shi Lin Wang	Shanghai Jiao Tong University, China
Wei Wang	Shenzhen MSU-BIT University, China
Weiqun Wang	Chinese Academy of Sciences, China
Xiaoyu Wang	Tokyo Institute of Technology, Japan
Xin Wang	Southwest University, China
Xin Wang	Southwest University, China
Yan Wang	Chinese Academy of Sciences, China
Yan Wang	Sichuan University, China
Yonghua Wang	Guangdong University of Technology, China
Yongyu Wang	JD Logistics, China
Zhenhua Wang	Northwest A&F University, China
Zi-Peng Wang	Beijing University of Technology, China
Hongxi Wei	Inner Mongolia University, China
Guanghui Wen	Southeast University, China
Guoguang Wen	Beijing Jiaotong University, China
Ka-Chun Wong	City University of Hong Kong, China
Anna Wróblewska	Warsaw University of Technology, Poland
Fengge Wu	Institute of Software, Chinese Academy of Sciences, China
Ji Wu	Tsinghua University, China
Wei Wu	Inner Mongolia University, China
Yue Wu	Shanghai Jiao Tong University, China
Likun Xia	Capital Normal University, China
Lin Xiao	Hunan Normal University, China

Contents – Part I

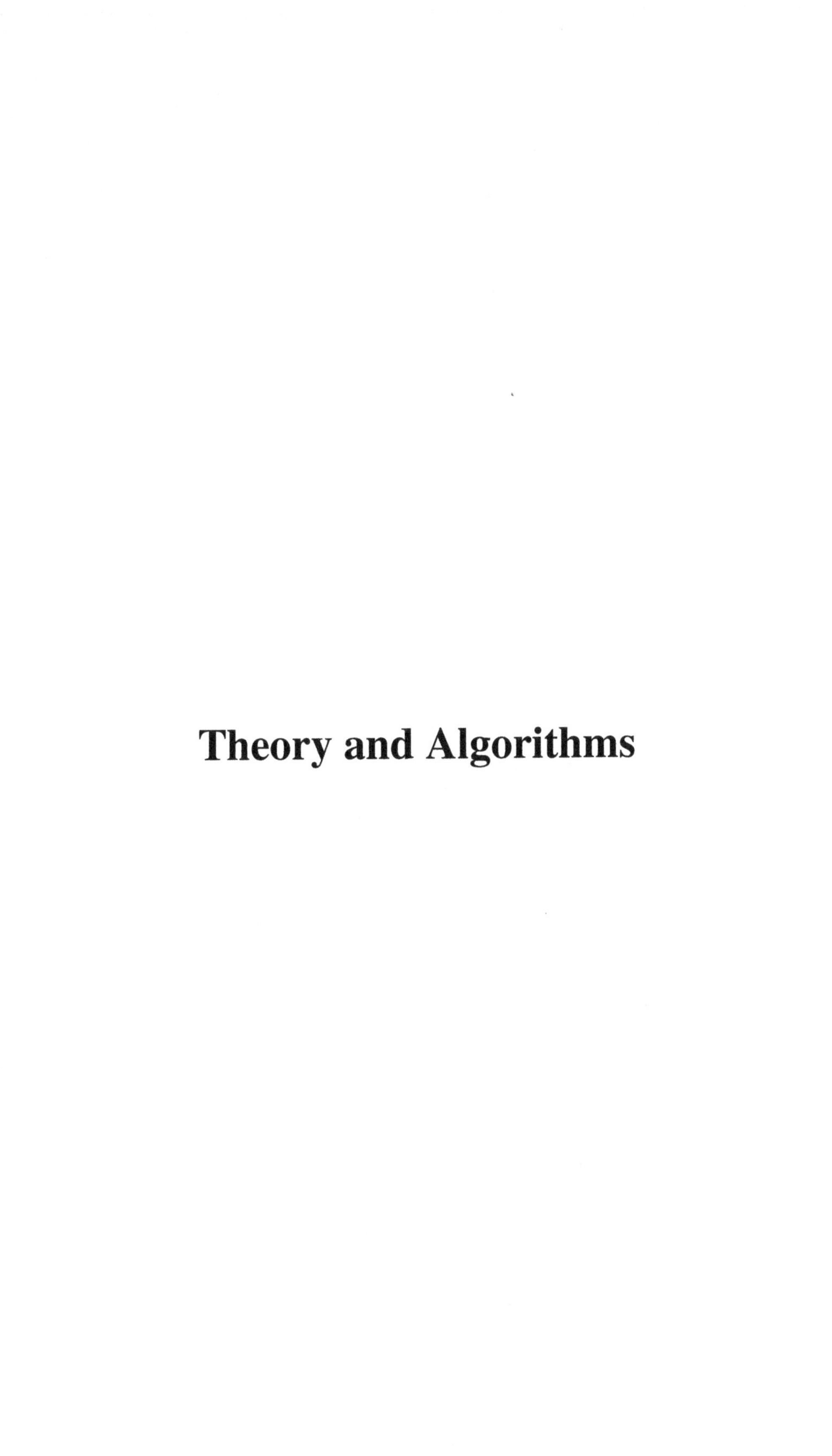

Theory and Algorithms

Single Feedback Based Kernel Generalized Maximum Correntropy Adaptive Filtering Algorithm

Jiaming Liu[1], Ji Zhao[1(✉)], Qiang Li[1(✉)], Lingli Tang[2], and Hongbin Zhang[3]

[1] School of Information Engineering, Southwest University of Science and Technology, Mianyang 621010, China
{zhaoji,liqiangsir}@swust.edu.cn

[2] Department of Social Sciences, Southwest University of Science and Technology, Mianyang 621010, China

[3] School of Information and Communication Engineering, University of Electronic Science and Technology of China, Chengdu 611731, China

Abstract. This paper presents a novel single feedback based kernel generalized maximum correntropy (SF-KGMC) algorithm by introducing a single delay into the framework of kernel adaptive filtering. In SF-KGMC, the history information implicitly existing in the single delayed output can enhance the convergence rate. Compared to the second-order statistics criterion, the generalized maximum correntropy (GMC) criterion shows better robustness against outliers. Therefore, SF-KGMC can efficiently reduce the influence of impulsive noise and avoids significant performance degradation. In addition, for SF-KGMC, the theoretical convergence analysis is also conducted. Simulation results on chaotic time-series prediction and real-world data applications validate that SF-KGMC achieves better filtering accuracy and a faster convergence rate.

Keywords: Kernel adaptive filtering · Generalized maximum correntropy · Single feedback · Impulsive noise

1 Introduction

Over the past decade, kernel methods have been widely used to solve nonlinear problems. Examples include time-series prediction, pattern classification, and system identification [1]. In the concept of the kernel method, the input space is mapped into a high dimensional (or even infinite dimension) reproducing kernel

This work is supported in part by the National Natural Science Foundation of China(Grant no. 62201478 and 61971100), in part by the Southwest University of Science and Technology Doctor Fund (Grant no. 20zx7119), in part by the Sichuan Science and Technology Program (Grant no. 2022YFG0148), and in part by the Heilongjiang Provincial Science and Technology Program (No. 2022ZX01A16).

B. Luo et al. (Eds.): ICONIP 2023, LNCS 14447, pp. 3–14, 2024.
https://doi.org/10.1007/978-981-99-8079-6_1

Hilbert space (RKHS) by the use of nonlinear mapping $\varphi(\cdot)$, which is induced by Mercer kernels [2,3]. Additionally, as a family of online learning methods, kernel adaptive filtering (KAF) algorithms can obtain the dynamic statistical characteristics of real-time data and make timely and efficient predictions. Moreover, in KAF algorithms, the inner-product operation in RKHS can be estimated using a kernel function $\kappa(\boldsymbol{u}(i), \boldsymbol{u}(j))$, where $\boldsymbol{u}(i)$ is the input signal at instance i. Therefore, the KAF is a powerful method to solve some nonlinear problems [4]. Some famous KAF algorithms include the kernel least mean square, the kernel affine projection algorithm, and the (extended) kernel recursive least squares. More details are in the well-known book [5].

However, the mentioned KAF algorithms generally suffer from filtering performance degradation in non-Gaussian noise environments, especially for impulsive noise. The main reason is that such KAF algorithms use the second-order statistics criterion as the cost function, which is only suitable for Gaussian noise. To overcome this problem, as an alternative solution, a large number of robust cost functions can resist the influence of outliers [6]. In these cost functions, the maximum correntropy criterion (MCC) induced from information-theoretic learning has received widespread attention in the field of kernel adaptive learning [7]. Examples include the kernel maximum correntropy algorithm [8], the kernel recursive maximum correntropy algorithm [9], and the parallel kernel data-reusing maximum correntropy algorithm [10]. Additionally, based on the definition of MCC, the kernel bandwidth as the only hyperparameter may limit the accuracy of local similarity between the prediction value and the real one. Therefore, [11] proposed a more powerful correntropy named generalized correntropy using the generalized Gaussian density function as a kernel function. And some more effective and robust KAF algorithms have been derived and studied, such as the generalized kernel maximum correntropy algorithm [12], and the kernel recursive generalized maximum correntropy algorithm [13].

In addition, several researchers have reported that the history information is a benefit for improving the performance of algorithms [14,15]. Therefore, based on single (or multiple) delay output, the feedback structures are introduced into the framework of KAF. Examples include the linear recurrent kernel online learning (LRKOL) algorithm [16], the kernel least mean square with single feedback (SF-KLMS) algorithm [17], the regularized KLMS algorithm with multiple-delay feedback [18], the LRKOL based on maximum correntropy criterion (LRKOL-MCC) and kernel recursive maximum correntropy with multiple feedback [19]. However, to our best knowledge, there are still few reports on kernel generalized maximum correntropy (KGMC) algorithm with the history information in KAF.

Therefore, from the perspective of computation efficiency and compact network structure, in this paper, by injecting the single output delay into KGMC, we have proposed a new KAF algorithm named the single feedback based KGMC (SF-KGMC) algorithm. Armed with the single output delay and the GMC criterion, SF-KGMC not only achieves a better filtering performance but is also robust against impulsive noise. For SF-KGMC, we also conduct the theoretical convergence analysis. Furthermore, without loss of generality, the coherence criterion (CC) method is used to combat the computational burden of the proposed algorithm [20]. Compared with related competitors, simulation results on Mackey-Glass time-series prediction and real-world data applications show that SF-KGMC achieves a smaller misadjustment and a faster convergence rate under an impulsive noise environment.

2 Kernel Generalized Maximum Correntropy Algorithm

In this part, after briefly introducing the concept of the generalized correntropy criterion, we review the derivation of KGMC from the perspective of maximizing the generalized correntropy criterion.

2.1 Generalized Correntropy Criterion

Given two random variables X and Y, the generalized correntropy is defined as

$$V_{\alpha,\beta}(X,Y) = \boldsymbol{E}\left[\kappa_{\alpha,\beta}(X,Y)\right] = \int \kappa_{\alpha,\beta}(x,y)dF_{XY}(x,y), \tag{1}$$

where $\boldsymbol{E}[\cdot]$ is an expectation operator; $\kappa_{\alpha,\beta}(x,y) = \xi_{\alpha,\beta}\exp(-\lambda\left|x-y\right|^{\alpha})$ is the generalized Gaussian density function, in which $\alpha > 0$ is the shape parameter, $\beta > 0$ is the scale factor, $\lambda = \beta^{-\alpha}$ is the kernel parameter, $\xi_{\alpha,\beta} = \frac{\alpha}{2\beta\Gamma(1/\alpha)}$ is the normalization constant, and $\Gamma(\cdot)$ is the Gamma function; and $F_{XY}(x,y)$ is the joint distribution function of (X,Y) [11]. In practice, the joint distribution of X and Y is commonly unknown, only a finite number of samples $\{(x_i,y_i)\}_{i=1}^{N}$ are available. Denote $e_i = x_i - y_i$, and thus (1) can be approximated as

$$\hat{V}_{\alpha,\beta}(X,Y) = \frac{1}{N}\sum_{i=1}^{N}\kappa_{\alpha,\beta}(e_i). \tag{2}$$

From (2), we get that the correntropy is only a special case of $\alpha = 2$, and thus the generalized correntropy inherits some properties in terms of symmetry, positive and bounded from correntropy [11]. Additionally, in the field of adaptive filters, a large number of works have demonstrated that one can maximize the generalized correntropy between filter output y_i and observation d_i to construct a robust adaptive filtering algorithm [12,13,21,22]. Hence, armed with the generalized maximum correntropy (GMC) criterion, a kernel adaptive algorithm updates by maximizing the following function

$$J_N = \frac{1}{N}\sum_{i=1}^{N}\kappa_{\alpha,\beta}(d_i - y_i). \tag{3}$$

2.2 Kernel Generalized Maximum Correntropy Algorithm

Let $\boldsymbol{u}(j) \in \mathbb{U}$ be a input vector at instance j, and $d(j) \in \mathbb{R}$ be the corresponding desired output. Under a given sequence of input-output pairs $\{\boldsymbol{u}(j), d(j)\}_{j=1}^{i}$, this work aims to build a continuous input-output mapping function $f:\mathbb{U}\rightarrow\mathbb{R}$ based on the input-output information. According to the kernel method, the function $f(\cdot)$ can be reconstructed in RKHS as the following linear form

$$f(\cdot)=\boldsymbol{\omega}(i)^T\varphi(\cdot) \quad \text{with} \quad \boldsymbol{\omega}(i)=\sum_{j=1}^{i}a_j(i)\varphi(\boldsymbol{u}(j)), \tag{4}$$

where $\boldsymbol{\omega}(i)$ is the weight coefficient vector in RKHS. Additionally, it can also be expressed as a linear combination of the transformed input vector $\varphi(\boldsymbol{u}(j))$ in RKHS and $a_j(i)$ is the expansion coefficient. Furthermore, based on the kernel trick, (4) can be rewritten as

$$f(\boldsymbol{u}(i))=\sum_{j=1}^{i}a_j(i)\varphi(\boldsymbol{u}(j))^T\varphi(\boldsymbol{u}(i))=\sum_{j=1}^{i}a_j(i)\kappa_h(\boldsymbol{u}(i),\boldsymbol{u}(j)), \tag{5}$$

where $\kappa_h(\cdot,\cdot)$ is a Gaussian kernel with kernel width $h>0$.

In KAF, various algorithms can be obtained using different cost functions and optimization strategies. In this work, based on GMC in (3) and the stochastic gradient ascent strategy, the kernel generalized maximum correntropy (KGMC) algorithm can be derived as

$$\boldsymbol{\omega}(i)=\boldsymbol{\omega}(i-1)+\frac{\mu}{\lambda\alpha}\frac{\partial J_1}{\partial\boldsymbol{\omega}(i-1)}=\boldsymbol{\omega}(i-1)+\mu p(e(i))\varphi(\boldsymbol{u}(i)), \tag{6}$$

where $\mu>0$ is a step-size; $p(e(i))=\exp(-\lambda\left|e(i)\right|^{\alpha})\left|e(i)\right|^{\alpha-1}sgn(e(i))$ and $sgn(\cdot)$ is the sign operator; $e(i)=d(i)-y(i)$ is a prediction error and $y(i)=\boldsymbol{\omega}(i-1)^T\varphi(\boldsymbol{u}(i))$ is the estimated output. Let $\boldsymbol{\omega}(i)=\mathbf{0}$, thus a more compact version of (6) can be represented as

$$\boldsymbol{\omega}(i)=\sum_{j=1}^{i}a_j(i)\varphi(\boldsymbol{u}(j)) \quad \text{with} \quad a_j(i)=\mu p(e(j)), \tag{7}$$

which leads to the estimated output as $y(i)=\sum_{j=1}^{i}a_j(i)\kappa_h(\boldsymbol{u}(i),\boldsymbol{u}(j))$.

3 Single Feedback Based KGMC

3.1 SF-KGMC

From the perspective of computation efficiency and compact network structure, a single delay $y(i-1)$ is introduced into KGMC, and Fig. 1 shows the structure of proposed algorithm, which yields the following output

$$y(i)=\boldsymbol{k}_h(i)^T\boldsymbol{a}(i)+\gamma(i)y(i-1), \tag{8}$$

where $\boldsymbol{k}_h(i)=\left[\kappa_h(\boldsymbol{u}(i),\boldsymbol{u}(C_{1i})),\ldots,\kappa_h(\boldsymbol{u}(i),\boldsymbol{u}(C_{li}))\right]^T$ denotes a kernel evaluation vector between input $\boldsymbol{u}(i)$ and dictionary $\boldsymbol{C}(i)=\{\boldsymbol{u}(C_{1i}),\boldsymbol{u}(C_{2i}),\ldots,$

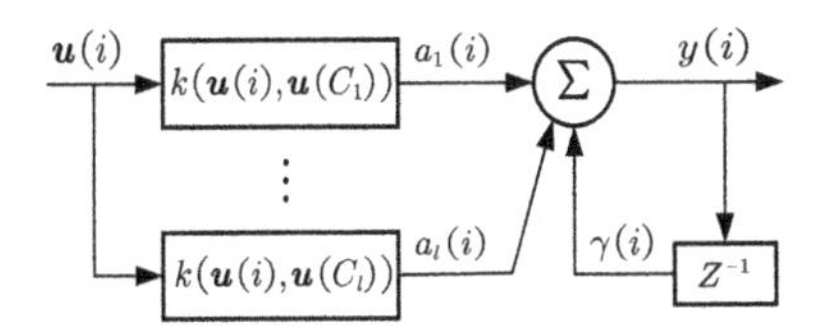

Fig. 1. The structure of SF-KGMC.

$\boldsymbol{u}(C_{li})\}$; $y(i-1)$ is the single delay output; $\boldsymbol{a}(i) = [a_1(i), a_2(i), \ldots, a_l(i)]^T$ is a feedforward vector (FFV) and $\gamma(i)$ is a feedback scalar (FBS).

To compute our proposed algorithm efficiently, we consider the CC sparsification method for constructing a compact dictionary [20], i.e.,

$$\max_{l \in \boldsymbol{C}(i)} (\kappa_h(\boldsymbol{u}(i), \boldsymbol{u}(\boldsymbol{C}_l(i)))) \leq \zeta, \tag{9}$$

where $0 \leq \zeta \leq 1$ is a threshold parameter balancing the filtering performance and network sparsity.

In the single feedback structure, we can define the optimal FFV as $\boldsymbol{a}^*$ and FBS as γ^*, then the desired output $d(i)$ can be expressed as $d(i) = \boldsymbol{k}_h(i)^T\boldsymbol{a}^* + \gamma^* y(i-1)$, and the prediction error $e(i) = d(i) - y(i) + \varepsilon(i)$, i.e.,

$$e(i) = \boldsymbol{k}_h(i)^T (\boldsymbol{a}^* - \boldsymbol{a}(i)) + y(i-1)(\gamma^* - \gamma(i)) + \varepsilon(i) \tag{10}$$

where $\varepsilon(i)$ is a disturbance error due to noise and will be used in later. Similar to the derivations of KGMC, the update equations of FFV and FBS can be obtained by maximizing the generalized correntropy as follows

$$\boldsymbol{a}(i+1) = \boldsymbol{a}(i) + \eta_a \nabla_{\boldsymbol{a}} J(\boldsymbol{a}(i), \gamma(i)), \ \gamma(i+1) = \gamma(i) + \eta_\gamma \nabla_\gamma J(\boldsymbol{a}(i), \gamma(i)), \tag{11}$$

where $J(\boldsymbol{a}(i), \gamma(i)) = \kappa_{\alpha,\beta}(e(i))$, and $\nabla_{\boldsymbol{a}} J(\boldsymbol{a}(i), \gamma(i))$ is the gradient of $J(\boldsymbol{a}(i), \gamma(i))$ with respective to $\boldsymbol{a}(i)$; η_a and η_γ are positive step-sizes of FFV and FBS, respectively. In detail, the weight update formula of FFV can be derived as

$$\boldsymbol{a}(i+1) = \boldsymbol{a}(i) + \frac{\eta_a(i)}{\rho_a(i)} p(e(i)) \nabla_{\boldsymbol{a}} y(i), \tag{12}$$

where $\rho_a(i)$ is a normalization factor to prevent the vanishing radius problem [16,17], and $p(e(i))$ has the same definition in (6). For simplicity, let $\boldsymbol{D}_{\boldsymbol{a}}(i) = \nabla_{\boldsymbol{a}} y(i)$ be the evaluated derivative of $y(i)$ with respect to FFV, and we have

$$\boldsymbol{D}_{\boldsymbol{a}}(i) = \frac{\partial y(i)}{\partial \boldsymbol{a}(i)} + \mu_a(i) \frac{\partial y(i)}{\partial y(i-1)} \frac{\partial y(i-1)}{\partial \boldsymbol{a}(i)} \approx \boldsymbol{k}_h(i) + \mu_a(i)\gamma(i)\boldsymbol{D}_{\boldsymbol{a}}(i-1), \tag{13}$$

where $\mu_a(i)$ is a parameter for adjusting the recurrent gradient information, i.e., $\boldsymbol{D}_{\boldsymbol{a}}(i-1)$. Combining (12) and (13), the update formula of feedforward part in SF-KGMC is obtained as

$$\boldsymbol{a}(i+1) = \boldsymbol{a}(i) + \frac{\eta_a(i)}{\rho_a(i)} p(e(i)) \boldsymbol{D}_a(i). \tag{14}$$

Additionally, armed with similar definitions, let $D_\gamma(i) = \nabla_\gamma y(i)$, and the weight update formula of FBS can be derived as

$$\gamma(i+1) = \gamma(i) + \frac{\eta_\gamma(i)}{\rho_\gamma(i)} p(e(i)) D_\gamma(i), \quad D_\gamma(i) \approx y(i-1) + \mu_\gamma(i)\gamma(i) D_\gamma(i-1). \quad (15)$$

Furthermore, to make sure the convergence of SF-KGMC, following update rules about parameters $\eta_\diamond(i), \rho_\diamond(i)$ and $\mu_\diamond(i)$ with $\diamond \in \{a, \gamma\}$ are listed.

- The parameter $\eta_\diamond(i)$ in (14) or (15) is determined by

$$\eta_\diamond(i) = \begin{cases} 1, & \text{if} \quad \Omega_\diamond(i) + \Omega_\diamond(i-1) \leq 2p(e(i))(e(i) - \bar{\varepsilon}_\diamond), \\ 0, & \text{otherwise}, \end{cases} \quad (16)$$

  where $\Omega_\diamond(i) = \frac{\chi_\diamond^2}{\rho_\diamond(i)}$, and $\chi_\diamond \in \{\|\boldsymbol{D}_\omega(i)\| |p(e(i))|, |D_\gamma(i) p(e(i))|\}$.
- As a normalization factor, $\rho_\diamond(i)$ in (14) or (15) is obtained as

$$\rho_\diamond(i) = \max\left(\rho_\diamond(i-1), \nu_\diamond \rho_\diamond(i-1) + \max\left(\bar{\rho}_\diamond, \chi_\diamond^2\right)\right), \quad (17)$$

  where $\nu_\diamond \in (0, 1)$ is a coefficient that decides the effect degree of the previous factor on the current one; and $\bar{\rho}_\diamond$ is a positive regulatory factor that avoids the initial step-size becoming too small [16].
- As a hybrid step-size, $\mu_\diamond(i)$ controls the recurrent information, and updates by

$$\mu_\diamond(i) = \begin{cases} sgn(\theta(i)), & \text{if} \quad |p(e(i))| (\delta + |\gamma(i)|) < |p(e(i-1))|, \\ 0, & \text{otherwise}, \end{cases} \quad (18)$$

  where $\theta(i) = \gamma(i) p(e(i)) p(e(i-1))$; and $\delta > 0$ is a regularization factor.

Remarks: Armed with equations (13) to (18), we have obtained the single-feedback based kernel generalized maximum correntropy (SF-KGMC) algorithm, which is summarized in **Algorithm 1**. Different from KGMC in (7), the weight parameters of SF-KGMC will be fully updated at each instant during the learning process. Such update operations has the benefit for filtering performance improvement. In addition, the relations between SF-KGMC and LRKOL-MCC are that: 1) When $\alpha = 2$, SF-KGMC becomes to KMC with single-feedback; 2) Although LRKOL-MCC contains a single-delay output, multiple-feedbacks are used in LRKOL-MCC. Therefore, SF-KGMC has not only a more compact network, but also enhances the filtering performance with flexible parameters setting. Moreover, similar to the momentum LMS algorithm [23], there exists a momentum term in the update of FFV as shown in the next subsection.

Algorithm 1 : The SF-KGMC Algorithm

Parameters: $\nu_a, \nu_\gamma, \bar{\rho}_a, \bar{\rho}_\gamma, \bar{\varepsilon}_a, \bar{\varepsilon}_\gamma, \alpha, \lambda, h$
Initialization: $\boldsymbol{C}(1) = \boldsymbol{u}(1), \boldsymbol{a}(1) = d(1), \gamma(1) = 0.$
Computation:
While $\{\boldsymbol{u}(i), d(i)\}$, $(i \geq 1)$ **do**
1) Compute the output $y(i)$:
$\boldsymbol{k}_h(i) = [\kappa_h(\boldsymbol{u}(i), \boldsymbol{u}(C_1)), \ldots, \kappa_h(\boldsymbol{u}(i), \boldsymbol{u}(C_l))]^T$, $y(i) = \boldsymbol{k}(i)^T \boldsymbol{a}(i) + \gamma(i) y(i-1)$
2) Compute the error: $e(i) = d(i) - y(i)$
3) Compute the parameters: $\eta_\diamond(i), \rho_\diamond(i), \mu_\diamond(i), \diamond \in \{a, \gamma\}$
4) Update the estimated derivative:
$\boldsymbol{D}_a(i) = \boldsymbol{k}_h(i) + \mu_\omega(i)\gamma(i)\boldsymbol{D}_a(i-1)$, $D_\gamma(i) = y(i-1) + \mu_\gamma(i)\gamma(i)D_\gamma(i-1)$
5) Update the FFV and FBS:
$\boldsymbol{a}(i+1) = \boldsymbol{a}(i) + \dfrac{\eta_a(i)}{\rho_a(i)} f(e(i))\boldsymbol{D}_a(i)$, $\gamma(i+1) = \gamma(i) + \dfrac{\eta_\gamma(i)}{\rho_\gamma(i)} f(e(i)) D_\gamma(i)$
End While

3.2 The Momentum Existed in SF-KGMC

In [23], the weight update equation of momentum LMS (MLMS) is

$$\mathbf{w}(i+1) = \mathbf{w}(i) + \eta \frac{\partial J(i)}{\partial \mathbf{w}(i)} + \vartheta \left(\mathbf{w}(i) - \mathbf{w}(i-1)\right), \tag{19}$$

where η is a positive step-size; and ϑ is the momentum coefficient. In general, when $|\vartheta| < 1$, MLMS is convergent and stable. And it realizes a faster convergence speed with positive ϑ value.

For SF-KGMC, based on (14), we have

$$\boldsymbol{D}_a(i-1) = \left(\boldsymbol{a}(i) - \boldsymbol{a}(i-1)\right) \frac{\rho_a(i-1)}{\eta_a(i-1)p(e(i-1))}, \tag{20}$$

and substituting (20) into (13), then (14) can be rewritten as

$$\boldsymbol{a}(i+1) = \boldsymbol{a}(i) + \frac{\eta_a(i)}{\rho_a(i)} p(e(i)) \frac{\partial y(i)}{\partial \boldsymbol{a}(i)} + \bar{\mu}_a(i)\left(\boldsymbol{a}(i) - \boldsymbol{a}(i-1)\right), \tag{21}$$

where $\bar{\mu}_a(i) = \dfrac{\eta_a(i)\rho_a(i-1)p(e(i))\gamma(i)\mu_a(i)}{\rho_a(i)\eta_a(i-1)p(e(i-1))}$, which can be treated as a momentum coefficient; and $\boldsymbol{a}(i) - \boldsymbol{a}(i-1)$ is the momentum term. When SF-KGMC converges, we have

$$\lim_{i \to \infty} \frac{\eta_a(i)\rho_a(i-1)}{\rho_a(i)\eta_a(i-1)} \approx 1. \tag{22}$$

Furthermore, according to (22) and the definition of $\mu_a(i)$ in (18), we obtain

$$|\bar{\mu}_a(i)| = \left| \frac{\eta_a(i)\rho_a(i-1)p(e(i))\gamma(i)\mu_a(i)}{\rho_a(i)\eta_a(i-1)p(e(i-1))} \right| \leq \left| \frac{\eta_a(i)\rho_a(i-1)}{\rho_a(i)\eta_a(i-1)} \right| \approx 1. \tag{23}$$

Therefore, an equivalent momentum term exists in the learning processing of SF-KGMC. Armed with the momentum during wight update, SF-KGMC can effectively improve the filtering performance by reusing the past information.

3.3 Convergence Analysis

In this part, for SF-KGMC, the convergence analysis is conducted. Subtracting the optimal FFV $\boldsymbol{a}^*$ from both sides of (14), we have

$$\hat{\boldsymbol{a}}(i+1) = \hat{\boldsymbol{a}}(i) + \frac{\eta_a(i)}{\rho_a(i)} p(e(i)) \boldsymbol{D}_a(i), \tag{24}$$

where $\hat{\boldsymbol{a}}(i) = \boldsymbol{a}(i) - \boldsymbol{a}^*$. Based on $\boldsymbol{D}_a(i)$ in (13), squaring both sides of (24) and denoting $\Delta(\hat{\boldsymbol{a}}(i)) = \|\hat{\boldsymbol{a}}(i+1)\|^2 - \|\hat{\boldsymbol{a}}(i)\|^2$, we have

$$\Delta(\hat{\boldsymbol{a}}(i)) = \frac{\eta_a(i)}{\rho_a(i)} \left(A_1(i) + A_2(i) + A_3(i)\right), \tag{25}$$

where $A_1(i) = \eta_a(i)\rho_a(i)^{-1}p(e(i))^2\|\boldsymbol{D}_a(i)\|^2$, $A_2(i) = 2p(e(i))\hat{\boldsymbol{a}}(i)^T\boldsymbol{k}_h(i)$, and $A_3(i) = 2p(e(i))\mu_a(i)\gamma(i)\hat{\boldsymbol{a}}(i)^T\boldsymbol{D}_a(i-1)$. Additionally, according to (24) and denoting $\xi_a(i) = \rho_a(i)^{-1}\eta_a(i)$, then $A_3(i)$ can be rewritten as

$$A_3(i) = H(i)\xi_a(i-1)^{-1}\left(\Delta(\hat{\boldsymbol{a}}(i-1)) + \xi_a(i-1)^2 p(e(i-1))^2\|\boldsymbol{D}_a(i-1)\|^2\right), \tag{26}$$

where $H(i)$ is defined as

$$H(i) = \begin{cases} \left|\dfrac{p(e(i))\gamma(i)}{p(e(i-1))}\right|, & \text{if } \ |p(e(i))| < \dfrac{|p(e(i-1))|}{\delta + |\gamma(i)|}, \\ 0, & \text{otherwise}, \end{cases} \tag{27}$$

which means that $0 \le H(i) \le 1$. Then, denoting $\Xi_{aH}(i) = \xi_a(i-1)\mathrm{H}(i)p(e(i-1))^2\|\boldsymbol{D}_a(i-1)\|^2$ and $\Xi_{aD}(i) = \xi_a(i)p(e(i))^2\|\boldsymbol{D}_a(i)\|^2$, then (25) is rewritten as

$$\Delta(\hat{\boldsymbol{a}}(i)) = A_4(i) + A_5(i)\Delta(\hat{\boldsymbol{a}}(i-1)), \tag{28}$$

where $0 \le A_5(i) = \xi_a(i)\xi_a(i-1)^{-1}H(i) \le 1$, and

$$\begin{aligned} A_4(i) &= \xi_a(i)\left\{\Xi_{aD}(i) + 2p(e(t))\hat{\boldsymbol{a}}(i)^T\boldsymbol{k}_h(i) + \Xi_{aH}(i)\right\} \\ &\overset{(b_1)}{\le} \xi_a(i)\left\{\Xi_{aD}(i) - 2p(e(t))(e(i) - \bar{\varepsilon}_a(i)) + \Xi_{aH}(i)\right\} \overset{(b_2)}{\le} 0, \end{aligned} \tag{29}$$

the first inequality (b_1) is based on the following fact

$$\begin{aligned} e(i) &= -\boldsymbol{k}_h(i)^T\hat{\boldsymbol{a}}(i) + y(i-1)\left(\gamma^* - \gamma(i)\right) + \varepsilon(i) \\ &= \varepsilon_a(i) - \boldsymbol{k}_h(i)^T\hat{\boldsymbol{a}}(i) \le \bar{\varepsilon}_a - \boldsymbol{k}_h(i)^T\hat{\boldsymbol{a}}(i) \end{aligned}, \tag{30}$$

where $\varepsilon_a(i) = y(i-1)\left(\gamma^* - \gamma(i)\right) + \varepsilon(i)$ is the equivalent disturbance, and $\bar{\varepsilon}_a = \max(|\varepsilon_a(i)|)$. And the second inequality (b_2) is derived by the definition of $\eta_a(i)$ in (16). With a initial condition, i.e., $\Delta(\hat{\boldsymbol{a}}(1)) \le 0$, based on $A_4(i) \le 0$ and $A_5(i) \ge 0$, for (28), we get $\Delta(\hat{\boldsymbol{a}}(i)) \le 0$. Which means $\|\hat{\boldsymbol{a}}(i+1)\|^2 \le \|\hat{\boldsymbol{a}}(i)\|^2$ leading to the convergence of FFV. Similarly, the convergence of FBS can also be ensured.

4 Simulation Results

In this section, to verify the filtering performance of SF-KGMC, a time-series prediction and two real-world applications are considered. In addition, for comparison, we consider some related algorithms, i.e., KGMC, LRKOL, LRKOL-MCC and SF-KLMS. For simplicity, we introduce a parameter vector $\mathbf{p} = [\mu, \alpha, \beta]^T$ for KGMC; two parameter vectors $\mathbf{w}_1 = [\bar{\varepsilon}_a, \bar{\rho}_a, \nu_a]^T$ and $\mathbf{r}_1 = [\bar{\varepsilon}_\gamma, \bar{\rho}_\gamma, \nu_\gamma]^T$ for LRKOL, LRKOL-MCC and SF-KLMS; and three parameter vectors $\mathbf{w}_2 = [\bar{\varepsilon}_a, \bar{\rho}_a, \nu_a]^T$, $\mathbf{r}_2 = [\bar{\varepsilon}_\gamma, \bar{\rho}_\gamma, \nu_\gamma]^T$ and $\mathbf{g}_2 = [\alpha, \beta]^T$ for SF-KGMC. To evaluate the filtering performance, the testing mean square error (TMSE) is adopted [5]. In addition, the non-Gaussian noise is modeled by the alpha-stable noise (α-SN) with parameter vector $\mathbf{A}_0 = [\alpha_0, \beta_0, \gamma_0, \delta_0]^T$, where $\alpha_0 \in (0, 2]$ is the characteristic factor, $\beta_0 \in [-1, 1]$ is the symmetry parameter, $\gamma_0 > 0$ is the dispersion parameter and $\delta_0 \in (-\infty, +\infty)$ is the location parameter [24]. Furthermore, without loss of generality, the CC sparsification method with sparsity coefficient $\zeta = 0.45$ is used to reduce the computational burden of these KAFs. In this simulation, all parameters are set to achieve similar initial convergence behaviors. And, all results are obtained by averaging over 50 independent runs.

4.1 Mackey-Glass Time-Series Prediction

Considering a benchmark example, i.e., Mackey-Glass (MK) time-series prediction [5], which is modeled as $dx(t)/dt = -0.1x(t) + 0.2x(t-30)/(1 + x(t-30)^{10})$. The time series is discretized at a sampling period of 6 s, and $\boldsymbol{u}(i) = [x(i-10), ..., x(i-1)]^T$ is used as the input to predict the current one $x(i)$. In this experiment, the training data size is 1500 and the testing data size is 100.

Here, for SF-KGMC-CC[1], we investigate the joint effect of α and β on the steady-state TMSE obtained by averaging over last 100 iterations. To this end, we set $\mathbf{w}_2 = [0.02, 0.3, 0.75]^T$, $\mathbf{r}_2 = [0.02, 150, 0.9]^T$, $h = 0.25$, and $\mathbf{A}_0 = [1.4, 0, 0.05, 0]^T$. The corresponding results are plotted in Fig. 2(a). From this figure, we get that SF-KGMC-CC achieves a better filtering accuracy when $\alpha = 1.6$ and a larger β, such $\beta = 12$. Then, SF-KGMC-CC is compared to these related KAFs. For all KAFs, the $h = 0.25$; for LRKOL-CC, LRKOL-MCC-CC, SF-KLMS-CC and SF-KGMC-CC, we set $\mathbf{w}_1 = \mathbf{w}_2 = [0.02, 0.3, 0.75]^T$,$\mathbf{r}_1 = \mathbf{r}_2 = [0.02, 150, 0.9]^T$, and $\mathbf{g}_2 = [1.6, 12]^T$; for LRKOL-MCC-CC, the kernel width of MCC is $\sigma = 3$; for KGMC-CC, we set $\mathbf{p} = [0.08, 1.6, 12]^T$. The corresponding learning curves are shown in Fig. 2(b). As expected, under the non-Gaussian noise, MCC and/or GMCC based KAFs are more robust than LRKOL-CC and SF-KLMS-CC. And SF-KGMC-CC achieves the best filtering performance. In addition, for SF-KGMC type algorithms, SF-KGMC-CC outperforms SF-KGMC in terms of convergence rate and steady-state misalignment. This is beyond our intuition, because we think that the sparsification should always decrease the

[1] SF-KGMC-CC stands for SF-KGMC sparsified by the CC sparsification method.

accuracy, and the final steady-state misalignment of SF-KGMC should be always smaller than that of SF-KGMC-CC. Such phenomena have also shown in [24,25].

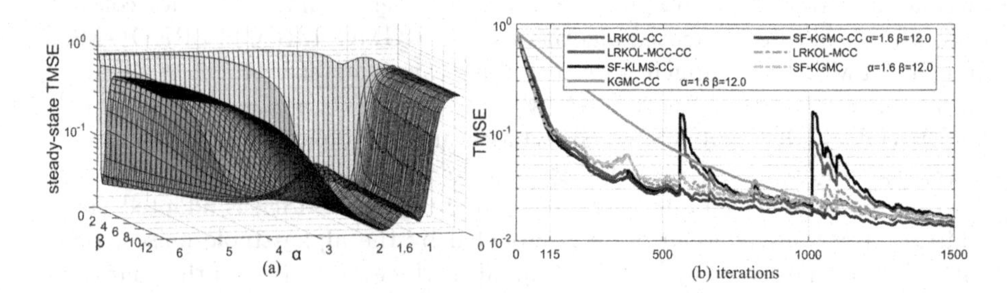

Fig. 2. For MK time-series prediction under α-SN; (a) The joint effect of α and β on the steady-state TMSE of SF-KGMC-CC; (b) The TMSE curves of various KAFs.

4.2 Real-World Data Predictions

In this experiment, two real-world data sets, i.e., the Internet traffic and Sunspot number, are considered to verify the superior performance of proposed SF-KGMC. The Internet traffic data with 14772 samples are from the homepage of Prof. Paulo Cortez[2], and we use 2000 samples as training data and another 100 samples as testing data. The Sunspot number with 3235 samples are recorded from the January 1749 to September 2018[3], and we use 1500 samples and other 100 samples as training and testing data, respectively. To eliminate the influence of the scale of the origin data, the zero-one normalization is used.

For Internet traffic prediction, we use the 7 previous points as input to predict the current one. For all KAFs, we set $h = 0.5$; for SF-KGMC-CC and KGMC-CC, we set $\mathbf{g}_2 = [4, 0.5]^T$ and $\mathbf{p} = [0.1, 4, 0.5]^T$; for feedback based KAFs, we set $\mathbf{w}_1 = \mathbf{w}_2 = [0.02, 0.8, 0.85]^T$ and $\mathbf{r}_1 = \mathbf{r}_2 = [0.02, 10, 0.9]^T$. The corresponding TMSE learning curves are plotted in Fig. 3(a), which shows that SF-KGMC-CC outperforms other competing KAFs in terms of convergence rate and filtering accuracy. And, for Sunspot number prediction, we use the 5 past points to predict the current one. For all KAFs, we set $h = 0.35$; for SF-KGMC-CC and KGMC-CC, we set $\mathbf{g}_2 = [4, 0.5]^T$ and $\mathbf{p} = [0.3, 4, 0.5]^T$; For feedback based KAFs, we set $\mathbf{w}_1 = \mathbf{w}_2 = [0.02, 0.7, 0.65]^T$, $\mathbf{r}_1 = \mathbf{r}_2 = [0.02, 10, 0.93]^T$. Figure 3(b) plots the corresponding results, and revels that SF-KGMC-CC can still realize better filtering accuracy and faster convergence rate.

[2] http://www3.dsi.uminho.pt/pcortez/series/.

[3] https://www.sidc.be/silso/datafiles.

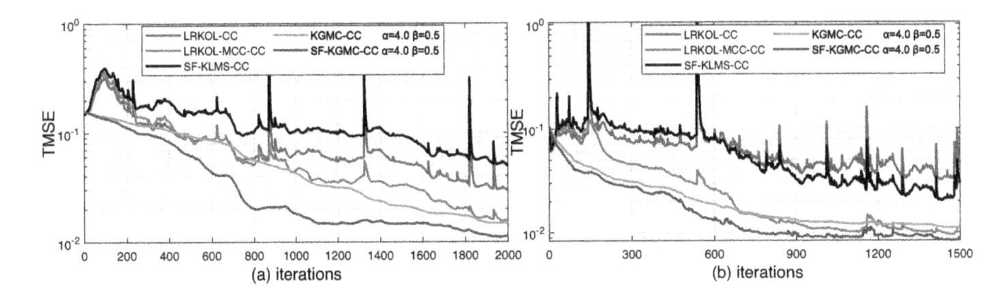

Fig. 3. Real world applications: (a) The TMSE curves of Internet traffic prediction with $\mathbf{A}_0 = [1.4, 0, 0.05, 0]^T$; (b) The TMSE curves of Sunspot number prediction with $\mathbf{A}_0 = [1.4, 0, 0.05, 0]^T$.

5 Conclusion

In this paper, based on the single feedback structure, a new robust kernel generalized maximum correntropy adaptive filtering algorithm (SF-KGMC) has been proposed. Compared to MSE based KAFs, KGMC is more robust against outliers. In addition, armed with proper shape parameter and scale factor, KGMC is more flexible to different kinds of noises. Therefore, SF-KGMC achieves better filtering performance than SF-KLMS, LRKOL, and LRKOL-MCC. Furthermore, for SF-KGMC, the convergence is also proved from the perspective of energy non-increase. Simulation results for a benchmark example and two real-world data predictions demonstrate that, in comparison with competitors, SF-KGMC achieves better filtering performances. However, due to some hyperparameters involved, selecting the appropriate ones becomes a challenging task, thus it is a worthwhile problem to study in future work.

References

1. Pillonetto, G., Dinuzzo, F., Chen, T., De Nicolao, G., Ljung, L.: Kernel methods in system identification, machine learning and function estimation: a survey. Automatica **50**(3), 657–682 (2014)
2. Takizawa, M., Yukawa, M.: Adaptive nonlinear estimation based on parallel projection along affine subspaces in reproducing Kernel Hilbert space. IEEE Trans. Signal Process. **63**(16), 4257–4269 (2015)
3. Ghasemi, M., Fardi, M., Ghaziani, R.K.: Numerical solution of nonlinear delay differential equations of fractional order in reproducing Kernel Hilbert space. Appl. Math. Comput. **268**, 815–831 (2015)
4. Xinghan, X., Ren, W.: Random fourier feature kernel recursive maximum mixture correntropy algorithm for online time series prediction. ISA Trans. **126**, 370–376 (2022)
5. Príncipe, J.C., Liu, W., Haykin, S.: Kernel Adaptive Filtering: A Comprehensive Introduction. Wiley (2011)
6. Kumar, K., Pandey, R., Karthik, M.L.N.S., Bhattacharjee, S.S., George, N.V.: Robust and sparsity-aware adaptive filters: a review. Signal Process. **189**, 108276 (2021)

7. Liu, W., Pokharel, P.P., Principe, J.C.: Correntropy: properties and applications in non-gaussian signal processing. IEEE Trans. Signal Process. **55**(11), 5286–5298 (2007)
8. Zhao, S., Chen, B., Principe, J.C.: Kernel adaptive filtering with maximum correntropy criterion. In: 2011 International Joint Conference on Neural Networks, pp. 2012–2017. IEEE (2011)
9. Wu, Z., Shi, J., Zhang, X., Ma, W., Chen, B., IEEE Senior Member: Kernel recursive maximum correntropy. Signal Process. **117**, 11–16 (2015)
10. Qishuai, W., Li, Y., Xue, W.: A parallel kernelized data-reusing maximum correntropy algorithm. IEEE Trans. Circuits Syst. II Express Briefs **67**(11), 2792–2796 (2020)
11. Chen, B., Xing, L., Zhao, H., Zheng, N., Prı, J.C., et al.: Generalized correntropy for robust adaptive filtering. IEEE Trans. Signal Process. **64**(13), 3376–3387 (2016)
12. He, Y., Wang, F., Yang, J., Rong, H., Chen, B.: Kernel adaptive filtering under generalized maximum correntropy criterion. In: 2016 International Joint Conference on Neural Networks (IJCNN), pp. 1738–1745. IEEE (2016)
13. Zhao, J., Zhang, H.: Kernel recursive generalized maximum correntropy. IEEE Signal Process. Lett. **24**(12), 1832–1836 (2017)
14. Fei, J., Liu, L.: Real-time nonlinear model predictive control of active power filter using self-feedback recurrent fuzzy neural network estimator. IEEE Trans. Industr. Electron. **69**(8), 8366–8376 (2021)
15. Wang, S., Takyi-Aninakwa, P., Fan, Y., Chunmei, Yu., Jin, S., Fernandez, C., Stroe, D.-I.: A novel feedback correction-adaptive kalman filtering method for the whole-life-cycle state of charge and closed-circuit voltage prediction of lithium-ion batteries based on the second-order electrical equivalent circuit model. Int. J. Electr. Power Energy Syst. **139**, 108020 (2022)
16. Fan, H., Song, Q.: A linear recurrent kernel online learning algorithm with sparse updates. Neural Netw. **50**, 142–153 (2014)
17. Zhao, J., Liao, X., Wang, S., Chi, K.T.: Kernel least mean square with single feedback. IEEE Signal Process. Lett. **22**(7), 953–957 (2014)
18. Wang, S., Zheng, Y., Ling, C.: Regularized kernel least mean square algorithm with multiple-delay feedback. IEEE Signal Process. Lett. **23**(1), 98–101 (2015)
19. Wang, S., Dang, L., Wang, W., Qian, G., Chi, K.T.: Kernel adaptive filters with feedback based on maximum correntropy. IEEE Access **6**, 10540–10552 (2018)
20. Richard, C., Bermudez, J.C.M., Honeine, P.: Online prediction of time series data with kernels. IEEE Trans. Signal Process. **57**(3), 1058–1067 (2008)
21. Zhao, J., Zhang, H., Wang, G.: Fixed-point generalized maximum correntropy: convergence analysis and convex combination algorithms. Signal Process. **154**, 64–73 (2019)
22. Ma, W., Duan, J., Chen, B., Gui, G., Man, W.: Recursive generalized maximum correntropy criterion algorithm with sparse penalty constraints for system identification. Asian J. Control **19**(3), 1164–1172 (2017)
23. Qian, N.: On the momentum term in gradient descent learning algorithms. Neural Netw. **12**(1), 145–151 (1999)
24. Zhao, J., Zhang, H., Wang, G., Zhang, J.A.: Projected kernel least mean p-power algorithm: convergence analyses and modifications. IEEE Trans. Circuits Syst. I Regul. Pap. **67**(10), 3498–3511 (2020)
25. Chen, B., Zhao, S., Zhu, P., Príncipe, J.C.: Quantized kernel least mean square algorithm. IEEE Trans. Neural Networks Learn. Syst. **23**(1), 22–32 (2011)

Application of Deep Learning Methods in the Diagnosis of Coronary Heart Disease Based on Electronic Health Record

Hanyang Meng(✉) and Xingjun Wang

Shenzhen International Graduate School, Tsinghua University, Shenzhen, China
{mhy20,wangxingjun}@mails.tsinghua.edu.cn

Abstract. With the development of Internet technology, the number of electronic health record data has surged. Also, artificial intelligence simulates the ability of human beings to solve problems and make decisions by conducting complex and fast calculation on a large amount of data. Based on the electronic health records of hypertension patients in Peking University Shenzhen Hospital, we proposed the use of deep learning methods to achieve intelligent coronary heart disease diagnosis for hypertensive patients. We firstly conducted statistical data analysis and effective feature selection experiments. Then, we established an intelligent diagnosis model for coronary heart disease based on the Transformer and contrastive learning. The model integrates multiple types of health record data such as patient's personal information, symptoms, concurrent diseases and test data, and the results proved that our model achieved the best classification performance and accuracy compared with CNN, RNN and LSTM, with AUC value reached 0.9349. In the future, this model can be extended to the diagnosis of general chronic diseases.

Keywords: electronic health record · coronary heart disease · disease diagnosis · deep learning

1 Introduction

Currently, there have been significant changes in the way of life and population structure in China. The aging process of the population is accelerating, and people's unhealthy lifestyles such as lack of exercise, high mental pressure, and alcoholism are becoming more and more common, which leads to the incidence rate of chronic diseases rising year by year, especially cardiovascular diseases. As the warning of COVID-19, if COVID-19 patients themselves suffer from other basic chronic diseases, they will face the threat of becoming more serious and dying of viruses or other complications. If cardiovascular diseases cannot be well prevented, it will cause a devastating blow to patients. At the same time, the increasing demand for chronic disease management has also increased the pressure on resource allocation and direct expenditure in China's healthcare system. Therefore, timely diagnosis and effective management of chronic cardiovascular

B. Luo et al. (Eds.): ICONIP 2023, LNCS 14447, pp. 15–26, 2024.
https://doi.org/10.1007/978-981-99-8079-6_2

diseases not only have significant implications for the patients' health, but also have a special impact on the allocation of resources and disease prevention and control strategies in China's health system.

With the development of Internet technology, the state has paid more attention to "Internet Medicine" and encouraged patients to carry out online chronic disease management. More and more hospitals have began to use standardized electronic health records to build information platforms and hospital information system, which makes the medical data foundation in China more and more solid. Electronic health records contain various data of patients during diagnosis, hospitalization and treatment, including patient personal information, diagnostic information, laboratory tests, examination descriptions, clinical record texts, etc. All of this information is stored electronically, which improves medical efficiency a lot [1]. The key challenge is how to extract the necessary knowledge and learn useful features from complex, high-dimensional, heterogeneous, and unstructured electronic health record data for intelligent healthcare [2–9]. In the period of machine learning, researchers often manually extracted features or some combination of features [10,11]; In deep learning, the model can actively learn its key features from complex data itself and transform them into efficient and meaningful vector representations. Therefore, some researchers began using architectures such as CNN and RNN (including LSTM and GRU) to extract effective representations of patient health records for disease prediction and diagnosis [12,13].

Through literature research, we found that there are many similarities between the representation of electronic health record data and natural language processing. Currently, Transformer is completely based on the self attention mechanism to calculate vector representations, and has shown excellent performance in processing natural language representations. And the development of unsupervised contrastive learning also brought directions for this article.

2 Materials and Methods

2.1 Data Resource

The electronic health record data used in this study is from the hypertension specialized disease management database established by Peking University Shenzhen Hospital. During this study, a total of 14389 electronic health records are collected, and each patient is assigned a unique identification number, each outpatient or admission record of the patient has a unique record number. The electronic health record data is divided into several parts: patient information, diagnostic records, test results and health record text. The records in different data tables are associated through patient identification number and record number.

2.2 Data Preprocessing

Based on the analysis of electronic health records at Peking University Shenzhen Hospital, we proposed targeted data preprocessing.

Firstly, due to the continuous development of information technology and the large amount of non-standard writing by doctors when recording diagnostic information, there were diversified diagnostic codes and names in diagnostic records, accompanied by a large number of missing data. In response to these issues, we firstly took the first three digits of ICD-10 codes in all diagnostic records, and standardized the diagnostic names referring to the writing method of disease names in ICD-10. For example, the diagnostic names corresponding to data with ICD-10 code I10-I15 were all standardized as "hypertension". In response to the issue of missing ICD code or diagnostic name data, we filled in the missing data based on the standardized correspondence between the two, and deleted the diagnostic records that both are missing.

Secondly, we processed the test data. The test data includes numerical data, categorical data, and textual descriptive data. For numerical test item values, non numerical symbols (such as ">", "<", "+", "-") were removed in this article; For categorical test item values, we used numbers to replace categories (such as 0 for negative and 1 for positive); Then we removed the inspection items for textual descriptions.

Next, for the record text, we hoped to further extract the necessary symptom element words from the patient's main complaint and case characteristics. We used the pkuseg toolkit to segment all recorded texts. After removing stop words and punctuation mark marks, we counted the frequency of all words, manually filtered out symptom words with high frequency, and normalized some symptom words. For negative element words, we explored the patterns of negative words appearing in health records and established certain rules to remove negative keywords.

After data preprocessing, we associated the above records by patient identification number and record number, and removed duplicate and isolated data. Finally, we obtained electronic health record data of 5583 hypertensive patients.

2.3 Feature Selection

Due to the complex types of feature variables in the original electronic health records, we further extracted features after data preprocessing. Based on existing researches on coronary heart disease risk, combined with clinical experience of doctors and available data, we preliminarily screened variables related to coronary heart disease risk in hypertensive patients from cleaned electronic health records, including patient information, existing diseases, symptom manifestations, and laboratory tests. We also removed features with missing values exceeding 80% from the sample.

Then, we conducted corresponding statistical analysis and testing on the correlation between the selected feature variables and coronary heart disease.

For the relationship between categorical variables and coronary heart disease labels, we used chi square test for correlation analysis, and the results were shown in the Table 1.

Table 1. Chi square test results.

Categorical Features	Chi square statistic	P-value
Chest pain	680.3209017	1.8629E-148
Hypertension	621.562044	3.4199E-137
Chest tightness	551.0963649	2.1426E-120
Syncope	272.352534	7.23434E-60
Hyperlipidemia	254.0319735	3.43132E-57
Diabetes	242.5328806	1.10265E-54
Renal failure	226.5973497	3.29175E-51
Smoking	223.975781	2.31354E-49
Short of breath	203.8826355	5.33886E-45
Dizziness	178.2809075	1.9355E-39
Arrhythmias	98.84584225	2.72943E-23
Rheumatic Heart Disease	90.97934981	1.45179E-21
Vomiting	75.15285681	4.79473E-17
Dyspnea	71.45462505	3.04667E-16
Palpitations	57.26950019	3.6651E-13
Nausea	56.78524832	4.66917E-13
Drinking	43.41531872	3.73666E-10
Sleep disorders	25.11956478	5.38834E-07
Hyperuricemia	22.01654133	2.70311E-06
Gout	19.72906771	8.92337E-06
Cough	13.46351	0.001192438
Cerebrovascular diseases	6.421557907	0.011274321
Headache	1.190593277	0.551398975
Fatigue	0.029709423	0.985255075

For the relationship between numerical variables and coronary heart disease labels, we used z-test for correlation analysis, and the results were shown in the Table 2.

Table 2. Z test results.

Numerical Features	Z statistic	P-value
Anti Human Immunodeficiency Virus Antibodies [HIV-Ab]	-18.3658131	6.1825E-73
High sensitivity troponin T [hs cTnT]	-18.32213962	5.84506E-72
#Triglycerides [TG]	-17.90743944	1.5991E-69
Free triiodothyronine [FT3]	-17.66028754	1.58206E-66

continued

Table 2. continued

Numerical Features	Z statistic	P-value
Free thyroxine [FT4]	-17.24956374	9.32025E-64
Jaundice Index	-16.68083119	3.21922E-59
Troponin I [cTnI]	-16.04125048	4.42355E-56
Hemolysis index	-15.760308	1.09682E-53
#Total cholesterol [TC]	-15.55488445	1.40363E-52
#Low density lipoprotein cholesterol [LDL-C]	-15.08063209	1.09621E-49
Lipid blood index	-14.95412537	4.34917E-49
#High-density lipoprotein cholesterol [HDL-C]	-13.92453633	1.03686E-42
Creatine kinase MB subtype activity [CK-MB]	-13.67837212	8.80199E-42
Conductivity (Cond.)	13.33744522	1.64203E-39
Fibrinogen [FIB]	-13.24489157	5.14741E-39
Glucose [GLU]	-12.76695711	1.13922E-36
PT activity [PT %]	-12.01535296	1.67091E-32
eGFR	11.00561677	1.28449E-27
Activated partial coagulation kinase time [APTT]	-10.68610011	3.344E-26
Thrombin coagulation time [TT]	-10.28584645	1.50896E-24
Lymphocyte percentage	9.940719566	6.21563E-23
Percentage of undamaged red blood cells (NLRBC %)	9.504007545	3.5864E-21
Uric acid [UA]	-9.459870264	5.44289E-21
Creatine kinase [CK]	-8.588640501	1.15562E-17
Neutrophil percentage	-6.818170711	1.08523E-11
Absolute value of neutrophils	-6.676517523	2.84676E-11
#Thyroid hormone [TSH]	-6.351358751	2.36029E-10
Absolute value of eosinophils	-6.177915781	7.08295E-10
Urea	6.040611851	1.6635E-09
Prothrombin time [PT]	-5.806503461	7.11297E-09
#White blood cell count	-5.761946566	9.03706E-09
Plasma prothrombin time ratio [PT-R]	-5.699611749	1.32957E-08
Monocyte percentage	4.932134231	8.56447E-07
Specific Gravity (SG)	4.232514707	2.36726E-05
Large platelet ratio	-4.214818229	2.56426E-05
Platelet volume distribution width	-4.191413154	2.84159E-05
Specific antibodies against Treponema pallidum [TP Ab]	-4.129219681	3.69855E-05
D-dimer [D-D]	-3.853700563	0.000118237
PT international standardized ratio [INR]	-3.745446741	0.000183778
#Albumin [Alb]	3.735625463	0.000190084
PH	3.534030259	0.000414293
Percentage of eosinophils	-3.474969397	0.000516497
#Total protein [TP]	3.267074401	0.001096691
#Sodium [Na]	-3.186152893	0.001456804
Calcium [Ca]	-3.090133296	0.002019238
Absolute value of basophils	-3.07824694	0.002094185
Mean platelet volume	-3.042358786	0.002365667
#Chlorine [Cl]	-2.88658788	0.003922595
#Anti hepatitis C virus antibody [HCV Ab]	-2.744812815	0.006074757
Globulin [Glb]	2.744524257	0.006089657
Absolute value of lymphocytes	2.675754759	0.007491823
Direct bilirubin [DB]	2.368488131	0.017951745
MUCUS	2.070749359	0.038470115
Percentage of nucleated red blood cells	1.971764759	0.048740517

continued

Table 2. continued

Numerical Features	Z statistic	P-value
RBC distribution width SD	1.944830077	0.051877991
RBC distribution width CV	1.832564678	0.066955665
Epithelial cells (EC)	1.75688779	0.079047189
White globulin ratio [A/G]	1.726195386	0.084397332
Urea/creatinine ratio [urea/Cr]	1.704241246	0.088429314
#Potassium [K]	-1.665604518	0.095894524
#Alanine aminotransferase [ALT]	-1.532740441	0.125417528
CAST	-1.477536342	0.139588785
Phosphorus [P]	-1.471301526	0.141288247
Platelet hematocrit	-1.410221265	0.15857348
Percentage of basophils	1.374996011	0.169208608
#Hemoglobin	-1.289730712	0.197227229
Crystallization (XTAL)	1.161317838	0.24565248
Absolute value of nucleated red blood cells	1.157267941	0.247267135
Average hemoglobin concentration	-0.972263625	0.330986298
Red blood cells (RBC)	-0.939725573	0.347399139
Yeast like bacteria (YLC)	-0.873397947	0.382483834
Average hemoglobin content	-0.866446567	0.38630478
Absolute value of monocytes	-0.864344268	0.387466993
Magnesium [Mg]	-0.816490787	0.414273502
Carbon dioxide binding capacity [CO2-CP]	0.747474949	0.454833374
White blood cells (WBC)	0.685003544	0.493412812
Hematocrit	-0.640502548	0.521887191
#Red blood cell count	-0.591937502	0.553930128
#Hepatitis B virus surface antigen [HBsAg]	-0.482800911	0.629266958
Hypersensitivity C-reactive protein	0.184316516	0.853776989
Average red blood cell volume	-0.133575801	0.893745734
Indirect bilirubin [IB]	0.09660348	0.923048141
Creatinine [Cr]	-0.057421729	0.954213301
#Platelet count	-0.022954961	0.981687664

We set p-value of$<$0.05 as having strong correlation. After the correlation analysis, a total of 78 characteristic variables with strong correlation with coronary heart disease were extracted, including patient gender, age, 12 symptoms, 10 related diseases, and 54 laboratory tests.

2.4 Model

In this part, we proposed EHR-former, a coronary heart disease diagnosis model for hypertensive patients based on Tranformer and contrastive learning. The model consists of four parts: embedding layer, semantic extraction layer, data augmentation layer, and diagnosis classification layer (see Fig. 1).

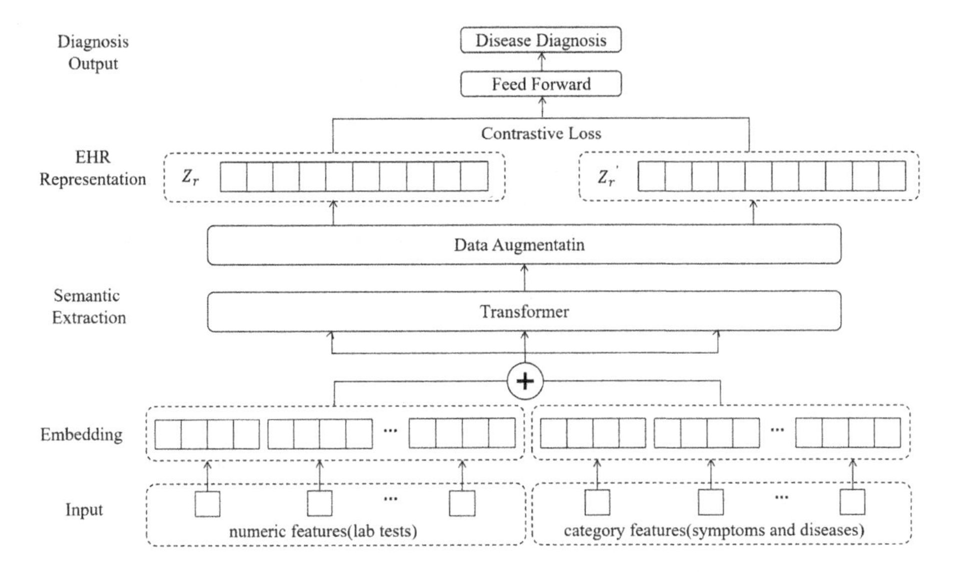

Fig. 1. EHR-former: a coronary heart disease diagnosis model for hypertensive patients based on Tranformer and contrastive learning

Embedding Layer. In the feature embedding layer, we aimed to map both numerical and categorical features to a fixed dimensional vector and embedded them into the same feature space. Here, we used different encoding methods to embed different types of data.

For the numerical features (test data), we first normalized the data and mapped it between [0,1]. Then, we used the Dense network to transform the data into model input features.

For categorical features (symptoms, concurrent diseases), we first converted the index of the feature categories into one hot encoding. In one hot encoding, only the data feature value of a certain dimension is 1, and the rest is 0. Next, we used the Dense network to further transform the vector into a fixed dimension.

After encoding the numerical and categorical features separately, we concatenated all feature vectors to obtain the original representation vector of the patient $z_r = (z_{numeric}, z_{category})$ as input to the model.

Semantic Extraction Layer. In the Semantic information extraction layer, we wanted to extract the semantic information of the original representation of patients' electronic health records, which can better represent the patients' health information.

There are many similarities between electronic health record data and natural language text. If we consider each patient's record as a natural language sentence, then each clinical data in it can be regarded as a word. Just as the complex grammatical structure of natural language determines the meaning of the sentences, these data features in a patient's health record largely determine the patient's current health status. Currently, the Transformer model has achieved better performance in almost all tasks in the field of natural language processing [14].

Therefore, we used the multi-head self attention mechanism in the Transformer model to extract the semantic information in the original health record representation z_r, which realized the fusion and extraction of the record data, and obtained high-dimensional vector representation. Attention here is the method of mapping a query (Q) and key value pair (K-V) to the output, where Q, K, and V are vectors, and the calculation is as formula (1).

$$Attention(K, Q, V) = softmax(\frac{QK^T}{\sqrt{d_k}})V \tag{1}$$

Here, d_k the dimension of Q, K, or V. Dividing by d_k is to prevent the product from being too large, causing the softmax function to enter the saturation region, thereby causing the gradient to be too small.

Data Augmentation Layer. Inspired by unsupervised contrastive learning [15,16], we used droupout to add noise to the data. Dropout is the process of randomly dropping connections of some neurons in a neural network. Due to the different connections of neurons, the final output of the model also varies. We encoded the same health record in the Transformer layer and input it twice into the dropout layer to obtain two different feature vectors, namely a positive case pair. The negative case pair represents other health records in the same batch, as shown in the Fig. 2.

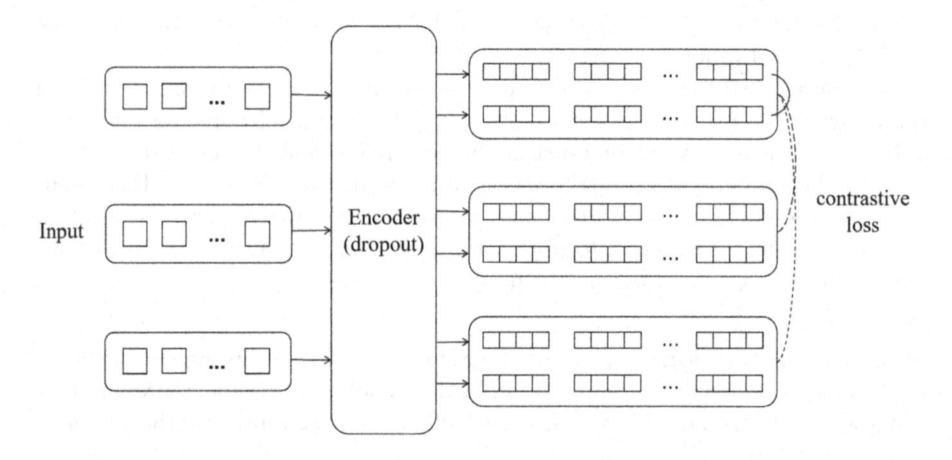

Fig. 2. Contrastive learning

The contrastive learning method usually optimizes the contrastive learning loss function to make the distance between positive cases in the feature space closer and the distance between negative cases farther. In this paper, we used cosine similarity to calculate the similarity between samples, as shown in equation (2). The similarity between positive sample pairs was taken as the numerator in the loss function, and the denominator was defined as the sum of the similarity between positive cases and all negative cases. Finally, the contrastive loss function was defined as formula (3).

$$sim(u, v) = u^T v / \|u\| \, \|v\| \tag{2}$$

$$l_{i,j} = -log \frac{exp(sim(z_i, z_j)/\tau)}{\sum_{k=1}^{2N} exp(sim(z_i, z_j)/\tau)} \tag{3}$$

Among them, z_i and z_j are the representation vectors of positive sample pairs, and τ represents the temperature parameter, whose size can control the model's attention to difficult samples. The final comparative loss $L_{contrast}$ can be expressed as the arithmetic mean value of all positive samples to the loss in a batch, as shown in formula (4).

$$L_{contrast} = \frac{1}{2N} \sum_{k=1}^{N} [l(2k-1, 2k) + l(2k, 2k-1)] \tag{4}$$

Diagnosis Classification Layer. Finally, the representation vector z_r of the patient record was fed into the classification layer to determine whether the patient has a certain disease. For the binary task of disease diagnosis, the final output layer was defined as equation(5):

$$O = \sigma(W z_r + b) \tag{5}$$

where, W matrix and b vector are trainable parameters, and σ represents sigmoid activation function.

In this paper, the loss of disease classification was defined as the loss of binary cross entropy, which is used to calculate the gap between the real label and the predicted label. See formula (6):

$$L_{cross\ entropy} = -[y \log O + (1-y) \log (1-O)] \tag{6}$$

Among them, $y \in \{0, 1\}$ is a binary label for disease diagnosis.

3 Experiments and Results

3.1 The Problem of Imbalanced Categories

After strict data preprocessing, we integrated 5583 records of hypertension patients, of which 3788 diagnoses included coronary heart disease, accounting for 67.8%. So it is necessary to address the issue of imbalanced sample categories before modeling.

Here, we used the SMOTE (Synthetic Minority Oversampling Technique) [17] method to upsample the samples. In the end, this article obtained 7576 patient records, of which 3788 were labeled as having coronary heart disease and 3788 were labeled as not having coronary heart disease.

3.2 Experimental Setup

We set the batch size to 64, and randomly divided all data into a training set(N=6016) and a testing set(N=1472) in a 4:1 ratio, discarding the data less than one batch size. The label distribution of the training set and testing set was shown in Table 3.

Table 3. Label distribution

	Not having coronary heart disease	Having coronary heart disease
Training set(N=6016)	3016	3000
Testing set(N=1472)	731	741

For missing features in the data, we used a 0 value to fill in. And during the model training process, we used the Adam optimizer and learning rate decay.

3.3 Results

We compared different deep learning methods such as CNN, RNN, LSTM, Transformer, etc. for extracting health record representations and disease diagnosis with the EHR_Former model proposed in this article. The diagnostic results of coronary heart disease on the test set were shown in Table 4.

Table 4. Test results of coronary heart disease diagnosis

Deep learning methods	Accuracy	Precision	Recall	F1	AUC
CNN	0.8383	0.8360	0.8428	0.8394	0.9166
RNN	0.8329	0.8320	0.8387	0.8353	0.9128
LSTM	0.8492	0.8403	0.8652	0.8526	0.9269
Transformer	0.8505	0.8529	0.8482	0.8505	0.9345
EHR_Former(Transformer+Contrastive learning)	0.8628	0.8664	0.8617	0.8641	0.9349

It can be seen that the Transformer model had better feature extraction performance, and after adding contrastive learning, the EHR_Former performed the best, with an AUC value of 0.9349. This proved the effectiveness of the coronary heart disease diagnosis model based on Transformer and contrastive learning.

4 Conclusion and Future Work

This article used deep learning methods to achieve intelligent diagnosis of coronary heart disease based on electronic health record data of hypertensive patients.

Firstly, we preprocessed the electronic health record data to solve the problems of unstructured, sparse, high-dimensional, and inconsistency in the original data. This improved the usability of the electronic health record data and laied the foundation for subsequent experiments and work.

Subsequently, we conducted feature selection through correlation analysis and determined the required feature variables for the coronary heart disease diagnosis model. These factors are likely to participate in the process of disease occurrence and can serve as biomarkers for early disease detection. So the screening and mining of these characteristic variables was very important for the prevention and diagnosis of coronary heart disease in hypertensive patients.

Finally, we built a disease diagnosis model based on Transformer and contrastive learning. And the final AUC value in the diagnosis task of coronary heart disease reached 0.9349, surpassing the model characterized only by CNN, RNN, LSTM, and Transformer.

Electronic health records are essential medical data in research. Currently, this article has achieved some results in feature extraction of electronic health records and their application in disease diagnosis research. However, there are still many problems worth further research. The prospects for future work are as follows:

(1) Due to the sparse data of some variables in the dataset of this study, potential cardiovascular risk factors such as diastolic blood pressure, body mass index, blood glucose values, etc., were not included in the model. Therefore, the selected feature variables in this study have certain bias and incompleteness. If this missing important variable data can be supplemented in future samples, the results of the model will definitely be greatly improved.

(2) In the future, with the continuous improvement and accumulation of electronic health record data, the coronary heart disease diagnosis model for hypertensive patients can be extended to more comprehensive chronic disease prevention and diagnosis, and a more comprehensive chronic disease management system for hypertensive patients can be established.

Acknowledgements. This work was supported in part by the Research and Development Program of Shenzhen under Grant KCXFZ202002011010487 and Grant WDZC20200818121348001. We would like to thank Peking University Shenzhen Hospital for their cooperation and help in electronic health record data extraction.

References

1. Shickel, B., Tighe, P.J., Bihorac, A., et al.: Deep EHR: a survey of recent advances in deep learning techniques for electronic health record (EHR) analysis. IEEE J. Biomed. Health Inform. **22**(5), 1589–1604 (2017)
2. Miotto, R., Wang, F., Wang, S., et al.: Deep learning for healthcare: review, opportunities and challenges. Brief. Bioinform. **19**(6), 1236–1246 (2018)
3. Zhang, J., Kowsari, K., Boukhechba, M., et al.: Sparse longitudinal representations of electronic health record data for the early detection of chronic kidney disease in diabetic patients. In: 2020 IEEE International Conference on Bioinformatics and Biomedicine (BIBM), pp. 885–892. IEEE (2020)

4. Jensen, P.B., Jensen, L.J., Brunak, S.: Mining electronic health records: towards better research applications and clinical care. Nat. Rev. Genet. **13**(6), 395–405 (2012)
5. Xie, F., Yuan, H., Ning, Y., et al.: Deep learning for temporal data representation in electronic health records: a systematic review of challenges and methodologies. J. Biomed. Inform. **126**, 103980 (2022)
6. Si, Y., Du, J., Li, Z., et al.: Deep representation learning of patient data from electronic health records (EHR): a systematic review. J. Biomed. Inform. **115**, 103671 (2021)
7. Rajkomar, A., Oren, E., Chen, K., et al.: Scalable and accurate deep learning with electronic health records. NPJ Digit. Med. **1**(1), 18 (2018)
8. Xiao, C., Choi, E., Sun, J.: Opportunities and challenges in developing deep learning models using electronic health records data: a systematic review. J. Am. Med. Inform. Assoc. **25**(10), 1419–1428 (2018)
9. Miotto, R., Li, L., Kidd, B.A., et al.: Deep patient: an unsupervised representation to predict the future of patients from the electronic health records. Sci. Rep. **6**(1), 1–10 (2016)
10. Wu, J., Roy, J., Stewart, W.F.: Prediction modeling using EHR data: challenges, strategies, and a comparison of machine learning approaches. Med. Care **48**, S106–S113 (2010)
11. Du, Z., Yang, Y., Zheng, J., et al.: Accurate prediction of coronary heart disease for patients with hypertension from electronic health records with big data and machine-learning methods: model development and performance evaluation. JMIR Med. Inform. **8**(7), e17257 (2020)
12. Nguyen, P., Tran, T., Wickramasinghe, N., et al.: Deepr: a convolutional net for medical records. arXiv:1607.07519 (2016)
13. Choi, E., Bahadori, M.T., Schuetz, A., et al.: Doctor AI: predicting clinical events via recurrent neural networks. In: Machine Learning for Healthcare Conference, pp. 301–318. PMLR (2016)
14. Vaswani, A., Shazeer, N., Parmar, N., et al.: Attention is all you need. In: Advances in Neural Information Processing Systems, vol. 30 (2017)
15. Chen, T., Kornblith, S., Norouzi, M., et al.: A simple framework for contrastive learning of visual representations. In: ICML 2020: Proceedings of the 37th International Conference on Machine Learning, pp. 1597–1607 (2020)
16. Gao, T., Yao, X., Chen, D.: SimCSE: simple contrastive learning of sentence embeddings. arXiv preprint arXiv:2104.08821 (2021)
17. Chawla, N.V., Bowyer, K.W., Hall, L.O., et al.: SMOTE: synthetic minority oversampling technique. J. Artif. Intell. Res. **16**, 321–357 (2002)

Learning Adaptable Risk-Sensitive Policies to Coordinate in Multi-agent General-Sum Games

Ziyi Liu(✉) and Yongchun Fang

College of Artificial Intelligence, NanKai University, Tianjin, China
liuzy@mail.nankai.edu.cn, fangyc@nankai.edu.cn

Abstract. In general-sum games, the interaction of self-interested learning agents commonly leads to socially worse outcomes, such as defect-defect in the iterated stag hunt (ISH). Previous works address this challenge by sharing rewards or shaping their opponents' learning process, which require too strong assumptions. In this paper, we observe that agents trained to optimize expected returns are more likely to choose a safe action that leads to guaranteed but lower rewards. To overcome this, we present Adaptable Risk-Sensitive Policy (ARSP). ARSP learns the distributions over agent's return and estimates a dynamic risk-seeking bonus to discover risky coordination strategies. Furthermore, to avoid overfitting training opponents, ARSP learns an auxiliary opponent modeling task to infer opponents' types and dynamically alter corresponding strategies during execution. Extensive experiments show that ARSP agents can achieve stable coordination during training and adapt to non-cooperative opponents during execution, outperforming a set of baselines by a large margin.

Keywords: Distributional Reinforcement Learning · Policy Adaptation · General-Sum Games · Multi-Agent Systems

1 Introduction

Most existing works in multi-agent reinforcement learning (MARL) focus on the fully cooperative [15,18] and competitive [17,21] settings. However, these settings only represent a fraction of potential real-world multi-agent environments. General-sum games, in which multiple self-interested learning agents optimize their own rewards independently and win-win outcomes are only achieved through coordination [12], describe many domains such as self-driving cars [20] and human-robot interactions [16].

Coordination is often coupled with risk. There are many scenarios where there is a safe action that leads to guaranteed but lower rewards, and a risky action that leads to higher rewards only if agents cooperate, such as alleviating traffic congestion [10], sustainably sharing limited resources [8], and altruistic human-robot interaction [16]. This kind of multi-agent coordination problem presents

B. Luo et al. (Eds.): ICONIP 2023, LNCS 14447, pp. 27–40, 2024.
https://doi.org/10.1007/978-981-99-8079-6_3

unique challenges that are not presented in single-agent and fully cooperative multi-agent reinforcement learning [24], and simply applying RL algorithms to train self-interested agents typically converge on unconditional mutual defection, which is the globally worst outcome [19]. Furthermore, the pre-trained cooperative agent in general-sum games should have the ability to adapt to non-cooperative opponents during execution in order to get a guaranteed reward.

To avoid such catastrophic outcomes, one set of approaches use explicit reward shaping to force agents to be prosocial, such as by making agents care about rewards of their partners [14,19], which can be viewed as shaping the risk degree of coordination strategies. However, this requires the strong assumption that agents involved are altruistic and can access other agents' reward function. Other works either treat partners as stationary [13] or take into account their learning steps in order to shape their policy [4]. By contrast, we focus on general settings where multiple decentralized, separately-controlled, and partially-observable agents interact in the environment and only care about maximizing their own rewards - while the objective is still to increase the probability of coordination. Furthermore, pre-trained cooperative agents should adapt to non-cooperative opponents during execution to avoid being exploited.

In this paper, we propose Adaptable Risk-Sensitive Policy (ARSP), which trains agents to achieve stable coordination during training, and can adapt to non-cooperative opponents during execution. To discover coordination strategies in general-sum games, ARSP first estimates a dynamic risk-seeking bonus which makes the agent be more risk tolerance and pursuing the maximum potential payoff. Next, ARSP learns an auxiliary self-supervised opponent modeling objective to assist decision making utilizing opponent's history behaviors. During execution, the agent can not access rewards but we can still optimize the opponent modeling objective to update it's policy, thus adapting to non-cooperative opponents during execution.

We evaluate ARSP in four different Markov Games: Iterated Stag Hunt (ISH) [23], Iterated Prisoners' Dilemma (IPD) [4,24], Monster-Hunt [19,27] and Escalation [14]. Compared with baseline methods, ARSP agents learn substantially faster, achieve stable mutual coordination during training, and can adapt to non-cooperative opponents during execution. To the best of our knowledge, ARSP is the first decentralized method that can achieve mutual coordination in IPD and ISH and adapt to non-cooperative opponents during execution.

2 Background

2.1 Stochastic Games

In this work, we consider multiple self-interested learning agents interacting with each other. We model the problem as a Partially-Observable Stochastic Game (POSG) [5], which consists of N agents, a state space $\mathcal{S}$ describing the possible configurations of all agents, a set of actions $\mathcal{A}^1, \ldots, \mathcal{A}^N$ and a set of observations $\mathcal{O}^1, \ldots, \mathcal{O}^N$ for each agent. At each time step, the agent i receives its own observation $o^i \in \mathcal{O}^i$, and selects an action $a^i \in \mathcal{A}^i$ based on a stochastic policy

$\pi^i : \mathcal{O}^i \times \mathcal{A}^i \mapsto [0, 1]$, which results in a joint action vector $\boldsymbol{a}$. The environment then produces a new state s' based on the transition function $P(s'|s, \boldsymbol{a})$. Each agent i obtains rewards as a function of the state and its action $R^i : \mathcal{S} \times \mathcal{A}^i \mapsto \mathbb{R}$. The initial states are determined by the distribution $\rho : \mathcal{S} \mapsto [0, 1]$. We treat the reward "function" R^i as a random variable to emphasize its stochasticity and use $Z^{\pi^i}(s, a^i) = \sum_{t=0}^{T} \gamma^t R^i(s_t, a_t^i)$ to denote the random variable of the cumulative discounted rewards, where $S_0 = s$, $A_0^i = a^i$,γ is a discount factor and T is the time horizon.

2.2 Distorted Expectation

Distorted expectation is a risk-weighted expectation of value distribution under a specific distortion function [25]. A function $g : [0, 1] \mapsto [0, 1]$ is a distortion function if it is non-decreasing and satisfies $g(0) = 0$ and $g(1) = 1$ [1]. The distorted expectation of Z under g is defined as $\Psi(Z) = \int_0^1 F_Z^{-1}(\tau) dg(\tau) = \int_0^1 g'(\tau) F_Z^{-1}(\tau) d\tau$, where F_Z^{-1} is the quantile function at $\tau \in [0, 1]$ for the random variable Z. We introduce two common distortion functions as follows:

- **CVaR** is the expectation of the lower or upper tail of the value distribution, corresponding to risk-averse or risk-seeking policy respectively. Its distortion function is $g(\tau) = \min(\tau/\alpha, 1)$ (risk-averse) or $\max(0, 1 - (1 - \tau)/\alpha)$ (risk-seeking), $\alpha \in (0, 1)$ denotes confidence level.
- **WT** is proposed by Wang [22]: $g_\lambda(\tau) = \Phi\left(\Phi^{-1}(\tau) + \lambda\right)$, where Φ is the distribution of a standard normal. The parameter λ is called the market price of risk and reflects systematic risk. $\lambda > 0$ for risk-averse and $\lambda < 0$ for risk-seeking.

CVaR_α assigns a 0-value to all percentiles below the α or above $1-\alpha$ significance level which leads to erroneous decisions in some cases [1]. Instead, WT is a complete distortion risk measure and ensures using all the information in the original loss distribution which makes training much more stable, and we will empirically demonstrate it in Sect. 4.

3 Methods

3.1 Motivation

Considering a classical matrix game in the game theory: Stag Hunt (SH). A group of hunters are tracking a big stag silently; now a hare shows up, each hunter should decide whether to keep tracking the stag or kill the hare immediately. This leads to the 2-by-2 matrix-form stag-hunt game in Table 1 with two actions for each agent, Stag (C) and Hare (D). There exists two pure strategy Nash Equilibrium (NE): the Stag NE, where both agents choose C and receive a high payoff a (e.g., a = 2), and the Hare NE, where both agents choose D and receive a lower payoff d (e.g., d = 1). The Stag NE is "risky" because if one agent defects, they still receive a decent reward b (e.g., b = 1) for eating the hare alone while the other agent with an S action may suffer from a big loss c for being hungry (e.g., c = −10). [19] demonstrates that when the risk is high, the probability that the Policy Gradient (PG) method converges to the Stag NE is low. It is noteworthy that the PG aims to maximize one's expected cumulative returns and ignores the complete information of the payoff distribution, e.g., the upper and lower tail information when the return distribution is asymmetric.

Figure 1 left part shows the quantile value distribution of hunting stag (Coop) and hunting hare (Defect) in the ISH learned by a risk-neutral policy. The expected return of Defect is higher than that of Coop, but the Coop distribution has a longer upper tail meaning that it has a higher potential payoff. However, the expectation unable to express this information. Figure 1 right part shows the WT weighted quantile distribution learned by a risk-seeking policy. The risk-seeking policy gives the upper tail higher weight when computing distorted expectations, making the agent forward-looking and pursuing the highest payoff.

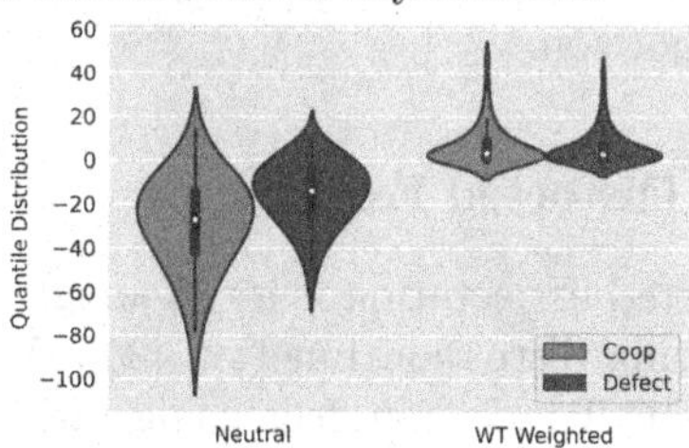

Fig. 1. Quantile value distribution of cooperation and defection in ISH weighted by WT compared with risk-neutral policy.

3.2 Risk-Seeking Bonus

Based on analysis above, we propose to use WT distortion function to weight the expectation of quantile distribution. By following [3], we first represent the return distribution of each agent i with policy π^i by a uniform mix of M supporting quantiles:

$$Z_\theta^{\pi^i}(o^i, a^i) \doteq \frac{1}{M}\sum_{k=1}^{M} \delta_{\theta_k^{\pi^i}(o^i,a^i)} \tag{1}$$

where δ_x denotes a Dirac Delta function at $x \in \mathbb{R}$, and each $\theta_k^{\pi^i}$ is an estimation of the quantile value corresponding to the quantile fraction $\hat{\tau}_k \doteq \frac{\tau_{k-1}+\tau_k}{2}$ with $\tau_k \doteq \frac{k}{M}$ for $0 < k \leq M$. The state-action value $Q^{\pi^i}(o^i, a^i)$ can then be approximated by $\frac{1}{M}\sum_{k=1}^{M} \theta_k^{\pi^i}(o^i, a^i)$.

Furthermore, we propose the risk-seeking bonus Ψ defined as:

$$\Psi(Z_\theta^{\pi^i}) = \int_0^1 g'_\lambda(\tau) F^{-1}_{Z_\theta^{\pi^i}}(\tau) d\tau \approx \frac{1}{M}\sum_{k=1}^{M} g'_\lambda(\hat{\tau}_k)\theta_k^i, \tag{2}$$

where $g'_\lambda(\tau)$ is the derivative of WT distortion function at $\tau \in [0, 1]$, and λ controls the risk-seeking level. Figure 1 right part shows the WT weighted quantile distribution in which the upper tail is extended and the lower tail is shrinked.

A naive approach for exploration would be to use the variance of the estimated distribution as a bonus. Specifically, the Right Truncated Variance tells about lower tail variability and the Left Truncated Variance tells about upper tail variability. For instantiating optimism in the face of uncertainty, the upper tail variability is more relevant than the lower tail, especially if the estimated distribution is asymmetric. So we adopt the Left Truncated Variance of quantile distribution for efficient exploration. The left truncated variance is defined as

$$\sigma_+^2 = \frac{1}{2M} \sum_{j=\frac{M}{2}}^{M} \left(\theta_{\frac{M}{2}} - \theta_j \right)^2 . \tag{3}$$

The index starts from the median, i.e., $M/2$, rather than the mean due to its well-known statistical robustness [7]. We anneal the two exploration bonuses dynamically so that in the end we produce unbiased policies. The anneal coefficients are defined as $c_{tj} = c_j \sqrt{\frac{\log t}{t}}, j = 1, 2$ which is the parametric uncertainty decay rate [9], and c_j is a constant factor. This approach leads to choosing the action such that

$$a^{i*} = \arg\max_{a^i \in \mathcal{A}^i} \left(Q^{\pi^i}(o^i, a^i) + c_{t1} \Psi(Z^{\pi^i}(o^i, a^i)) + c_{t2} \sqrt{\sigma_+^2(o^i, a^i)} \right) \tag{4}$$

These quantile estimates are trained using the Huber [3,6] quantile regression loss. The loss of the quantile value network of each agent i at time step t is then given by

$$\mathcal{J}\left(o_t^i, a_t^i, r_t^i, o_{t+1}^i; \theta^i\right) = \frac{1}{M} \sum_{k=0}^{M-1} \sum_{j=0}^{M-1} \rho_{\hat{\tau}_k}^{\kappa}\left(\delta_{kj}^{ti}\right), \tag{5}$$

where $\delta_{kj}^{ti} \doteq r_t^i + \gamma \theta_j^i \left(o_{t+1}^i, \pi^i\left(o_{t+1}^i\right)\right) - \theta_k^i(o_t^i, a_t^i)$, and $\rho_{\hat{\tau}_k}^{\kappa}(x) \doteq |\hat{\tau}_k - \mathbb{I}\{x < 0\}| \frac{\mathcal{L}_\kappa(x)}{\kappa}$ where $\mathbb{I}$ is the indicator function and $\mathcal{L}_\kappa(x)$ is the Huber loss:

$$\mathcal{L}_\kappa(x) \doteq \begin{cases} \frac{1}{2}x^2 & \text{if } x \leq \kappa \\ \kappa\left(|x| - \frac{1}{2}\kappa\right) & \text{otherwise} \end{cases} \tag{6}$$

3.3 Auxiliary Opponent Modeling Task

In order to alter the agent's strategies under different opponents, we share parameters between policy and auxiliary opponent modeling task. Specifically, we split the Q value network into two parts: feature extractor $\mathcal{E}_\phi$ and decision maker $\mathcal{D}_{\psi_a}$. The parameters of the Q value network Q_{θ^i} for agent i are sequentially divided into ϕ^i and ψ_a^i, i.e., $\theta^i = (\phi^i, \psi_a^i)$. The auxiliary opponent modeling task shares a common feature extractor $\mathcal{E}_{\phi^i}$ with the value network. We can update the parameters of $\mathcal{E}_{\phi^i}$ during execution using gradients from the auxiliary opponent modeling task, such that π_{θ^i} can generalize to different opponents. The

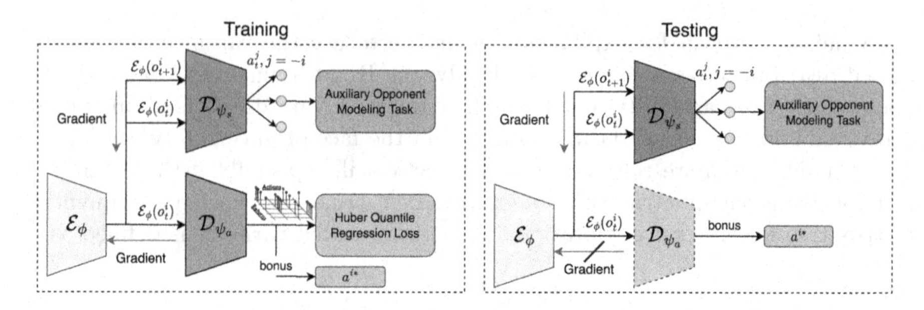

Fig. 2. Left: Diagram of ARSP architecture during training. Outputs of $\mathcal{E}_\phi$ are fed into $\mathcal{D}_{\psi_a}$ and $\mathcal{D}_{\psi_s}$, so features are shared between policy and auxiliary opponent modeling. The prediction head $\mathcal{D}_{\psi_s}$ outputs other agents' actions. **Right:** Test-Time policy adaptation. The agent can not receive environment rewards during testing, so we only optimize the auxiliary opponent modeling objective.

supervised prediction head and its specific parameters are $\mathcal{D}_{\psi_s^i}$ with ψ_s^i. The details of our network architecture are shown in Fig. 2.

During training, the agent i can collect a set of transitions $\{(o_t^i, o_{t+1}^i, \mathbf{a}_t^{-i})\}_{t=0}^T$ where $\mathbf{a}_t^{-i}$ indicates the joint actions of other agents except i at time step t. We use the embeddings of agent i's observations o_t^i and o_{t+1}^i to predict the joint actions $\mathbf{a}_t^{-i}$, i.e., the $\mathcal{D}_{\psi_s^i}$ is a multi-head neural network whose outputs are actions of other agents, and the objective function of the auxiliary opponent modeling task can be formulated as

$$\mathcal{L}\left(o_t^i, o_{t+1}^i, \mathbf{a}_t^{-i}; \phi^i, \psi_s^i\right) \tag{7}$$

$$= \frac{1}{N-1} \sum_{j=1, j\neq i}^{N} \ell\left(a_t^j, \mathcal{D}_{\psi_s^i}\left(\mathcal{E}_{\phi^i}\left(o_t^i\right), \mathcal{E}_{\phi^i}\left(o_{t+1}^i\right)\right)^j\right), \tag{8}$$

where $\ell(\cdot)$ is the cross-entropy loss for discrete action space or mean squared error for continuous action space. The strategies of opponents will change constantly during the procedure of multi-agent exploration and thus various strategies will emerge. The agent can leverage them to gain some experience about how to make the best response by jointly optimizing the auxiliary opponent modeling task and quantile value distribution. The joint training problem is therefore

$$\min_{\phi^i, \psi_s^i, \psi_a^i} \mathcal{J}\left(o_t^i, a_t^i, r_t^i, o_{t+1}^i; \phi^i, \psi_a^i\right) + \mathcal{L}\left(o_t^i, o_{t+1}^i, \mathbf{a}_t^{-i}; \phi^i, \psi_s^i\right) \tag{9}$$

During execution, we can not optimize $\mathcal{J}$ anymore since the reward is unavailable, but we assume the agent can observe actions made by its opponents during execution, then we can continue optimizing $\mathcal{L}$ to update parameters of the shared feature extractor $\mathcal{E}_\phi$. Learning from opponents' behaviors at test time makes the agent adapt to non-cooperative opponents efficiently. This can be formulated as

$$\min_{\phi^i, \psi_s^i} \mathcal{L}\left(o_t^i, o_{t+1}^i, \mathbf{a}_t^{-i}; \phi^i, \psi_s^i\right) \tag{10}$$

4 Experimental Setup

4.1 Environments

Iterated Games. We consider two kinds of iterated matrix games: Iterated Stag Hunt (ISH) and Iterated Prisoners' Dilemma (IPD) with a constant episode length of 10 time steps. At each time step, agents can choose either cooperation or defection. If both agents choose to cooperate simultaneously, each of them gets a bonus of 2. However, if only one agent chooses to cooperate, he gets a penalty of -10 in ISH and -1 in IPD, and the other agent gets a bonus of 1 and 3, respectively. If both agents choose to defect, they get a bonus of 1 in ISH and 0 in IPD. The one-step payoff matrix of the SH and PD is shown in Tables 1 and 2. The optimal strategy in ISH and IPD is to cooperate at each time step, and the highest global payoffs are 40, i.e., 20 for each of them.

Table 1. Payoff Matrix for the SH

	C	D
C	(2, 2)	(−10, 1)
D	(1, −10)	(1, 1)

Table 2. Payoff Matrix for the PD.

	C	D
C	(2, 2)	(−1, 3)
D	(3, −1)	(0, 0)

Monster-Hunt. The environment is a 5×5 grid world, consisting of two agents, two apples, and one monster. The apples are static while the monster keeps moving toward its closest agent. When a single agent meets the monster, he gets a penalty of -10. If two agents catch the monster together, they both get a bonus of 5. If a single agent meets an apple, he gets a bonus of 2. Whenever an apple is eaten or the monster meets an agent, the entity will respawn randomly. The optimal strategy, i.e., both agents move towards and catch the monster, is a risky coordination strategy since an agent will receive a penalty if the other agent deceives.

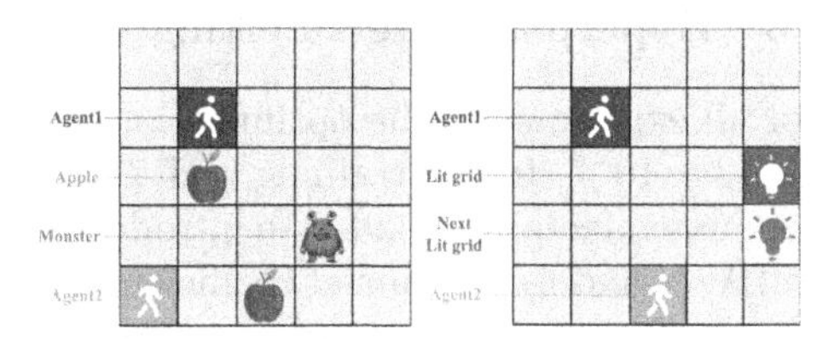

Fig. 3. Monster-Hunt.

Escalation. Escalation is a 5×5 grid-world with sparse rewards, consisting of two agents and a static light. If both agents step on the light simultaneously, they receive a bonus of 1, and then the light moves to a random adjacent grid. If only one agent steps on the light, he gets a penalty of $1.5L$, where L denotes the latest consecutive cooperation steps, and the light will respawn randomly. To maximize their individual payoffs and global rewards, agents must coordinate to stay together and step on the light grid. For each integer L, there is a corresponding coordination strategy where each agent follows the light for L steps and then simultaneously stops coordination.

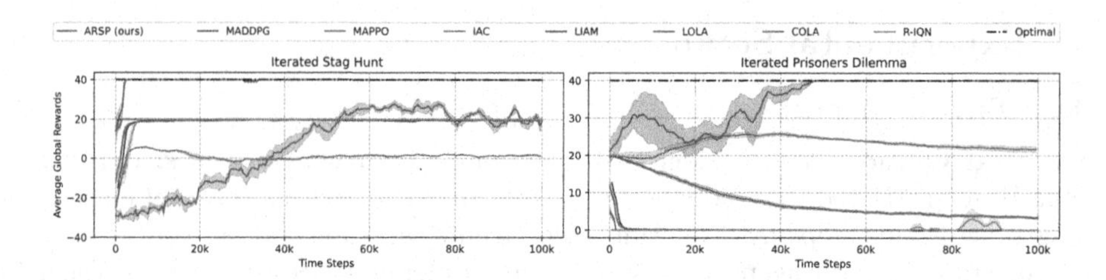

Fig. 4. Mean evaluation returns for ARSP, MADDPG, MAPPO, IAC, LIAM and LOLA on two iterated matrix games. The average global rewards equal to 40 means that all agents have learned coordination strategy, i.e., cooperating at each time step.

4.2 Baseline Comparisons

We compare our approach with several baselines in three categories: 1) centralized training and decentralized execution methods, including MADDPG [11] and MAPPO [26]. 2) fully decentralized learning methods, including Independent Actor-Critic (IAC) and Risk-Sensitive Implicit Quantile Networks (R-IQN) with risk-seeking policy [2]. 3) opponent modeling methods, including LIAM [13], LOLA [4] and COLA [24].

4.3 Hyperparameters Setting

For all experiments, the optimization is conducted using Adam with a learning rate 3×10^{-4} during training and 7×10^{-3} during policy adaptation, and the discounted factor γ is 0.99. We introduce three important hyperparameters: c_1, c_2 and λ. c_1 and c_2 are anneal coefficients for risk-seeking bonus and left truncated variance separately, and λ controls the risk-seeking level. We detail (c_1, c_2, λ) of our methods in Table 3.

Table 3. Hyperparameters setting of ARSP for all tasks. $*$ denotes the hyperparameter used during execution.

Hyperparameters	ISH	IPD	MH	Escalation
c_1	20	10	80	10
c_1^*	5	5	5	5
c_2	50	100	80	100
c_2^*	0	0	0	0
λ	-0.9	-1.0	-0.9	-1.75
λ^*	0.2	0.2	-0.5	-0.5

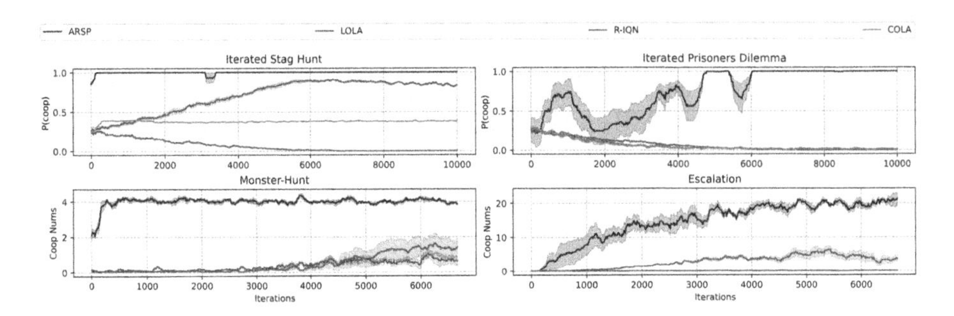

Fig. 5. The probability of agents cooperate with each other in ISH and IPD (first row) during training and the number of mutual cooperation steps in one episode in Monster-Hunt and Escalation (second row) during training.

5 Results

In this subsection, we evaluate all methods on four multi-agent environments, and the returns are averaged on 5 different initial seeds. We pause training every 50 episodes and run 30 independent episodes to evaluate the average performance of each method. We only compare with COLA in iterated games since it can not scale to high dimensional settings [24].

5.1 Iterated Games

Figure 4 shows the average global rewards, i.e., the summation of all agents' average returns evaluated in ISH and IPD. The shadowed part represents a 95% confidence interval. The average global rewards equal to 40 means that all agents' have discovered the coordination strategy, i.e., cooperating at each time step. ARSP significantly outperforms all baseline methods in IPD and ISH. Notably, it is the only algorithm that discovers the optimal coordination strategy with high sample efficiency in two iterated games. To our best knowledge, ARSP is the first method to achieve this result. COLA outperforms other baselines in IPD but can not learn any meaningful strategies in ISH since it is designed explicitly for prisoner's dilemma games. R-IQN is a risk-seeking method similar to ARSP but failed to discover coordination strategies. We suspect that this happens because R-IQN implicitly achieves risk-seeking by sampling quantile fractions from the distorted distribution, lacking of a particular mechanism to guide agents to discover coordination strategies. In contrast, ARSP explicitly constructs risk-seeking exploration bonuses to encourage agents to discover coordination strategies more efficiently. The first row of Fig. 5 shows the probability of agents cooperating with each other in ISH and IPD during training. ARSP agents can converge to coordination strategies efficiently and stably in probability one. Although LOLA and COLA agents can discover coordination strategies, they can not achieve stable coordination, leading to much more bigger losses and lower global rewards.

5.2 Grid-Worlds

We further show the effectiveness of ARSP in two high-dimensional grid-world games - Monster Hunt (MH) and Escalation [19]. Both of them have multiple NEs with different payoffs.

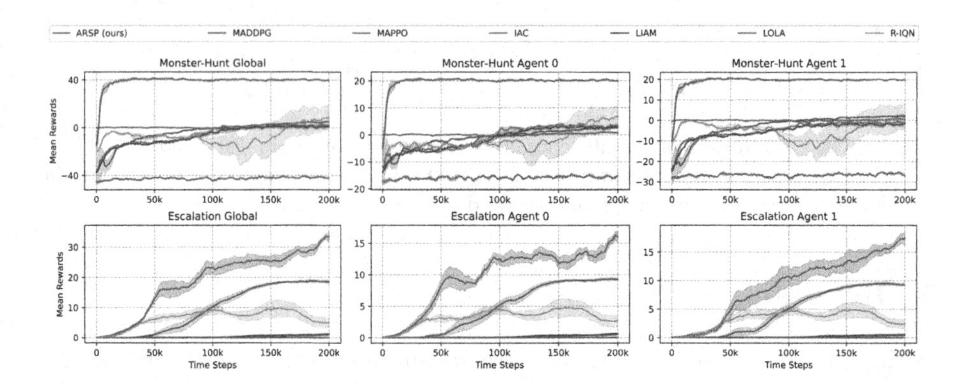

Fig. 6. Mean evaluation returns for ARSP, MADDPG, MAPPO, IAC, LIAM and LOLA on Monster-Hunt and Escalation. Global rewards are summation of both agents rewards.

The first row of Fig. 6 is the evaluation results of ARSP and other baselines in MH. In MH, ARSP agents can rapidly discover the high payoff NE where all agents stay together and wait for the monster. Besides, ARSP can converge to the coordination strategy in less than 25k time steps, which shows the superiority of our method in sample efficiency. Most other baseline methods can only converge to some guaranteed but lower payoff NEs, i.e., elude the stag and eat apples alone. Though LOLA agents can achieve around one-time cooperation as shown in 5, most of the time they suffer from punishment because of the opponent's defection. We suspect that this happens because LOLA agents assume their opponents are naive learners, leading to unstable coordination. R-IQN performs better than other baselines in MH, demonstrating that the risk-seeking policy plays an important role in discovering coordination strategies. Similar results are also achieved in the Escalation environment, where ARSP significantly outperforms other baselines both in asymptotic performance and sample efficiency. Although with the increase of cooperation times, the punishment suffered by agents for betrayal is also increasing in the Escalation, ARSP agents can still achieve stable coordination. MADDPG agents perform well in Escalation since they can observe others' policies, i.e., observations and actions, and cooperation is the only source of rewards. While MAPPO agents only have access to opponents' observations. ARSP does not make these assumptions but still achieves more cooperation times, as shown in Fig. 5.

5.3 Adaptation Study

This subsection investigates the ability of the pre-trained ARSP agent to adapt to different opponents during execution. The cooperative opponents are trained by the ARSP method because ARSP is the only method that can produce cooperative strategies, and non-cooperative opponents are MADDPG agents. During the evaluation, random seeds of four environments are different from that during training, and hyperparameters of the ARSP, e.g., risk-seeking level λ, are same and fixed between different opponents. Furthermore, the pre-trained agent can no longer use environment rewards to update its policy and it must utilize the auxiliary opponent modeling task to adapt to different opponents.

Table 4. Mean evaluation return of GRSP with and without auxiliary opponent modeling task on four multi-agent environments.

Oppo: Coop(Defect)	ISH	IPD	M-H	Escalation
GRSP-No-Aom	20(−100)	20(−5)	20.62(−15.03)	9.45(−0.545)
GRSP-Aom	**20(0.65)**	**20(−1.08)**	**21.36(−12.07)**	**11.3(0.175)**

Table 4 shows the mean evaluation return of the ARSP agent with and without auxiliary opponent modeling (Aom) task on four multi-agent environments when interacting with different opponents. All returns are averaged on 100 episodes. ARSP-No-Aom means that the agent is trained without Aom task. CC indicates the opponent is a cooperative agent while CD means the opponent will defect. In ISH and IPD, the ARSP agent chooses to cooperate at the first time step. After observing its opponent's decision, it will choose to keep cooperating or defecting, depending on its opponent's strategies. So the ARSP agent can avoid being exploited by its opponent and gets guaranteed rewards, while ARSP-No-Aom can not, as shown in Table 4. Similar results also appear in M-H and Escalation. Furthermore, when interacting with cooperative opponents, ARSP agent can adapt to their policies and receive higher individual rewards than ARSP-No-Aom, e.g., 21.36 in M-H and 11.3 in Escalation. The experiment results also demonstrate that policies learned independently can overfit other agents' policies during training, and our Aom method provides a way to tackle this problem.

5.4 Ablations

In this subsection, we perform an ablation study to examine the components of ARSP to better understand our method. ARSP is based on QR-DQN and has three components: the risk-seeking exploration bonus, the left truncated variance (Tv) and the auxiliary opponent modeling task (Aom). We design and evaluate six different ablations of ARSP in two grid-world environments, as show in Fig. 7. The evaluation return of ARSP-No-Aom which we ablate the Aom module and

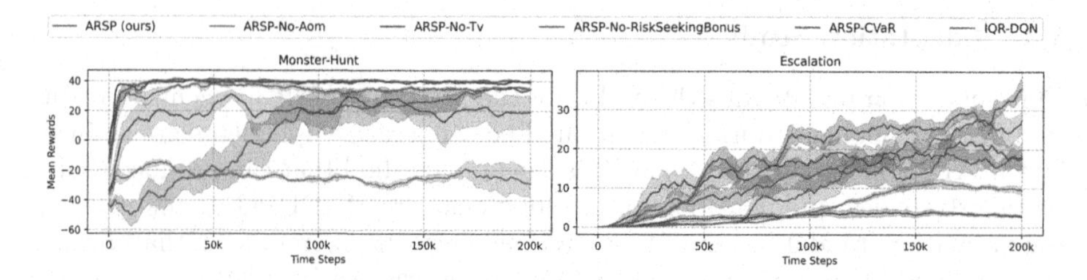

Fig. 7. Mean evaluation return of ARSP compared with other ablation methods in two grid-world multi-agent environments.

retain all other features of our method is a little lower than that of ARSP but has a much higher variance, indicating that learning from opponent's behaviors can stable training and improve performance. Moreover, the ARSP-No-Aom is a completely decentralized method whose training without any opponent information, and the ablation results of ARSP-No-Aom indicate that our risk-seeking bonus is the determining factor for agents to achieve mutual coordination in our experiments. We observe that ablating the left truncated variance module leads to a lower return than ARSP in the Escalation but no difference in the Monster-Hunt. Furthermore, ablating the risk-seeking bonus increases the training variance, leads to slower convergence and worse performance. It is noteworthy that the Escalation is a sparse reward and hard-exploration multi-agent environment because two decentralized agents can not get any reward until they navigate to and step on the light simultaneously and constantly. These two ablations indicate that the exploration ability of left truncated variance is important for our method and the risk-seeking bonus can encourage agents to coordinate with each other stably and converge to high-risky cooperation strategies efficiently. We also implement our risk-seeking bonus with CVaR instead of WT, and the results are shown as ARSP-CVaR. The ARSP-CVaR performs worse than our method and has a higher training variance. Finally, we ablate all components of the ARSP and use ϵ-greedy policy for exploration which leads to the IQR-DQN algorithm. As shown in Fig. 7, IQR-DQN can not learn effective policies both in the Monster-Hunt and the Escalation.

6 Conclusion

In this paper, we propose a novel decentralized learning method, Adaptable Risk-Sensitive Policy (ARSP), which can achieve mutual coordination in a class of general-sum games and adapt to non-cooperative opponents efficiently during execution. By estimating the risk-seeking exploration bonus, ARSP agents can efficiently converge to risky coordination strategies and receive socially optimal payoffs. To avoid overfitting training opponents, we propose the auxiliary opponent modeling task, which leverages the opponent's past behaviors to infer its strategies and alter the ego agent's policy dynamically during execution, thus adapting to non-cooperative opponents. Extensive experiments in iterated games and high-dimensional settings show that ARSP significantly outperforms other baseline methods, demonstrating the superiority and scalability of our method.

References

1. Balbás, A., Garrido, J., Mayoral, S.: Properties of distortion risk measures. Methodol. Comput. Appl. Probab. **11**(3), 385–399 (2009)
2. Dabney, W., Ostrovski, G., Silver, D., Munos, R.: Implicit quantile networks for distributional reinforcement learning. In: International Conference on Machine Learning, pp. 1096–1105. PMLR (2018)
3. Dabney, W., Rowland, M., Bellemare, M., Munos, R.: Distributional reinforcement learning with quantile regression. In: Proceedings of the AAAI Conference on Artificial Intelligence, vol. 32 (2018)
4. Foerster, J.N., Chen, R.Y., Al-Shedivat, M., Whiteson, S., Abbeel, P., Mordatch, I.: Learning with opponent-learning awareness. arXiv preprint arXiv:1709.04326 (2017)
5. Hansen, E.A., Bernstein, D.S., Zilberstein, S.: Dynamic programming for partially observable stochastic games. In: AAAI, vol. 4, pp. 709–715 (2004)
6. Huber, P.J.: Robust estimation of a location parameter. In: Kotz, S., Johnson, N.L. (eds.) Breakthroughs in Statistics. Springer Series in Statistics. Springer (1992). https://doi.org/10.1007/978-1-4612-4380-9_35
7. Huber, P.J.: Robust statistics. In: Lovric, M. (eds.) International Encyclopedia of Statistical Science. Springer, Heidelberg (2011). https://doi.org/10.1007/978-3-642-04898-2_594
8. Hughes, E., et al.: Inequity aversion improves cooperation in intertemporal social dilemmas. In: Advances in Neural Information Processing Systems 31 (2018)
9. Koenker, R., Hallock, K.F.: Quantile regression. J. Econ. Persp. **15**(4), 143–156 (2001)
10. Lazar, D.A., Bıyık, E., Sadigh, D., Pedarsani, R.: Learning how to dynamically route autonomous vehicles on shared roads. Transport. Res. Part C: Emerg. Technol. **130**, 103258 (2021)
11. Lowe, R., Wu, Y.I., Tamar, A., Harb, J., Pieter Abbeel, O., Mordatch, I.: Multi-agent actor-critic for mixed cooperative-competitive environments. In: Advances in Neural Information Processing Systems 30 (2017)
12. Matignon, L., Laurent, G.J., Le Fort-Piat, N.: Independent reinforcement learners in cooperative Markov games: a survey regarding coordination problems. Knowl. Eng. Rev. **27**(1), 1–31 (2012)
13. Papoudakis, G., Christianos, F., Albrecht, S.: Agent modelling under partial observability for deep reinforcement learning. In: Advances in Neural Information Processing Systems 34 (2021)
14. Peysakhovich, A., Lerer, A.: Prosocial learning agents solve generalized stag hunts better than selfish ones. arXiv preprint arXiv:1709.02865 (2017)
15. Rashid, T., Samvelyan, M., Schroeder, C., Farquhar, G., Foerster, J., Whiteson, S.: QMIX: monotonic value function factorisation for deep multi-agent reinforcement learning. In: International Conference on Machine Learning, pp. 4295–4304. PMLR (2018)
16. Shirado, H., Christakis, N.A.: Locally noisy autonomous agents improve global human coordination in network experiments. Nature **545**(7654), 370–374 (2017)

17. Silver, D., et al.: Mastering the game of go without human knowledge. Nature **550**(7676), 354–359 (2017)
18. Son, K., Kim, D., Kang, W.J., Hostallero, D.E., Yi, Y.: QTRAN: learning to factorize with transformation for cooperative multi-agent reinforcement learning. In: International Conference on Machine Learning, pp. 5887–5896. PMLR (2019)
19. Tang, Z., et al.: Discovering diverse multi-agent strategic behavior via reward randomization. arXiv preprint arXiv:2103.04564 (2021)
20. Toghi, B., Valiente, R., Sadigh, D., Pedarsani, R., Fallah, Y.P.: Social coordination and altruism in autonomous driving. arXiv preprint arXiv:2107.00200 (2021)
21. Vinyals, O., et al.: Grandmaster level in StarCraft II using multi-agent reinforcement learning. Nature **575**(7782), 350–354 (2019)
22. Wang, S.S.: A class of distortion operators for pricing financial and insurance risks. J. Risk Insurance **67**, 15–36 (2000)
23. Wang, W.Z., Beliaev, M., Bıyık, E., Lazar, D.A., Pedarsani, R., Sadigh, D.: Emergent prosociality in multi-agent games through gifting. arXiv preprint arXiv:2105.06593 (2021)
24. Willi, T., Letcher, A.H., Treutlein, J., Foerster, J.: COLA: consistent learning with opponent-learning awareness. In: International Conference on Machine Learning, pp. 23804–23831. PMLR (2022)
25. Wirch, J.L., Hardy, M.R.: Distortion risk measures: coherence and stochastic dominance. In: International congress on Insurance: Mathematics and Economics, pp. 15–17 (2001)
26. Yu, C., Velu, A., Vinitsky, E., Wang, Y., Bayen, A., Wu, Y.: The surprising effectiveness of PPO in cooperative, multi-agent games. arXiv preprint arXiv:2103.01955 (2021)
27. Zhou, Z., Fu, W., Zhang, B., Wu, Y.: Continuously discovering novel strategies via reward-switching policy optimization. In: Deep RL Workshop NeurIPS 2021 (2021)

Traffic Data Recovery and Outlier Detection Based on Non-negative Matrix Factorization and Truncated-Quadratic Loss Function

Linfang Yu[1,2], Hao Wang[1,2](✉), Yuxin He[3], and Yang Wen[1,2]

[1] Guangdong Multimedia Information Service Engineering Technology Research Center, Shenzhen University, Shenzhen, China

[2] Guangdong Key Laboratory of Intelligent Information Processing, Shenzhen, China
haowang@szu.edu.cn

[3] College of Urban Transportation and Logistics, Shenzhen Technology University, Shenzhen, China

Abstract. Intelligent Transportation System (ITS) plays a critical role in managing traffic flow and ensuring safe transportation. However, the presence of missing and corrupted traffic data may undermine the accuracy and reliability of the system. The problem of recovering traffic data can often be transformed into a low-rank matrix factorization problem by exploiting the intrinsic low-rank characteristics of the traffic matrix. While many existing methods demonstrate excellent recovery performance under the assumption of noiseless or Gaussian noise, they often exhibit suboptimal performance in the presence of outliers. In this paper, we propose a novel method for recovering traffic data using non-negative matrix factorization with a truncated-quadratic loss function. Although the objective function in our model is non-convex and non-smooth, we convert it to a convex formulation using half-quadratic theory. Then, a solver based on block coordinate descent is developed. Our experiments on real-world traffic datasets demonstrate superior performance compared to state-of-the-art methods.

Keywords: Traffic data recovery · Matrix factorization · Outliers

1 Inrodution

Traffic data collected from road networks plays a crucial role in Intelligent Transportation System (ITS), and the scale of traffic data has increased significantly with the rapid development of distributed sensor systems. However, due to sensor damage or network outages, traffic data is often missing or contaminated, making it difficult to process. Therefore, it is of great significance for ITS to achieve accurate and robust traffic data recovery through the observed data.

Traffic data is usually characterized by an inherently low-rank structure. Therefore, many researchers have focused on utilizing this attribute to recover

B. Luo et al. (Eds.): ICONIP 2023, LNCS 14447, pp. 41–52, 2024.
https://doi.org/10.1007/978-981-99-8079-6_4

missing entries. There are two general categories for traffic matrix completion based on the low-rank nature: rank minimization and matrix factorization. Rank minimization-based approaches often transform the traffic data recovery problem into a traffic matrix rank minimization problem. However, the rank minimization problem is NP-hard and is typically handled by seeking a surrogate function to replace the rank function through equivalent transformation or appropriate relaxation. The nuclear norm is a widely used convex surrogate for the rank function and involves iterative updating with full singular value decomposition (SVD), but it has limitations due to its expensive computation and sensitivity to parameter dependency. To avoid the full SVD, some non-convex functions have been proposed to serve as surrogates for the nuclear norm. In [1], a truncated nuclear norm (TNN) is used to further improve the recovery accuracy as well as reduce the computational cost. In [2], Schatten p-norm is proposed to replace the rank function with the slackness to algebraic rank controlled by the value of p. Combining the truncated nuclear norm and Schatten p-norm, a truncated tensor Schatten p-norm is designed in [3].

The main idea behind matrix factorization is to seek low-rank matrices of the incomplete data [4]. The sparsity regularized matrix factorization (SRMF) in [5] decomposes the incomplete matrix into low-rank matrices that may contain negative elements, and the spatial and temporal constraints on the reconstructed matrices are introduced in this approach. Unlike the former, the temporal and adaptive spatial constrained low rank (TAS-LR) method in [6] imposes spatial and temporal constraints on the low-rank matrices separately. The spatiotemporal non-negative matrix factorization (ST-NMF) method in [7] decomposes the incomplete matrix into non-negative matrices and introduces an ℓ_1-norm term. Joint low-rank and fundamental diagram constraints (JLFDC) [8] models the incomplete matrix as the sum of low-rank normal matrices and sparse noise matrices. While these methods demonstrate good accuracy in traffic data recovery under the assumption of noise-free or Gaussian noise, their performance degrades in the presence of outliers.

In this paper, we propose a novel approach for addressing the traffic data recovery problem. Specifically, we formulate it as a non-negative matrix factorization (NMF) problem and introduce spatial and temporal constraints to the factor matrices. A truncated-quadratic function is used to handle outliers. The contributions of this work are summarized as follows:

1. A truncated-quadratic function is introduced to mitigate the impact of noise and outliers.
2. With the truncated-quadratic function, our model is non-convex and non-smooth. The half-quadratic optimization theory (HO) [9] is leveraged to convert the objective function into an easy-to-solve form, and an efficient solver based on block coordinate descent (BCD) is developed [10].
3. The superior recovery performance of our model is demonstrated on real-world traffic datasets with variant missing scenarios and different levels of missing rates.

The rest of this paper is organized as follows. In Sect. 2, we provide notations and some background information. In Sect. 3, we develop the proposed model and present its numerical solution. In Sect. 4, we conduct the experiment on real-world datasets. The conclusions are drawn in Sect. 5.

2 Preliminaries

2.1 Traffic Matrices

The traffic state data of N road links during a period of M consecutive time intervals can be described by a traffic matrix $\mathbf{X} = [\mathbf{x}_1, \mathbf{x}_2, \ldots, \mathbf{x}_j, \ldots, \mathbf{x}_M] = [\mathbf{x}_1^\top; \mathbf{x}_2^\top; \ldots; \mathbf{x}_i^\top; \ldots; \mathbf{x}_N^\top]$, where the row $\mathbf{x}_i^\top$ represents the time series of ith road link, the column $\mathbf{x}_j$ denotes the road traffic status at jth time interval, and matrix element X_{ij} is the traffic status of ith road link at jth time interval. To distinguish between the observed data and missing data, the incomplete matrix $\mathbf{X}_{\boldsymbol{\Omega}}$ is

$$\mathbf{X}_{\boldsymbol{\Omega}} = \mathbf{X} \odot \boldsymbol{\Omega} = \begin{cases} X_{ij}, & \text{if } X_{ij} \text{ is available,} \\ 0, & \text{otherwise,} \end{cases} \tag{1}$$

where $\boldsymbol{\Omega} \in \mathbb{R}^{N \times M}$ denotes a binary position matrix consisting of 0 and 1.

2.2 Background

Traffic Matrix Completion: The traffic data recovery problem can be transformed into a traffic matrix completion problem based on NMF. In this context, the recovered matrix $\hat{\mathbf{X}}$ can be factorized into a basic matrix $\mathbf{W}$ and a coefficient matrix $\mathbf{H}$, such that $\hat{\mathbf{X}} = \mathbf{WH}$. The typical NMF is formulated using the Frobenius norm [11] as:

$$\min_{\mathbf{W},\mathbf{H}} \|\mathbf{X}_{\boldsymbol{\Omega}} - (\mathbf{WH})_{\boldsymbol{\Omega}}\|_F^2, \quad \text{s.t. } \mathbf{W} \geq 0, \mathbf{H} \geq 0. \tag{2}$$

Given the spatiotemporal nature of traffic data, the recovery performance can be improved by mining the correlations in the rows and columns of traffic matrix $\mathbf{X}_{\boldsymbol{\Omega}}$. One approach to achieving this is to utilize factor matrices that incorporate spatial correlation and temporal smoothing constraints, as proposed in [7]. The corresponding formulation is given by

$$\begin{aligned} \min_{\mathbf{W},\mathbf{H},\mathbf{A}} \ & \|\mathbf{X}_{\boldsymbol{\Omega}} - (\mathbf{WH})_{\boldsymbol{\Omega}} - \mathbf{A}_{\boldsymbol{\Omega}}\|_F^2 + \alpha_1 \operatorname{tr}(\mathbf{W}^\top \mathbf{L} \mathbf{W}) + \alpha_2 \|\mathbf{HT}\|_F^2 + \alpha_3 \|\mathbf{A}_{\boldsymbol{\Omega}}\|_1, \\ \text{s.t. } \ & \mathbf{W} \geq 0, \mathbf{H} \geq 0. \end{aligned} \tag{3}$$

where $\mathbf{L}$ and $\mathbf{T}$ represent the spatial and temporal constraint matrices respectively, α_1, α_2 and α_3 are trade-off parameters, and $\mathbf{A}_{\boldsymbol{\Omega}}$ denotes the outlies in the observation of traffic data. The ℓ_1-norm is used to alleviate the influence of anomalies. Since $\mathbf{L}$ denotes the Laplacian matrix of traffic matrix $\mathbf{X}_{\boldsymbol{\Omega}}$, the regularizaition $\operatorname{tr}(\mathbf{W}^\top \mathbf{L} \mathbf{W})$ can be used to measure the spatial similarity. $\mathbf{T}$ is a differential matrix which satisfies $T_{ii} = -1$ and $T_{(i+1)i} = 1$ for $i \in [1, M-1]$. The differential matrix $\mathbf{T}$ induces similarity between adjacent columns in $\mathbf{H}$, thereby conforming to the temporal smoothness of the actual traffic state.

Half-Quadratic Optimization: Given an objective function $\phi(x)$, there exist a quadratic function $g(x, y)$ and a convex *dual* function $\varphi(y)$ that holds the following equation:

$$\phi(x) = \inf_{y} g(x, y) + \varphi(y) = \inf_{y} \frac{(x-y)^2}{2} + \varphi(y) \tag{4}$$

where $y \in \mathbb{R}$ is auxilariy variable.

3 Methodology

3.1 TMC-TQ Model

Although NMF has been widely used to solve the traffic data recovery problem, its performance is limited by the assumption that the noise follows a Gaussian distribution. Real-world traffic matrices are prone to outliers due to traffic accidents and sensor faults, making it challenging to accurately recover the missing data. To address this issue, we propose a novel traffic matrix completion approach that incorporates a truncated-quadratic function into the model. This function helps to mitigate the impact of gross noise in the data, improving the accuracy of the recovered traffic matrix. The definition of the truncated-quadratic function is given by

$$\|\mathbf{N}\|_{\phi} = \phi(\mathbf{N}) = \sum_{i,j} \phi(N_{ij}) = \sum_{i,j} \frac{1}{2} \min\left\{N_{ij}^2, e^2\right\} \tag{5}$$

where $e > 0$ is an upper bound to suppress outliers, and $\mathbf{N} := \mathbf{X} - \mathbf{WH}$ denotes the residual error between the truth traffic matrix and the recovered traffic matrix. By incorporating the truncated-quadratic function, we reformulate the traffic data recovery problem as follows:

$$\min_{\mathbf{W},\mathbf{H}} \|\mathbf{X}_{\Omega} - (\mathbf{WH})_{\Omega}\|_{\phi} + \alpha_1 \mathrm{tr}(\mathbf{W}^{\top}\mathbf{L}\mathbf{W}) + \alpha_2 \|\mathbf{HT}\|_F^2, \text{ s.t. } \mathbf{W} \geq 0, \mathbf{H} \geq 0, \tag{6}$$

where $\alpha_1 \geq 0$ and $\alpha_2 \geq 0$ are trade-off parameters. Similar to the ST-NMF model in equation (3), the two regularization terms induce temporal and spatial attributes on the factor matrices. These attributes are advantageous for improving the effectiveness and stability of the traffic data recovery model. The model presented in Eq. (6) is referred to as Traffic Matrix Completion with Truncated-Quadratic function (TMC-TQ). The truncated-quadratic function can not only reduce the influence of general Gaussian noise, but also restrain the negative impact of outliers. However, it is worth noting that the objective function in (6) has non-convexity and non-smoothness. Therefore, in the following subsection, we will explore the solution method for this model.

3.2 The Solution of TMC-TQ

Based on HO theory in (4), we introduce an auxiliary variable $\mathbf{A}$ to model the outlies of residual error and rewrite (6) as:

$$\begin{aligned} \min_{\mathbf{W},\mathbf{H},\mathbf{A}} \ & \|\mathbf{X}_{\Omega} - (\mathbf{W}\mathbf{H})_{\Omega} - \mathbf{A}_{\Omega}\|_F^2 + \varphi(\mathbf{A}_{\Omega}) + \alpha_1 \mathrm{tr}(\mathbf{W}^\top \mathbf{L}\mathbf{W}) + \alpha_2 \|\mathbf{H}\mathbf{T}\|_F^2, \\ \text{s.t. } & \mathbf{W} \geq 0, \mathbf{H} \geq 0, \end{aligned} \tag{7}$$

where the derivation of the *dual* function $\varphi(\mathbf{A}_{\Omega}) = \sum_{(i,j)|\Omega_{ij}=1} \varphi(A_{ij})$ is provided in Appendix A. Using the BCD framework, problem (7) with three variables can be decomposed into subproblems, which can be solved iteratively.

Updating W: By fixing the variables $\mathbf{H}$ and $\mathbf{A}$ and omitting terms that do not involve $\mathbf{W}$, the factor matrix $\mathbf{W}$ can be solved using the following formula:

$$\mathbf{W}^{k+1} = \arg\min_{\mathbf{W} \geq 0} \|\mathbf{X}_{\Omega} - (\mathbf{W}\mathbf{H}^k)_{\Omega} - \mathbf{A}_{\Omega}^k\|_F^2 + \alpha_1 \mathrm{tr}(\mathbf{W}^\top \mathbf{L}\mathbf{W}), \tag{8}$$

which can be decoupled with respect to $\mathbf{w}_j^\top$ for $j \in [1, \dots, N]$, resulting in:

$$(\mathbf{w}_j^\top)^{k+1} = \arg\min_{\mathbf{w}_j^\top \geq 0} \|(\mathbf{x}_j^\top - \mathbf{a}_j^\top)_{\mathbf{o}_j^\top} - (\mathbf{w}_j^\top \mathbf{H}^k)_{\mathbf{o}_j^\top}\|_2^2 + \alpha_1 \sum_{i=1}^{N} L_{ji} \mathbf{w}_i^\top \mathbf{w}_j, \tag{9}$$

where $\mathbf{w}_j^\top$, $\mathbf{a}_j^\top$ and $\mathbf{o}_j^\top$ represent the jth row of $\mathbf{W}^k$, $\mathbf{A}_{\Omega}^k$ and $\boldsymbol{\Omega}$, respectively. By applying the *Majorization-Minimization* (MM) algorithm [12], we can deduce that the equation (9) has a closed-form solution, given as follows:

$$\mathbf{w}_j^{t+1} = \mathcal{J}(\mathbf{q}^t) = \begin{cases} -\frac{q_i^t}{2}, & \text{if } q_i^t < 0, i = 1, \dots, R, \\ 0, & \text{otherwise}, \end{cases} \tag{10}$$

and

$$\mathbf{q}^t = \frac{2}{\lambda_1} \{(\mathbf{w}_j^t)^\top [\mathbf{H}^k (\mathbf{H}^k)^\top - \lambda_1 \mathbf{I}_R] - [(\mathbf{x}_j^\top - \mathbf{a}_j^\top)_{\mathbf{o}_j^\top}](\mathbf{H}^k)^\top + \frac{\alpha_1}{2} \sum_{i=1}^{N} L_{ji} \mathbf{w}_i^\top\}, \tag{11}$$

where $\mathbf{I}_R \in \mathbb{R}^{R \times R}$ is an identity matrix, λ_1 is the largest eigenvalue of $\mathbf{H}^k (\mathbf{H}^k)^\top$ and the superscript t denotes the iteration number of MM algorithm. When $\mathbf{w}_j^{t+1}$ converges, it becomes the optimal solution to (9). The derivation of (10) and (11) is provided in Appendix B.

Updating H: By fixing variables $\mathbf{W}$ and $\mathbf{A}$ and leaving out the terms without $\mathbf{H}$, the coefficent matrix $\mathbf{H}$ can be updated by:

$$\mathbf{H}^{k+1} = \arg\min_{\mathbf{H} \geq 0} \|\mathbf{X}_{\Omega} - (\mathbf{W}^{k+1}\mathbf{H})_{\Omega} - \mathbf{A}_{\Omega}^k\|_F^2 + \alpha_2 \|\mathbf{H}\mathbf{T}\|_F^2, \tag{12}$$

The expression can be decoupled with respect to $\mathbf{h}_j$ for $j \in [1, \dots, M]$, resulting in:

$$\mathbf{h}_j^{k+1} = \arg\min_{\mathbf{h}_j \geq 0} \|(\mathbf{x}_j - \mathbf{a}_j)_{\mathbf{o}_j} - (\mathbf{W}^{k+1}\mathbf{h}_j)_{\mathbf{o}_j}\|_2^2 + \alpha_2 \sum_{i=1}^{M} \hat{T}_{ji} \mathbf{h}_i^\top \mathbf{h}_j \tag{13}$$

where $\mathbf{h}_j$, $\mathbf{x}_j$ and $\mathbf{o}_j$ represent the column of $\mathbf{H}^k$, $\mathbf{X}_\Omega$ and $\mathbf{\Omega}$, respectively. Then, the Eq. (13) has the colsed-form solution given by:

$$\mathbf{h}_j^{t+1} = \mathcal{J}\left(\mathbf{p}^t\right) = \begin{cases} -\frac{p_i^t}{2}, & \text{if } p_i^t < 0,\ i = 1, \ldots, R \\ 0, & \text{otherwise} \end{cases} \tag{14}$$

and

$$\mathbf{p}^t = \frac{2}{\lambda_2}\{[(\mathbf{W}^{k+1})^\top \mathbf{W}^{k+1} - \lambda_2 \mathbf{I}_R]\mathbf{h}_j^t - (\mathbf{W}^{k+1})^\top[(\mathbf{x}_j - \mathbf{a}_j)_{\mathbf{o}_j}] + \frac{\alpha_2}{2}\sum_{i=1}^{M}\hat{T}_{ji}\mathbf{h}_i\} \tag{15}$$

where λ_2 is the largest eigenvalue of $(\mathbf{W}^{k+1})^\top \mathbf{W}^{k+1}$ and $\hat{T}_{ji}$ represents the element of $\hat{\mathbf{T}} = \mathbf{T}\mathbf{T}^\top$. When $\mathbf{h}_j^{t+1}$ converges, it becomes the optimal solution to (13). The derivation is similar to (10) and (11) in Appendix B.

Updating A: Given $\mathbf{W}^{k+1}$ and $\mathbf{H}^{k+1}$, let the residual error $\mathbf{N}_\mathbf{\Omega}^{k+1} = \mathbf{X}_\Omega - (\mathbf{W}^{k+1}\mathbf{H}^{k+1})_\Omega$ and then the auxiliary variable $\mathbf{A}$ can be updated by:

$$\mathbf{A}^{k+1} = \arg\min_{\mathbf{A}} \|\mathbf{N}_\mathbf{\Omega}^{k+1} - \mathbf{A}_\Omega\|_F^2 + \varphi(\mathbf{A}_\Omega) \tag{16}$$

According to the derivation in Appendix A, the solution of the problem in (16) is given by

$$\mathbf{A}^{k+1} = \delta(\mathbf{N}^{k+1}) = N_{ij}\text{sign}(|N_{ij}| - \min\{|N_{ij}|, e\}),\ i \in [1, N],\ j \in [1, M]. \tag{17}$$

It is important to note that the value of $\mathbf{A}$ depends on the parameter e. This implies that larger values of e may not effectively filter out outliers, while smaller values of e can lead to the truncation of normal residuals during the updating iterations, resulting in performance degradation. Therefore, choosing a dynamic adaptive value of e is crucial for the proposed method. Assuming that outliers have large amplitudes, they are expected to be well-separated from normal errors. To achieve this, a non-linear Laplacian kernel function is used in this paper. This function exhibits a decreasing trend as the variable increases, particularly for larger values of the variable where the function tends towards zero. The Laplacian kernel function for the observed entries of $\mathbf{N}^{k+1}$ is formulated as follows:

$$K(N_{ij}) = \exp\left(-\frac{|N_{ij}|}{\sigma}\right), \quad \{(i,j) \mid \Omega_{ij} = 1\}, \tag{18}$$

where σ is the kernel width and set by Silverman's rule [13] defined as:

$$\sigma = 1.06 \times \min(\hat{\sigma}, IQR/1.34) \times (\|\mathbf{\Omega}\|_1)^{-0.2}, \tag{19}$$

$\hat{\sigma}$ and IQR represent the standard deviation and interquartile range of the observed entries of $\mathbf{N}^{k+1}$, respectively. We can obtain the outlier group Φ using (18):

$$\Phi = \{(i,j) \mid K(N_{ij}) < \epsilon\} \tag{20}$$

As a result, the value of e is updated by the following strategy:

$$e^{k+1} = \min\{|N_{ij}|, e^k\}, \quad (i, j) \in \Phi. \tag{21}$$

Termination Condition: The stopping criterion is suggested as the following:

$$\max\left\{ \frac{\|\mathbf{W}^{k+1} - \mathbf{W}^k\|_\infty}{\|\mathbf{W}^k\|_\infty}, \frac{\|\mathbf{H}^{k+1} - \mathbf{H}^k\|_\infty}{\|\mathbf{H}^k\|_\infty}, \frac{\|\mathbf{A}^{k+1} - \mathbf{A}^k\|_\infty}{\|\mathbf{A}^k\|_\infty} \right\} \leq 10^{-3}. \tag{22}$$

If the stopping criterion is satisfied, the iteration terminates. The overall optimization steps are summarized in Algorithm 1.

Algorithm 1. TMC-TQ

Input: Traffic matrix $\mathbf{X}_{\mathbf{\Omega}}$, position matrix $\mathbf{\Omega}$, α_1, α_2, R, $T = 500$.
  Initialize: Randomly initialize $\mathbf{W}^0 \in \mathbb{R}^{N \times R}$, $\mathbf{H}^0 \in \mathbb{R}^{R \times M}$.
  while not converged and $k \leq T$ **do**
    Update $\mathbf{w}_j^{k+1}$, j=1, ..., N via (10) and (11)
    Update $\mathbf{h}_j^{k+1}$, j=1, ..., M via (14) and (15)
    Update e^{k+1} via (21)
    Update $\mathbf{A}^{k+1}$ via (17)
    Check termination condition via (22)
    $k = k + 1$.
  end while
Output: $\hat{\mathbf{X}} = \mathbf{WH}$.

4 Experiment and Result Analysis

4.1 Dataset and Missing Data Scenario

We conducted experiments using three publicly available traffic datasets.

- **Guangzhou dataset**: Guangzhou dataset records traffic speed data from 214 urban road links at 10-minute intervals over two months (August and September 2016) in Guangzhou, China. The size of the input traffic matrix is (214 × 8784).
- **Portal dataset**: Portal dataset records traffic speed data from 1003 highways at 5-minute intervals over a week (March 6 to March 12, 2022) in the Portland-Vancouver metropolitan area. The size of the input traffic matrix is (1003 × 2016).
- **Los dataset**: Los dataset records traffic speed data from 207 road links at 5-minute intervals over a week (March 1 to March 7, 2012) in Los Angeles highways, and the size of the input traffic matrix is (207 × 2016).

To evaluate the performance of our proposed algorithm, we consider three missing data scenarios that can be described as:

1. Random missing scenario (RM): This scenario simulates the situation where missing data occurs randomly and instantaneously, which may go undetected in real-world traffic data but is common in experiments. In this scenario, certain entries are randomly and independently selected and masked as missing data.
2. Time missing scenario (TM): This scenario simulates the situation where missing data occurs successively on the part of road links for a continuous period of time, which may be caused by traffic sensors failure. In this scenario, successive entries in the row of the traffic matrix are randomly selected and masked as missing data.
3. Space missing scenario (SM): This scenario simulates the situation where missing data occurs simultaneously on certain road links, which may be caused by software for the data collection center breaking down. In this scenario, the columns of the traffic matrix are randomly selected, and certain entries among them are simultaneously masked as missing data.

The latter two missing scenarios are more common in ITS and demonstrate structured missing generated by events.

4.2 Experiment Setting

For the assessment of traffic matrix completion performance, Normalized Mean Absolute Error (NMAE) is used to measure the accuracy. It is defined as follows:

$$NMAE = \frac{\sum_{(i,j)|\Omega_{ij}=0} |X_{ij} - \hat{X}_{ij}|}{\sum_{(i,j)|\Omega_{ij}=0} |X_{ij}|}, \tag{23}$$

where X_{ij} and $\hat{X}_{ij}$ respectively denote the ground truth and the recovered value of entry (i, j).

We choose five baseline models for comparison, namely, NMF for matrix completion [11], SRSVD [5], SRMF [5], TAS-LR [6], and ST-NMF [7]. The hyper-parameters of these models are set according to the recommended values in the original papers. For our proposed approach, we set the trade-off parameters $\alpha_1 = 0.1$ and $\alpha_2 = 0.1$. For the latent rank of all models, we set $R = 8$ in the following experiments.

4.3 Experimental Results and Analysis

The traffic data recovery results are presented in Fig. 1, 2, and 3. Our proposed model outperforms other baseline models in terms of traffic matrix completion for the three traffic datasets under three missing data scenarios with varying missing rates. We observe that traffic matrix completion under the time missing (TM) and space missing (SM) scenarios is more challenging for all models compared to the random missing (RM) scenario. Besides, TAS-LR, ST-NMF, and our model perform better in diverse scenarios and across the three datasets. They all utilize temporal and spatial characteristics in the process of data recovery. Therefore, incorporating information in the spatial structure using the Laplacian Eigenmaps (LE) and capturing time-variant patterns through the difference operator in the proposed model is reasonable.

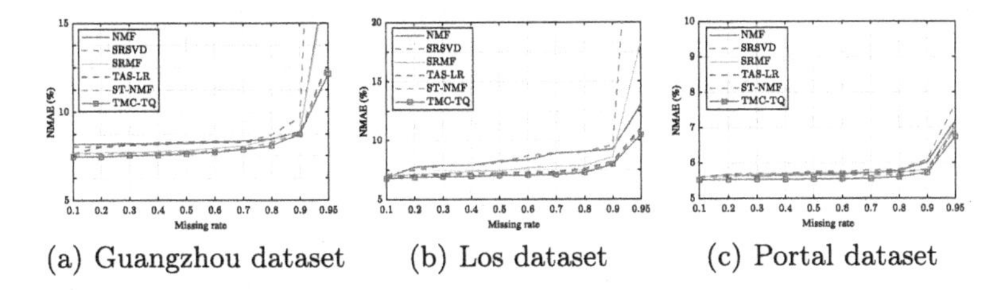

(a) Guangzhou dataset (b) Los dataset (c) Portal dataset

Fig. 1. Recovery performance under RM scenario with variant missing rates

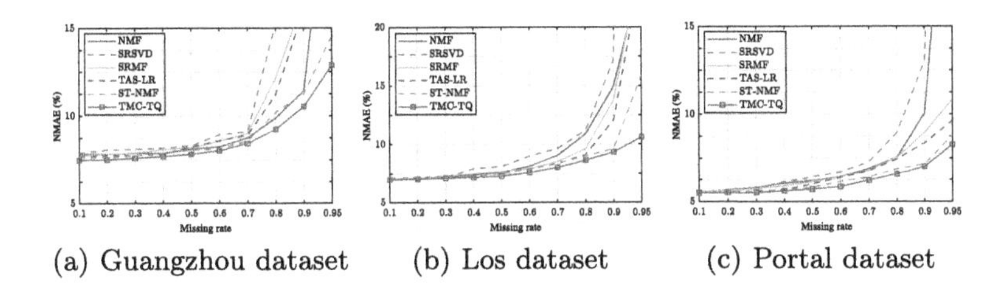

(a) Guangzhou dataset (b) Los dataset (c) Portal dataset

Fig. 2. Recovery performance under TM scenario with variant missing rates

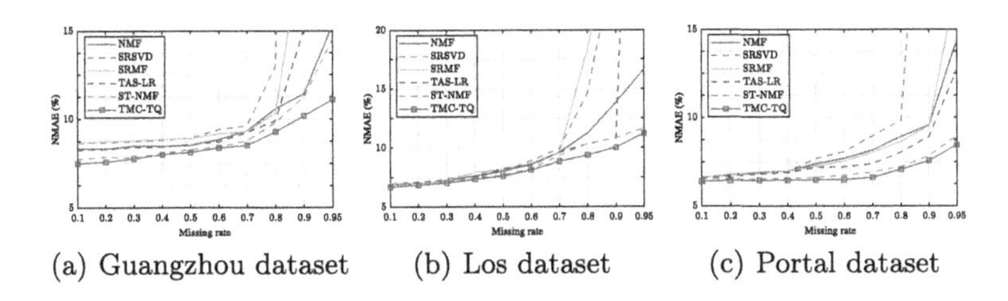

(a) Guangzhou dataset (b) Los dataset (c) Portal dataset

Fig. 3. Recovery performance under SM scenario with variant missing rates

Furthermore, we observed that our proposed method consistently achieves promising performance with lower values of NMAE across varying levels of missing data. In contrast, most of the baseline models experience a rapid degradation in performance as the rate of missing data increases, particularly at high levels of missing data. At low levels of missing data, there is a sufficient amount of truly useful data available, enabling all models to perform well. However, as the percentage of missing data increases, the availability of truly useful data decreases rapidly. This trend suggests that the presence of outliers in the data may have a more significant impact as the proportion of missing data increases, ultimately resulting in inaccuracies in data recovery. In contrast, our proposed method utilizes the truncated-quadratic function to detect outliers in the residuals, enabling us to extract more useful data. Compared with other baseline models, our approach has a higher tolerance for outliers. By effectively addressing outliers, our method enhances the accuracy and reliability of traffic data recovery, even in scenarios with high levels of missing and corrupted data.

Fig. 4 provides visual example of traffic data recovery through our approach on the Guangzhou dataset under the TM scenario with a missing rate of 0.5. It can be observed that the speed at the point on Day 2 experiences a sharp drop from approximately 35 km/h to around 8 km/h. Despite the fluctuation in the average speed ranging from 30 km/h to 40 km/h for most of the time, this specific instance is considered an outlier and is not well-fitted by our model.

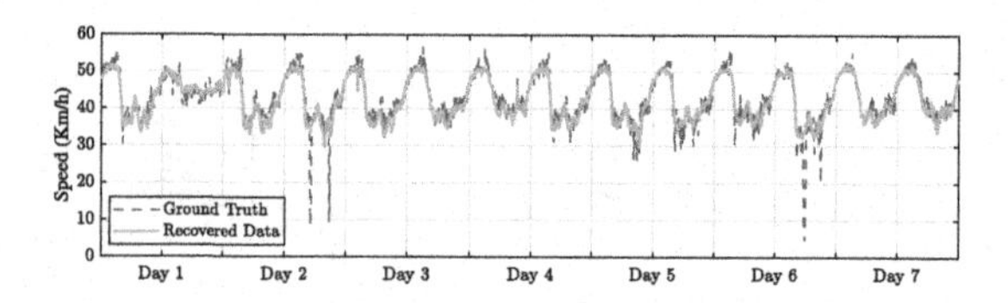

Fig. 4. Visualization example of data recovery under TM scenario with missing rate of 0.5

5 Conclusion

In this paper, we propose a robust method for traffic data recovery and outliers detection based on non-negative matrix factorization. The local spatiotemporal similarity of traffic data is utilized in model. To handle outliers in the traffic matrix, we adopt the truncated-quadratic function to model the residuals between the observed and recovered data. Experimental results on real-world traffic datasets under three missing scenarios with varying missing rates demonstrate that our method outperforms several state-of-the-art traffic matrix completion approaches. In the future, we will focus on reorganizing the traffic speed data into tensor form to better capture the feature structure of the data.

Acknowledgements. The work was supported by the National Natural Science Foundation of China (Grant No. 62206178) and Guangdong Basic and Applied Basic Research Foundation (Grant No. 2021A1515110731).

Appendix

A Derivation of the Dual Function and the Minimizer Function

The *dual* function $\varphi(A_{ij})$ of the truncated-quadratic function $\phi(N_{ij})$ in (5) can be obtained by

$$\begin{aligned}\varphi(A_{ij}) =& \sup_{N_{ij}} -\frac{(N_{ij}-A_{ij})^2}{2} + \phi(N_{ij}) = \sup_{N_{ij}} \frac{1}{2}\left[-(N_{ij}-A_{ij})^2 + \min\{N_{ij}^2, e^2\}\right] \\ =& \frac{1}{2}\left[-(\min\{|A_{ij}|, e\} - e)^2 + e^2\right].\end{aligned} \tag{24}$$

Substituting (5) and (24) into (4) leads to:

$$\begin{aligned}\phi(N_{ij}) =& \inf_{A_{ij}} \frac{(N_{ij}-A_{ij})^2}{2} + \varphi(A_{ij}) \\ =& \inf_{A_{ij}} \frac{1}{2}\left[(N_{ij}-A_{ij})^2 - (\min\{|A_{ij}|, e\} - e)^2 + e^2\right].\end{aligned} \tag{25}$$

Hence, let $\delta(N_{ij})$ denotes optimal solution of function (25), thus

$$\begin{aligned}\delta(N_{ij}) :=& \arg\inf_{A_{ij}} \frac{(N_{ij}-A_{ij})^2}{2} + \varphi(A_{ij}) \\ =& \arg\inf_{A_{ij}} \frac{1}{2}\left[(N_{ij}-A_{ij})^2 - (\min\{|A_{ij}|, e\} - e)^2 + e^2\right] \\ =& N_{ij}\mathrm{sign}(|N_{ij}| - \min\{|N_{ij}|, e\}).\end{aligned} \tag{26}$$

B Derivation of the Closed-Form Solution of (9) and (13)

Both (9) and (13) have the same form as

$$f(\mathbf{x}) = \|\mathbf{y} - \mathbf{Q}\mathbf{x}\|_2^2 + \mathbf{P}\mathbf{x}. \tag{27}$$

Then, the term without $\mathbf{x}$ is dropped, leading to:

$$f(\mathbf{x}) = \mathbf{x}^\top\mathbf{Q}^\top\mathbf{Q}\mathbf{x} - 2(\mathbf{Q}^\top\mathbf{y})^\top\mathbf{x} = \mathbf{x}^\top\mathbf{L}\mathbf{x} - 2(\mathbf{Q}^\top\mathbf{y})^\top\mathbf{x} \tag{28}$$

where $\mathbf{L} = \mathbf{Q}^\top\mathbf{Q}$. We adopt to MM algorithm and use a surrogate function to approximate the quadratic term $\mathbf{x}^\top\mathbf{L}\mathbf{x}$ in order to obtain the closed-form solution of (28). According to [14], the surrogate function of $\mathbf{x}^\top\mathbf{L}\mathbf{x}$ is formulated as:

$$\mathbf{x}^\top\mathbf{M}\mathbf{x} + 2\mathbf{x}^\top(\mathbf{L}-\mathbf{M})\mathbf{x}^t + (\mathbf{x}^t)^\top(\mathbf{M}-\mathbf{L})\mathbf{x}^t = \lambda_{max}\mathbf{x}^\top\mathbf{x} + 2[(\mathbf{L}-\mathbf{M})\mathbf{x}^t]^\top\mathbf{x} \tag{29}$$

where $\mathbf{M} = \lambda_{max}\mathbf{I}_R$, λ_{max} denotes the largest eigenvalue of $\mathbf{L}$ and the superscript t represents the iteration number of the MM algotithm. Then, we obtain the surrogate function $g(\mathbf{x} \mid \mathbf{x}^t)$ of (28) as:

$$\begin{aligned}g(\mathbf{x} \mid \mathbf{x}^t) =& \lambda_{max}\mathbf{x}^\top\mathbf{x} + 2[(\mathbf{L}-\mathbf{M})\mathbf{x}^t - \mathbf{Q}^\top\mathbf{y}]^\top\mathbf{x} \\ =& \lambda_{max}\{\mathbf{x}^\top\mathbf{x} + \frac{2}{\lambda_{max}}[(\mathbf{L}-\mathbf{M})\mathbf{x}^t - \mathbf{Q}^\top\mathbf{y}]^\top\mathbf{x}\} = \lambda_{max}\left[\mathbf{x}^\top\mathbf{x} + (\mathbf{q}^t)^\top\mathbf{x}\right]\end{aligned} \tag{30}$$

where $\mathbf{q}^t = \frac{2}{\lambda_{max}}[(\mathbf{L} - \mathbf{M})\mathbf{x}^t - \mathbf{Q}^\top \mathbf{y}]$. According to MM and KKT condition, $\mathbf{x}^{t+1}$ can be updated as:

$$\mathbf{x}^{t+1} = \arg\min_{\mathbf{x}\geq 0} g(\mathbf{x} \mid \mathbf{x}^t) = \arg\min_{\mathbf{x}\geq 0} \mathbf{x}^\top \mathbf{x} + (\mathbf{q}^t)^\top \mathbf{x} = \begin{cases} -\frac{q_i^t}{2}, & q_i^t < 0 \\ 0, & \text{otherwise.} \end{cases} \tag{31}$$

As a result, $\mathbf{x}^{k+1} = \mathbf{x}^{t+1}$ when $\mathbf{x}^{t+1}$ is convergent.

References

1. Chen, X., Yang, J., Sun, L.: A nonconvex low-rank tensor completion model for spatiotemporal traffic data imputation. Trans. Res. Part C: Emerg. Technol. **117**, 102673 (2020)
2. Yu, J., Stettler, M.E.J., Angeloudis, P., Hu, S., Chen, X.M.: Urban network-wide traffic speed estimation with massive ride-sourcing gps traces. Trans. Res. Part C: Emerg. Technol. **112**, 136–152 (2020)
3. Nie, T., Qin, G., Sun, J.: Truncated tensor schatten p-norm based approach for spatiotemporal traffic data imputation with complicated missing patterns. Trans. Res. Part C: Emerg. Technol. **141**, 103737 (2022)
4. Gao, Y., Liu, J., Yan, X., Lan, M., Liu, Yu.: A spatiotemporal constraint non-negative matrix factorization model to discover intra-urban mobility patterns from taxi trips. Sustainability **11**(15), 4214 (2019)
5. Roughan, M., Zhang, Y., Willinger, W., Qiu, L.: Spatio-temporal compressive sensing and internet traffic matrices (extended version). IEEE/ACM Trans. Netw. **20**(3), 662–676 (2012)
6. Wang, Y., Zhang, Y., Piao, X., Liu, H., Zhang, K.: Traffic data reconstruction via adaptive spatial-temporal correlations. IEEE Trans. Intell. Transp. Syst. **20**(4), 1531–1543 (2019)
7. Wang, Y., Zhang, Y., Wang, L., Yongli, H., Yin, B.: Urban traffic pattern analysis and applications based on spatio-temporal non-negative matrix factorization. IEEE Trans. Intell. Transp. Syst. **23**(8), 12752–12765 (2022)
8. Wang, Y., Zhang, Y., Qian, Z., Wang, S., Yongli, H., Yin, B.: Multi-source traffic data reconstruction using joint low-rank and fundamental diagram constraints. IEEE Intell. Transp. Syst. Mag. **11**(3), 221–234 (2019)
9. Nikolova, M., Chan, R.H.: The equivalence of half-quadratic minimization and the gradient linearization iteration. IEEE Trans. Image Process. **16**(6), 1623–1627 (2007)
10. Yangyang, X., Yin, W.: A block coordinate descent method for regularized multiconvex optimization with applications to nonnegative tensor factorization and completion. SIAM J. Imag. Sci. **6**(3), 1758–1789 (2013)
11. Sun , D.L., Mazumder, R.: Non-negative matrix completion for bandwidth extension: a convex optimization approach. In 2013 IEEE International Workshop on Machine Learning for Signal Processing (MLSP), pp. 1–6. IEEE (2013)
12. Sun, Y., Babu, P., Palomar, D.P.: Majorization-minimization algorithms in signal processing, communications, and machine learning. IEEE Trans. Signal Process. **65**(3), 794–816 (2016)
13. Silverman, B.W.: Density estimation for statistics and data analysis, vol. 26. CRC Press (1986)
14. Song, J., Babu, P., Palomar, D.P.: Optimization methods for designing sequences with low autocorrelation sidelobes. IEEE Trans. Signal Process. **63**(15), 3998–4009 (2015)

ADEQ: Adaptive Diversity Enhancement for Zero-Shot Quantization

Xinrui Chen, Renao Yan, Junru Cheng, Yizhi Wang, Yuqiu Fu, Yi Chen, Tian Guan(✉), and Yonghong He

Shenzhen International Graduate School, Tsinghua University, Shenzhen, China
guantian@sz.tsinghua.edu.cn

Abstract. Zero-shot quantization (ZSQ) is an effective way to compress neural networks, especially when real training sets are inaccessible because of privacy and security issues. Most existing synthetic-data-driven zero-shot quantization methods introduce diversity enhancement to simulate the distribution of real samples. However, the adaptivity between the enhancement degree and network is neglected, i.e., whether the enhancement degree benefits different network layers and different classes, and whether it reaches the best match between the inter-class distance and intra-class diversity. Due to the absence of the metric for class-wise and layer-wise diversity, maladaptive enhancement degree run the vulnerability of mode collapse of the inter-class inseparability. To address this issue, we propose a novel zero-shot quantization method, ADEQ. For layer-wise and class-wise adaptivity, the enhancement degree of different layers is adaptively initialized with a diversity coefficient. For inter-class adaptivity, an incremental diversity enhancement strategy is proposed to achieve the trade-off between inter-class distance and intra-class diversity. Extensive experiments on the CIFAR-100 and ImageNet show that our ADEQ is observed to have advanced performance at low bit-width quantization. For example, when ResNet-18 is quantized to 3 bits, we improve top-1 accuracy by 17.78% on ImageNet compared to the advanced ARC. Code at https://github.com/dangsingrue/ADEQ.

Keywords: Zero-shot Quantization · Diversity Enhancement · Class-wise Adaptability · Layer-wise Adaptability · Inter-class Separability

1 Introduction

Quantization is a promising method for neural network compression, improving the computational efficiency of edge devices by compressing data storage precision [1]. For example, when moving the storage precision of weights and activation tensor in neural networks from 32-bit to 8-bit, the memory overhead is reduced by a factor of 4, while the computational overhead of matrix multiplication is reduced by a factor of 16. Thanks to the robustness of neural networks, appropriate quantization bit width has little impact on model accuracy [2,3].

B. Luo et al. (Eds.): ICONIP 2023, LNCS 14447, pp. 53–64, 2024.
https://doi.org/10.1007/978-981-99-8079-6_5

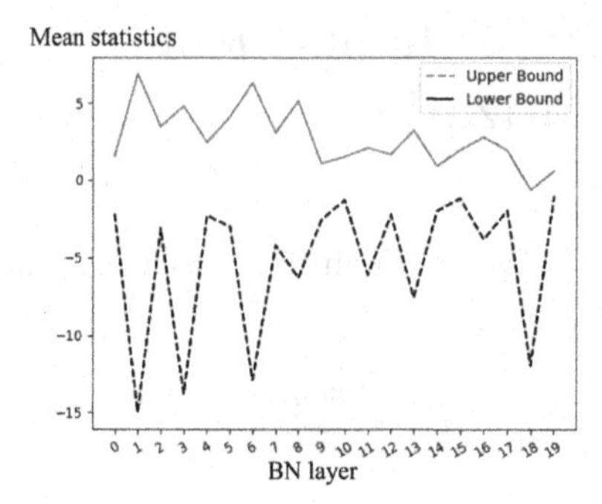

Fig. 1. The upper and lower bounds of the mean statistics in different BN layers. Taking one image from the ImageNet training set as the input of ResNet18.

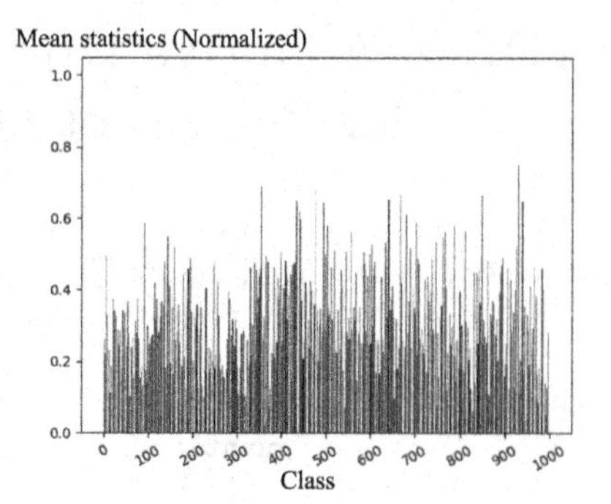

Fig. 2. The range of the mean statistics in the same BN layer. Taking images of 1000 classes from ImageNet as the input of ResNet18.

Research on neural network quantization has focused on quantization-aware training (QAT) [4–7] and post-training quantization (PTQ) [8–11]. The former relies on the complete training set for model fine-tuning, while the latter relies on a small part of the training set to optimize the quantization parameters. Although PTQ reduces data dependency, its drawback also comes from unavoidable access to real data. Due to privacy and security constraints, access to certain real datasets are prohibited, such as medical images, confidential business information, etc. In this context, zero-shot quantization (ZSQ) is naturally proposed to circumvent the limitations. Synthetic datasets have been introduced in many advanced studies and proven to be effective. Most zero-shot quantization methods synthesize samples by fitting statistics in full-precision models, which tends to trigger homogeneous synthetic data, limiting the performance of quantized models. Therefore, sample diversity enhancement methods have become a hot research topic, which avoids homogeneous samples by relaxing constraints [12], boundary constraints [13], and noise interference [14]. However, these methods adopt constant and network-wise enhancement degrees by empirical settings, neglecting the layer-wise and class-wise adaptivity of diversity enhancement. Specifically, there are several key questions: Does the network-wise enhancement degree benefit samples from all network layers and classes? Does the intra-class diversity enhancement reduce the inter-class distance and thus destroy the inter-class separability?

Unfortunately, we observe the maladaptivity of existing methods with the same enhancement degree in layer-wise, class-wise, and inter-class separability. For layer-wise, as shown in Fig. 1, the range of data distribution within different BN layers varies. Thus, constant enhancement degrees for different layers can easily lead to over-enhancement in deep layers and under-enhancement in shallow layers. For class-wise, as shown in Fig. 2, the data distribution of different classes in the same BN layer varies. For inter-class separability, previous studies have shown that real data are characterized by large inter-class distances and high intra-class diversity, as in Fig. 3b. However, inter-class distances and intra-class diversity are relatively contradictory, and blindly increasing intra-class diversity will lead to inter-class inseparability, i.e., over-enhancement, as shown in Fig. 3c. While conservative enhancement degree limits sample diversity, i.e., under-enhancement, as in Fig. 3a.

How to set the enhancement degree for different layers and classes? How to adjust the enhancement degree to maintain intra-class diversity and inter-class separability? To answer the above questions, a novel zero-shot quantization method named ADEQ is proposed. We introduce a diversity coefficient to initialize the enhancement degree for different layers and classes, achieving layer-wise and class-wise adaptability. Then, the incremental diversity enhancement strategy is proposed to maintain the inter-class separability and intra-class diversity. With the above methods, the proposed method achieves advanced performance in low bit-width quantization. Our contributions are:

- The maladaptivity of existing diversity enhancement methods in layer-wise, class-wise, and inter-class separability is highlighted. To the extent of our knowledge, this work is the first to propose adaptive diversity enhancement for zero-shot network quantization.
- Diversity coefficient is introduced to compensate the absence of the metric for class-wise and layer-wise diversity.
- With the proposed adaptive enhancement degree initialization and incremental diversity enhancement strategy, we achieve a 17.78% top-1 accuracy increase compared to the state-of-the-art on large-scale ImageNet, when ResNet-18 is quantized to 3 bits.

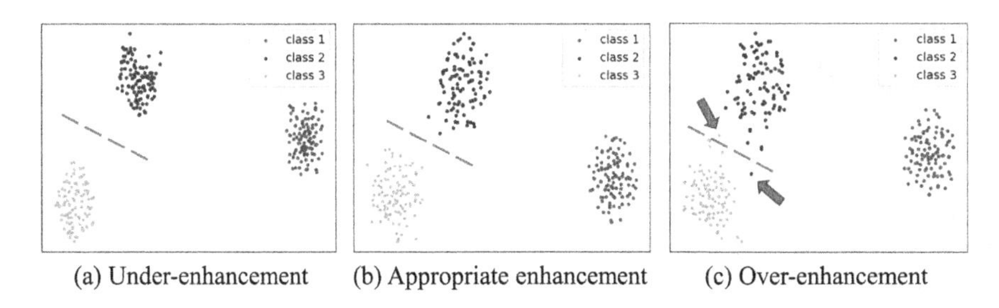

(a) Under-enhancement (b) Appropriate enhancement (c) Over-enhancement

Fig. 3. Effect of different enhancement degrees on inter-class separability. (a) Insufficient enhancement degree results in low intra-class diversity. (b) Appropriate enhancement degree trades off intra-class diversity and inter-class distance. (c) Excessive enhancement degree leads to inter-class inseparability.

2 Related Work

ZSQ avoids accessing to real data and extracts information from full precision networks. Most studies recover quantization error from two perspectives: quantization parameter design [15–18] and sample synthesizing [12,13,19,20].

The first perspective is to design quantization parameters with module properties and model statistics without requiring any data. For example, Nagel et al. [15] proposed a scale-equivariance property of activation functions to equalize the weight ranges of the network. Meller et al. [18] highlighted an inversely proportional factorization of convolutional neural networks to decrease the degradation caused by quantization. The above methods design simple parameters to accomplish zero-shot quantization but suffer from unrecoverable quantization errors at low bit widths. For example, DFQ achieves only a 0.11% top-1 accuracy when quantizing MobileNet V2 to 4-bit on ImageNet. Therefore, more ZSQ methods resort to synthetic samples in the hope of obtaining the distribution of the real ones.

The second perspective generates synthetic samples to fine-tune the model or knowledge distillation. ZeroQ [19] adopted an optimization method to generate synthetic samples and introduce batch normalization statistics (BNS) alignment loss. Similarly, GDFQ [20] proposes a generator to produce labeled synthetic data with cross entropy loss and BNS alignment loss. However, tight BNS alignment causes homogenized synthetic samples, and many studies have been conducted on the enhancement of sample diversity. DSG [12] relaxed the constraint of the BNS alignment to avoid homogenized synthetic data. IntraQ [13] highlighted the property of intra-class heterogeneity and devised three methods including a local object reinforcement, a marginal distance constraint, and a soft inception loss. ClusterQ [14] enhanced sample diversity by introducing Gaussian noise at the cluster center of the feature distribution. These studies provide us with effective methods for sample diversity enhancement, yet they neglect the adaptivity between the enhancement degree and the network.

3 Methodology

3.1 Preliminaries

Quantizer. Following GDFQ [20], asymmetric quantization is adopted. Given a floating-point value x_f (weights or activations) and quantization bit width BW, the asymmetric quantizer can be defined as:

$$\begin{cases} S = \frac{2^{BW}-1}{x_{f\,\max}-x_{f\,\min}} \\ ZP = x_{f\,\min} \cdot S \\ x_q = \lfloor x_f \cdot S - ZP \rceil \end{cases}, \quad (1)$$

where x_q denotes the quantized fixed-point number, S denotes the scaling factor mapping the floating-point number to a fixed-point integer, ZP denotes a zero point mapping the floating-point minimum to the fixed-point zero, $\lfloor input \rceil$

rounds its input to the nearest integer. With the quantized network, computational costs can be reduced significantly.

The dequantized value x_d can be obtained as:

$$x_d = \frac{x_q + ZP}{S}. \tag{2}$$

Data Synthesis. To compensate for the lack of real dataset $D = \{(x, y)\}$, most ZSQ methods introduce synthetic dataset $\overline{D} = \{(\overline{x}, \overline{y})\}$ to fine-tune the model. Many existing studies synthesize samples that match the batch normalization statistics (BNS) in the full-precision model F. The BNS alignment loss is expressed as:

$$\mathcal{L}_{BNS} = \sum_{l=1}^{L} \left\|\mu_l(\overline{x}) - {\mu_l}^F\right\|_2 + \left\|\sigma_l(\overline{x}) - {\sigma_l}^F\right\|_2, \tag{3}$$

where L is the total number of BN layers, ${\mu_l}^F$ and ${\sigma_l}^F$ are the running mean and variance stored in the l-th BN layer in F. The mean and variance of synthetic image batch $\overline{x}$ are given by $\mu_l(\overline{x})$ and $\sigma_l(\overline{x})$, respectively.

To generate label-oriented synthetic samples, the cross entropy loss is commonly used to constrain the output of F and the given label, as defined in Eq. 4. For the generator (G), given a random white noise z and a target label $\overline{y}$, the synthetic sample can be obtained with $\overline{x} = G\left(z|\overline{y}\right)$. During data synthesis, F is fixed and G is updated by backpropagation.

$$\mathcal{L}_{CE} = \mathbb{E}_{(x,y)\sim\{(\overline{x},\overline{y})\}}\left[\text{Cross Entropy}\left(F(\overline{x}),\overline{y}\right)\right]. \tag{4}$$

3.2 Adaptive Diversity Enhancement

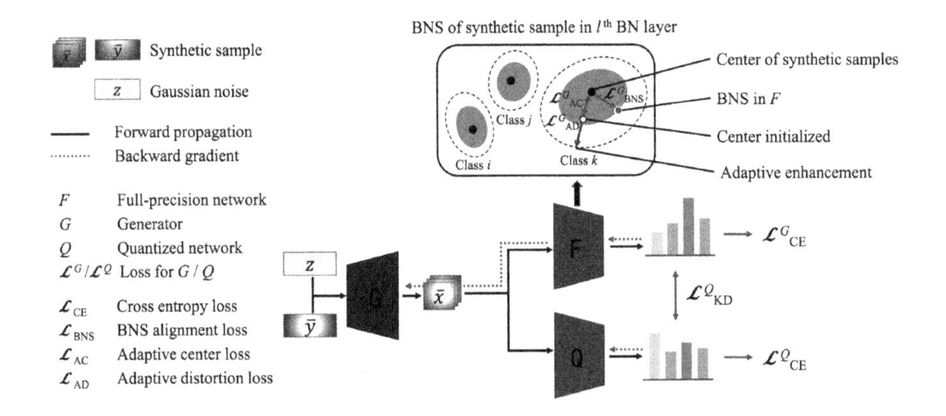

Fig. 4. The proposed ADEQ framework.

In this section, the framework of adaptive diversity enhancement for zero-shot network quantization (ADEQ) is proposed, as shown in Fig. 4. We first extract the centers of synthetic samples and constrain G to fit them with adaptive center loss. Next, adaptive distortion loss is introduced to interfere with the sample centers and enhance diversity. Specifically, we introduce the diversity coefficient and implement it to conduct adaptive initialization of enhancement degrees for different layers and classes, and use the incremental diversity enhancement strategy to maintain inter-class adaptivity.

Diversity Coefficient. To measure the diversity degree of layer-wise and class-wise, we introduce the concept of diversity coefficient for synthetic samples, as shown in Eq. 5. It represents the range of the mean or variance tensor in the l-th BN layer and class k when taking synthetic samples as the input of F, indicating the diversity of distribution within a class or layer.

$$\begin{cases} {d_{l,k}}^{\mu} = \max_i \mu_{l,i}\left(\overline{x}|\overline{y}=k\right) - \min_i \mu_{l,i}\left(\overline{x}|\overline{y}=k\right) \\ {d_{l,k}}^{\sigma} = \max_i \sigma_{l,i}\left(\overline{x}|\overline{y}=k\right) - \min_i \sigma_{l,i}\left(\overline{x}|\overline{y}=k\right) \end{cases}, \tag{5}$$

where $\mu_{l,i}\left(\overline{x}|\overline{y}=k\right)$ and $\sigma_{l,i}\left(\overline{x}|\overline{y}=k\right)$ are the mean and variance in the l-th BN layer of F with input of synthetic sample, respectively. i denotes the i-th element in the tensor.

With the proposed metric, we take the ImageNet training set as an example to illustrate the existing maladaptive properties. When class k is fixed, various range of the statistics in different layers can be observed in Fig. 1. When BN layer l is fixed, the inter-class diversity coefficients differ in different classes, as shown in Fig. 2. Thus, same enhancement degree for different layers and classes is maladaptive. Next, we propose the following two strategies.

Adaptive Enhancement Degree Initialization. Adaptive initialization strategy is proposed to improve the adaptability in layer-wise and class-wise. First, we use the generator G to synthesize N samples for each class and extract their mean centers and variance centers in F for each layer and class with Eq. 6.

$$\begin{cases} {\mu_{l,k}}^{C} = \frac{1}{N} \sum_{n=1}^{N} \mu_l\left(\overline{x_n}|\overline{y}=k\right) \\ {\sigma_{l,k}}^{C} = \frac{1}{N} \sum_{n=1}^{N} \sigma_l\left(\overline{x_n}|\overline{y}=k\right). \end{cases} \tag{6}$$

Then, we introduce adaptive center loss (${\mathcal{L}^G}_{AC}$), as defined in Eq. 7, constraining the synthetic samples to fit the centers of the corresponding layers and classes.

$${\mathcal{L}^G}_{AC} = \sum_{\lceil L/2 \rceil - 2}^{L} \left\|\mu_l(\overline{x}|\overline{y}=k) - {\mu_{l,k}}^{C}\right\|_2 + \left\|\sigma_l(\overline{x}|\overline{y}=k) - {\sigma_{l,k}}^{C}\right\|_2. \tag{7}$$

To enhance sample diversity, adaptive distortion loss (${\mathcal{L}^G}_{AD}$) is proposed, as defined in Eq. 8, which provides layer-wise and class-wise adaptive white noise interference to avoid overfitting the center and intra-class homogenization.

$$\begin{aligned} {\mathcal{L}^G}_{AD} = \sum_{\lceil L/2 \rceil - 2}^{L} & \left\| \left(\mu_l\left(\overline{x}|\overline{y}=k\right) - N\left({\mu_{l,k}}^C, {\nu_{l,k}}^{\mu}\right)\right) \right\|_2 \\ & + \left\| \left(\sigma_l\left(\overline{x}|\overline{y}=k\right) - N\left({\sigma_{l,k}}^C, {\nu_{l,k}}^{\sigma}\right)\right) \right\|_2, \end{aligned} \tag{8}$$

where ${\nu_{l,k}}^{\mu}$ and ${\nu_{l,k}}^{\sigma}$ are enhancement degree of mean and variance, which are initialized with Eq. 9 after warm-up for w epochs. λ=0.4 is observed to perform best.

$$\begin{cases} {\nu_{l,k}}^{\mu}|^{epoch=w} &= \lambda \cdot {d_{l,k}}^{\mu} \\ {\nu_{l,k}}^{\sigma}|^{epoch=w} &= \lambda \cdot {d_{l,k}}^{\sigma}. \end{cases} \tag{9}$$

Incremental Diversity Enhancement Strategy. After initializing the enhancement degree, it is necessary to consider how to adjust the enhancement degree to balance intra-class diversity and inter-class separability. However, we do not have any access to the diversity coefficients in real samples, which is essentially an unsupervised self-adaptivity problem. For this reason, we borrow the idea of early stop and propose the incremental diversity enhancement strategy. Specifically, the enhancement degree is linearly increased as Eq. 10 until the accuracy of the quantized model Q no longer raise in continuous 20 epochs.

$$\begin{cases} {\nu_{l,k}}^{\mu}|^{epoch=w+e} &= {\nu_{l,k}}^{\mu}|^{epoch=w} \cdot \theta^e \\ {\nu_{l,k}}^{\sigma}|^{epoch=w+e} &= {\nu_{l,k}}^{\sigma}|^{epoch=w} \cdot \theta^e. \end{cases} \tag{10}$$

3.3 Quantization Process

Synthetic Sample Generation. To synthesize samples, cross entropy loss and BNS alignment loss are used to fit the BNS and label in F. The total loss used by the generator G can be summarized as:

$$\mathcal{L}^G = {\mathcal{L}^G}_{CE} + \alpha_1 \cdot {\mathcal{L}^G}_{BNS} + \alpha_2 \cdot {\mathcal{L}^G}_{AC} + \alpha_3 \cdot {\mathcal{L}^G}_{AD}, \tag{11}$$

where the α_1, α_2 and α_3 are the trade-off parameters.

Quantized Model Fine-Tuning. Taking the synthetic data as the input of Q, the cross entropy loss in Eq. 12 is applied to fine-tune Q.

$${\mathcal{L}^Q}_{\text{CE}} = \mathbb{E}_{(x,y)\sim\overline{D}} \left[\text{Cross Entropy}\left(Q(\overline{x}),\overline{y}\right)\right]. \tag{12}$$

Subsequently, the synthetic data are used as the input of F. The Kullback Leibler distance in Eq. 13 is applied to transfer the output of F to Q.

$${\mathcal{L}^Q}_{KD} = \mathbb{E}_{(x,y)\sim\overline{\text{D}}} \left[\text{Kullback Leibler}\left(Q(\overline{x}),F(\overline{x})\right)\right]. \tag{13}$$

At this point, the overall loss for fine-tuning Q can be summarized as:

$$\mathcal{L}^Q = \mathcal{L}^Q{}_{CE} + \alpha_4 \cdot \mathcal{L}^Q{}_{KD}, \tag{14}$$

where the α_4 is the trade-off parameter.

Table 1. Results of ResNet-20 on CIFAR-100. WBAB indicates the weights and activations are quantized to B-bit.

Bit-width	Method	Top-1 Accuracy (%)
W32A32	Full-precision	70.33
W5A5	GDFQ [20] (ECCV 2020)	67.52
	ARC [21] (IJCAI 2021)	68.40
	Qimera [22] (NeurIPS 2021)	69.02
	ZAQ [23] (CVPR 2021)	68.70
	ARC+AIT [24] (CVPR 2022)	68.40
	ADEQ (Ours)	**69.34**
W4A4	GDFQ [20] (ECCV 2020)	63.75
	ARC [21] (IJCAI 2021)	62.76
	Qimera [22] (NeurIPS 2021)	65.10
	ZAQ [23] (CVPR 2021)	60.42
	IntraQ [13] (CVPR 2022)	64.98
	ARC+AIT [24] (CVPR 2022)	61.05
	ADEQ (Ours)	**66.35**
W3A3	GDFQ [25] (ECCV 2020)	47.61
	ARC [21] (IJCAI 2021)	40.15
	Qimera [25] (NeurIPS 2021)	46.13
	IntraQ [13] (CVPR 2022)	48.25
	ARC+AIT [24] (CVPR 2022)	41.34
	ADEQ (Ours)	**54.33**

4 Experiments

4.1 Experimental Settings and Details

We report top-1 accuracy on the validation set of CIFAR100 [26] and ImageNet [27]. The quantized networks selected include ResNet-20 [28] for CIFAR100 and ResNet-18 [28] for ImageNet. Note that real training sets are not used in quantization. All experiments are implemented with Pytorch [29] via the code of FDDA [30], and run on an NVIDIA GeForce GTX 3090 GPU together with an AMD(R) EPYC 7542 @ 2.90GHz CPU.

To generate synthetic images, we adopt the generator proposed by GDFQ [20]. The initial learning rate is set to 1e-3 and multiplied by 0.1 every 100 epochs. To the quantized network, the batch size for fine-tuning is 128 for CIFAR-100 and 16 for ImageNet. The initial learning rate is 3e-6, adjusted by cosine annealing. SGD with Nesterov is adopted as an optimizer for the quantized network. Before formal training, we warm up the generator G for 100 epochs, and then use a total of 300 epochs to update the generator G and the quantized model Q. Based on existing studies and experience [20,30], the hyperparameters from α_1 to α_4 are respectively set to 0.4, 0.02. 4, and 20.

Table 2. Results of ResNet-18 on ImageNet. WBAB indicates the weights and activations are quantized to B-bit.

Bit-width	Method	Top-1 Accuracy (%)
W32A32	Full-precision	71.47
W5A5	GDFQ [20] (ECCV 2020)	68.49
	ARC [21] (IJCAI 2021)	68.88
	Qimera [22] (NeurIPS 2021)	69.29
	ZAQ [23] (CVPR 2021)	64.54
	IntraQ [13] (CVPR 2022)	69.94
	ARC+AIT [24] (CVPR 2022)	70.28
	ADEQ (Ours)	**69.90**
W4A4	GDFQ [20] (ECCV 2020)	60.60
	ARC [21] (IJCAI 2021)	61.32
	Qimera [22] (NeurIPS 2021)	63.84
	ZAQ [23] (CVPR 2021)	52.64
	IntraQ [13] (CVPR 2022)	66.47
	ARC+AIT [24] (CVPR 2022)	65.73
	ADEQ (Ours)	**66.32**
W3A3	GDFQ [25] (ECCV 2020)	20.23
	ARC [21] (IJCAI 2021)	23.37
	Qimera [25] (NeurIPS 2021)	1.17
	ADEQ (Ours)	**41.15**

4.2 Performance Comparison

CIFAR-100. We first compare with advanced ZSQ methods on CIFAR-100 with the quantized network ResNet-20. Table 1 summarizes our comparison results.

Specifically, our method achieves advanced performance at different quantization bit widths. Among them, we improve the top-1 accuracy by 6.08% in the

3-bit case compared to the advanced IntraQ and 1.25% in the 4-bit case compared to the advanced Qimera. In particular, the proposed method achieves a low accuracy loss of 0.99% in the 5-bit case, indicating that our proposed method fits the distribution of real data well.

ImageNet. To demonstrate the advancement of our method, we further conduct comparison on the challenging ImageNet with the quantized network ResNet-18. Results are presented in Table 2.

In low-bit width cases, most of the existing methods suffer from large accuracy loss, while our method maintains generalization capability. Specifically, in the 3-bit case, the top-1 accuracy is improved by 17.78% compared to the advanced ARC, reaching an outstanding performance of 41.15%.

5 Ablation Study

In this section, we perform ablation studies on the key components of ADEQ, including adaptive initialization and incremental diversity enhancement strategy in Sect. 3.2. All the experiments are designed on quantizing ResNet-18 to 3-bit on ImageNet. Results are presented in Table 3. The control group in the bottom column uses the traditional enhancement method without considering adaptivity. It can be observed that discarding one or two key components of ADEQ leads to a significant performance decrease (up to 3.58%).

Table 3. Ablations on key components of our ADEQ. "AEDI" denotes the adaptive enhancement degree initialization, and "IDES" denotes the incremental diversity enhancement strategy. We report the top-1 accuracy of 3-bit ResNet-18 on ImageNet.

AEDI	IDES	Top-1 Accuracy (%)
✓	✓	**41.15**
✓		39.79
	✓	39.17
		37.57

6 Conclusion

In this paper, we emphasize the adaptivity of enhancement degree at the layer-wise, class-wise as well as class separability, and propose a novel zero-shot quantization method, ADEQ. To achieve layer-wise and class-wise adaptability, we introduce diversity coefficients and initialize the enhancement degree adaptively for different layers and different classes. To achieve inter-class adaptivity, the incremental diversity enhancement strategy is proposed to achieve a trade-off between inter-class distance and intra-class diversity. Extensive experiments on ImageNet and CIFAR-100 show that our ADEQ is observed to have advanced performance in low bit-width quantization.

Acknowledgements. This work is jointly supported by the National Natural Science Foundation of China (NSFC) (61975089), the Science and Technology Research Program of Shenzhen City (KCXFZ20201221173207022, WDZC20200821141349001), and the Jilin Fuyuan Guan Food Group Co., Ltd.

References

1. Jacob, B., et al.: Quantization and training of neural networks for efficient integer-arithmetic-only inference. In: Proceedings of the IEEE Conference on Computer Vision and Pattern Recognition, pp. 2704–2713 (2018)
2. Nagel, M., Fournarakis, M., Amjad, R.A., Bondarenko, Y., Baalen, M.V., Blankevoort, T.: A white paper on neural network quantization (2021). arXiv preprint arXiv:2106.08295
3. Krishnamoorthi, R.: Quantizing deep convolutional networks for efficient inference: A whitepaper (2018). arXiv preprint arXiv:1806.08342
4. Kim, D., Lee, J., Ham, B.: Distance-aware quantization. In Proceedings of the IEEE/CVF International Conference on Computer Vision, pp. 5271–5280 (2021)
5. Nahshan, Y., et al.: Loss aware post-training quantization. Mach. Learn. **110**(11–12), 3245–3262 (2021)
6. Bai, H., Cao, M., Huang, P., Shan, J.: BatchQuant: quantized-for-all architecture search with robust quantizer. Adv. Neural. Inf. Process. Syst. **34**, 1074–1085 (2021)
7. Zhao, S., Yue, T., Hu, X.: Distribution-aware adaptive multi-bit quantization. In: Proceedings of the IEEE/CVF Conference on Computer Vision and Pattern Recognition, pp. 9281–9290 (2021)
8. Fang, J., Shafiee, A., Abdel-Aziz, H., Thorsley, D., Georgiadis, G., Hassoun, J.H.: Post-training piecewise linear quantization for deep neural networks. In: Vedaldi, A., Bischof, H., Brox, T., Frahm, J.-M. (eds.) ECCV 2020. LNCS, vol. 12347, pp. 69–86. Springer, Cham (2020). https://doi.org/10.1007/978-3-030-58536-5_5
9. Jeon, Y., Lee, C., Cho, E., Ro, Y.: Mr. BiQ: post-training non-uniform quantization based on minimizing the reconstruction error. In: Proceedings of the IEEE/CVF Conference on Computer Vision and Pattern Recognition, pp. 12329–12338 (2022)
10. Banner, R., Nahshan, Y., Soudry, D.: Post training 4-bit quantization of convolutional networks for rapid-deployment. In: Advances in Neural Information Processing Systems, vol. 32 (2019)
11. Finkelstein, A., Fuchs, E., Tal, I., Grobman, M., Vosco, N., Meller, E.: QFT: post-training quantization via fast joint finetuning of all degrees of freedom. In: Karlinsky, L., Michaeli, T., Nishino, K. (eds.) Computer Vision – ECCV 2022 Workshops. ECCV 2022. LNCS, vol. 13807. Springer, Cham (2023). https://doi.org/10.1007/978-3-031-25082-8_8
12. Zhang, X., et al.: Diversifying sample generation for accurate data-free quantization. In: Proceedings of the IEEE/CVF Conference on Computer Vision and Pattern Recognition, pp. 15658–15667 (2021)
13. Zhong, Y., et al.: IntraQ: learning synthetic images with intra-class heterogeneity for zero-shot network quantization. In: Proceedings of the IEEE/CVF Conference on Computer Vision and Pattern Recognition, pp. 12339–12348 (2022)

14. Gao, Y., Zhang, Z., Hong, R., Zhang, H., Fan, J., Yan, S.: Towards feature distribution alignment and diversity enhancement for data-free quantization. In: 2022 IEEE International Conference on Data Mining (ICDM), pp. 141–150. IEEE (2022)
15. Nagel, M., Baalen van, M., Blankevoort, T., Welling, M.: Data-free quantization through weight equalization and bias correction. In: Proceedings of the IEEE/CVF International Conference on Computer Vision, pp. 1325–1334 (2019)
16. Yvinec, E., Dapogny, A., Cord, M., Bailly, K.: SPIQ: data-free per-channel static input quantization. In: Proceedings of the IEEE/CVF Winter Conference on Applications of Computer Vision, pp. 3869–3878 (2023)
17. Guo, C., et al.: SQuant: On-the-fly data-free quantization via diagonal hessian approximation (2022). arXiv preprint arXiv:2202.07471
18. Meller, E., Finkelstein, A., Almog, U., Grobman, M.: Same, same but different: recovering neural network quantization error through weight factorization. In: International Conference on Machine Learning, pp. 4486–4495. PMLR (2019)
19. Cai, Y., Yao, Z., Dong, Z., Gholami, A., Mahoney, M.W., Keutzer, K.: ZeroQ: a novel zero shot quantization framework. In: Proceedings of the IEEE/CVF Conference on Computer Vision and Pattern Recognition, pp. 13169–13178 (2020)
20. Xu, S., et al.: Generative low-bitwidth data free quantization. In: Vedaldi, A., Bischof, H., Brox, T., Frahm, J.-M. (eds.) ECCV 2020. LNCS, vol. 12357, pp. 1–17. Springer, Cham (2020). https://doi.org/10.1007/978-3-030-58610-2_1
21. Zhu, B., Hofstee, P., Peltenburg, J., Lee, J., Alars, Z.: AutoReCon: neural architecture search-based reconstruction for data-free. In: International Joint Conference on Artificial Intelligence (2021)
22. Choi, K., Hong, D., Park, N., Kim, Y., Lee, J.: Qimera: data-free quantization with synthetic boundary supporting samples. Adv. Neural. Inf. Process. Syst. **34**, 14835–14847 (2021)
23. Liu, Y., Zhang, W., Wang, J.: Zero-shot adversarial quantization. In: Proceedings of the IEEE/CVF Conference on Computer Vision and Pattern Recognition, pp. 1512–1521 (2021)
24. Choi, K., et al.: It's all in the teacher: zero-shot quantization brought closer to the teacher. In: Proceedings of the IEEE/CVF Conference on Computer Vision and Pattern Recognition, pp. 8311–8321 (2022)
25. Qian, B., Wang, Y., Hong, R., Wang, M.: Adaptive data-free quantization. arXiv e-prints, pages arXiv-2303 (2023)
26. Krizhevsky, A., Hinton, G., et al.: Learning multiple layers of features from tiny images (2009)
27. Russakovsky, O., et al.: ImageNet large scale visual recognition challenge. Int. J. Comput. Vision **115**, 211–252 (2015)
28. He, K., Zhang, X., Ren, S., Sun, J.: Deep residual learning for image recognition. In: Proceedings of the IEEE Conference on Computer Vision and Pattern Recognition, pp. 770–778 (2016)
29. Paszke, A., et al.: PyTorch: an imperative style, high-performance deep learning library. In: Advances in Neural Information Processing Systems, vol. 32 (2019)
30. Zhong, Y., et al.: Fine-grained data distribution alignment for post-training quantization. In: Avidan, S., Brostow, G., Cissé, M., Farinella, G.M., Hassner, T. (eds.) Computer Vision – ECCV 2022. ECCV 2022. LNCS, vol. 13671. Springer, Cham (2022). https://doi.org/10.1007/978-3-031-20083-0_5

ASTPSI: Allocating Spare Time and Planning Speed Interval for Intelligent Train Control of Sparse Reward

Haotong Zhang[1] and Gang Xian[2](✉)

[1] College of Information Science and Engineering, Chongqing Jiaotong University, Chongqing, China
tongzh@mails.cqjtu.edu.cn

[2] College of Computer, National University of Defense Technology, Changsha, China
xiangang@nudt.edu.cn

Abstract. When using deep reinforcement learning (DRL) to solve train operation control in urban railways, encounter complex and dynamic environments with sparse rewards. Therefore, it is crucial to alleviate the negative impact of sparse rewards on finding the optimal trajectory. This paper introduces a novel algorithm called Allocating Spare Time and Planning Speed Intervals (ASTPSI), which can reduce the blindness of exploration dramatically of intelligent train agents under sparse rewards when using DRL and significantly improve their learning efficiency and operation quality. The ASTPSI can generate real-time train trajectories that meet the requirements by combining different DRL algorithms. To evaluate the algorithm's performance, we verified the convergence rate of the ASTPSI-DRL to optimize train trajectories in the face of sparse rewards on a real track. ASTPSI-DRL has better performance and stability than genetic algorithms and original DRL algorithms in reducing train energy consumption, punctuality, and accurate stopping.

Keywords: Train trajectory optimization · ASTPSI · Sparse reward · Deep reinforcement learning · Energy efficiency

1 Introduction

One of the urban railways' essential functions of the automatic train operation (ATO) system is to generate a recommended train speed profiles before train departure, which serves as the control target for the train [5]. This process is referred to as Train Trajectory Optimization (TTO) issue, a complex optimal control issue that will face challenges in finding the optimal speed profile under complex predetermined constraints such as travel time, slope, safety, etc.

Many researchers and engineers have been developing effective methods to solve the TTO issue in recent years. Traditional computational methodologies

B. Luo et al. (Eds.): ICONIP 2023, LNCS 14447, pp. 65–77, 2024.
https://doi.org/10.1007/978-981-99-8079-6_6

such as PMP [4], dynamic programming and heuristic algorithms can fall into problems such as local optima and dimension explosion. These algorithms have certain limitations and cannot generate truly optimal train trajectories. Regarding TTO calculation, deep reinforcement learning (DRL) has become one of the current research hotspots. DRL can learn the mapping from the current state to the decision sequence without knowing the final result and guide learning based on reward signals. Therefore, DRL algorithms can solve the dimension explosion issue and local optima and have more flexible adaptability and intelligence. However, in the practical application of the TTO issue, the agent struggles to receive rewards, making it difficult to explore the correct strategies. Therefore, alleviating the issue of sparse rewards in DRL models has become an important and challenging research direction. We propose a novel method called Allocating Spare Time and Planning Speed Intervals (ASTPSI) to overcome the difficulty of enormous solution space and generate practical solutions under complicated constraints such as speed limits, gradients, and predetermined running time. We combine DDPG and TD3 with ASTPSI, respectively, to validate the ability of the algorithm to alleviate the negative impact of sparse rewards. Testing the ASTPSI algorithm on a real railway indicates that our approach effectively reduces the search space of the deep reinforcement algorithm, avoiding the generation of unreasonable or infeasible trajectories, alleviating the learning difficulties in sparse reward, and significantly improving the learning efficiency and operation quality of its intelligent agent.

The remainder of this paper is structured as follows: Sect. 2 introduces the issue description and the ASTPSI algorithm. Section 3 elaborates on the design of the reinforcement learning framework. Section 4 establishes numerical simulation experiments and analyzes the performance of the proposed algorithm. Section 5 reviews related work, and Sect. 6 provides a summary of this paper.

2 Issue Description and Solution

To simulate train operation and generate a set of practicable train speed profiles, make the following assumptions: This paper aims to minimize traction energy consumption in driving the train to its destination station without factoring in regenerative braking energy as part of the synthesized energy. Regenerative braking converts and stores the train's kinetic energy during braking instead of converting it into useless heat, and adding it increases the braking speed and the coasting time later [3]. Therefore, implicitly discouraging regenerative braking energy from compensating for unnecessary traction, the assumption is made that regenerative braking energy is not considered.

2.1 Train Dynamics Model

In general, a train can be simplified as a single particle, and when calculating the additional resistance, is considered a multi-particle. Subject to various forces during operation, and the train's motion can be expressed as follows:

$$\frac{dv(x)}{dx} = \frac{M(x) \cdot F(v) - C(v) - S(x)}{\rho \cdot M \cdot v(x)} \tag{1}$$

where $F(v)$ is the maximum traction or braking effort at speed v, and $M(x)$ represents the percentage of traction or braking effort applied at point x. ρ is the rotating mass factor. M is the train weight. $C(v)$ represents the basic resistances, including frictional and air resistance, expressed as a quadratic velocity function according to the Davis equation [10]. $S(x)$ represents the additional resistance at the train's position x, including the gradient, curve, and tunnel resistance. The elapsed time $t(x)$ satisfies a differential equation:

$$\dot{t}(x) = \frac{dt(x)}{dx} = \frac{1}{v(x)} \tag{2}$$

In order to ensure safe, train speed must be within the speed limit along the track. Hence, the train speed should satisfy the speed limit constraint:

$$v(x) \leq v_{lim}(x) \tag{3}$$

2.2 Allocating Spare Time and Planning Speed Intervals

Allocating Spare Time. The train's punctuality is critical in evaluating the train system model. Only one trajectory can achieve the minimum energy consumption of the countless train trajectories leading to the destination within a given scheduled time. Based on this, the remaining spare running time of the train can be calculated by subtracting the minimum running time under speed limit conditions from the scheduled running time. This remaining spare running time can be allocated to different sections of the train route to achieve punctual arrival and reduce energy consumption during operation. As the energy consumed by the train during acceleration is proportional to the square of the speed, time should be allocated first to the higher-speed sections from the remaining spare running time to achieve a reduction in energy consumption [14]. Therefore, this paper proposes a novel algorithm for allocating the remaining spare running time.

To begin with, we use the entire stretch of the road divided into N segments s_i based on the speed limit. The trajectory with the shortest running time is used to obtain the minimum running time t_i^{min} for each segment. After that, each segment's scheduled running time t_i is initialized as the minimum running time t_i^{min}. The total spare running time T_{ts} can be calculated using the following formula:

$$T_{ts} = T_p - \sum_{t=0}^{N} t_i^{min} \tag{4}$$

where T_p is the scheduled running time for the route, and now we define the average speed $\overline{v}_i^s$ for segment s_i, calculated as follows:

$$\overline{v}_i^s = \frac{l_i}{t_i} = 1, 2, ..., N \tag{5}$$

where l_i is the length of segment s_i. Subsequently, J different blocks b_j are divided based on each segment's average speed $\overline{v}_i^s$. Segments with the same average speed $\overline{v}_i^s$ are grouped into the same block, and sorted by average speed, with block b_0 having the fastest average speed. It should be noted that segments within a block may not necessarily be adjacent. The average speeds $\overline{v}_0$ and $\overline{v}_1$ of blocks b_0 and b_1 are selected. The remaining spare time T_{rs} for block b_0 is calculated as follows:

$$T_{rs} = (\frac{1}{\overline{v}_1} - \frac{1}{\overline{v}_0}) \cdot L \tag{6}$$

where L is the length of block b_0. If the remaining spare running time T_{rs} is less than the total spare running time T_{ts}, then $T_{rs} = T_{ts}$. The segment's scheduled running time t_i are allocated to the segments s_i within block b_0, and the running time for each segment is updated, calculated as follows:

$$t_i = t_i + \frac{l_i}{L} \cdot T_{rs} \tag{7}$$

where l_i is the length of segment s_i. Meanwhile, update the total spare running time T_{ts}, calculated as follows:

$$T_{ts} = T_{ts} - T_{rs} \tag{8}$$

Finally, the new average speeds $\overline{v}_i^s$ for the segments of block b_0 are recalculated, and blocks set B are updated. The spare running time continues to be allocated until the total spare time $T_{ts} = 0$. Block b_0 will merge with block b_1 to form a new block b_0, with the new average speed of block b_0 becoming the average speed v_1 of the original block b_1. Based on the above definitions and derivations, each segment is assigned a running time and average speed to ensure the punctuality of trains and reduce energy consumption.

Planning Speed Intervals. To reduce a large number of infeasible state spaces, we propose an algorithm that basically satisfies the optimal control sequence derived from PMP. The algorithm first generates a suboptimal train trajectory, upon which it plans a speed interval to allow the agent to learn meticulously within this filtered space and obtain the optimal train trajectory. Based on the Allocating Spare Time (AST) calculated from $\overline{v}_i^s$ and t_i, a suboptimal train trajectory is found using three operating modes: full traction, cruising, and full braking mode.

The first and the last segment adopt modes t and b, respectively. The other segments are separated from the middle and divided into two parts, the first half implements mode t, and the second half implements mode b. Two operating modes, full Traction and Cruising (Tr-Cr) mode t and Cruising and full Braking (Cr-Br) mode b, are provided for it.

The starting point x_s and the ending point x_f of the segment are defined. Then, $v(x)$ is calculated. Firstly, the low point x_l of the s_k segment is set to x_s, the high point x_h is set to x_f, and the midpoint x_m is set to $\frac{x_s+x_f}{2}$. The operating mode is assigned to the train based on the segment position, as shown below:

$$m(x_i) = \begin{cases} Tr, & if\ x_i < x_m, mode = t, \\ Br, & if\ x_i > x_m, mode = b, \\ Cr, & otherwise. \end{cases} \quad (9)$$

Based on the operating mode, calculate the speed profiles as follows:

$$v(x_i) = \sqrt{v(x_{i-1})^2 + 2 \cdot \frac{m(x_i) \cdot F(v(x_{i-1})) - C(v(x_{i-1})) - S(x_i)}{\rho \cdot M \cdot v(x_i)} \cdot \Delta l} \quad (10)$$

The actual running time t_i^a is calculated as follows:

$$t_i^a = \int_{x_{i-1}}^{x_i} \frac{dx}{v(x)} \quad (11)$$

If the actual running time of segment t_i^a is less than the scheduled running time of segment t_i, then $x_h = x_m - 1$. Otherwise, $x_l = x_m + 1$. The midpoint x_m is recalculated as follows:

$$x_m = (x_l + x_h)2 \quad (12)$$

Update the set of running times until $x_h = x_l$ to continue end the loop. Then continue calculate $v(x)$ until all segments have been traversed, where the starting speed is equal to the ending speed of the previous speed profiles.

There are two-speed profiles for $v(x)$, namely the Tr-Cr profiles or the Cr-Br profiles. Therefore, it is evident that $v(x)$ has monotonicity, so the running time t_i^a of the segment also has monotonicity and has only one solution. Through the above method, a suboptimal train trajectory $v(x)$ is calculated, and the upper bound $v_{ub}(x)$ and the lower bound $v_{lb}(x)$ of the planned speed interval are defined, ensuring that the speed profiles are always below the Automatic Train Protection (ATP) curve. The definition of planned speed interval is as follows.

$$v_{ub}(x_i) = \begin{cases} v(x_i) \cdot w_{pf}, & if\ v(x_i) > v_{min}(x_i), \\ v_{min}(x_i), & otherwise. \end{cases} \quad (13)$$

$$v_{lb}(x_k) = v(x_k) \cdot w_{nf} \quad (14)$$

where w_{pf} and w_{nf} are positive weights. Finally, the energy consumed by $V(x)$ will be used as the evaluation indicators of the train energy efficiency E_{astpsi} for the next section, and $v(x)$ will be Gaussian blurred to provide initial data for early experience replay memory. The planned speed interval (PSI) is generated based on the above definitions and derivations. Since this space contains both the optimal solution and many suboptimal solutions, the agent can quickly learn energy-saving, punctuality, and accurate stopping skills in this filtered space.

3 Principles of Deep Reinforcement Learning

3.1 State and Environment

This paper is based on the model described in the Sect. 2.1, a simulation environment is established that is based on the dynamic train model and considers speed restrictions, gradients, and train parameters. Based on input actions, the environment can provide feedback on correct acceleration, speed, time, and energy consumption.

The upper and lower bounds of the PSI are calculated in Sect. 2.2. They are used as state information for the agent, allowing it to know the boundary of the search space to mitigate the negative effects of sparse rewards and accelerate convergence.

The train operation states involve the current speed V_i, the remaining distance to the destination D_i, the remaining scheduled running time T_i, and the upper bounds V_i^{ub} and lower bounds V_i^{lb} of the PSI at the current point. Additionally, research [18] shows that utilizing layer normalization for the state information can increase the agent's stability and slightly improve the convergence speed.

3.2 Agent and Action

The agent corresponds to the driver of the train in DRL and makes appropriate actions according to the environment to drive the train. We define the action of the agent as follows: $A = [M(x), C(x)],\ M(x), C(x) \in \{-1, 1\}$. The action A is a continuous action space, where $M(x)$ is the output percentage of the maximum traction/braking effort. $M(x) > 0$ represents the traction effort output and vice versa for the braking effort. The coasting action is taken when $C(x) > 0$. Based on the action input in the current state, a new state is generated and returned to the agent. Additionally, when the speed exceeds the PSI, the ASTPSI algorithm controls the speed to remain within the boundary until the current action can return to the PSI.

3.3 Reward Function

The main factors that need to be considered in the TTO issue are energy saving, punctuality, and accurate stopping [17]. Therefore, these three evaluation indicators of the train's operation trajectory are considered factors for the reward function. The actual running time T of the train's trajectory $v(x)$ from the starting point x_s to the ending point x_f can be described as follows:

$$T = \int_{x_s}^{x_f} \frac{dx}{v(x)} \tag{15}$$

where $T(v)$ is the maximum traction effort at speed v, and $A_t(x)$ is the action of traction at point x. The actual energy consumption E of the train from the starting point x_s to the ending point x_f can be described as follows:

$$E = \int_{x_s}^{x_f} T(v) \cdot A_t(x)dx \tag{16}$$

In order to align with the platform doors, the train's parking accuracy should be within 0.3 m [11]. The accurate stopping A can be described by the actual running distance x_a as follows:

$$A = |x_f - x_a| \tag{17}$$

The reward function R comprises the following three parts: accurate stopping reward, punctuality arrival reward, and energy efficiency reward.

Regarding the accurate stopping reward, it can be described as follows:

$$r_a = \begin{cases} C_m - C_a \cdot A^2, & if\ r_a < C_m, \\ C_m, & otherwise. \end{cases} \tag{18}$$

Regarding the punctuality reward, it can be described as follows:

$$r_t = \begin{cases} C_m - C_t \cdot \frac{(T_p - T)^2}{T_p}, & if\ r_t < C_m, \\ C_m, & otherwise. \end{cases} \tag{19}$$

Regarding the energy efficiency reward, it can be described as follows:

$$r_e = \begin{cases} C_m - C_e \cdot \frac{(E_{astpsi} - E)^2}{E_{astpsi}}, & if\ r_e \leq C_m \\ C_m, & otherwise. \end{cases} \tag{20}$$

where C_m is a relatively large positive weight that prevents any one reward from being too dominant. C_a, C_t, and C_e are positive weights to balance the three objectives. T_p is the planned running time for the line, and E_{astpsi} is the train energy consumption calculated using the ASTPSI algorithm.

4 Experiment

This section presents the adaptability of the train trajectory simulation environment and demonstrates the performance of ASTPSI-DRL through a series of experiments. The experiments were conducted on an actual section of the Chongqing Rail Transit Line 2, which is approximately 1.1 km long and has various speed limits and more significant gradients. Table 1 shows the gradient and speed limit data. Actual experiments were conducted on a straight track to determine the characteristics of the train in traction mode. Based on the line data and train characteristics, the minimum running time was calculated to be 74.2 s, while the planned running time was set to 93 s.

Table 1. Speed Limits and Gradients Data.

Item	Value	Segment(m)
Speed Limits (km/h)	50	[0,120]
	75	[120,920]
	50	[920,1056]
Gradients (‰)	-3	[0,400]
	27.8	[400,641]
	-12.439	[641,971]
	0	[971,1056]

4.1 ASTPSI-DRL with Scheduled Running Time

In this experiment, we explore the effect of ASTPSI-DRL on TTO by comparing ASTPSI and ASTPSI-TD3. The agent outputs specific actions to achieve train operation based on the PSI generated by the ASTPSI algorithm and the calculated energy weighting E_{astpsi}. Figure 1(a) compares train trajectories between the ASTPSI-TD3 and ASTPSI methods. Figure 1(b) corresponds to energy/time consumption between the ASTPSI-TD3 and ASTPSI methods, where the dotted line represents the energy consumption curve and the solid line represents the time consumption curve.

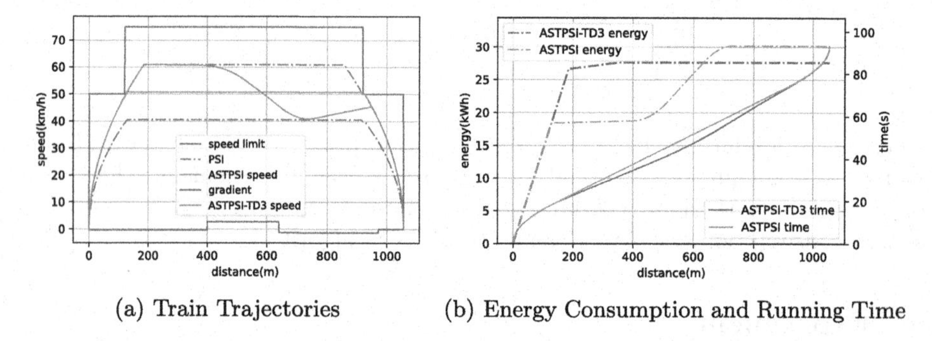

(a) Train Trajectories (b) Energy Consumption and Running Time

Fig. 1. Train Trajectories and Energy Consumption and Practical Running Time from the ASTPSI-TD3 and ASTPSI.

Figure 1 demonstrates that ASTPSI-TD3 and ASTPSI employ the maximum power output during the traction phase, overlapping the speed, energy consumption, and running time profiles. Subsequently, ASTPSI-TD3 adopts the maximum traction up to the upper limit of the PSI and maintains it for a certain distance in this range. The time saved by running at a higher speed is compensated in the following coasting phase to ensure the arrival punctuality of the train. Additionally, although ASTPSI-TD3 consumes more energy in the early stages, it utilizes the downhill sections of 641m to 971m to accelerate through coasting, thereby offsetting the energy consumption during the early traction phase and ultimately consuming less energy than ASTPSI.

4.2 Performance Comparison

In this experiment, we use two algorithms, DDPG and TD3, to explore the learning robustness of ASTPSI-DRL when facing sparse rewards in the context of TTO. The relevant parameters of these reinforcement learning algorithms trained are shown in Table 2. During the training process, when the current speed deviates from the PSI, the ASTPSI algorithm executes actions until the agent's actions return to the PSI. When the agent reaches the destination point, the training is stopped and reset to the initial state for further training.

Table 2. Parameters in the Deep Reinforcement Learning.

Item	DDPG	TD3
Training episode e	[0,40000]	[0,40000]
learning rate for critic network α_c	0.005	0.005
learning rate for actor network α_a	0.005	0.005
Soft replacement factor τ	0.01	0.01
Reward discount factor γ	0.99	0.99
Replay memory size $\mathcal{M}$	10^6	10^6
Batch size $\mathcal{B}$	512	512
Structure of critic network	$256 \times 256 \times 1$	$256 \times 256 \times 1$
Structure of actor network	$256 \times 256 \times 2$	$256 \times 256 \times 2$

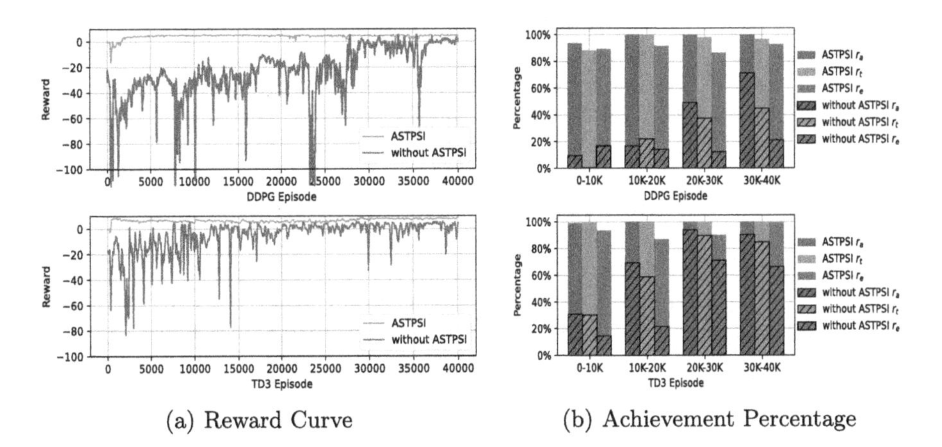

(a) Reward Curve (b) Achievement Percentage

Fig. 2. Performance Comparison of DDPG and TD3 Before and After Using SATRSI.

Figure 2(a) shows the reward curve for DDPG and TD3 before and after using the ASTPSI algorithm for training. The agent cannot escape local optimal solutions because of sparse rewards when ASTPSI is unused. As a result, the reward curve exhibits long periods of fluctuation, leading to slow and unstable convergence. As a result, DDPG and TD3 still generate many infeasible solutions even after completing their training. When ASTPSI is used, feasible solutions generated by using ASTPSI actions along with Gaussian blur are added to the experience replay memory from approximately 0 to 1000 episodes, which helps the agent accelerate its training. After experiencing reward fluctuations for approximately 500 and 3000 episodes, TD3 and DDPG gradually train from almost all infeasible solutions to mostly feasible ones. After this stage, the reward curve gradually stabilizes and continues to produce more and more valuable actions. During the deceleration stopping stage, where the speed exceeds the PSI, the ASTPSI algorithm takes control. Figure 2(b) displays the distribution of the percentage of accurate stopping, punctuality, and energy efficiency completion achieved by DDPG and TD3 every episode before and after using the ASTPSI algorithm. Compared to not using the ASTPSI algorithm, the agent can quickly learn the skill of accurate parking and produce feasible solutions more stably.

Table 3. Practical Energy Efficiency, Accurate Stopping and Punctuality of GA, ASTPSI, and ASTPSI-DRL Under Scheduled Running Time.

Item	ASTPSI	GA	DDPG	ASTPSI-DDPG	TD3	ASTPSI-TD3
Energy (kWh)	30.135	30.343	32.425	**28.846**	29.726	**27.644**
Energy Saving P_A (%)	–	–0.690	–7.599	**4.277**	1.357	**8.266**
Energy Saving P_G (%)	0.685	–	–6.862	**4.934**	2.033	**8.895**
Time (s)	93.097	91.342	96.352	**91.670**	94.871	**93.435**
Time Error T_E (%)	0.104	1.751	3.785	**1.430**	2.012	**0.462**
Accurate Stopping (m)	0.00	0.21	0.25	**0.04**	0.14	**0.01**

Energy Saving P_A = (Actual energy - Energy of ASTPSI) / Energy of ASTPSI
Energy Saving P_G = (Actual energy - Energy of GA) / Energy of GA
Time Error T_E = |Scheduled time - Actual time| / Scheduled time

Table 3 provides a detailed comparison of two reinforcement learning methods with and without the ASTPSI algorithm, and the runtime and energy savings achieved by ASTPSI and the Genetic Algorithm (GA) within the current planned operating time (93 s), as well as the accuracy of parking. ASTPSI-DRL performs better than the Genetic Algorithm and conventional DRL algorithms in reducing train energy consumption, improving accurate stopping, and achieving punctuality.

5 Related Work

PMP-Based Method: PMP aims to solve the optimal control differential equations and identify the shift points and optimal state sequence; it is a commonly used method for solving the TTO issue. Howlett et al. [6] theoretically demonstrated, based on the PMP, that the optimal energy-saving operation sequence for trains is "maximum traction-cruising-coasting-maximum braking." Liu et al. [9] provided a numerical algorithm for solving the transition points between energy-saving operation conditions. Howlett et al. [7] discovered that trains on tracks with steep slopes should implement maximum traction before entering the uphill section. Albrecht et al. [1] proved that the optimal switching points in each slope region of the track are uniquely defined, thus deriving the uniqueness of the globally optimal strategy. Cao et al. [2] approximated all non-linear constraints using piecewise-affine (PWA) functions and transformed the trajectory optimization issue into a mixed-integer linear programming (MILP) issue, which can be solved using existing solvers.

DRL-Based Method: Many researchers are attempting to use DRL to optimize train trajectories for practical use. However, applying DRL to solve the TTO issue presents a challenge due to the issue of sparse rewards. In such cases, the agent's learning process becomes complicated and slow. This paper will introduce some recent research progress in this area. Ladosz et al. [8] provided a comprehensive overview of existing solutions to address the issue of sparse rewards, such as rewarding novel states, diversified behavior, and goal-based methods. Zhu et al. [17] proposed an eco-driving method based on Q-learning to determine the optimal energy allocation strategy, effectively reducing the space complexity of the state method. Ning et al. [12] proposed an algorithm that allocates running time, selects running modes to improve train punctuality, and provides a series of discrete running modes as recommended actions for the model. Additionally, Zhang et al. [15] proposed a train operation method based on ERL (DRTO) and designed a dynamic incentive system that provides motivating rewards for the agent. Shang et al. [13] proposed a DRL method with a reference system (DRL-rs) to check and correct the agent's learning progress, avoiding getting further away from the correct path due to the difficulty of sparse rewards. Zhou et al. [16] proposed two intelligent train operation algorithms by combining expert knowledge with reinforcement learning algorithms, namely, the STO algorithm (STOD) based on deep deterministic policy gradients and the STON algorithm based on normalized advantage functions.

6 Conclusion

In this paper, we propose the ASTPSI algorithm, which allocates the remaining spare time to sections with higher average speeds and generates a planned speed interval, thereby reducing the unreasonable and infeasible solution space and alleviating the learning difficulty caused by sparse rewards. It significantly improves the agent learning efficiency with a more reasonable trajectory, which can reduce the blindness of exploration dramatically.

We conduct experiments on a section of Chongqing Rail Transit Line 2, and the results show that the trained agents significantly improve their convergence rate and operation quality and can generate excellent train speed curves under given scheduled running times. Based on the research results of this paper, our future research will further improve the ASTPSI algorithm and the PSI algorithm of upper and lower bounds programming to reduce the size of the search space further, and study the TTO issue of high-speed trains with even sparser rewards.

Acknowledgment. This work was supported by the Graduate Student Research Innovation Program of Chongqing Jiaotong University (CYS23515).

References

1. Albrecht, A.R., Howlett, P.G., Pudney, P.J., Vu, X.: Energy-efficient train control: from local convexity to global optimization and uniqueness. Automatica **49**(10), 3072–3078 (2013)
2. Cao, Y., Zhang, Z., Cheng, F., Shuai, S.: Trajectory optimization for high-speed trains via a mixed integer linear programming approach. IEEE Trans. Intell. Transp. Syst. **23**(10), 17666–17676 (2022)
3. Chen, J., et al.: Integrated regenerative braking energy utilization system for multi-substations in electrified railways. IEEE Trans. Industr. Electron. **70**(1), 298–310 (2022)
4. Deng, K., et al.: An adaptive PMP-based model predictive energy management strategy for fuel cell hybrid railway vehicles. eTransportation **7**, 100094 (2021)
5. Dong, H., Ning, B., Cai, B., Hou, Z.: Automatic train control system development and simulation for high-speed railways. IEEE Circuits Syst. Mag. **10**(2), 6–18 (2010)
6. Howlett, P.: An optimal strategy for the control of a train. ANZIAM J. **31**(4), 454–471 (1990)
7. Howlett, P.G., Pudney, P.J., Vu, X.: Local energy minimization in optimal train control. Automatica **45**(11), 2692–2698 (2009)
8. Ladosz, P., Weng, L., Kim, M., Oh, H.: Exploration in deep reinforcement learning: a survey. Inf. Fusion **85**, 1–22 (2022)
9. Liu, R.R., Golovitcher, I.M.: Energy-efficient operation of rail vehicles. Transp. Res. Part A: Policy Pract. **37**(10), 917–932 (2003)
10. Liu, W., Shuai, S., Tang, T., Wang, X.: A DQN-based intelligent control method for heavy haul trains on long steep downhill section. Transp. Res. Part C: Emerg. Technol. **129**, 103249 (2021)
11. Lu, M., Ou, D., Hua, Z., Gu, L.: Analysis of stopping accuracy deviation of urban rail transit train in ATO driving Mode. In: Qin, Y., Jia, L., Liang, J., Liu, Z., Diao, L., An, M. (eds.) Proceedings of the 5th International Conference on Electrical Engineering and Information Technologies for Rail Transportation (EITRT) 2021. EITRT 2021. LNEE, vol. 868. Springer, Singapore (2022). https://doi.org/10.1007/978-981-16-9913-9_72

12. Ning, L., Zhou, M., Hou, Z., Goverde, R.M.P., Wang, F.-Y., Dong, H.: Deep deterministic policy gradient for high-speed train trajectory optimization. IEEE Trans. Intell. Transp. Syst. **23**(8), 11562–11574 (2021)
13. Shang, M., Zhou, Y., Fujita, H.: Deep reinforcement learning with reference system to handle constraints for energy-efficient train control. Inf. Sci. **570**, 708–721 (2021)
14. Xiao, Z., Wang, Q., Sun, P., You, B., Feng, X.: Modeling and energy-optimal control for high-speed trains. IEEE Trans. Transp. Electrification **6**(2), 797–807 (2020)
15. Zhang, L., Zhou, M., Li, Z., et al.: An intelligent train operation method based on event-driven deep reinforcement learning. IEEE Trans. Industr. Inf. **18**(10), 6973–6980 (2021)
16. Zhou, K., Song, S., Xue, A., You, K., Hui, W.: Smart train operation algorithms based on expert knowledge and reinforcement learning. IEEE Trans. Syst. Man Cybern. Syst. **52**(2), 716–727 (2020)
17. Zhu, Q., Shuai, S., Tang, T., Liu, W., Zhang, Z., Tian, Q.: An eco-driving algorithm for trains through distributing energy: a Q-learning approach. ISA Trans. **122**, 24–37 (2022)
18. Zhuang, D., Gan, V.J.L., Tekler, Z.D., Chong, A., Tian, S., Shi, X.: Data-driven predictive control for smart HVAC system in IoT-integrated buildings with time-series forecasting and reinforcement learning. App. Energy **338**, 120936 (2023)

Amortized Variational Inference via Nosé-Hoover Thermostat Hamiltonian Monte Carlo

Zhan Yuan(✉), Chao Xu, Zhiwen Lin, and Zhenjie Zhang

Unit 91977 of PLA, Beijing, China
Yuanzhan202306@163.com

Abstract. Sampling latents from the posterior distribution efficiently and accurately is a fundamental problem for posterior inference. Markov chain Monte Carlo (MCMC) is such a useful tool to do that but at the cost of computational burden since it needs many transition steps to converge to the stationary distribution for each datapoint. Amortized variational inference within the framework of MCMC is thus proposed where the learned parameters of the model are shared by all observations. Langevin autoencoder is a newly proposed method that amortizes inference in parameter space. This paper generalizes the Langevin autoencoder by utilizing the stochastic gradient Nosé-Hoover Thermostat Hamiltonian Monte Carlo to conduct amortized updating of the parameters of the inference distribution. The proposed method improves variational inference accuracy for the latent by subtly dealing with the noise introduced by stochastic gradient without estimating that noise explicitly. Experiments benchmarking our method against baseline generative methods highlight the effectiveness of our proposed method.

Keywords: variational inference · Hamiltonian Monte Carlo · generative model · autoencoder · neural network

1 Introduction

Capturing the underlying statistical structure from real world data in high dimension space to produce new reasonable samples is a challenging task for generative models, since the distributions to be modeled in the framework of Bayesian sampling are generally intractable in most cases. As a kind of generative model, variational autoencoders (VAEs) [1] have received much attention [2–8] since it provides a principled framework for learning inference model (encoder) and generative model (decoder) simultaneously. Especially, VAEs share variational parameters across observations and needn't iteratively optimize the inference model for each datapoint, and thus, the cost of variational inference can be amortized. In practice, VAEs are suitable to deal with the problems with large data-sets.

Generally, the success of generative models including VAEs greatly depends on how closely the parameterized inference posterior distribution approximate the true yet intractable posterior distribution [9, 10], due to the fact that the intractable evidence

B. Luo et al. (Eds.): ICONIP 2023, LNCS 14447, pp. 78–90, 2024.
https://doi.org/10.1007/978-981-99-8079-6_7

lower bound (ELBO) is usually used as the proxy objective function to optimize the model parameters rather than the intractable marginal likelihood of the observations. Hence, in order to reduce the gap between the ELBO and the marginal likelihood, we should make sure that the inference posterior distribution be accurate enough, while being computationally efficient and easy to sample from. These requirements place strict limitation on the choise of the family of inference posterior distribution and often lead to use of simple distributions such as Gaussian posterior. Unfortunately, these simple distributions generally lack the flexibility to approximate non-Gaussian and multi-modal posteriors. As a result, generative models with these inference distributions produce sub-optimal latents and thus generate unrealistic new samples.

To improve the flexibility and the accuracy of the inference posterior, resent work like Hamiltonian Variational Auto-Encoder [11] and Monte Carlo Variational Auto-Encoders [12] has sought to combine Markov Chain Monte Carlo (MCMC) with variational inference [13, 14], given the guarantee of asymptotically exacting convergence of MCMC to the true posterior. Although achieving superior performance over ordinary VAEs, MCMC-based methods take a long time to perform stochastic transition operator for each data-point, bringing about high computation cost for the approximation procedure. Inspired by the amortized strategy used in VAE, the Langevin autoencoder (LAE) was proposed [15], where the Langevin dynamics is exploited to generate latent samples that are converged to the true posterior distribution. The main difference from other MCMC-based methods is that LAE updates the parameters of the variational inference distribution iteratively rather than data-point updating to generate latents.

Following LAE, in this paper, we introduce an alternative approach where the more sophisticated stochastic gradient Nosé-Hoover Thermostat Hamiltonian Monte Carlo [16, 17] is exploited to learn the parameters of the variational inference distribution, which we refer to as the Nosé-Hoover Thermostat autoencoder (HMCAE). Compared with Langevin dynamics, stochastic gradient Nosé-Hoover Thermostat Hamiltonian Monte Carlo can fully explore the modality embedded in data space. By doing so, our proposed method can improve approximation capacity of the inference model and bring the ELBO closer to the true marginal likelihood objective. Experiments on real world data-set conformed the effectiveness of our proposed method.

2 Background

2.1 The Variational Autoencoder

Given a dataset $\{x^{(i)}\}_{i=1}^{N}$ observed from the real world, we hope to construct a probability model $p_\theta(x)$ with learnable parameter θ to fit the observed data. Generally, the parameter θ is optimized by maximizing the following log-likelihood function:

$$\theta^* = argmax_\theta E_{p_{data}(x)} logp_\theta(x). \quad (1)$$

Under the generative model framework, the observed samples are assumed to be generated by some latent variables z representing the underlying structure of the real world data, and we have:

$$p_\theta(x) = \int p_\theta(x, z)dz = \int p_\theta(x|z)p(z)dz, \quad (2)$$

where $p(z)$ is the prior distribution, $p_\theta(x|z)$ is the generative distribution.

Since true data distribution $p_\theta(x)$ is generally unavailable in practice, the integration in Eq. (2) is usually intractable and the parameter θ is hard to be estimated. To deal with this optimization problem, various approximate methods are proposed. Variational AutoEncoders (VAE) introduced by [1] gives an Evidence Lower Bound (ELBO) to the log-likelihood as follows:

$$\begin{aligned}\mathcal{L}_{\theta,\phi}(x) &= \int log\left(\frac{p_\theta(x,z)}{q_\phi(z|x)}\right)q_\phi(z|x)dz \\ &= E_{q_\phi(z|x)}(logp_\theta(x|z)) - D_{KL}\big(q_\phi(z|x)||p(z)\big) \\ &= logp_\theta(x) - D_{KL}\big(q_\phi(z|x)||p_\theta(z|x)\big) \\ &\leq logp_\theta(x),\end{aligned} \tag{3}$$

where $q_\phi(z|x)$ is the variational inference distribution which approximates the true posterior distribution $p(z|x)$, and ϕ is the learnable distribution parameter. By optimizing the ELBO, we can infer the joint parameters $\{\phi, \theta\}$ under certain assumptions on the encoder $q_\phi(z|x)$ and the decoder $p_\theta(x|z)$ which are parameterized by different neural networks respectively. Generally speaking, we hope $q_\phi(z|x)$ can approximate p(z|x) accurately enough to make the learned latent space close to the data manifold, as shown in (3), in terms of Kullback-Leibler (KL) divergence, the better $q_\phi(z|x)$ approximates $p(z|x)$, the smaller the gap between the marginal likelihood and the ELBO. However, for the sake of easier sampling and lower computation, simple parametric variational distribution families are considered, and the produced latent variables may fail to capture the true statistical property.

2.2 Langevin Autoencoder

In generative model, calculating the loss function generally requires sampling the latent variable z from the posterior distribution $q_\phi(z|x)$, which is generally realized by the MCMC method. In the framework of auto-encoder, we could evolve the latent variable z or the posterior distribution parameter ϕ according to the random process established by MCMC whose stationary distribution is the interesting posterior distribution. Langevin Autoencoder adopted the later way by which the posterior distribution parameter ϕ is shared by observation samples. To obtain latent variables that approach to the true posterior distribution, Langevin Autoencoder utilizes Langevin MCMC method to establish a sampling chain that converges to the true posterior distribution. It also constructs an amortized framework where the updates of the encoder neural network in parameter space replace the data-point MCMC iterations in latent space through the Langevin dynamics. By doing so, the latent is implicitly updated via the encoder. Interestingly, as proved in [15], diffusing only the last linear layer of the encoder $f_{z|x}$ can get the variational posterior distribution $q(Z|X)$ induced by Eq. (2) converge to the stationary distribution, i.e.,

$$q(Z|X) := \prod\nolimits_{i=1}^{n} q\left(z^{(i)}|x^{(i)}\right) \propto exp\left(-\sum\nolimits_{i=1}^{n} U\left(x^{(i)}, z^{(i)}\right)\right). \tag{4}$$

In the following, we denote the encoder network as $f_{z|x} = \phi g(x; \psi)$, where $\phi \in R^l$ represents l-dim weight parameter vector of the last layer of the encoder, ψ is the parameter set of the remaining layers of the encoder. In LAE, ψ is updated by gradient ascent criterion and ϕ diffuses according to an SDE defined as follows:

$$d\phi_t = h(\phi_t, t)dt + \sigma(\phi_t, t)dB_t, t \in [0, T], \quad (5)$$

where $h(\phi_t, t)$ and $\sigma(\phi_t, t)$ are the drift coefficient and the diffusion coefficient of the SDE respectively, $\{B_t\}$ is thestandard Wiener process. In LAE, $\sigma(\phi_t, t) = \sqrt{2d}$ is a scalar, $h(\phi_t, t)$ is the summation of the scores of $p_\theta(x, z)$ over minibatch $\{x^{(i)}\}_{i=1}^n$:

$$h(\phi_t, t) := -\nabla_{\phi_t} \sum_{i=1}^{n} U\left(x^{(i)}, f_{z|x}\left(x^{(i)}; \phi, \psi\right); \theta\right), \quad (6)$$

$$U(x, z; \theta) = -logp_\theta(x, z), \quad (7)$$

where $f_{z|x}$ is the deterministicencoder. The Euler–Maruyama discretization method is utilized to solve the SDE (5) and we run M steps of Langevin MCMC to update the parameters of the encoder:

$$\phi_{m+1} = \phi_m - \eta \sum_{i=1}^{n} \nabla_\phi U\left(x^{(i)}, z^{(i)} = f_{z|x}\left(x^{(i)}; \phi_m, \psi\right); \theta\right) + \sqrt{2\eta}\varepsilon_m, m = 0, 1, \ldots M, \quad (8)$$

where $\eta = \frac{T}{M}$ is the step size, and ε_m is standard normal. Every step in Eq. (8) is followed by a Metropolis-Hastings (MH) rejection step to decider whether to accept $\phi\prime$ or not. The acceptance rate is calculated as follows:

$$\alpha_t = min\left\{1, \frac{exp(-V(\phi_{t+1})q(\phi_t|\phi_{t+1}))}{exp(-V(\phi_t)q(\phi_{t+1}|\phi_t))}\right\}. \quad (9)$$

Once $\phi\prime$ is updated for M times, for each data $x^{(i)}$ in minibatch $\{x^{(i)}\}_{i=1}^n$, its corresponding latent $z^{(i)}$ is updated by the encoder with new parameters $\phi\prime$. In this sense, the computation of inferring latent from the variational posterior distribution $q_\phi(z|x)$ for each $x^{(i)}$ is amortized by inferring ϕ which are shared by all data points in a minibatch.

Similar idea is presented in [19], where only the last linear layer of an L-layer neural network is updated by the Laplace approximation, and the feature extractor defined by the first L-1 layers remain to be estimated by maximum a-posterior.

3 Method

Inspired by [15–17], we propose an alternative approach to LAE which generalizes the Langevin dynamics in LAE to more sophisticated Hamiltonian Langevin dynamics, i.e., stochastic gradient Nosé-Hoover Thermostat dynamics. Figure 1 shows its diagram. In summary, the introduction of stochastic gradient Nosé-Hoover Thermostat dynamics can bring at least two benefits for our generative model: implicitly eliminating error introduced by the noise when calculating the gradient based on minibatch and increasing the acceptance rate due to the energy-preserving property of the Hamiltonian Langevin dynamics.

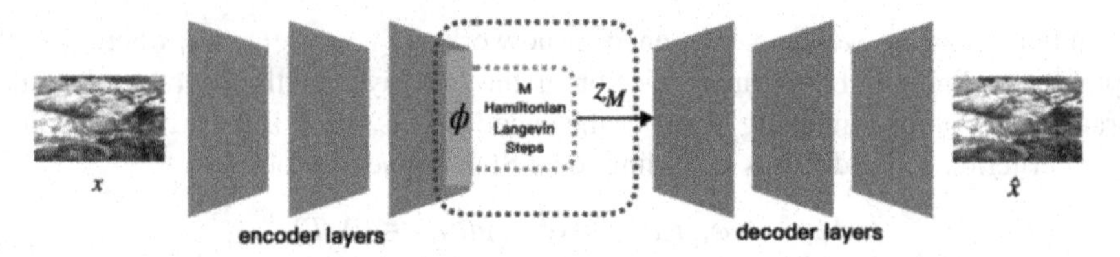

Fig. 1. A diagram showing HMCAE. The difference between traditional VAE and HMCAE is that the parameters of the last layer of the encoder are optimized by stochastic gradient Nosé-Hoover Thermostat Hamiltonian Monte Carlo, while the remaining parameters of the encoder and the decoder are optimized via gradient descent.

3.1 Amortized Variational Inference via Nosé-Hoover Thermostat Dynamics

In our proposed method, the parameter of the last layer of the encoder is updated by the Hamiltonian-like dynamics as shown in Fig. 1. By appropriately updating the parameter of the encoder, we could sample from the encoder to obtain accurate latent variables which could be used to generate new samples. There are different ways to realize that depending on the choise of the dynamics [20]. For traditional Hamiltonian Langevin dynamics, an auxiliary velocity variable ρ and the interesting position variable ϕ corresponding to the parameters of the last layer of the encoder in our model constitute a Hamiltonian dynamics $H(\phi, \rho)$ which is described by the following stochastic differential equations:

$$d\rho_t = -\xi \rho_t dt - \nabla_\phi V(\phi_t)dt + \sqrt{2\zeta} dB_{t,} \tag{10}$$

$$d\phi_t = \rho_t dt, \tag{11}$$

where ξ is the friction coefficient, ζ is the intensity of the external random force. Under mild assumptions on the function $V(\phi)$, the Markov process (ϕ_t, ρ_t) has a unique stationary distribution which is proportional to $exp(-H(\phi, \rho))$, where $H(\phi, \rho) = V(\phi) + K(\rho)$ is the Hamiltonian energy that is conserved during the Markov process, $V(\phi)$ is the potential energy, $K(\rho) = \frac{1}{2\sigma^2}\|\rho\|^2$ is the kinetic energy. Compared to the Langevin dynamics (3), HMC dynamics introduce the velocity variable for the parameters of the inference model and, consequently, the complex learning tasks are conducted in an augmented space instead of the parameter space, which makes the exploring of different distribution landscapes more efficient [21]. In that sense, the update of parameters of the inference model and the momentum variable coupled in the joint parameters-velocity space helps to capture the posterior distribution more efficiently and smoothly. Due to its properties of energy-preservation and reversibility, HMC has been received wide application to generate samples from target distribution.

However, in order to be scaled to the problems with very large data-sets, the gradients of the potential energy in Eq. (10) are evaluated using mini-batches in various models in lieu of full-batch, resulting in the so-called stochastic gradient method. Stochastic gradient brings run-time efficiency but also unavoidably introduces stochastic noise to the estimation of the gradient in Eq. (10) causing inference samples from the converged chain will not follow the target distribution. However, accurate estimation of the stochastic noise is very hard. To reduce the negative influence of the stochastic gradient noise, we sought to find more appropriate HMC methods to perform variational inference.

In this paper, we make use of the so-called Nosé-Hoover thermostat dynamics [16–18] to replace the Langevin dynamics in LAE, which extends the traditional Hamiltonian Langevin dynamics by introducing a thermostat. This thermostat plays an important role in controlling thermal equilibrium that makes the dynamics adaptive to the noise introduced by the stochastic gradient without having to estimate it, which could result in more accurate estimation of the gradient of the potential energy. Concretely, in Nosé-Hoover thermostat dynamics, a fixed temperature T is introduced. The variables ξ and ζ in Eq. (10) have the following relation, $\zeta = \xi K_B T$, where K_B is the Boltzmann constant. The probability of the states (ϕ, ρ) in a canonical ensemble instead follows the canonical distribution $exp(-H(\phi, \rho)/K_B T)$. To keep thermal equilibrium and make sure the dynamics could correctly simulate the canonical ensemble, the following condition should be satisfied:

$$\frac{K_B T}{2} = \frac{1}{n} E[K(\rho)]. \tag{12}$$

Equation (12) means that the kinetic energy is constant and equal to equilibrium value.

Since this noise couldn't be estimated accurately and explicitly, the thermal equilibrium condition breaks, and the dynamics (10) may drift away from thermal equilibrium and produce inaccurate samples. In order to generate correct samples, Nosé-Hoover thermostat dynamics makes the friction coefficient in Eq. (10) called thermostat change its value according to the ρ variable in order to adaptively control the mean kinetic energy. We adapt the SDE for thermostat presented in [16]. The whole Nosé-Hoover thermostat dynamics equation is as follows:

$$d\phi_t = \rho_t dt, \tag{13}$$

$$d\rho_t = -\nabla_\phi V(\phi_t) dt - \xi_t \rho_t dt + \sqrt{2\zeta} dB_t, \tag{14}$$

$$d\xi_t = \left(\frac{1}{n}\rho_t^T \rho_t - 1\right) dt, \tag{15}$$

where $\nabla_\phi V(\phi) = \frac{1}{n}\sum_{i=1}^{n} \nabla_\phi U\left(x^{(i)}, z^{(i)} = f_{z|x}\left(x^{(i)}; \phi, \psi\right); \theta\right)$. The dynamics presented by Eqs. (13)–(15) generalize the model in [16] to deep learning model where the important ingredient in the model, i.e., the potential energy, is realized and estimated by a neural network.

Euler–Maruyama discretization of Eqs. (13)–(15) gives rise to:

$$\rho_{m+1} = \rho_m - \xi_m \eta \rho_m - \eta \nabla_\phi V(\phi_m) + \sqrt{2\zeta\eta}\varepsilon_m, \tag{16}$$

$$\phi_{m+1} = \phi_m + \eta \rho_m, \tag{17}$$

$$\xi_{m+1} = \xi_m + \eta\left(\frac{1}{n}\rho_m^T \rho_m - 1\right), m = 0, 1 \ldots M. \tag{18}$$

Typically, due to the discretization error, each iteration of Eqs. (16)–(18) is followed by a MH rejection step. But in our proposed method, we choose to omit the MH step, mainly considering the thermostat variable can adaptively control the Hamiltonian energy, in which case we don't reject the proposal value.

3.2 Learning the Model Parameters

In the proposed method, we first learn ϕ using the discretization Eqs. (16)–(18), and then optimize the log-likelihood function to learn the remaining parameters ψ and θ for the encoder and decoder. These parameters are generally estimated using stochastic gradient, i.e., a minibatch of the observations $\{x^{(i)}\}_{i=1}^{n}$ drawn randomly from the entire data set are used to approximate the expectation in Eq. (1), which gives rise to:

$$\mathcal{L}(\psi,\theta) = \frac{1}{n}\sum\nolimits_{i=1}^{n} logp_{\psi,\theta}(x_i). \tag{19}$$

ψ and θ are updated by gradient descent:

$$\psi \leftarrow \psi - \lambda\nabla_{\psi}\mathcal{L}(\psi,\theta) = \psi - \lambda\frac{1}{n}\sum\nolimits_{i=1}^{n} \nabla_{\psi} logp_{\psi,\theta}\left(x^{(i)}\right), \tag{20}$$

$$\theta \leftarrow \theta - \lambda\nabla_{\theta}\mathcal{L}(\psi,\theta) = \theta - \lambda\frac{1}{n}\sum\nolimits_{i=1}^{n} \nabla_{\theta} logp_{\psi,\theta}\left(x^{(i)}\right), \tag{21}$$

where λ is the learning rate. The gradient of the marginal log-likelihood w.r.t. ψ,θ for each observation x is computed as follows:

$$\begin{aligned}
\nabla_{\psi,\theta} logp_{\psi,\theta}(x) &= \frac{1}{p_{\psi,\theta}(x)}\nabla_{\psi,\theta} p_{\psi,\theta}(x) = \frac{1}{p_{\psi,\theta}(x)}\nabla_{\psi,\theta}\int p_{\psi,\theta}(x,z)dz \\
&= \int \frac{\nabla_{\psi,\theta} p_{\psi,\theta}(x,z)}{p_{\psi,\theta}(x)}dz = \int q_{\phi,\psi}(z|x)\frac{\nabla_{\psi,\theta} p_{\psi,\theta}(x,z)}{p_{\psi,\theta}(x,z)}dz \\
&= E_{q_{\phi,\psi}(z|x)}\left[\nabla_{\psi,\theta} p_{\psi,\theta}(x,z)\right] \\
&= E_{q_{\phi,\psi}(z|x)}\left[\nabla_{\psi,\theta}(logp(z) + logp_{\theta}(x|z))\right].
\end{aligned} \tag{22}$$

Expectation in Eq. (22) is approximated by sampling z from $q_{\phi,\psi}(z|x)$:

$$\begin{aligned}
\nabla_{\psi,\theta} logp_{\psi,\theta}(x) &\approx \frac{1}{M}\sum\nolimits_{m=1}^{M} \nabla_{\psi,\theta} logp_{\theta}(x,z_m) \\
&= \frac{1}{M}\sum\nolimits_{m=1}^{M} \nabla_{\psi,\theta}(logp(z_m) + logp_{\theta}(x|z_m)),
\end{aligned} \tag{23}$$

where $z_m \sim q_{\phi,\psi}(z|x)$, M is the iteration number of the Nosé-Hoover thermostat dynamics.

Algorithm below summarizes the training procedure. It is worth noting that compared with LAE, the MH rejection step is not included in HMCAE, and the performance of our method is not effected as a consequence, as shown in our experiments.

Algorithm Nosé-Hoover Thermostat Autoencoder

Input: *Data-set $\mathcal{D}$, minibatch size n, number of HMC steps M, step-size η, learning rate λ.*

Randomly initialize the encoder network parameters $\{\phi, \psi\}$, the decoder network parameter θ, the momentum variable ρ, and thermostat variable ξ.

***while** not converged:*

select a minibatch $\{x^{(i)}\}_{i=1}^{n}$ from the data-set $\mathcal{D}$.

for** m=1 **to** M **do

$$V(\phi_m, \psi, \theta) = \frac{1}{n}\sum_{i=1}^{n} logp\left(x^{(i)}, z^{(i)} = f_{z|x}\left(x^{(i)}; \phi_m, \psi\right); \theta\right)$$

$$\rho_{m+1} = \rho_m - \xi_m \eta \rho_m - \eta \nabla_\phi V(\phi_m, \psi, \theta) + \sqrt{2\zeta\eta}\varepsilon_m$$

$$\phi_{m+1} = \phi_m + \eta \rho_m$$

$$\xi_{m+1} = \xi_m + \eta(\frac{1}{n}\rho_m^T \rho_m - 1)$$

end for

$$\psi \leftarrow \psi - \lambda \frac{1}{n} \nabla_{\psi,\theta} \sum_{i=1}^{n} logp_{\psi,\theta}(x^{(i)}) = \psi - \lambda \frac{1}{M} \nabla_\psi \sum_{m=1}^{M} V(\phi_m, \psi, \theta)$$

$$\theta \leftarrow \theta - \lambda \frac{1}{n} \nabla_{\psi,\theta} \sum_{i=1}^{n} logp_{\psi,\theta}(x^{(i)}) = \theta - \lambda \frac{1}{M} \nabla_\theta \sum_{m=1}^{M} V(\phi_m, \psi, \theta)$$

end while

return ϕ, ψ, θ

4 Experiments

In this section, we present a set of experiments to compare the proposed method with a variety of generative methods including VAE [1], DLGM [22], FLOW [23] and LAE [15]. We conducted experiments on binarized MNIST handwritten digit data-set, consisting of a training set and a validation set, at the present stage, and experiments evaluating our model on other complex data-set will be conducted in the future.

In our experiments, most of the settings of the hyper-parameters are consistent with [15] for fair comparison of various methods. The inference distribution and the generative distribution are modeled by fully connected deep neural networks. Since our work is to evaluate the efficiency of the propose method, the complex network architecture with regularization operations is not used in the experiments. The main hyper-parameters set in these experiments are given in Table 1. Two steps of iteration seem enough to obtain satisfactory results in our experiments. Similar conclusion appeared in [12] and [15], i.e., most of the improvement of the MCMC method benefits from the first two iterations. Of course, how many steps should we choose depends on the problems we study.

As we have noted that the minibatch size relates to the noise introduced by stochastic gradient, and in turn affect the sampling accuracy of latents, we set the minibatch size to 32, 64 and 128 to observe the change of the performance of different methods. Figures 2(a)–(c) show the evolution of the training loss for various methods during training with different batch sizes. We observe that our proposed method consistently achieves

Table 1. The main hyper-parameters set in our experiments

parameter	*values*
epochs	*40*
step size η	*1e-4*
batch size	*32, 64, 128*
latent dimensionality	*8*
learning rate	*1e-4*
HMC steps	*2*

lower training loss compared to baseline methods on training set in case of different batch sizes.

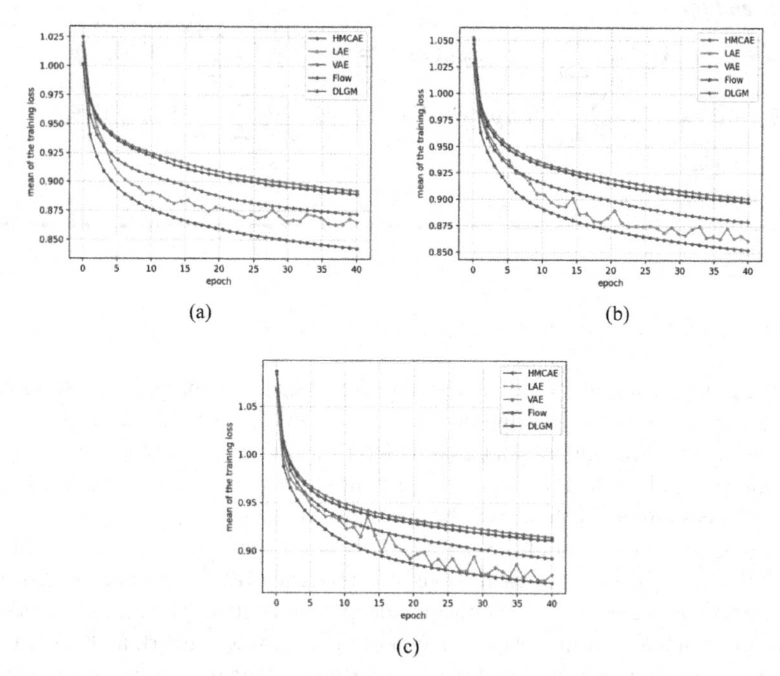

Fig. 2. Evaluation of the training loss for HMCAE, LAE, VAE, DLGM and FLOW with different minibatch sizes on the number of epoch. (a): batch size = 32, (b): batch size = 64, (c): batch size = 128.

Since we are mainly concerned with the accuracy of posterior inference in this paper, to verify that, we quantitatively compare the reconstruction error between the reconstructed image and the test image using mean square error. The minibatch size is set to 64 and 128. The reconstructed images are generated from the latents sampled from the approximated posterior distribution. Approximating the true posterior more accurately means lower reconstruction error and higher sample quality. Figures 3(a)–(b) show the results over numbers of training epochs on validation set. Tables 2–3 report the values measured by reconstruction error at different stages of the training. On the whole, our method outperforms other methods in terms of the effectiveness of posterior inference. Note that when the minibatch size increases (from 64 to 128 in our experiments), LAE and HMCAE achieve comparable reconstruction error since the larger the minibatch size, the more sufficiently the noises are averaged, which brings about more accurate latents. Figures 4–5 show the reconstruction error results for different generative methods with boxplot. We note that HMCAE achieves higher reconstruction precision and lower reconstruction variance than other generative methods.

Table 2. Results of reconstruction error of the different methods on MNIST (minibatch size = 64).

reconstruction error on test image									
number of epochs		5	10	15	20	25	30	35	40
Methods	VAE	2.531	1.507	1.48	1.253	1.269	1.16	1.314	1.151
	FLOW	2.556	1.529	1.436	1.259	1.181	1.113	1.133	1.087
	DLGM	**2.137**	1.522	1.426	1.311	1.254	1.276	1.175	1.063
	LAE	2.604	2.039	1.15	**1.056**	1.215	1.078	1.04	0.987
	HMCAE	**1.385**	**1.102**	1.087	**1.081**	**0.963**	**0.938**	**0.902**	**1.385**

Table 3. Results of reconstruction error of the different methods on MNIST (minibatch size = 128).

reconstruction error on test image									
number of epochs		5	10	15	20	25	30	35	40
Methods	VAE	6.894	3.615	3.014	3.025	2.686	2.621	2.537	2.419
	FLOW	5.733	3.186	3.006	3.293	2.556	2.51	2.432	2.35
	DLGM	6.667	3.61	3.207	2.795	2.766	2.564	2.332	2.403
	LAE	5.678	3.13	2.671	2.627	2.166	**2.083**	2.071	1.975
	HMCAE	**4.985**	**3.084**	**2.449**	**2.606**	**2.094**	2.164	**1.991**	**1.921**

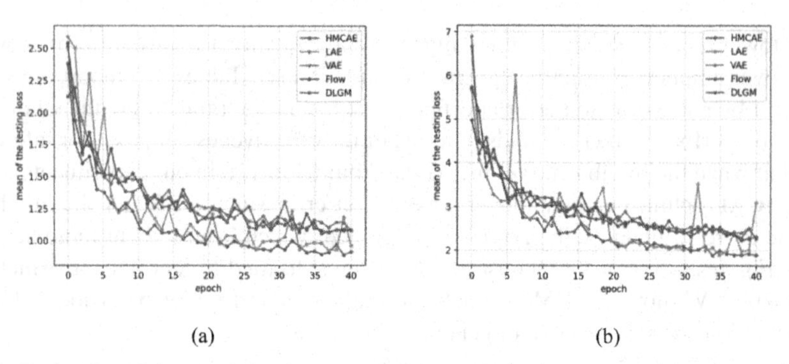

Fig. 3. Evaluation of the reconstruction error for HMCAE, LAE, VAE, DLGM and FLOW with different minibatch sizes on the number of epoch. (a): batch size = 64, (b): batch size = 128.

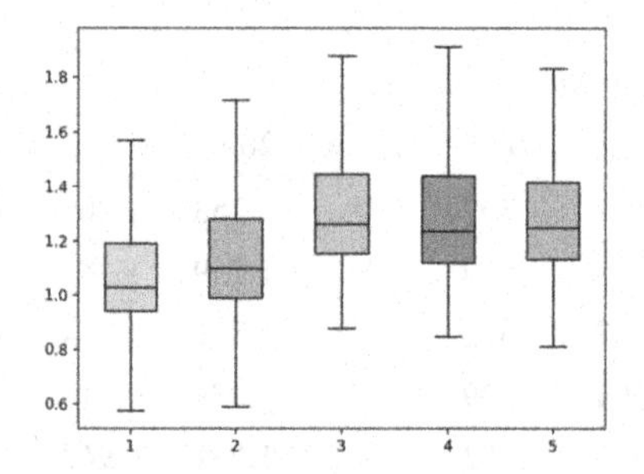

Fig. 4. Boxplot of the reconstruction error for different generative methods with minibatch size set to 64. From left to right: HMCAE, LAE, VAE, DLGM and FLOW.

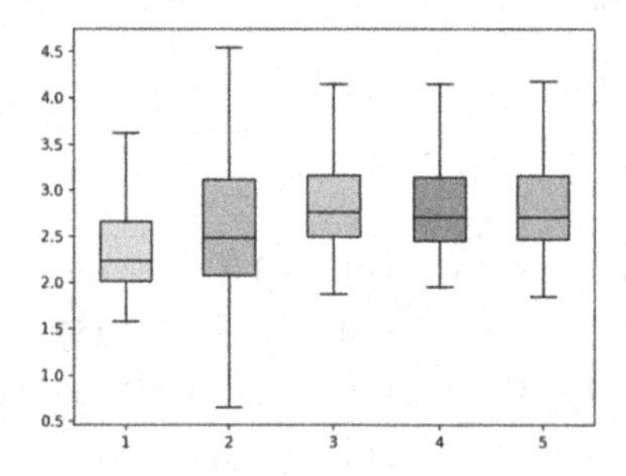

Fig. 5. Boxplot of the reconstruction error for different generative methods with minibatch size set to 128. From left to right: HMCAE, LAE, VAE, DLGM and FLOW.

5 Conclusions

In this work, we use Nosé-Hoover thermostat dynamics instead of the Langevin dynamics within the framework of LAE method to sample from the posterior distribution, and thus, a new generative model is designed. Benefitting from the use of a thermostat variable, the noises caused by stochastic gradient and the errors generated by discretizing the dynamics, encountered often by other models, are remedied properly. Experiments demonstrated that our proposed method gives the state-of-the-art performance over baseline generative methods in terms of ELBO estimation and reconstruction error. We also note that our proposed method increases accepting rate in per iteration and achieves more accurate posterior inference than LAE dose even without using the MH rejection step. Hence, our proposed method achieves a good balance between computational-cost and precision. A possible avenue for future work would be to utilize a neural network to estimate the thermostat at each iteration. The input to this network is the momentum variable generated at current step and the output is the estimated thermostat that is utilized to update the momentum variable at next step. We hope to learn a better estimate of the thermostat in this way to get better approximation of variational inference.

References

1. Kingma, D.P., Welling, M.: Auto-encoding variational bayes. arXiv preprint arXiv:1312.6114 (2013)
2. Gulrajani, I., et al.: PixelVAE: a latent variable model for natural image. arXiv:1611.05013 (2016)
3. Makhzani, A., Shlens, J., Jaitly, N., Goodfellow, I., Frey, B.: Adversarial autoencoder. arXiv: 1511.05644 (2016)
4. Higgins, I., et al.: Beta-VAE: learning basic visual concepts with a constrained variational framework. In: 5th International Conference on Learning Representations, Toulon, France (2017)
5. Burgess, C.P., et al.: Understanding disentangling in β-VAE. arXiv:1804.03599 (2017)
6. Rezaabad, A.L., Vishwanath, S.: Learning representation by maximizing Mutual Information in VAE. arXiv:1912.13361 (2019)
7. Tolstikhin, I., Bousquet, O., Gelly, S., Schoelkopf, B.: Wasserstein auto-encoders. arXiv: 1711.01558 (2019)
8. Xiao, Z., Kreis, K., Kautz, J., Vahdat, A.: VAEBM: a symbiosis between variation autoencoders and energy-based models. arXiv:2010.00654 (2021)
9. Ganguly, A., Earp, S.W.F.: An Introduction to Variational Inference. arXiv:2108.13083 (2021)
10. Blei, D.M., Kucukelbir, A., McAuliffe, J.D.: Variational inference: a review for statisticians. arXiv:1601.00670 (2017)
11. Caterini, A.L., Doucet, A., Sejdinovic, D.: Hamiltonian variational auto-encoder. arXiv:1805. 11328v2 (2018)
12. Thin, A., Kotelevskii, N., Durmus, A., Panov, M., Moulines, E., Doucet, A.: Monte carlo variational auto-encoders. arXiv:2106.15921 (2021)
13. Salimans, T., Kingma, D.P., Welling, M.: Markov chain monte carlo and variational inference: bridging the gap. arXiv:1410.6460 (2015)
14. Wolf, C., Karl M., van der Smagt, P.: Variational Inference with Hamiltonian Monte Carlo. arXiv:1609.08203 (2016)
15. Taniguchi, S., Iwasawa, Y., Kumagai, W., Matsuo, Y.: Langevin autoencoders for learning deep latent variable models. arXiv:2209.07036 (2022)
16. Ding, N., Fang, Y.H., Babbush, R., Chen, C.Y., Skeel, R.D., Neven, H.: Bayesian sampling using stochastic gradient thermostats. In: Advances in Neural Information Processing Systems (2014)
17. Luo, R., Wang, J.H., Yang, Y.D., Zhu, Z.X., Wang, J.: Thermostat-assisted continuously-tempered Hamiltonian Monte Carlo for Bayesian learning. arXiv:1711.11511 (2018)
18. Leimkuhler B., Reich, S.: A metropolis adjusted Nose-Hoover thermostat. ESAIM: Math. Modelling Num. Anal. **43**, 743–755 (2009)

19. Daxberger, E.: Laplace redux-effortless Bayesian deep learning. arXiv:2106.14806 (2021)
20. Chen, T.Q., Fox, E.B., Guestrin, C.: Stochastic Gradient Hamiltonian Monte Carlo. arXiv: 1402.4102 (2014)
21. Dockhorn, T., Vahdat, A., Kreis, K.: Score-based generative modeling with critically-damped Langevin diffusion. arXiv:2112.07068 (2022)
22. Hoffman, M.D.: Learning deep latent gaussian models with Markov chain monte carlo. In: 34th International Conference on Machine Learning, pp.1510–1519. Sydney, NSW, Australia (2017)
23. Rezende, D., Mohamed, S.: Variational inference with normalizing flows. In: 32th International Conference on Machine Learning, Lille, France, pp. 1530–1538 (2015)

AM-RRT*: An Automatic Robot Motion Planning Algorithm Based on RRT

Peng Chi[1], Zhenmin Wang[1], Haipeng Liao[1], Ting Li[1], Jiyu Tian[1], Xiangmiao Wu[1], and Qin Zhang[2(✉)]

[1] School of Mechanical and Automotive Engineering, South China University of Technology, WuShan Street, Guangzhou 510641, China
{mechipeng,melt}@mail.scut.edu.cn,
{wangzhm,tianjiyu,xmwu}@scut.edu.cn

[2] School of Computer Science and Engineering, South China University of Technology, Xiaoguwei Street, Guangzhou 511400, China
cszhangq@scut.edu.cn

Abstract. Motion planning is a very important part of robot technology, where the quality of planning directly affects the energy consumption and safety of robots. Focusing on the shortcomings of traditional RRT methods such as long, unsmooth paths, and uncoupling with robot control system, an automatic robot motion planning method was proposed based on Rapid Exploring Random Tree called AM-RRT* (automatic motion planning based on RRT*). First, the RRT algorithm was improved by increasing the attractive potential fields of the target points of the environment, making it more directional during the sampling process. Then, a path optimization method based on a dynamic model and cubic B-spline curve was designed to make the planned path coupling with the robot controller. Finally, an RRT speed planning algorithm was added to the planned path to avoid dynamic obstacles in real time. To verify the feasibility of AM-RRT*, a detailed comparison was made between AM-RRT* and the traditional RRT series algorithms. The results showed that AM-RRT* improved the shortcomings of RRT and made it more suitable for robot motion planning in a dynamic environment. The proposal of AM-RRT* can provide a new idea for robots to replace human labor in complex environments such as underwater, nuclear power, and mines.

Keywords: Motion planning · Sampling-based algorithms · RRT · Speed planning

Supported by Guangdong Provincial Science and Technology Plan Project (Grant number 2021B1515420006, Grant number 2021B1515120026); Guangdong Province Marine Economic Development Special Fund Project(Six Major Marine Industries) (GDNRC [2021]46); National Natural Science Foundation of China (Grant number U2141216, Grant number 51875212); Shenzhen Technology Research Project (JSGG20201201100401005, JSGG20201201100400001).

B. Luo et al. (Eds.): ICONIP 2023, LNCS 14447, pp. 91–103, 2024.
https://doi.org/10.1007/978-981-99-8079-6_8

1 Introduction

Motion planning are important components of mobile robots, which can be used to guide robots to work autonomously in complex environments and avoid obstacles [1]. How to plan a short, smooth, and controllable path is a major challenge. The methods of mobile robot motion planning and their improvements can be divided into classic methods and intelligent methods [1].

The classic methods include environment-modeling-based methods [2–6], graph-search-based methods, sampling-based methods, and curve-based methods. There are three most commonly used environmental modeling methods: grid map [2], road map [3–5], and artificial potential field [6]. The grid map can be divided into two parts: obstacle and free, which relies on the accuracy of the grid. The road map used a limited number of nodes to represent the environment and the nodes were connected to form edges. The more representative road map methods are Visibility Graph [3], Voronoi Diagram [4], and Probabilistic maps [5]. The artificial potential field method [6] was designed by a potential function, which included the attractive field of the target point and the repulsive field of the obstacles. The disadvantage is that it is easy to fall into the local minimum. The graph-search-based methods relied on known maps and obstacle information in the map to construct a feasible path from the start point to the end point [6–9]. However, these methods required a large amount of computation and were not suitable for path planning in dynamic scenes. The sampling-based path planning methods performed random sampling in the environmental space [5,10,11]. The advantage of these methods is that it does not need to model the environmental space. But the planning process is random and the optimal path is not obtained. The most representative random sampling methods include Probabilistic Roadmap Method (PRM) [5] and Rapidly Exploring Random Tree (RRT) [10]. The curve-based method referred to the method of constructing or inserting a new data set within a known data set [12,13]. The curve can fit the previous set of path points to a smoother trajectory [14,15]. Overall, the classic methods tend to favor static path planning and have the disadvantage of easily falling into local minimum, making them unsuitable for dynamic path planning.

The path planning problem is not only a search problem but a reasoning problem, so many intelligent methods can also be used in path planning problems. The most typical intelligent algorithm is the neural-network-based algorithm [16], which simulates the behavior of animal neural networks and performs distributed information processing. The genetic algorithm is a search optimal solution algorithm that simulates the evolution of Darwinian organisms [17], which can realize synchronous planning and tracking. The ant colony algorithm is derived from the exploration of ant colony foraging behavior, which is effective and robust [18]. It can find relatively good paths but is suitable for real-time planning. At present, intelligent planning algorithms are not yet mature and require high hardware devices, making them unsuitable for real-time path planning solutions.

Given the above problems of existing planning methods, a new motion planning method suitable for robot control was proposed in this paper called AM-RRT*. The main contributions are as follows:

- A new RRT path planning algorithm based on elliptical region and potential field constraints was proposed, which had the shottest path length compared to traditional RRT algorithms.
- A path optimization method based on robot dynamic model constraints and cubic B-spline curves was proposed to ensure that the planned path was smooth enough and tightly coupled with the robot controller.
- A speed planning method based on RRT was proposed, which endowed robots with the ability to autonomously control speed in dynamic environments to save energy consumption and time.

2 Related Work

The AM-RRT* proposed in this paper is based on the RRT algorithm. The RRT has been proposed for many years, and many researchers have proposed numerous improved algorithms for it.

The RRT was proposed to solve complex obstacle constraints and a high degree of freedom path planning problems [10]. The idea is to gradually build a tree, attempting to quickly and uniformly explore configuration or state space, providing benefits similar to probabilistic roadmap methods, but applicable to a wider range of problems [11]. The RRT* algorithm maintained the tree structure like RRT while maintaining the asymptotic optimality [19]. Its basic idea was the fusion between the sampling-based motion planning algorithm and the random geometric graph theory. Informed RRT* (IRRT*) [20] was proposed for centralized search by directly sampling the subset to address the issue of low efficiency and inconsistency with their single query properties. Q-RRT* [21] considered more vertices during the optimization process, thereby expanding the possible set of parent vertices, resulting in a path with lower cost than RRT*. IB-RRT* [22] utilized the bidirectional tree method and introduced intelligent sample insertion heuristic to quickly converge to the optimal path solution [23]. P-RRT* was proposed to combine the artificial potential field algorithm in RRT*, which allowed for a significant reduction in the number of iterations. PQ-RRT* [24] combined the advantages of P-RRT* and Q-RRT*, further improving operating speed while possessing superiority. These improvement methods are aimed at optimizing the performance of path. However, robots have the characteristic of multiple modules tightly coupled.

Shi et al. [25] improved and optimized the RRT algorithm by establishing a vehicle steering model to increase vehicle steering angle constraints and solve the problems of high randomness, slow convergence speed, and large deviation. However, the applied model is relatively simple and the optimization effect is not significant enough. Chen et al. [26] incorporated mobility reliability into mission planning and proposed a reliability-based path smoothing algorithm to solve the suboptimal problem. The method has high robustness, but low universality.

Therefore, it is necessary to design a RRT-based path planning method that is short in length, smooth, and coupled with the robot motion control, and to form an autonomous robot motion planning system with the speed planning method.

3 Methodology

The overview of AM-RRT* is shown in Fig. 1. The input of the system is from the environment perception of a robot, which includes three parts: map information, static obstacles information, and trajectory of dynamic obstacles. First, an improved RRT algorithm for robot path planning was utilized, which included the elliptical sampling area of IRRT* and the attractive potential field of Voronoi points. Then, the planned path was optimized to make it smoother and more suitable for robot motion control, including robot dynamic constraints and optimization of cubic spline curves. Finally, the improved RRT algorithm mentioned above was applied to the speed planning of robots, where the extension was used in the distance-time (s-t) image to obtain real-time speed.

3.1 Path Planning Based on RRT

According to the shortcomings of the globssal sampling of the RRT algorithm, the elliptical area was defined and restricted to the sampling area to improve its sampling efficiency [20]. An elliptical region C with the shortest path length is defined as:

$$C = \{x \in \chi \,|\, \|x - x_{init}\| + \|x - x_{goal}\| < length(\sigma_{\min})\} \tag{1}$$

where $\sigma_{\min}$ is the length of the shortest path. x_{init}, x_{goal} is the initial point and the goal point of the path. x is the current point. Since the average planned path of RRT is about 1.3–1.5 times the shortest path, two kinds of attractive potential fields were added to produce a total attractive potential field. The whole attractive potential field is defined as:

$$U_{art}(x) = U_{goal}(x) + U_{vor}(x) \tag{2}$$

where $U_{goal}(x)$ is the attractive potential field of the target point. $U_{vor}(x)$ is the attractive potential field of the Voronoi diagram of the environment.

The $U_{goal}(x)$ is defined as:

$$U_{goal}(x)=\frac{1}{2}k_{goal}(x - x_{goal})^2 \tag{3}$$

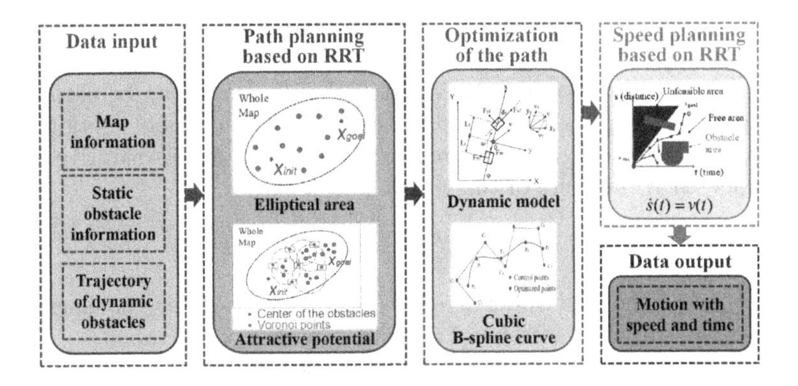

Fig. 1. Overview of the proposed AM-RRT* motion planning system.

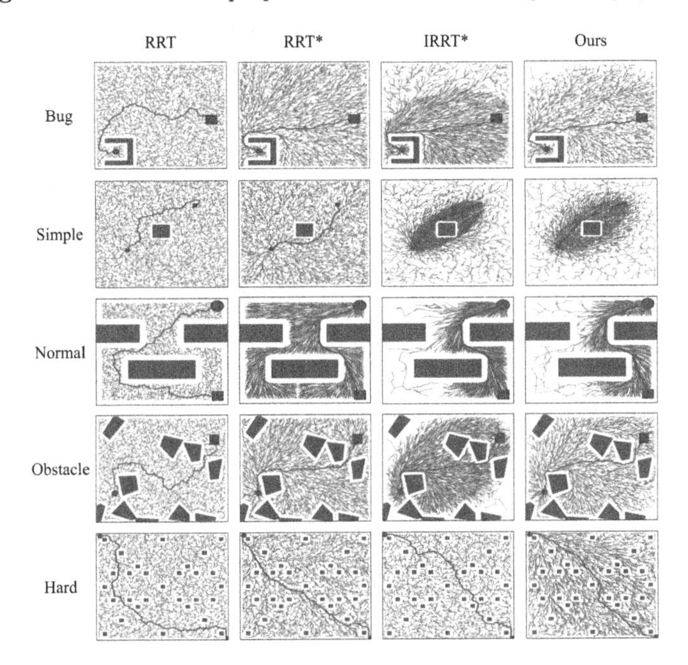

Fig. 2. Path planning results of RRT [10], RRT* [19], IRRT* [20], and our proposed method in five different scenarios.

where k_{goal} is a constant weighting factor. The $U_{vor}(x)$ is defined as:

$$U_{vor}(x)=\frac{1}{2}k_{vor}\sum\nolimits_{x_{vor}\in\chi_{vor}}(x-x_{vor})^2 \tag{4}$$

where k_{vor} is the constant weight coefficient. χ_{vor} is the set of points that make up the Voronoi points. Since Voronoi points are the circumscribed center of the three obstacles, the problem that when the obstacles are too close, random sampling can not find the result can be solved.

3.2 Optimization of the Path

Constraints based on robot kinematics and dynamic model can be used to optimize the path, making the planned path more suitable for robot motion control [25,27]. A 5-DOF vehicle dynamic model was selected for comparative analysis in this paper [27]. The state of the vehicle is defined by:

$$x = (x_g, y_g, \varphi, v_y, \omega)^T \tag{5}$$

where x_g, y_g represent the robot's center of gravity. φ is the yaw angle of the robot. v_y is the speed of lateral and ω is the yaw rate of the robot.

According to the second Newton's law, the fundamental law of dynamics can be defined as:

$$\begin{cases} m(\dot{v}_x - v_y\omega) = -F_{xf}\cos\delta_f - F_{yf}\sin\delta_f - F_{xr} \\ m(\dot{v}_y + v_x\omega) = F_{yf}\cos\delta_f - F_{xf}\sin\delta_f + F_{yr} \\ I_z\dot{\omega} = L_f(F_{yf}\cos\delta_f - F_{xf}\sin\delta_f) - L_rF_{yr} \end{cases} \tag{6}$$

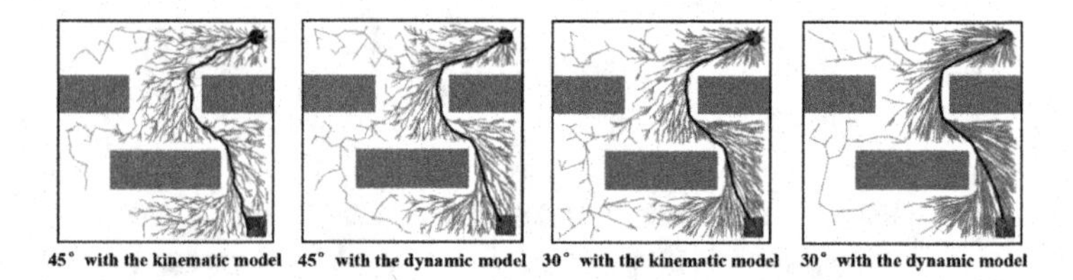

Fig. 3. Optimization results of RRT path planning based on dynamic models in Normal scene.

where m is the mass of the robot. v_x is the speed of longitude. $F_{xf}, F_{yf}, F_{xr}, F_{yr}$ are the longitudinal and lateral force of the front and rear wheels. δ_f is the front wheel angle. I_z is the moment of inertia around the z-axis. L_f, L_r are the distance from the center of gravity to the front and rear wheel.

The traditional RRT algorithms are not smooth and often require post-processing to make the path smoother. The cubic B-spline curve [15] was selected as the optimization method. A B-spline curve is defined by a series of control points. These points consist of all nodes in the suboptimal path searched by RRT. At the same time, each point of the generated path is only related to the four nearby control points. The planned path can be defined as the x, y coordinates Φ_x, Φ_y:

$$\begin{cases} \Phi_x = [\phi_{x,-1}\phi_{x,0}\cdots\phi_{x,m-1}] \\ \Phi_y = [\phi_{y,-1}\phi_{y,0}\cdots\phi_{y,m-1}] \end{cases} \tag{7}$$

where $\phi_{x,i}$ $\phi_{y,i}$ are the control points. Define the setting $s \in [0,1]$ is the percentage of the path length that has been traveled, the final path can be expressed as:

$$\begin{cases} x(s) = \sum_{k=0}^{3} f_k(t) \cdot \phi_{x,k+l} \\ y(s) = \sum_{k=0}^{3} f_k(t) \cdot \phi_{y,k+l} \end{cases} \tag{8}$$

where $[\cdot]$ is the round down function. t and l are used to adjust the control points:

$$\begin{cases} t = s \cdot \sigma - [s \cdot \sigma] \\ l = [s \cdot \sigma] - 1 \end{cases} \tag{9}$$

where σ is the length of planned path. $\{f_k(t)\}_{k=[0,3]}$ is the basic function of the B-spline, defined as:

$$\{f_k(t)\}_{k=[0,3]} = \begin{cases} f_0(t) = \frac{1}{6}(1-t)^3 \\ f_1(t) = \frac{1}{6}(3t^3 - 6t^2 + 4) \\ f_2(t) = \frac{1}{6}(-3t^3 + 3t^2 + 3t + 1) \\ f_3(t) = \frac{1}{6}t^3 \end{cases} \tag{10}$$

To optimize the path length and smoothness under the premise of ensuring the safety of the path without collision, the definition of the optimization function consists of two parts: path length and potential field energy along the path. The potential field energy is composed of the obstacle's repulsive potential field:

$$E(\Phi) = L(\Phi) + \lambda \cdot P(\Phi) \tag{11}$$

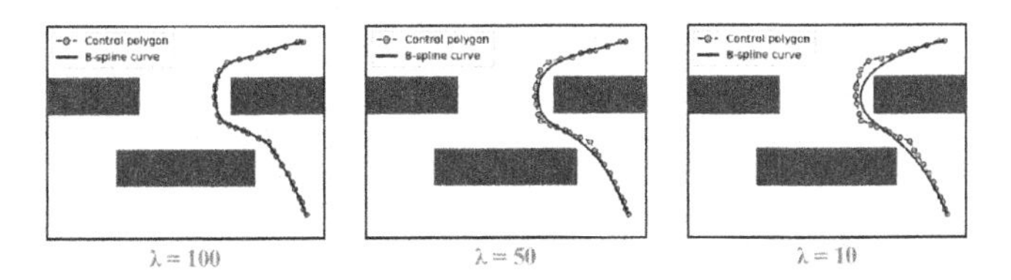

Fig. 4. Optimization results of RRT path planning based on cubic spline curve in Normal scene.

where $L(\Phi)$ is the path length and $P(\Phi)$ is the potential field energy. λ represents the weight between path length and potential. If the value of λ is larger, the final path will be farther away from the obstacles but the path length may be longer.

The cost of the path length is defined as:

$$L = \int_0^1 [\dot{x}^2(s) + \dot{y}^2(s)]ds \tag{12}$$

The cost of the potential field energy is defined as:

$$P = \int_0^1 P(x(s), y(s))ds \tag{13}$$

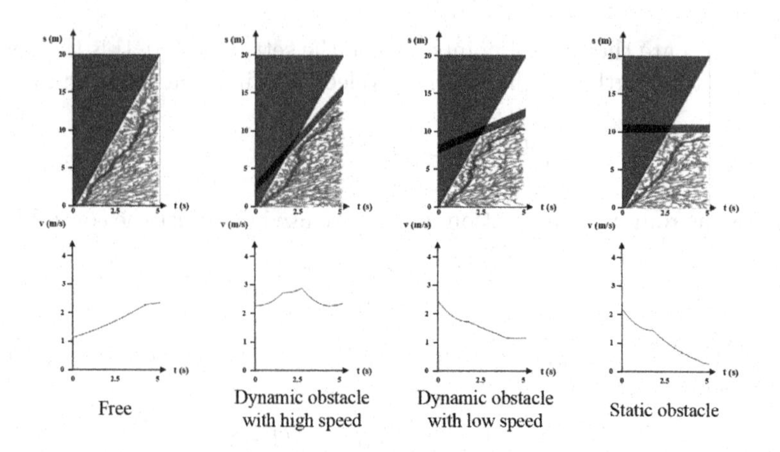

Fig. 5. Speed planning results for four different scenarios, including free, dynamic obstacle with low speed, dynamic obstacle with high speed, and static obstacle.

3.3 Speed Planning Based on RRT

For speed planning, the RRT-based planning method is still used. Assuming that the robot's pitching motion is not considered, the motion generally includes three parameters: the coordinates of the robot on the path $\rho_x(s), \rho_y(s)$ and the heading angle of the robot along the path $\rho_\theta(s)$:

$$\rho(s) = [\rho_x(s), \rho_y(s), \rho_\theta(s)] \tag{14}$$

The trajectory of the dynamic obstacles is defined as:

$$obs_i(t)_{i \in [1,M]} = [x_{obs,i}(t) y_{obs,i}(t)] \tag{15}$$

where t is time. M is the number of obstacles and $x_{obs,i}(t)$,$y_{obs,i}(t)$ are the coordinates of the obstacle i at time t. Given the reference path $\rho(s)$ and the initial state of the robot, the task of speed planning is to generate a longitudinal motion function $s(t)(t \in [0, T_{forward}])$, where the speed can be expressed as $v(t) = \dot{s}(t)$. The robot status can be described as $x = (s, t)$ and the s-t motion space χ can be described as the state space within the range $x(s \geq s_0, t \geq t_0)$. The s_0 and t_0 represent the initial path length and time of the robot. Assume that the set of states that will collide with obstacles in s-t space is the obstacle space χ_{obs}:

$$\chi_{obs} = \{x \in \chi | \, \|\rho(s) - obs_i(t)\| < d_{danger}\} \tag{16}$$

where d_{danger} is the threshold of the dangerous distance from the robot to the obstacles.

The state in the s-t space should satisfy the following constraints: $|\dot{s}| \leq v_{\max}$, $a_{\min} \leq \ddot{s} \leq a_{\max}$, $|\ddot{s}| \cdot \kappa \leq \omega_{\max}$, where $v_{\max}$ is the maximum speed of the robot. $[a_{\min}, a_{\max}]$ are the acceleration limit of the robot. κ and $\omega_{\max}$ represent the

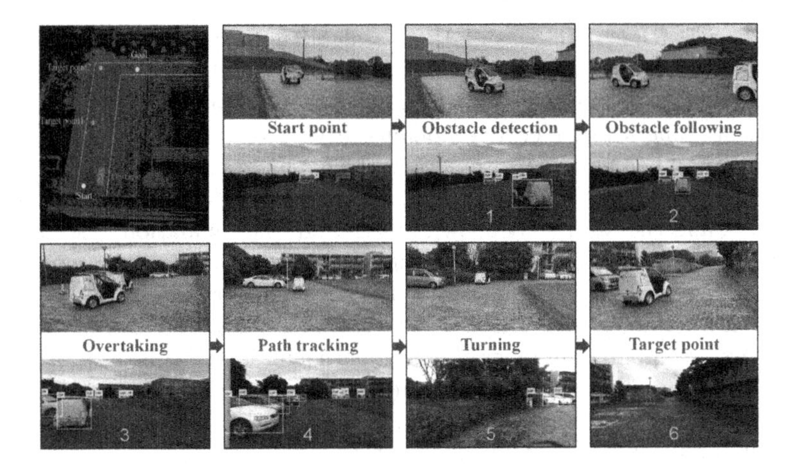

Fig. 6. On-site testing scenarios and distribution of the stages.

curvature of the reference path and the maximum angular velocity of the robot. To narrow the search range and improve the search efficiency, the kinematics unfeasible space χ_{unf}:

$$\chi_{unf} = \{x \in \chi \,|t_0 \leq t \leq t_{\max}, 0 \leq s - s_0 \leq v_{\max} \cdot (t - t_0)\} \tag{17}$$

The free area χ_{free} can be defined as:

$$\chi_{free} = \{x \in \chi|!(\chi_{obs} \cup \chi_{unf})\} \tag{18}$$

The extension of RRT will be extended in the free area to calculate the speed of the robot.

4 Experimental Results

To verify the effectiveness and robustness of our proposed AM-RRT* method, simulation experiments and on-site tests were conducted. First, AM-RRT* was

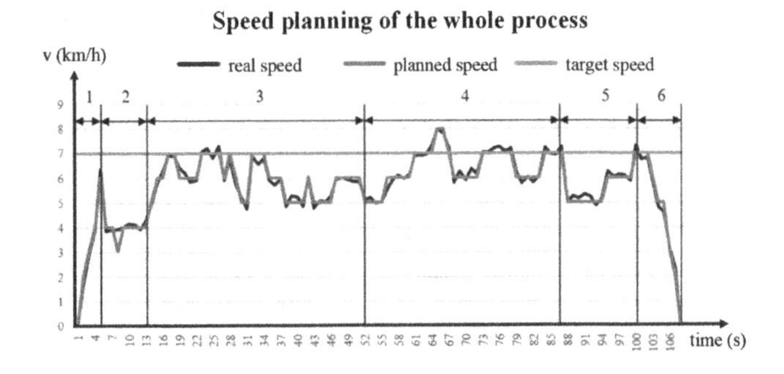

Fig. 7. Speed planning results of the whole on-site testing process.

compared with RRT, RRT*, and IRRT* algorithms. Subsequently, the path optimization method and speed planning method were tested in the normal scene. In the on-site testing section, AM-RRT* was conducted in the actual environment using a modified autonomous driving vehicle. The results showed that our method had good performance in the number of nodes and path length.

First, the proposed path planning algorithm was compared with RRT, RRT*, and IRRT*, and the results are shown in Fig. 2. In the simulation, 5 scenes were designed: Bug, Simple, Normal, Obstacle, and Hard. These scenes comprehensively demonstrate the performance of the algorithm, and our method yields smoother paths compared to IRRT*. For a more rigorous comparison, three perspectives were compared: number of nodes, path length, and time cost. As shown in Tab. 1, Our method has significant advantages in the number of nodes and path length, and with the addition of optimized methods (RRT* and IRRT*), we also have advantages in terms of time cost.

In the path optimization, to verify the effectiveness of the model constraints, a kinematics model [25] was used for comparative verification. As shown in Fig. 3, two models and steering constraints were used for planning results in Normal scenarios, respectively. Under the 30° constraint, the results of both models are better than those under the 45° constraint, but the 5-degree of freedom model we used has a smoother path. Table 2 shows the performance of the four models at 5000 iterations. In addition to time cost, the dynamic model has significant advantages in terms of node number, path length, and maximum angle.

The improved B-spline curve method was also tested under the Normal scene, and the comparison was made by changing the weight value λ. As shown in Fig. 4, the weight values were set to 100, 50, and 10, respectively. As can be seen from the results that when the weight value becomes smaller, the path length will become smaller, but it is much closer to the obstacles. At the same time, due to the use of a fitting B-spline curve, the optimized curve will not pass through all control points. Considering the error caused by the sensor and path tracking in actual situations, it is necessary to set a reasonable weight value.

To verify the feasibility of the speed planning algorithm, four typical scenes were selected for simulation verification, which is shown in Fig. 5. In the simulation process, it is assumed that the maximum speed of the robot is 4 m/s, the expected speed is 2.5 m/s, the acceleration is limited to $\pm 1\,\mathrm{m/s^2}$ and the forward-looking time is 5 s, so the maximum forward-looking distance is 20 m. In the s-t diagram, the blue area is the defined unfeasible area and the dynamic obstacle area, the gray lines are the RRT search tree for speed planning, and the red line is the final selection speed path for the robot.

The on-site test of an automatic driving vehicle was tested to verify its feasibility under vehicle conditions. For the hardware platform, the VLP-16 LIDAR and ZED binocular cameras were used to perceive the surrounding environment and detect the position of obstacles. As shown in Fig. 6, LiDAR-based mapping work was conducted on a closed road and set the starting point and some target points. Another vehicle was used to simulate dynamic obstacles. The speed planning results for the process are shown in Fig. 7, with the target speed set at

Table 1. Comparison of the number of nodes, path length, and time cost of RRT [10], RRT* [19], IRRT* [20], and our proposed method in 5 scenarios.

Environment	methods	No. of nodes	Path length (m)	Time cost (s)
Bug (12*12) 10000 iterations	RRT	8361	17.26	**15.24**
	RRT*	7180	12.88	30.02
	IRRT*	6164	13.03	42.78
	Ours	**5615**	**12.49**	30.83
Simple (16*16) 10000 iterations	RRT	9158	15.67	**16.91**
	RRT*	9166	12.97	37.54
	IRRT*	8738	10.89	81.89
	Ours	**8456**	**10.65**	68.57
Normal (6*6) 10000 iterations	RRT	5001	12.86	**9.85**
	RRT*	5060	6.74	40.69
	IRRT*	2614	6.72	31.31
	Ours	**2324**	**6.47**	25.26
Obstacle (12*12) 10000 iterations	RRT	7102	19.61	**13.64**
	RRT*	7099	14.1	33.89
	IRRT*	7611	13.53	73.19
	Ours	**6333**	**13.02**	39.72
hard (16*16) 10000 iterations	RRT	7893	30.27	**14.95**
	RRT*	7719	24.11	35.12
	IRRT*	7268	27.39	46.02
	Ours	**7141**	**23.89**	30.62

Table 2. Comparison of number of nodes, path length, time cost and maximum angle of kinematics model [25] and dynamics model [27] in Normal scene.

Environment	Methods	Time cost (s)	No. of nodes	Path length (m)	Maximum angle (°)
Normal (6*6) 5000 iterations	45° kinematic	**5.73**	1619	6.95	23.5
	45° dynamic	7.03	1653	6.58	20.2
	30° kinematic	8.38	1307	6.46	14.4
	30° dynamic	12.35	**1194**	**6.36**	**11.6**

7 km/h. First, a dynamic obstacle was detected after the path planning at the start point, as shown in stage 1. Then, in stage 2, the motion planning performed the obstacle following operation and gradually reduced the speed to 4 km/h. In stage 3, with the cooperation of the environment perception, the motion planning executed the overtaking action and increased the speed. After completing the overtaking operation, the action planner re-planned the path and brought the speed closer to the target speed, as shown in stage 4. Finally, as shown in stages 5 and 6, after turning and reaching the goal point, the vehicle slowed down and stopped. From the results, the robot motion planning system AM-RRT* proposed in this paper can be applied to the robot in the real environment through the coupling control of path planning, path optimization, and speed planning.

5 Conclusion and Future Work

This paper proposed an automatic RRT-based motion planning method suitable for robots called AM-RRT*. First, to solve the problem of long paths of traditional RRT, an improved scheme for elliptical regions and attractive potential fields was designed. Then, to solve the problem of insufficient smoothness and large angles in traditional path planning, an optimization method based on a 5-degree-of-freedom dynamic model and a cubic spline curve was proposed. Finally, in response to the shortcomings of traditional path planning not being coordinated with speed control, an RRT-based speed planning method was proposed. Multiple simulation experiments and on-site testing results have demonstrated the effectiveness and robustness of our method. In future work, we will be committed to developing robot motion planning solutions under special conditions, such as magnetic adsorption robots, underwater robots, etc., to enable robots to automatically and intelligently replace humans in hazardous work.

References

1. Hichri, B., Gallala, A., Giovannini, F., Kedziora, S.: Mobile robots path planning and mobile multirobots control: a review. Robotica **40**(12), 4257–4270 (2022)
2. Tae, N., Jae, S., Young, C.: A 2.5d map-based mobile robot localization via cooperation of aerial and ground robots. Sensors **17**(12), 2730 (2017)
3. Lozano-Perez, T., Wesley, M.: An algorithm for planning collision-free paths among polyhedral obstacles. Commun. ACM **22**, 560–570 (1979)
4. Takahashi, O., Schilling, R.: Motion planning in a plane using generalized voronoi diagrams. IEEE Trans. Robot. Autom. **5**, 143–150 (1989)
5. Kavraki, L.E., Svestka, P.: Probabilistic roadmaps for path planning in high-dimensional configuration spaces. IEEE Trans. Rob. Autom. **12**(4), 566–580 (1996)
6. Khatib, O.: Real-time obstacle avoidance for manipulators and mobile robots. In: 1985 IEEE International Conference on Robotics and Automation (Cat. No. 85CH2152-7), pp. 500–505 (1984)
7. Hwang, J., Kim, J., Lim, S., Park, K.: A fast path planning by path graph optimization. IEEE Trans. Syst. Man Cybern. Part A-Syst. Humans **33**(1), 121–128 (2003)
8. Kushleyev, A., Likhachev, M.: Time-bounded lattice for efficient planning in dynamic environments. In: ICRA: 2009 IEEE International Conference Robotics and Automation, vol. 1–7. pp. 4303–4309 (2009)
9. Likhachev, M., Gordon, G., Thrun, S.: Ara*: anytime a* with provable bounds on sub-optimality. In: Thrun, S., Saul, K., Scholkopf, B. (eds.) Advances in Neural Information Processing Systems, vol. 16, pp. 767–774 (2004)
10. Lavalle, S.M.: Rapidly-exploring random trees: a new tool for path planning. Research Report (1999)
11. LaValle, S., Kuffner, J.: Rapidly-exploring random trees: progress and prospects. In: Donald, B., Lynch, K., Rus, D. (eds.) Algorithmic and Computational Robotics: New Directions, pp. 293–308 (2001)
12. Reeds, J.A., Shepp, R.A.: Optimal paths for a car that goes both forward and backward. Pac. J. Math. **145** (1991)

13. Brezak, M., Petrovic, I.: Real-time approximation of clothoids with bounded error for path planning applications. IEEE Trans. Rob. **30**(2), 507–515 (2014)
14. Jolly, K.G., Kumar, R.S., Vijayakumar, R.: A bezier curve based path planning in a multi-agent robot soccer system without violating the acceleration limits. Rob. Auton. Syst. **57**(1), 23–33 (2009)
15. Koyuncu, E., Inalhan, G.: A probabilistic b-spline motion planning algorithm for unmanned helicopters flying in dense 3d environments. In: 2008 IEEE/RSJ International Conference on Robots and Intelligent Systems, Conference Proceedings, vol. 1–3, pp. 815–821 (2008)
16. Luo, M., Hou, X., Yang, S.X.: A multi-scale map method based on bioinspired neural network algorithm for robot path planning. IEEE Access **7**, 142682–142691 (2019)
17. Xin, J., Zhong, J., Sheng, J., Li, P., Cui, Y.: Improved genetic algorithms based on data-driven operators for path planning of unmanned surface vehicle. Int. J. Rob. Autom. **34**(6), 713–722 (2019)
18. Luo, Q., Wang, H., Zheng, Y., He, J.: Research on path planning of mobile robot based on improved ant colony algorithm. Neural Comput. Appl. **32**(6SI), 1555–1566 (2020)
19. Karaman, S., Frazzoli, E.: Incremental sampling-based algorithms for optimal motion planning. Rob. Sci. Syst. **104**, 267–274 (2010)
20. Gammell, J.D., Srinivasa, S.S., Barfoot, T.D.: Informed rrt*: optimal sampling-based path planning focused via direct sampling of an admissible ellipsoidal heuristic. In: 2014 IEEE/RSJ International Conference on Intelligent Robots and Systems (IROS 2014), pp. 2997–3004 (2014)
21. Jeong, I.B., Lee, S.J., Kim, J.H.: Quick-rrt*: triangular inequality-based implementation of rrt* with improved initial solution and convergence rate. Expert Syst. Appl. **123**(JUN.), 82–90 (2019)
22. Qureshi, A.H., Ayaz, Y.: Intelligent bidirectional rapidly-exploring random trees for optimal motion planning in complex cluttered environments. Rob. Auton. Syst. **68**, 1–11 (2015)
23. Jordan, M., Perez, A.: Optimal bidirectional rapidly-exploring random trees (2013)
24. Li, Y., Wei, W., Gao, Y., Wang, D., Fan, Z.: Pq-rrt*: an improved path planning algorithm for mobile robots. Expert Syst. Appl. **152** (2020)
25. Shi, Y., Li, Q., Bu, S., Yang, J., Zhu, L.: Research on intelligent vehicle path planning based on rapidly-exploring random tree. Math. Prob. Eng. **2020** (2020)
26. Jiang, C., et al.: R2-rrt*: reliability-based robust mission planning of off-road autonomous ground vehicle under uncertain terrain environment. IEEE Trans. Autom. Sci. Eng. **19**(2), 1030–1046 (2022)
27. Pepy, R., Lambert, A., Mounier, H.: Path planning using a dynamic vehicle model. In: 2006 2nd International Conference on Information & Communication Technologies, vol. 1, pp. 781–786 (2006)

MS3DAAM: Multi-scale 3-D Analytic Attention Module for Convolutional Neural Networks

Yincong Wang, Shoubiao Tan, and Chunyu Peng(✉)

Anhui University, Hefei 230601, Anhui, China
p21201013@stu.ahu.edu.cn, tsb@ustc.edu, cyupeng@ahu.edu.cn

Abstract. In this paper, we propose a compact and effective module, called multi-scale 3-D analytic attention module (MS3DAAM) to address this challenge. We significantly reduce model complexity by developing a decoupling-and-coupling strategy. Firstly, we factorize the regular attention along channel, height and width directions and then efficiently encode the information via 1-D convolutions, which greatly saves the computational power. Secondly, we multiply the weighted embedding results of the three direction vectors to regain a better 3-D attention map, which allocates an independent weight to each neuron, thus developing a unified measurement method for attention. Furthermore, multi-scale method is introduced to further strengthen our module capability in locating by capturing both the inter-channel relationships and long-range spatial interactions from different receptive fields. Finally, we develop a structural re-parameterization technique for multi-scale 1-D convolutions to boost the inference speed. Extensive experiments in classification and object detection verify the superiority of our proposed method over other state-of-the-art counterparts. This factorizing-and-combining mechanism with the beauty of brevity can be further extended to simplify similar network structures.

1 Introduction

Attention mechanisms tell models "what" and "where" to place emphases on, but they all require more or less computational power. In recent years, various ingenious attention methods are proposed to push the performance of CNNs, thus narrowing the gap between lightweight CNNs [6,12] and deep CNNs [4,14]. However, some complex attention mechanisms inevitably bring a lot of computational overhead and lag in training and inference speed, so how to overcome the paradox of trade-off between complexity and performance becomes a problem.

Because of its simplicity and effectiveness, the Squeeze-and-Excitation (SE) net [8] is commonly used. ECANet [15], pointing out that the nonlinear relationship obtained by dimensionality reduction in SENet may cause side-effect, seeks to emphasize the inter-channel relationships by introducing lightweight 1-D convolution rather than multi-layer perceptron (MLP).

CBAM [16] further exploits positional information by pooling along channel dimension and then computing spatial attention by convolutions. However,

B. Luo et al. (Eds.): ICONIP 2023, LNCS 14447, pp. 104–117, 2024.
https://doi.org/10.1007/978-981-99-8079-6_9

convolutions in CBAM may ignore long-range dependencies. To capture the long-range spatial interaction with precise position information, CA [5] factorizes the global pooling into two directional pooling operations. Two generated feature maps are then respectively encoded to capture dependencies along certain spatial direction and thus preserve the positional information. However, these attention mechanisms heavily relying on traditional 2-D convolution inevitably increase computational overhead. Moreover, due to the fixed kernel size, feature extraction is limited to a definite receptive field.

The receptive field reflects how far the convolutions in the network can probe. Due to the fixed kernel size, feature extraction is limited to a definite receptive field in some networks. So, some network architectures that utilize convolutions of different kernel sizes are proposed to simultaneously obtain receptive fields of different scales [3].

To leverage the advantages of multi-scale convolutions without substantial computational cost, we propose MS3DAM to factorize the attention along three directions of channel, width and height. This novel module calculates channel attention and spatial attention in parallel, so that the two are intrinsically linked. First, our module globally aggregates features along the vertical and horizontal directions to get two directional feature maps acting as a bridge between channel attention and spatial attention. Then, our module exploits an average pooling operation to aggregate 1-D attention from any directional feature map along a certain spatial direction, while another two average pooling operations are used to yield two direction vectors from the two feature maps. These two direction vectors embedded with direction-aware information are then respectively encoded into two directional attention maps by multi-scale 1-D convolutions. Therefore, precise positional information can be regained by multiplying the two directional attention vectors. Similarly, by multiplying the results of channel attention and spatial attention, a 3-D attention map with the same size of the input feature map can be generated, which can reflect the importance of each pixel in the input feature map.

The main contributions of this work are summarized as follows:

- We present a compact attention module using a decoupling-and-coupling strategy to regain a better 3-D attention map with precise positional information.
- We utilize multi-scale 1-D convolutions to efficiently encode information and develop a structural re-parameterization method for 1-D convolutions to further boost the inference speed.
- We conduct extensive experiments on standard benchmarks to show that our method outperforms other counterparts without any architecture search-based method or hyperparameter adjustment.

2 Related Work

2.1 Attention Mechanisms

Attention mechanisms are widely used to facilitate models to focus on more prominent features. According to the different dimensions of the generated attention map, attention mechanisms can be divided into the following types.

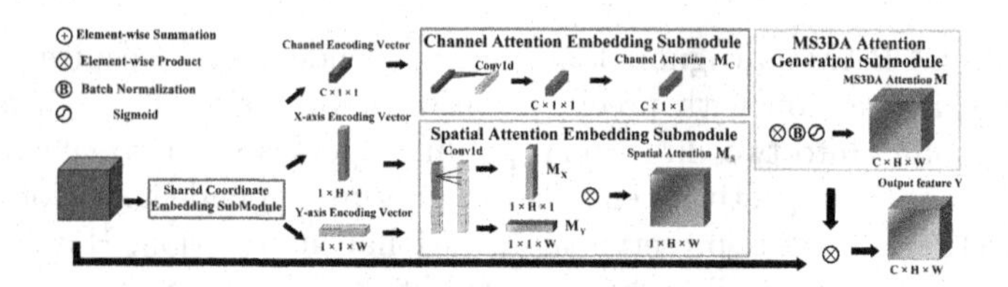

Fig. 1. The overview of MS3DA attention module.

1-D Attention Mechanisms. 1-D attention refers to channel attention, which treats neurons in different channels differently. SENet uses global average pooling to squeeze spatial information and then establish a connection between each channel by a shared MLP. The channel attention module (CAM) of CBAM adopts both maximum and average pooling operations to reduce the loss of information caused by global pooling. SKNet [9] has two channel attention branches and uses multi-scale 2-D convolutions to better capture the inter-channel relationships at high costs.

2-D Attention Mechanisms. 2-D attention refers to spatial attention, which treats neurons in different locations differently. The spatial attention module (SAM) of CBAM exploits positional information by convolutions with large kernels. Later, different spatial attention methods are adopted in the works like GENet [7].

3-D Attention Mechanisms. However, 1-D and 2-D treat neurons in each channel or spatial location equally. To solve this problem, 3-D attention simultaneously captures channel and spatial information. Classic attention methods like CBAM [16] and BAM [11], advance this idea by connecting channel and spatial attention in serial or parallel way. TAM [10] further extends this idea by computing through pairwise combination. Parameter-Free attention SimAM [18] proposes an energy function to achieve this measurement, but the hyperparameters in which need to be adjusted through extensive experiments according to different datasets and tasks.

2.2 Structural Re-Parameterization

Structural re-parameterization means that the model adopts a multi-branch structure in the training phase and fuses parameters into one branch in the inference phase to boost the speed. DBB [2] is a module able to lift the performance of the network, which can replace the traditional convolution. RepVGG [3] designs a VGG-like network using the structural re-parameterization technique.

3 MS3DA Attention

Our MS3DA attention mechanism can globally emphasize salient features and enhance the representation ability of the network. For a given feature map $\mathbf{X} \in$

$\mathbb{R}^{C\times H\times W}$ as input, we can obtain the refined feature map $\mathbf{Y} \in \mathbb{R}^{C\times H\times W}$ by multiplying $\mathbf{X}$ and the 3-D attention map $\mathbf{M} \in \mathbb{R}^{C\times H\times W}$ generated by our attention module.

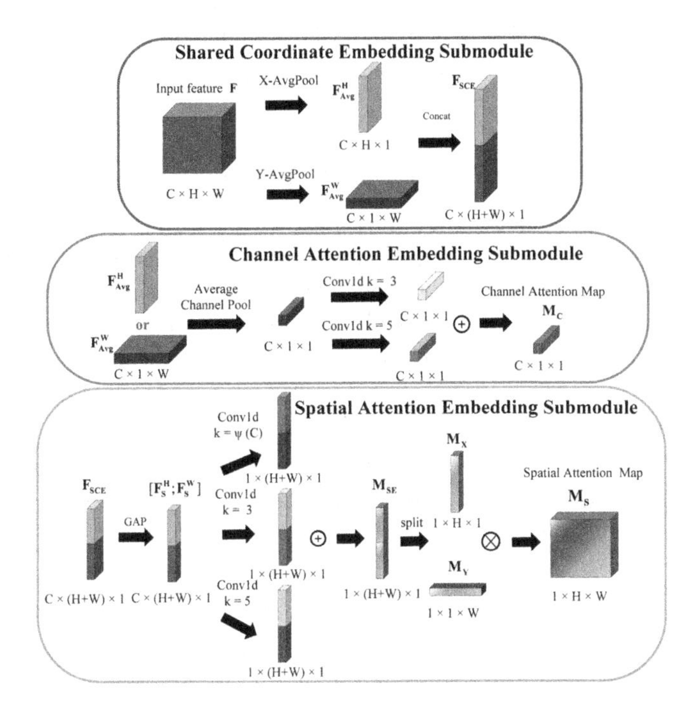

Fig. 2. Diagram of detailed MS3DA attention submodules.

3.1 Multi Scale 3-Dimension Analytic Attention Module

Using traditional 2-D convolutions to directly compute $\mathbf{M}$ brings a large computational overhead, so we reduce the cost by factorizing the attention into channel attention $\mathbf{M_C} \in \mathbb{R}^C$, X-axis attention $\mathbf{M_X} \in \mathbb{R}^H$, and Y-axis attention $\mathbf{M_Y} \in \mathbb{R}^W$ along the channel, width, and height directions, where $\mathbf{M_X}$ and $\mathbf{M_Y}$ can be further composed of spatial attention by element-wise multiplication. With these three directional attention maps, the 3-D attention map can be computed as:

$$\mathbf{M} = \sigma(BN(\mathbf{M_C} \otimes \mathbf{M_S})) = \sigma(BN(\mathbf{M_C} \otimes (\mathbf{M_X} \otimes \mathbf{M_Y}))). \tag{1}$$

Here, $\otimes$ refers to element-wise product, σ denotes sigmoid activation function and BN is batch normalization.

Through this way of factorizing-and-combining, our proposed MS3DA attention can simultaneously focus on the channel relationships and the direction-aware dependencies with precise positional information, and further assigns an independent weight to each neuron of the input feature map.

The overview of MS3DAM is shown in Fig. 1, which consists of four submodules. The shared coordinate embedding (SCE) submodule acts as a bridge connecting the channel attention embedding (CAE) submodule and the spatial attention embedding (SAE) submodule, which are then used to respectively generate 1-D channel attention map and 2-D spatial attention map. Finally, the multi-scale 3-D attention generation submodule is utilized to compute the 3-D attention map.

3.2 Shared Coordinate Embedding Submodule

The structure of the SCE submodule is illustrated above in Fig. 2. Specifically, for a given input $\mathbf{F} \in \mathbb{R}^{C\times H\times W}$, we use two global average pooling operations with kernels of $(H, 1)$ or $(1, W)$ to respectively encode spatial information along the horizontal coordinate and vertical direction, yielding $\mathbf{F}^{\mathbf{H}}_{\text{Avg}} \in \mathbb{R}^{C\times H}$ and $\mathbf{F}^{\mathbf{W}}_{\text{Avg}} \in \mathbb{R}^{C\times W}$. The output in channel of c and height of h can be computed as:

$$F_c^h(h) = \frac{1}{W}\sum_{0\leq i<W} x_c(h,i). \tag{2}$$

The output in channel of c and width of w can be written as:

$$F_c^w(w) = \frac{1}{H}\sum_{0\leq j<H} x_c(j,w). \tag{3}$$

Then we can obtain $\mathbf{F_{SCE}} \in \mathbb{R}^{H+W}$ by concatenating the above two tensors. The overall SCE process is written as:

$$\mathbf{F_{SCE}}(\mathbf{F}) = [X - AvgPool(\mathbf{F}); Y - AvgPool(\mathbf{F})] = [\mathbf{F}^{\mathbf{H}}_{\text{Avg}}; \mathbf{F}^{\mathbf{W}}_{\text{Avg}}], \tag{4}$$

where, $[*; *]$ refers to the concatenation operation. Embedding structure like this can capture long-range dependencies of input feature map and preserve more precise spatial information. In addition, the SCE submodule connects the CAE submodule and the SAE submodule to ensure that the channel attention and the spatial attention can share the same embedded information.

3.3 Channel Attention Embedding Submodule

The structure of the CAE submodule is shown in the middle of Fig. 2. The CAE submodule further aggregates information from the SCE submodule along the other spatial direction to generate a spatial context descriptor. Since the global average pooling process can be computed as:

$$F_c = \frac{1}{H\times W}\sum_{i=1}^{H}\sum_{j=1}^{W} x_c(i,j) = \frac{1}{H}\sum_{i=1}^{H} F_c^h(h) = \frac{1}{W}\sum_{j=1}^{W} F_c^w(w), \tag{5}$$

and the height equals the width of the input feature map, any one of the outputs of the SCE submodule ($\mathbf{F}^{\mathbf{H}}_{\text{Avg}}$ or $\mathbf{F}^{\mathbf{W}}_{\text{Avg}}$) can be aggregated to get a same output through global average pooling, which can be written as $\mathbf{F}_{\mathbf{C}}^{\mathbf{H}} \in \mathbb{R}^C$ or $\mathbf{F}_{\mathbf{C}}^{\mathbf{W}} \in \mathbb{R}^C$. Subsequently, we can use multi-scale 1-D convolutions $f_C^{\psi(C)}$, f_C^3 , and f_C^5 to efficiently encode this feature map, then all the encoding results are added to yield the channel attention map $\mathbf{M_C} \in \mathbb{R}^C$. The process of channel attention encoding can be computed by either of the following two formulas:

$$\mathbf{M_C}(\mathbf{F}_{\mathbf{Avg}}^{\mathbf{H}}) = f_C^{\psi(C)}(\mathbf{F}_{\mathbf{C}}^{\mathbf{H}}) + f_C^3(\mathbf{F}_{\mathbf{C}}^{\mathbf{H}}) + f_C^5(\mathbf{F}_{\mathbf{C}}^{\mathbf{H}}), \tag{6}$$

$$\mathbf{M_C}(\mathbf{F}_{\mathbf{Avg}}^{\mathbf{H}}) = f_C^{\psi(C)}(\mathbf{F}_{\mathbf{C}}^{\mathbf{W}}) + f_C^3(\mathbf{F}_{\mathbf{C}}^{\mathbf{W}}) + f_C^5(\mathbf{F}_{\mathbf{C}}^{\mathbf{W}}). \tag{7}$$

An adaptive method of ECANet is used to determine the convolution kernel size $\psi(C)$, which can be calculated by:

$$k = \psi(C) = \left| \frac{\log_2(C)}{\gamma} + \frac{b}{\gamma} \right|_{odd}, \tag{8}$$

where $|t|_{odd}$ denotes rounding the number t up to the nearest odd number, γ is set to 2 and b is set to 1. Through extensive experiments (see Sect. 4.1), the kernel sizes in CAE module are determined to 3, 5, and $\psi(C)$.

3.4 Spatial Attention Embedding Submodule

The diagram of the SAE submodule is shown below in Fig. 2. The output of the SCE submodule, as the input of SAE submodule, is sent to a max pooling operation to yield $[\mathbf{F}_{S}^{\mathbf{H}}; \mathbf{F}_{S}^{\mathbf{W}}] \in \mathbb{R}^{C+H}$ along the channel direction. Then we use multi-scale 1-D convolutions f_S^3 and f_S^5 to further encode the concatenated direction vectors from different scales of receptive fields to simultaneously grasp the relationships between two groups of orthogonal directions, obtaining:

$$\begin{aligned}\mathbf{M_{SE}}(\mathbf{F_{SCE}}) &= f_S^3(CAP([\mathbf{F}_{\mathbf{Avg}}^{H}; \mathbf{F}_{\mathbf{Avg}}^{W}])) + f_S^5(CAP([\mathbf{F}_{\mathbf{Avg}}^{\mathbf{H}}; \mathbf{F}_{Avg}^{\mathbf{W}}])) \\ &= f_S^3([\mathbf{F}_{\mathbf{S}}^{\mathbf{H}}; \mathbf{F}_{\mathbf{S}}^{\mathbf{W}}]) + f_S^5([\mathbf{F}_{\mathbf{S}}^{\mathbf{H}}; \mathbf{F}_{\mathbf{S}}^{\mathbf{H}}]),\end{aligned} \tag{9}$$

where CAP is an average pooling operation along channel.

The encoding result $\mathbf{M_{SE}}$ is then divided into X-axis coordinate attention $\mathbf{M_X} \in \mathbb{R}^H$ and Y-axis coordinate attention $\mathbf{M_Y} \in \mathbb{R}^W$. We obtain a 2-D spatial attention map by sending the two direction vectors to an element-wise product operation, which can be written as:

$$\begin{aligned}\mathbf{M_S}(\mathbf{F}_{\mathbf{Avg}}^{\mathbf{H}}, \mathbf{F}_{\mathbf{Avg}}^{\mathbf{W}}) &= (f_S^3(\mathbf{F}_{\mathbf{S}}^{\mathbf{H}}) + f_S^5(\mathbf{F}_{\mathbf{S}}^{\mathbf{H}})) \otimes (f_S^3(\mathbf{F}_{\mathbf{S}}^{\mathbf{W}}) + f_S^5(\mathbf{F}_{\mathbf{S}}^{\mathbf{W}})) \\ &= \mathbf{M_X} \otimes \mathbf{M_Y}.\end{aligned} \tag{10}$$

The kernel sizes of multi-scale convolutions in the SAE module are determined to 3 and 5 (see Sect. 4.1).

3.5 MS3DA Attention Generation Submodule

After we obtain channel attention map and spatial attention map from the CAE and the SAE submodules, the two attention maps are directly multiplied in this submodule to highlight representations more important to network.

Given the input $\mathbf{F} \in \mathbb{R}^{C \times H \times W}$, the final MS3DA attention map $\mathbf{M}$ and the output $\mathbf{Y}$ can be represented as:

$$\begin{aligned}\mathbf{M}(\mathbf{F}) &= \sigma(BN(\mathbf{M_C}(\mathbf{F}^{\mathbf{H}}_{\mathbf{Avg}}) \otimes \mathbf{M_S}(\mathbf{F}^{\mathbf{H}}_{\mathbf{Avg}}, \mathbf{F}^{\mathbf{W}}_{\mathbf{Avg}})) \\ &= \sigma(BN(\mathbf{M_C}(\mathbf{F}^{\mathbf{W}}_{\mathbf{Avg}}) \otimes \mathbf{M_S}(\mathbf{F}^{\mathbf{H}}_{\mathbf{Avg}}, \mathbf{F}^{\mathbf{W}}_{\mathbf{Avg}}))),\end{aligned} \tag{11}$$

$$\mathbf{Y} = \mathbf{M}(\mathbf{F}) \otimes \mathbf{F}. \tag{12}$$

Here, σ denotes the sigmoid activation function.

3.6 Re-Param for Multi-scale 1-D Convolutions

In the CAE submodule and the SAE submodule, the information is encoded into two direction vectors. We use multi-scale 1-D convolutions to encode the direction vectors, and pad the input with 0 to ensure that the size of the output is unchanged. According to the rules of convolution calculation, it is obvious that when we pad the input direction vector and the 1-D convolution kernel with 0 at the same time, the result of convolution keeps consistent. Taking advantage of this, in the inference phase, we can use the structural re-parameterization method to fuse multi-scale convolutions to boost the inference speed and save memory.

We use $\mathbf{F_{In}} \in \mathbb{R}^{C}$ to denote the input direction vector and $\mathbf{W_a}$, $\mathbf{W_b}$ to denote the 1-D convolutions with kernel sizes a and b, respectively. Assuming the kernels sizes are 3 and 5, the process of re-parameterization can be illustrated in Fig. 3. During the transformation, $\mathbf{W'_a}$ and $\mathbf{W'_b}$ denotes the padded convolution and the fused convolution, respectively.

Due to the property of 1-D convolution, it is possible to pad the smaller convolution kernel to a larger one, which will not change the final calculation result. Then, we can obtain a merged convolution kernel by adding the corresponding padded parameters one by one. In the inference phase, the fused convolution can completely replace the original two convolutions. Considering the more general case, assuming that a is smaller than b, the re-parameterization process can be written as:

$$\begin{aligned}\mathbf{F_{Out}} &= \mathbf{F_{In}} * \mathbf{W_a} + \mathbf{F_{In}} * \mathbf{W_b} = \sum_{i=0}^{C-1} \mathbf{W_a}(i) \star \mathbf{F_{In}}(i) + \sum_{j=0}^{C-1} \mathbf{W_b}(j) \star \mathbf{F_{In}}(j) \\ &= \sum_{k=0}^{C-1} (\mathbf{W_a}'(k) + \mathbf{W_b}(k)) \star \mathbf{F_{In}}(k) \\ &= \sum_{k=0}^{C-1} \mathbf{W_b}'(k) \star \mathbf{F_{In}}(k) = \mathbf{F_{In}} * \mathbf{W'_b},\end{aligned} \tag{13}$$

where $*$ and $\star$ respectively denote the convolution and autocorrelation operation. If the two kernels are equal in size, the parameters can be added directly without padding. When we have three convolution kernels, we can pad the two smaller kernels with 0 to be equal to the largest kernel and then add the parameters of all three together.

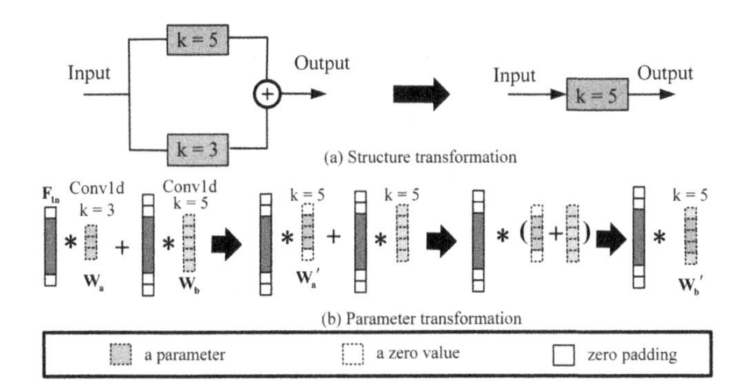

Fig. 3. Diagram of structure and parameter transformations during the process of structural re-parameterization. k and $*$ stand for 1-D convolution kernel size and convolution operation, respectively.

4 Experiments

In this section, we demonstrate the effectiveness of MS3DA attention module on standard benchmarks including image classification and object detection. First, we conduct extensive ablation experiments to determine the structure of our module. Then, we reproduce many state-of-the-art networks in the PyTorch framework and verify that our method outperforms various counterparts, demonstrating the general applicability of our module.

4.1 Ablation Studies on CIFAR-100

In this subsection, we perform extensive ablation experiments on ResNets [4] to determine the combining strategies and the kernel sizes of multiscale 1-D convolutions in our attention module. We evaluate Top-1 accuracy in CIFAR-100 classification. Learning strategies and image enhancement methods are the same as [1].

The Combination of Channel Attention and Spatial Attention. Table 1a summarizes the results of experiments, which verify that each branch of channel and spatial attention contributes to the improvement of performance. And we observe significant boosting in accuracy when the two branches are grouped in an appropriate way, indicating that best-combining strategy can further make full use of inter-channel relationships and spatial dependencies to yield a more precise object localization.

Table 1. ResNets classification to determine the kernel sizes and combining strategies of multi-scale 1-D convolutions in MS3DAAM. (a) Ablation studies on the combining methods of MS3DA attention. $\mathbf{M_C}$, $\mathbf{M_S}$, $\mathbf{M_X}$, and $\mathbf{M_Y}$ represent channel attention, spatial attention, X-axis attention, and Y-axis attention, respectively. $\oplus$ and $\otimes$ respectively denote element-wise summation operation and element-wise multiplication operation. (b) Impact of simultaneously using different kernel sizes of 1-D convolutions in the CAE submodule and the SAE submodules. k_C and k_S refer to the convolution kernel sizes in the CAE and the SAE submodules, respectively.

(a)

Settings	Base	A	B	C	D	E	F	+Ours
$\mathbf{M_C}$		✓			✓	✓	✓	✓
$\mathbf{M_S}$			✓	✓	✓	✓	✓	✓
$\mathbf{M_C} \oplus \mathbf{M_S}$					✓	✓		
$\mathbf{M_C} \otimes \mathbf{M_S}$							✓	✓
$\mathbf{M_X} \oplus \mathbf{M_Y}$			✓		✓		✓	
$\mathbf{M_X} \otimes \mathbf{M_Y}$				✓		✓		✓
ResNet18 [4]	77.8	78.5	78.2	78.2	78.6	78.2	78.4	**78.8**
ResNet50 [4]	80.0	80.7	80.4	80.1	80.6	80.5	80.5	**80.8**

(b)

Kernel Size	Base	A	B	C	D	E	F	G	H	+Ours
$k_S = 3$		✓		✓	✓	✓	✓	✓	✓	✓
$k_S = 5$			✓	✓	✓	✓	✓	✓	✓	✓
$k_C = \psi(C)$		✓	✓	✓				✓	✓	✓
$k_C = 3$					✓		✓	✓	✓	
$k_C = 5$						✓	✓		✓	✓
ResNet18 [4]	77.8	78.4	78.4	78.4	78.2	78.2	78.3	78.5	78.5	**78.8**
ResNet50 [4]	78.0	80.4	80.4	80.4	80.2	80.3	80.7	80.7	80.7	**80.8**

Combining Methods of Multi-scale 1-D Convolutions. MS3DA attention module generates attention maps along three directions, so there are many combining strategies. In Table 1a, we also compare four different combining strategies on ResNet18 and ResNet50: the combination strategy between $\mathbf{M_C}$ and $\mathbf{M_S}$, $\mathbf{M_X}$ and $\mathbf{M_Y}$ can be either element-wise summation or element-wise product.

According to the results shown in Table 1a, we observe that the network achieves the best performance by using element-wise product in both CAE and SAE submodules. It is significantly different from BAM which prefers element-wise summation. The possible reasons are diversified. Firstly, from the spatial point of view, the SAM submodule concatenates the X-axis direction vector and Y-axis direction vector and then uses the same set of multi-scale convolutions for encoding, which largely avoids the phenomenon of uneven gradient distribution in the backward phase caused by multiplication mentioned in BAM. Secondly, by using element-wise product, the regained spatial attention map can further widen the chasm between the weights of the spatial attention, resulting in emphases on positional information. Thirdly, the sigmoid functions provide a more unified measurement for the combination. Finally, the 3-D attention map calculated by multiplication gives greater weight to the region of interest, which enables the network to highlight more prominent features and capture the precise locations.

The Selection of the Kernel Size of 1-D Convolution. We introduce multi-scale convolutions to push network performance. We conduct experiments on ResNet18 and ResNet50 to determine the appropriate 1-D convolution kernel sizes. Since it is difficult to determine the kernel sizes of the CAE submodule and the SAE submodule at the same time, we first fix the kernel sizes of the CAE submodule to $\psi(C)$ Considering the images of the CIFAR-100 and ImageNet datasets are respectively resized to 32×32 and 224×224 during preprocessing, the corresponding convolution kernel sizes are determined to 3 and 5, calculated by Eq. 8. Table 1b shows that when the kernel size of the CAE submodule is fixed, the network achieves the best performance by using multi-scale convolutions that can further boost the performance of the network.

After we determine appropriate k_S from the experiment above, we fix it and then investigate the impact of k_C on following experiments. As shown in Table 1b, the results illustrate that when k_S is fixed, using multi-scale convolutions can also yield better results. In addition, adding new branches contributes to the lift of performance. In particular, when k_C is set to $\psi(C)$, 3, and 5, we observe a boosting in accuracy, which further verifies the effectiveness of multi-scale 1-D convolutions.

Table 2. Classification results on CIFAR-100 and ImageNet-1K. We compute the average inference frame rate by forwarding images of validation sets one by one on an Intel(R) Core i7-12700 CPU and a GTX 3080Ti GPU. †: The inference speed is evaluated before and after re-parameterization. The number to the right of the plus sign indicates the speed gain.

(a) CIFAR-100 classification results

Settings	Param. (M)	FLOPs (G)	Speed (FPS)	Top-1 Acc.	Top-5 Acc.
ResNet18 [4]	11.22	0.56	309	77.81	93.86
+SE [8]	11.31	0.56	230	78.30	93.97
+CBAM [16]	11.31	0.56	149	78.01	93.85
+CA [5]	11.30	0.56	145	78.44	94.31
+ECA [15]	11.22	0.56	224	77.88	94.36
+TAM [10]	11.23	0.56	148	78.00	94.47
+SimAM [18]	11.22	0.56	218	78.63	**94.48**
+Ours	11.22	0.56	134+35†	**78.83**	**94.48**
ResNet50 [4]	23.71	1.31	152	79.95	95.34
+SE [8]	26.22	1.32	113	80.28	95.25
+CBAM [16]	26.22	1.32	70	80.32	95.01
+CA [5]	25.62	1.32	72	79.92	94.94
+ECA [15]	23.71	1.31	115	80.33	95.13
+TAM [10]	23.71	1.34	73	79.22	94.99
+SimAM [18]	23.71	1.31	114	80.47	95.31
+Ours	23.74	1.32	71+14†	**80.77**	95.17

(b) ImageNet-1K classification results

Settings	Param. (M)	FLOPs (G)	Speed (FPS)	Top-1 Acc.	Top-5 Acc.
ResNet50 [4]	25.56	4.13	126	75.01	92.25
+SE [8]	28.07	4.14	98	75.73	92.74
+CBAM [16]	28.07	4.15	63	75.89	92.74
+CA [5]	27.47	4.17	65	75.56	92.48
+ECA [15]	25.56	4.14	99	75.48	92.52
+TAM [10]	25.56	4.18	62	75.38	92.31
+SimAM [18]	25.56	4.13	97	75.24	92.49
+Ours	25.59	4.17	62+14†	**76.04**	**92.91**
ResNeXt50 [17]	25.03	4.29	150	75.68	92.50
+SE [8]	27.54	4.29	114	76.25	92.93
+CBAM [16]	27.55	4.30	68	76.27	93.07
+CA [5]	26.95	4.33	70	76.09	92.67
+ECA [15]	25.03	4.29	113	75.55	92.54
+TAM [10]	25.03	4.33	70	75.55	92.45
+SimAM [18]	25.03	4.29	113	76.00	92.72
+Ours	25.06	4.32	67+15†	**76.42**	**92.97**
MobileNetV2 [12]	3.51	0.33	165	68.91	88.81
+SE [8]	4.07	0.33	119	70.33	89.59
+CBAM [16]	4.07	0.33	71	69.93	89.49
+CA [5]	3.95	0.34	74	70.87	**89.92**
+ECA [15]	3.51	0.33	124	70.81	89.74
+TAM [10]	3.51	0.33	74	70.16	89.40
+SimAM [18]	3.51	0.33	123	69.13	88.88
+Ours	3.52	0.33	98+14†	**70.97**	89.90

4.2 Comparision with State-of-the-Arts

In Table 2, we compare the implementation of our module with various attention mechanisms in ResNets. Due to its ingenuity, MS3DA attention achieves the best results in the Top-1 accuracy with minimal parameters and computational costs compared to other attention methods.

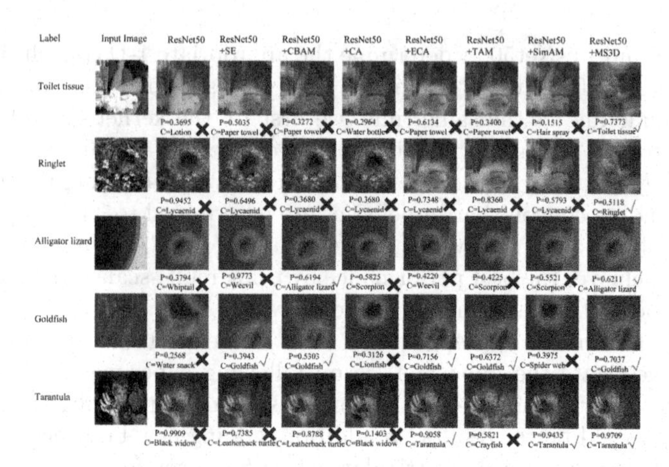

Fig. 4. Visualization of the feature maps in ResNet50 [4] with various attention mechanisms by Grad-CAM [13]. The grad-CAM visualization is computed for the last convolutional layer. C and P denote the predicted class and the corresponding softmax score, respectively. ✓ means that the predicted category and label are consistent, while ✓ indicates that they are inconsistent. Better to zoom in.

In addition to accuracy, the diversified results further prove the superiority of MS3DA attention module. Firstly, in terms of the number of parameters, MS3DA attention is close to the recently proposed parameter-free SimAM as well as some near-parameter-free attention methods such as ECANet, TAM, and NAM and significantly outperforms SE, CBAM and CA. Secondly, in complexity, it is preferable to TAM and CA and close to the rest. Finally, in inference speed, our proposed module runs much faster than CBAM, CA, and TAM after re-parameterization.

We further verify that MS3DAAM is beneficial to ImageNet classification. Learning strategies and image enhancement methods are the same as ResNets.

The experiments in Table 2 verify that on ResNet50 [4] and ResNeXt50 [17], our proposed MS3DA attention outperforms other counterparts in terms of Top-1 classification accuracy with virtually no extra computational cost. Compared with SE, CBAM, and CA using traditional 2-D convolutions, MS3DA attention has nearly no parameters. In inference speed, our proposed module runs faster than CBAM, CA, and TAM. It is worth noting that, compared with CA, our method has less computation and parameters and faster inference speed.

On MobileNetV2 [12], our proposed MS3DA attention achieves 2.06% performance gain in top-1 accuracy. And our method outperforms the classical SE and CBAM as well as many lightweight methods such as ECANet and SimAM. Compared with CA, our method achieves competitive performance with less computational overhead and faster inference speed, which is especially suitable for lightweight networks. The re-parameterization method we develop for 1-D convolutions greatly boost the inference speed, with a speedup of 14.28%.

4.3 Visualization with Grad-CAM

In order to show the effect of attention on the network, we used Grad-CAM [13] to visualize the feature maps of models with different attention mechanisms. The visualization results in Fig. 4 obviously demonstrate the superiority of our proposed MS3DA attention over other methods in locating the objects of interest. We note that when the baseline network fails to classify an object correctly, adding an appropriate attention mechanism can guide the network to make the correct judgment. In addition, through the visualization results, we suppose that channel attention is able to emphasize prominent features, and the 1-D attention mechanisms using fully connected or shared MLP directly strengthen the features of interest along the channel dimension. However, in some cases, the network is easy to be misled by some salient features, so as to make wrong judgments including misclassification of quite-different categories and misjudgment of alike classes (see the classification results of toilet paper and ringlet in Fig. 4, respectively). Therefore, channel attention has to be guided by spatial attention, which can focus on the precise positional locations.

The results of the classification of toilet tissue and ringlet further validate our idea: excessive attention to salient features may sometimes provide higher score, but it may lead to misjudgment of alike categories. The ringlet and lycaenid are morphologically very similar, with the exception that the ringlet has more eye spots on both front and back wings. The visualization results show that MS3DA attention can emphasize the inter-channel relationships and the spatial dependencies, and better capture subtle features by establishing the connection between channel and space, thus prompting the model to make the correct judgment. Furthermore, the classification results of the alligator lizard, goldfish, and tarantula also verify that our attention can guide the network to make correct judgments in a considerable number of cases and always provide higher confidence score.

Table 3. Object detection results on MS COCO and VOC. For MS COCO, we use SSD320 as detector and ResNet50 [4] as backbone. For VOC, we adopt SSDLite320 detector and MobileNetV2 [12] backbone.

(a) Object detection results on MS COCO

Method	Backbone	Param. (M)	FLOPs (G)	AP (%)	AP_50 (%)	AP_75 (%)	AP_S (%)	AP_M (%)	AP_L (%)
SSD320	ResNet50 [4]	20.30	2.03	23.4	40.3	23.8	6.9	25.1	37.8
SSD320	ResNet50 + SE [8]	24.10	2.03	23.6	40.8	23.7	6.8	25.7	38.5
SSD320	ResNet50 + CA [5]	23.60	2.03	23.8	41.0	23.9	6.7	25.5	38.3
SSD320	ResNet50 + Ours	22.90	2.03	**24.2**	**41.9**	**24.2**	**7.1**	**25.9**	**39.3**

(b) Object detection results on VOC

Method	Backbone	Param. (M)	FLOPs (G)	AP (%)
SSDLite320	MobileNetV2 [12]	3.43	0.74	65.81
SSDLite320	MobileNetV2 + SE [8]	3.99	0.74	66.18
SSDLite320	MobileNetV2 + Ours	3.44	0.77	**67.31**

4.4 Object Detection Results on COCO and VOC

We conduct object detection on MS COCO and VOC datasets, according to [4,12,17].

As shown in Table 3, the results further demonstrate the superiority of MS3DA attention over other counterparts. For MS COCO, our MS3DA attention outperforms other attention methods and boosts the performance in the mean AP by 0.8% with a minimal number of parameters. Obviously, classical attention methods fail to push the performance of small object detection which is a challenging task for networks. On the contrary, they may bring side effects. It is worth noting that our proposed method lifts the performance of small object detection by 0.4%, which further verifies the transferable capability and capability in locating of our method. On VOC dataset, we also observe a significant improvement of 1.5%.

5 Conclusion

In this paper, we present MS3DA attention, a novel lightweight attention mechanism, which can advance the localization of objects of interest by emphasizing directional features along three dimensions: channel, height and width. The classification results on CIFAR-100 and ImageNet and object detection results on MS COCO prove the effectiveness of MS3DAAM. We also verify the superiority of our proposed method by network visualization. We hope that our attention mechanism could not only boost the performance of more CNN architectures but also greatly simplify their structures.

References

1. DeVries, T., Taylor, G.W.: Improved regularization of convolutional neural networks with cutout. arXiv preprint arXiv:1708.04552 (2017)
2. Ding, X., Zhang, X., Han, J., Ding, G.: Diverse branch block: building a convolution as an inception-like unit. In: Proceedings of the IEEE/CVF Conference on Computer Vision and Pattern Recognition, pp. 10886–10895 (2021)
3. Ding, X., Zhang, X., Ma, N., Han, J., Ding, G., Sun, J.: Repvgg: making vgg-style convnets great again. In: Proceedings of the IEEE/CVF Conference on Computer Vision and Pattern Recognition, pp. 13733–13742 (2021)
4. He, K., Zhang, X., Ren, S., Sun, J.: Deep residual learning for image recognition. In: Proceedings of the IEEE Conference on Computer Vision and Pattern Recognition, pp. 770–778 (2016)
5. Hou, Q., Zhou, D., Feng, J.: Coordinate attention for efficient mobile network design. In: Proceedings of the IEEE/CVF Conference on Computer Vision and Pattern Recognition, pp. 13713–13722 (2021)
6. Howard, A., et al.: Searching for mobilenetv3. In: Proceedings of the IEEE/CVF International Conference on Computer Vision, pp. 1314–1324 (2019)
7. Hu, J., Shen, L., Albanie, S., Sun, G., Vedaldi, A.: Gather-excite: exploiting feature context in convolutional neural networks. Adv. Neural Inf. Process. Syst. **31** (2018)
8. Hu, J., Shen, L., Sun, G.: Squeeze-and-excitation networks. In: Proceedings of the IEEE Conference on Computer Vision and Pattern Recognition, pp. 7132–7141 (2018)
9. Li, X., Wang, W., Hu, X., Yang, J.: Selective kernel networks. In: Proceedings of the IEEE/CVF Conference on Computer Vision and Pattern Recognition, pp. 510–519 (2019)
10. Misra, D., Nalamada, T., Arasanipalai, A.U., Hou, Q.: Rotate to attend: convolutional triplet attention module. In: Proceedings of the IEEE/CVF Winter Conference on Applications of Computer Vision, pp. 3139–3148 (2021)

11. Park, J., Woo, S., Lee, J.Y., Kweon, I.S.: Bam: bottleneck attention module. arXiv preprint arXiv:1807.06514 (2018)
12. Sandler, M., Howard, A., Zhu, M., Zhmoginov, A., Chen, L.C.: Mobilenetv 2: inverted residuals and linear bottlenecks. In: Proceedings of the IEEE Conference on Computer Vision and Pattern Recognition, pp. 4510–4520 (2018)
13. Selvaraju, R.R., Cogswell, M., Das, A., Vedantam, R., Parikh, D., Batra, D.: Grad-cam: visual explanations from deep networks via gradient-based localization. In: Proceedings of the IEEE International Conference on Computer Vision, pp. 618–626 (2017)
14. Simonyan, K., Zisserman, A.: Very deep convolutional networks for large-scale image recognition. arXiv preprint arXiv:1409.1556 (2014)
15. Wang, Q., Wu, B., Zhu, P., Li, P., Zuo, W., Hu, Q.: ECA-NET: efficient channel attention for deep convolutional neural networks. In: Proceedings of the IEEE/CVF Conference on Computer Vision and Pattern Recognition, pp. 11534–11542 (2020)
16. Woo, S., Park, J., Lee, J.Y., Kweon, I.S.: CBAM: convolutional block attention module. In: Proceedings of the European conference on computer vision (ECCV), pp. 3–19 (2018)
17. Xie, S., Girshick, R., Dollár, P., Tu, Z., He, K.: Aggregated residual transformations for deep neural networks. In: Proceedings of the IEEE Conference on Computer Vision and Pattern Recognition, pp. 1492–1500 (2017)
18. Yang, L., Zhang, R.Y., Li, L., Xie, X.: Simam: a simple, parameter-free attention module for convolutional neural networks. In: International Conference on Machine Learning, pp. 11863–11874. PMLR (2021)

Nonlinear Multiple-Delay Feedback Based Kernel Least Mean Square Algorithm

Ji Zhao[1(✉)], Jiaming Liu[1], Qiang Li[1(✉)], Lingli Tang[2], and Hongbin Zhang[3]

[1] School of Information Engineering, Southwest University of Science and Technology, Mianyang 621010, China
{zhaoji,liqiangsir}@swust.edu.cn

[2] Department of Social Sciences, Southwest University of Science and Technology, Mianyang 621010, China

[3] School of Information and Communication Engineering, University of Electronic Science and Technology of China, Chengdu 611731, China

Abstract. In this paper, a novel algorithm called nonlinear multiple-delay feedback kernel least mean square (NMDF-KLMS) is proposed by introducing a nonlinear multiple-delay into the framework of multikernel adaptive filtering. The proposed algorithm incorporates the nonlinear multiple-delay to enhance the filtering performance in comparison with the kernel adaptive filtering algorithm using linear feedback. Furthermore, for NMDF-KLMS, the theoretical mean-square convergence analyses is also conducted. Simulation results under chaotic time-series prediction and real-world data applications show that NMDF-KLMS achieves a faster convergence rate and superior filtering accuracy.

Keywords: Multikernel · Nonlinear feedback · Multiple-delay · Kernel adaptive filtering

1 Introduction

In recent years, the kernel method has gained widespread applications in machine learning and signal processing. The kernel method can transform the input data into a reproducing kernel Hilbert spaces (RKHS) with high or even infinite dimensions. By doing so, it becomes feasible to find linear solutions to nonlinear problems of input space. Therefore, it is valuable for learning tasks, such as pattern classification, system identification, time-series prediction, and other related applications [1,2].

Use the kernel method for linear adaptive filtering leading to the well-known kernel adaptive filtering (KAF) algorithms, which are efficient to solve nonlinear signal processing [3]. And, some celebrated examples include the kernel

This work is supported in part by the National Natural Science Foundation of China(Grant no. 62201478 and 61971100), in part by the Southwest University of Science and Technology Doctor Fund (Grant no. 20zx7119), in part by the Sichuan Science and Technology Program (Grant no. 2022YFG0148), and in part by the Heilongjiang Provincial Science and Technology Program (No. 2022ZX01A16).

B. Luo et al. (Eds.): ICONIP 2023, LNCS 14447, pp. 118–129, 2024.
https://doi.org/10.1007/978-981-99-8079-6_10

least mean square (KLMS) algorithm [4], the kernel affine projection (KAP) algorithm [5], and the kernel recursive least squares (KRLS) algorithm [6]. To further improve the filtering performance of KAF algorithms, feedback frameworks are introduced into these systems resulting in the family of recurrent KAF. Examples include the linear recurrent kernel online learning (LRKOL) algorithm [7], the single feedback KLMS (SF-KLMS) algorithm [8], the regularized KMLS with multiple-delay feedback (RKLMS-MDF) algorithm [9], and the KRLS with multiple feedback algorithm [10]. Simulation results have demonstrated that these recurrent KAF algorithms outperform their counterparts. However, in these recurrent KAFs, only linear feedback information is considered.

To remedy such limitation, it's better to consider the nonlinear contribution of feedback information [11,12]. The literature has proposed a nonlinear recurrent kernel normalized LMS (NR-KNLMS) algorithm [13], which is developed using a multikernel adaptive filtering framework to solve nonlinear autoregressive systems [14–16]. In NR-KNLMS, only a single-delay desired output is used to construct a recurrent structure. However, some references have validated that multiple-delay leads to more filtering performance enhancement than that of single-delay [9,10]. Therefore, in this work, inspired by the multikernel adaptive filtering, a new KAF called nonlinear multiple-delay feedback kernel least mean square (NMDF-KLMS) is proposed by considering a nonlinear multiple-delay estimated output. For NMDF-KLMS, some theoretical analyses in terms of mean square convergence and steady-state performance are also conducted. Furthermore, without loss of generality, we apply the coherence criterion (CC) method to mitigate computational burdens of NMDF-KLMS [17]. Simulation results under a time-series prediction and four real-world applications demonstrate the superiority of the proposed algorithm over competing algorithms.

The rest of this paper is organized as follows. In Sect. 2, we briefly introduce the concept of multikernel adaptive filtering. In Sect. 3, we propose the nonlinear multiple-delay feedback kernel least mean square (NMDF-KLMS) algorithm and conduct convergence analysis. Section 4 presents the simulation results. Finally, concluding remarks are given in Sect. 5.

2 Multikernel Least Mean Square Algorithm

In general, a number of real-world problems can be modeled as the estimation of a nonlinear function, i.e., $f(\cdot) : \mathbb{U} \to \mathbb{D}$, where $\mathbb{U} \subseteq \mathbb{R}^n$ and $\mathbb{D} \subseteq \mathbb{R}$ are input space and output space, respectively. In the standard KLMS algorithm [3], the regressors are single kernel evaluations, and the weight update formula is

$$\boldsymbol{\Omega}(i) = \eta \sum_{j=1}^{i} e(j)\boldsymbol{\varphi}(j), \tag{1}$$

where $\eta > 0$ is a step-size; $\boldsymbol{\varphi}(\cdot)$ is an unknown mapping induced by a Mercer kernel $\kappa(\cdot,\cdot)$; $\boldsymbol{\Omega}(i) \in \mathcal{H}$ stands for the weight vector, and $\mathcal{H}$ is a real (or complex) valued RKHS and satisfies $\{\varphi(\boldsymbol{u}) : \mathbb{U} \rightarrow \mathcal{H}, \boldsymbol{u} \in \mathbb{U}\}$; $e(i) = d(i) - \boldsymbol{\Omega}(i-1)^T \varphi(i)$ is an estimation error, and $d(i)$ denotes the desired output. When KLMS convergences, the nonlinear function is identified as $f(i) = \boldsymbol{\kappa}(i)^T \boldsymbol{\omega}(i)$, where $\boldsymbol{\kappa}(i) = [\kappa(\boldsymbol{u}_1, \boldsymbol{u}_i), \ldots, \kappa(\boldsymbol{u}_i, \boldsymbol{u}_i)]^T$ and $\boldsymbol{\omega}(i) = \eta[e(1), \ldots, e(i)]^T$.

However, the single kernel used in KLMS has the limitation making full use of the opportunities offered in the feature space. Therefore, the benefits of kernel-based adaptive filtering can be maximized considering multikernel strategy [16]. To this end, consider the multidimensional mapping given by

$$\boldsymbol{\Phi} : \mathbb{U} \rightarrow \mathbb{H}_L \quad \Rightarrow \quad \boldsymbol{u} \mapsto \boldsymbol{\Phi}(\boldsymbol{u}) = [\varphi_1(\boldsymbol{u}), \ldots, \varphi_l(\boldsymbol{u}), \ldots, \varphi_L(\boldsymbol{u})]^T, \tag{2}$$

where $\varphi_l(\boldsymbol{u}) \in \mathcal{H}_l$; and $\mathbb{H}_L$ is a vector-valued RKHS, L is the total number of monokernel. Using the feature elements $\boldsymbol{\Phi}(\boldsymbol{u})$, the output of the multikernel LMS (MKLMS) can be expressed as

$$y = \langle \boldsymbol{\Omega}, \boldsymbol{\Phi}(\boldsymbol{u}) \rangle, \tag{3}$$

where $\boldsymbol{\Omega} \in \mathbb{H}_L$ is the weight vector coefficient. Similar to that of the KLMS algorithm, the update of $\boldsymbol{\Omega}$ is obtained utilizing the stochastic gradient as follows

$$\boldsymbol{\Omega}(i) = \boldsymbol{\Omega}(i-1) + \eta e(i) \boldsymbol{\Phi}(\boldsymbol{u}(i)) = \eta \sum_{j=1}^{i} e(j) \boldsymbol{\Phi}(\boldsymbol{u}(j)). \tag{4}$$

Substituting (4) into (3), and according to the reproducing property of vector-valued Hilbert space [16], we have

$$y(i) = \left\langle \eta \sum_{j=1}^{i-1} e(j) \boldsymbol{\Phi}(\boldsymbol{u}(j)), \boldsymbol{\Phi}(\boldsymbol{u}(i)) \right\rangle = \eta \sum_{j=1}^{i-1} e(j) \langle \boldsymbol{\Phi}(\boldsymbol{u}(j)), \boldsymbol{\Phi}(\boldsymbol{u}(i)) \rangle. \tag{5}$$

Similar to the KLMS algorithm utilizing the kernel trick to express the inner product via a kernel evaluation, for MKLMS, inner product in (5) is obtained as

$$\langle \boldsymbol{\Phi}(\boldsymbol{u}(j)), \boldsymbol{\Phi}(\boldsymbol{u}(i)) \rangle = \sum_{l=1}^{L} \langle \varphi_l(\boldsymbol{x}(j)), \varphi_l(\boldsymbol{x}(i)) \rangle = \sum_{l=1}^{L} \kappa_l(\boldsymbol{u}(j), \boldsymbol{u}(i)), \tag{6}$$

where $\kappa_l(\cdot, \cdot)$ is the l-th kernel function. And thus, for MKLMS, the output estimate is derived as

$$y(i) = \sum_{j=1}^{i-1} \sum_{l=1}^{L} a_j(i) \kappa_l((\boldsymbol{u}(j), \boldsymbol{u}(i))) \tag{7}$$

with $a_j(i) = \eta e(j)$. Equation (7) can be regarded as a generalization of KLMS in (1), which not only obtains benefits of the kernels κ_l within MKLMS, but also extracts the multimodality information of input signal.

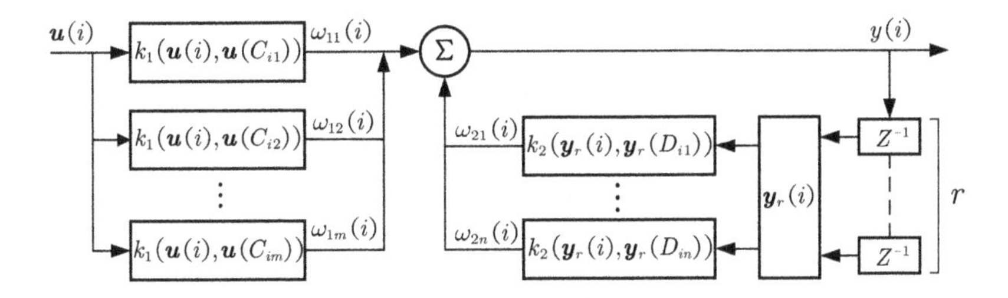

Fig. 1. The structure of NMDF-KLMS.

3 Nonlinear Multiple-Delay Feedback Based KLMS

3.1 NMDF-KLMS

Motivated from the idea of MKLMS, in this work, we propose a novel estimated output delay feedback kernel adaptive filtering algorithm, and show the structure of proposed algorithm in Fig. 1. In which, the estimated output passes r delay operations z^{-1}, and produces a multiple-delay vector $\boldsymbol{y}_r$. Then we treat $\boldsymbol{y}_r$ as a kind of new input data, and thus, from Fig. 1, the output of our proposed algorithm is given as

$$\begin{aligned} y(i) &= \sum_{t=1}^{m} \omega_{1t}(i)\kappa_1(\boldsymbol{u}(C_{it}), \boldsymbol{u}(i)) + \sum_{t=1}^{n} \omega_{2t}(i)\kappa_2(\boldsymbol{y}_r(D_{it}), \boldsymbol{y}_r(i)) \\ &= \boldsymbol{\kappa}_1^T(i)\boldsymbol{\omega}_1(i) + \boldsymbol{\kappa}_2^T(i)\boldsymbol{\omega}_2(i), \end{aligned} \tag{8}$$

where $\boldsymbol{\kappa}_1(i) = [\kappa_1(\boldsymbol{u}(C_{i1})), \ldots, \kappa_1(\boldsymbol{u}(C_{im}))]^T$ represents a kernel evaluation vector computed from input $\boldsymbol{u}(i)$ and dictionary $\boldsymbol{C}(i) = [\boldsymbol{u}(C_{i1}), \ldots, \boldsymbol{u}(C_{im})]^T$; $\boldsymbol{\kappa}_2(i)$ is an other kernel evaluation vector between $\boldsymbol{y}_r(i) = [y(i-1), y(i-2), \ldots, y(i-r)]^T$ and $\boldsymbol{D}(i) = [\boldsymbol{y}_r(D_{i1}), \ldots, \boldsymbol{y}_r(D_{in})]^T$; $\boldsymbol{\omega}_1(i) = [\omega_{11}(i), \ldots, \omega_{1m}(i)]^T$ is the feedforward vector, and $\boldsymbol{\omega}_2(i) = [\omega_{21}(i), \ldots, \omega_{2n}(i)]^T$ is the feedback vector. Although, a variety of kernel functions have been reported in the literature, the universal Gaussian kernel function is a usually selection. Therefore, in this work, two Gaussian kernels with different positive kernel-bandwidths h_1 and h_2 are chosen, i.e., $\kappa_1(\boldsymbol{u}_1, \boldsymbol{u}_2) = \exp(-\|\boldsymbol{u}_1 - \boldsymbol{u}_2\|^2/2h_1^2)$ and $\kappa_2(\boldsymbol{y}_{r_1}, \boldsymbol{y}_{r_2}) = \exp(-\|\boldsymbol{y}_{r_1} - \boldsymbol{y}_{r_2}\|^2/2h_2^2)$.

For our proposed algorithm to be computationally efficient, we consider the following coherence criterion sparsification method [17] for constructing compact dictionaries

$$\mathrm{Coh}_1 = \max_{t \in \boldsymbol{C}(i)} (\kappa_1(\boldsymbol{u}(i), \boldsymbol{u}(\boldsymbol{C}_{it}))) \leq \delta_1, \tag{9}$$

$$\mathrm{Coh}_2 = \max_{t \in \boldsymbol{D}(i)} (\kappa_2(\boldsymbol{y}_r(i), \boldsymbol{y}_r(\boldsymbol{D}_{it}))) \leq \delta_2, \tag{10}$$

where $0 \leq \delta_1, \delta_2 \leq 1$ are the threshold parameters that balance the filtering performance and network sparsity.

Similar to that of the MKLMS algorithm, the weight update formula of our proposal can be derived as follows

$$\boldsymbol{\omega}_1(i) = \boldsymbol{\omega}_1(i-1) + \eta e(i)\boldsymbol{\kappa}_1(i), \quad \boldsymbol{\omega}_2(i) = \boldsymbol{\omega}_2(i-1) + \eta e(i)\boldsymbol{\kappa}_2(i). \tag{11}$$

Observe (11), a compact version can be obtained as

$$\boldsymbol{\omega}(i) = \boldsymbol{\omega}(i-1) + \eta e(i)\boldsymbol{\kappa}(i) \quad \text{with} \quad e(i) = d(i) - \boldsymbol{\kappa}^T(i)\boldsymbol{\omega}(i-1) + v(i), \tag{12}$$

where $\boldsymbol{\omega}(i) = [\boldsymbol{\omega}_1^T(i), \boldsymbol{\omega}_2^T(i)]^T$ and $\boldsymbol{\kappa}(i) = [\boldsymbol{\kappa}_1^T(i), \boldsymbol{\kappa}_2^T(i)]^T$; $e(i)$ is the prediction error, and $v(i)$ is an additive noise.

Remark: Equation (12) is call as *nonlinear multiple-delay feedback based KLMS* (NMDF-KLMS) algorithm, which is summarized in **Algorithm** 1. In comparison with MKLMS, NMDF-KLMS treats multiple-delay feedback as another input signal, while MKLMS only uses the original input data as the kernel centers for multikernels. In addition, compared to the existing monokernel adaptive filters with linear feedback structure, i.e., LRKOL, SF-KLMS, and RKLMS-MDF, the NMDF-KLMS algorithm constructs the feedback information into the framework of multikernel adaptive filtering, so that resulting in a superior filtering performance. Furthermore, there are two differences from the NR-KNLMS algorithm: 1) NMDF-KLMS uses the estimated output of filter as feedback information instead of the desired signal, which skips the requirement to possess prior knowledge of the desired signal; 2) NMDF-KLMS uses the multiple-delay feedback ($r \geq 1$) that considers more past information leading to improved convergence rate and filtering accuracy.

3.2 Mean Square Convergence Analysis

We first derive the energy conservation relation for NMDF-KLMS. By defining the optimal weight vector as $\boldsymbol{\omega}^o$, the prediction error $e(i)$ can be rewritten as

$$e(i) = \boldsymbol{\kappa}^T(i)\tilde{\boldsymbol{\omega}}(i-1) + v(i) = e_a(i) + v(i), \tag{13}$$

where $\tilde{\boldsymbol{\omega}}(i-1) = \boldsymbol{\omega}^o - \boldsymbol{\omega}(i-1)$ is a weight vector error, and $e_a(i) = \boldsymbol{\kappa}^T(i)\tilde{\boldsymbol{\omega}}(i-1)$ is *a priori* error. Subtracting $\boldsymbol{\omega}^o$ from both sides of (12), yields

$$\tilde{\boldsymbol{\omega}}(i) = \tilde{\boldsymbol{\omega}}(i-1) - \eta e(i)\boldsymbol{\kappa}(i). \tag{14}$$

Let $e_p(i) = \boldsymbol{\kappa}^T(i)\tilde{\boldsymbol{\omega}}(i)$ be the *a posteriori* error, and multiply both sides of (14) by $\boldsymbol{\kappa}^T(i)$, yields

$$e_p(i) = e_a(i) - \eta e(i)\|\boldsymbol{\kappa}(i)\|^2. \tag{15}$$

Substituting (15) into (14), we have

$$\tilde{\boldsymbol{\omega}}(i) = \tilde{\boldsymbol{\omega}}(i-1) + \frac{e_p(i) - e_a(i)}{\|\boldsymbol{\kappa}(i)\|^2}\boldsymbol{\kappa}(i). \tag{16}$$

Furthermore, squaring both sides of (16), an energy equality can be obtained

$$\|\tilde{\boldsymbol{\omega}}(i)\|^2 + \frac{e_a^2(i)}{\|\boldsymbol{\kappa}(i)\|^2} = \|\tilde{\boldsymbol{\omega}}(i-1)\|^2 + \frac{e_p^2(i)}{\|\boldsymbol{\kappa}(i)\|^2}, \tag{17}$$

Algorithm 1 : The NMDF-KLMS Algorithm

Parameters: $\delta_1, \delta_2, h_1, h_2, \eta$.
Initialization: $\boldsymbol{C}(1) = \boldsymbol{u}(1), \boldsymbol{\omega}_1(1) = d(1), \boldsymbol{D}(1) = \mathbf{0}, \boldsymbol{\omega}_2(1) = 0$.
Computation:
While $\{\boldsymbol{u}(i), d(i)\}$, $(i \geq 1)$ **do**
1) Compute the output $y(i) = \boldsymbol{\kappa}^T(i)\boldsymbol{\omega}(i)$.
2) Compute the error $e(i) = d(i) - y(i) + v(i)$.
3) Compute the Coh_1 and Coh_2 by (9) and (10) .
4) Update $\boldsymbol{\omega}_1(i)$ and $\boldsymbol{\omega}_2(i)$:
If $\text{Coh}_1 \leq \delta_1$, $\text{Coh}_2 \leq \delta_2$ **Then**
$\boldsymbol{\omega}_1(i-1) = [\boldsymbol{\omega}_1^T(i-1), 0]^T$, $\boldsymbol{\omega}_2(i-1) = [\boldsymbol{\omega}_2^T(i-1), 0]^T$,
$\boldsymbol{\omega}(i) = \boldsymbol{\omega}(i-1) + \eta e(i)\boldsymbol{\kappa}(i)$
Else If $\text{Coh}_1 \leq \delta_1$, $\text{Coh}_2 > \delta_2$ **Then**
$\boldsymbol{\omega}_1(i-1) = [\boldsymbol{\omega}_1^T(i-1), 0]^T$, $\boldsymbol{\omega}_2(i-1) = \boldsymbol{\omega}_2(i-1)$,
$\boldsymbol{\omega}(i) = \boldsymbol{\omega}(i-1) + \eta e(i)\boldsymbol{\kappa}(i)$
Else If $\text{Coh}_1 > \delta_1$, $\text{Coh}_2 \leq \delta_2$ **Then**
$\boldsymbol{\omega}_1(i-1) = \boldsymbol{\omega}_1(i-1)$, $\boldsymbol{\omega}_2(i-1) = [\boldsymbol{\omega}_2^T(i-1), 0]^T$,
$\boldsymbol{\omega}(i) = \boldsymbol{\omega}(i-1) + \eta e(i)\boldsymbol{\kappa}(i)$
Else
$\boldsymbol{\omega}(i) = \boldsymbol{\omega}(i-1) + \eta e(i)\boldsymbol{\kappa}(i)$
End If
End While

where $\|\tilde{\boldsymbol{\omega}}(i)\|^2 = \tilde{\boldsymbol{\omega}}(i)^T\tilde{\boldsymbol{\omega}}(i)$ is the weight vector error power. To obtain insight into the mean square convergence behavior of NMDF-KLMS, we inject (15) into (17) and take expectations of both sides that yield

$$\mathbf{E}[\|\tilde{\boldsymbol{\omega}}(i)\|^2] = \mathbf{E}[\|\tilde{\boldsymbol{\omega}}(i-1)\|^2] - 2\eta\mathbf{E}[e_a(i)e(i)] + \eta^2\mathbf{E}[e^2(i)\|\boldsymbol{\kappa}(i)\|^2]. \tag{18}$$

To facilitate the mean square convergence analysis, a set of hypothesis are made as follows:

H1: The noise $v(i)$ is zero-mean with variance $\xi_v^2 = \mathbf{E}[v^2(i)]$.
H2: The noise $v(i)$ is independent, identically distributed (i.i.d.).
H3: The noise $v(i)$ is independent of $\boldsymbol{\kappa}(i)$.
Combining (18), **H1**, **H2**, and **H3**, we have

$$\mathbf{E}[\|\tilde{\boldsymbol{\omega}}(i)\|^2] = \mathbf{E}[\|\tilde{\boldsymbol{\omega}}(i-1)\|^2] - 2\eta\mathbf{E}[e_a^2(i)] + \eta^2\mathbf{E}[\|\boldsymbol{\kappa}(i)\|^2](\mathbf{E}[e_a^2(i)] + \xi_v^2), \tag{19}$$

which leads to

$$\begin{aligned}\mathbf{E}[\|\tilde{\boldsymbol{\omega}}(i)\|^2] \leq \mathbf{E}[\|\tilde{\boldsymbol{\omega}}(i-1)\|^2] &\Leftrightarrow \eta^2\mathbf{E}[\|\boldsymbol{\kappa}(i)\|^2](\mathbf{E}[e_a^2(i)] + \xi_v^2) \leq 2\eta\mathbf{E}[e_a^2(i)] \\ &\Leftrightarrow \eta \leq \frac{2\mathbf{E}[e_a^2(i)]}{\mathbf{E}[\|\boldsymbol{\kappa}(i)\|^2](\mathbf{E}[e_a^2(i)] + \xi_v^2)}.\end{aligned} \tag{20}$$

Therefore, when η satisfies (20), $\mathbf{E}[\|\tilde{\boldsymbol{\omega}}(i)\|^2]$ is non-increasing, and NMDF-KLMS is theoretically convergent.

3.3 Steady-State Performance

In this part, the steady-state filtering accuracy is obtained by analyzing the limiting behavior of transient performance. For (19), when $i \to \infty$, yields

$$\begin{aligned}\lim_{i\to\infty} \mathbf{E}[\|\tilde{\boldsymbol{\omega}}(i)\|^2] = \lim_{i\to\infty} \mathbf{E}[\|\tilde{\boldsymbol{\omega}}(i-1)\|^2] - \lim_{i\to\infty} 2\eta\mathbf{E}[e_a^2(i)] \\ + \lim_{i\to\infty} \eta^2\mathbf{E}[\|\boldsymbol{\kappa}(i)\|^2](\mathbf{E}[e_a^2(i)] + \xi_v^2).\end{aligned} \tag{21}$$

When NMDF-KLMS converges, we have

$$\begin{aligned}&\lim_{i\to\infty} \mathbf{E}[\|\tilde{\boldsymbol{\omega}}(i)\|^2] = \lim_{i\to\infty} \mathbf{E}[\|\tilde{\boldsymbol{\omega}}(i-1)\|^2] \\ &\Leftrightarrow \lim_{i\to\infty} 2\eta\mathbf{E}[e_a^2(i)] = \lim_{i\to\infty} \eta^2\mathbf{E}[\|\boldsymbol{\kappa}(i)\|^2](\mathbf{E}[e_a^2(i)] + \xi_v^2),\end{aligned} \tag{22}$$

which results in the extra mean-square-error (EMSE), $\varDelta = \lim_{i\to\infty} \mathbf{E}[e_a^2(i)]$, as

$$\varDelta = \frac{\eta\xi_v^2\mathrm{Tr}(\boldsymbol{R}_k)}{2-\eta\mathrm{Tr}(\boldsymbol{R}_k)} = \frac{\xi_v^2}{2\left(\eta\mathrm{Tr}(\boldsymbol{R}_k)\right)^{-1} - 1}, \tag{23}$$

where $\mathrm{Tr}(\boldsymbol{R}_k) = \mathrm{Tr}(\mathbf{E}[\boldsymbol{\kappa}(i)\boldsymbol{\kappa}^T(i)]) = \mathbf{E}[\|\boldsymbol{\kappa}(i)\|^2]$, and $\mathrm{Tr}(\cdot)$ is the trace operator. From (23) we can observe that steady-state EMSE of NMDF-KLMS will increase as the step-size and the noise power increase.

4 Simulation Results

In this section, for NMDF-KLMS, we conduct some simulations in time-series predictions and real-world data applications to test the filtering performance. And the testing mean-square-error (TMSE) is used as the performance measure, i.e., $\mathrm{TMSE} = N^{-1}\sum_{i=1}^{N}(d(i)) - y(i))^2$, where N is the number of testing data. In addition, we also consider some related competing algorithms, such as KNLMS, LRKOL, SF-KLMS, RKLMS-MDF, and NR-KNLMS, in which the first one is KAF based on feedforward network, the second three are KAFs based on linear feedback and the last one is derived from a nonlinear single-delay feedback network. And all simulation results are averaged over 100 runs.

4.1 Mackey-Glass Time-Series Prediction

In this experiment, we consider the Mackey-Glass short time-series prediction, which is generated as follows

$$\frac{dx(t)}{dt} = -bx(t) + \frac{ax(t-\tau)}{1+x(t-\tau)^{10}}, \tag{24}$$

where $b = 0.1$, $a = 0.2$, and τ=30. The time-series is sampled at a sampling period of 6 seconds, and $\boldsymbol{u}(i) = [x(i-10), x(i-9), ..., x(i-2), x(i-1)]$ is used

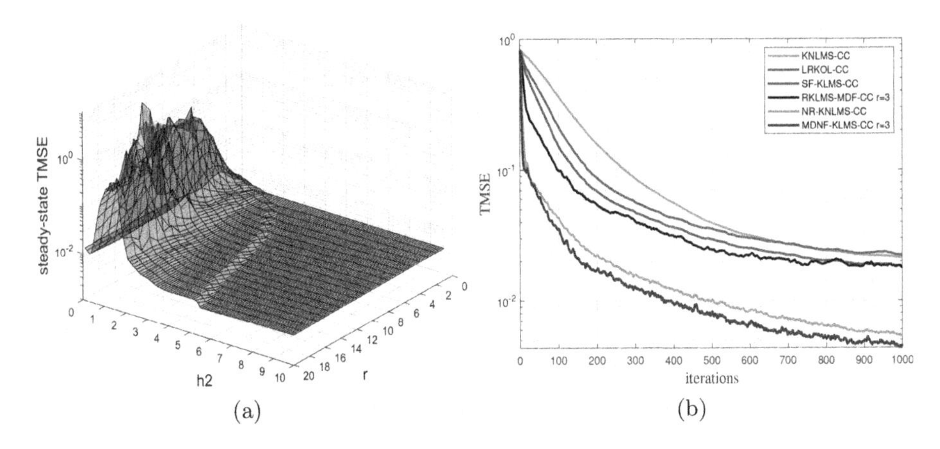

Fig. 2. (a)The relation between steady-state TMSE and delay number r of NMDF-KLMS-CC. (b)Learning curves of different KAFs in Mackey-Glass time series prediction.

as the input to predict the current one $d(i) = x(i)$ [3]. In this trial, the training data size is 1000 and the testing data size is 100.

Firstly, for NMDF-KLMS-CC[1], we investigate the joint effect of delay number r and the feedback kernel width h_2 on the steady-state TMSE, which is obtained by averaging over the last 100 iterations. To this end, we set $\delta_1 = 0.5$, $\delta_2 = 0.7$, $h_1 = 0.3$, $\eta = 0.2$, $h_2 \in [0.1, 10]$ and $r \in [0, 20]$. The corresponding results are plotted in Fig. 2(a). From this figure, we can get that, when $h_2 \geq 2$, NMDF-KLMS-CC achieves a better filtering accuracy.

Then, NMDF-KLMS-CC is compared to these related KAFs. For all algorithms, the parameters are set to realize a similar initial convergence rate, and Table 1 lists the parameters setting. Figure 2(b) plots the corresponding TMSE learning curves, which show that: 1) KAFs with nonlinear feedback outperform these based on linear feedback or without feedback in terms of convergence speed and filtering accuracy; 2) NMDF-KLMS-CC achieves the best filtering performance. Furthermore, Table 2 summarizes the simulation results in terms of the steady-state TMSE (STMSE), the feedforward network size (FWNS), and the feedback network size (FBNS), respectively.

4.2 Real-World Data Applications

In this trial, we also consider some real-world data applications including the combined cycle power plant (CCPP) dataset [18], the sunspot number dataset[2], and the wine quality dataset [19,20]. For the sake of simplification, we summarize

[1] NMDF-KLMS-CC stands for NMDF-KLMS sparsified by the CC sparsification method.

[2] http://www.sidc.be/silso/datafiles.

Table 1. Parameters for different KAFs in Mackey-Glass time-series prediction

Algorithm	Parameters
KNLMS-CC	$h_1 = 0.3$; $\delta_1 = 0.5$; $\eta = 0.05$; $\epsilon = 0.001$
LRKOL-CC or SF-KLMS-CC	$h_1 = 0.3$; $\delta_1 = 0.5$; $[\nu_\alpha, \nu_\lambda] = [0.9, 0.9]$; $[\bar{\rho}_\alpha, \bar{\rho}_\lambda] = [1, 100]$; $\varepsilon_m^\alpha = \varepsilon_m^\lambda = 0.03$
RKLMS-MDF-CC	$h_1 = 0.3$; $\delta_1 = 0.5$; $[\lambda_1, \lambda_2] = [0.01, 0.99]$; $[\theta_1, \theta_2] = [0.55, 0.9]$; $[\alpha_1, \alpha_2] = [0.55, 0.9]$; $[\mu_1, \mu_2] = [0.25, 0.01]$; $P_1 = P_2 = 1$; $r = 3$
NR-KNLMS-CC	$[h_1, h_2] = [0.3, 4]$; $[\delta_1, \delta_2] = [0.5, 0.7]$; $\eta = 0.2$; $\epsilon = 0.001$
NMDF-KLMS-CC	$[h_1, h_2] = [0.3, 4]$; $[\delta_1, \delta_2] = [0.5, 0.7]$; $\eta = 0.2$; $r = 3$

Table 2. Simulation results in Mackey-Glass time-series prediction

Algorithm	STMSE	FWNS	FBNS
KNLMS-CC	0.0213	86	–
LRKOL-CC	0.0221	86	–
SF-KLMS-CC	0.0183	86	–
RKLMS-MDF-CC	0.0176	86	–
NR-KNLMS-CC	0.0054	86	1
NMDF-KLMS-CC	**0.0046**	86	1

the parameters setting and the corresponding filtering results of these KAFs in Table 3 and Table 4, respectively.

The CCPP dataset contains 9568 data points collected from a Combined Cycle Power Plant over 6 years (2006-2011) when the power plant was set to work with a full load. Features consist of hourly average ambient variables, i.e., Temperature (T), Ambient Pressure (AP), Relative Humidity (RH), and Exhaust Vacuum (V), to predict the net hourly electrical energy output (EP) of the plant. Inhere, we use the 4 quantitative variables as input vector, and 1500 input-output pairs for training, the other 100 pairs for testing.

The Sunspot number dataset spans from January of 1749 to September of 2018, with a total of 3235 data points. In this part, we use the 8 preceding data points to predict the current one, where 1500 samples are allocated for training and the other 100 points for testing.

The wine quality dataset incorporates two distinct data sets that pertain to the red and white variations of the Portuguese "Vinho Verde" wine. The red wine data set contains 1599 input-output pairs, whereas the white wine data set consists of 4898 pairs. In addition, both datasets incorporate 11 attributes as input information. For red wine prediction, the first 1500 pairs are utilized as training samples, and the following 99 pairs are used for testing samples. For white wine prediction, 1000 samples are used for training and the other 100 for testing.

Table 3. Parameters for different KAFs in real-world data applications

Algorithm	Parameter Symbols	Parameter Values			
		CCPP	Sunspot Number	White Wine	Red Wine
LRKOL-CC or SF-KLMS-CC	$[h_1, \delta_1]$	$[0.25, 0.5]$	$[0.15, 0.4]$	$[0.25, 0.45]$	$[0.2, 0.45]$
	$[\nu_\alpha, \nu_\alpha]$	[0.92.0.99]	[0.95.0.98]	[0.92.0.95]	[0.93.0.95]
	$[\bar{\rho}_\alpha, \bar{\rho}_\alpha]$	$[1, 10]$	$[1.5, 10]$	$[2, 10]$	$[2, 10]$
	$[\varepsilon_m^\alpha, \varepsilon_m^\lambda]$	0.02	0.03	0.02	0.02
RKLMS-MDF-CC	$[h_1, \delta_1]$	$[0.25, 0.5]$	$[0.15, 0.4]$	$[0.25, 0.45]$	$[0.2, 0.45]$
	$[\lambda_1, \lambda_2]$	[0.01.0.09]	[0.01.0.09]	[0.01.0.09]	[0.01.0.09]
	$[\theta_1, \theta_2]$	$[0.35, 0.99]$	$[0.1, 0.1]$	$[0.45, 0.1]$	$[0.1, 0.1]$
	$[\alpha_1, \alpha_2]$	$[0.65, 0.65]$	$[0.1, 0.1]$	$[0.1, 0.1]$	$[0.1, 0.1]$
	$[\mu_1, \mu_2]$	$[0.35, 0.01]$	$[0.35, 0.02]$	$[0.12, 0.02]$	$[0.22, 0.02]$
	$[P_1, P_2]$	1	1	1	1
	r	3	3	6	4
NR-KNLMS-CC	$[h_1, h_2]$	$[0.25, 1]$	$[0.15, 0.75]$	$[0.25, 1]$	$[0.2, 0.75]$
	$[\delta_1, \delta_2]$	[0.5.0.6]	[0.4.0.4]	[0.45.0.65]	[0.45.0.55]
	$[\eta, \epsilon]$	$[0.12, 0.001]$	$[0.035, 0.001]$	$[0.05, 0.001]$	$[0.03, 0.001]$
NMDF-KLMS-CC	$[h_1, h_2]$	$[0.25, 1]$	$[0.15, 0.75]$	$[0.25, 1]$	$[0.2, 0.75]$
	$[\delta_1, \delta_2]$	[0.5.0.6]	[0.4.0.4]	[0.45.0.65]	[0.45.0.55]
	$[\eta, r]$	$[0.12, 3]$	$[0.035, 3]$	$[0.05, 6]$	$[0.03, 4]$

Table 4. Simulation results in in real-world data applications

Algorithm	CCPP			Sunspot Number			White Wine			Red Wine		
	STMSE	FWNS	FBNS	STMSE	FWNS	FBNS	STMSE	FWNS	FBNS	STMSE	FWNS	FBNS
LRKOL-CC	0.0114	25	-	0.0202	127	-	0.0323	61	-	0.0316	127	-
SF-KLMS-CC	0.0098	25	-	0.0177	127	-	0.0288	61	-	0.0289	127	-
RKLMS-MDF-CC	0.0093	25	-	0.0169	127	-	0.0263	61	-	0.0244	127	-
NR-KNLMS-CC	0.0065	25	1	0.0113	127	1	0.0185	61	1	0.0159	127	1
NMDF-KLMS-CC	**0.0061**	25	2	**0.0108**	127	1	**0.0172**	61	3	**0.0146**	127	2

Figure 3 plots the corresponding TMSE learning curves of various KAFs. From this figure, we can observe that the nonlinear feedback KAFs outperform their linear feedback counterparts in terms of convergence speed and filtering accuracy. Moreover, the proposed NMDF-KLMS achieves the best filtering performance among the evaluated algorithms.

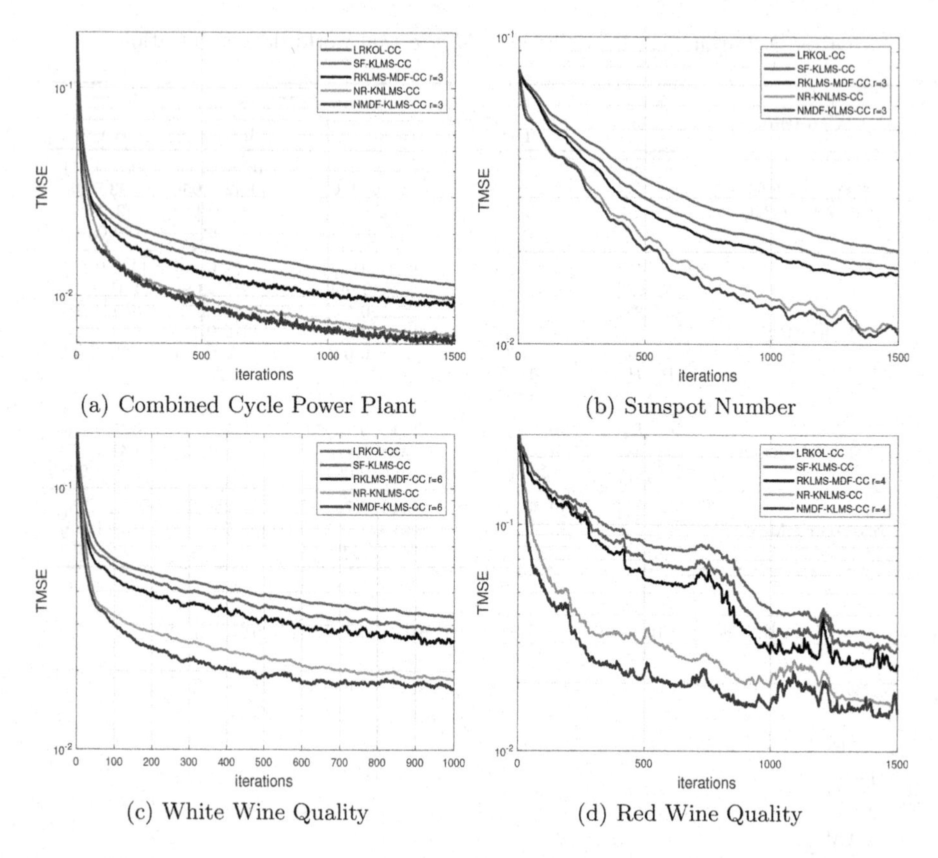

(a) Combined Cycle Power Plant

(b) Sunspot Number

(c) White Wine Quality

(d) Red Wine Quality

Fig. 3. The learning curves of different KAFs algorithms in real-world data applications.

5 Conclusion

In this paper, based on the multiple-delay framework, a new kernel adaptive filtering algorithm has been proposed, i.e., the nonlinear multiple-delay feedback based KLMS (NMDF-KLMS) algorithm. We also conduct some theoretical analysis in terms of mean square convergence and steady-state performance. The effectiveness and advantage of NMDF-KLMS are demonstrated by Mackey-Glass short time-series prediction and four real-world data applications. The results of this work suggest that the proposed algorithm has significant potential for enhancing filtering performance in time-series predictions. This paper employs the Mackey-Glass short time-series prediction as a benchmark example for conducting performance simulation testing. In future works, additional differential equations, such as Chua's chaotic time-series and Lorenz's chaotic time-series prediction, can also be utilized for performance simulation testing.

References

1. Apsemidis, A., Psarakis, S., Moguerza, J.M.: A review of machine learning kernel methods in statistical process monitoring. Comput. Indust. Eng. **142** 106376 (2020)
2. Rojo-Álvarez, J.L. Martínez-Ramón, M., Muñoz-Marí, J., Camps-Valls, G.: Kernel Feature Extraction in Signal Processing (2018)
3. Príncipe, J.C., Liu, W., Haykin, S.: Kernel Adaptive Filtering: A Comprehensive Introduction. John Wiley & Sons (2011)
4. Liu, W., Pokharel, P.P., Principe, J.C.: The kernel least-mean-square algorithm. IEEE Trans. Signal Process. **56**(2), 543–554 (2008)
5. Liu, W., Príncipe, J.C.: Kernel affine projection algorithms. EURASIP J. Adv. Signal Process. **2008**, 1–12 (2008)
6. Engel, Y., Mannor, S., Meir, R.: The kernel recursive least-squares algorithm. IEEE Trans. Signal Process. **52**(8), 2275–2285 (2004)
7. Fan, H., Song, Q.: A linear recurrent kernel online learning algorithm with sparse updates. Neural Netw. **50**, 142–153 (2014)
8. Zhao, J., Liao, X., Wang, S., Chi, K.T.: Kernel least mean square with single feedback. IEEE Signal Process. Lett. **22**(7), 953–957 (2014)
9. Wang, S., Zheng, Y., Ling, C.: Regularized kernel least mean square algorithm with multiple-delay feedback. IEEE Signal Process. Lett. **23**(1), 98–101 (2015)
10. Wang, S., Wang, W., Duan, S., Wang, L.: Kernel recursive least squares with multiple feedback and its convergence analysis. IEEE Trans. Circuits Syst. II Express Briefs **64**(10), 1237–1241 (2017)
11. Wang, Y., Jin, X., Yin, Y.: Using nonlinear feedback control to improve aircraft nose landing gear shimmy performance. Meccanica **57**(9), 2395–2411 (2022)
12. Gao, S., Zhang, X.: Course keeping control strategy for large oil tankers based on nonlinear feedback of swish function. Ocean Eng. **244**, 110385 (2022)
13. Fu, D., Gao, W., Song, M., Zhang, L.: Nonlinear recurrent kernel normalized LMS algorithm for nonlinear autoregressive system. In: 2021 IEEE International Conference on Signal Processing, Communications and Computing (ICSPCC), pp. 1–6. IEEE (2021)
14. Yukawa, M.: Nonlinear adaptive filtering techniques with multiple kernels. In: 2011 19th European Signal Processing Conference, pp. 136–140. IEEE (2011)
15. Yukawa, M.: Multikernel adaptive filtering. IEEE Trans. Signal Process. **60**(9), 4672–4682 (2012)
16. Tobar, F.A., Kung, S.Y., Mandic, D.P.: Multikernel least mean square algorithm. IEEE Trans. Neural Netw. Learn. Syst. **25**(2), 265–277 (2013)
17. édric Richard, C., Bermudez, J.M.C., Honeine, P.: Online prediction of time series data with kernels. IEEE Trans. Signal Process. **57**(3), 1058–1067 (2008)
18. Tüfekci, P.: Prediction of full load electrical power output of a base load operated combined cycle power plant using machine learning methods. Int. J. Electrical Power Energy Syst. **60**, 126–140 (2014)
19. Cortez, P., Cerdeira, A., Almeida, F., Matos, T., Reis, J.: Modeling wine preferences by data mining from physicochemical properties. Decis. Support Syst. **47**(4), 547–553 (2009)
20. Zhao, J., Zhang, H., Liao, X.: Variable learning rates kernel adaptive filter with single feedback. Digital Signal Process. **83**, 59–72 (2018)

AGGDN: A Continuous Stochastic Predictive Model for Monitoring Sporadic Time Series on Graphs

Yucheng Xing[1(✉)], Jacqueline Wu[2], Yingru Liu[1], Xuewen Yang[1,3], and Xin Wang[1]

[1] Stony Brook University, Stony Brook, NY 11790, USA
yucheng.xing@stonybrook.edu
[2] New York University, New York, NY 10012, USA
[3] InnoPeak Technology, Inc., Palo Alto, CA 94303, USA

Abstract. Monitoring data of real-world networked systems could be sparse and irregular due to node failures or packet loss, which makes it a challenge to model the continuous dynamics of system states. Representing a network as graph, we propose a deep learning model, *Adversarial Graph-Gated Differential Network (AGGDN)*. To accurately capture the spatial-temporal interactions and extract hidden features from data, AGGDN introduces a novel module, *dynDC-ODE*, which empowers Ordinary Differential Equation (ODE) with learning-based Diffusion Convolution (DC) to effectively infer relations among nodes and parameterize continuous-time system dynamics over graph. It further incorporates a Stochastic Differential Equation (SDE) module and applies it over graph to efficiently capture the underlying uncertainty of the networked systems. Different from any single differential equation model, the ODE part also works as a control signal to modulate the SDE propagation. With the recurrent running of the two modules, AGGDN can serve as an accurate online predictive model that is effective for either monitoring or analyzing the real-world networked objects. In addition, we introduce a soft masking scheme to capture the effects of partial observations caused by the random missing of data from nodes. As training a model with SDE component could be challenging, Wasserstein adversarial training is exploited to fit the complicated distribution. Extensive results demonstrate that AGGDN significantly outperforms existing methods for online prediction.

Keywords: Graph Sequence Prediction · Sporadic Time Series · Continuous Model · Stochastic Model

1 Introduction

Many systems, such as social networks, vehicle networks, communication networks, and power grids, are networked and can be represented as graphs. Although a practical system operates continuously, its states are normally collected periodically at discrete time instants. Due to practical constraints such as

B. Luo et al. (Eds.): ICONIP 2023, LNCS 14447, pp. 130–146, 2024.
https://doi.org/10.1007/978-981-99-8079-6_11

cost, unreliable communications, device malfunction or failure, the observations of systems are often sporadic, which are sparse and irregular in both spatial and temporal domains. In order to timely and effectively control the systems for reliable and intelligent operations, it is important to accurately predict the system states based on the collected observations. The aim of this paper is to attack the challenge of modeling the graph dynamics where signals of nodes evolve continuously but the observations are sporadic.

It is highly non-trivial to learn the structured dynamics from sporadic observations on a graph. The challenges mainly come from three sources. First, the signals from nodes are time varying, and it is hard to model the spatial-temporal interaction across the whole graph. Second, only partial dynamics can be observed from sporadic data, which makes it difficult to model the underlying process. Last, as the output signals of networked systems may contain both process uncertainty (e.g. the distributed energy resource control signal of a microgrid is influenced by the uncertainty of its input.) and measurement noise, the distribution of data can be complicated and difficult to estimate.

With the rapid development of Graph Neural Network (GNN) [17,38,44], there are considerable number of studies on learning the dynamics of time series on graphs [16,46,48]. However, existing efforts mostly consider discrete-time dynamics, assuming systems are fully observed with complete data taken periodically. Although data missing is considered in [4,39] for graph prediction, the scheme still considers discrete data samples and can not depict the ground-true dynamics of the continuous-time network systems in real world. Recently, several ODE-based models are proposed to learn the continuous dynamics of the graph time series [2,8,12,31,35], but ODE is only applied to learn the deterministic dynamics. Neglecting the process uncertainty of the system, they are unable to capture the complicated distribution of observations in real-world systems.

We propose a continuous-time graph-based recurrent neural network, *Adversarial Graph-Gated Differential Network (AGGDN)*, to capture the underlying dynamics on the graph structure from sporadic observations of node signals. AGGDN models the stochastic process of dynamic states of networked systems with a novel compounded infrastructure with ODE modulated SDE running continuously over a graph. On the one hand, SDE module can supplement the deterministic ODE module with a stochastic variation; on the other hand, ODE part can work as a control signal to modulate the SDE propagation. In addition, in order to address the challenge of learning the complete system states with incomplete data samples, we first propose a soft-masking scheme that can better extract disentangled hidden features of partial observations to well explore the spatial-temporal relation in the graph. Furthermore, to more accurately parameterize the ODE module, we enhance diffusion convolution [20] with the learning of impacts of nodes in the graph and call it as *dynamic Diffusion Convolution (dynDC)*. The SDE component is incorporated to efficiently capture the process uncertainty of the underlying system dynamics with a flexible tracking of data distribution. However, optimizing an SDE model is difficult, so we further exploit Wasserstein adversarial training to efficiently train AGGDN.

The contributions of this paper can be summarized as follows:

1. We propose a novel AGGDN architecture to accurately predict system states of a networked system, given sparse and irregular data samples and system uncertainty. It accurately models the continuous-time process over graph with a compounded ODE-SDE structure, where SDE tracks the stochastic variation of states to capture system uncertainties and ODE provides a modulation for the whole model propagation.
2. We design a dynDC-ODE module with ODE empowered with the dynamic diffusion convolution to learn different impacts of nodes, that our model can well adapt to the dynamic changes of graph.
3. We introduce soft-masking that AGGDN can better learn from the partial observations of a networked system to more effectively infer missing data.
4. We employ Wasserstein adversarial training for our continuous-time AGGDN model, rather than taking variational inference methods (assumed by most SDE models) that are inefficient for high-dimensional data.

The rest of this paper is organized as follows. The problem formulation and related works are introduced in Sect. 2. The detailed architectures and training method of our model are proposed in Sect. 3. Extensive experiments and analysis are given in Sect. 4 and conclusions are made in Sect. 5. The source code and data are available at https://github.com/SBU-YCX/AGGDN.

2 Background

2.1 Problem Formulation

Notations. We denote a graph with time-varying signals at nodes by $\mathbb{G} = \{\mathcal{V}, \mathcal{E}, \{\mathcal{X}_{t_n}, \mathcal{M}_{t_n}, t_n\}_{n=0}^{N}\}$, where $\mathcal{V} = \{v_i\}_{i=1}^{|\mathcal{V}|}$ is the set of nodes and $\mathcal{E} = \{(v_i, v_j)|v_i, v_j \in \mathcal{V}\}$ is the set of edges between nodes. The cardinalities $|\mathcal{V}|$ and $|\mathcal{E}|$ denote the number of elements in $\mathcal{V}$ and $\mathcal{E}$. We further denote the 0-1 adjacent matrix of a graph as A. In a given network $\{\mathcal{V}, \mathcal{E}\}$, the dynamic states are described by a sequence of N frames. A frame contains a multi-variate signal $\mathcal{X}_{t_n} \in \mathbb{R}^{|\mathcal{V}| \times d}$ captured at the discrete time $t_n \in \mathbb{R}_+$, where d is the corresponding dimension of the signal. Since sensing or transmission problems in real-world systems often cause sample missing, in each frame, a mask $\mathcal{M}_{t_n} = \{0, 1\}^{|\mathcal{V}| \times d}$ is used to indicate if there exists a signal in the corresponding dimension. Therefore, the actual observation sequence $\mathcal{O} = \{\mathcal{X}_{t_n} \odot \mathcal{M}_{t_n}\}_{n=0}^{N}$ fed into the model is a sporadic time series with irregular data in both temporal and spatial domains. $\mathcal{O}_{t_n} = \mathcal{X}_{t_n} \odot \mathcal{M}_{t_n}$ denotes the input data at time t_n, where $\odot$ represents the element-wise multiplication between two matrices.

Objective. Given a collection of data $D = \{\mathbb{G}^{(k)}\}_{k=1}^{|D|}$, where $\mathbb{G}^{(k)}$ is a data sequence introduced in Notation part above and $|D|$ is the total number of such sequences in the dataset, our goal is to learn a continuous-time recurrent predictive model $\mathcal{G}$ to maximize the masked log-likelihood:

$$\mathcal{L}_{ll}(\mathcal{G}) = \mathbb{E}_{\mathbb{G}^{(k)} \in D} \sum_{n=1}^{N} \mathcal{M}_{t_n} \otimes \log P_{\mathcal{G}}(\mathcal{X}_{t_n} | \mathcal{O}_{t_0:t_{n-1}}, A), \tag{1}$$

where $\otimes$ is defined as the sum of element-wise product of two matrices, $P_{\mathcal{G}}(\cdot)$ denotes the probability density of each element in the feature matrices. We want to emphasize that the log-likelihood is only evaluated on the observed training data indicated by the binary masks instead of full data. Therefore, the objective described in (1) can be regarded as an unsupervised one.

2.2 Related Works

Graph Convolution. It is the core operation of Graph Convolutional Network (GCN) [6,17,20]. To explore the relation among multi-hop neighbors in a graph, diffusion convolution was proposed in [20]. Rather than only using a binary adjacency matrix to indicate if there exist edges between nodes, we are interested in actively learning the relation among nodes for the more accurate modeling and thus more accurate prediction of dynamic system states. Our model, however, does not depend on the choice of graph convolutional operators and can be straight-forwardly adapted for other graph convolution methods.

Graph Recurrent Networks. Most existing sequential graph models are proposed for discrete-time data [7,28,29,36,47,48], and assume that the data sequence is fully observed. Recurrent models have also been applied to non-sequential data on static graphs, for applications such as graph generation [3,22,23,37,45] and feature learning [13,14,21,40]. DynGEM [10] and dyngraph2vec [9] are deep learning models for tracking the structure evolution of the graph topology. Our study, however, focuses mainly on modeling the stochastic process of the signals on the graph [11,43], especially under partial observations.

Neural Differential Equations (NeuralODE). NeuralODE [1] incorporates a deep learning module to parameterize nonlinear ordinary differential equation (ODE). In order to better model the continuous-time process of the vectorized data sequence, the structure of NeuralODE is further extended [5,34] by introducing a recurrent component to efficiently integrate the data information into the feature trajectories. Some ODE-based studies [2,8,12,31,35] are made to learn the continuous-time dynamics of node features on a graph, assuming that the underlying dynamics are deterministic and neglecting the process uncertainty existing in many real-world systems. These models can merely parameterize a simplified data distribution, not to say capture the complicated stochastic process of the systems with both process and measurement uncertainties.

To better model the randomness of data, neural stochastic differential equation (NeuralSDE) is proposed to bridge the gap between nonlinear SDE and deep learning models [18,19,25,30,41,42]. In [18,25,30], neural network components are introduced into SDE to define more robust and accurate deep learning architectures to solve supervised learning problems such as image classification. A scalable method is proposed in [19] to compute the gradients for optimizing NeuralSDE. To the best of our knowledge, most NeuralSDE models [19,24,26,27] are proposed for vector or matrix data but not for representing time series on graphs. Our focus, however, is on accurately predicting system states under dynamics and data missing through sequential learning over graphs.

We develop a compound model with the integration of NeuralODE and NeuralSDE into one infrastructure. It not only extends over a graph to capture the spatial interaction of data in multiple hops but also simultaneously track the deterministic dynamics and process uncertainty through our proposed dynamic Diffusion Convolution (dynDC). In a single ODE or SDE model, the propagation within an interval between two time stamps only relies on the preceding observed values and make an update at any time point directly. Instead, our model first gets an approximate prediction through the ODE part, which is then used as a control signal to modulate the latent state thus the final data predicted have more continuous changes. The natural cubic spline interpolation method [2,15] makes a three-degree polynomial assumption about the curve between two observations, but needs a full time series to construct the underlying process approximation. Instead, ODE module allows creating the corresponding control signal as time evolves. We further introduce a learning-based soft-masking scheme to work with our dynDC-ODE components that AGGDN can adapt to spatially irregular observations for a more accurate modeling and apply the Wasserstein adversarial training to efficiently train our model.

3 Methodology

We propose a flexible continuous-time recurrent model, Adversarial Graph-Gated Differential Network (AGGDN), which is capable of learning the continuous graph dynamics from spatial-temporal irregular observation sequences. We first introduce its architecture, then provide the details of the key components. To effectively train the stochastic part, we adopt the adversarial training strategy and will discuss the training details at the end of this section.

3.1 Architecture

As shown in Fig. 1, our model consists of two trajectories: one hidden feature trajectory to integrate the topological relation in the graph and one latent state trajectory to capture the system uncertainties. As a continuous-time model to predict the system states at any time $t \in \mathbb{R}_+$, these two trajectories need to evolve continuously and simultaneously in the temporal domain. The two trajectories are realized with two major components, an ordinary differential equation (ODE) module parameterized with dynamic Diffusion Convolution, named as *dynDC-ODE*, and a stochastic differential equation (SDE) module. The former

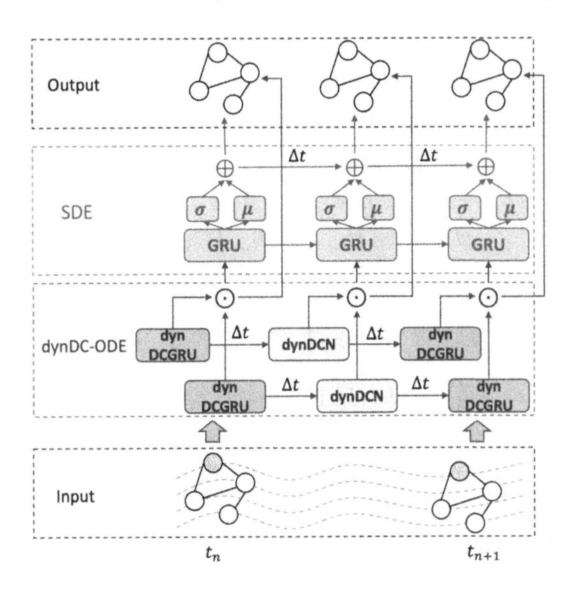

Fig. 1. The architecture of our proposed AGGDN.

one encodes the stable parts and the topological information into the hidden features $H_t \in \mathbb{R}^{|\mathcal{V}| \times d_h} (t \in \mathbb{R}_+)$ while the latter one embeds the system uncertainties into the stochastic latent states $Z_t \in \mathbb{R}^{|V| \times d_z} (t \in \mathbb{R}_+)$. Although a single ODE or SDE is a continuous model, the observations are discrete, and the propagation within the interval between two observations always rely on the preceding observed data. As a result, the hidden state trajectory only makes an update at an observation point, and thus changes abruptly. To address this issue, in AGGDN, the ODE module first gives an approximation of H_t about the underlying hidden state, which is used to modulate the SDE module so that the latent state Z_t has a continuous dependency on the data, which in turn further refines the coarse hidden features H_t to get the final accurate output.

3.2 Component–Improved DynDC-ODE Module

Given an observation at a time instant t_n, our dynDC-ODE module extracts the topological relations of the graph and the stable properties of signals, which are embed into the continuous-time features H_{t_n}. The dynDC-ODE module has two functions: one is to employ a nonlinear encoder $G(\cdot)$ to directly integrate the information from data $\mathcal{O}_{t_n}$ observed at t_n, and the other is to apply a propagation function $F(\cdot)$ to update the features H_t within the interval (t_{n-1}, t_n).

Specifically, for the encoder $G(\cdot)$, we adopt the Diffusion Convolution Gated Recurrent Unit (DCGRU) [20] to integrate information from multi-hop neighbors for each node in the graph, which can be expressed as

$$H_{t_n} = G(H_{t_n - \delta t}, \mathcal{O}_{t_n}, A), \tag{2}$$

where A is the binary adjacency matrix of the graph. To capture the different impacts of neighbors during the integration, we learn another weight matrix W_A during the training and use the element-wise multiplication of two matrices $W_A \odot A$ to replace A in (2), forming our *dynamic Diffusion Convolution Unit (dynDCGRU)*. Therefore, at the observation time t_n, we can rewrite (2) as

$$H_{t_n} = G(H_{t_n - \delta t}, \mathcal{O}_{t_n}, W_A \odot A), \tag{3}$$

For the propagation function $F(\cdot)$, we parameterize the dynamics of hidden features during the interval (t_{n-1}, t_n) using ODEs:

$$\begin{aligned} \frac{dH_t}{dt} &= F(H_t, W_A \odot A), \\ H_t &= H_{t_{n-1}} + \int_{t_{n-1}}^{t} F(H_\tau, W_A \odot A) d\tau, \end{aligned} \tag{4}$$

where the first-order derivative $F(\cdot)$ is formed through our *dynamic Diffusion Convolution Networks (dynDCN)*. For simplicity, we compute the integration in (4) by Euler Method, which can be expressed as

$$H_t = H_{t-\delta t} + F(H_{t-\delta t}, W_A \odot A)\delta t, \tag{5}$$

where δt is the propagation step size of the ODE module during the interval.

The observable data $\mathcal{O}_{t_n} = \mathcal{X}_{t_n} \odot \mathcal{M}_{t_n}$ are often irregular in the spatial domain due to the sample missing caused by sensor malfunction or transmission loss frequently appearing in the real-world networked system. To make our model adaptive to different missing locations (i.e., data loss from different nodes), we further introduce a *soft-masking function* into our dynDC-ODE module. The hidden feature H_t is then composed of a feature factor $H_{t,f}$ to extract the information and properties of the network, and a masking factor $H_{t,m} \in (0,1)^{|\mathcal{V}| \times d_h}$ to modulate the values in the feature. So the final feature would be

$$H_t = \rho(H_{t,m} W_m + b_m) \odot H_{t,f}, \tag{6}$$

where $\rho(\cdot)$ denotes the nonlinear activation function, W_m and b_m are the weight and bias parameters of a feed-forward network. Both $H_{t,f}$ and $H_{t,m}$ are learnt from the observed input and processed with (3) and (5). During the interval (t_{n-1}, t_n), the two factors will be updated by dynDC-ODEs as

$$\begin{aligned} H_{t,f} &= H_{t-\delta t,f} + F_f(H_{t-\delta t,f}, W_A \odot A)\delta t, \\ H_{t,m} &= H_{t-\delta t,m} + F_m(H_{t-\delta t,m}, W_A \odot A)\delta t, \end{aligned} \tag{7}$$

At an observation time t_n, the new information from input data will be integrated by encoders as

$$\begin{aligned} H_{t_n,f} &= G_f(H_{t_n-\delta t,f}, \mathcal{O}_{t_n}, W_A \odot A), \\ H_{t_n.m} &= G_m(H_{t_n-\delta t,m}, \mathcal{O}_{t_n}, W_A \odot A). \end{aligned} \tag{8}$$

3.3 Component–SDE Module

In many real-world networked systems, the observations are influenced by the process uncertainties. For example, in the case of a migrogrid network, the uncertainty in the control signal of distributed energy resource (DER) will cause oscillation in the current flows within the microgrid. To capture such uncertainties, we introduce a random latent state Z_t and parameterize it with a nonlinear SDE on the graph:

$$dZ_t = \mu(Z_t, H_{\leq t})dt + \sigma(H_{\leq t})dB_t,$$
$$Z_t = Z_{t_{n-1}} + \int_{t_{n-1}}^{t} \mu(Z_\tau, H_{\leq \tau})d\tau + \int_{t_{n-1}}^{t} \sigma(H_{\leq \tau})dB_\tau, \tag{9}$$

where μ and σ are the drift and diffusion functions respectively, B_t represents the standard Brownian motion. We define $\mu(\cdot)$ as the function of the current latent state Z_t and historical hidden features $H_{\leq t}$. However, $\sigma(\cdot)$ is only a function of the historical ODE features $H_{\leq t}$, as including Z_t also into $\sigma(\cdot)$ will bring additional noise term into the gradient computation [26] and make the training more difficult. Similar to the computation in dynDC-ODE, we compute the integration in (9) using Euler-Maruyama Method for simplicity:

$$Z_t = Z_{t-\delta t} + \mu(Z_{t-\delta t}, H_{\leq t})\delta t + \sqrt{\delta t}\sigma(H_{\leq t})\epsilon_t, \tag{10}$$

where δt is the step size and $\epsilon_t \in \mathcal{N}(0, 1)$ is the standard Gaussian noise. In implementation, since the hidden features H_t extracted through dynDC-ODE have already captured the spatial topological relations within the network graph, we simply use a Gated Recurrent Unit (GRU) to integrate the historical information $H_{\leq t}$ of the ODE feature trajectory. Similarly, the drift function $\mu(\cdot)$ and the diffusion function $\sigma(\cdot)$ in (10) are also parameterized with Dense Neural Networks (DNNs) on the GRU features.

After computing hidden features H_t and latent states Z_t, the dynamic system states can be predicted with a trajectory of data dynamics and a residual term

$$\hat{\mathcal{X}}_t = N_{\hat{X}}(H_t) + N_{\hat{X}}^{(res)}(H_t, Z_t), \tag{11}$$

where $N_{\hat{X}}(\cdot)$ is the function of H_t to predict the smooth and stable trend of the signals while $N_{\hat{X}}^{(res)}(\cdot)$ incorporates the latent state to estimate the residual stochastic variations. Both $N_{\hat{X}}(\cdot)$ and $N_{\hat{X}}^{(res)}(\cdot)$ are implemented by simple DNNs.

3.4 Training Details

Since our model incorporates the latent states to model the system uncertainty, the log-likelihood in (1) of a single data instance $\mathcal{O}$, under an adjacency matrix A, can be rewritten as

$$\begin{aligned}\mathcal{L}_{ll}(\mathcal{O}) &= \sum_{n=1}^{N} \mathcal{M}_{t_n} \otimes \log P_{\mathcal{G}}(\mathcal{X}_{t_n}|\mathcal{O}_{t_0:t_{n-1}}, A) \\ &= \sum_{n=1}^{N} \mathcal{M}_{t_n} \otimes \log \int P_{\mathcal{G}}(\mathcal{X}_{t_n}|Z_{t_n}, \mathcal{O}_{t_0:t_{n-1}}, A) \\ &\qquad \times P_{\mathcal{G}}(Z_{t_n}|\mathcal{O}_{t_0:t_{n-1}}, A) dZ_{t_n},\end{aligned} \tag{12}$$

where $P_{\mathcal{G}}(Z_{t_n}|\mathcal{O}_{t_0:t_{n-1}}, A)$ is the conditional distribution of the latent states induced by SDE module while $P_{\mathcal{G}}(\mathcal{X}_{t_n}|Z_{t_n}, \mathcal{O}_{t_0:t_{n-1}}, A)$ is the conditional distribution of the observation.

In general, $P_{\mathcal{G}}(Z_{t_n}|\mathcal{O}_{t_0:t_{n-1}}, A)$ does not have a closed-form solution as Z_t is computed by the integration of a nonlinear SDE, thus $\mathcal{L}_{ll}(\cdot)$ does not have a closed-form solution either. However, after we synthesize a trajectory of the latent states $\{Z_t\}$ by (10), an alternative is to simplify the log-likelihood in (12) as the logarithm of the conditional distribution of observations:

$$\mathcal{L}_{con}(\mathcal{O}|\{Z_t\}) = \sum_{n=1}^{N} \mathcal{M}_{t_n} \otimes \log P_{\mathcal{G}}(\mathcal{X}_{t_n}|Z_{t_n}, \mathcal{O}_{t_0:t_{n-1}}, A), \tag{13}$$

where only one latent trajectory $\{Z_t\}$ is enough for estimating $P_{\mathcal{G}}(Z_{t_n}|\mathcal{O}_{t_0:t_{n-1}}, A)$ by Monte-Carlo Method. However, there is still a large difference between (13) and (12), which will compromise the training quality of our model. Conventionally, a state-space model with SDE is usually trained with the variational inference, where an evidence lower bound of the log-likelihood is given through an auxiliary inference model. The performance of variational inference is highly dependent on the accuracy of the inference model, and it often requires a Monte-Carlo Method with multiple samples to reduce the variance of the evidence lower bound. But in many real-world applications, if there is a large number of nodes and signals, an accurate inference model is hard to define and a Monte-Carlo Method running over a large number of samples is computationally expensive.

To efficiently train our model and avoid the drawbacks of variational inference method, we utilize the Wasserstein adversarial training objective:

$$\begin{aligned}\mathcal{L}_{adv}(\mathcal{G}, \mathcal{F}) &= \mathbb{E}_{\{\mathcal{X}_{t_n}, \mathcal{M}_{t_n}\} \in D}[\mathcal{F}(\{\mathcal{X}_{t_n} \odot \mathcal{M}_{t_n}\}) \\ &\quad - \mathbb{E}_{\epsilon_t \in \mathcal{N}(0,1)} \mathcal{F}(\{\hat{\mathcal{X}}_{t_n} \odot \mathcal{M}_{t_n}\})],\end{aligned} \tag{14}$$

where $\mathcal{G}$ is our proposed model and $\mathcal{F}$ is the discriminator in the adversarial training. D is the dataset while $\{\hat{\mathcal{X}}_{t_n}\}$ is the set of predicted states given by our model, and $\{\epsilon_t\}$ is the uncertainty term in the SDE module. Based on the adversarial training, the discriminator is optimized by maximizing (14) while our model is optimized by minimizing the combination of the conditional log-likelihood in (13) and the adversarial loss in (14), i.e.

$$\mathcal{G}_* = arg \min_{\mathcal{G}} \Big(\lambda \mathcal{L}_{adv}(\mathcal{G}, \mathcal{F}) - \mathbb{E}_{\mathcal{O} \in D} \mathcal{L}_{con}(\mathcal{O}|\{Z_t\}) \Big), \tag{15}$$

$$\mathcal{F}_* = arg \max_{\mathcal{F}} \Big(\mathcal{L}_{adv}(\mathcal{G}, \mathcal{F}) \Big), \tag{16}$$

where λ is the weight coefficient.

Discriminator: The discriminator in our design is a function to convert the sporadic observation sequence, including the original one $\{\mathcal{O}_{t_n} = \mathcal{X}_{t_n} \odot \mathcal{M}_{t_n}\}$ and the predicted one $\{\hat{\mathcal{O}}_{t_n} = \hat{\mathcal{X}}_{t_n} \odot \mathcal{M}_{t_n}\}$, into a scalar in $\mathbb{R}$. Actually, our proposed model is independent of the choice of the discriminator, but for simplicity, we incorporate a DCGRU to extract features from inputs. The extracted feature in the last frame is applied to compute the scalar for each load in the microgrid. The output of the discriminator is given as the summation of all these scalar values. To meet the K-Lipschitz requirement in the Wasserstein adversarial objective, we apply Spectral Normalization (SN) to all the weight parameters in DCGRU and the output layer in the discriminator.

4 Experiments

We demonstrate the effectiveness of each component in our proposed model and show its robustness under different conditions through experiments on several datasets.

4.1 Experimental Setups

Datasets. We first introduce the benchmarks and the pre-processing of data as follows:

IEEE33-Nodes: As a typical microgrid instance, the IEEE-33 Bus System contains 5 electricity sources and 28 load nodes. The 2-dimentional DQ current signals going through each node are generated using a hardware-based tool RTDS [33]. We collect 50 trajectories with signals sampled at the interval of 3 milliseconds for 21 different network topologies. For each trajectory, we split the sequence of samples into segments of 100 frames for training and testing. Before fed into the model, all data are normalized with the mean and the standard derivation in the temporal domain of the corresponding node.

METR-LA [20]: This traffic dataset records the speeds of vehicles on the highway of Los Angeles County. The data are collected by 207 sensors every 5 minutes. We split the data samples into segments of 36 frames and normalize the samples with the global means and standard derivations of all sensors in the temporal domain.

PEMS-BAY [20]: Similar to METR-LA, this is also a traffic dataset collected by 325 sensors every 5 minutes in the Bay Area. The pre-processing is the same as we did to METR-LA.

Evaluation Metrics. We compare the values predicted with our proposed model based on sporadic observations and fully observable ground-true data using the metrics Mean Absolute Error (MAE, ↓), Root-Mean-Square Error (RMSE, ↓), and Mean Absolute Percentage Error (MAPE, ↓).

Implementation Details. Our model consists two modules: one dynDC-ODE module and one SDE module. In the dynDC-ODE module, we use a dynDCGRU to encode the input signals into 32-dimensional hidden features and a 1-layer dynDCN with random-walk range $K = 3$ to parameterize the ODE. In the SDE module, a simple GRU cell and a 2-layer DNN are used to convert the hidden features from dynDC-ODE to the corresponding 2-dimensional latent states. The propagation step of both dynDC-ODE and SDE is $\delta t = 0.1\Delta T$, where ΔT is the sampling interval of input data. For each trajectory in the datasets, we synthesize 25 sporadic observation sequences with the random selection, which we will give details in the following part. After that, all data are split into training/validation/test sets with ratio 0.8/0.1/0.1. All models are trained by the ADAM optimizer with the learning rates $[10^{-2}, 10^{-3}, 10^{-4}]$, where 100 epochs are trained for each rate. λ in the adversarial training objective of (15) is 1.0.

Table 1. Testing Performance of different models on various datasets ($p_t = 0.5, p_s = 0.8$)

Datasets	IEEE33-Nodes			METR-LA			PEMS-BAY		
	MAE	RMSE	MAPE	MAE	RMSE	MAPE	MAE	RMSE	MAPE
Discrete									
STGCN [47]	0.0812	0.1273	18.18%	0.2290	0.4375	37.58%	0.2499	0.4468	49.63%
Graph-GRU [36,48]	0.0349	0.0643	9.65%	0.1978	0.4226	33.45%	0.1925	0.3725	41.17%
Continuous									
Graph-ODE-RNN [32]	0.0306	0.0578	8.36%	0.1918	0.4217	32.73%	0.1774	0.3605	37.34%
Graph-GRU-ODE [5]	0.0313	0.0607	8.49%	0.1947	0.4322	32.90%	0.1726	0.3488	37.44%
Ours									
AGGDN	**0.0243**	**0.0457**	**7.03%**	**0.1612**	**0.3480**	**30.41%**	**0.1489**	**0.2739**	**35.32%**

4.2 Experiment Results

Overall Performance. To synthesize the scenario of the sporadic observations, we randomly select a ratio p_t of the data frames in the temporal dimension as observed data. For each selected frame, we further assume only signals from p_s of the nodes are observed. Therefore, the data fed into our model has only $p_t \times p_s$ values remained, which can be regarded as sparse and also irregular in both temporal and spatial domains due to the random selection. Table 1 shows the performance comparison between our method and some other representative literature works on ($p_t = 0.5, p_s = 0.8$) case. From the table, we can see that our proposed method has a better performance on all datasets.

Ablation Study. We perform experiments on the IEEE33-Nodes dataset to demonstrate the effectiveness of each component in our proposed model, including the *continuous-time modeling*, the *dynamic diffusion convolution* and the *soft-masking function* in the dynDC-ODE module, the *usage of the SDE module* and the *adversarial training strategy.*

Continuous-Time Modeling: From Table 1, we can see that no matter our proposed model or other continuous-time models are superior to traditional discrete-time models, which only update features at the observation time instants. Continuous models make the updates $\Delta T/\delta t$ times between two neighboring observations, where the propagation step δt is much smaller than the sampling interval ΔT (we set $\delta t = 0.1\Delta T$ in the experiment). When used with graph convolutions, the integrated information from adjacent nodes will continuously help correct and update the states of the current node. Therefore, even though we use the simplest Euler-Method to approximately calculate the integration, the predicted trajectory fits the original one better than the trajectory provided by discrete models. Moreover, since the propagation step can be arbitrary small, the continuous-time model can provide predictions at any time to enable timely control, rather than only discretely on observation time points.

Dynamic Diffusion Convolution: Unlike most literature works using Graph Convolution (GC), Diffusion Convolution (DC) in our model integrates the information from neighbors within K hops (we set $K = 3$). This is especially helpful when data are sparse and superior to traditional graph convolution, which only collect and share information among 1-hop neighbors. Besides, our model learns the impacts of different neighbors and puts in weights in the aggregation, forming the dynamic Diffusion Convolution (dynDC), so that the information integration is more effective and accurate. To illustrate the effectiveness, we conduct the corresponding experiments and the results are shown in Table 2.

Table 2. Performance comparison among models using graph convolution (GC), diffusion convolution (DC) and dynamic diffusion convolution (dynDC) on IEEE33-Nodes ($p_t = 0.5, p_s = 0.8$).

	GC [17]			DC [20]			dynDC		
	MAE	RMSE	MAPE	MAE	RMSE	MAPE	MAE	RMSE	MAPE
Graph-GRU	0.0349	0.0643	9.65%	0.0341	0.0622	9.64%	**0.0319**	**0.0571**	**9.13%**
Graph-ODE-RNN	0.0306	0.0578	8.36%	0.0279	0.0522	7.83%	**0.0264**	**0.0504**	**7.49%**
Graph-GRU-ODE	0.0313	0.0607	8.49%	0.0298	0.0567	8.18%	**0.0272**	**0.0519**	**7.65%**

Soft-Masking Function: As described in Sec. 3, we introduce a soft-masking function to modulate the hidden features in our dynDC-ODE module for better adapting to the missing data cases. The comparison results in Table 3 have proved the role of such a design.

SDE Module: To better capture uncertainties existing in all the real-world systems, the stochastic modules, i.e. SDE, is included in our model. Different from

Table 3. Performance comparison of our dynDC-ODE module with & without the soft-masking function on IEEE33-Nodes ($p_t = 0.5, p_s = 0.8$).

	MAE	RMSE	MAPE
AGGDN (w/o soft-masking)	0.0264	0.0504	7.49%
AGGDN (w/ soft-masking)	**0.0250**	**0.0471**	**7.21%**

ODE models which make strong Gaussian assumptions about the data distribution, the real distribution is learnt by Monte-Carlo sampling process within the SDE propagation. We also implement a simplified version, AGGDN(ODE), that does not have the SDE part for comparison. The experiments results based on both schemes running on the same data are shown in Table 4, and the improvement brought by the SDE module is obvious.

Table 4. Performance comparison of our full model and the simplified model without SDE on IEEE33-Nodes ($p_t = 0.5, p_s = 0.8$).

	MAE	RMSE	MAPE
AGGDN (ODE)	0.0250	0.0471	7.21%
AGGDN (full)	**0.0243**	**0.0457**	**7.03%**

Adversarial Training: Since we incorporate stochastic terms in our model, the training process becomes more difficult. To better train our model, we utilize an adversarial training strategy to avoid the possible gradient explosion problem. For reference, we also attach the comparison results before and after using the adversarial training in Table 5.

Table 5. Performance comparison of our model on IEEE33-Nodes ($p_t = 0.5, p_s = 0.8$) before & after using adversarial training.

	MAE	RMSE	MAPE
AGGDN (w/o adversarial training)	0.0250	0.0473	7.20%
AGGDN (w/ adversarial training)	**0.0243**	**0.0457**	**7.03%**

Robustness Study. To test the generalization capability of our model, we also conduct extra experiments using the models trained by IEEE33-Node data with the observation ratio ($p_t = 0.4, p_s = 0.6$), and further evaluate the performance on data by varying the observation ratios. Besides our AGGDN, our dynamic

Diffusion Convolution (dynDC) also helps improve the baselines and the results are plotted in Fig. 2. We can see that, compared with continuous-time models, the Graph-GRU cannot adapt well to different observation ratios and its performance even becomes worse when more data are observed. Compared with other continuous-time models, our AGGDN consistently performs better, especially when the observation ratio is small. When $p_s = 1.0$, the simplified version AGGDN(ODE) is a little bit better than the full version AGGDN(full) since the effect of uncertainties brought by data missing is reduced and all the data properties and topological relation can be directly inferred from data.

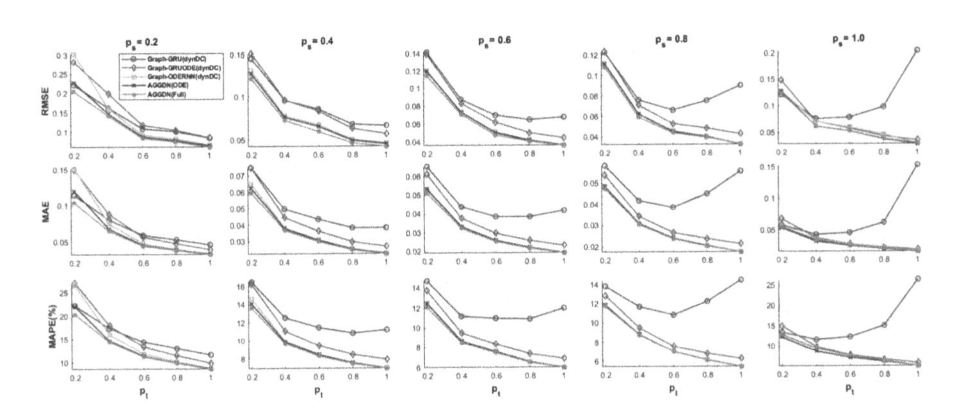

Fig. 2. Performance of various models with respect to different ratios of observations.

5 Conclusion

We propose a novel continuous-time stochastic model called AGGDN to model the dynamics of real-world networked systems from sporadic observations. Our model adopts a compounded ODE-SDE structure to capture the topological information and the signal properties on the graph while taking the underlying uncertainties in the system into consideration. The ODE component provides an approximate estimation of the signal, which further modulates SDE to refine the ODE result to provide a more accurate output. To better model the interactions among nodes, we propose a dynDC-ODE module with enhanced diffusion convolutions to learn the impact of different nodes during information integration. Besides, we introduce a soft-masking function to make our model adapt to the sparse and irregular data cases. To address the challenge of training with SDE module, the Wasserstein adversarial objective is incorporated. Experimental evaluations demonstrate that our model is effective in the state prediction and outperforms previous works on partial-observable networked systems such as microgrid and traffic networks.

Acknowledgments. This work was supported in part by NSF under the award number ITE 2134840 and the U.S. DOE's Office of EERE under the Solar Energy Technologies Office Award Number 38456.

References

1. Chen, R.T.Q., Rubanova, Y., Bettencourt, J., Duvenaud, D.K.: Neural ordinary differential equations. In: Advances in Neural Information Processing Systems 31 (2018)
2. Choi, J., Choi, H., Hwang, J., Park, N.: Graph neural controlled differential equations for traffic forecasting (2021)
3. Chu, H., et al.: Neural turtle graphics for modeling city road layouts. In: Proceedings of the IEEE/CVF International Conference on Computer Vision (ICCV) (October 2019)
4. Cui, Z., Lin, L., Pu, Z., Wang, Y.: Graph markov network for traffic forecasting with missing data. Trans. Res. Part C: Emerging Technol. **117** (2020)
5. De Brouwer, E., Simm, J., Arany, A., Moreau, Y.: GRU-ODE-Bayes: continuous modeling of sporadically-observed time series. In: Advances in Neural Information Processing Systems 32, pp. 7379–7390 (2019)
6. Defferrard, M., Bresson, X., Vandergheynst, P.: Convolutional neural networks on graphs with fast localized spectral filtering. In: Advances in Neural Information Processing Systems 29 (2016)
7. Diao, Z., Wang, X., Zhang, D., Liu, Y., Xie, K., He, S.: Dynamic spatial-temporal graph convolutional neural networks for traffic forecasting. In: Proceedings of the 33rd AAAI Conference on Artificial Intelligence (AAAI 2019) (Feb 2019)
8. Fang, Z., Long, Q., Song, G., Xie, K.: Spatial-temporal graph ODE networks for traffic flow forecasting. In: Proceedings of the 27th ACM SIGKDD Conference on Knowledge Discovery & amp Data Mining. ACM (Aug 2021). https://doi.org/10.1145/3447548.3467430
9. Goyal, P., Chhetri, S.R., Canedo, A.: dyngraph2vec: capturing network dynamics using dynamic graph representation learning. arXiv abs/ arXiv: 1809.02657 (2018)
10. Goyal, P., Kamra, N., He, X., Liu, Y.: Dyngem: deep embedding method for dynamic graphs. arXiv abs/ arXiv: 1805.11273 (2018)
11. Hajiramezanali, E., Hasanzadeh, A., Duffield, N., Narayanan, K.R., Zhou, M., Qian, X.: Variational graph recurrent neural networks. CoRR abs/ arXiv: 1908.09710 (2019)
12. Huang, Z., Sun, Y., Wang, W.: Learning continuous system dynamics from irregularly-sampled partial observations. In: Advances in Neural Information Processing Systems 33 (2020)
13. Ioannidis, V.N., Marques, A.G., Giannakis, G.B.: A recurrent graph neural network for multi-relational data. In: 2019 IEEE International Conference on Acoustics, Speech and Signal Processing (ICASSP), pp. 8157–8161 (2019)
14. Jin, Y., JáJá, J.F.: Learning graph-level representations with gated recurrent neural networks. arXiv abs/ arXiv: 1805.07683 (2018)
15. Kidger, P., Morrill, J., Foster, J., Lyons, T.: Neural controlled differential equations for irregular time series (2020)
16. Kipf, T., Fetaya, E., Wang, K.C., Welling, M., Zemel, R.: Neural relational inference for interacting systems. In: Proceedings of the 35th International Conference on Machine Learning, pp. 2688–2697 (2018)

17. Kipf, T.N., Welling, M.: Semi-supervised classification with graph convolutional networks. In: International Conference on Learning Representations (ICLR) (2017)
18. Kong, L., Sun, J., Zhang, C.: SDE-Net: equipping deep neural network with uncertainty estimates. In: Proceedings of the 37th International Conference on Machine Learning (2020)
19. Li, X., Wong, T.K.L., Chen, R.T.Q., Duvenaud, D.: Scalable gradients for stochastic differential equations. In: 23rd International Conference on Artificial Intelligence and Statistics, pp. 3870–3882 (Aug 2020)
20. Li, Y., Yu, R., Shahabi, C., Liu, Y.: Diffusion convolutional recurrent neural network: data-driven traffic forecasting. In: International Conference on Learning Representations (ICLR) (2018)
21. Li, Y., Tarlow, D., Brockschmidt, M., Zemel, R.S.: Gated graph sequence neural networks. In: International Conference on Learning Representations (ICLR) (2016)
22. Li, Y., Vinyals, O., Dyer, C., Pascanu, R., Battaglia, P.W.: Learning deep generative models of graphs. arXiv abs/ arXiv: 1803.03324 (2018)
23. Liao, R., et al.: Efficient graph generation with graph recurrent attention networks. In: Advances in Neural Information Processing Systems 32, pp. 4255–4265 (2019)
24. Liu, X., Xiao, T., Si, S., Cao, Q., Kumar, S., Hsieh, C.J.: Neural sde: stabilizing neural ode networks with stochastic noise (2019). https://doi.org/10.48550/ARXIV.1906.02355, https://arxiv.org/abs/1906.02355
25. Liu, X., Xiao, T., Si, S., Cao, Q., Kumar, S., Hsieh, C.J.: How does noise help robustness? explanation and exploration under the neural sde framework. In: IEEE/CVF Conference on Computer Vision and Pattern Recognition (CVPR) (June 2020)
26. Liu, Y., et al.: Learning continuous-time dynamics by stochastic differential networks. ArXiv abs/ arXiv: 2006.06145 (2020)
27. Liu, Y., et al.: Continuous-time stochastic differential networks for irregular time series modeling. In: Mantoro, T., Lee, M., Ayu, M.A., Wong, K.W., Hidayanto, A.N. (eds.) ICONIP 2021. CCIS, vol. 1516, pp. 343–351. Springer, Cham (2021). https://doi.org/10.1007/978-3-030-92307-5_40
28. Manessi, F., Rozza, A., Manzo, M.: Dynamic graph convolutional networks. Pattern Recogn. **97** (2020)
29. Pareja, A., et al.: EvolveGCN: evolving graph convolutional networks for dynamic graphs. In: Proceedings of the 34th AAAI Conference on Artificial Intelligence (AAAI 2020) (2020)
30. Peluchetti, S., Favaro, S.: Infinitely deep neural networks as diffusion processes. In: 23rd International Conference on Artificial Intelligence and Statistics, vol. 108, pp. 1126–1136 (Aug 2020)
31. Poli, M., Massaroli, S., Park, J., Yamashita, A., Asama, H., Park, J.: Graph neural ordinary differential equations. arXiv abs/ arXiv: 1911.07532 (2019)
32. Poli, M., et al.: Continuous-depth neural models for dynamic graph prediction (2021). https://doi.org/10.48550/ARXIV.2106.11581, https://arxiv.org/abs/2106.11581
33. RTDS-Technologies-Inc.: Power hardware-in-the-loop (phil) (2022). https://www.rtds.com/applications/power-hardware-in-the-loop/
34. Rubanova, Y., Chen, T.Q., Duvenaud, D.K.: Latent ordinary differential equations for irregularly-sampled time series. In: Advances in Neural Information Processing Systems 32, pp. 5321–5331 (2019)
35. Sanchez-Gonzalez, A., Bapst, V., Cranmer, K., Battaglia, P.W.: Hamiltonian graph networks with ODE integrators. arXiv abs/ arXiv: 1909.12790 (2019)

36. Seo, Y., Defferrard, M., Vandergheynst, P., Bresson, X.: Structured sequence modeling with graph convolutional recurrent networks. In: The 25th International Conference on Neural Information Processing, pp. 362–373 (2018)
37. Shrivastava, H., et al.: Glad: learning sparse graph recovery. In: International Conference on Learning Representations (ICLR) (2020)
38. Simonovsky, M., Komodakis, N.: Dynamic edge-conditioned filters in convolutional neural networks on graphs. In: Proceedings of the IEEE/CVF Conference on Computer Vision and Pattern Recognition (CVPR) (July 2017)
39. Sun, C., Karlsson, P., Wu, J., Tenenbaum, J.B., Murphy, K.: Predicting the present and future states of multi-agent systems from partially-observed visual data. In: International Conference on Learning Representations (ICLR) (2019)
40. Taheri, A., Gimpel, K., Berger-Wolf, T.: Learning graph representations with recurrent neural network autoencoders. In: KDD 2018 Deep Learning Day (2018)
41. Tzen, B., Raginsky, M.: Neural stochastic differential equations: deep latent gaussian models in the diffusion limit. ArXiv abs/ arXiv: 1905.09883 (2019)
42. Tzen, B., Raginsky, M.: Theoretical guarantees for sampling and inference in generative models with latent diffusions. In: 32nd Annual Conference on Learning Theory, vol. 99, pp. 3084–3114 (Jun 2019)
43. Yan, T., Zhang, H., Li, Z., Xia, Y.: Stochastic graph recurrent neural network (2020). https://doi.org/10.48550/ARXIV.2009.00538, https://arxiv.org/abs/2009.00538
44. Ying, R., You, J., Morris, C., Ren, X., Hamilton, W.L., Leskovec, J.: Hierarchical graph representation learning with differentiable pooling. In: Advances in Neural Information Processing Systems 31, pp. 4805–4815 (2018)
45. You, J., Ying, R., Ren, X., Hamilton, W., Leskovec, J.: GraphRNN: generating realistic graphs with deep auto-regressive models. In: Proceedings of the 35th International Conference on Machine Learning, pp. 5708–5717 (2018)
46. Yu, B., Li, M., Zhang, J., Zhu, Z.: 3D graph convolutional networks with temporal graphs: a spatial information free framework for traffic forecasting arXiv: 1903.00919 (2019)
47. Yu, B., Yin, H., Zhu, Z.: Spatio-temporal graph convolutional networks: a deep learning framework for traffic forecasting. In: Proceedings of he 27th International Joint Conference on Artificial Intelligence (IJCAI) (2018)
48. Yu, B., Yin, H., Zhu, Z.: ST-UNet: a spatio-temporal u-network for graph-structured time series modeling. arXiv abs/ arXiv: 1903.05631 (2019)

Attribution Guided Layerwise Knowledge Amalgamation from Graph Neural Networks

Yunzhi Hao[1,2], Yu Wang[1], Shunyu Liu[1], Tongya Zheng[1](✉), Xingen Wang[1], Xinyu Wang[1], Mingli Song[1], Wenqi Huang[3], and Chun Chen[1]

[1] Zhejiang University, Hangzhou 310058, China
{ericohyz,yu.wang,liushunyu,tyzheng,newroot, wangxinyu,brooksong,chenc}@zju.edu.cn

[2] Zhejiang University-China Southern Power Grid Joint Research Centre on AI, Hangzhou 310058, China

[3] Digital Grid Research Institute, China Southern Power Grid, Guangzhou 510663, China
huangwq@csg.cn

Abstract. Knowledge Amalgamation (KA), aiming to transfer knowledge from multiple well-trained teacher networks to a multi-talented and compact student, is gaining attention due to its crucial role in resource-constrained scenarios. Previous literature on KA, although exhibiting promising results, is primarily geared toward Convolutional Neural Networks (CNNs). However, when transferred to Graph Neural Networks (GNNs) with non-grid data, KA techniques face new challenges that can be difficult to overcome. Moreover, the layerwise aggregation of GNNs produces significant noise as they progress from a shallow to a deep level, which can impede KA students' deep-level semantic comprehension. This work aims to overcome this limitation and propose a novel strategy termed LAyerwIse Knowledge Amalgamation (LaiKA). It involves Hierarchical Feature Alignment between the teachers and the student, which enables the student to directly master the feature aggregation rules from teacher GNNs. Meanwhile, we propose a Selective Attribution Transfer (SAT) module that identifies task-relevant topological substructures to assist the capacity-limited student in mitigating noise and enhancing performance. Extensive experiments conducted on six datasets demonstrate that our proposed method equips a single student GNN to handle tasks from multiple teachers effectively and achieve comparable or superior results to those of the teachers without human annotations.

Keywords: Knowledge Amalgamation · Knowledge Transfer · Attribution Graph · Graph Neural Networks

1 Introduction

Recent years have witnessed the unprecedented progress of Deep Neural Networks (DNNs) in various fields ranging from computer vision to natural language

B. Luo et al. (Eds.): ICONIP 2023, LNCS 14447, pp. 147–160, 2024.
https://doi.org/10.1007/978-981-99-8079-6_12

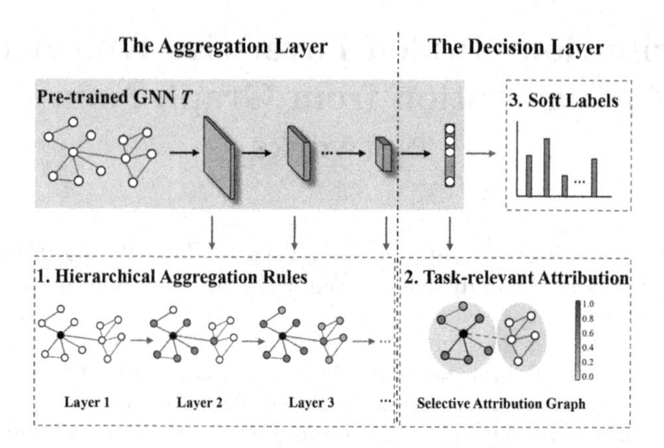

Fig. 1. Illustrations of three parts of knowledge available for amalgamation from a pre-trained GNN T: 1) Hierarchical aggregation rules from the aggregation (intermediate) layer; 2) Task-relevant topological attribution from the decision layer, and colors encode the importance of each edge; 3) Soft labels produced by T.

processing. Numerous pre-trained DNNs have been released online [8,22]. As these well-trained models have consumed a large number of computing resources to mine the knowledge underlying the raw data, they are of great potential to reduce the cost of addressing new tasks if reused properly. Various studies have been conducted to explore the ways to reuse pre-trained DNNs [16]. One approach is transfer learning [15], which adapts pre-trained DNNs to new tasks. Another technique is Knowledge Distillation (KD) where a compressed student is created by mimicking the outputs of a pre-trained teacher [9], which has shown exceptional outcomes. More recently, several attempts have been made at learning a compact and multi-talented student from multiple teachers that specialize in different tasks [2,17,29], which is termed *Knowledge Amalgamation* (KA).

Unlike KD which focuses on training a student for the same task as the teacher, KA aims to train a multitasking student that can handle various tasks of all teachers. Despite the encouraging results achieved, most of them are tailored for Convolution Neural Networks (CNNs), restricted to grid data. Recently, Graph Neural Networks (GNNs) [7,12,24], owing to the large power on non-grid data in graphs, have gained increasing popularity in various domains including social networks, gene regulatory networks, and knowledge graphs, *etc.* Different from CNNs, GNNs learn graph topology by propagating neighbors' information. Unfortunately, existing CNN-based KA methods are not feasible for accomplishing graph tasks since they overlook the semantics in interconnected datasets.

Several attempts apply KD to GNNs [5,28]. However, limited research has been devoted to the GNN-based KA task, with the only known exception being the work of AmalgamateGNN [10]. The authors amalgamated knowledge from multiple teachers at the decision layer and introduce a topological attribution map to highlight structural saliency. Nevertheless, since only the decision layer is employed, AmalgamateGNN does not provide sufficient guidance since students need to make a gradual alignment from shallow to deep layers of teachers. Additionally, AmalgamateGNN renders the student a passive recipient of knowledge, hindering its ability to filter task-irrelevant noise from pre-trained GNNs.

Motivated by the above issues, we propose a *LAyerwIse Knowledge Amalgamation* scheme (LaiKA) to fully transfer knowledge from multiple teachers. As depicted in Fig. 1, we split the knowledge from a pre-trained GNN T into three parts: 1) Aggregation rules from intermediate layers; 2) Task-relevant knowledge from the decision layer; 3) Soft labels produced by T. LaiKA first leverages *Hierarchical Feature Alignment* to enable the student to capture feature aggregation rules underlying intermediate layers. By training on unlabeled graphs, the student generates the initial amalgamated features at each layer. These features are then projected onto teachers' domains to derive the transferred features. Finally, LaiKA computes an alignment loss with the features from corresponding layers.

As the student is capacity-limited compared to the ensemble of the teachers, we introduce *Selective Attribution Transfer* (SAT) that identifies task-relevant substructures to facilitate the compact student capturing task-relevant features. Specifically, this is achieved by generating a selective attribution graph, which explicitly constructs task-relevant attributions. To avoid redundancy from teachers, a selective adjacency matrix is generated with graph clustering to allocate task-relevant weight for closely related nodes. By optimizing the similarity between the student's selective attribution graph and the ones of the teachers, we can transfer comprehensive and task-relevant knowledge to the student. Extensive experiments are performed to assess the efficacy of the proposed LaiKA. Results show that the student in LaiKA achieves comparable or even superior performance to those of the teachers without any human annotations. In summary, the contributions of this work are listed as follows.

- We propose a layerwise GNN-based KA technique that enables the training of a compact, yet versatile student model capable of addressing cross-domain teachers' tasks without requiring any human annotation.
- The hierarchical feature alignment and selective attribution transfer are devised to aid the student in acquiring feature aggregation rules and task-relevant attribution from multiple teachers.
- Extensive experiments are conducted and the results indicate that the student in LaiKA provides comparable or even superior performance compared to its teachers and outperforms the previous KA approaches.

2 Related Work

(1) Knowledge Amalgamation: Knowledge amalgamation (KA) is originally developed from the knowledge distillation (KD) proposed by [9], which aims to train a petite and shallow student that performs comparably to its teacher [13,27,30]. Unlike KD which focuses on training a student for the same task as the teacher, KA aims to introduce a multitasking, lightweight student that can handle various tasks of all teachers. For instance, [17] leverages prior knowledge via lateral connections to previous features. And [21] proposes a pilot strategy that comprises feature amalgamation and parameter learning. Despite the encouraging progress, existing KA works are primarily limited to methods tailored for CNNs, which are unsuitable for GNNs with non-grid data.

(2) Graph Neural Networks: Graph neural networks (GNNs) encode the graph structures with complex relations between objects [18,25,26]. GCN is one of the pioneers that follow the neighbor aggregation scheme [12]. While GraphSAGE [7] generates embeddings by aggregating features from local structures. GAT [24] introduces an attention mechanism to specify fine-grained weights on neighbors. Despite the superior effectiveness, only a few works investigate reusing GNNs. The work of [28], as the first attempt, generalizes KD to GNNs. [5] proposes graph-free KD by modeling topology with multivariate Bernoulli distribution. As for KA on GNNs, to our knowledge, only [10] amalgamates knowledge from heterogeneous GNNs by topological attribution maps. However, [10] is limited by sparse and insufficient amalgamation as only the decision layer is employed, and the progressive learning of GNNs could result in task-irrelevant noise, posing challenges for students to learn attribution.

(3) Attribution Graph: Attribution graph, which belongs to knowledge transferability learning [32], refers to assigning importance scores to the inputs for a specific output [23]. The probe data is first passed through the pre-trained model to obtain features after a forward pass. Backward passes then produce the attributions. The back-propagation rule depends on the attribution methods adopted [1]. By calculating the similarity between corresponding attribution graphs, we can directly evaluate the knowledge transferability of the models. Since the explicit representations of teachers and compact student knowledge are important in KA, we maximize the similarity of the attribution graphs between the student and teachers, we can transfer task-relevant information from teachers and train a versatile student without human annotations.

3 Problem Statement

We aim to solve the GNN-based KA task as follows. Assume that we are given N pre-trained GNNs $\{T_1, T_2, ..., T_n, ..., T_N\}$, where each T_n corresponding to a graph $G_n = (V_n, X_n, E_n)$ with its adjacency matrix A_n, respectively. Our objective is to construct a single multi-talented and lightweight student GNN S that amalgamates the knowledge of all the teachers and handles their tasks, with no annotated training samples. The information accumulation strategy of the teachers can be identical or different, with no specific constraints. The student S is anticipated to be more robust in tackling comprehensive classification tasks and more practical and resource economic than the ensemble of the teachers.

4 Method

4.1 Overview

This study presents a **La**yerwise **K**nowledge **A**malgamation strategy (LaiKA) for addressing the GNN-based KA problem. The overall KA process, which

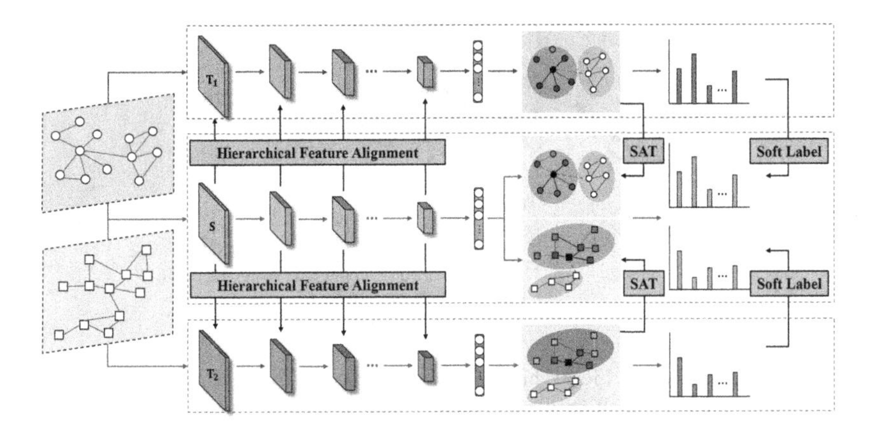

Fig. 2. The overall workflow of the proposed LaiKA in the two-teacher case. LaiKA mainly consists of two parts: hierarchical feature alignment and Selective Attribution Transfer (SAT). The student learns feature aggregation rules in the hierarchical feature alignment module and then imitates the selective attribution graphs of the teachers in SAT module. Finally, Soft label is applied as additional supervision.

involves two pre-trained teachers, is illustrated in Fig. 2. Admittedly, we assume for now that teachers share the same number of layers. This assumption might be arguably strong but it does hold in many cases, where most GNNs have three-layer architectures to aggregate neighbors within three hops. Inspired by [10], the proposed LaiKA firstly sets the input channel number of the student adaptively based on the maximum value of the dimensions of all input nodes. Specifically, adaptive graph convolutional layers replace the student layers to prevent compatibility problems. Aligning the merged embeddings of nodes with various teachers at each level permits the student to capture several feature aggregation rules simultaneously in the hierarchical feature alignment module. Then the selective attribution transfer is applied by generating task-relevant substructures for both teachers and the student to further enhance the performance. Finally, LaiKA utilizes soft predictions generated by multiple teachers as additional supervised information and optimizes the student with the loss of the above three parts. The details of each module are given as follows.

4.2 Hierarchical Feature Alignment

Teacher GNNs adopt a neighbor aggregation approach to encode topological and semantic information in graphs, in contrast to traditional convolutional networks for grid data. After l aggregations, the representation of node v aggregates the node features within its l-hop neighbors. Formally, the l-th layer of a teacher GNN can be defined as:

$$\mathbf{a}_v^l = \text{AGGREGATE}^l(\mathbf{e}_u^{l-1} | u \in \mathcal{N}_v \cup v), \tag{1}$$

$$\mathbf{e}_v^l = \text{COMBINE}^l(\mathbf{e}_v^{l-1}, \mathbf{a}_v^l), \tag{2}$$

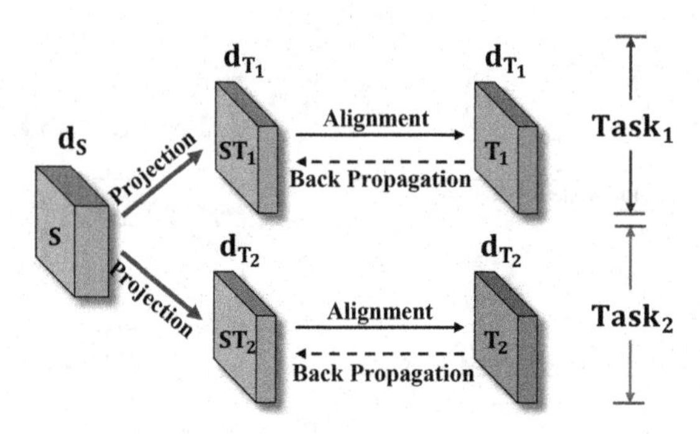

Fig. 3. Illustrations of the hierarchical feature alignment. Node embeddings from student S are projected to each task and aligned with the corresponding teacher output.

where $\mathbf{a}_v^l$ is the aggregated features of the l-hop neighbors, $\mathcal{N}_v$ denotes the the neighborhood of node v . The choices of the AGGREGATE and COMBINE operation are crucial to the performance, such as graph convolutional layers [12] and graph attention layers [24]. Such aggregation layers encode feature aggregation rules within node neighborhoods and are essential to GNNs. However, existing GNN-based KA methods concentrate mainly on extracting knowledge from the decision layer, thereby appearing to overlook the aggregation process.

To bridge this gap, we introduce a *hierarchical feature alignment* module, shown in Fig. 3, which enables the transfer of encoded aggregation rules from teacher to student by imitating the neighborhood features in a layerwise manner. By first training on the unlabeled graphs, the student can learn initial node embeddings at each layer. Considering that the student S contains amalgamated knowledge of multiple tasks, an adaption layer is employed to project the amalgamated node embeddings onto each of the teachers' respective domains. Thus the task-wise embeddings of the student can be aligned with each teacher at the current layer.

To achieve a student with fewer parameters, we integrate the semantics of multiple tasks into compact node embeddings with fewer dimensions. Here we use an adaption layer equipped with a node embedding projection function to isolate distinct tasks from the student. We use $\mathbf{E}_S^l$ and $\mathbf{E}_{T_n}^l$ to denote the outputs of student S and teacher T_n at l-th layer respectively and use $\mathbf{E}_{ST_n}^l$ to denote the projected student embeddings processed by the adaption layer. The transferred node embeddings of the student share the same task domain as the corresponding teacher T_n. The adaption layer is formed as:

$$\mathbf{E}_{ST_n}^l = f_n^l(\mathbf{E}_S^l), \tag{3}$$

where f_n^l is the proposed linear mapping function at the l-th layer for S and each teacher T_n. After generating the task-wise transferred student node embeddings $\mathbf{E}_{ST_n}^l$, the Mean Squared Error (MSE) is utilized to compute the hierarchical

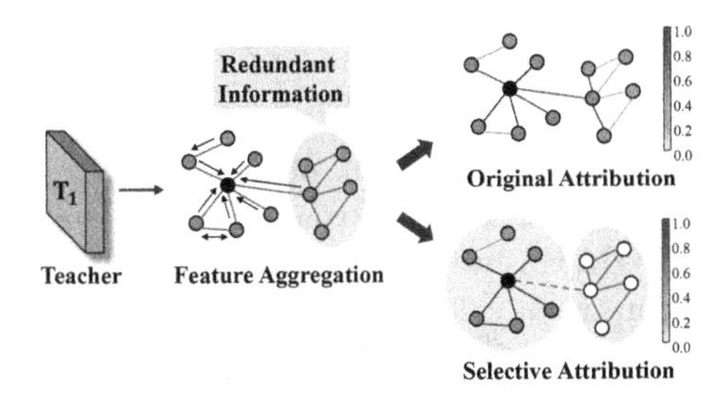

Fig. 4. Illustrations of the selective attribution graph of teacher T_1. Compare with the original attribution graph, SAT can filter out significant information in T_1.

alignment loss with $\mathbf{E}^l_{T_n}$ and $\mathbf{E}^l_{ST_n}$, and further enforce the neighbor aggregation rules of the student to imitate those of multiple teachers at each layer.

4.3 Selective Attribution Transfer

After amalgamating the aggregation rules underlying the intermediate layers, the compact student is also expected to capture task-relevant knowledge from the decision layer to further enhance the performance. Therefore, representing task-relevant knowledge encoded in teachers plays a crucial role in the KA task. We notice that attribution graph [33], referring to assigning importance scores to the inputs for a task-specified output, has been proposed to quantify the knowledge transferability in CNNs and made remarkable progress [23]. Given two pre-trained deep networks C_1 and C_2, the knowledge transferability between them can be measured by the similarity of their corresponding attribution graphs.

Though the notable advancements in CNN-based knowledge transferability, existing methods present certain challenges when applied to GNNs. Unlike CNNs, the training loss in GNNs (such as node classification loss) cannot be decomposed into individual terms for each sample because the loss term on GNNs' nodes relies on multiple neighbors [4]. Moreover, traditional GNNs suffer from over-smoothing due to redundant aggregation of distant neighbors, which leads to degradation in performance. Figure 4 illustrates that pre-trained GNN T_1 redundantly aggregates the distant neighborhood on the right subgraph (shaded in grey). Hence the corresponding attribution graph encodes the redundant information, adversely affecting the student during the KA task.

To solve this dilemma, we introduce Selective Attribution Transfer (SAT) to explicitly and selectively transfer task-relevant substructures with pre-trained teachers. The attribution transfer process of teacher T_n is given as follows. To mitigate the effects of irrelevant neighbors, we first adopt graph clustering, such as Metis [11] and Graclus [6], to construct partitions over nodes in G_n such that within-clusters edges are much more than between-cluster edges. Initially, each node and its neighbors are located in one cluster; after the clustering process, neighbors with tight relations remain in the same cluster, whereas those with sparse edges are separated into different clusters. Once the node clusters are derived, a selective adjacency matrix $\overline{A}_n$ of $\overline{G}_n$ can be generated with c clusters:

$$\overline{A}_n = \{A_n - \beta \cdot \sum A_{n,pq} | p, q \in 1, ...c\}, \tag{4}$$

where A_n is the original adjacency matrix of G_n, and $\{A_{n,pq} | p, q \in 1, ...c\}$ contains the between-cluster edges. $\beta = [0, 1]$ is a parameter that controls the weight of between-cluster edges. In this way, we can learn a more intensive distribution of task-relevant attention when calculating the attribution graph of T_n.

Specifically, the selective attribution graph SAg_n is produced for each teacher by computing the gradients of the output label scores from the decision layer with respect to the selective adjacency matrix $\overline{A}_n$. After the predicted class scores Sc_n are derived by T_n with the input graph G_n, the selective attribution graph SAg_n of T_n can be computed as:

$$SAg_n = \left. \frac{\partial Sc_n}{\partial \overline{A}_n} \right|_{\overline{G}_n}. \tag{5}$$

As shown in the lower right of Fig. 4, we construct two clusters and decrease the weight of the between-cluster edges, then the selective attribution graph appears to focus on closely related neighbors. With the selective attribution graphs of both the student and teachers generated by Eqn. (5), we can selectively transfer task-relevant semantics from different teachers to the student.

4.4 Model Optimization

As mentioned above, there are three parts of alignment losses, termed hierarchical alignment loss, selective attribution alignment loss, and soft label loss.

Loss of Hierarchical Alignment. We weigh all hierarchical losses between each teacher and the student and write the aggregation alignment loss as:

$$\mathcal{L}_{align} = \sum_{n=1}^{N} \sum_{l=1}^{L-1} \alpha_n^l \cdot \left\| \mathbf{E}_{ST_n}^l - \mathbf{E}_{T_n}^l \right\|, \tag{6}$$

where $\alpha_n^l = \frac{\left\| \mathbf{E}_{ST_n}^l - \mathbf{E}_{T_n}^l \right\|}{\sum_{l=1}^{L-1} \left\| \mathbf{E}_{ST_n}^l - \mathbf{E}_{T_n}^l \right\|}$ is the weight of alignment loss at l-th layer.

Loss of Selective Attribution Transfer. Similarly, the knowledge transferability between the student and teachers is measured by quantifying the distance between their corresponding attribution graphs:

$$\mathcal{L}_{att} = \sum_{n=1}^{N} \left\| \frac{\partial Sc_n}{\partial \overline{A}_n} - \frac{\partial Sc_s}{\partial \overline{A}_s} \right\|. \tag{7}$$

The node features are multiplied with $\overline{A}_n$ to calculate the gradient. And the hierarchical weight of $\mathcal{L}_{att}$ is computed together with the $\mathcal{L}_{align}$ after normalization.

Table 1. Statistics and splitting protocols. The unit of PPI for splitting is graph.

Datasets	PPI	Cora	Citeseer	Pubmed	WiKi	Amazon
Nodes	56,944	2,708	3,327	19,717	11,701	13,572
Edges	818,716	5,429	4,732	44,338	216,123	574,418
Features	50	1,433	3,703	500	300	767
Labels	121	7	6	3	10	10
Train	20	1,208	1,827	18,217	1,170	1,357
Valid	2	500	500	500	1,170	1,357
Test	2	1,000	1,000	1,000	9,361	10,858

Loss of Soft Label. $\mathcal{L}_{soft}$ is computed as the MSE loss among the soft predictions from the student and multiple teachers. Finally, we incorporate the above loss terms into our final loss function:

$$\mathcal{L} = \mathcal{L}_{align} + \mathcal{L}_{att} + \mathcal{L}_{soft}. \tag{8}$$

By optimizing this loss function, the student GNN is generated with the amalgamation of multiple teachers without human annotations.

5 Experiments

In this section, we demonstrate the efficiency and robustness of the proposed LaiKA on a list of publicly available benchmarks across different tasks and provide implementation details, comparative results, and ablation study as follows.

5.1 Datasets

We evaluate LaiKA on three sets of six real-world graph datasets across various domains. Table 1 provides statistics and splitting protocol of these graphs.

Biological Graphs. We use Protein-protein interaction (PPI) dataset [35] for multi-label node classification. We follow the same dataset splitting protocol with [28] and divide PPI into two subsets for training two teachers: 61 labels for T_1 and the other 60 labels for T_2.

Citation Graphs. For the single-label node classification task, we primarily use three citation graphs: Cora, Citeseer, and Pubmed [19], which have distinct channel numbers for each node. We construct three teachers corresponding to the above three graphs respectively and follow the same splitting protocol in [3].

Cross-Domain Graphs. Further, we adopt the reference network Wiki-CS [14] and the computer co-purchase network Amazon-Computers [20] form Wikipedia and Amazon respectively. We follow the splitting protocol as in [34]. We pretrain two teachers with input from different domains and expect the student to combine their advanced knowledge.

5.2 Baselines

We compare LaiKA with AmalgamateGNN [10] and the multi-task learning (MTL) with attention transfer (AT) [31]. We also combine the local structure preserving (LSP) module from [28] and the MTL scheme to develop a KA approach.

For implementations, there are two different teachers GCN [12] and GAT [24] with three-layer architectures. For the proposed LaiKA, we adopt Metis for graph clustering and set the cluster number of PPI as 50, Pubmed as 10, and the rest as 5. The parameters and network architectures of all models are the same.

5.3 Node Classification

We perform semi-supervised node classification which aims to predict the most probable label for target nodes with GCN and GAT.

Table 2. Results of amalgamating knowledge from all combinations of teacher models on PPI, in terms of micro-averaged F1 score (F1).

Models	Teachers		Students			
	T_1(GCN)	T_2(GCN)	MTL+AT	MTL+LSP	AmalgateGNN	**LaiKA**
Size	362.7K	361.8K	14.22M	14.22M	14.22M	**14.22M**
F1	69.48	63.62	68.05/61.23	69.01/62.16	69.32/62.21	**69.83/63.09**
Models	T_1(GCN)	T_2(GAT)	MTL+AT	MTL+LSP	AmalgateGNN	**LaiKA**
Size	362.7K	11.24M	14.22M	14.22M	14.22M	**14.22M**
F1	69.48	98.62	68.99/97.37	69.34/97.89	69.71/97.61	**70.46/98.35**
Models	T_1(GAT)	T_2(GCN)	MTL+AT	MTL+LSP	AmalgateGNN	**LaiKA**
Size	11.30M	361.8K	14.22M	14.22M	14.22M	**14.22M**
F1	98.73	63.62	97.41/62.19	97.77 / 62.36	97.84/62.74	**98.69/63.47**
Models	T_1(GAT)	T_2(GAT)	MTL+AT	MTL+LSP	AmalgateGNN	**LaiKA**
Size	11.30M	11.24M	14.22M	14.22M	14.22M	**14.22M**
F1	98.73	98.62	97.03/96.99	97.27/97.22	98.02/97.95	**98.84/98.73**

Amalgamation from PPI. Table 2 shows our results on PPI. T_1 is shorthand for T{PPI_1}. We observe that the GAT student with LaiKA consistently outperforms all the baselines while maintaining a compact model size. Generally speaking, KA models (AmalgamateGNN, LaiKA) yield better results than MTL-based models. This demonstrates the necessity of aligning topological information. MTL+LSP performs less effectively as it lacks task-relevant knowledge. Compared with the best competitor, LaiKA performs better on both PPI_1 and PPI_2, and is sometimes superior to the teachers, which demonstrates the effectiveness of hierarchical feature alignment and selective attribution transfer.

Amalgamation from Citation Graphs. To assess the results comprehensively, we employ various combinations of citation graphs to train multiple teachers and observe the performance of students. We utilize GAT as teachers for efficiency. Table 3 illustrates that LaiKA consistently outperforms AmalgateGNN and achieves 2% improvements on Citeseer and Pubmed. Moreover, the student with LaiKA demonstrates competitive performance compared with teachers and even outperforms the teacher on Citeseer, which emphasizes the selective transfer of task-relevant knowledge. As the hierarchical alignment of LaiKA can identify the feature aggregation rules of various graphs, the student stores more inclusive classification knowledge than a single teacher.

Table 3. Results of all combinations of citation graphs. Both teachers and students adopt GAT models. The accuracy (Acc.) and the model size are reported as follows.

Teachers	Size	**Acc.**	Students	Size	Acc.	Students	Size	Acc.
T{Cora}	739.6K	87.92	AmalgateGNN	1.43M	87.20	**LaiKA**	**1.43M**	**87.93**
T{Citeseer}	1.901M	78.87			76.80			**78.90**
T{Cora}	739.6K	87.92	AmalgateGNN	1.43M	86.90	**LaiKA**	**1.43M**	**87.45**
T{Pubmed}	259.8K	85.70			83.40			**85.36**
T{Citeseer}	1.901M	78.87	AmalgateGNN	1.43M	76.57	**LaiKA**	**1.43M**	**79.07**
T{Pubmed}	259.8K	85.70			83.60			**85.44**
T{Cora}	739.6K	87.92	AmalgateGNN	1.43M	86.65	**LaiKA**	**1.43M**	**87.52**
T{Citeseer}	1.901M	78.87			76.69			**78.63**
T{Pubmed}	259.8K	85.70			83.20			**85.20**

Amalgamation from Different Domains. Based on the above experiments, we proceeded to verify our approach on different domains. Edges in Wiki-CS are hyperlinks between computer science articles, whereas edges in Amazon-Computers indicate computer co-purchase relations. Despite the scenario differences, they share similar aggregation rules, thereby inspiring the learning potential of hierarchical feature alignment. As shown in Table 4, the student of LaiKA also delivers comparable results, demonstrating the potential of LaiKA on cross-domain KA tasks.

Table 4 displays the results conducted on PPI and citation graphs. AmalgamateGNN experiences evident decreases in performance on citation graphs (-4.52%, -6.27%, -4.50%). In contrast, the student of LaiKA is competent to simultaneously handle all tasks and delivers competitive performance on 5 datasets, with a performance drop of at most 2.98% compared to the teachers.

Table 4. Results of amalgamating knowledge from different domains. Both teachers and students are GATs. The F1/Acc. of the node classification task is reported.

Teachers	T{WiKi}	T{Amazon}	T{PPI_1}	T{PPI_2}	T{Cora}	T{Citeseer}	T{Pubmed}
F1/Acc.	77.71	87.23	98.73	98.62	87.92	78.87	85.70
Students	AmalgamateGNN{WA}		AmalgamateGNN{PPI, Citation}				
F1/Acc.	74.16	84.21	98.20	98.17	83.40	72.60	81.20
Students	**LaiKA{WA}**		**LaiKA{PPI, Citation}**				
F1/Acc.	**76.87**	**86.35**	**98.47**	**98.42**	**85.56**	**75.89**	**84.08**

5.4 Ablation Study

To verify the effectiveness of the hierarchical feature alignment and the selective attribution transfer, three variant models are designed as follows:

LaiKA w/o align is a simplified version of LaiKA without the hierarchical feature alignment. We simply use the decision layer to amalgamate knowledge.

LaiKA w/o att trains the student by only calculating the hierarchical alignment loss, and the selective attribution transfer module is removed.

LaiKA w/o cluster removes the graph clustering preprocessing module and generates the attribution graph with the original adjacency matrix.

Figure 5 depicts the results on PPI and citation graphs. Evidently, the original LaiKA yields the best overall performance, indicating the effectiveness of imitating the layerwise aggregation rules and selective attribution graphs. LaiKA w/o att performs better than LaiKA w/o cluster, further demonstrating that unessential information fails to influence student performance positively.

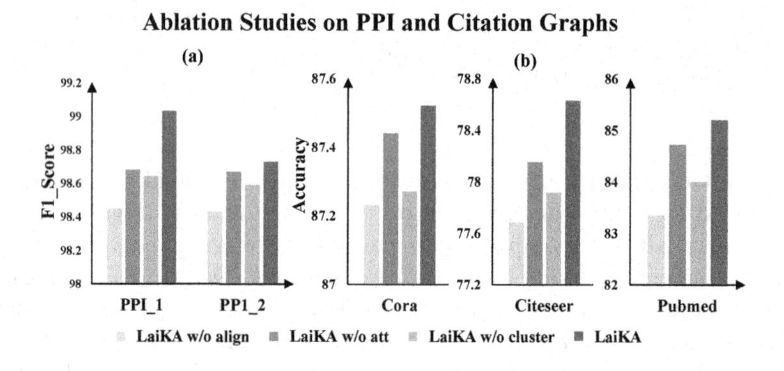

Fig. 5. Ablation studies on PPI and Citation graphs with GAT models.

6 Conclusion

In this paper, we present LaiKA, a novel KA scheme to train a versatile and compact student from pre-trained GNNs specializing in heterogeneous tasks. LaiKA functions by hierarchical feature alignment, where the student imitates feature aggregation rules gradually from the simple to the complex. In addition, a novel selective attribution graph is proposed to enhance student performance by explicitly encoding task-relevant semantics. To validate LaiKA's effectiveness, we conduct node classification experiments on six datasets from different domains and find that it achieves comparable or superior results to those of the teachers. Ablation studies demonstrate the validity of two essential components of LaiKA. We plan to explore an automatic method of generating dense subgraphs in future work to further decrease the graph clustering preprocessing step.

Acknowledgements. This work is funded by the National Key Research and Development Project (Grant No: 2022YFB2703100), National Natural Science Foundation of China (61976186, U20B2066), Zhejiang Province High-Level Talents Special Support Program"Leading Talent of Technological Innovation of Ten-Thousands Talents Program" (No. 2022R52046), the Starry Night Science Fund of Zhejiang University Shanghai Institute for Advanced Study (Grant No. SN-ZJU-SIAS-001), the Fundamental Research Funds for the Central Universities (2021FZZX001-23, 226-2023-00048), Shanghai Institute for Advanced Study of Zhejiang University, and ZJU-Bangsun Joint Research Center.

References

1. Ancona, M., Ceolini, E., Öztireli, C., Gross, M.: Towards better understanding of gradient-based attribution methods for deep neural networks. arXiv preprint arXiv:1711.06104 (2017)
2. de Carvalho, M.V., Pratama, M., Zhang, J., San, Y.: Class-incremental learning via knowledge amalgamation. In: ECML/PKDD (2022)
3. Chen, J., Ma, T., Xiao, C.: Fastgcn: fast learning with graph convolutional networks via importance sampling. arXiv preprint arXiv:1801.10247 (2018)
4. Chiang, W.L., Liu, X., Si, S., Li, Y., Bengio, S., Hsieh, C.J.: Cluster-gcn: an efficient algorithm for training deep and large graph convolutional networks. In: Proceedings of the 25th ACM SIGKDD International Conference on Knowledge Discovery & Data Mining, pp. 257–266 (2019)
5. Deng, X., Zhang, Z.: Graph-free knowledge distillation for graph neural networks. In: The 30th International Joint Conference on Artificial Intelligence (2021)
6. Dhillon, I.S., Guan, Y., Kulis, B.: Weighted graph cuts without eigenvectors a multilevel approach. IEEE Trans. Pattern Anal. Mach. Intell. **29**(11), 1944–1957 (2007)
7. Hamilton, W.L., Ying, R., Leskovec, J.: Inductive representation learning on large graphs. In: Proceedings of the 31st International Conference on Neural Information Processing Systems, pp. 1025–1035 (2017)
8. He, K., Zhang, X., Ren, S., Sun, J.: Deep residual learning for image recognition. In: Proceedings of the IEEE Conference on Computer Vision and Pattern Recognition, pp. 770–778 (2016)
9. Hinton, G., Vinyals, O., Dean, J.: Distilling the knowledge in a neural network. arXiv preprint arXiv:1503.02531 (2015)
10. Jing, Y., Yang, Y., Wang, X., Song, M., Tao, D.: Amalgamating knowledge from heterogeneous graph neural networks. In: Proceedings of the IEEE/CVF Conference on Computer Vision and Pattern Recognition, pp. 15709–15718 (2021)
11. Karypis, G., Kumar, V.: A fast and high quality multilevel scheme for partitioning irregular graphs. SIAM J. Sci. Comput. **20**(1), 359–392 (1998)
12. Kipf, T.N., Welling, M.: Semi-supervised classification with graph convolutional networks. arXiv preprint arXiv:1609.02907 (2016)
13. Liu, Y., et al.: Knowledge distillation via instance relationship graph. In: 2019 IEEE/CVF Conference on Computer Vision and Pattern Recognition (CVPR), pp. 7089–7097 (2019)
14. Mernyei, P., Cangea, C.: Wiki-cs: a wikipedia-based benchmark for graph neural networks (2020)
15. Pan, S.J., Yang, Q.: A survey on transfer learning. IEEE Trans. Knowl. Data Eng. **22**(10), 1345–1359 (2010)

16. Romero, A., Ballas, N., Kahou, S.E., Chassang, A., Gatta, C., Bengio, Y.: Fitnets: hints for thin deep nets. arXiv preprint arXiv:1412.6550 (2014)
17. Rusu, A.A., et al.: Progressive neural networks. arXiv preprint arXiv:1606.04671 (2016)
18. Scarselli, F., Gori, M., Tsoi, A.C., Hagenbuchner, M., Monfardini, G.: The graph neural network model. IEEE Trans. Neural Netw. **20**(1), 61–80 (2008)
19. Sen, P., Namata, G., Bilgic, M., Getoor, L., Galligher, B., Eliassi-Rad, T.: Collective classification in network data. AI Mag. **29**(3), 93–93 (2008)
20. Shchur, O., Mumme, M., Bojchevski, A., Günnemann, S.: Pitfalls of graph neural network evaluation. arXiv preprint arXiv:1811.05868 (2018)
21. Shen, C., Wang, X., Song, J., Sun, L., Song, M.: Amalgamating knowledge towards comprehensive classification. In: Proceedings of the AAAI Conference on Artificial Intelligence, vol. 33, pp. 3068–3075 (2019)
22. Simonyan, K., Zisserman, A.: Very deep convolutional networks for large-scale image recognition. arXiv preprint arXiv:1409.1556 (2014)
23. Song, J., et al.: Depara: deep attribution graph for deep knowledge transferability. In: Proceedings of the IEEE/CVF Conference on Computer Vision and Pattern Recognition, pp. 3922–3930 (2020)
24. Veličković, P., Cucurull, G., Casanova, A., Romero, A., Lio, P., Bengio, Y.: Graph attention networks. arXiv preprint arXiv:1710.10903 (2017)
25. Wu, Z., Pan, S., Chen, F., Long, G., Zhang, C., Philip, S.Y.: A comprehensive survey on graph neural networks. IEEE Trans. Neural Netw. Learn. Syst. **32**(1), 4–24 (2020)
26. Xu, K., Hu, W., Leskovec, J., Jegelka, S.: How powerful are graph neural networks? arXiv preprint arXiv:1810.00826 (2018)
27. Yang, C., Liu, J., Shi, C.: Extract the knowledge of graph neural networks and go beyond it: An effective knowledge distillation framework. In: Proceedings of the Web Conference 2021, WWW 2021, pp. 1227–1237. Association for Computing Machinery, New York (2021)
28. Yang, Y., Qiu, J., Song, M., Tao, D., Wang, X.: Distilling knowledge from graph convolutional networks. In: Proceedings of the IEEE/CVF Conference on Computer Vision and Pattern Recognition, pp. 7074–7083 (2020)
29. Ye, J., Ji, Y., Wang, X., Ou, K., Tao, D., Song, M.: Student becoming the master: Knowledge amalgamation for joint scene parsing, depth estimation, and more. In: Proceedings of the IEEE/CVF Conference on Computer Vision and Pattern Recognition, pp. 2829–2838 (2019)
30. Yuan, F., Shou, L., Pei, J., Lin, W., Gong, M., Fu, Y., Jiang, D.: Reinforced multi-teacher selection for knowledge distillation (2020)
31. Zagoruyko, S., Komodakis, N.: Paying more attention to attention: improving the performance of convolutional neural networks via attention transfer. arXiv preprint arXiv:1612.03928 (2016)
32. Zamir, A., Sax, A., Shen, W., Guibas, L., Malik, J., Savarese, S.: Taskonomy: disentangling task transfer learning (2018)
33. Zamir, A.R., Sax, A., Shen, W.B., Guibas, L.J., Malik, J., Savarese, S.: Taskonomy: disentangling task transfer learning. In: IEEE Conference on Computer Vision and Pattern Recognition (CVPR), IEEE (2018)
34. Zhu, Y., Xu, Y., Yu, F., Liu, Q., Wu, S., Wang, L.: Graph contrastive learning with adaptive augmentation. In: Proceedings of the Web Conference 2021 (Apr 2021)
35. Zitnik, M., Leskovec, J.: Predicting multicellular function through multi-layer tissue networks. Bioinformatics **33**(14), i190–i198 (2017)

Distributed Neurodynamic Approach for Optimal Allocation with Separable Resource Losses

Linhua Luan[1], Yining Liu[1], Sitian Qin[1](✉), and Jiqiang Feng[2]

[1] Harbin Institute of Technology, Weihai 264200, China
qinsitian@163.com
[2] Shenzhen University, Shenzhen 518060, China
fengjq@szu.edu.cn

Abstract. To solve the optimal allocation problem with separable resource losses, this paper proposes a neurodynamic approach based on multi-agent system. By using KKT condition, the nonlinear coupling equality constraint in the original problem is equivalently transformed into a convex coupling inequality constraint. Then, with the help of finite-time tracking technology and fixed-time projection method, a neurodynamic approach is designed and its convergence is strictly proved. Finally, simulation results verify the effectiveness of the proposed neurodynamic approach.

Keywords: Distributed neurodynamic approach · Optimal allocation problem · Separable resource losses

1 Introduction

Solving optimal allocation problems plays a pivotal role in achieving optimal resource utilization and load balancing. Rational allocation of resources holds the potential to enhance the efficiency of extensive systems, including power distribution networks [8], integrated energy system [16], and sensor network [20], thereby reducing overall costs while ensuring equitable resource allocation among nodes.

In general, for the network equipped with n nodes, it is responsible to allocate a certain amount of resources and achieve the objective of minimizing the overall production cost of the entire system. According to the actual situation, the production cost incurred by each node is determined by its own device configuration and the local amount of resource allocation. Considering the resource constraints, the optimal allocation problem can be mathematically formulated as follows:

$$\begin{aligned} \min \ & \sum_{i=1}^{n} f_i(x_i) \\ \text{s.t.} \ & \sum_{i=1}^{n} x_i = D,\ x_i \in \mathcal{X}_i \end{aligned} \tag{1}$$

B. Luo et al. (Eds.): ICONIP 2023, LNCS 14447, pp. 161–172, 2024.
https://doi.org/10.1007/978-981-99-8079-6_13

where $x_i \in \mathbb{R}$ denotes the actual allocation of node i, $f_i : \mathbb{R} \to \mathbb{R}$ is the production cost function of node i, $D \in \mathbb{R}$ is the given amount of resources, and $\mathcal{X}_i$ is the local constraints imposed on node i. Recently, as an optimized parallel computing model, the brain-like nonlinear dynamic system in the recurrent neural network, the so-called neurodynamic approach, has been known. With the help of the parallel computing and information interaction ability of multi-agent systems, which are not available in traditional optimization numerical methods, designing distributed neurodynamic approaches based on multi-agent systems has become a prevalent method for optimal allocation problems Eq. (1), such as [2,3,6,10,19].

In the optimal allocation problem Eq. (1), the equality constraint is related to the resources of all nodes in the network, so it is actually a global constraint, which is essentially different from local constraints. The coupling of global constraints usually makes it difficult to design distributed neurodynamic approaches. For example, the neurodynamic approaches which can effectively solve the distributed optimization problem with only local constraints [1,12] cannot be directly used to solve the optimal allocation problem Eq. (1). Additionally, global inequality constraints often exist in practical applications, which is more challenging to maintain the distributed manner of neurodynamic approaches. Some related studies have been given in [11,13].

Although the aforementioned results advance the technology of neurodynamic approaches, there are few neurodynamic approaches that can deal with resource losses. Resource loss usually destroys the convexity of coupling equality constraint, which brings great difficulties to search the solution of resource allocation problems. To handle this challenge, a neurodynamic approach is proposed to solve the optimal allocation problem with separable resource losses by combining projection operator and symbolic function in this paper. Compared with the existing literature, the main contributions of this paper are as follows:

- In contrast to [2,3,6,8,10,11,13], the optimal allocation problems in this paper takes the resource losses into account. Thus, the discussed optimal allocation problem is more general and challenging.
- The states of agents along the proposed neurodynamic approach finally reach at the exact optimal solution to the the optimal allocation problem rather than the optimal solution set (see [9,18]), which ensures a better convergence property.

2 Preliminaries and Problem Description

2.1 Preliminaries

Graph Theory. In the paper, the distributed neurodynamic approach is based on the multi-agent system, and the system operation is closely related to the communication network. The n agents in the system can be regarded as n nodes on the network, and the communication network can be represented by the topology graph $\mathcal{G}(\mathcal{V}, \mathcal{E}, \mathcal{A})$, which is composed of node set $\mathcal{V} = \{1, 2, \cdots, n\}$, edge set $\mathcal{E} = \{(i,j) \subseteq \mathcal{V} \times \mathcal{V} : a_{ij} > 0\}$, and adjacency matrix $\mathcal{A} = [a_{ij}]_{n \times n}$. For the multi-agent system, $(i,j) \in \mathcal{E}$ means that agent i and agent j can exchange information. If the adjacency matrix $\mathcal{A}$ is symmetric, the graph $\mathcal{G}$ is said to be undirected. A undirected graph $\mathcal{G}$ is said to be connected if there are links between every two different nodes. More details can be found in [7].

Convex Analysis. A subset $\Omega \subseteq \mathbb{R}^n$ is called a convex set if for any $x \in \Omega$, $y \in \Omega$, and $\lambda \in [0,1]$, it has $(1-\lambda)x + \lambda y \in \Omega$. For the convex set Ω, if $f((1-\lambda)x+\lambda y) \leqslant (1-\lambda)f(x) + \lambda f(y)$ holds for all x, $y \in \Omega$ and $\lambda \in [0,1]$, then the function $f : \Omega \to \mathbb{R}$ is called a convex function. Furthermore, if there is $w > 0$ such that $f((1-\lambda)x+\lambda y) \leqslant (1-\lambda)f(x) + \lambda f(y) - \frac{w}{2}\lambda(1-\lambda)\|x-y\|^2$ holds for all x, $y \in \Omega$ and $\lambda \in [0,1]$, then the function $f : \Omega \to \mathbb{R}$ is called a w-strongly convex function. For a point x outside the convex set Ω, the distance from the projected point $\mathcal{P}_\Omega(x)$ to x is the minimum distance from x to Ω. Its mathematical definition is as follows

$$\mathcal{P}_\Omega(x) := \arg\min_{y \in \Omega} \|x - y\|.$$

The projection of x on a convex set Ω is unique and satisfies $\langle x - \mathcal{P}_\Omega(x), y - \mathcal{P}_\Omega(x)\rangle \leqslant 0$ for all $y \in \Omega$.

2.2 Problem Description

In the field of power grid, there are usually various forms of losses. For example, according to the operating conditions, the copper loss or core loss of the generator can reach one-tenth of the power generation [4], which can not be ignored. In addition, the transmission and distribution of power will also cause losses, which will greatly reduce the overall efficiency. Following the symbol of the optimal allocation problem Eq. (1), the optimal allocation problem with separable resource losses is

$$\begin{aligned} \min \ & \sum_{i=1}^{n} f_i(x_i) \\ \text{s.t.} \ & \sum_{i=1}^{n} x_i = D + \Psi(x), \ \ x_i \in \mathcal{X}_i \end{aligned} \tag{2}$$

where $\Psi(x) = \sum_{i=1}^{n} \Psi_i(x_i)$ is the separable resource losses, $\mathcal{X}_i = [x_i^L, x_i^U]$ is the local box constraint for node i, which is defined with the lower bound x_i^L and the upper bound x_i^U.

Assumption 1. The functions in the optimal allocation problem Eq. (2) are defined as $f_i(x_i) = \alpha_i x_i^2 + \beta_i x_i + \gamma_i$ and $\Psi_i(x_i) = a_i x_i^2 + b_i x_i$ with $\alpha_i > 0$, $\beta_i > 0$, and $a_i > 0$ for all $i \in \mathcal{V} := \{1, 2, \cdots, n\}$.

Assumption 2. For all $i \in \mathcal{V}$, it holds $1 - b_i - 2a_i x_i^U > 0$ and the total amount D of resource satisfies $\sum_{i=1}^{n}(x_i^L - \Psi_i(x_i^L)) \leqslant D \leqslant \sum_{i=1}^{n}(x_i^U - \Psi_i(x_i^U))$.

Assumption 3. The communication graph $\mathcal{G}$ of the multi-agent system is undirected and connected.

Remark 1. It should be note that Assumption 1 implies the objective function and loss function of the optimal allocation problem Eq. (2) are both strongly convex. But the optimal allocation problem Eq. (2) is not necessarily a convex optimization problem, because the equality constraint is not affine. Assumption 2 guarantees that the feasible region of the optimal allocation problem Eq. (2) is nonempty and $1 - b_i - 2a_i x_i^U > 0$ follows $1 - \frac{\mathrm{d}\Psi(x)}{\mathrm{d}x_i} > 0$, which refers to the resource losses cannot exceed total resource supply.

3 Main Results

3.1 Approach Description

In view of the nonlinearity of equality constraints, it is impossible to directly use the related theory of convex optimization to solve the optimal solution of the optimal allocation problem Eq. (2). Denote $D = \sum_{i=1}^{n} d_i$, the following conclusion shows that solving the optimal allocation problem Eq. (2) can be transformed into solving a convex optimization problem with global inequality constraints.

Theorem 1. *Under Assumptions 1 and 2, the optimal solution of the following problem*

$$\begin{aligned} &\min\ F(x) = \sum_{i=1}^{n} f_i(x_i) \\ &s.t.\ \sum_{i=1}^{n} (d_i + \Psi_i(x_i) - x_i) \leqslant 0,\ x_i \in \mathcal{X}_i \end{aligned} \tag{3}$$

is the optimal solution of the optimal allocation problem Eq. (2).

Proof. Let $x^* = (x_1^*, x_2^*, \cdots, x_n^*)^{\mathrm{T}}$ be the optimal solution of the problem Eq. (3), then there are optimal Lagrange multipliers $\theta > 0$, $u^* = (u_1^*, u_2^*, \cdots, u_n^*)^{\mathrm{T}}$, and $v^* = (v_1^*, v_2^*, \cdots, v_n^*)^{\mathrm{T}}$ satisfies

$$\begin{aligned} &\nabla f_i(x_i^*) - \theta(1 - \nabla \Psi_i(x_i^*)) + u_i^* - v_i^* = 0 \\ &\theta \sum_{i=1}^{n} (d_i + \Psi_i(x_i^*) - x_i^*) = 0,\ \theta > 0 \\ &\sum_{i=1}^{n} (d_i + \Psi_i(x_i^*) - x_i^*) \leqslant 0 \\ &u_i^*(x_i^* - x_i^U) = 0,\ u_i^* \geqslant 0,\ x_i^* - x_i^U \leqslant 0 \\ &v_i^*(x_i^L - x_i^*) = 0,\ v_i^* \geqslant 0,\ x_i^L - x_i^* \leqslant 0 \end{aligned} \tag{4}$$

for all $i \in \mathcal{V}$.

Next, we prove that $\sum_{i=1}^{n}(d_i + \Psi_i(x_i^*) - x_i^*) = 0$. Suppose that this equation is not hold, then it implies $\theta = 0$ and $\sum_{i=1}^{n}(d_i + \Psi_i(x_i^*) - x_i^*) < 0$ from Eq. (4). According to Assumption 1, one has $\nabla f_i(x_i^*) = 2\alpha_i x_i^* + \beta_i > 0$ for all $i \in \mathcal{V}$. Hence, for any $i \in \mathcal{V}$,

$$v_i^* = \nabla f_i(x_i^*) - \theta(1 - \nabla\Psi_i(x_i^*)) + u_i^* = \nabla f_i(x_i^*) + u_i^* > 0.$$

This reveals that $x_i^* = x_i^L$ and $\sum_{i=1}^{n}(d_i + \Psi_i(x_i^L) - x_i^L) = \sum_{i=1}^{n}(d_i + \Psi_i(x_i^*) - x_i^*) < 0$, which contradicts $\sum_{i=1}^{n} x_i^L - \Psi(x_i^L) \leqslant D \leqslant \sum_{i=1}^{n} x_i^U - \Psi(x_i^U)$ in Assumption 2. It can be obtained that $D = \sum_{i=1}^{n} x_i^* - \Psi(x^*)$, that is, x^* is a feasible solution of the problem Eq. (2). Combining with the definition of x^*, we get that x^* is a optimal solution of problem Eq. (2).

Based on the Lagrangian multiplier method, here comes to a useful lemma.

Lemma 1. *[17] Under Assumptions 1 and 2, $x^* = (x_1^*, x_2^*, \cdots, x_n^*)^{\mathrm{T}}$ is an optimal solution to the problem Eq. (3) if and only if there exists $x^* \in \mathbb{R}^n$ and $y^* \in \mathbb{R}$ such that*

$$\begin{aligned} x^* &= \mathcal{P}_{\mathcal{X}}(x^* - \nabla F(x^*) - \nabla g(x^*) y^*) \\ y^* &= [y^* + g(x^*)]^+ \end{aligned} \tag{5}$$

where $\mathcal{X} = \mathcal{X}_1 \times \cdots \times \mathcal{X}_n$, $\nabla F(x^*) = (\nabla f_1(x_1^*), \nabla f_2(x_2^*), \cdots, \nabla f_n(x_n^*))^{\mathrm{T}}$, $\nabla g(x^*) = (\nabla g_1(x_1^*), \nabla g_2(x_2^*), \cdots, \nabla g_n(x_n^*))^{\mathrm{T}}$, and $g(x^*) = \sum_{i=1}^{n} g_i(x_i^*)$.

To tackle the convex optimization problem Eq. (3), we introduce a multi-agent system composed with n agents to represent the n nodes in the network, and for each agent $i \in \mathcal{V}$, design the following neurodynamic approach

$$\begin{cases} \dot{x}_i(t) = -x_i(t) + P_{\mathcal{X}_i}\Big(x_i(t) - \nabla f_i(x_i(t)) - \nabla g_i(x_i(t))[y_i(t) + z_i(t)]^+\Big) \\ \dot{y}_i(t) = J_i(t) + |J_i(t)|\mathrm{sign}(K_i(t)) - \mathrm{sig}(K_i(t))^{\mu} - \mathrm{sig}(K_i(t))^{\nu} \\ \dot{z}_i(t) = k_1 \sum\limits_{j \in \mathcal{N}_i} \mathrm{sign}\big(z_j(t) - z_i(t)\big) + n\nabla g_i(x_i(t))^{\mathrm{T}} \dot{x}_i(t) \\ J_i(t) = k_2 \sum\limits_{j \in \mathcal{N}_i} \mathrm{sign}\big(y_j(t) - y_i(t)\big) - \dfrac{1}{2} y_i(t) + \dfrac{1}{2}[y_i(t) + z_i(t)]^+ \\ K_i(t) = [y_i(t)]^+ - y_i(t) \end{cases} \tag{6}$$

with initial values $x_i(0) \in \mathcal{X}_i$ and $z_i(0) = ng_i(x_i(0))$, where $g_i(x_i) = d_i + \Psi_i(x_i) - x_i$, $[s]^+ = \max\{s, 0\}$, $\mathrm{sig}(s)^{\mu} = \mathrm{sign}(s)|s|^{\mu}$, and $\mathcal{N}_i$ is the neighbor set of agent i. In the neurodynamic approach Eq. (6), control parameters $k_1 > 0$, $k_2 > 0$, $0 < \mu < 1$, and $\nu > 1$ are usually

Remark 2. It is worth noting that the existence of the solution of the neurodynamic approach Eq. (6) has been discussed in [5,13]. Furthermore, although the total number n of agents is needed in the neurodynamic approach Eq. (6), it can be easily determined distributively by employing the finite-time tracking technique introduced in [14].

3.2 Convergence Analysis

In this section, the finite-time tracking technique and the fixed-time projection method constructed in the neurodynamic approach Eq. (6) are used to make distributed estimates of the required global information, and the finite-time consistency of the auxiliary variables and the convergence of the proposed neurodynamic approach are discussed.

Lemma 2. *Under Assumptions 1–3, starting from $x_i(0) \in \mathcal{X}_i$, there is $T_1 > 0$ such that $y_i(t) \geqslant 0$ and $x_i(t)$ is bounded.*

Proof. **Step 1.** We prove that there is T_1 such that $y_i(t) \geqslant 0$ when $t \geqslant T_1$. Consider the following Lyapunov function $V_1(t) = (y_i - [y_i]^+)^2$. Clearly, the derivative of V_1 along the neurodynamic approach Eq. (6) is

$$\begin{aligned}\dot{V}_1 &= 2(y_i - [y_i]^+)^{\mathrm{T}}\dot{y}_i \\ &= 2(y_i - [y_i]^+)^{\mathrm{T}}J_i(t) - 2|J_i(t)||y_i - [y_i]^+| - 2|y_i - [y_i]^+|^{1+\mu} - 2|y_i - [y_i]^+|^{1+\nu} \\ &\leqslant -2V_1^{\frac{1+\mu}{2}} - 2V_1^{\frac{1+\nu}{2}}.\end{aligned} \tag{7}$$

Therefore, from Lemma 1 in [15], there is $T_1 \leqslant \frac{2}{2(1-\mu)} + \frac{1}{2(\nu-1)}$ such that $y_i(t) \geqslant 0$ when $t \geqslant T_1$.

Step 2. Let us show that $x_i(t) \in \mathcal{X}_i$ for all $t \geqslant 0$. Denote $p_i(t) = \mathcal{P}_{\mathcal{X}_i}(x_i(t) - \nabla f_i(x_i(t)) - \nabla g_i(x_i(t))[y_i(t) + z_i(t)]^+)$, then $p_i(t) \in \mathcal{X}_i$. From Eq. (6), we have

$$x_i(t) = \mathrm{e}^{-t}x_i(0) + (1 - \mathrm{e}^{-t})\int_0^t p_i(s)\frac{\mathrm{e}^s}{\mathrm{e}^t - 1}\mathrm{d}s.$$

Since $\int_0^t \frac{\mathrm{e}^s}{\mathrm{e}^t - 1}\mathrm{d}s = 1$, $x_i(0) \in \mathcal{X}_i$ and $\mathcal{X}_i$ is convex set, it holds that $x_i(t) \in \mathcal{X}_i$, for any $t \geqslant 0$ and $i \in \mathcal{V}$. Hence, $x_i(t)$ is bounded from the boundedness of $\mathcal{X}_i$.

According to the conclusion of Lemma 2, we assume that M satisfies $\|x_i(t)\| \leqslant M$, and then the neurodynamic approach Eq. (6) is reduced as

$$\begin{cases}\dot{x}_i(t) = -x_i(t) + \mathcal{P}_{\mathcal{X}_i}\Big(x_i(t) - \nabla f_i(x_i(t)) - \nabla g_i(x_i(t))[y_i(t) + z_i(t)]^+\Big) \\ \dot{y}_i(t) = k_2\sum\limits_{j\in\mathcal{N}_i}\mathrm{sign}\big(y_j(t) - y_i(t)\big) - \dfrac{1}{2}y_i(t) + \dfrac{1}{2}[y_i(t) + z_i(t)]^+ \\ \dot{z}_i(t) = k_1\sum\limits_{j\in\mathcal{N}_i}\mathrm{sign}\big(z_j(t) - z_i(t)\big) + n\nabla g_i(x_i(t))^{\mathrm{T}}\dot{x}_i(t)\end{cases} \tag{8}$$

when $t \geqslant T_1$. Since $\nabla g_i(x_i)$ $(i \in \mathcal{V})$ are bounded based on Assumption 1, there is $M_1 > M$ such that $n^2\|\nabla g_i(x_i(t))^{\mathrm{T}}\dot{x}_i(t)\| \leqslant M_1$. Thus, from the Lemma 3.2 in [13], if $k_1 > 2M_1$, then we have $T_2 > T_1$ satisfying $z_i(t) = \frac{1}{n}\sum_{j=1}^n z_j(t)$ $\forall i \in \mathcal{V}$ when $t \geqslant T_2$. From $z_i(0) = ng_i(x_i(0))$, it follows that

$$z_i(t) = \sum_{i=1}^n g_i(x_i(t))$$

for all $i \in \mathcal{V}$ when $t \geqslant T_2$. Thus, it is induced that $z_i(t)$ is bounded due to the definition of g_i in Assumption 1. Similarly, we can find $M_2 > 0$ such that $\| -\frac{1}{2}y_i(t) + \frac{1}{2}[y_i(t) + z_i(t)]^+\| \leqslant \frac{1}{2}\|z_i\| \leqslant M_2$ when $t \geqslant T_2$, which implies that if $k_2 > 2M_2$, there exists $T_3 > T_2$ holding

$$y_i(t) = \sum_{j=1}^{n} y_j(t)$$

for all $i \in \mathcal{V}$ when $t \geqslant T_3$.

From Lemma 2 and the above discuss, one can get that the values of the control parameters μ and ν in the neurodynamic approach Eq. (6) determine the upper bound of the time for $y_i(t)$ greater than zero, and the difference of k_1 and k_2 will also change the maximum time for $z_i(t)$ and $y_i(t)$ to reach consensus. Therefore, the value of the control parameters will affect the convergence rate of the neurodynamic approach Eq. (6) to some extent. In practical application, the value of the control parameter is usually selected and adjusted according to the parameter adjustment experience.

Theorem 2. *Under Assumptions 1–3, starting from initial values $x_i(0) \in \mathcal{X}_i$ and $z_i(0) = ng_i(x_i(0))$, if $k_l > 2M_l$ $(l = 1, 2)$, then the trajectories $x_i(t)$ of neurodynamic approach Eq. (6) asymptotically converges to the optimal solution to the problem Eq. (2).*

Proof. Let $x = (x_1, x_2, \cdots, x_n)^{\mathrm{T}}$, $y = \sum_{i=1}^{n} y_i$, and $z = \sum_{i=1}^{n} z_i$, then when $t \geqslant T_3$, $z(t) = g(x(t))$ and the neurodynamic approach Eq. (8) can be rewritten as

$$\begin{cases} \dot{x}(t) = -x(t) + \mathcal{P}_{\mathcal{X}}\Big(x(t) - \nabla F(x(t)) - \nabla g(x(t))[y(t) + g(x(t))]^+\Big) \\ \dot{y}(t) = -\dfrac{1}{2}y(t) + \dfrac{1}{2}[y(t) + g(x(t))]^+ \end{cases} \tag{9}$$

Define $\Theta = \{(x^*, y^*) \in \mathbb{R}^{n+1} : (x^*, y^*) \text{ satisfies } (5)\}$, then according to Lemma 1, $(x^*, y^*) \in \Theta$ is the equilibrium point of neurodynamic approach (9), and x^* is the optimal solution to the problem (3).

Letting $h(x, y) = F(x) + \frac{1}{2}\|[y + g(x)]^+\|^2$, it is obvious that $h(x, y)$ is convex with respect to $(x, y) \in \mathbb{R}^n \times \mathbb{R}$. For $t \geqslant T_3$, consider a Lyapunov function

$$V(x, y) = h(x, y) - h(x^*, y^*) - (x - x^*, y - y^*)^{\mathrm{T}} \nabla h(x^*, y^*) + \frac{1}{2}\|x - x^*\|^2 + \frac{1}{2}\|y - y^*\|^2$$

with $(x^*, y^*) \in \Theta$, then we have $y^* = [y^* + g(x^*)]^+$ and $V(x, y) \geqslant \frac{1}{2}\|x - x^*\|^2 + \frac{1}{2}\|y - y^*\|^2$. According to $x^* = \mathcal{P}_{\mathcal{X}}(x^* - \nabla F(x^*) - \nabla g(x^*)y^*)$, denote $\hat{y} = [y + g(x)]^+$ and $\hat{y}^* = [y^* + g(x^*)]^+$, it gets

$$\begin{aligned} \dot{V} = &-\langle \nabla F(x) + \nabla g(x)\hat{y} - \nabla F(x^*) - \nabla g(x^*)\hat{y}^* + x - x^*, \\ &x - x^* + x^* - \mathcal{P}_{\mathcal{X}}(x - \nabla F(x) - \nabla g(x)\hat{y})\rangle + \frac{1}{2}\langle \hat{y} - y, \hat{y} + y - 2y^*\rangle \\ = &-\langle x - x^*, \nabla F(x) + \nabla g(x)\hat{y} - \nabla F(x^*) - \nabla g(x^*)\hat{y}^*\rangle - \|x - x^*\|^2 \\ &- \langle \nabla F(x) + \nabla g(x)\hat{y} - \nabla F(x^*) - \nabla g(x^*)\hat{y}^* + x - x^* \\ &x^* - \mathcal{P}_{\mathcal{X}}(x - \nabla F(x) - \nabla g(x)\hat{y})\rangle - \frac{1}{2}\|\hat{y} - y\|^2 + \langle \hat{y} - y, \hat{y} - y^*\rangle \end{aligned} \tag{10}$$

Since

$$\begin{aligned}\|x-x^*\|^2 &= \|x-\mathcal{P}_{\mathcal{X}}(x-\nabla F(x)-\nabla g(x)\hat{y})+\mathcal{P}_{\mathcal{X}}(x-\nabla F(x)-\nabla g(x)\hat{y})-x^*\|^2\\ &= \|x-\mathcal{P}_{\mathcal{X}}(x-\nabla F(x)-\nabla g(x)\hat{y})\|^2+\|\mathcal{P}_{\mathcal{X}}(x-\nabla F(x)-\nabla g(x)\hat{y})-x^*\|^2\\ &\quad+2\langle x-\mathcal{P}_{\mathcal{X}}(x-\nabla F(x)-\nabla g(x)\hat{y}),\mathcal{P}_{\mathcal{X}}(x-\nabla F(x)-\nabla g(x)\hat{y})-x^*\rangle\end{aligned} \tag{11}$$

then

$$\begin{aligned}\dot{V} =& -\langle x-x^*,\nabla F(x)+\nabla g(x)\hat{y}-\nabla F(x^*)-\nabla g(x^*)\hat{y}^*\rangle\\ &-\|x-\mathcal{P}_{\mathcal{X}}(x-\nabla F(x)-\nabla g(x)\hat{y})\|^2-\langle x^*-\mathcal{P}_{\mathcal{X}}(x-\nabla F(x)-\nabla g(x)\hat{y}),\\ &\quad\mathcal{P}_{\mathcal{X}}(x-\nabla F(x)-\nabla g(x)\hat{y})-x+\nabla F(x)+\nabla g(x)\hat{y}\rangle\\ &-\langle\mathcal{P}_{\mathcal{X}}(x-\nabla F(x)-\nabla g(x)\hat{y})-x^*,\nabla F(x^*)+\nabla g(x^*)\hat{y}^*\rangle.\end{aligned} \tag{12}$$

From the property of projection operators and x^* is the optimal solution to the problem Eq. (3), we can get

$$\langle x^*-\mathcal{P}_{\mathcal{X}}(x-\nabla F(x)-\nabla g(x)\hat{y}),\mathcal{P}_{\mathcal{X}}(x-\nabla F(x)-\nabla g(x)\hat{y})-x+\nabla F(x)+\nabla g(x)\hat{y}\rangle\geqslant 0.$$

Since $F(x)$ and $g(x)$ are convex, then $\langle\nabla F(x)-\nabla F(x^*),x-x^*\rangle\geqslant 0$ and $\langle\mathcal{P}_{\mathcal{X}}(x-\nabla F(x)-\nabla g(x)\hat{y})-x^*,\nabla F(x^*)+\nabla g(x^*)\hat{y}^*\rangle\geqslant 0$. Thus,

$$\begin{aligned}&\langle x-x^*,\nabla F(x)+\nabla g(x)\hat{y}-\nabla F(x^*)-\nabla g(x^*)\hat{y}^*\rangle\\ \geqslant&\langle x-x^*,\nabla g(x)\hat{y}-\nabla g(x^*)\hat{y}^*\rangle\\ =&\langle x-x^*,\nabla g(x)\hat{y}-\nabla g(x^*)y^*\rangle.\end{aligned} \tag{13}$$

Furthermore, from the definition of $\hat{y}$, it has $y-\hat{y}=[y+g(x)]^- -g(x)$. Due to $\hat{y}[y+g(x)]^-=0$, $\hat{y}g(x^*)\leqslant 0$, $\hat{y}^*[y+g(x)]^-\leqslant 0$, and $\hat{y}^*g(x^*)=0$, then

$$\begin{aligned}&-\langle x-x^*,\nabla g(x)\hat{y}-\nabla g(x^*)y^*\rangle-\langle y-\hat{y},\hat{y}-y^*\rangle\\ =&-(\nabla g(x)\hat{y})^{\mathrm{T}}(x-x^*)+(\nabla g(x^*)y^*)^{\mathrm{T}}(x-x^*)\\ &-\hat{y}([y+g(x)]^- -g(x))+\hat{y}^*([y+g(x)]^- -g(x))\\ =&-\hat{y}([y+g(x)]^- -g(x)+\nabla g(x)^{\mathrm{T}}(x-x^*))\\ &+y^*([y+g(x)]^- -g(x)+\nabla g(x^*)^{\mathrm{T}}(x-x^*))\leqslant 0.\end{aligned} \tag{14}$$

Therefore, by combining the above inequalities, it has

$$\dot{V}\leqslant-\|x-\mathcal{P}_{\mathcal{X}}(x-\nabla F(x)-\nabla g(x)\hat{y})\|^2-\|\hat{y}-y\|^2=-\|\dot{x}\|^2-2\|\dot{y}\|^2\leqslant 0. \tag{15}$$

Next, we show that there is an increasing sequence $\{t_k\}$ such that

$$\lim_{k\to\infty}\|\dot{x}(t_k)\|^2+2\|\dot{y}(t_k)\|^2=0.$$

If not, there exists $r>0$ satisfying $\liminf_{t\to\infty}\|\dot{x}(t_k)\|^2+2\|\dot{y}(t_k)\|^2=r$, which means that there is $T>T_3$ such that $\|\dot{x}(t_k)\|^2+2\|\dot{y}(t_k)\|^2\geqslant\frac{r}{2}$ for all $t\geqslant T$. As a result, it can obtain that $\dot{V}(x,y)\leqslant-\frac{r}{2}$, for all $t\geqslant T$. Integrating the formula, one can obtain that $V(x(t), y(t)) \leqslant V(x(T), y(T)) - \frac{r}{2}(t - T)$. Then we have $\lim_{t\to\infty} V(x(t), y(t)) = -\infty$. This contradicts the fact that $V \geqslant 0$. Thus, we have

$$\lim_{k\to\infty} \|x(t_k) - \mathcal{P}_{\mathcal{X}}(x(t_k) - \nabla F(x(t_k)) - \nabla g(x(t_k))\hat{y}(t_k))\| = 0,$$

$$\lim_{k\to\infty} \|\hat{y}(t_k) - y(t_k)\| = 0.$$

According to Eq. (15) and $V(x, y) \geqslant \frac{1}{2}\|x - x^*\|^2 + \|y - y^*\|^2$, it deduces that $x(t)$ and $y(t)$ are bounded. In addition, from Lemma 1, there is a convergent subsequence (still denoted) $\{t_k\}$, and there are $(x_0, y_0) \in \Theta$ such that $\lim_{k\to\infty} x(t_k) = x_0$ and $\lim_{k\to\infty} y(t_k) = y_0$. Letting $x^* = x_0$ and $y^* = y_0$, similar to above analysis, we have

$$\lim_{t\to\infty} V(x(t), y(t)) = V(x^*, y^*) = 0,$$

which follows $\lim_{t\to\infty} \|x(t) - x^*\| = 0$ and $\lim_{t\to\infty} \|y(t) - y^*\| = 0$. Hence, the trajectories $x_i(t)$ of neurodynamic approach Eq. (6) asymptotically converges to the optimal solution to the problem Eq. (3). According to Theorem 1, the conclusion is obviously valid.

4 Simulation Studies

In this section, we display the effectiveness of the neurodynamic approach Eq. (6) with twelve-agent system for the optimal allocation problem Eq. (2). The communication network of the multi-agent system is a undirected connected graph.

Table 1. Variables of the optimal allocation problem Eq. (2).

agent i	G_1	G_2	G_3	G_4	G_5	G_6	G_7	G_8	G_9	G_{10}	G_{11}	G_{12}
α_i	0.4	1.2	3.2	2.8	0.4	0.4	0.4	0	1.6	2.8	2.8	2
β_i	0.3	1.8	0.3	0.3	0.3	0.3	0.6	0.6	0.9	0.9	0.6	0.9
γ_i	1.8	2.1	1	3.6	2.6	2.2	3	4	1	4	3.7	2
a_i	0.1	0.24	0.07	0.06	0.8	0.14	0.02	0.11	0.08	0.2	0.04	0.06
b_i	-0.05	-0.05	-0.03	-0.01	-0.23	-0.12	-0.11	-0.07	-0.07	-0.1	-0.1	-0.2
x_i^L	0	0	0	0	0	0	0	0	0	0	0	0
x_i^U	1	1	1	2	2	2	3	3	3	4	4	4

To achieve the optimal allocation problem Eq. (2) with $D = 5.5$ and the parameters in Table 1 in this simulation, we apply the neurodynamic approach Eq. (6) by taking $k_1 = 1500$, $k_2 = 1250$, $\mu = 0.5$, and $\nu = 2$. The trajectories of $x_i(t)$ $(i = 1, 2, \cdots, 12)$ are shown in Fig. 1. Visibly, Fig. 1 illustrates that the neurodynamic approach Eq. (6) is able to get the optimal solution

$$x^* = [0.85, 0, 0.13, 0.14, 0.42, 0.85, 0.76, 2.42, 0.09, 0.05, 0.11, 0.11]$$

of the problem Eq. (3) and shows the convergence of the neurodynamic approach Eq. (6). It can be seen from Fig. 2 that the states $y_i(t)$ and $z_i(t)$ reach consensus within a finite time, which is consistent with the theoretical results in this paper. Furthermore, Fig. 3(a) describes the evaluations of the total cost value, and Fig. 3(b) shows that the global equation constraint of the problem Eq. (3) with resource losses can be satisfied. To sum up, these numerical results verify that the neurodynamic approach designed in this paper is effective for solving the optimal allocation problem with resource losses.

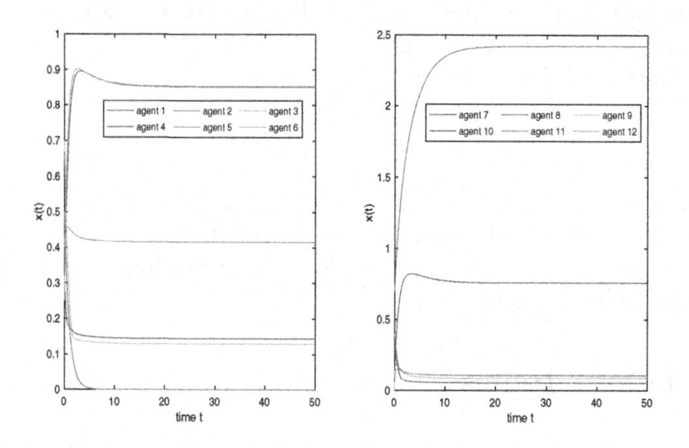

Fig. 1. The trajectories of x_i generated by neurodynamic approach Eq. (6).

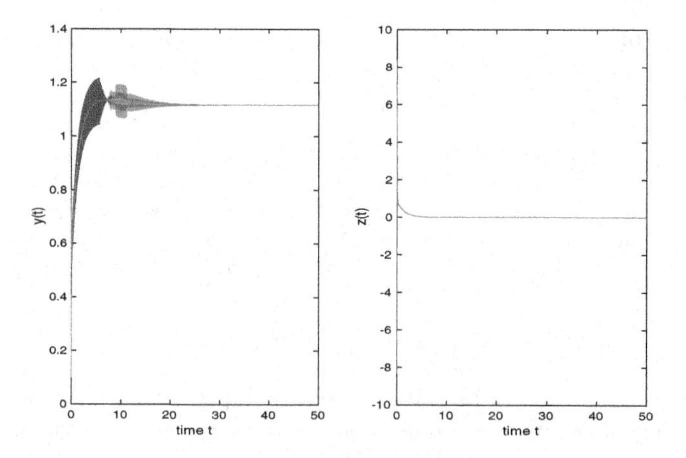

Fig. 2. The trajectories of y_i and z_i generated by neurodynamic approach Eq. (6).

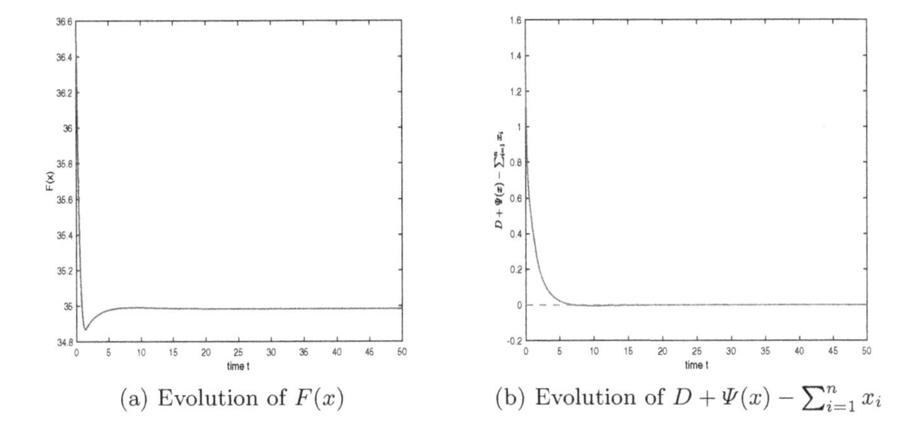

(a) Evolution of $F(x)$ (b) Evolution of $D + \Psi(x) - \sum_{i=1}^{n} x_i$

Fig. 3. Evolution of total cost and global constraint with losses in the optimal allocation problem Eq. (2).

5 Conclusion

In this paper, we investigated the distributed optimal allocation problem with separable resource losses. Through applying finite-time tracking technology and the properties of the projection operator, a distributed neurodynamic approach based on multi-agent system was designed and analyzed. Moreover, we showed that the states of the proposed neurodynamic approach can converge to the optimal solution of the considered problem both theoretically and numerically.

Acknowledgements. This paper is supported by the National Natural Science Foundation of China (No. 62176073, No. 12271127), Taishan Scholars of Shandong Province (No. tsqn202211090), and in part by the Natural Scientific Research Innovation Foundation in Harbin Institute of Technology.

References

1. Chen, G., Li, Z.: A fixed-time convergent algorithm for distributed convex optimization in multi-agent systems. Automatica **95**, 539–543 (2018)
2. Chen, G., Li, Z.: Distributed optimal resource allocation over strongly connected digraphs: a surplus-based approach. Automatica **125**, 109459 (2021)
3. Deng, Z., Liang, S., Hong, Y.: Distributed continuous-time algorithms for resource allocation problems over weight-balanced digraphs. IEEE Trans. Cybern. **48**(11), 3116–3125 (2018)
4. Dutta, R., Chong, L., Rahman, F.M.: Analysis and experimental verification of losses in a concentrated wound interior permanentmagnet machine. Progr. Electromagnet. Res. **48**, 221–248 (2013)
5. Filippov, A.F.: Differential Equations With Discontinuous Right-Hand Sides. Kluwer Academic, The Netherlands (1988)

6. Jia, W., Liu, N., Qin, S.: An adaptive continuous-time algorithm for nonsmooth convex resource allocation optimization. IEEE Trans. Autom. Control **67**(11), 6038–6044 (2022)
7. Knauer, U., Knauer, K.: Algebraic Graph Theory. De Gruyter (2011)
8. Li, C., Yu, X., Yu, W., Huang, T., Liu, Z.: Distributed event-triggered scheme for economic dispatch in smart grids. IEEE Trans. Industr. Inf. **12**(5), 1775–1785 (2016)
9. Li, Q., Liu, Y., Zhu, L.: Neural network for nonsmooth pseudoconvex optimization with general constraints. Neurocomputing **131**, 336–347 (2014)
10. Lian, M., Guo, Z., Wang, X., Wen, S., Huang, T.: Adaptive exact penalty design for optimal resource allocation. IEEE Trans. Neural Netw. Learn. Syst. **34**(3), 1430–1438 (2023)
11. Liang, S., Zeng, X., Hong, Y.: Distributed nonsmooth optimization with coupled inequality constraints via modified Lagrangian function. IEEE Trans. Autom. Control **63**(6), 1753–1759 (2017)
12. Liu, S., Qiu, Z., Xie, L.: Continuous-time distributed convex optimization with set constraints. IFAC Proc. Volumes **47**(3), 9762–9767 (2014)
13. Luan, L., Li, H., Qin, S.: Neurodynamic approaches to multiple constrained distributed resource allocation with planned or self-regulated demand. IEEE Trans. Industr. Inf. (2023). https://doi.org/10.1109/TII.2023.3262340
14. Martin, S., Hendrickx, J.M.: Continuous-time consensus under non-instantaneous reciprocity. IEEE Trans. Autom. Control **61**(9), 2484–2495 (2016)
15. Polyakov, A.: Nonlinear feedback design for fixed-time stabilization of linear control systems. IEEE Trans. Autom. Control **57**(8), 2106–2110 (2012)
16. Qin, Y., et al.: Optimal operation of integrated energy systems subject to coupled demand constraints of electricity and natural gas. CSEE J. Power Energy Syst. **6**, 444–457 (2020)
17. Ruszczyński, A.: Nonlinear Optimization. Princeton University Press, New Jersey (2006)
18. Wei, B., Ma, L., Qin, S., Xue, X.: Neural network for nonsmooth pseudoconvex optimization with general convex constraints. Neural Netw. **101**, 1–14 (2018)
19. Yi, P., Hong, Y., Liu, F.: Initialization-free distributed algorithms for optimal resource allocation with feasibility constraints and its application to economic dispatch of power systems. Automatica **74**, 259–269 (2016)
20. Zhang, Y., Lou, Y., Hong, Y., Xie, L.: Distributed projection based algorithms for source localization in wireless sensor networks. IEEE Trans. Wireless Commun. **14**(6), 3131–3142 (2015)

Multimodal Isotropic Neural Architecture with Patch Embedding

Hubert Truchan, Evgenii Naumov, Rezaul Abedin, Gregory Palmer, and Zahra Ahmadi(✉)

L3S Research Center, Leibniz University Hannover, Hannover, Germany
{truchan,naumov,abedin,gpalmer,ahmadi}@L3S.de

Abstract. Patch embedding has been a significant advancement in Transformer-based models, particularly the Vision Transformer (ViT), as it enables handling larger image sizes and mitigating the quadratic runtime of self-attention layers in Transformers. Moreover, it allows for capturing global dependencies and relationships between patches, enhancing effective image understanding and analysis. However, it is important to acknowledge that Convolutional Neural Networks (CNNs) continue to excel in scenarios with limited data availability. Their efficiency in terms of memory usage and latency makes them particularly suitable for deployment on edge devices. Expanding upon this, we propose Minape, a novel multimodal isotropic convolutional neural architecture that incorporates patch embedding to both time series and image data for classification purposes. By employing isotropic models, Minape addresses the challenges posed by varying data sizes and complexities of the data. It groups samples based on modality type, creating two-dimensional representations that undergo linear embedding before being processed by a scalable isotropic convolutional network architecture. The outputs of these pathways are merged and fed to a temporal classifier. Experimental results demonstrate that Minape significantly outperforms existing approaches in terms of accuracy while requiring fewer than 1M parameters and occupying less than 12 MB in size. This performance was observed on multimodal benchmark datasets and the authors' newly collected multi-dimensional multimodal dataset, Mudestreda, obtained from real industrial processing devices[1]([1]Link to code and dataset: https://github.com/hubtru/Minape).

Keywords: Multimodal Classification · Isotropic Architecture · Patch Embedding · Time Series

1 Introduction

Humans perceive the world through multiple senses, such as vision and hearing, leading to a multimodal understanding. This combination of data from different modalities offers augmented and complementary information, enabling more robust inference. In recent years, deep learning approaches have made remarkable progress in leveraging data from various modalities, resulting in improved performance across classical problems, including action recognition [14] and semantic

B. Luo et al. (Eds.): ICONIP 2023, LNCS 14447, pp. 173–187, 2024.
https://doi.org/10.1007/978-981-99-8079-6_14

segmentation [1]. Despite these advancements, effectively integrating information from multiple modalities remains a fundamental challenge in multimodal learning. Several research efforts have focused on designing fusion paradigms to fuse multimodal data [12,33]. However, these approaches often require manual design and are specific to certain tasks and modalities, primarily focusing on image and text as the common modalities [31]. Yet, in applications such as intelligent production and healthcare, time series and audio data serve as the major source of information. For instance, device state recognition is a vital problem in intelligent production and healthcare systems, demanding precise monitoring due to the dynamic nature of the device's physical and environmental properties. Maintenance reports show that these devices exhibit symptoms of changing conditions and damage before experiencing complete failure during operation. These symptoms, including distinctive sounds, can serve as indicators for identifying the state of the device. Traditional methods like single modality approaches [22] or self-supervised [18] are impractical in these scenarios due to the complexity and variability of data. Such applications often suffer from noise and conflicts between modalities, which can significantly impact prediction accuracy. Moreover, the limited availability of training samples further complicates the development of efficient models. An ideal algorithm should address these challenges by being robust to noise, selectively leveraging strong modalities, and effectively capturing complementary information among modalities.

In this paper, we present **(Minape)**, a novel multimodal isotropic neural architecture with patch embedding that integrates time series and image data as input. Minape is based on convolutional neural networks (CNNs) and uses a patch-based representation to learn local features from the data. One key advantage of Minape is its isotropic nature, which allows it to handle inputs of varying sizes and aspect ratios and to be trained with fewer labelled examples compared to existing methods. Minape can be employed in numerous applications, including healthcare systems, intelligent production, and real-time monitoring of device states. Through extensive experiments, we demonstrate that Minape achieves significantly higher accuracy compared to the state-of-the-art approaches on both public multimodal classification datasets and our newly introduced multimodal device state dataset (**Mudestreda**) collected from industrial processing devices. In addition, Minape requires significantly less memory to store its much smaller model and can be trained on small and medium size multimodal datasets. Overall, the contributions of this paper are as follows:

- The Minape framework, a novel multimodal isotropic convolutional architecture with patch embedding that requires less than 12 MB size and less than 1M Parameters, which yields the inference output over 90 images/s (results reported for Nvidia 1080TI) for audiovisual data,
- An alternate solution to transformer-based fusion models that can be effectively trained on the small and medium size multimodal datasets with less than 1k instances and capable of operating on edge devices and being deployed in real standalone systems,

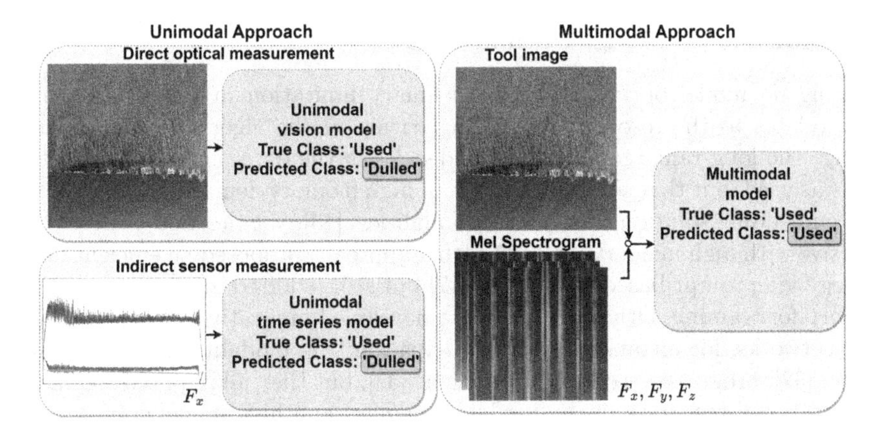

Fig. 1. A sample instance from Mudestreda that shows the importance of leveraging multimodal data for device state prediction.

- Minape yields significantly *higher accuracy* compared to the state-of-the-art approaches. Figure 1 intuitively demonstrates that this augmentation leads to a better predictive performance by an example from Mudestreda, emphasizing its real-world application readiness,
- Minape exhibits *scalability* concerning model pathways and depth dimension. It permits flexible model scaling based on data size and task complexity,
- Finally, we share our collected data, Mudestreda, which is publicly available and aims for multimodal device state recognition.

2 Related Work

2.1 Multimodal Fusion

The focus of multimodal fusion has been primarily on exploring different architectures and techniques for effectively integrating different data modalities to improve the performance of the model. In particular, an Action Segmentation model (ASPnet) based on ASFormer disentangles hidden features into modality-shared components and projects them into independent FC layers [1]. Similarly, a Transformer-based approach [13] creates a single shared representation space using multiple types of image-paired data. [24] also utilized a Transformer-based architecture to leverage audiovisual context in action localization. Other works introduced diverse multimodal fusion strategies involving semantic fusion [31], temporal sequence-based fusion [36], and cross-attention mechanism [18]. Our approach simplifies the fusion process by using isotropic convolutional neural architectures, making it computationally efficient and suitable for smaller datasets.

2.2 Multimodal Model Optimisation

Focusing on model optimization, the primary limitation in the existing literature relates to the resource-demanding nature of the algorithms. In order to manage the long-range temporal dependencies in the data, an attention bottleneck was used, but that solution resulted in high memory demand [1]. Similarly, a joint embedding solution reduced the modalities [13], yet the model is resource-intensive with high memory usage. Another approach proposed an efficient fusion strategy via prompt-based techniques [17] but still required more than 12 GB of memory for training. Other methods also developed innovative ways to optimize their networks, for instance, feature anticipation [36], modality-wise L_2 normalization [32], progressive reduction of tokens [37], but they all required significant computational resources, including large memory and high model size.

Additionally, recent works applied SWIN-transformer [22], TimeSformer [18], VGG-M [27], 3D-Resnet-18 [26] and all reported models with parameters exceeding tens of millions, far above the size of our proposed model. Similarly, recent research showcased innovative techniques with pyramid cross-fusion Transformer [35], contrastive-based alignment training [14], modality dropout training [12], single channel version of ViT [14], knowledge distillation [2], cross-modal prototypical loss [33] and injecting trainable parameters into a frozen ViT [19], but these methods were computationally expensive and required large-scale multimodal datasets for pre-training. Furthermore, recent works presented history-aware [5] and weakly-supervised parsing [20], but these are not feasible for small-to medium-sized multimodal datasets. In contrast, our work proposes a novel, lightweight isotropic convolutional architecture that performs well on limited datasets, requiring less than 1M parameters and less than 12 MB size which is 24M parameter and 88 MB less than the lightest available model.

3 Minape: A Multimodal Isotropic Neural Framework

We consider a target process that generates a sequence of M-modal multi-dimensional data points, $X(1), X(2), \ldots, X(l)$, consisting of time series and images and expressed as the set of tensors $T_{1,1}, T_{1,2}, ..., T_{L,M}$, where $X \in \mathbb{R}^D$ and the mth tensor is $D_{l,m}^{h \times b \times d}$ dimensional in $\mathbb{R}^D$, where $h \times b \times d$ describes tensor dimension, as height, bright and depth respectively, M is the number of auxiliary sensors, and l is the number of data points. Tensors are grouped into g groups according to their modality, where $T_{i,G_1}, i = 1..L, G_1 \in M$ is the set of the tensors belonging to modality type one of the size g_1.

Our objective is to train a classifier $f : \mathbb{R}^D \rightarrow y$ where y is the class of data point $X(i)$. This is measured by sparse categorical cross-entropy in our experiments. We assume the samples are drawn from an ergodic and stationary process. Multiple sensors are grouped in a cluster to reduce the Wasserstein distance between multimodal feature distributions [34].

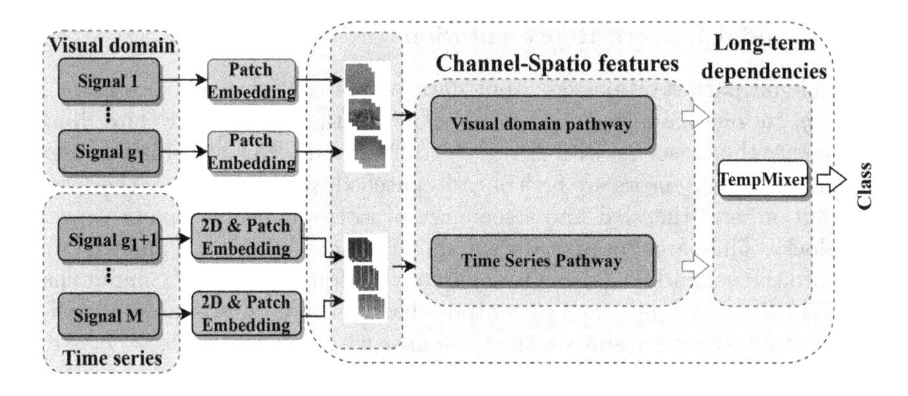

Fig. 2. The Minape general architecture.

To benefit from the synergy of the available M multimodal tensors and enhance the manifolds mapping abilities, we propose deep concatenation of the latent spaces of the linearly embedded two-dimensional representation of input samples from $f_{1:g}$ classifiers to train a classifier $f' : \mathbb{R}^{D_u} \rightarrow y$, where D_u represents the dimension of the deep concatenation layer, and g is the number of modalities of the same type.

3.1 The Minape Architecture Overview

Minape (Fig. 2) is a two-pathway convolutional-based model with visual and time-series feature extractions to obtain a joint representation of the multimodal data. The shared multimodal representation is then passed to the recurrent neural layer to capture the long-term dependencies of the data sequence. The data points comprise M tensors, which are initially grouped into g groups based on their modality type and input into representative $f_i, i = (1 \dots g)$ subnetworks that share parameters across tensors in the same modality group. Then, the learned representations of modalities $u_i, i = (1 \dots g)$ are merged using the concatenation layer, resulting in vector u:

$$u = [f_1(H_1), \dots f_i(H_i) \dots f_g(H_g)], f_i : \mathbb{R}^{D_{H_i}} \rightarrow \mathbb{R}^{D_{u_i}} (i = 1 \dots g), u \in \mathbb{R}^{D_u}. \tag{1}$$

The merged representation, u, serves as the input of the TempMixer block, representing the temporal convolutional network to model the temporal dependencies. This is followed by fully connected and softmax layers:

$$u' = f_{temp}(u), \; f_{temp} : \mathbb{R}^{D_u} \rightarrow \mathbb{R}^{D'_u},$$

$$v = W \cdot u' + b, P(y(i)) = \frac{e^{v_i}}{\sum_{j=1}^{y} e^{v_j}} \; for \; i = 1, 2, \dots, y. \tag{2}$$

3.2 Unimodal Feature Representation

The unimodal pathway takes as input data points consisting of g_v tensors T_j, $j = 1 \ldots g_v$, of one modality, and stacks them together along their third dimension to create the tensor T_v with $D^{h_v \times b_v \times d_v}$ dimensions. To assure high-precision representation, the tensors are first linearly patched, which decreases their internal resolution, and then fed into a sequence of isotropic channel-point convolutional blocks. The patching procedure and linear embedding are seamlessly integrated into the network [30] using a single convolution layer with d_{g_v} input channels, kernel size k, stride p, and c_{g_v} output channels. This step enables the effective representation and results in the Q_v tensor with the depth of $c_v = g_v \times d_{g_v}$:

$$Q_v = BN(\sigma\{Conv(X_{i,j}, stride = p, kernelSize = k)\}) : D^{h_{g_v}/p \times b_{g_v}/p \times c_v}. \quad (3)$$

The downsampled input is then processed by the Gaussian Error Linear Unit (σ) activation function, which applies nonlinearity by weighting the input by their percentile and can be considered as a smoothed version of the ReLU activation function. The fast version of the GELU [15] is used, which exhibits a sufficient trade-off between the latency and the degree of approximation:

$$H_v = BN(Q_v \cdot \frac{1}{2}[1 + erf(\frac{Q_v}{\sqrt{2}}]). \quad (4)$$

The activation is followed by a batch normalization (BN) layer, which normalizes each channel across mini-batch samples, helping to decrease the susceptibility to variations throughout the data. The channel-point convolutional block consists of the repeated sequence of grouped convolutions and pointwise convolutions with the skip connection that fosters information propagation.

The outcome of the convolution process of each group is combined independently as the dot product of the input and the filters sliding vertically and horizontally across the input field and a sum of the bias term. The grouped convolution with groups equal to the number of channels (c_v) is used. Thus, it allows for a channel-wise separable (depth-wise separable) convolution:

$$H_v^{'} = BN(\sigma\{DepthConv(H_v), groups = c_v\}) + H_v. \quad (5)$$

The depthwise convolution is followed by pointwise convolution, defined as the convolution layer with a kernel size of 1×1 and a number of filters equal to the number of channels output by the patch embedding block that allows for linear combinations across channels:

$$H_{v+1} = BN(\sigma\{PointConv(H_v^{'}), kenel = 1 \times 1\}). \quad (6)$$

The last layer is a two-dimensional global average pooling layer that performs downsampling by calculating the average of the vertical and horizontal dimensions of the input volume.

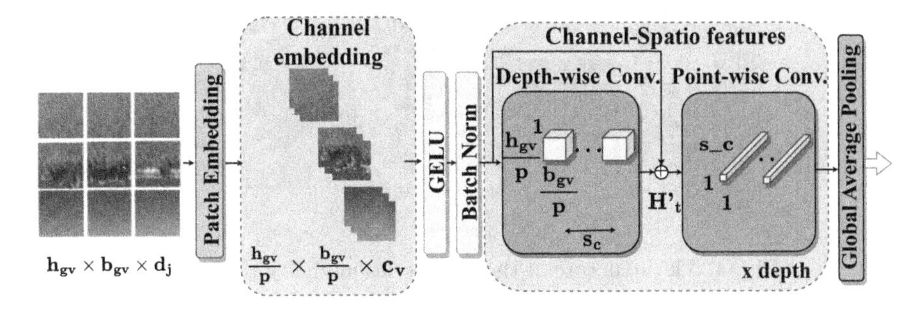

Fig. 3. The structure of the visual modality pathway using patch embeddings.

Visual Domain Pathway: In our architecture, the first pathway groups the tensors in the visual domain into G_v and normalizes their spatial dimensions to meet the subnetwork dimension requirements. These tensors are stacked together along their third dimension to create the tensor T_v with the depth of d_{g_v}:

$$\forall i = 1 \dots L, \forall j \in G_v \quad norm_v(T_{i,j}) : D^{h_j \times b_j \times d_j} \rightarrow D^{h_{g_v} \times b_{g_v} \times d_{g_v}}. \tag{7}$$

Following the above procedure, as depicted in Fig. 3, we patch and linearly embed the input tensor T_v using a single convolution layer. The outcome of Eq. 3 is then normalized using the fast version of GELU, followed by batch normalization as described in Eq. 4. The resulting tensor H_v serves as input to the sequence of isotropic depth-wise and point-wise convolutions that perform the operations in Eqs. 5 and 6, respectively. The outcome of the visual domain pathway is the vector u_v. The adjustable depth of the isotropic architecture ensures scalability according to the task complexity.

Time Series Pathway: The second pathway (Fig. 4) groups time series tensors into G_s, transforms them into two-dimensional embeddings, normalizes and spatially aligns them to ensure temporal correlations, applies patch embedding, and extracts signal representations with the channel-point convolutions block. The time series sequences of tensor T_j are first resampled to n_{g_s}, and the dimension h_j is trimmed or padded to the h_{g_s} length:

$$\forall i = 1 \dots L, \forall j \in G_s \quad norm_s(T_{i,j}) : D^{h_j \times b_j \times d_j} \rightarrow D^{h_{g_s} \times b_j}. \tag{8}$$

The two-dimensional embedding is defined as the logarithmically scaled amplitudes of the Mel spectrogram, which performs a windowing transformation of the signal to create a local frequency analysis.

First, the window used for the STFT calculation with a size of $n_{stft} \leq n_{g_s}$ is defined. Due to zero padding and varying amplitudes of T_s, the Hann windowing function is used to normalize the signals. The Mel filter bank is defined as $mel(sr, n_{stft}, n_{mels})$, where sr is the sampling rate of the incoming signal, n_{stft} is the length of the STFT window, and n_{mels} is the number of Mel bands to generate. The output is a two-dimensional array $\mathbb{R}^{n_{mels} \times (1+n_s tft/2)}$.

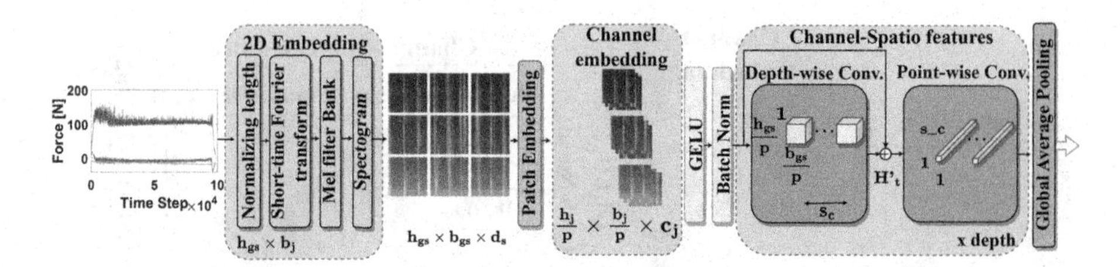

Fig. 4. The structure of the time series modality pathway.

Table 1. Summary of the characteristics of multimodal datasets.

Dataset	Task	Modalities	Types of Modalities	Instances	Classes
Visuo-Haptic [4]	Object recognition	4	Image, Pressure, Texture, Proximity	305	63
MeX [29]	Activity recognition	4	Image, Acceleration, Proximity, Pressure	710	7
BAUM-1a [10]	Emotion recognition	2	Image, Audio	273	9
emoFBVPs [25]	Emotion recognition	2	Image, Audio	1380	23
HA4M [7]	Assembly task	6	RGB images, Depth maps, IR images, RGB-to-Depth-Alignments, Skeleton data	2604	12
Energy [11]	Consumption classification	5	photoplethysmography (PPG), electrocardiography (ECG), Accelerometer, a fraction of oxygen in expired breath (VO2), Gyroscope	1192	3
Mudestreda	Device State recognition	4	Image, Dynamometer data	512	3

The final two-dimensional embedding is obtained using the following formula:

$$2D_{emb} = log_{10}\{mel(STFT(Input_{seq}, stride, hann(n_{stft})))\}, \quad (9)$$

where $hann(n_{stft})$ refers to the Hanning window function applied with a window length of n_{stft} for the short-time Fourier transformation ($stft$). The stride size determines the sliding window step used for STFT, and the Mel filterbank downsamples the signal to obtain the Mel spectrogram. The obtained two-dimensional representations of tensors grouped in G_s are stacked, resulting in the tensor T_s with $d_s = \sum_{j \in G_s} b_j$ channels:

$$\forall j \in G_s, \forall k = 1 \ldots b_j \;\; T_s = 2D_{emb}[T_{:,j}(:, k)] : D^{h_{g_s} \times b_{g_s} \times d_s}. \quad (10)$$

Next, the spectrograms are patched and linearly embedded using Eq. 3, followed by the channel-point convolutional block in Eqs. 5 and 6.

4 Experiments

4.1 Experimental Setup

We test the performance of the Minape framework on several open-source multimodal benchmarks (Table 1)[1]. All of them contain visual and time series data: Visuo-Haptic object recognition for robots contains time series (pressure, texture, proximity) and images, MeX human activity recognition contains time series data (acceleration, proximity, pressure) and images, BAUM-1a face mimic recognition and emoFBVPs physiological signals recognition contain images and audio signals, HA4M assembly task recognition contains images (RGB, depth maps, infra-red images, RGB-to-Depth alignments) and time series data (skeleton data), and Energy consumption estimation contains images (photoplethysmography) and time series data (electrocardiography, accelerometer, fraction of oxygen in expired breath, gyroscope).

We present two versions of multimodal fusion neural network architecture: one that uses isotropic neural architecture with patch embedding **Minape**, and the second which uses patch embedding with efficientnetv2-m for intermediate feature representation (**Minape-S**). The TempMixer block consists of four one-dimensional convolution layers with 64 filters each. The model is trained according to an improved training procedure presented in timm [28] that combines best practices for training, such as novel optimization and data augmentation[2]. We have enriched this procedure with the AdamW [23] optimizer and AutoAugment [8]. The results of Minape and Minape-S are reported for depth size = 16, 3×10^3 iterations, learning rate = 0.001, kernel size = 13, embedding dimension = 1, *globalAveragePooling* layer, and the GELU activation function.

In addition, we varied the batch size, number of epochs, steps per epoch, the pooling layer (globalMaxPooling2dLayer, globalAveragePooling), and the activation function (GELU, ReLu). We compare Minape with two well-established unimodal algorithms, two recent multimodal versions, and transformers-based architectures. The selected algorithms allow for testing a broad range of fusion methods, e.g., multiplicative combination, embracement, averaging, concatenation, and fusion methods:

- The Temporal Convolutional Network (TCN)[3] leverages dilated causal convolutions and residual blocks for efficiently processing long input sequences and learning complex temporal patterns [3].
- EfficientNetv2-m is a convolution-based image learning method that combines training-aware neural architecture search and scaling, which adaptively adjusts regularization along with image size.
- Multiplicative Multimodal network (Mulmix)[4] uses multiplicative modality mixture combination by first additively creating mixture candidates and then

[1] All experiments were performed on a single Nvidia GTX1080Ti 12 GB GPU.
[2] https://github.com/martinsbruveris/tensorflow-image-models.
[3] https://github.com/locuslab/TCN.
[4] https://github.com/skywaLKer518/MultiplicativeMultimodal.

Table 2. The average accuracy percentage of our proposed framework (Minape-S and Minape with primary learner) compared to other state-of-the-art methods on seven multimodal benchmarks. The results are calculated on five trials and the ranks are reported.

Dataset	TCN*	EfficientNetv2-m**	Mulmix	EmbraceNet	ViT	Minape-S	Minape
Visuo-Haptic	71.9 ± 2.0 (5)	70.2 ± 3.8 (6)	67.6 ± 2.2 (7)	73.3 ± 3.1 (4)	78.1 ± 2.7 (3)	80.7 ± 1.7 (2)	**85.4** ± 2.3 **(1)**
MeX	72.9 ± 3.7 (7)	85.0 ± 4.3 (5)	81.5 ± 3.5 (6)	90.8 ± 3.1 (2)	87.7 ± 3.2 (3)	86.2 ± 4.8 (4)	**93.4** ± 1.9 **(1)**
BAUM-1a	31.3 ± 5.5 (7)	41.8 ± 5.1 (6)	44.0 ± 3.5 (5)	48.2 ± 4.5 (3)	45.6 ± 2.9 (4)	51.5 ± 2.5 (2)	**57.7** ± 3.2 **(1)**
emoFBVPs	81.8 ± 7.2 (7)	83.2 ± 8.0 (6)	86.6 ± 6.0 (4)	83.4 ± 5.8 (5)	89.3 ± 1.8 (2)	88.1 ± 3.3 (3)	**92.7** ± 3.1 **(1)**
HA4M	57.5 ± 9.0 (7)	61.3 ± 6.5 (6)	65.1 ± 4.8 (5)	68.9 ± 6.2 (4)	74.8 ± 1.4 (2)	72.7 ± 3.7 (3)	**76.5** ± 2.9 **(1)**
Energy	65.2 ± 6.1 (7)	67.8 ± 6.6 (6)	72.3 ± 5.7 (5)	74.1 ± 3.9 (4)	77.3 ± 2.7 (3)	78.9 ± 2.1 (2)	**81.6** ± 1.8 **(1)**
Mudestreda	71.4 ± 3.5 (7)	79.9 ± 2.8 (6)	87.1 ± 2.5 (5)	91.3 ± 2.0 (4)	93.7 ± 2.3 (3)	94.7 ± 1.5 (2)	**98.2** ± 1.0 **(1)**
Average rank	6.7	5.9	5.3	3.7	2.9	2.6	1.0

* - trained only on time series, ** - trained only on the visual modality.

selecting useful modality mixtures with multiplicative combination procedure [21].
- EmbraceNet[5] is a deep learning method that leverages cross-modal correlations during the training phases [6]. It uses the *embracement* process to probabilistically select subsets of information from each modality to model the inter-modal correlations.
- Visual Transformer (ViT)[6] uses the Transformer layer with self-attention applied to image patch sequences [9].

4.2 Experimental Results

For increased reliability, we performed five independent trials using ten-fold cross-validation by partitioning each dataset into training (80%), validation (10%), and testing (10%) subsets. The outcomes of this comprehensive comparison are detailed in Table 2. For each method, the result of the best hyperparameter setting is reported. All results are in terms of accuracy percentage, and the mean and standard deviation of ten repeats are reported.

The tested datasets have varied sensor sequence lengths ranging from 276 in Visuo-Haptic to 1.5×10^5 in Mudestreda. We change the depth of the Minape architectures from four with a width of 128 in Visuo-Haptic to 16 with a width of 256 in BAUM-1a. We observe that wider networks and successively downsampled convolutional network designs yield better results, even when trained for fewer epochs. Moreover, changing the kernel size from 5 in Mex to 13 in Mudestreda indicates that larger kernels result in better performance. The kernels in the depthwise layer define the size of the receptive field. Thus, large kernels mix arbitrarily distant spatial locations, allowing them to capture spatial dependencies more comprehensively.

Our experiments indicate that the best results belong to patch embedding and isotropic feature extraction architectures. In all tested datasets, Minape

[5] https://github.com/idearibosome/embracenet.
[6] https://github.com/keras-team/keras-io/blob/master/examples/vision/vit_small_ds.py.

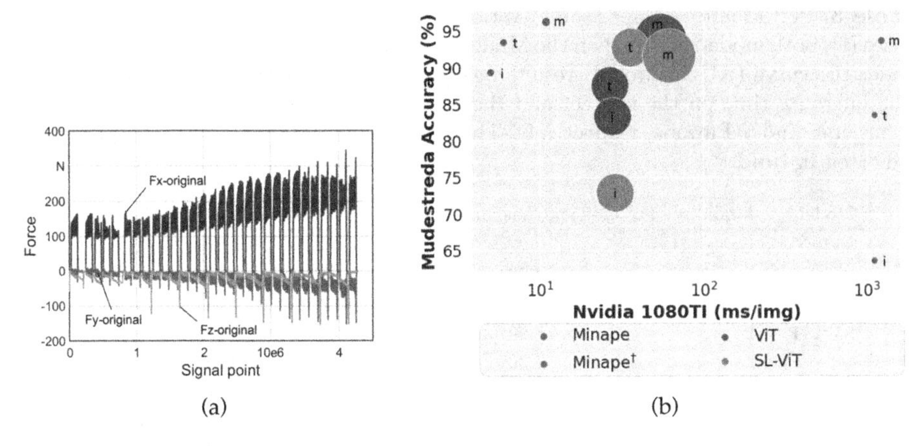

Fig. 5. (a) Example of force signals (Fx, Fy, Fz) from tool nr. 1 (T1) for 30 milling phases. (b) Comparison of multimodal models on Mudestreda. Various modalities are represented with i: image, t: time series, and m: multimodal. Minape† is a model with no patches. Minape (m) is 12 MB and SL-ViT (m) is 863 MB.

achieves significantly better results compared to other methods of comparison. Furthermore, we should note that the corresponding number of subnetworks does not grow exponentially with the number of multimodalities, making the solution scalable, even to high-dimensional spaces. Both the patch embedding and the large kernel size preserve locality well, suggesting that the spatial representation is sensitive to the relative embedding dimensions and filter size and generates equivalent results even without downsampling the representation in subsequent layers. Given that Minape yields better results than Minape-S and the fact that their difference lies only in the convolutional layer architecture, it can be concluded that the performance is more related to the kernel dimension and internal network resolution than to the depth of the classifier.

4.3 Mudestreda: A Multimodal Device State Recognition Use Case

In addition to the existing benchmarks, we study a real-case application of the milling process and propose a new multimodal industrial device state recognition dataset called **Mudestreda**. Mudestreda comprises 512 four-dimensional multimodal observations consisting of three force signal sequences and one RGB image of the shaft milling tool over five weeks from the Production Centrum. The three dimensions of time series modality are forces recorded in three axes, Fx, Fy, and Fz, with a frequency of 10 kHz. After each milling phase, a picture of the tool is taken, and the flank wear is measured to assign each observation to the respective class based on the defined metric: class-1 $[0, 71)\mu m$ (sharp), class-2 $[71, 110)\mu m$ (used), and class-3 $[110, +\infty)\mu m$ (dulled). Figure 5a shows an example of the collected signal, where a strong correlation between the tool wear and the force amplitudes can be observed. It indicates the smallest amplitudes for the sharp tool increase with tool wear.

Table 3. Performance comparison of various models and modalities (image (i), time-series (t), both modalities (m)) on the Mudestreda dataset in terms of accuracy and area under the curve (AUC). Model† results belong to a model without patch embedding. The inference time on the test set and the number of model parameters are reported as ms/img and #Params, respectively. The base model without a primary learner is indicated in **bold**.

Model	Modality	Fusion-Type	ms/img	Size(MB)	#Params	AUC	Accuracy
ViT	i	-	27	421	36.3M	0.89	87.5 ± 3.1
	t	-	28	422	36.4M	0.88	83.3 ± 2.7
	m	Concat	55	844	72.8M	0.98	93.7 ± 2.3
SL-ViT	i	-	29	428	36.5M	0.84	72.9 ± 2.2
	t	-	36	434	36.4M	0.97	92.7 ± 1.8
	m	Concat	62	863	73.7M	0.95	91.6 ± 1.5
Minape	i	-	5	6	0.4M	0.95	**89.3** ± 1.9
	t	-	6	6	0.4M	0.97	**93.4** ± 1.6
	m	**Concat**	**11**	**12**	**0.9M**	**0.98**	**96.2** ± 1.2
Minape †	i	-	1,118	3	0.2M	0.83	63.6 ± 2.6
	t	-	1,122	3	0.2M	0.89	83.5 ± 1.9
	m	Concat	1,239	6	0.4M	0.97	93.7 ± 2.2

Table 3 provides a detailed analysis of the impact of different modalities for Minape and ViT as the second-best comparison method in Table 2. In addition to ViT, we report the results for SL-ViT[7], which is a modified version of the ViT with Shifted Patch Tokenization (SPT) and Locality Self-Attention (LSA) addressing the problem of locality inductive bias and allowing training a small-size datasets [16]. The SL-ViT parameters are set as ViT with flags spt = true, lsa = true.

Table 3 shows that considering the combination of various modalities significantly enhances model accuracy across all examined models. Patches in Minape effectively improve the model's accuracy while reducing inference times. However, transformer models (ViT, SL-ViT), despite their comparable accuracy, have longer inference times and higher memory demands, which limit their applicability. The Minape model stands out with the highest accuracy, lowest memory usage, and shortest latency, making it a top choice for practical applications, given its balanced performance metrics. Figure 5b illustrates that Minape significantly outperforms other models in terms of inference latency and model size (MB). Minape exhibits a model parameter size reduction of more than 80 times compared to the multimodal ViT.

We scrutinized the sensitivity of Minape on the Mudestreda dataset, as shown in Table 4. The results demonstrate that the base model, when utilizing a concatenation fusion strategy, significantly outperforms other fusion types. The key advantage of concatenation fusion is its simplicity, as it does not introduce additional hyperparameters or necessitate complex computations. This not only

[7] https://github.com/keras-team/keras-io/blob/master/examples/vision/vit_small_ds.py.

Table 4. The results for sensitivity analysis of Minape on the Mudestreda dataset. The best model is indicated in **bold**.

Minape	Fusion-Type	Prime Learner	ms/img	Size(MB)	#Param(M)	Accuracy	Δ(%)
Fusion	Add	×	10	6	1.3	87.3 ± 2.8	−8.9
	Multiply	×	12	6	1.3	91.6 ± 1.7	−4.6
	Average	×	11	6	1.3	87.5 ± 2.5	−8.7
Learner	Concat	GRU	10	13	1.9	96.3 ± 1.2	+0.1
	Concat	LSTM	10	16	2.1	94.3 ± 1.6	−1.9
	Concat	**TCN**	**11**	**39**	**4.0**	**98.2** ± 1.0	**+2**
Attention ✓	Concat	×	15	8	2.0	95.1 ± 1.4	−1.1
Attention ✓	Concat	TCN	16	17	2.8	94.9 ± 1.8	−1.3

ensures efficient memory utilization but also meets the computational requirements of the system. Among our evaluated primary learners, which include GRU, LSTM, and TCN, the TCN model enhanced the baseline performance by 2% without any impact on the inference time. Our study also underscores the relevance of the attention mechanism in this workflow. Further experiments show that the attention layer does not improve the model performance[8].

5 Conclusion

We proposed Minape, a novel multimodal isotropic neural architecture with patch embedding, to improve multimodal learning for time series and image data. We observed that the patch representation and wider networks that successively down-sample the convolutional network size yield better results than deeper models and even ViT. It uses less memory and has faster inference allowing the use of the model on stand-by devices. The empirical results demonstrate that the modality fusion with the patch embedding yields higher accuracy than state-of-the-art and baseline methods on six multimodal test benchmarks and Mudestreda, our newly introduced real-case multimodal device state recognition dataset.

References

1. van Amsterdam, B., Kadkhodamohammadi, A., Luengo, I., Stoyanov, D.: Aspnet: action segmentation with shared-private representation of multiple data sources. In: IEEE/CVF Conference on Computer Vision and Pattern Recognition (CVPR), pp. 2384–2393 (2023)
2. Aslam, M.H., Zeeshan, M.O., Pedersoli, M., Koerich, A.L., Bacon, S., Granger, E.: Privileged knowledge distillation for dimensional emotion recognition in the wild. In: IEEE/CVF Conference on Computer Vision and Pattern Recognition (CVPR), pp. 3337–3346 (2023)

[8] Further ablation studies on the impact of hyperparameters can be found at https://github.com/hubtru/Minape.

3. Bai, S., Kolter, J.Z., Koltun, V.: An empirical evaluation of generic convolutional and recurrent networks for sequence modeling. arXiv:1803.01271 (2018)
4. Bonner, L.E.R., Buhl, D.D., Kristensen, K., Navarro-Guerrero, N.: Au dataset for visuo-haptic object recognition for robots. arXiv preprint arXiv:2112.13761 (2021)
5. Chen, S., Guhur, P.L., Schmid, C., Laptev, I.: History aware multimodal transformer for vision-and-language navigation. Adv. Neural Inform. Process. Syst. (NeurIPS) **34**, 5834–5847 (2021)
6. Choi, J.H., Lee, J.S.: Embracenet: a robust deep learning architecture for multimodal classification. Inform. Fusion **51**, 259–270 (2019)
7. Cicirelli, G., et al.: The ha4m dataset: multi-modal monitoring of an assembly task for human action recognition in manufacturing. Sci. Data **9**(1), 745 (2022)
8. Cubuk, E.D., Zoph, B., Mane, D., Vasudevan, V., Le, Q.V.: Autoaugment: Learning augmentation policies from data. arXiv preprint arXiv:1805.09501 (2018)
9. Dosovitskiy, A., et al.: An image is worth 16x16 words: transformers for image recognition at scale. In: International Conference on Learning Representation (ICLR) (2021)
10. Eroglu Erdem, C., Turan, C., Aydin, Z.: Baum-2: a multilingual audio-visual affective face database. Multimed. Tools Appl. **74**(18), 7429–7459 (2015)
11. Gashi, S., Min, C., Montanari, A., Santini, S., Kawsar, F.: A multidevice and multimodal dataset for human energy expenditure estimation using wearable devices. Sci. Data **9**(1), 537 (2022)
12. Geng, T., Wang, T., Duan, J., Cong, R., Zheng, F.: Dense-localizing audio-visual events in untrimmed videos: A large-scale benchmark and baseline. In: IEEE/CVF Conference on Computer Vision and Pattern Recognition (CVPR), pp. 22942–22951 (2023)
13. Girdhar, R., et al.: Imagebind: one embedding space to bind them all. In: IEEE/CVF Conference on Computer Vision and Pattern Recognition (CVPR), pp. 15180–15190 (2023)
14. Gong, X., et al.: MMG-ego4D: multimodal generalization in egocentric action recognition. In: IEEE/CVF Conference on Computer Vision and Pattern Recognition (CVPR), pp. 6481–6491 (2023)
15. Hendrycks, D., Gimpel, K.: Gaussian error linear units (gelus). arXiv preprint arXiv:1606.08415 (2016)
16. Lee, S.H., Lee, S., Song, B.C.: Vision transformer for small-size datasets. arXiv preprint arXiv:2112.13492 (2021)
17. Li, Y., Quan, R., Zhu, L., Yang, Y.: Efficient multimodal fusion via interactive prompting. In: IEEE/CVF Conference on Computer Vision and Pattern Recognition (CVPR), pp. 2604–2613 (2023)
18. Lialin, V., Rawls, S., Chan, D., Ghosh, S., Rumshisky, A., Hamza, W.: Scalable and accurate self-supervised multimodal representation learning without aligned video and text data. In: IEEE/CVF Winter Conference on Applications of Computer Vision (WACV), pp. 390–400 (2023)
19. Lin, Y.B., Sung, Y.L., Lei, J., Bansal, M., Bertasius, G.: Vision transformers are parameter-efficient audio-visual learners. In: IEEE/CVF Conference on Computer Vision and Pattern Recognition (CVPR), pp. 2299–2309 (2023)
20. Lin, Y.B., Tseng, H.Y., Lee, H.Y., Lin, Y.Y., Yang, M.H.: Exploring cross-video and cross-modality signals for weakly-supervised audio-visual video parsing. Adv. Neural Inform. Process. Syst. (NeurIPS) **34**, 11449–11461 (2021)
21. Liu, K., Li, Y., Xu, N., Natarajan, P.: Learn to combine modalities in multimodal deep learning. arXiv preprint arXiv:1805.11730 (2018)

22. Liu, X., Lu, H., Yuan, J., Li, X.: Cat: causal audio transformer for audio classification. In: IEEE International Conference on Acoustics, Speech and Signal Processing (ICASSP), pp. 1–5 (2023)
23. Loshchilov, I., Hutter, F.: Decoupled weight decay regularization. In: International Conference on Learning Representation (ICLR) (2018)
24. Ramazanova, M., Escorcia, V., Caba, F., Zhao, C., Ghanem, B.: Owl (observe, watch, listen): Audiovisual temporal context for localizing actions in egocentric videos. In: IEEE/CVF Conference on Computer Vision and Pattern Recognition (CVPR), pp. 4879–4889 (2023)
25. Ranganathan, H., Chakraborty, S., Panchanathan, S.: Multimodal emotion recognition using deep learning architectures. In: IEEE winter conference on Applications of Computer Vision (WACV), pp. 1–9 (2016)
26. Ryan, F., Jiang, H., Shukla, A., Rehg, J.M., Ithapu, V.K.: Egocentric auditory attention localization in conversations. In: IEEE/CVF Conference on Computer Vision and Pattern Recognition (CVPR), pp. 14663–14674 (2023)
27. Senocak, A., Kim, J., Oh, T.H., Li, D., Kweon, I.S.: Event-specific audio-visual fusion layers: a simple and new perspective on video understanding. In: IEEE/CVF Winter Conference on Applications of Computer Vision (WACV), pp. 2237–2247 (2023)
28. Wightman, R., Touvron, H., Jégou, H.: Resnet strikes back: an improved training procedure in timm. arXiv preprint arXiv:2110.00476 (2021)
29. Wijekoon, A., Wiratunga, N., Cooper, K.: Mex: multi-modal exercises dataset for human activity recognition. arXiv preprint arXiv:1908.08992 (2019)
30. Wu, H., et al.: Cvt: introducing convolutions to vision transformers. In: IEEE/CVF Conference on Computer Vision and Pattern Recognition (CVPR), pp. 22–31 (2021)
31. Xiao, Y., Ma, Y., Li, S., Zhou, H., Liao, R., Li, X.: Semanticac: semantics-assisted framework for audio classification. In: IEEE International Conference on Acoustics, Speech and Signal Processing (ICASSP), pp. 1–5 (2023)
32. Xu, R., Feng, R., Zhang, S.X., Hu, D.: Mmcosine: multi-modal cosine loss towards balanced audio-visual fine-grained learning. In: IEEE International Conference on Acoustics, Speech and Signal Processing (ICASSP), pp. 1–5 (2023)
33. Xue, Z., Marculescu, R.: Dynamic multimodal fusion. In: IEEE/CVF Conference on Computer Vision and Pattern Recognition (CVPR), pp. 2574–2583 (2023)
34. Zhang, X., Tang, X., Zong, L., Liu, X., Mu, J.: Deep multimodal clustering with cross reconstruction. In: Pacific-Asia Conference on Knowledge Discovery and Data Mining (PAKDD), pp. 305–317 (2020)
35. Zhang, Z., et al.: Abaw5 challenge: a facial affect recognition approach utilizing transformer encoder and audiovisual fusion. In: IEEE/CVF Conference on Computer Vision and Pattern Recognition (CVPR), pp. 5724–5733 (2023)
36. Zhong, Z., Schneider, D., Voit, M., Stiefelhagen, R., Beyerer, J.: Anticipative feature fusion transformer for multi-modal action anticipation. In: IEEE/CVF Winter Conference on Applications of Computer Vision (WACV), pp. 6068–6077 (2023)
37. Zhu, W., Omar, M.: Multiscale audio spectrogram transformer for efficient audio classification. In: IEEE International Conference on Acoustics, Speech and Signal Processing (ICASSP), pp. 1–5 (2023)

Determination of Local and Global Decision Weights Based on Fuzzy Modeling

Bartłomiej Kizielewicz[1], Jakub Więckowski[1], Bartosz Paradowski[2], Andrii Shekhovtsov[1], and Wojciech Sałabun[1](✉)

[1] Research Team on Intelligent Decision Support Systems, Department of Artificial Intelligence and Applied Mathematics, Faculty of Computer Science and Information Technology, West Pomeranian University of Technology in Szczecin, ul. Żołnierska 49, 71-210 Szczecin, Poland
{bartlomiej-kizielewicz,jakub-wieckowski,andrii-shekhovtsov, wojciech.salabun}@zut.edu.pl

[2] National Institute of Telecommunications, Szachowa 1, 04-894 Warsaw, Poland
b.paradowski@il-pib.pl

Abstract. An essential challenge in multi-criteria decision analysis (MCDA) is the determination of criteria weights. These weights map the decision maker's preferences for decision problems in determining the importance of criteria. However, these values are not necessarily constant in the whole domain. Although many approaches are related to their determination, some MCDA models can have local weights that are difficult to map in global spaces. This paper focuses on an approach in which we determine global and local weights from the Characteristic Objects METhod (COMET) by using linear regression. Moreover, obtained linear models are compared with COMET and Technique for Order of Preference by Similarity to Ideal Solution (TOPSIS) models to answer how similar they are. Then, the relationships between the obtained global and local weights are analyzed based on a simple case study. The results demonstrate the high sensitivity of the COMET method and the applicability of the proposed approach for determining global and local weights. The most useful contribution is the proposed approach to identify local weights that can be used for deeper decision analysis.

Keywords: Global weights · Local weights · Linear regression · MCDA · COMET · TOPSIS

1 Introduction

Multi-Criteria Decision Analysis (MCDA) methods are a popular tool used in Decision Support Systems (DSS) [24]. Their purpose is to determine the preference values of the alternatives considered in the decision problem. In addition, they indicate the quality of each alternative [29]. It, in turn, proves to be helpful for the expert, especially with a significant dimensionality of the problem

B. Luo et al. (Eds.): ICONIP 2023, LNCS 14447, pp. 188–200, 2024.
https://doi.org/10.1007/978-981-99-8079-6_15

and a large number of decision variants considered. Many methods are used for multi-criteria analysis, which have different paradigms [30]. The use of different approaches in calculating preference values usually results in disparities in the rankings obtained [27].

An additional difference noticed among existing multi-criteria techniques is the set of parameters necessary for proper operation. Most of the methods use a vector of criterion weights [7]. It is used to determine the importance of factors describing the attractiveness of an alternative. In other words, the vector of weights indicates to what extent a given criterion has contributed to the indicated evaluation [10]. Consequently, it directly affects the calculated preference value and the ranking position of a given decision alternative. Existing studies show that the choice of the method used to calculate the weights in a problem results in differences in the final rankings [37]. Therefore, it demonstrates that this decision should be accompanied by a detailed analysis of the problem specification so that the results obtained are as close to reality as possible.

However, some methods do not rely on a vector of criterion weights to determine the preference of alternatives. One of them is Characteristic Objects Method (COMET) [23]. Instead, the attractiveness of decision alternatives is defined in a pairwise comparison of Characteristic Objects (COs). First, the expert determines relations between COs so that the desired values are preferable. Then, the Matrix of Expert Judgment (MEJ) matrix, the Summed Judgement (SJ) vector, and the alternatives' final preference value are determined based on the relations [22].

The performance of the COMET method is therefore characterized by the non-explicit indication of the vector of criterion weights. It causes the expert not to receive information about the influence of individual criteria on the final result. That creates a research gap to be filled. To date, some research has been made that aims to reduce the dimensionality of problems solved by the COMET method [33] or determine how to define characteristic values most effectively [26]. However, no attempt has been made to indicate to what extent the criteria influence the obtained evaluation of the decision options considered [13]. Knowing this allows the selection of a given alternative to be justified comprehensively and reliably. Furthermore, it provides the possibility to justify the relationship between the criteria and the evaluation in a detailed way. It is crucial in real-life problems, where in addition to the results obtained, it is also worth indicating what and how influenced their outcome.

In this paper, we face the problem of weights identification in multi-criteria COMET models. The used approach contains establishing a functional relationship between the importance of model weights. For this purpose, simulated decision-making examples have been used. The obtained results have been averaged to generalize them and examine the principle of operation in many research cases. The study aimed to determine hypersurfaces from the resulting Characteristic Objects and linear models using a linear regression function. These surfaces were then used to determine the differences between them to verify the correctness of the results obtained. The proper matching of the vector of criteria weights to the problems in the COMET method is a valuable contribution to the practical application of this method in real multi-criteria problems, equipping the expert with knowledge of the influence of each criterion on the outcome.

The rest of the paper is organized as follows. Section 2 presents a literature review on the identification of multi-criteria weights. Section 3 presents descriptions of the selected methods for the study. Section 4 presents a case study related to the research object. Section 6 provides conclusions and future research directions.

2 Literature Review

Identifying weights in multi-criteria models is a critical problem accompanying the use of these methods in complex systems [18]. Particular attention should be paid to selecting appropriate methods for determining the weights of criteria, as they directly impact the results obtained. This vector determines to what extent a given criterion influences the outcome in problems solved by MCDA methods. Often, weighting methods based on the variation of data in the decision matrix are used to ensure an objective approach [6]. A review of existing studies shows that techniques such as entropy, standard deviation, or Gini index are popular among researchers [5,8,34].

Most methods from the MCDA family are based on their calculations on an explicitly indicated vector of weights, based on which the evaluation of alternatives is determined [11]. It makes the expert aware from the beginning to what extent a given criterion influences the assessment [28]. However, some methods do not rely explicitly on the vector of criteria weights in the multi-criteria evaluation process [32]. Furthermore, it is necessary to develop techniques to determine the criterion weights from the results obtained.

The literature review shows that the approaches used to identify the weights mainly consider establishing the relevance of the criteria needed to determine the inputs in MCDA models. For example, in 2008, Renaud set the weights of the criteria by parametric identification, considering the weights were not fixed by criterion but set according to utility level [19]. Using random weights in identifying models using neural networks is also a popular approach [35,36]. These studies have shown generally satisfactory results, with a narrower range of solutions having more desirable outcomes. The work was also directed towards adaptive weight identification. Multiple machine learning conducted on the available data proved to be an effective solution for determining the significance of parameters. Furthermore, it allowed the vector of weights to be matched to the problems being solved [9,14].

Additionally, some approaches use linear models to identify weights. Some studies focus on using the Best-Worst Method (BWM) to evaluate alternatives based on the weights just determined by the linear models [1]. The results show that the combination of these techniques and their use in complex problems can be effectively used [20,21]. Linear models have been used to identify input weights also for the VIekriterijumsko KOmpromisno Rangiranje

(VIKOR), the Preference Ranking Organization Method for Enrichment Evaluation (PROMETHEE), or The Technique for Order of Preference by Similarity to Ideal Solution (TOPSIS) methods [2,4,17]. However, no studies are dedicated to this issue using the COMET method. In the case of this multi-criteria technique, the problem is to backwardly determine the significance of the weights based on the results obtained, rather than in other methods where the weights are identified as input values to get the most reliable results. It accounts for the novel approach to the problem of identifying the weights, aiming to justify in detail the influence of individual criteria on the results obtained by the COMET method.

3 Descriptions of the Selected Methods

3.1 Linear Regression

Linear regression allows determining the linear relationship between related and unrelated variables [25], which is described by the following Equation (1).

$$E\left[Y \mid X_1=x_1, X_2=x_2, \ldots, X_K=x_K\right]=\phi\left(x_1, x_2, \ldots, x_K\right) \tag{1}$$

where Y is the response variable and the explanatory variables are $X_1, X_2, \ldots, X_K$.

It enables to solve the problem of estimating the expected value [12]. Many linear regression models take a different approach to the expected value estimation. Their choice is determined by the type of problem we face and the accuracy we want to achieve. The generalized linear regression equation is represented as follows (2):

$$y=ax+b \tag{2}$$

where a is the coefficient of the independent variable, and b is constant term [3].

3.2 The COMET Method

The Characteristic Objects Method belongs to the rule-based MCDA methods group [31]. The alternative preference is obtained by conducting the pairwise comparison of determined Characteristic Objects (COs). The main advantage of this technique is that it is completely free of the ranking reversal phenomenon [23]. Moreover, complex problems can be split into smaller subproblems, reducing the dimensionality and calculation time. The COMET effectiveness has been verified many times and proved to be reliable tool. Its main assumptions should be shortly recalled.

Step 1. Define the Space of the Problem – the expert determines the dimensionality of the problem by selecting the number r of criteria, $C_1, C_2, ..., C_r$. Then, a set of fuzzy numbers is selected for each criterion C_i, e.g., $\{\tilde{C}_{i1}, \tilde{C}_{i2}, \ldots, \tilde{C}_{ic_i}\}$ (3):

$$\begin{aligned} C_1 &= \left\{\tilde{C}_{11}, \tilde{C}_{12}, \ldots, \tilde{C}_{1c_1}\right\} \\ C_2 &= \left\{\tilde{C}_{21}, \tilde{C}_{22}, \ldots, \tilde{C}_{2c_2}\right\} \\ &\cdots \\ C_r &= \left\{\tilde{C}_{r1}, \tilde{C}_{r2}, \ldots, \tilde{C}_{rc_r}\right\} \end{aligned} \tag{3}$$

where $C_1, C_2, \ldots, C_r$ are the ordinals of the fuzzy numbers for all criteria.

Step 2. Generate Characteristic Objects – The characteristic objects (CO) are obtained by using the Cartesian Product of fuzzy numbers cores for each criteria as follows (4):

$$CO = C(C_1) \times C(C_2) \times ... \times C(C_r) \tag{4}$$

Step 3. Rank the Characteristic Objects – the expert determines the MEJ. It is a result of pairwise comparison of the COs by the problem expert. The MEJ matrix contains results of comparing characteristic objects by the expert, where α_{ij} is the result of comparing CO_i and CO_j by the expert. The function f_{exp} denotes the mental function of the expert. It depends solely on the knowledge of the expert and can be presented as (5). Afterwards, the vertical vector SJ is obtained as follows (6).

$$\alpha_{ij} = \begin{cases} 0.0, \; f_{exp}(CO_i) < f_{exp}(CO_j) \\ 0.5, \; f_{exp}(CO_i) = f_{exp}(CO_j) \\ 1.0, \; f_{exp}(CO_i) > f_{exp}(CO_j) \end{cases} \tag{5}$$

$$SJ_i = \textstyle\sum_{j=1}^{t} \alpha_{ij} \tag{6}$$

Finally, values of preference are approximated for each characteristic object. As a result, the vertical vector P is obtained, where $i-th$ row contains the approximate value of preference for CO_i.

Step 4. The Rule Base – each characteristic object and value of preference is converted to a fuzzy rule as follows (7):

$$IF \; C(\tilde{C}_{1i}) \; AND \; C(\tilde{C}_{2i}) \; AND \; ... \; THEN \; P_i \tag{7}$$

In this way, the complete fuzzy rule base is obtained.

Step 5. Inference and Final Ranking – each alternative is presented as a set of crisp numbers (e.g., $A_i = \{a_{1i}, a_{2i}, ..., a_{ri}\}$). This set corresponds to criteria $C_1, C_2, ..., C_r$. Mamdani's fuzzy inference method is used to compute preference of $i - th$ alternative. The rule base guarantees that the obtained results are unequivocal. The bijection makes the COMET a completely rank reversal free.

3.3 The TOPSIS Method

The Technique for Order Preference by Similarity to an Ideal Solution (TOPSIS) was first introduced by Chen and Hwang [16]. The principle of it is based on logic related to reference points, where alternatives are evaluated according to their distance from them. It uses two reference points a positive ideal solution and a negative ideal solution. Its entire procedure can be presented as follows:
Step 1. Normalization of the defined decision matrix X based on Equation (8).

$$\begin{gathered}\textbf{Profit: } r_{ij} = \frac{x_{ij} - \min_j(x_{ij})}{\max_j(x_{ij}) - \min_j(x_{ij})} \quad \textbf{Cost: } r_{ij} = \frac{\max x_j(x_{ij}) - x_{ij}}{\max x_j(x_{ij}) - \min_j(x_{ij})} \\ i = 1, \ldots, n \quad j = 1, \ldots, J\end{gathered} \tag{8}$$

Step 2. Calculation of a weighted normalized decision matrix (9):

$$v_{ij} = w_i \cdot r_{ij}, \quad i = 1, \ldots, n \quad j = 1, \ldots, J \tag{9}$$

Step 3. Identification of the Positive Ideal Solution (A^*) and Negative Ideal Solution (A^-) for a defined decision-making problem (10):

$$\begin{aligned} A^* &= \{v_1^*, \ldots, v_n^*\} = \left\{\left(\max_j v_{ij} \mid i \in I^P\right), \left(\min_j v_{ij} \mid i \in I^C\right)\right\} \\ A^- &= \{v_1^-, \ldots, v_n^-\} = \left\{\left(\min_j v_{ij} \mid i \in I^P\right), \left(\max_j v_{ij} \mid i \in I^C\right)\right\} \end{aligned} \tag{10}$$

where I^P stands for profit type criteria and I^C for cost type.

Step 4. Calculation of the Positive and Negative Distances using the n-dimensional Euclidean distance with Equation (11).

$$\begin{aligned} D_j^* &= \sqrt{\sum_{i=1}^n \left(v_{ij} - v_i^*\right)^2}, \quad j = 1, \ldots, J \\ D_j^- &= \sqrt{\sum_{i=1}^n \left(v_{ij} - v_i^-\right)^2}, \quad j = 1, \ldots, J \end{aligned} \tag{11}$$

Step 5. Calculation of the relative closeness to the Ideal Solution (12):

$$C_j^* = \frac{D_j^-}{\left(D_j^* + D_j^-\right)}, \quad j = 1, \ldots, J \tag{12}$$

4 Study Case

In the case study under consideration, we focus on how to extract global weights from the COMET method. For this purpose, we will use linear regression to obtain a linear model based on the preferences of the characteristic objects. Furthermore, uniformly distributed points in a given space will be used to measure the similarity of the two models. This study determined that points will be generated using a Cartesian product of 100 values for all criteria uniformly distributed in the range [0, 1]. An additional aspect addressed is the comparison of the two

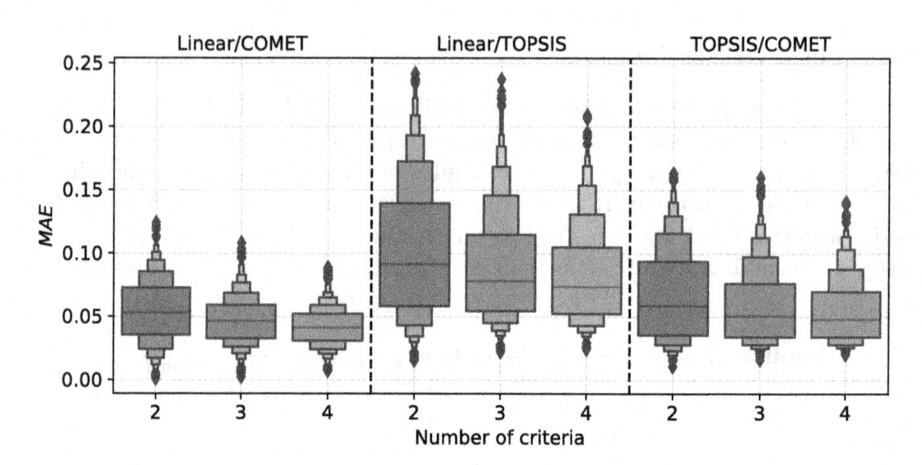

Fig. 1. Mean absolute error (MAE) for the models considered against the number of criteria.

models with the TOPSIS model to show the difference between the models. This study uses the pymcdm library version 1.0, available through PyPi [15].

A study on model similarity comparison was performed for 1000 randomly generated COMET models. An additional aspect addressed was the effect of increasing the number of criteria on the similarity of these models. Therefore, the study was done for 2, 3, and 4 criteria. With the help of illustration 1 the obtained results from the model comparisons using mean absolute error (MAE) are shown. The effect of the number of criteria on the similarity of the models can be seen here, whereas the number of criteria increases, and the similarity of the model's increases. As a result, the most similar models were found to be linear models with COMET models. On the other hand, the least similar models turned out to be linear models with TOPSIS models. The slightest error value was obtained for comparisons between linear and COMET models and was 0.00121. In contrast, the most significant error value was obtained between linear and TOPSIS models and was 0.24203.

After investigating the similarity between linear and COMET models, the similarity of local weights with global weights was examined. The global weights were determined using the regression equation coefficients for the mapped COMET model. The COMET model, on the other hand, was randomly generated. A set formed from a single alternative was used to determine the local weights, whose values for each criterion were incremented and decremented by the value of $\epsilon = 0.1$. The Cartesian product of such values then produced a set used to identify the local weights. The local weights were determined in similar steps as the global weights. For the study, a set consisting of 10 alternatives and 10 criteria was used, represented by a Table 1. The criteria values for all alternatives were randomly generated from the interval [0, 1]. The alternatives were evaluated using the COMET method. Their preferences were then used to identify the linear model.

Table 1. The set of alternatives used to identify the linear model.

A_i	C_1	C_2	C_3	C_4	C_5	C_6	C_7	C_8	C_9	C_{10}
A_1	0.3004	0.8723	0.8864	0.5351	0.8915	0.1970	0.8368	0.3999	0.2645	0.7461
A_2	0.1449	0.5784	0.3394	0.4672	0.6048	0.2943	0.3890	0.5251	0.8964	0.2078
A_3	0.8006	0.7790	0.1542	0.5660	0.4963	0.5561	0.2813	0.5995	0.6032	0.5120
A_4	0.2226	0.2842	0.7842	0.4466	0.1461	0.7277	0.5816	0.4601	0.6923	0.8310
A_5	0.5113	0.3422	0.8482	0.4814	0.2766	0.8102	0.3111	0.4188	0.2956	0.5398
A_6	0.4143	0.1478	0.7171	0.4793	0.8966	0.4139	0.3229	0.5952	0.6057	0.2790
A_7	0.7767	0.3694	0.1276	0.7097	0.1548	0.2058	0.6632	0.3685	0.1775	0.3314
A_8	0.5169	0.4446	0.7346	0.3349	0.8398	0.3588	0.5548	0.1832	0.3783	0.6351
A_9	0.4630	0.6246	0.8179	0.5870	0.1031	0.5879	0.4567	0.2255	0.7692	0.5151
A_{10}	0.2943	0.2576	0.4256	0.1049	0.8356	0.3342	0.4144	0.2147	0.1410	0.7799

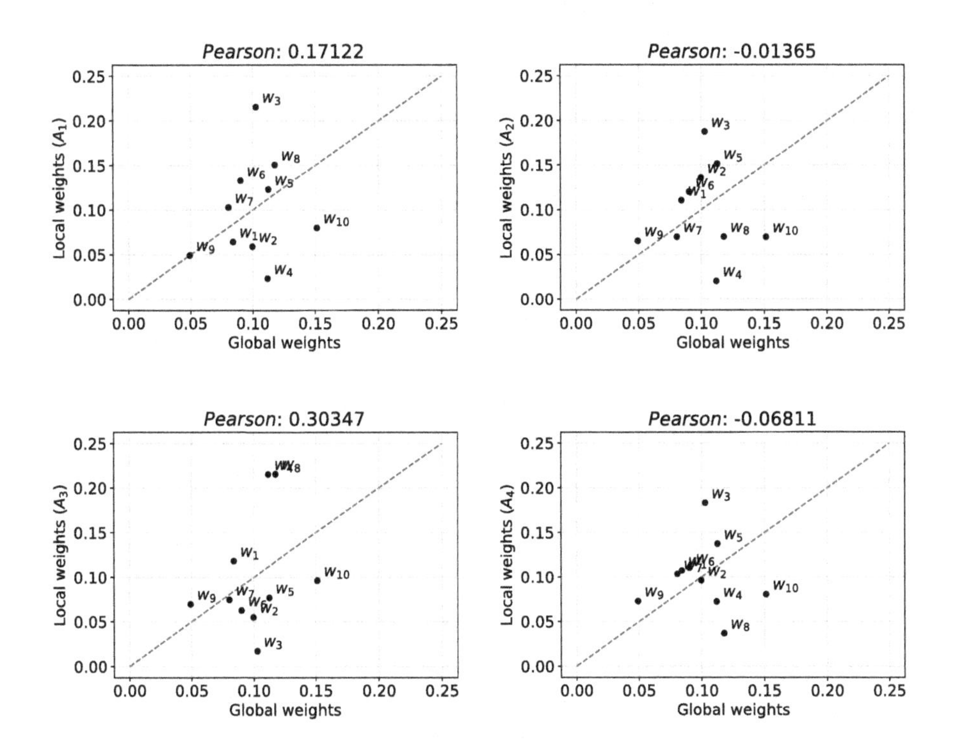

Fig. 2. Relationship between local weights and global weights.

In Fig. 2, charts of the relationship between the global and local weights for the first four alternatives of the considered set are presented. As can be seen in the graphs, the obtained weights differ significantly from each other. The minor differences among the considered weights are noticeable for weights w_7, w_9, which took similar results for both approaches, i.e., local and global weights. The most significant differences between the values of local and global weights are observed for the weight w_3. In the case of local weights for the alternatives A_1, A_2, A_4, the weight w_3 has a much higher local value than the global one. On the other hand, for alternative A_3, it took a smaller value. Moreover, all comparisons shown in the graphs are characterized by a low value of Pearson's coefficient, which comes from the range [-0.07, 0.30]. This indicates the lack of correlation between the considered approaches to determining the weights.

Table 2 shows the global and local weights obtained using the crowned approach for the alternatives presented above. The last row in the Table shows the Pearson coefficient value for the comparisons of a given column of weights with the global weights. The highest similarity was achieved between the local weights derived from the created set based on the alternatives A_5, A_{10}, and the global weights. Pearson's coefficient for these combinations was 0.72863 and 0.78541, respectively. On the other hand, the least correlated local weights with global weights were obtained from the created sets from the alternatives A_2, A_4 and were -0.01365 and -0.06811, respectively. Most of the obtained local weights have very low similarity with global weights.

Table 2. Overview of the obtained local and global weights.

w_i	Global	A_1	A_2	A_3	A_4	A_5	A_6	A_7	A_8	A_9	A_{10}
w_1	0.0840	0.0641	0.1106	0.1180	0.1072	0.0988	0.0634	0.1094	0.0668	0.0539	0.0889
w_2	0.0997	0.0590	0.1358	0.0549	0.0961	0.1304	0.0584	0.0509	0.1323	0.1408	0.0413
w_3	0.1028	0.2154	0.1877	0.0171	0.1831	0.0883	0.1704	0.0159	0.2240	0.1810	0.0902
w_4	0.1120	0.0233	0.0203	0.2151	0.0723	0.0182	0.2288	0.1830	0.0895	0.1943	0.1620
w_5	0.1127	0.1229	0.1513	0.0767	0.1374	0.1352	0.0816	0.1498	0.1721	0.1459	0.0770
w_6	0.0902	0.1328	0.1201	0.0625	0.1105	0.0523	0.1315	0.1189	0.0717	0.0564	0.0965
w_7	0.0804	0.1028	0.0697	0.0744	0.1034	0.0622	0.0791	0.0886	0.0833	0.0672	0.0560
w_8	0.1180	0.1507	0.0699	0.2155	0.0368	0.1803	0.0362	0.1618	0.0382	0.0308	0.1622
w_9	0.0492	0.0490	0.0651	0.0695	0.0727	0.0290	0.0717	0.0413	0.0388	0.0628	0.0412
w_{10}	0.1512	0.0799	0.0696	0.0963	0.0804	0.2054	0.0790	0.0804	0.0833	0.0671	0.1848
Pearson	-	0.17122	−0.01365	0.30347	−0.06811	0.72863	0.09894	0.32809	0.23553	0.18739	0.78541

5 Disccusion

The application of linear regression in the context of the described study of COMET models is an important aspect that contributes to understanding and evaluating the effectiveness of these models. Linear regression is a powerful data analysis tool with many advantages in this study's context. Among the key advantages is its interpretability. In studies of COMET models, where it is essential to understand the impact of individual criteria on the outcome, linear regression allows us to gain precise knowledge of this. Indeed, for example, experts can determine exactly which criteria have the most significant impact on the results of COMET models, which can provide valuable guidance.

It is also worth noting that linear regression is a valuable tool for comparing different models, including COMET models. We can assess how well a given model reflects reality and what criteria are essential in decision-making. This is crucial, especially for COMET models designed to support decision-making processes.

The potential for future research is also worth noting. While this study focused on comparing COMET models with linear models and TOPSIS, future research could consider comparisons with other popular techniques such as Reference Ideal Method (RIM), Stable Preference Ordering Towards Ideal Solution (SPOTIS), or Multi-Attributive Border Approximation area Comparison (MABAC) [15]. This will give us a better understanding of which of these techniques performs best in different decision-making contexts. Also, considering the differences in test results depending on the number of criteria is essential, as it can influence the choice of the appropriate method in a particular case.

6 Conclusion

The study demonstrated the feasibility of using linear regression to obtain global weights from the COMET model. Another aspect was to compare the linear models, COMET and TOPSIS, to each other to determine how similar they are. The global weights obtained were then compared with the local weights to examine the similarity of the linear model with the COMET model and show the proposed approach's effectiveness.

The highest similarity was between the linear and the COMET models, which was the desired effect for extracting weights from the identified linear models. A slightly lower similarity was obtained between TOPSIS and COMET models, demonstrating the validity of choosing linear models to derive global weights from COMET models. In the correlation analysis between global weights and local weights, low correlation is visible, which proves the high sensitivity of the COMET method.

Future research should propose a new approach for determining global weights that give highly correlated results with local weights. In addition, an approach would need to be developed whereby the accuracy of the identified models can be compared.

References

1. Abadi, F., Sahebi, I., Arab, A., Alavi, A., Karachi, H.: Application of best-worst method in evaluation of medical tourism development strategy. Decis. Sci. Lett. **7**(1), 77–86 (2018)
2. Alsalem, M., et al.: Multiclass benchmarking framework for automated acute Leukaemia detection and classification based on BWM and group-VIKOR. J. Med. Syst. **43**(7), 1–32 (2019)

3. Badillo, S., et al.: An introduction to machine learning. Clin. Pharmacol. Therapeutics **107**(4), 871–885 (2020)
4. Balugani, E., Lolli, F., Butturi, M.A., Ishizaka, A., Sellitto, M.A.: Logistic regression for criteria weight elicitation in PROMETHEE-based ranking methods. In: Ahram, T., Karwowski, W., Vergnano, A., Leali, F., Taiar, R. (eds.) IHSI 2020. AISC, vol. 1131, pp. 474–479. Springer, Cham (2020). https://doi.org/10.1007/978-3-030-39512-4_74
5. Bhunia, S.S., Das, B., Mukherjee, N.: EMCR: routing in WSN using multi criteria decision analysis and entropy weights. In: Fortino, G., Di Fatta, G., Li, W., Ochoa, S., Cuzzocrea, A., Pathan, M. (eds.) IDCS 2014. LNCS, vol. 8729, pp. 325–334. Springer, Cham (2014). https://doi.org/10.1007/978-3-319-11692-1_28
6. Branke, J.: MCDA and multiobjective evolutionary algorithms. In: Greco, S., Ehrgott, M., Figueira, J.R. (eds.) Multiple Criteria Decision Analysis. ISORMS, vol. 233, pp. 977–1008. Springer, New York (2016). https://doi.org/10.1007/978-1-4939-3094-4_23
7. de Brito, M.M., Almoradie, A., Evers, M.: Spatially-explicit sensitivity and uncertainty analysis in a MCDA-based flood vulnerability model. Int. J. Geogr. Inf. Sci. **33**(9), 1788–1806 (2019)
8. Chen, W., Zhang, S., Li, R., Shahabi, H.: Performance evaluation of the GIS-based data mining techniques of best-first decision tree, random forest, and naïve Bayes tree for landslide susceptibility modeling. Sci. Total Environ. **644**, 1006–1018 (2018)
9. Cunha, M., Marques, J., Creaco, E., Savić, D.: A dynamic adaptive approach for water distribution network design. J. Water Resour. Plan. Manag. **145**(7), 04019026 (2019)
10. Groothuis-Oudshoorn, C.G.M., Broekhuizen, H., van Til, J.: Dealing with uncertainty in the analysis and reporting of MCDA. In: Marsh, K., Goetghebeur, M., Thokala, P., Baltussen, R. (eds.) Multi-Criteria Decision Analysis to Support Healthcare Decisions, pp. 67–85. Springer, Cham (2017). https://doi.org/10.1007/978-3-319-47540-0_5
11. Hatefi, S., Torabi, S.: A common weight MCDA-DEA approach to construct composite indicators. Ecol. Econ. **70**(1), 114–120 (2010)
12. Jalal, H., Goldhaber-Fiebert, J.D., Kuntz, K.M.: Computing expected value of partial sample information from probabilistic sensitivity analysis using linear regression metamodeling. Med. Decis. Making **35**(5), 584–595 (2015)
13. Jeong, J.S., García-Moruno, L., Hernández-Blanco, J., Sánchez-Ríos, A.: Planning of rural housings in reservoir areas under (mass) tourism based on a fuzzy DEMATEL-GIS/MCDA hybrid and participatory method for Alange. Spain. Habitat Int. **57**, 143–153 (2016)
14. Jia, J., Ruan, Q., An, G., Jin, Y.: Multiple metric learning with query adaptive weights and multi-task re-weighting for person re-identification. Comput. Vis. Image Underst. **160**, 87–99 (2017)
15. Kizielewicz, B., Shekhovtsov, A., Sałabun, W.: pymcdm-The universal library for solving multi-criteria decision-making problems. SoftwareX **22**, 101368 (2023)
16. Liu, Q.: TOPSIS model for evaluating the corporate environmental performance under intuitionistic fuzzy environment. Int. J. Knowl.-Based Intell. Eng. Syst. **26**(2), 149–157 (2022)
17. Olson, D.L.: Comparison of weights in TOPSIS models. Math. Comput. Model. **40**(7–8), 721–727 (2004)
18. Reid, S.G.: Acceptable risk criteria. Prog. Struct. Mat. Eng. **2**(2), 254–262 (2000)

19. Renaud, J., Levrat, E., Fonteix, C.: Weights determination of OWA operators by parametric identification. Math. Comput. Simul. **77**(5–6), 499–511 (2008)
20. Rezaei, J.: Best-worst multi-criteria decision-making method: some properties and a linear model. Omega **64**, 126–130 (2016)
21. Rezaei, J., van Roekel, W.S., Tavasszy, L.: Measuring the relative importance of the logistics performance index indicators using Best Worst Method. Transp. Policy **68**, 158–169 (2018)
22. Sałabun, W., Piegat, A., Wątróbski, J., Karczmarczyk, A., Jankowski, J.: The COMET method: the first MCDA method completely resistant to rank reversal paradox. European Working Group Series 3 (2019)
23. Sałabun, W., Ziemba, P., Wątróbski, J.: The rank reversals paradox in management decisions: the comparison of the AHP and COMET methods. In: Czarnowski, I., Caballero, A.M., Howlett, R.J., Jain, L.C. (eds.) Intelligent Decision Technologies 2016. SIST, vol. 56, pp. 181–191. Springer, Cham (2016). https://doi.org/10.1007/978-3-319-39630-9_15
24. Santos, J., Bressi, S., Cerezo, V., Presti, D.L.: SUP&R DSS: a sustainability-based decision support system for road pavements. J. Clean. Prod. **206**, 524–540 (2019)
25. Seber, G.A., Lee, A.J.: Linear regression analysis. John Wiley & Sons (2012)
26. Shekhovtsov, A., Więckowski, J., Kizielewicz, B., Sałabun, W.: effect of criteria range on the similarity of results in the COMET method. In: 2021 16th Conference on Computer Science and Intelligence Systems (FedCSIS), pp. 453–457. IEEE (2021)
27. Shekhovtsov, A., Więckowski, J., Wątróbski, J.: Toward reliability in the MCDA rankings: comparison of distance-based methods. In: Czarnowski, I., Howlett, R.J., Jain, L.C. (eds.) Toward Reliability in the MCDA Rankings: Comparison of Distance-Based Methods. SIST, vol. 238, pp. 321–329. Springer, Singapore (2021). https://doi.org/10.1007/978-981-16-2765-1_27
28. Steele, K., Carmel, Y., Cross, J., Wilcox, C.: Uses and misuses of multicriteria decision analysis (MCDA) in environmental decision making. Risk Anal. Int. J. **29**(1), 26–33 (2009)
29. Stewart, T.J.: Dealing with uncertainties in MCDA. In: Dealing with uncertainties in MCDA. ISORMS, vol. 78, pp. 445–466. Springer, New York (2005). https://doi.org/10.1007/0-387-23081-5_11
30. Uhde, B., Andreas Hahn, W., Griess, V.C., Knoke, T.: Hybrid MCDA methods to integrate multiple ecosystem services in forest management planning: a critical review. Environ. Manage. **56**(2), 373–388 (2015)
31. Wątróbski, J., Jankowski, J., Ziemba, P., Karczmarczyk, A., Zioło, M.: Generalised framework for multi-criteria method selection. Omega **86**, 107–124 (2019)
32. Wątróbski, J., Sałabun, W., Karczmarczyk, A., Wolski, W.: Sustainable decision-making using the COMET method: an empirical study of the ammonium nitrate transport management. In: 2017 Federated Conference on Computer Science and Information Systems (FedCSIS), pp. 949–958. IEEE (2017)
33. Więckowski, J., Dobryakova, L.: A fuzzy assessment model for freestyle swimmers-a comparative analysis of the MCDA methods. Procedia Comput. Sci. **192**, 4148–4157 (2021)
34. Yariyan, P., Ali Abbaspour, R., Chehreghan, A., Karami, M., Cerdà, A.: GIS-based seismic vulnerability mapping: a comparison of artificial neural networks hybrid models. Geocarto International, pp. 1–24 (2021)
35. Yu, W., Pacheco, M.: Impact of random weights on nonlinear system identification using convolutional neural networks. Inf. Sci. **477**, 1–14 (2019)

36. Zhang, Y., Ma, W., Hou, R., Rong, D., Qin, X., Cheng, Y., Wang, H.: Spectroscopic profiling-based geographic herb identification by neural network with random weights. Spectrochimica Acta Part A: Molecular and Biomolecular Spectroscopy, p. 121348 (2022)
37. Zyoud, S.H., Fuchs-Hanusch, D.: A bibliometric-based survey on AHP and TOPSIS techniques. Expert Syst. Appl. **78**, 158–181 (2017)

Binary Mother Tree Optimization Algorithm for 0/1 Knapsack Problem

Wael Korani(✉)

University of North Texas, Denton, USA
wael.korani@unt.edu

Abstract. The knapsack problem is a well-known strongly NP-complete problem where the profits of collection of items in knapsack is maximized under a certain weight capacity constraint. In this paper, a novel Binary Mother Tree Optimization Algorithm (BMTO) and Knapsack Problem Framework (KPF) are proposed to find an efficient solution for 0/1 knapsack problem in a short time. The proposed BMTO method is built on the original MTO and a binary module to solve an optimization problem in a discrete space. The binary module converts a set of real numbers equal to the dimension of the knapsack problem to a binary number using a threshold and the sigmoid function. In fact, the KPF makes the implementation of a metaheuristic algorithm to solve the knapsack problem much simpler. In order to assess the performance of the proposed solutions, extensive experiments are conducted. In this regard, several statistical analyses on the resulting solution are evaluated when solved for two sets of knapsack instances (small and large scale). The results demonstrate that BMTO can produce an efficient solution for knapsack instances of different sizes in a short time, and it outperforms two other algorithms Binary Particle Swarm Optimization (BPSO) and Binary Bacterial Foraging (BBF) algorithms in terms of best solution and time. In addition, the results of BPSO and BBF show the effectiveness of KPF compared to the results in the literature.

Keywords: Binary Mother Tree Algorithm · Knapsack Problem · Knapsack Problem Framework

1 Introduction

The Knapsack problem is a well known NP-complete combinatorial problem [16]. There are a few knapsack problems with different item types [9]: the bounded knapsack problem, the unbounded knapsack problem, and the 0/1 knapsack problem. In this paper, the proposed solutions are used to solve the 0/1 knapsack problem, which is called a binary knapsack, and it is defined as follows:

$$Maximize\ f = \sum_{i=1}^{n} P_i x_i,\ x_i \in \{0, 1\} \tag{1a}$$

$$subject\ to\ \sum_{i=1}^{n} w_i x_i \leq MaxCap,\ x_i \in \{0, 1\} \tag{1b}$$

B. Luo et al. (Eds.): ICONIP 2023, LNCS 14447, pp. 201–213, 2024.
https://doi.org/10.1007/978-981-99-8079-6_16

where n is a set of items that are contained by knapsack, and each item has a weight w_i and a profit p_i. $MaxCap$ is the maximum weight capacity of knapsack, and x_i is a binary variable to select an item in the knapsack. The knapsack problem arises in many practical applications such as: decision making [12], resource sharing and allocation [15], and project management and selection [10].

To solve the knapsack problem, there are many methods, and they are categorized into two categories: exact and metaheuristic methods. Exact methods can produce an exact solution, but they take longer when the size of the problem increases such as branch and bound [14] and linear programming [13]. However, metaheuristic methods produce approximate solutions in a reasonable time frame [6]. Metaheuristic includes swarm intelligence (SI) methods and evolutionary algorithms (EAs), and many of those SI methods and EAs were developed in the last few decades to solve large scale 0/1 knapsack problems to overcome the exact methods shortage.

In [17], the authors introduced a genetic algorithm variant based on adaptive evolution in dual population called DPAGA to solve the 0/1 knapsack problem; however, the authors implemented the proposed method on just two instances in order to evaluate the performance of the method; however, the average error was a little high. In [2], a variant of the differential evolution algorithm (DEA) was introduced to solve the 0/1 knapsack problem, and the results were compared to several differential evolution variants. The results showed that the proposed method could solve the small size instances, and the performance degrades to solve large instances and run time significantly increases.

In [4], few variants of Firefly Algorithm (FA) were introduced to solve the 0/1 knapsack problem; however, the running time of all variants were too long, and the authors assessed the variants on just small scale instances. In [11], a variant of Binary Bacteria Foraging (BF) was introduced and the results were compared to the original BF and Binary Particle Swarm Optimization (BPSO). The results showed that the average error of the proposed method was high specially when the size of instances increases.

The goal of the current paper is to introduce an efficient and fast method to solve different sizes of the knapsack problem and overcome the shortage in the previous methods. In addition, a framework is introduced to facilitate the evaluation of other metaheuristic methods to solve knapsack problem. The rest of the paper is organized as follows. Section 2 presents the proposed BMTO method. Section 3 introduces a new framework for knapsack problem. Section 4 shows the experimentation and results to evaluate the performance of the proposed BMTO and the framework. Finally, Sect. 5 lists concluding remarks and ideas for feature works.

2 Proposed Binary Mother Tree Algorithm

The Mother Tree Optimization (MTO) algorithm is a swarm intelligence algorithm that is built on the symbiotic relationship between Douglas fir trees and mycorrhizal fungi network [7]. The heart of MTO is the Fixed-Offspring (FO) topology, which facilitates the communication between agents in the population. Agents in the FO topology are divided into three groups: Top Mother Tree (TMT); Partially Connected Trees (PCTs) group; Fully Connected Trees (FCTs) group. In addition, the PCTs group is divided into First-PCTs and Last-PCTs. Agents' positions are updated according to the group they belong to.

2.1 Top Mother Tree (TMT)

The TMT performs two levels of search. It takes a random direction with a step size δ. The updated position is computed as follows:

$$P_1(x_{k+1}) = P_1(x_k) + \delta R(d),\ where\ R(d) = \frac{R}{\sqrt{R \cdot R^{\mathsf{T}}}}, \tag{2}$$

where $R = 2\ (round(rand\ (d, 1)) - 1)\ rand(d, 1)$, δ is a step size and $R(d)$ is a random vector, and d is the dimension of the problem. At the second level, the TMT's position is updated using a different random direction with step size Δ.

$$P_1(x_{k+1}) = P_1(x_k) + \Delta R(d). \tag{3}$$

2.2 Partially Connected Trees (PCTs)

Agents of ***First-PCTs*** group are in range $[2 : \frac{\mathrm{N_T}}{2} - 1]$, and their positions are updated as follows:

$$P_n(x_{k+1}) = P_n(x_k) + \sum_{i=1}^{n-1} \frac{1}{n-i+1}(P_i(x_k) - P_n(x_k)), \tag{4}$$

where $\mathrm{N_T}$ refers to the population size, $P_n(x_{k+1})$ refers to the updated position of agent n, $P_n(x_k)$ refers to the current position of agent n, and $P_i(x_k)$ refers to the current position of agent i. However, if any member of this group uses the defense mechanism, it updates the position using

$$P_n(x_{k+1}) = P_n(x_k) + \phi R(d), \tag{5}$$

where ϕ is a small deviation from the current position. The user has the option to enable or disable the defense mechanism.

Agents of ***Last-PCTs*** group are in range $[\frac{\mathrm{N_T}}{2} + 3\text{: } \mathrm{N_T}]$, and their positions are updated as follows:

$$P_n(x_{k+1}) = P_n(x_k) + \sum_{i=n-N_{os}}^{N_T - N_{os}} \frac{1}{n-i+1}(P_i(x_k) - P_n(x_k)). \tag{6}$$

2.3 Fully Connected Trees (FCTs)

Agents of FCTs group are in range $[\frac{\mathrm{N_T}}{2} : \frac{\mathrm{N_T}}{2} + 2]$, and agents' positions are updated as follows:

$$P_n(x_{k+1}) = P_n(x_k) + \sum_{i=n-N_{os}}^{n-1} \frac{1}{n-i+1}(P_i(x_k) - P_n(x_k)). \tag{7}$$

In [5], the authors introduced a variant of MTO called DMTO to solve a discrete problem called Traveling Salesman Problem. The DMTO was built on a swap concept operation. In this paper, a novel variant of MTO called Binary-MTO (BMTO) is introduced to solve the binary knapsack problem as shown in Algorithm 1.

Algorithm 1. The BMTO algorithm

1: **Inputs:**
   $N_T, P_T, d, K_{rs}, Profits, Weights,$ and $MaximumCap$
2: N_T: The population size
3: P_T: The position of agent
4: d: The dimension of the knapsack problem
5: K_{rs}: The number of iterations
6: **Initialize:**
   Distribute T agents uniformly over the search space $(P_1, \ldots, P_T)$
7: **Evaluate:**
   [suggestedPosition, fintess(T), Maxweight]= $knapsackFunction(Binary(P_1 \ldots P_T))$
8: $S = Sort(fitness(1) \ldots fitness(T))$, fitnesses represent the profits
9: The sorted positions with the same rank of S stored in array $A = (P_1 \ldots P_T)$
10: **for** $k_{rs} = 1$ to K_{rs} **do**
11: Use equations (2)–(7) to update the position of each agent in A
12: Evaluate the fitness (profits) of the updated positions
13: Sort solutions in descending order and store them in S
14: Update A
15: **end for**
16: $S = Sort(S_1 \ldots S_T)$
17: $Best\ profit = Max(S)$
18: **Output:**
   Suggested Solution, Max profit, Max weight

3 The Knapsack Problem Framework

In the last few decades, many SI methods and EAs were proposed to find a fast and efficient solution for the knapsack problem and its applications. In some cases, a proposed method with the same set of parameters behaves in a different way to tackle a specific problem. The main reason for this aforementioned behavior is how the proposed method was implemented. In this paper, I propose an efficient framework for the knapsack problem called KPF that accelerates the implementation of a metaheuristic algorithm and achieves better results.

The framework makes the process of evaluating a new method for the knapsack problem very simple. The KPF is evaluated using a set of knapsack instances along with three SI methods PSO, BF, and MTO.

The KPF consists of three modules: the optimizer, binary, and knapsack modules as shown in Fig. 1. In general, the optimizer module can be a meta-heuristic algorithm and in particular can be a SI or EA. In fact, a continuous optimization algorithm can handle just real numbers. The KPF idea is to convert the real numbers to binary string using the binary module, evaluate the output binary string and suggest a new solution using the knapsack module.

The binary module limits the real value to a certain threshold and then uses the sigmoid function

$$f(x) = \frac{1}{1 + exp(-x)}, \tag{8}$$

to convert the real number to a real number between [0, 1]. Then, a random number is generated between [0, 1], if the output of the sigmoid function is greater than generated number then the value will be set to be one, if not the value will be set to zero as shown in Fig. 2-a. The knapsack module is the heart of the KPF to compute the maximum profit and keep the maximum capacity under or equal to a certain value. The main idea is to use the binary number that was generated from the binary module and check the selected items and keep the weights under a given constraint capacity as shown in Fig. 2-b. Finally, a new solution is suggested by the knapsack module to the optimizer module that satisfies the maximum profit under the weight constraint. The maximum profit will be assigned to the associated agent in the optimizer module as a fitness value as shown in Fig. 1.

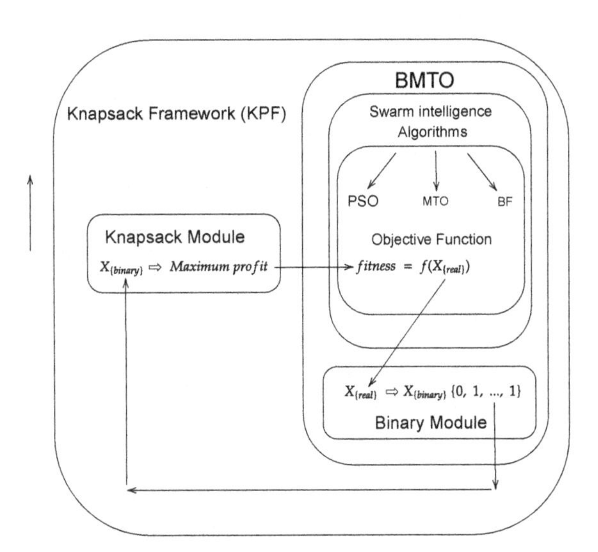

Fig. 1. Knapsack proposed framework.

Binary Module

```
input : x (real number)
initialize : i, thres

if x > thresh
    x = thresh
elseif x < -thresh
    x = -thresh
end

while i < size(x) do {

    temp(i) = sigmoid(i)

    if temp(i) > rand

        temp(i) = 1
    else
        temp = 0
    end

    i = i +1

}
output : temp (binary number)
```

(a) Convert real number to Binary

Knapsack Module

```
input : (P, W, MaxCap, X_binary)
initialize : MaxP, MaxW, Sol, i

while i < size(P) do {

    if X_binary(i) = 1

        MaxW = MaxW + W(i)

        if MaxW > MaxCap
            MaxW = MaxW - W(i)
            continue
        end

        MaxP = MaxP + P(i)
        sol(i) = 1
    else
        sol(i) = 0
    end
    i = i + 1
}
output : Sol, MaxP, MaxW
```

(b) Compute max-profit under weight-constraint.

Fig. 2. Binary and knapsack modules

4 Experimentation and Results

In this paper, two experiments are conducted. In the first experiment, the number of function evaluations (No. FEs) is set low to show the effectiveness of the proposed KPF. The No. FEs is the number of times a cost function is evaluated. In the second experiment, the No. FEs in terms of the number of iterations is increased to find the optimal solution for hard large scale instances.

The two sets of instances are listed as shown in Table 1: small scale instances and large scale instances. The small scale instances set includes eight binary knapsack instances P01-P08 that have different sizes (number of items) ranging from five to 24 items, and it is available on [1]. The aforementioned set was used to evaluate the performance of a DEA [2] and a few variants of FA [4]. In fact, the difficulty level of the knapsack problem increases when the size of the problem increases. The large scale set includes six instances that have different sizes ranging from 10 to 100 as shown in Table 2. The large scale set was used to evaluate the performance of BBF and BPSO algorithms [11]. In addition, a partial part of the large scale set was used to evaluate DEA [2], Schema-Guiding Evolutionary Algorithm (SGEA) [8], and a hybrid of HS and Jaya algorithms [3].

The three algorithms PSO, BF, and MTO are coded in MATLAB R2022a and are performed on a MacBook Pro with Apple M1 Max and 32 GB memory. Each algorithm is executed 30 times for each instance with different seed values from 1 to 30 to avoid bias in the results. Then, the maximum, minimum, average, and average error percentage (AE%) from global solution

$$AE_i\% = \frac{Optimal_i - Mean_i}{Mean_i} * 100 \quad (9)$$

values are recorded for comparison. In addition, the time is recorded for the run that achieves the highest profit.

The BPSO, BBF, and BMTO algorithms are evaluated in terms of time and accuracy using the two aforementioned sets, and the results of the two experiments are compared to other algorithms in the literature. The performance of the proposed KPF is evaluated on the three algorithms. The results are compared to the same algorithms in the literature to show the effectiveness and efficiency of the proposed framework.

Table 1. Two sets of knapsack instances.

Small scale instances			Large scale instances		
Instance	No. of items	Optimal	Instance	No. of items	Optimal
P01	10	309	*P*09	10	295
P02	5	51	*P*10	20	1042
P03	6	150	*P*11	15	481.0694
P04	7	107	*P*12	23	9767
P05	8	900	*P*13	50	3169
P06	7	1735	*P*14	100	15170
P07	15	1458	–	–	–
P08	24	13549094	–	–	–

4.1 Parameter Settings

The parameters' values of SI algorithms have a significant effect on their performance. The parameters of all algorithms are recorded in Table 3, and (Ped = 0.25 for BF). All parameters' values for PSO and BF are recommended in the literature.

4.2 Results

The results of the first experiment are recorded in Table 4 and Table 5. The results of the small scale set are recorded in Table 4. The results show that BMTO and BPSO are relatively fast compared to BBF and the duration to reach the global solution ranges from 0.0362 to 0.0935 fraction of sec. The BMTO is faster in three instances out of eight. The three algorithms achieve the optimal solution in all cases; however, BBF could not reach the global solution for P08.

Comparing results to other studies, in [4], the authors reported the results of the original FA and three variants. Although the authors used more No. FEs, all methods could not find the optimal solution for P07 and P08 instances during 10 runs. In addition, the run time for these two instances was indeed long and when the authors increased the number of iterations the run time for one variant was approximately 70 secs per run. To evaluate the effectiveness of KPF, a similar experiment is conducted for just the original FA and similar results obtained in a shorter time. In [2], the authors reported the AE% for several variants of DEA. The AE% of BMTO and BPSO is much lower than all other DEAs, for instance P08.

Table 2. Binary knapsack instances.

Large scale instances	
Inst.	Parameters (Profit (P), Wights (W) and Max Capacity (C)
*P*09	P = (55, 10, 47, 5, 4, 50, 8, 61, 85, 87), W =(95, 4, 60, 32, 23, 72, 80, 62, 65, 46), C = 269
*P*10	P= (92, 4, 43, 83, 84, 68, 92, 82, 6, 44, 32, 18, 56, 83, 25, 96, 70, 48, 14, 58(,w = (44, 46, 90, 72, 91, 40, 75, 35, 8, 54, 78 40, 77, 15, 61, 17, 75, 29, 75,63(, C = 878
*P*11	P=(0.125126, 19.330424, 58.500931,35.029145, 82.284005, 17.410810, 71.050142, 30.399487, 9.140294, 14.731285, 98.852504, 11.908322, 0.891140, 53.166295, 60.176397), W= (56.358531 , 80.874050, 47.987304, 89.596240, 74.660482, 85.894345, 51.353496, 1.498459, 36.445204, 16.589862, 44.569231, 0.466933, 37.788018, 57.118442,60.716575), C = 375
*P*12	P= (981, 980, 979, 978, 977,976, 487, 974, 970, 485, 485, 970, 970,484, 484, 976, 974, 482, 962, 961, 959,958, 857), W = (983, 982, 981, 980, 979, 978, 488,976, 972, 486, 486, 972, 972, 485, 485,969, 966, 483, 964, 963, 961, 958, 959), C = 10000
*P*13	P=(220, 208, 198, 192, 180, 180, 165, 162, 160, 158, 155, 130, 125, 122, 120, 118, 115, 110, 105, 101, 100, 100, 98, 96, 95, 90, 88, 82, 80, 77, 75, 73, 72, 70, 69, 66, 65, 63, 60, 58, 56, 50, 30, 20, 15, 10, 8, 5, 3, 1), W = (80, 82, 85, 70, 72, 70, 66, 50, 55, 25, 50, 55, 40, 48, 50, 32, 22, 60, 30, 32, 40, 38, 35, 32, 25, 28, 3, 22, 50, 30, 45, 30, 60, 50, 20, 65, 20, 25, 30, 10, 20, 25, 15, 10, 10, 10, 4, 4, 2, 1), C= 1000
*P*14	P= (297, 295, 293, 292, 291, 289, 284, 284, 283, 283, 281, 280, 279, 277, 276, 275, 273, 264, 260, 257, 250, 236, 236, 235, 235, 233, 232, 232, 228, 218, 217, 214, 211, 208, 205, 204, 203, 201, 196, 194, 193, 193, 192, 191, 190, 187, 187, 184, 184, 184, 181, 179, 176, 173, 172, 171, 160, 128, 123, 114, 113, 107, 105, 101, 100, 100, 99, 98, 97, 94, 94, 93, 91, 80, 74, 73, 72, 63, 63, 62, 61, 60, 56, 53, 52, 50, 48, 46, 40, 40, 35, 28, 22, 22, 18, 15, 12, 11, 6, 5), W= (54, 95, 36, 18, 4, 71, 83, 16, 27, 84, 88, 45, 94, 64, 14, 80, 4,23, 75, 36, 90, 20, 77, 32, 58, 6, 14, 86, 84, 59, 71, 21, 30, 22, 96, 49, 81, 48, 37, 28, 6, 84, 19, 55, 88, 38, 51, 52, 79, 55, 70, 53, 64, 99, 61, 86, 1, 64, 32, 60, 42, 45, 34, 22,49, 37, 33, 1, 78, 43, 85, 24, 96, 32, 99, 57, 23, 8, 10, 74, 59, 89, 95, 40, 46, 65, 6, 89, 84, 83, 6, 19, 45, 59, 26, 13, 8, 26, 5, 9), C = 3820

Table 3. Parameter setting of PSO, BF, MTO.

PSO parameters					
c_1	c_1	w	Population size	Iterations	No. of FEs
2	2	0.9 - 0.6	20	200	4000
BF parameters					
N_c	N_{re}	N_{ed}	N_s	Population size	No. of FE
25	4	2	4	20	8000
MTO parameters					
ϕ	δ	Δ	Population size	Iterations	No. of FE
5.0	12.5	5.0	20	200	4000

Table 4. Results of the small scale instances.

Instance	Method	For all runs					For best run
		Best	Worse	Avg.	Std. dev.	AE%	time (sec)
P01	BPSO	309	309	309	0	0	0.0754
	BBF	309	309	309	0	0	0.2361
	BMTO	309	309	309	0	0	**0.0404**
P02	BPSO	51	51	51	0	0	0.0356
	BBF	51	51	51	0	0	0.1864
	BMTO	51	51	51	0	0	**0.0351**
P03	BPSO	150	150	150	0	0	**0.0362**
	BBF	150	150	150	0	0	0.1571
	BMTO	150	150	150	0	0	0.0364
P04	BPSO	107	107	107	0	0	**0.0385**
	BBF	107	107	107	0	0	0.1680
	BMTO	107	107	107	0	0	0.0498
P05	BPSO	900	900	900	0	0	**0.0424**
	BBF	900	900	900	0	0	0.1734
	BMTO	900	898	899.9333	0.3651	0.007	0.0512
P06	BPSO	1735	1735	1735	0	0	**0.0388**
	BBF	1735	1735	1735	0	0	0.1644
	BMTO	1735	1682	1731.8	11.04	0.184	0.0581
P07	BPSO	1458	1451	1455.9	2.2854	0.205	**0.0691**
	BBF	1458	1448	1452.6	2.7114	0.370	0.1862
	BMTO	1458	1438	1450.5	5.5096	0.514	0.0774
P08	BPSO	13549094	13458871	1.3521e+07	13007	0.002	0.0969
	BBF	13524340	13380730	1.3441e+07	30525	0.8	0.2262
	BMTO	13549094	13351612	1.3481e+07	4.7535	0.005	**0.0935**

Table 5. Results of the large scale instances.

Instance	Method	For all runs					For the best run
		Best	Worse	Avg.	Std. dev.	AE%	time (sec)
P09	BPSO	295	295	295	0	0	**0.0500**
	BBF	295	294	294.9333	0.2537	0.022	0.1774
	BMTO	295	252	291.233	9.90	1.276	0.0817
P10	BPSO	1042	1037	1041.7	1.2685	0.0287	0.1095
	BBF	1042	1007	1033.8	10.0519	0.7869	0.2023
	BMTO	1042	1013	1034.3	7.9522	0.7389	**0.1073**
P11	BPSO	481.0694	481.0694	481.0694	0	0	**0.0680**
	BBF	481.0694	427.9031	446.2197	17.7844	7.246	0.2050
	BMTO	481.0694	428.3151	471.3368	18.0757	2.00	0.1087
P12	BPSO	9767	9767	9767	0	0	**0.0982**
	BBF	9767	9759	9764.1	1.7604	0.0296	0.2217
	BMTO	9767	9762	9766.2	1.5625	0.008	0.1262
P13	BPSO	3138	3094	3115.3	10.6555	1.695	0.1936
	BBF	3122	3047	3074.8	17.4199	2.997	0.3117
	BMTO	3166	3089	3134.6	21.2434	1.08	**0.1563**
P14	BPSO	13953	12784	13369	242.5096	11.872	0.3828
	BBF	14991	14472	14781	102.5481	1.17	2.564
	BMTO	14913	12895	14209	549.4847	6.33	**0.2764**

The results of the large scale set are recorded in Table 5, and show that BSPO and BMTO are faster than BBF in all cases. The BMTO is faster than BPSO in three instances out of six. In addition, BMTO performs faster and achieves lower average error when the size of the knapsack problem increases. Although BPSO has the lowest average error, the average errors for all algorithms are close as shown in Fig. 3-a. The results show that none of the algorithms could find the optimal solution for P13 and P14 instances using low No. FEs. In the second experiment, the No. FEs is increased to 14,000 (700 iterations) for P13 and 80,000 FE (4000 iterations) for P14. The results show that BMTO achieves the lowest AE%, and is the only algorithm to find the global solution in a short time as shown in Table 6 and Fig. 3-b. Thus, BMTO can produce a solution for both small and large instances in a reasonable time frame.

Comparing the results to other studies, in [11], the authors used 30,000 FEs for instances (P09, P10, P11) and 50,000 FEs for instances (P12, P13, P14) for 30 runs. However, after calculating the average error for their results and comparing with PKF, the results show that PKF achieves lower average error, which shows the effectiveness of the proposed framework.

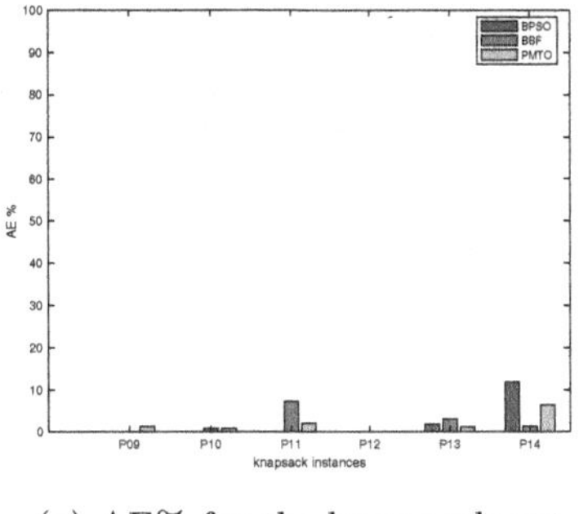

(a) AE% for the large scale set.

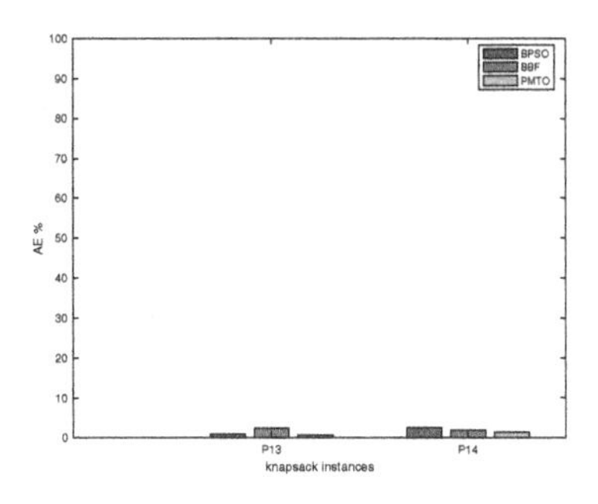

(b) AE% for P13 and P14

Fig. 3. A comparison of the AE% for all algorithms

In [8], the authors used (P10, P13, P14) to evaluate the performance of SGEA. The average errors of their results are higher than BMTO and the run time is longer than BMTO. In [2], the authors used (P10, P13, P14) again to evaluate different variants of DEA. The authors reported the run time for P10

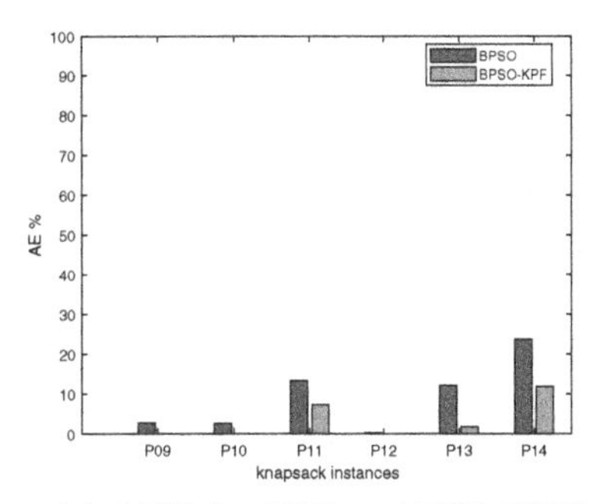

(a) AE% for PSO and PSO-KPF.

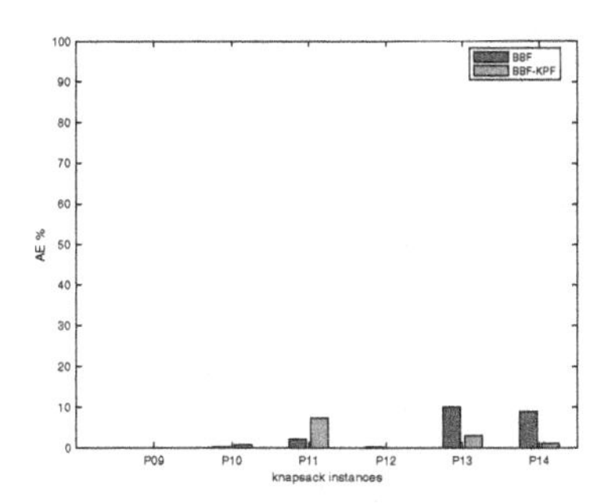

(b) AE% for BF and BF-KPF

Fig. 4. AE% for PSO and BF without and with KPF

Table 6. Results of the P13 and P14 instances.

Instance	Method	For all runs				For the best run	
		Best	Worse	Avg.	Std. dev.	AE%	time (sec)
P13	BPSO	3160	3118	3142.6	10.9815	0.833	0.6702
	BBF	3122	3056	3095.3	15.2631	2.325	1.2718
	BMTO	**3169**	3122	3149.1	13.3332	**0.627**	**0.5457**
P14	BPSO	14805	14350	14556	120.1742	2.4	4.047
	BBF	15061	14781	14893	60.7566	1.825	10.2787
	BMTO	**15170**	14461	15013	193.6572	**1.349**	**5.5548**

and P13. The results show that BMTO is much faster for both instances. In addition, BMTO achieves better value when compared to B-DE (the best DEA variant). In [3], the authors used a hybrid of HS and Jaya algorithms, and the results for (P09, P11, P12) were better than the reported results using KPF (Fig. 4).

5 Conclusion and Future Works

In this paper, a novel variant of MTO called BMTO, and a knapsack problem framework are developed. BMTO is a binary MTO method to solve the knapsack problem. This framework makes it simple to evaluate a new metaheuristic variant for the solving knapsack problem. In order to validate the effectiveness and efficiency of the proposed solutions, two experiments are conducted using fourteen knapsack instances. The results show that BMTO has the capability to solve small and large knapsack problems quickly and efficiently. The results of PSO and BBF are compared to the results in the literature to show the reliability of the proposed framework.

For further research, BMTO is used as a core part of a feature selection technique, and the performance is evaluated. The results show BMTO's effectiveness in achieving promising accuracy for high dimensional feature extraction problems.

References

1. KNAPSACK_01 data for the 01 knapsack problem. https://people.sc.fsu.edu/~jburkardt/datasets/knapsack_01/knapsack_01.html. Accessed 17 Aug 2014
2. Ali, I.M., Essam, D., Kasmarik, K.: An efficient differential evolution algorithm for solving 0–1 knapsack problems. In: 2018 IEEE Congress on Evolutionary Computation (CEC), pp. 1–8. IEEE (2018)
3. Alomoush, A., Alsewari, A.A., Alamri, H.S., Zamli, K.Z.: Solving 0/1 knapsack problem using hybrid HS and Jaya algorithms. Adv. Sci. Lett. **24**(10), 7486–7489 (2018)
4. David, D., Ronny, R., Widayanti, T., et al.: Modification of attractiveness and movement of the firefly algorithm for resolution to knapsack problems. In: 2022 4th International Conference on Cybernetics and Intelligent System (ICORIS), pp. 1–5. IEEE (2022)
5. Korani, W., Mouhoub, M.: Discrete mother tree optimization for the traveling salesman problem. In: Yang, H., Pasupa, K., Leung, A.C.-S., Kwok, J.T., Chan, J.H., King, I. (eds.) ICONIP 2020. LNCS, vol. 12533, pp. 25–37. Springer, Cham (2020). https://doi.org/10.1007/978-3-030-63833-7_3
6. Korani, W., Mouhoub, M.: Review on nature-inspired algorithms. Oper. Res. Forum **2**(3), 1–26 (2021). https://doi.org/10.1007/s43069-021-00068-x
7. Korani, W., Mouhoub, M., Spiteri, R.J.: Mother tree optimization. In: 2019 IEEE International Conference on Systems, Man and Cybernetics (SMC), pp. 2206–2213. IEEE (2019)

8. Liu, Y., Liu, C.: A schema-guiding evolutionary algorithm for 0–1 knapsack problem. In: 2009 International Association of Computer Science and Information Technology-Spring Conference, pp. 160–164. IEEE (2009)
9. Locatelli, A., Iori, M., Cacchiani, V.: Optimization methods for knapsack and tool switching problems (2023)
10. Mavrotas, G., Diakoulaki, D., Kourentzis, A.: Selection among ranked projects under segmentation, policy and logical constraints. Eur. J. Oper. Res. **187**(1), 177–192 (2008)
11. Niu, B., Bi, Y.: Binary bacterial foraging optimization for 0/1 knapsack problem. In: 2014 IEEE Congress on Evolutionary Computation (CEC), pp. 647–652. IEEE (2014)
12. Peeta, S., Salman, F.S., Gunnec, D., Viswanath, K.: Pre-disaster investment decisions for strengthening a highway network. Comput. Oper. Res. **37**(10), 1708–1719 (2010)
13. Senju, S., Toyoda, Y.: An approach to linear programming with 0–1 variables. Manag. Sci. **15**, B196–B207 (1968)
14. Shih, W.: A branch and bound method for the multiconstraint zero-one knapsack problem. J. Oper. Res. Soc. **30**(4), 369–378 (1979)
15. Vanderster, D.C., Dimopoulos, N.J., Parra-Hernandez, R., Sobie, R.J.: Resource allocation on computational grids using a utility model and the knapsack problem. Futur. Gener. Comput. Syst. **25**(1), 35–50 (2009)
16. Wojtczak, D.: On strong NP-completeness of rational problems. In: Fomin, F.V., Podolskii, V.V. (eds.) CSR 2018. LNCS, vol. 10846, pp. 308–320. Springer, Cham (2018). https://doi.org/10.1007/978-3-319-90530-3_26
17. Yan, T.S., Guo, G.Q., Li, H.M., He, W.: A genetic algorithm for solving knapsack problems based on adaptive evolution in dual population. In: Advanced Materials Research, vol. 756, pp. 2799–2802. Trans Tech Publ (2013)

Distributed State Estimation for Multi-agent Systems Under Consensus Control

Yan Li[1], Jiazhu Huang[1,2], Yuezu Lv[3](✉), and Jialing Zhou[3]

[1] School of Automation, Beijing Institute of Technology, Beijing, China
{yanl,jzh}@bit.edu.cn

[2] MIIT Key Laboratory of Complex-field Intelligent Sensing, Yangtze Delta Region Academy of Beijing Institute of Technology, Jiaxing, China

[3] Advanced Research Institute of Multidisciplinary Science, Beijing Institute of Technology, Beijing, China
{yzlv,jlzhou}@bit.edu.cn

Abstract. Distributed state estimation and consensus control for linear time-invariant multi-agent systems under strongly connected directed graph are addressed in this paper. The distributed output tracking algorithm and the local state estimator are designed for each agent to estimate the output and state of the entire multi-agent system, despite having access only to local output measurements that are insufficient to directly reconstruct the entire state. The consensus control protocol is further designed based on each agent's own entire state estimation. Neither distributed state estimation nor consensus control protocol design requires state information from neighboring agents, eliminating the transmission of the values of state estimations during the whole process. The theoretical analysis demonstrates that the realization of distributed output tracking and state estimation. Moreover, all agents achieve consensus. Finally, numerical simulations are worked out to show the effectiveness of the proposed algorithm.

Keywords: Multi-agent systems · Distributed output tracking · Distributed state estimation · Consensus control

1 Introduction

In recent years, multi-agent systems (MASs) have been widely applied in engineering fields, such as power engineering [1], market simulation [2], smart grid [3], and online monitoring [4], to solve problems that cannot be addressed by a single agent due to its limited capability or knowledge. Limited by the ability of agents to perceive global information, the state estimation of multi-agent systems has become a research hotspot. For large-scale multi-agent systems, distributed state estimation algorithms highlight a greater advantage compared to

This work was supported by the National Natural Science Foundation of China under Grants 62088101 and 62273045, and the Fellowship of China Postdoctoral Science Foundation under Grants 2021TQ0039 and 2022M710384.

B. Luo et al. (Eds.): ICONIP 2023, LNCS 14447, pp. 214–225, 2024.
https://doi.org/10.1007/978-981-99-8079-6_17

centralized state estimation algorithms in the sense of requiring less network communication and consuming fewer system resources.

Distributed state estimation algorithms are mainly classified into Kalman-filter-based techniques and observer-based approaches. The distributed Kalman filtering method based on the consensus algorithm is a novel distributed estimation algorithm coupling Kalman filtering and multi-agents, which is an efficient class of distributed state estimation algorithms using only the local information of nodes without fusion centers. For instance, [5] proposed a two-stage Kalman-consensus filtering approach consisting of a standard Kalman filtering update and a consensus update to solve the distributed state estimation problem under random faults in communication links. To reduce the communication effort, distributed event-triggered Kalman filtering consensus algorithms were proposed in [6–8]. In addition, the robustness stability margin of the distributed Kalman filtering algorithm based on the consensus theory was analyzed in [9], and robust distributed state estimator for uncertain systems was designed in [10].

On the other hand, observer-based distributed state estimation methods consider the joint observability or detectability of the system and construct local observers for the nodes in the network to estimate the entire state of the system in [11–15]. Reference [16] designed observers to estimate the entire state based on the consensus theory by using the local measured output of the system. To reduce the amount of communication in the network, reference [17] performed an observability decomposition, where a local Luenberger observer was designed for the observable part and an observer based on the consensus theory for the unobservable part was designed to jointly estimate the entire state of the system.

A limitation of existing research is the assumption that the system is autonomous or that the input information is available globally for all agents. In addition, few scholars have considered both state estimation and control for multi-agent systems with heterogeneous output [18–21]. Different from previous researches, this paper considers a class of linear time-invariant multi-agent systems with control inputs that are not globally known, and proposes an integrated protocol for distributed state estimation and consensus control under a strongly connected graph.

The details are as follows: 1) design an observer to estimate the overall output of the multi-agent system according to the local output measurement; 2) design a local observer to estimate the entire state of the multi-agent system according to the output estimation; 3) design a local consensus control protocol to realize consensus based on the state estimation.

The remainder of this paper is structured as follows. The Sect. 2 introduces the necessary basic knowledge and states the problem studied in this paper. The main results are shown in Sect. 3. The numerical simulation is illustrated in Sect. 4 and the conclusion is given in Sect. 5.

Notation: I_n stands for the $n \times n$ identity matrix. $\mathbf{1}_n$ is an all-ones vector. $\mathbb{R}^n$ denotes the set of n-dimensional real vectors, and the symbol $diag\,(s_1, s_2, \cdots, s_n)$ denotes the diagonal matrix with diagonal elements s_i that s_i can be either as a number or as a matrix. The symbol $col\,(s_1, s_2, \cdots, s_n)$ is equivalent to

$[s_1^T, s_2^T, \cdots s_n^T]^T$. $\otimes$ stands for the Kronecker product. $\sigma_{\max}(A)$ represents the maximum singular value of the matrix A, and $\lambda_{\max}(B)$ represents the maximum eigenvalue of the symmetric matrix B.

2 Preliminaries and Problem Description

2.1 Graph Theory

The communication network of the MASs containing N agents is described by an unweighted graph $\mathcal{G} = (\mathcal{V}, \mathcal{E}, \mathcal{A})$, where $\mathcal{V} = \{1, 2, \cdots, N\}$ is the set of nodes, $\mathcal{E} \subseteq \mathcal{V} \times \mathcal{V}$ is the set of edges (responding to the communication relationship between agents), and $\mathcal{A} = (a_{ij})_{N\times N}$ denotes the adjacency matrix. Define $a_{ii} = 0$, and $a_{ij} = 1$ when the node i can receive information from the node j (the node j is a neighbor of the node i), otherwise $a_{ij} = 0$ ($a_{ij} = a_{ji}$ for the undirected graph). The Laplacian matrix $\mathcal{L} = (l_{ij})_{N\times N}$ of the graph $\mathcal{G}$ is defined as

$$l_{ij} = \begin{cases} -a_{ij} & i \neq j, \\ \sum_{j=1}^{N} a_{ij} & i = j. \end{cases}$$

For a complete graph, the Laplacian matrix $\mathcal{L}_o = \left(l'_{ij}\right)_{N\times N}$ is given by

$$l'_{ij} = \begin{cases} -1 & i \neq j, \\ N-1 & i = j. \end{cases}$$

A strongly connected directed graph has a directed path between any two nodes, that is, there exists some nodes $k_1, k_2, \cdots, k_m$ between nodes i to node j such that $(i, k_1), (k_1, k_2), \cdots (k_m, j) \in \mathcal{E}$. An undirected graph is connected if there exists a connected path between any two nodes.

Assumption 1. *The communication graph $\mathcal{G}$ among the N agents is directed and strongly connected.*

2.2 Problem Description

The dynamics of a linear multi-agent system consisting of N agents without noise are described as

$$\dot{x}_i = Ax_i + Bu_i, i \in \mathcal{V}, \tag{1}$$

where $x_i \in \mathbb{R}^n$ represents the state vector, and $u_i \in \mathbb{R}^r$ is local control input. Let $x = col(x_1, x_2, \cdots, x_N)$ and $u = col(u_1, u_2, \cdots, u_N)$, obviously, we get

$$\dot{x} = (I_N \otimes A)\, x + (I_N \otimes B)\, u. \tag{2}$$

The output measurement $y_i \in \mathbb{R}^{p_i}$ of each agent is associated with the global state information of the multi-agent system, described as

$$y_i = C_i x, \tag{3}$$

where $C_i \in \mathbb{R}^{p_i \times Nn}$. Let $C = col(C_1, C_2, \cdots, C_N)$ and denote $y = col(y_1, y_2, \cdots, y_N)$ as the output measurement of all agents. Consider the MASs as a sensor network system with the dynamics of

$$\begin{aligned} \dot{x} &= (I_N \otimes A)\, x + (I_N \otimes B)\, u, \\ y &= Cx. \end{aligned} \tag{4}$$

Assumption 2. *The pair $(I_N \otimes A, C)$ is observable.*

Assumption 2 is the joint observability condition, which is widely used in distributed state estimation.

Define $\mathcal{D}_i = diag(a_{i1} I_{p1}, a_{i2} I_{p2}, \cdots, a_{iN} I_{pN})$ and $\mathcal{D} = diag(\mathcal{D}_1, \mathcal{D}_2, \cdots, \mathcal{D}_N)$, the following lemmas hold.

Lemma 1 ([14]). *Suppose the Assumption 1 holds, $\mathcal{L} \otimes I_p + \mathcal{D}$ is a non-singular M-matrix, where $p = \sum_{i=1}^{N} p_i$.*

Lemma 2 ([22]). *For any non-singular M matrix $\mathcal{M}$, there exists a diagonally positive definite matrix G such that $G\mathcal{M} + \mathcal{M}^T G > 0$.*

Lemma 3 ([23, Chap. 2, Th. 1]). *For the MASs (1) under a complete graph, the controller is designed as $\hat{u}_i = cK \sum_{j=1}^{N} (x_i - x_j)$, where $c > \frac{1}{N}$, $K = -B^T Q^{-1}$, Q is a positive definite matrix satisfying the linear matrix inequality (LMI) $AQ + QA^T - 2BB^T < 0$, then the MASs of (1) can achieve consensus.*

Remark 1. Lemma 3 illustrates the consensus control problem of MASs that the communication topology graph is a complete graph. For a general communication topology graph, the consensus control protocol can be designed as $\hat{u}_i = cK \sum_{j=1}^{N} a_{ij} (x_i - x_j)$, where a_{ij} corresponds to the elements in the adjacency matrix $\mathcal{A}$. If local agents have access to the global state information, then they can directly adopt the information of their neighbors when designing the consensus control protocol, avoiding the information interaction in the communication network. In this paper, the entire state of the multi-agent system will be estimated, meaning that all the agents' state estimation can be utilized in consensus protocol design for each agent. As a result, one can introduce the protocol under the complete graph, as is given in Lemma 3.

The goal of this paper is to design distributed observers by using local information to make each agent obtain global information (entire measurement output y and entire system state x), and to design consensus control protocol to achieve consensus.

3 Main Results

In this section, a framework for distributed output estimation, local state estimation and consensus control protocol is designed, and the convergence of estimation error and consensus error is proved.

Firstly, denote $\hat{x}_i = col(\hat{x}_{i1}, \hat{x}_{i2}, \cdots, \hat{x}_{iN})$ as the estimation of the entire state x by agent i, where $\hat{x}_{ij}$ is the state estimation of j-th agent by i-th agent. The consensus control protocol of the agent i is designed as

$$u_i = cK \sum_{j=1}^{N} (\hat{x}_{ii} - \hat{x}_{ij}), \tag{5}$$

where c and K are characterized in Lemma 3. Under control input (5), (2) converts to

$$\dot{x} = \bar{A}x + \bar{B}(\hat{x} - \mathbf{1}_N \otimes x), \tag{6}$$

where $\bar{A} = I_N \otimes A + c\mathcal{L}_o \otimes BK$, $\bar{B} = c\mathcal{L}^o \otimes BK$, $\mathcal{L}^o = diag(\mathcal{L}^1, \cdots, \mathcal{L}^N)$ that $\mathcal{L}^i$ is the i-th row of $\mathcal{L}_o$, and $\hat{x} = col(\hat{x}_1, \hat{x}_2, \cdots, \hat{x}_N)$.

The distributed output tracking protocol is proposed:

$$\dot{\hat{y}}_i = -\mu \left[\sum_{j=1}^{N} a_{ij} (\hat{y}_i - \hat{y}_j) + \mathcal{D}_i (\hat{y}_i - y) \right] + C\bar{A}\hat{x}_i, \tag{7}$$

where μ is a large positive constant and $\hat{y}_j$ is the estimation of y by the j-th agent.

Remark 2. According to the definition of $\mathcal{D}_i$, (7) is a distributed output tracking algorithm, since it uses only its own estimation of the entire state and output, and its neighbors' information of output estimation and measurement output.

The dynamics of the local state estimation are designed as

$$\dot{\hat{x}}_i = \bar{A}\hat{x}_i - F(C\hat{x}_i - \hat{y}_i), \tag{8}$$

where $F = P^{-1}C^T$, P and W are positive definite matrices satisfying the following equation:

$$\begin{aligned} -W &= (I_N \otimes \mathrm{P})(I_N \otimes \bar{A} - \mathbf{1}_N \otimes \bar{B}) \\ &\quad + (I_N \otimes \bar{A} - \mathbf{1}_N \otimes \bar{B})^T (I_N \otimes \mathrm{P}) - 2I_N \otimes C^T C \\ &< 0. \end{aligned} \tag{9}$$

Denote $\hat{y} = col(\hat{y}_1, \hat{y}_2, \cdots, \hat{y}_N)$, the dynamics of $\hat{x}$ and $\hat{y}$ are given by

$$\begin{aligned} \dot{\hat{y}} &= -\mu (\mathcal{L} \otimes I_p + \mathcal{D})(\hat{y} - \mathbf{1}_N \otimes y) + (I_N \otimes C\bar{A})\hat{x}, \\ \dot{\hat{x}} &= \left[I_N \otimes (\bar{A} - FC) \right] \hat{x} + (I_N \otimes F)\hat{y}. \end{aligned} \tag{10}$$

The distributed output tracking error is defined as $\widehat{y} = \hat{y} - \mathbf{1}_N \otimes y$, and the local state estimation error is defined as $\widehat{x} = \hat{x} - \mathbf{1}_N \otimes x$. Thus,

$$\begin{aligned} \dot{\widehat{y}} &= \dot{\hat{y}} - \mathbf{1}_N \otimes \dot{y} \\ &= -\mu \left(\mathcal{L} \otimes I_p + \mathcal{D}\right)\left(\hat{y} - \mathbf{1}_N \otimes y\right) + \left(I_N \otimes C\bar{A}\right)\hat{x} \\ &- \mathbf{1}_N \otimes \left(C\bar{A}x\right) - \mathbf{1}_N \otimes C\bar{B}(\hat{x} - \mathbf{1}_N \otimes x) \\ &= -\mu \left(\mathcal{L} \otimes I_p + \mathcal{D}\right)\widehat{y} + \left(I_N \otimes C\bar{A} - \mathbf{1}_N \otimes C\bar{B}\right)\widehat{x}, \end{aligned} \tag{11}$$

$$\begin{aligned} \dot{\widehat{x}} &= \dot{\hat{x}} - \mathbf{1}_N \otimes \dot{x} \\ &= \left[I_N \otimes \left(\bar{A} - FC\right)\right]\hat{x} + \left(I_N \otimes F\right)\hat{y} \\ &- \mathbf{1}_N \otimes \left(\bar{A}x + \bar{B}(\hat{x} - \mathbf{1}_N \otimes x)\right) \\ &= \left[I_N \otimes \left(\bar{A} - FC\right) - \mathbf{1}_N \otimes \bar{B}\right]\widehat{x} + \left(I_N \otimes F\right)\widehat{y}. \end{aligned} \tag{12}$$

Theorem 1. *Suppose that the Assumption 1 and Assumption 2 hold. If the constant μ satisfies*

$$\begin{aligned} \mu &\geq \mu_0 \\ &= \frac{1}{\lambda_0}\left[\frac{4\lambda_{\max}^2(G)\,\sigma_{\max}^2\left(I_N \otimes C\bar{A} - \mathbf{1}_N \otimes C\bar{B}\right)}{\lambda(W)} + \frac{4\sigma_{\max}^2(C)}{\lambda(W)}\right] \end{aligned} \tag{13}$$

where $\lambda_0 > 0$ is the minimum eigenvalue of $G\left(\mathcal{L} \otimes I_p + \mathcal{D}\right) + \left(\mathcal{L} \otimes I_p + \mathcal{D}\right)^T G$, and G is the diagonal positive definite matrix described in Lemma 2, then $\hat{y}_i$ and $\hat{x}_i$ can asymptotically converge to the overall measured output y and the entire state x respectively. Meanwhile, MASs (1) *can realize consensus with the control protocol* (5).

Proof. The proof of Theorem 1 contains two parts: one is the asymptotic convergence of the output tracking protocol and local state estimation proposed in this paper, and the other is that the N agents described in (1) achieve consensus in the sense of $\lim_{t\to\infty} \|x_i(t) - x_j(t)\| = 0, \forall i, j = 1, 2, \cdots, N$ under the control protocol (5) based on state estimation.

Firstly, the error system is shown to converge asymptotically. Consider the Lyapunov function

$$V = \widehat{y}^T G \widehat{y} + \widehat{x}^T \left(I_N \otimes P\right)\widehat{x}, \tag{14}$$

where G is defined in Lemma 2. According to lemma 2, the matrix $G > 0$ does exist. The time derivative of V can be written as

$$\begin{aligned} \dot{V} &= -\mu \widehat{y}^T \left[G\left(\mathcal{L} \otimes I_p + \mathcal{D}\right) + \left(\mathcal{L} \otimes I_p + \mathcal{D}\right)^T G\right]\widehat{y} \\ &+ 2\widehat{y}^T G\left[I_N \otimes C\bar{A} - \mathbf{1}_N \otimes C\bar{B}\right]\widehat{x} \\ &- \widehat{x}^T W \widehat{x} + 2\widehat{x}^T \left(I_N \otimes C^T\right)\widehat{y}. \end{aligned} \tag{15}$$

By Yong's inequality, we note that

$$\begin{aligned} & 2\widehat{y}^T G\left[I_N \otimes C\bar{A} - \mathbf{1}_N \otimes C(c\mathcal{L}^o \otimes BK)\right]\widehat{x} \\ & \leq \frac{4\lambda_{\max}^2(G)\,\sigma_{\max}^2\left(I_N \otimes C\bar{A} - \mathbf{1}_N \otimes C\bar{B}\right)}{\lambda(W)}\widehat{y}^T\widehat{y} + \frac{1}{4}\widehat{x}^T W\widehat{x}, \end{aligned} \tag{16}$$

and

$$2\widehat{x}^T\left(I_N \otimes C^T\right)\widehat{y} \leq 4\frac{\sigma_{\max}^2(C)}{\lambda(W)}\widehat{y}^T\widehat{y} + \frac{1}{4}\widehat{x}^T W\widehat{x}. \tag{17}$$

By Lemma 2, we have,

$$\begin{aligned} & -\mu\widehat{y}^T\left[G(\mathcal{L} \otimes I_p + \mathcal{D}) + (\mathcal{L} \otimes I_p + \mathcal{D})^T G\right]\widehat{y} \\ & \leq -\mu\lambda_0\widehat{y}^T\widehat{y}. \end{aligned} \tag{18}$$

From (16), (17) and (18), we can conclude that

$$\begin{aligned} \dot{V} & \leq -\lambda_0(\mu - \mu_0)\widehat{y}^T\widehat{y} - \frac{1}{2}\widehat{x}^T W\widehat{x} \\ & \leq 0. \end{aligned} \tag{19}$$

Therefore, V is bounded, so are $\widehat{y}$ and $\widehat{x}$. If $V \equiv 0$, then $\widehat{x} \equiv 0$ and $\widehat{y} \equiv 0$. It implies that $\widehat{x}$ and $\widehat{y}$ asymptotically converges to zero, meaning that the output tracking $\hat{y}_i$ asymptotically converge to y, and the local state estimation $\hat{x}_i$ asymptotically converge to entire state x.

Secondly, we show the MASs (1) can realize consensus under the control input (5). Since the $\hat{x}_i$ asymptotically convergence to x, then $u_i = cK\sum_{j=1}^{N}(\hat{x}_{ii} - \hat{x}_{ij})$ asymptotically convergence to $\hat{u}_i = cK\sum_{j=1}^{N}(x_i - x_j)$. It is clear that the N agents described in (1) can achieve consensus according to Lemma 3.

Remark 3. If the measurement output of the i-th agent fails, meaning $C_i = C_i' = \mathbf{0}$. Denote $C' = col\left(C_1, \cdots C_i', \cdots, C_N\right)$, as long as the pair $\left(I_N \otimes A, C'\right)$ is observable, the distributed state estimation and consensus control problem of the MASs can still be solved by utilizing the estimation algorithm presented in equations (7) and (8), along with the control input (5).

4 Simulation

In this section, numerical simulation results are provided to illustrate the theoretical results. Consider the MASs containing four agents with

$$A = \begin{bmatrix} 0.6 & 2 \\ -0.4 & -1.5 \end{bmatrix}, B = \begin{bmatrix} 2 \\ 1 \end{bmatrix},$$

and

$$C = \begin{bmatrix} C_1 \\ C_2 \\ C_3 \\ C_4 \end{bmatrix} = \begin{bmatrix} 2 & -2 & 3 & -5 & 1 & 5 & -6 & 0 \\ 0 & 1 & -1 & 0 & 0 & 2 & 0 & -3 \\ \hdashline 2 & -2 & 3 & 5 & 0 & 4 & -2 & 0 \\ 0 & 1 & 2 & -3 & 1 & 2 & 6 & -5 \\ 1 & -1 & 6 & 0 & -2 & 3 & 8 & 2 \end{bmatrix}.$$

The strongly connected directed graph $\mathcal{G}$ among four agents is depicted in Fig. 1. Then the adjacency matrix of the graph $\mathcal{G}$ is given by

$$\mathcal{A} = \begin{bmatrix} 0 & 0 & 0 & 1 \\ 1 & 0 & 0 & 0 \\ 1 & 1 & 0 & 0 \\ 0 & 0 & 1 & 0 \end{bmatrix},$$

and the Laplacian matrix of $\mathcal{G}$ is written as

$$\mathcal{L} = \begin{bmatrix} 1 & 0 & 0 & -1 \\ -1 & 1 & 0 & 0 \\ -1 & -1 & 2 & 0 \\ 0 & 0 & -1 & 1 \end{bmatrix}.$$

The initial state of each agent is chosen as $x_0 = \begin{bmatrix} 2 & 1 & -2 & 2 & 1 & -1 & 0 & 3 \end{bmatrix}^T$. Let $c = \frac{1}{4}$ and $\mu = 20$, solving the LMI $AQ + QA^T - 2BB^T < 0$ gives

$$Q = \begin{bmatrix} 6.3404 & -1.5213 \\ -1.5213 & 2.5514 \end{bmatrix},$$

then we obtain the feedback gain matrix K = [-0.4778 -0.6769]. Solving the equation (9) obtains $P = \begin{bmatrix} P_{11} & P_{12} \\ P_{12}^T & P_{22} \end{bmatrix}$, where

$$P_{11} = \begin{bmatrix} 1.2126 & -2.2152 & 2.1781 & -4.4453 \\ -2.2152 & 12.4962 & -6.0325 & 13.1961 \\ 2.1781 & -6.0325 & 14.2988 & -25.8695 \\ -4.4453 & 13.1961 & -25.8695 & 67.1919 \end{bmatrix},$$

$$P_{12} = \begin{bmatrix} 0.4354 & 1.3471 & -1.2243 & 2.0791 \\ -0.5903 & -4.6399 & 3.5227 & -8.8796 \\ -2.4920 & 10.4634 & 20.1462 & -37.1236 \\ 3.8313 & -19.5699 & -38.8462 & 81.4727 \end{bmatrix},$$

$$P_{22} = \begin{bmatrix} 2.9029 & -1.5245 & -8.0773 & 12.2416 \\ -1.5245 & 26.5315 & 15.3802 & -32.8396 \\ -8.0773 & 15.3802 & 59.3377 & -105.2542 \\ 12.2416 & -32.8396 & -105.2542 & 293.3230 \end{bmatrix}.$$

Thus, we can get $F = \begin{bmatrix} -0.7119 & 1.7700 & 7.2219 & -0.5964 & 1.9576 \\ 0.7373 & -0.1484 & -0.9192 & 0.1915 & -0.0896 \\ 1.2738 & -0.9522 & -1.3061 & 0.3571 & 0.6787 \\ -0.1457 & 0.0276 & 1.0093 & 0.0042 & 0.5702 \\ -0.1418 & -0.2145 & -0.5493 & 1.1534 & -0.3327 \\ 0.1604 & 0.1495 & 0.1371 & 0.0215 & -0.0134 \\ -0.8244 & 0.2437 & 0.8281 & 0.2611 & 0.4337 \\ -0.0429 & -0.0423 & -0.1892 & 0.0850 & 0.0858 \end{bmatrix}$.

The output tracking error is shown in Fig. 2(a), and the local state estimation error is depicted in Fig. 2(b). It is clearly that the output tracking and entire state estimation can asymptotically be realized. The dashed lines in Figs. 3(a) and Fig. 3(b) respectively depict the estimation of the state component one and two for each agent, and the solid lines are the actual dynamic curves of each agent, which reflect the efficiency of the consensus control of the multi-agent system. Obviously, under the control protocol (5), all agents achieve consensus. The consensus control protocol (5) is depicted in Fig. 4.

Next, we analyze the scenario where output measurement failure exists in MASs. Assuming that the output measurement of the 4-th agent fails, we have $C_4 = \mathbf{0}_{1\times 8}$ and $y_4 = \mathbf{0}$. It is evident that the MASs still satisfies the condition of joint observability. Furthermore, the initial state of each agent is set to x_0. Figure 5(a) and Fig. 5(b) show the real trajectories (solid line) and estimation trajectories (dashed line) of two state components, respectively, demonstrating that the distributed state estimation and consensus control method proposed in this paper still effective even in the presence of failures in the measurement output of agents.

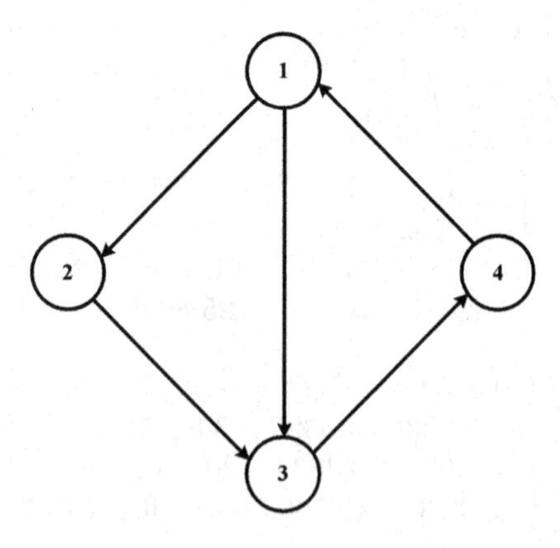

Fig. 1. The strongly connected communication graph.

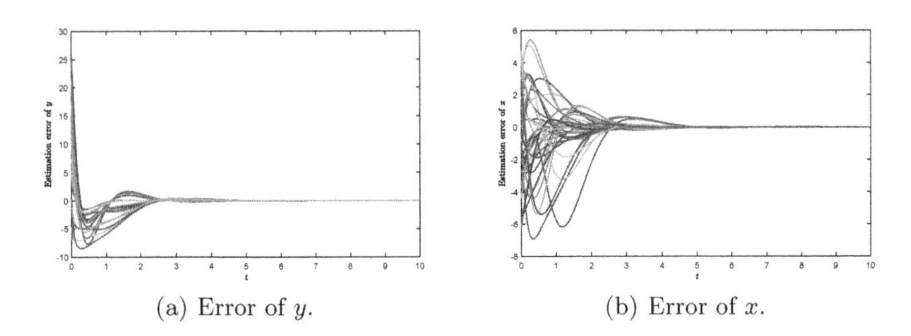

(a) Error of y. (b) Error of x.

Fig. 2. The estimation error.

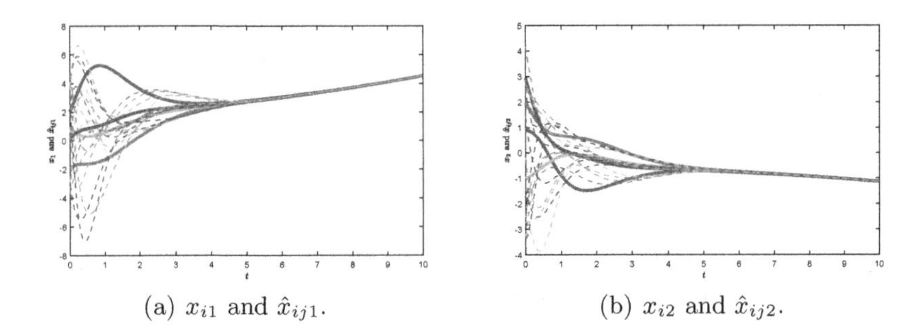

(a) x_{i1} and $\hat{x}_{ij1}$. (b) x_{i2} and $\hat{x}_{ij2}$.

Fig. 3. The real state x_i and the estimation $\hat{x}_{ij}$.

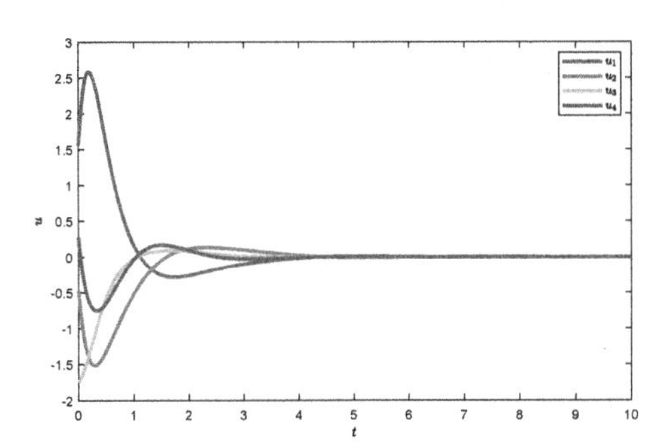

Fig. 4. The control input u.

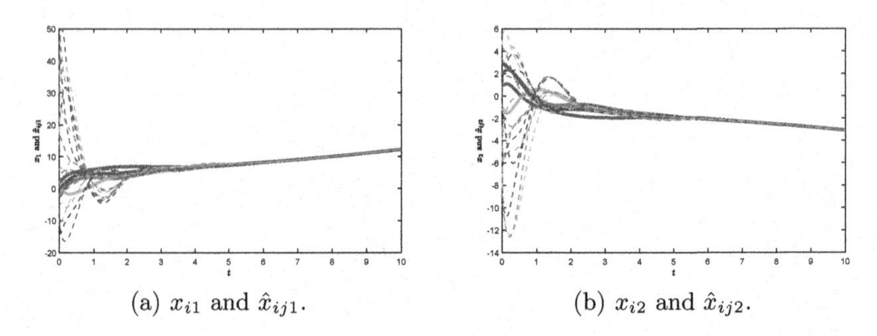

(a) x_{i1} and $\hat{x}_{ij1}$.

(b) x_{i2} and $\hat{x}_{ij2}$.

Fig. 5. The real state x_i and the estimation $\hat{x}_{ij}$ under a failure of output measurement.

5 Conclusion

In this paper, distributed state estimation for linear time-invariant MASs considering consensus control design was addressed, where each agent's local output is related to the entire state of the MASs and the information among the agents is exchanged via a strongly connected directed graph. The distributed output tracking observers are designed to estimate the entire output measurement of the MASs for each agent, and the local state estimators based on own output estimation that can avoid local state estimation to transmit among neighboring agents are proposed to make each agent obtain the entire state. Meanwhile, the distributed consensus protocol only relying on entire state estimation is introduced to ensure all agents achieve consensus. The main feature of this paper is to solve the problem of distributed state estimation and consensus control of MASs when each agent only knows its own output measurement related to all agents' states.

References

1. McArthur, S.D., et al.: Multi-agent systems for power engineering applications-part I: concepts, approaches, and technical challenges. IEEE Trans. Power Syst. **22**(4), 1743–1752 (2007)
2. Karpe, M., Fang, J., Ma, Z., Wang, C.: Multi-agent reinforcement learning in a realistic limit order book market simulation. In: Proceedings of the First ACM International Conference on AI in Finance, pp. 1–7. New York, NY, USA (2021)
3. Kar, S., Hug, G., Mohammadi, J., Moura, J.M.: Distributed state estimation and energy management in smart grids: a consensus + innovations approach. IEEE J. Sel. Top. Sig. Process. **8**(6), 1022–1038 (2014)
4. Xie, J., Liu, C.C., Sforna, M., Bilek, M., Hamza, R.: On-line physical security monitoring of power substations. Int. Trans. Electr. Energy Syst. **26**(6), 1148–1170 (2016)
5. Liu, Q., Wang, Z., He, X., Zhou, D.: On Kalman-consensus filtering with random link failures over sensor networks. IEEE Trans. Autom. Control **63**(8), 2701–2708 (2018)

6. Su, H., Li, Z., Ye, Y.: Event-triggered Kalman-consensus filter for two-target tracking sensor networks. ISA Trans. **71**, 103–111 (2017)
7. Liang, Y., Li, Y., Chen, S., Qi, G., Sheng, A.: Event-triggered Kalman consensus filter for sensor networks with intermittent observations. Int. J. Adapt. Control Sig. Process. **35**(8), 1478–1497 (2021)
8. Rezaei, H., Mahboobi Esfanjani, R., Akbari, A., Sedaaghi, M.H.: Event-triggered distributed Kalman filter with consensus on estimation for state-saturated systems. Int. J. Robust Nonlinear Control **30**(18), 8327–8339 (2020)
9. Lian, B., Lewis, F.L., Hewer, G.A., Estabridis, K., Chai, T.: Robustness analysis of distributed Kalman filter for estimation in sensor networks. IEEE Trans. Cybern. **52**(11), 12479–12490 (2022)
10. Duan, P., Lv, Y., Duan, Z., Chen, G.: Distributed state estimation for uncertain linear systems with a recursive architecture. IEEE Trans. Netw. Sci. Eng. **9**(3), 1163–1174 (2022)
11. Zhou, J., Lv, Y., Wen, G., Yu, X.: Resilient consensus of multiagent systems under malicious attacks: appointed-time observer-based approach. IEEE Trans. Cybern. **52**(10), 10187–10199 (2022)
12. Kim, T., Lee, C., Shim, H.: Completely decentralized design of distributed observer for linear systems. IEEE Trans. Autom. Control **65**(11), 4664–4678 (2020)
13. Xu, W., He, W., Ho, D.W., Kurths, J.: Fully distributed observer-based consensus protocol: adaptive dynamic event-triggered schemes. Automatica **139**, 110188 (2022)
14. Li, Y., Lv, Y., Duan, P., Zhou, J.: Output tracking based distributed state estimation. In: 2021 IEEE International Conference on Unmanned Systems (ICUS), pp. 146–150. IEEE, Beijing, China (2021)
15. Duan, P., Qian, J., Wang, Q., Duan, Z., Shi, L.: Distributed state estimation for continuous-time linear systems with correlated measurement noise. IEEE Trans. Autom. Control **67**(9), 4614–4628 (2022)
16. Han, W., Trentelman, H.L., Wang, Z., Shen, Y.: A simple approach to distributed observer design for linear systems. IEEE Trans. Autom. Control **64**(1), 329–336 (2019)
17. Silm, H., Efimov, D., Michiels, W., Ushirobira, R., Richard, J.P.: A simple finite-time distributed observer design for linear time-invariant systems. Syst. Control Lett. **141**, 104707 (2020)
18. Zhou, J., Wu, X., Lv, Y., Wen, G.: Terminal-time synchronization of multiple vehicles under discrete-time communication networks with directed switching topologies. IEEE Trans. Circ. Syst. II Express Briefs **67**(11), 2532–2536 (2020)
19. Wang, Q., Duan, Z., Wang, J.: Distributed optimal consensus control algorithm for continuous-time multi-agent systems. IEEE Trans. Circ. Syst. II Express Briefs **67**(1), 102–106 (2020)
20. Zhou, J., Wu, X., Lv, Y., Li, X., Liu, Z.: Recent progress on the study of multi-vehicle coordination in cooperative attack and defense: an overview. Asian J. Control **24**(2), 794–809 (2022)
21. Wang, Q., Duan, Z., Wang, J., Wang, Q., Chen, G.: An accelerated algorithm for linear quadratic optimal consensus of heterogeneous multiagent systems (2022)
22. Plemmons, R.J.: M-matrix characterizations. I-nonsingular m-matrices. Linear Algebra Appl. **18**(2), 175–188 (1977)
23. Li, Z., Duan, Z.: Cooperative Control of Multi-Agent Systems: A Consensus Region Approach. CRC Press (2014)

Integrated Design of Fully Distributed Adaptive State Estimation and Consensus Control for Multi-agent Systems

Jiazhu Huang[1,2], Yan Li[1], and Yuezu Lv[3](✉)

[1] School of Automation, Beijing Institute of Technology, Beijing, China
jzh@bit.edu.cn, yanl@bit.edu.cn
[2] MIIT Key Laboratory of Complex-Field Intelligent Sensing, Yangtze Delta Region Academy of Beijing Institute of Technology, Jiaxing, China
[3] Advanced Research Institute of Multidisciplinary Science, Beijing Institute of Technology, Beijing, China
yzlv@bit.edu.cn

Abstract. In this paper, the problem of fully distributed adaptive state estimation and consensus control for linear multi-agent systems is investigated. By designing fully distributed adaptive output tracking observers, the entire output information is available to each agent, and local state estimators based on the estimated output is constructed to estimate the overall state of multi-agent systems. The consensus control protocol based on state estimation for multi-agent systems is designed to ensure the agents achieve consensus. The proposed control input of each agent relies on its own estimation of the entire state. Theoretical analysis proves the effectiveness of the algorithm and practical applications are given by simulation.

Keywords: Multi-agent systems · Consensus control · Distributed state estimation · Adaptive output tracking

1 Introduction

In recent years, multi-agent systems have received more and more attention due to their in-depth research and applications in many fields such as energy systems [1], sensor networks [2], and formation flight of unmanned aerial vehicles [3]. The task types of multi-agent systems can be divided into consensus control, formation control, leader-following control, cluster control, and output regulation [4–6]. Among these problems, consensus control is a hot topic in research [7–12]. State feedback is a common approach used to complete the task, but not all state of the system can be directly obtained due to measurement limits. Therefore, state estimation algorithms are necessary to estimate state based on available output measurement and interaction among agents.

This work was supported by the National Natural Science Foundation of China under Grants 62088101 and 62273045, and the Fellowship of China Postdoctoral Science Foundation under Grants 2021TQ0039 and 2022M710384.

B. Luo et al. (Eds.): ICONIP 2023, LNCS 14447, pp. 226–237, 2024.
https://doi.org/10.1007/978-981-99-8079-6_18

When designing a consensus control protocol for multi-agent systems, it is generally first necessary to use the dynamics of the agents to obtain a feedback gain matrix, and then determine the parameters related to the communication topology. Some consensus control protocols in the above papers are not fully distributed since they need to use some global information, such as the non-zero eigenvalues of the Laplacian matrix. Adaptive gain can be introduced to automatically adjust the parameters to avoid knowing the global information of the topology graph, indicating that the consensus control protocols are fully distributed [13–15].

Distributed state estimation is a commonly used method for estimating the state of multi-agent systems. It can be broadly categorized into two types: Kalman-filter-based methods and observer-based approaches. In 2005, distributed Kalman filtering was first proposed and has since been rapidly evolving [16–18]. On the other hand, a distributed observer consists of multiple local observers, each responsible for estimating the state of a local subsystem [19–21]. The local observers are exchanged via communication to improve the estimation accuracy of the global state. The distributed observer is capable of handling large-scale systems and is robust to failures in individual agents or sensors. However, when performing distributed state estimation for multi-agent systems, it is generally assumed that the agents do not have control input or that the input is globally known, which may not always be reasonable in practice. This paper considers the case of locally knowable inputs and heterogeneous outputs, which poses a more challenging state estimation problem.

Both state estimation and consensus control of multi-agent systems have abundant results, but few consider these two aspects simultaneously. In this paper, one proposes an integrated design framework for distributed state estimation and consensus control. Firstly, the fully distributed adaptive output tracking algorithm is proposed such that each agent can estimate the overall output measurement, and the adaptive gain avoids using global information. Then the local state observer that relies on its own estimation of the overall output is constructed to estimate the entire state of multi-agent systems. Finally, the consensus control input is designed based on its own estimation of the overall state, and all agents realize consensus when the state estimation converges asymptotically. The issue of sensor output failures is considered for the agents.

The rest of this paper is structured as follows. Section 2 illustrates the problem studied in this paper. Section 3 contains the algorithms for consensus control and state estimation. The simulation results are given in Sect. 4. Finally, this paper is concluded in Sect. 5.

2 Problem Statement

Conside a multi-agent system consisting of N agents, the dynamics of each agent can be described by

$$\dot{x}_i = Ax_i + Bu_i, i = 1, \cdots, N, \tag{1}$$

where $x_i \in \mathbb{R}^n$ and $u_i \in \mathbb{R}^p$ are the state and control input of i-th agent, respectively; $A \in \mathbb{R}^{n\times n}$ and $B \in \mathbb{R}^{n\times p}$ are the known state and input matrices. Let $x = \left[x_1^T, \cdots, x_N^T\right]^T$, $u = \left[u_1^T, \cdots, u_N^T\right]^T$, one can get

$$\dot{x} = (I_N \otimes A)\, x + (I_N \otimes B)\, u, \tag{2}$$

where I_N represents the $N \times N$ identity matrix and $\otimes$ stands for the Kronecker product. The output measurement of the system is given by

$$y_i = C_i x, i = 1, \cdots, N, \tag{3}$$

where $y_i \in \mathbb{R}^{q_i}$ is the measured output of i-th agent and $C_i \in \mathbb{R}^{q_i \times Nn}$ is the output matrix. Let $y = \left[y_1^T, \cdots, y_N^T\right]^T$, $C = \left[C_1^T, \cdots, C_N^T\right]^T$, one can obtain

$$y = Cx. \tag{4}$$

Assumption 1. *The matrix pair $(I_N \otimes A, C)$ is observable.*

The communication topology among the N agents of the multi-agent system is described by a directed graph $\mathcal{G} = (\mathcal{V}, \mathcal{E}, \mathcal{A})$, where $\mathcal{V} = \{v_1, \cdots, v_N\}$ denotes the set of nodes, $\mathcal{E} \subseteq \mathcal{V} \times \mathcal{V}$ is the set of edges and $\mathcal{A} = [a_{ij}]_{N\times N}$ represents the adjacency matrix. $(v_i, v_j) \in \mathcal{E}$ means that node v_j can obtain information from node v_i, and v_i is also called a neighbor of v_j. For node v_{i_1} and node v_{i_k}, there is a directed path between v_{i_1} and v_{i_k} if there exists a series of adjacent edges $\left(v_{i_l}, v_{i_{l+1}}\right), l = 1, \cdots, k-1$. A directed graph is said to be strongly connected if there is a directed path from every node to all other nodes. Suppose there is no self-loop in the communication topology graph, the diagonal elements of the adjacency matrix $\mathcal{A}$ are represented by $a_{ii} = 0$. If $(v_j, v_i) \in \mathcal{E}$ then $a_{ij} = 1$, otherwise $a_{ij} = 0$. The degree matrix of the directed graph can be given by the diagonal matrix $\mathcal{D}$ with $d_{ii} = \sum_{j=1}^{N} a_{ij}$. Then the Laplacian matrix $\mathcal{L}$ corresponding to the graph can be defined as $\mathcal{L} = \mathcal{D} - \mathcal{A}$.

Assumption 2. *The directed graph $\mathcal{G}$ of the multi-agent system is strongly connected.*

Lemma 1 ([22]). *For the Laplacian matrix $\mathcal{L}$ and the adjacency matrix $\mathcal{A}$ of the strongly connected graph $\mathcal{G}$, let $\mathcal{A}^j = diag\left(a_{1j}, \cdots, a_{Nj}\right), j = 1, \cdots, N$, then $\mathcal{L}^j = \mathcal{L} + \mathcal{A}^j$ is a non-singular M-matrix.*

Lemma 2 ([23]). *For a non-singular M-matrix $\mathcal{M}$, there exists a diagonal matrix $G > 0$ such that $G\mathcal{M} + \mathcal{M}^T G > 0$.*

The purpose of this paper is to design fully distributed state estimators and output tracking observers so that each agent can estimate the state and the output of the entire system, i.e., $\lim_{t\to\infty} \|\hat{x}_i(t) - x(t)\| = 0$, $\lim_{t\to\infty} \|\hat{y}_i(t) - y(t)\| = 0$, $i = 1, \cdots, N$, and to devise the consensus protocol based on state estimation so that the agents in the system can achieve consensus, i.e., $\lim_{t\to\infty} \|x_i(t) - x_j(t)\| = 0$, $i, j = 1, \cdots, N$.

3 Main Results

3.1 Consensus Control

Each agent will perform state estimation for the N agents in the system, which means that every agent can use the entire state information of all agents in designing the consensus control protocol. The communication topology graph can be considered as an arbitrary known strongly connected graph $\mathcal{G}_o$, whose Laplacian matrix is denoted as $\mathcal{L}_o$ and adjacency matrix is represented by $\mathcal{A}_o$. A control protocol is illustrated as following to make the N agents in the system achieve consensus.

For N agents in (1), the control protocol is designed to be

$$\hat{u}_i = cK \sum\nolimits_{j=1}^{N} a_{ij_o} (x_i - x_j), \tag{5}$$

where a_{ij_o} is the element of the adjacency matrix $\mathcal{A}_o$, the parameters c and the matrix K in the control protocol are selected as $c > \frac{1}{\mathrm{Re}(\lambda_2)}$, where λ_2 is the the minimum non-zero eigenvalue of the Laplacian matrix $\mathcal{L}_o$ and $\mathrm{Re}(\lambda_2)$ represents its real part, and $K = -B^T P^{-1}$, where $P > 0$ is a solution of the linear matrix inequality $AP + PA^T - 2BB^T < 0$.

Lemma 3 ([24]). *Suppose Assumption 2 holds. Under the control protocol in* (5)*, the N agents in multi-agent system* (1) *can achieve consensus.*

Remark 1. If the control protocol is designed as $\hat{u}_i = cK \sum_{j=1}^{N} a_{ij} (x_i - x_j)$, then each agent needs non-zero eigenvalues of the Laplacian matrix $\mathcal{L}$ to calculate the parameter c. In such case, the control design is not fully distributed. Since each agent will estimate the state of all the agents in the system, the communication topology in the control input design can be chosen as a known strongly connected directed graph which is able to make the agents achieve consensus in a fully distributed manner.

3.2 Distributed Adaptive Output Tracking and State Estimation

Denote the state estimation of i-th agent as $\hat{x}_i = \left[\hat{x}_{i1}^T, \cdots, \hat{x}_{iN}^T\right]^T$. State estimation is required for consensus control of the system because the state is not directly available. The estimation of the control input is designed as follows:

$$u_i = cK \sum_{j=1}^{N} a_{ij_o} (\hat{x}_{ii} - \hat{x}_{ij}). \tag{6}$$

Substituting (6) to (2), the dynamics of the system can be expressed as

$$\begin{aligned} \dot{x} &= \hat{A}x + \mathcal{B}(\hat{x} - \mathbf{1}_N \otimes x), \\ y &= Cx, \end{aligned} \tag{7}$$

where $\hat{A} = I_N \otimes A + c\mathcal{L}_o \otimes BK$, $\mathcal{B} = c\mathcal{L}^o \otimes BK$, $\mathcal{L}^o = diag\left(\mathcal{L}_o^1, \cdots, \mathcal{L}_o^N\right)$, and $\mathcal{L}_o^i$ is i-th row of the Laplacian matrix $\mathcal{L}_o$. $\mathbf{1}_N$ is an N-dimensional column vector whose elements are all 1.

For the system described in (7), one needs to design distributed state estimators for each agent to estimate the state x. The control protocol will update based on the estimated state in real time to achieve consensus. All agents cannot directly use the entire outputs to build state estimators since they only have access to the output information of their own and neighbors. In view of this, the adaptive distributed output tracking observers that enable each agent to estimate the entire output are proposed, then state estimators are designed that are based on output estimation and enable each agent to achieve the estimation of the state.

Following above idea, the distributed adaptive output tracking observers and state estimators are proposed:

$$\begin{aligned} \dot{\hat{y}}_{ij} &= -\left(\varphi_{ij} + \psi_{ij}\right)\left[\sum_{k=1}^{N} a_{ik}\left(\hat{y}_{ij} - \hat{y}_{kj}\right) + a_{ij}\left(\hat{y}_{ij} - y_j\right)\right] + C_j \hat{A}\hat{x}_i, \\ \dot{\hat{x}}_i &= \left(\hat{A} - FC\right)\hat{x}_i + F\hat{y}_i, \end{aligned} \tag{8}$$

where $\hat{y}_{ij}$ represents the output estimation of i-th agent to j-th agent, $\hat{y}_i = \left[\hat{y}_{i1}^T, \cdots, \hat{y}_{iN}^T\right]^T$, φ_{ij} is the adaptive gain with an initial positive value, and its dynamic representation is $\dot{\varphi}_{ij} = \psi_{ij} = \left\|\sum_{k=1}^{N} a_{ik}\left(\hat{y}_{ij} - \hat{y}_{kj}\right) + a_{ij}\left(\hat{y}_{ij} - y_j\right)\right\|^2$, and F is the feedback gain matrix to be determined.

Remark 2. The introduction of adaptive gain φ_{ij} avoids the utilization of the global graph information, which means the proposed algorithm is fully distributed. φ_{ij} can automatically adjust its value, and when each agent has achieved an estimate of the overall output, the adaptive gain will converge to a certain positive constant.

Define $\tilde{y}_{ij} = \hat{y}_{ij} - y_j$ the error of output estimation of agent j by agent i, and $\tilde{x}_i = \hat{x}_i - x$ the estimation error of the state, then $\tilde{y}_{ij}$ and $\tilde{x}_i$ satisfy the following dynamics:

$$\begin{aligned} \dot{\tilde{y}}_{ij} &= -\left(\varphi_{ij} + \psi_{ij}\right)\left[\sum_{k=1}^{N} a_{ik}\left(\tilde{y}_{ij} - \tilde{y}_{kj}\right) + a_{ij}\tilde{y}_{ij}\right] + C_j\hat{A}\tilde{x}_i - C_j\mathcal{B}\tilde{x}, \\ \dot{\varphi}_{ij} &= \psi_{ij} = \left\|\sum_{k=1}^{N} a_{ik}\left(\tilde{y}_{ij} - \tilde{y}_{kj}\right) + a_{ij}\tilde{y}_{ij}\right\|^2, \\ \dot{\tilde{x}}_i &= \left(\hat{A} - FC\right)\tilde{x}_i - \mathcal{B}\tilde{x} + F\tilde{y}_i. \end{aligned} \tag{9}$$

Denote $\tilde{y} = \left[\tilde{y}_1^T, \cdots, \tilde{y}_N^T\right]^T$ and $\tilde{y}^j = \left[\tilde{y}_{1j}^T, \cdots, \tilde{y}_{Nj}^T\right]^T$, it is easy to get $\tilde{y}^T\tilde{y} = \sum_{j=1}^{N}\left(\tilde{y}^j\right)^T\tilde{y}^j$. The dynamics of $\tilde{y}^j$ are presented by

$$\dot{\tilde{y}}^j = -\left[\Gamma^j\mathcal{L}^j \otimes I_{q_j}\right]\tilde{y}^j + \left(I_N \otimes C_j\hat{A} - \mathbf{1}_N \otimes C_j\mathcal{B}\right)\tilde{x}, \tag{10}$$

where $\Gamma^j = \Phi^j + \Psi^j$, $\Phi^j = diag(\varphi_{1j}, \cdots, \varphi_{Nj})$, $\Psi^j = diag(\psi_{1j}, \cdots, \psi_{Nj})$, and $\mathcal{L}^j = \mathcal{L} + \mathcal{A}^j$ is a non-singular M-matrix defined in Lemma 1.

Let $\eta_{ij} = \sum_{k=1}^{N} a_{ik} (\tilde{y}_{ij} - \tilde{y}_{kj}) + a_{ij}\tilde{y}_{ij}$, and $\eta^j = \left[\eta_{1j}^T, \cdots, \eta_{Nj}^T\right]^T$, then one can get $\eta^j = \left(\mathcal{L}^j \otimes I_{q_j}\right) \tilde{y}^j$. The dynamics of η^j are written as

$$\begin{aligned} \dot{\eta}^j &= -\left[\mathcal{L}^j \Gamma^j \otimes I_{q_j}\right] \eta^j + \left(\mathcal{L}^j \otimes C_j \hat{A} - \alpha_j \otimes C_j \mathcal{B}\right) \tilde{x}, \\ \dot{\varphi}_{ij} &= \psi_{ij} = \eta_{ij}^T \eta_{ij}, \end{aligned} \tag{11}$$

where $\alpha_j = [a_{1j}, \cdots, a_{Nj}]^T$.

Theorem 1. *Suppose Assumption 1 and Assumption 2 hold. The feedback gain matrix F is chosen to be $F = Q^{-1}C^T$, where Q is a positive definite matrix satisfying the following linear matrix inequality*

$$\begin{aligned} -W = &(I_N \otimes Q) \left[I_N \otimes \left(\hat{A} - FC\right) - \mathbf{1}_N \otimes \mathcal{B}\right] \\ &+ \left[I_N \otimes \left(\hat{A} - FC\right) - \mathbf{1}_N \otimes \mathcal{B}\right]^T (I_N \otimes Q) < 0. \end{aligned} \tag{12}$$

Then $\hat{y}_{ij}$ can asymptotically track the output y_j of j-th agent, $\hat{x}_i$ can asymptotically estimate the state x of N agents, and the adaptive gain φ_{ij} is bounded. Meanwhile, the N agents in (7) *can achieve consensus under the control protocol* (6).

Proof. The Lyapunov function is chosen as

$$V = \mu \tilde{x}^T (I_N \otimes Q) \tilde{x} + \sum_{j=1}^{N} V_j, \tag{13}$$

where Q is the positive-definite matrix defined in (12), and V_j is chosen as $V_j = \sum_{i=1}^{N} \frac{g_{ij}}{2} \left[(2\varphi_{ij} + \psi_{ij}) \psi_{ij} + \left(\varphi_{ij} - \nu^j\right)^2\right]$, g_{ij} are the diagonal elements of the diagonal positive definite matrix G^j satisfying $G^j \mathcal{L}^j + \left(\mathcal{L}^j\right)^T G^j > 0$. μ and ν^j are positive constants that are to be determined later.

The time derivative of V is

$$\dot{V} = -\mu \tilde{x}^T W \tilde{x} + 2\mu \tilde{x}^T (I_N \otimes QF) \tilde{y} + \sum_{j=1}^{N} \dot{V}_j. \tag{14}$$

According to Young's inequality, one can get

$$\begin{aligned} 2\mu \tilde{x}^T (I_N \otimes QF) \tilde{y} &\leq \frac{\mu}{4} \tilde{x}^T W \tilde{x} + 4\mu \tilde{y}^T (I_N \otimes QF)^T W^{-1} (I_N \otimes QF) \tilde{y} \\ &\leq \frac{\mu}{4} \tilde{x}^T W \tilde{x} + \frac{4\mu \sigma_{\max}^2 (I_N \otimes QF)}{\lambda_{\min}(W)} \sum_{j=1}^{N} \left(\tilde{y}^j\right)^T \tilde{y}^j \\ &\leq \frac{\mu}{4} \tilde{x}^T W \tilde{x} + \sum_{j=1}^{N} \frac{4\mu \sigma_{\max}^2 (I_N \otimes QF)}{\lambda_{\min}(W) \sigma_{\min}^2 (\mathcal{L}^j)} \left(\eta^j\right)^T \eta^j, \end{aligned} \tag{15}$$

where $\sigma_{\max}(Z)$ represents the maximum singular value of the matrix Z, and the time derivative of V_j is

$$
\begin{aligned}
\dot{V}_j &= \sum_{i=1}^{N} g_{ij}\left[(\varphi_{ij}+\psi_{ij})\,\dot{\psi}_{ij}+\left(\varphi_{ij}+\psi_{ij}-\nu^j\right)\dot{\varphi}_{ij}\right] \\
&= -(\eta^j)^T\left[\Gamma^j\left(G^j\mathcal{L}^j+(\mathcal{L}^j)^T G^j\right)\Gamma^j\otimes I_{q_j}\right]\eta^j \\
&+2(\eta^j)^T\left[\Gamma^j G^j\otimes I_{q_j}\right]\left(\mathcal{L}^j\otimes C\hat{A}-\alpha_j\otimes C_j\mathcal{B}\right)\tilde{x} \\
&+(\eta^j)^T\left[\Gamma^j G^j\otimes I_{q_j}\right]\eta^j-\nu^j(\eta^j)^T\left(G^j\otimes I_{q_j}\right)\eta^j \\
&\le -\lambda_{0j}(\eta^j)^T\left[(\Gamma^j)^2\otimes I_{q_j}\right]\eta^j-\nu^j\lambda_{\min}\left(G^j\right)(\eta^j)^T\eta^j \\
&+2(\eta^j)^T\left[\Gamma^j G^j\otimes I_{q_j}\right]\left(\mathcal{L}^j\otimes C\hat{A}-\alpha_j\otimes C_j\mathcal{B}\right)\tilde{x} \\
&+(\eta^j)^T\left[\Gamma^j G^j\otimes I_{q_j}\right]\eta^j,
\end{aligned}
\tag{16}
$$

where λ_{0j} is the smallest eigenvalue of the positive definite matrix $G^j\mathcal{L}^j + (\mathcal{L}^j)^T G^j$. According to Young's inequality, it is obtained that

$$
\begin{aligned}
&2(\eta^j)^T\left[\Gamma^j G^j\otimes I_{q_j}\right]\left(\mathcal{L}^j\otimes C\hat{A}-\alpha_j\otimes C_j\mathcal{B}\right)\tilde{x} \\
&\le \frac{\lambda_{0j}}{2}(\eta^j)^T\left[(\Gamma^j)^2\otimes I_{q_j}\right]\eta^j \\
&+\frac{2\sigma_{\max}^2\left(G^j\mathcal{L}^j\otimes C\hat{A}-G^j\alpha_j\otimes C_j\mathcal{B}\right)}{\lambda_{0j}\lambda_{\min}(W)}\tilde{x}^T W\tilde{x}.
\end{aligned}
\tag{17}
$$

The constants μ and ν^j are given by $\mu = \sum_{j=1}^{N}\frac{8\sigma_{\max}^2\left(G^j\mathcal{L}^j\otimes C\hat{A}-G^j\alpha_j\otimes C_j\mathcal{B}\right)}{\lambda_{0j}\lambda_{\min}(W)}$, $\nu^j = \frac{4\mu\sigma_{\max}^2(I_N\otimes QF)}{\lambda_{\min}(G^j)\lambda_{\min}(W)\sigma_{\min}^2(\mathcal{L}^j)} + 8\frac{\lambda_{\max}^2(G^j)}{\lambda_{0j}\lambda_{\min}(G^j)}$ respectively. It is easy to get

$$
\begin{aligned}
&-\frac{\lambda_{0j}}{2}(\eta^j)^T\left[(\Gamma^j)^2\otimes I_{q_j}\right]\eta^j-8\frac{\lambda_{\max}^2\left(G^j\right)}{\lambda_{0j}}(\eta^j)^T\eta^j \\
&\le -2(\eta^j)^T\left[\Gamma^j G^j\otimes I_{q_j}\right]\eta^j.
\end{aligned}
\tag{18}
$$

Substituting (15)-(18) into (14) yields

$$
\begin{aligned}
\dot{V} &\le -\frac{\alpha}{2}\tilde{x}^T W\tilde{x}-\sum_{j=1}^{N}(\eta^j)^T\left[\Gamma^j G^j\otimes I_{m_j}\right]\eta^j \\
&\le -\frac{\alpha}{2}\tilde{x}^T W\tilde{x}-\sum_{j=1}^{N}(\eta^j)^T\left[\Phi^j(0)\,G^j\otimes I_{m_j}\right]\eta^j \\
&\le 0.
\end{aligned}
\tag{19}
$$

According to (13) and (19), one can get $V \ge 0$ and $\dot{V} \le 0$, thus $V(t)$ is bounded, and it means that $\tilde{x}$, η^j, φ_{ij} are also bounded. If $\dot{V} \equiv 0$, then $\tilde{x} \equiv 0$

and $\eta^j \equiv 0$. One can have $\lim_{t\to\infty}\tilde{x}(t) = 0$, $\lim_{t\to\infty}\eta^j(t) = 0$ owing to LaSalle's Invariant Principle [25]. Moreover, one can get $\lim_{t\to\infty}\tilde{y}(t) = 0$. Therefore the designed distributed adaptive tracker $\hat{y}_{ij}$ can asymptotically track to y_j at $t \to \infty$ and the state estimation $\hat{x}_i$ can asymptotically converge to the state x of the system. Then $\hat{u}_i$ will converge to u_i, and according to Lemma 3, the N agents in the system can achieve consensus.

Remark 3. When the sensors of i-th agent fail, the output of that agent cannot be measured, i.e., $y_i = [0, \cdots, 0]^T$, which means that all elements in C_i are set to zero. If the state estimation is performed only based on the agent's own output information, the system state can never be obtained by this agent when the measurement output fails. By using the state estimation equation (8) designed in the paper, when the agent's sensors fail, all elements in C_i are changed to zero, and the C in equation (4) is modified to $C = \left[C_1^T, \cdots, 0^{q_i \times Nn}, \cdots, C_N^T\right]^T$. If assumption 1 holds, the agent can still estimate the system state.

4 Simulation

In this section, the proposed algorithm is simulated and verified.

Consider five agents whose communication topology graph $\mathcal{G}$ is shown in Fig. 1. The adjacency matrix $\mathcal{A}$ of the graph $\mathcal{G}$ is

$$\mathcal{A} = \begin{bmatrix} 0&1&0&0&0 \\ 0&0&1&0&0 \\ 0&0&0&1&0 \\ 0&0&0&0&1 \\ 1&0&0&0&0 \end{bmatrix}.$$

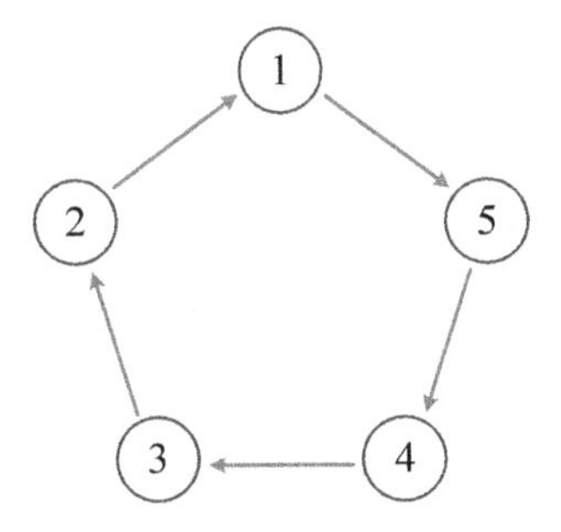

Fig. 1. The strongly connected communication graph.

The known matrices in the agent dynamics are chosen as

$$A = \begin{bmatrix} -1.5 & -8 \\ 0.1 & 0.6 \end{bmatrix}, B = \begin{bmatrix} 1 \\ 1 \end{bmatrix},$$

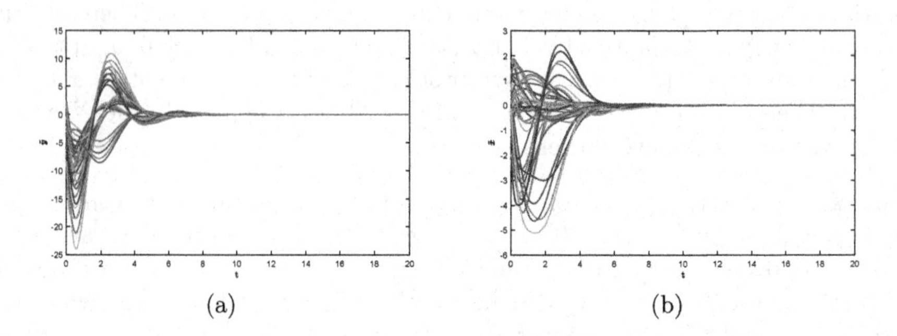

(a) (b)

Fig. 2. The output tracking error and state estimation error.

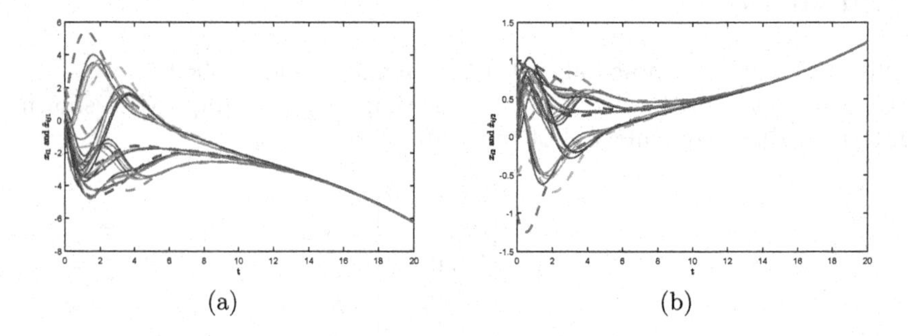

(a) (b)

Fig. 3. Trajectories of the state x and the local state estimation $\hat{x}$ under (8).

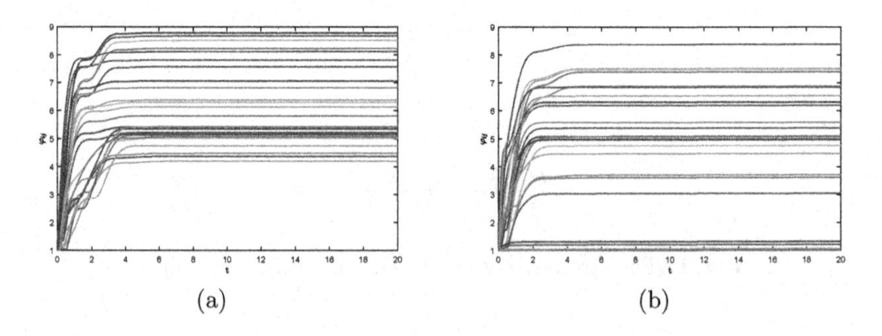

(a) (b)

Fig. 4. Trajectories of the adaptive gain φ_{ij}.

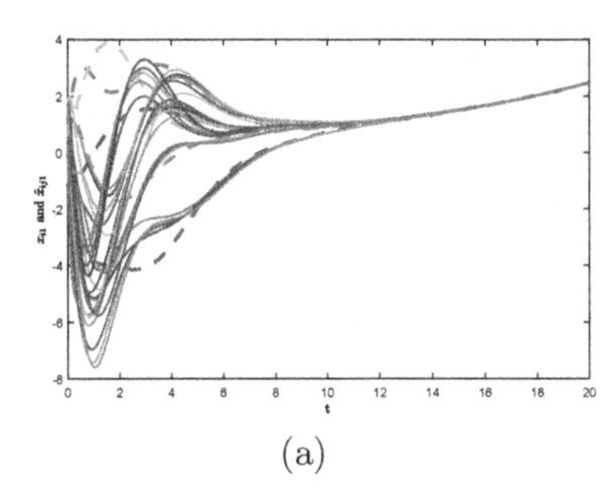

(a)

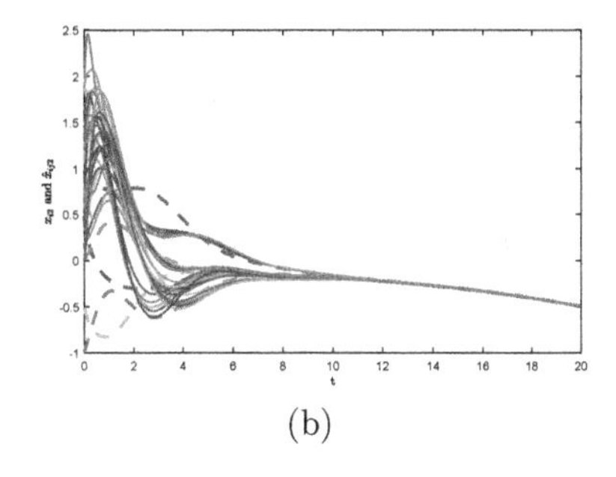

(b)

Fig. 5. Trajectories of the state x and the local state estimation $\hat{x}$ under (8) with failure.

and

$$C = \begin{bmatrix} C_1 \\ C_2 \\ C_3 \\ C_4 \\ C_5 \end{bmatrix} = \begin{bmatrix} 2 & -1 & 3 & -5 & 1 & 5 & -6 & 0 & 0 & -1 \\ \hdashline 2 & -2 & 3 & 5 & 0 & 4 & -2 & 0 & 3 & -2 \\ \hdashline 0 & 1 & 2 & -3 & 1 & 2 & -6 & -3 & 2 & 1 \\ \hdashline 1 & -1 & 6 & 0 & -2 & 3 & 8 & 2 & 2 & -2 \\ \hdashline 0 & 2 & -1 & 1 & 3 & 2 & -1 & 1 & 0 & -3 \\ 3 & 0 & 1 & -2 & 0 & -2 & 1 & 4 & 0 & -1 \end{bmatrix},$$

where $C_1, C_2, C_3, C_4 \in \mathbb{R}^{1\times 10}$ and $C_5 \in \mathbb{R}^{2\times 10}$.

According to the linear matrix inequality in Lemma 3, one can get

$$P = \begin{bmatrix} 35.5939 & -3.2511 \\ -3.2511 & 1.2582 \end{bmatrix},$$

then the feedback gain matrix in the consensus control is $K = \begin{bmatrix} -0.1318 & -1.1353 \end{bmatrix}$. The known strongly connected graph $\mathcal{G}_o$ is selected as a complete graph, whose Laplacian matrix is

$$\mathcal{L}_o = \begin{bmatrix} 4 & -1 & -1 & -1 & -1 \\ -1 & 4 & -1 & -1 & -1 \\ -1 & -1 & 4 & -1 & -1 \\ -1 & -1 & -1 & 4 & -1 \\ -1 & -1 & -1 & -1 & 4 \end{bmatrix}.$$

The coupling gain is taken as $c = 0.4$. The initial value of the adaptive gain φ_{ij} is set to $\varphi_{ij}(0) = 1$. According to the linear matrix inequality (12), one can get Q and F.

The output tracking error and the state estimation error are given in Fig. 2(a) and Fig. 2(b), which imply that the output tracking and state estimation can asymptotically converge to the true output and state, respectively. The state of the system and the trajectories of the state estimation are depicted in Fig. 3, both revealing that the system can achieve consensus. Figure 4(a) shows the adaptive gain φ_{ij}, which implies that it is bounded.

Then, one assumes that the output of 4th agent in the system fails, i.e., $y_4 = 0_{1\times 10}$. One can verify that assumption 1 still holds, and the system state can be estimated despite the failure. The system state and the estimation of each agent's state are shown in Fig. 5, indicating that the agents are still able to achieve consensus and each agent completes the estimation of the system state with the presence of failure. φ_{ij} is depicted in Fig. 4(b), demonstrating that the adaptive gain remains bounded.

5 Conclusion

In this paper, one proposes an integrated approach for distributed state estimation and consensus control for linear multi-agent systems. The fully distributed adaptive output tracking algorithm is proposed to make each agent eatimate the entire output measurement of the system, while the local state estimators enable each agent to obtain the entire state without information interaction between communication topological networks. The distributed output measurement tracking protocol, which introduces adaptive gain to avoid using global algebraic connectivity information, is indeed a distributed algorithm. Moreover, the consensus control protocol that relies only on state estimation is proposed. The estimation contains the state information of all agents in the system, thus the consensus control input no longer needs the state information of neighbors. Then the consensus problem of multi-agent systems under the general strongly connected graph can transform into the consensus problem under a strongly connected directed graph or an undirected connected graph. Future work can be done to develop an integrated design approach of distributed state estimation and consensus control for multi-agent systems with noise or nonlinear.

References

1. Azzollini, I.A., Yu, W., Yuan, S., Baldi, S.: Adaptive leader-follower synchronization over heterogeneous and uncertain networks of linear systems without distributed observer. IEEE Trans. Autom. Control **66**(4), 1925–1931 (2020)
2. Schenato, L., Fiorentin, F.: Average timeSynch: a consensus-based protocol for clock synchronization in wireless sensor networks. Automatica **47**(9), 1878–1886 (2011)
3. Sun, J., Geng, Z., Lv, Y.: Adaptive output feedback consensus tracking for heterogeneous multi-agent systems with unknown dynamics under directed graphs. Syst. Control Lett. **87**, 16–22 (2016)
4. Lv, Y., Fu, J., Wen, G., Huang, T., Yu, X.: Fully distributed anti-windup consensus protocols for linear mass with input saturation: the case with directed topology. IEEE Trans. Cybern. **51**(5), 2359–2371 (2021)
5. Jiang, W., Wen, G., Peng, Z., Huang, T., Rahmani, A.: Fully distributed formation-containment control of heterogeneous linear multiagent systems. IEEE Trans. Autom. Control **64**(9), 3889–3896 (2019)
6. Wang, Q., Duan, Z., Wang, J.: Distributed optimal consensus control algorithm for continuous-time multi-agent systems. IEEE Trans. Circuits Syst. II Express Briefs **67**(1), 102–106 (2020)

7. Lv, Y., Wen, G., Huang, T.: Adaptive protocol design for distributed tracking with relative output information: a distributed fixed-time observer approach. IEEE Trans. Control Netw. Syst. **7**(1), 118–128 (2020)
8. Qian, Y.Y., Liu, L., Feng, G.: Distributed event-triggered adaptive control for consensus of linear multi-agent systems with external disturbances. IEEE Trans. Cybern. **50**(5), 2197–2208 (2020)
9. Zhou, J., Wu, X., Lv, Y., Wen, G.: Terminal-time synchronization of multiple vehicles under discrete-time communication networks with directed switching topologies. IEEE Trans. Circ. Syst. II Express Briefs **67**(11), 2532–2536 (2020)
10. Wang, Q., Duan, Z., Wang, J., Wang, Q., Chen, G.: An accelerated algorithm for linear quadratic optimal consensus of heterogeneous multiagent systems. IEEE Trans. Autom. Control **67**(1), 421–428 (2022)
11. Zhou, J., Lv, Y., Wen, G., Yu, X.: Resilient consensus of multiagent systems under malicious attacks: appointed-time observer-based approach. IEEE Trans. Cybern. **52**(10), 10187–10199 (2021)
12. Lv, Y., Zhou, J., Wen, G., Yu, X., Huang, T.: Fully distributed adaptive NN-based consensus protocol for nonlinear mass: an attack-free approach. IEEE Trans. Neural Netw. Learn. Syst. **33**(4), 1561–1570 (2022)
13. Lv, Y., Li, Z., Duan, Z.: Distributed adaptive consensus protocols for linear multi-agent systems over directed graphs with relative output information. IET Control Theor. Appl. **12**(5), 613–620 (2018)
14. Lv, Y., Li, Z., Duan, Z., Chen, J.: Distributed adaptive output feedback consensus protocols for linear systems on directed graphs with a leader of bounded input. Automatica **74**, 308–314 (2016)
15. Li, Z., Ren, W., Liu, X., Fu, M.: Consensus of multi-agent systems with general linear and lipschitz nonlinear dynamics using distributed adaptive protocols. IEEE Trans. Autom. Control **58**(7), 1786–1791 (2012)
16. Hu, C., Qin, W., Li, Z., He, B., Liu, G.: Consensus-based state estimation for multi-agent systems with constraint information. Kybernetika **53**(3), 545–561 (2017)
17. Olfati-Saber, R.: Distributed Kalman filter with embedded consensus filters. In: Proceedings of the 44th IEEE Conference on Decision and Control, pp. 8179–8184. IEEE (2005)
18. Battilotti, S., Cacace, F., d'Angelo, M., Germani, A.: Asymptotically optimal consensus-based distributed filtering of continuous-time linear systems. Automatica **122**, 109189 (2020)
19. He, C., Huang, J.: Adaptive distributed observer for general linear leader systems over periodic switching digraphs. Automatica **137**, 110021 (2022)
20. Yang, G., Barboni, A., Rezaee, H., Parisini, T.: State estimation using a network of distributed observers with unknown inputs. Automatica **146**, 110631 (2022)
21. Yang, G., Rezaee, H., Alessandri, A., Parisini, T.: State estimation using a network of distributed observers with switching communication topology. Automatica **147**, 110690 (2023)
22. Li, Y., Lv, Y., Duan, P., Zhou, J.: Output tracking based distributed state estimation. In: 2021 IEEE International Conference on Unmanned Systems (ICUS), pp. 146–150. IEEE, Beijing, China (2021)
23. Zhang, H., Li, Z., Qu, Z., Lewis, F.L.: On constructing Lyapunov functions for multi-agent systems. Automatica **58**, 39–42 (2015)
24. Li, Z., Duan, Z.: Cooperative Control of Multi-Agent Systems: A Consensus Region Approach. CRC Press, Boca Raton, FL, USA (2014)
25. Khalil, H.K.: Nonlinear Systems. Patience Hall, Englewood Cliffs, NJ, USA (2002)

Computer Simulations of Applying Zhang Inequation Equivalency and Solver of Neurodynamics to Redundant Manipulators at Acceleration Level

Ji Lu[1], Min Yang[2], Ning Tan[1], Haifeng Hu[2], and Yunong Zhang[1](✉)

[1] School of Computer Science and Engineering, Sun Yat-sen University, Guangzhou 510006, People's Republic of China
luji3@mail2.sysu.edu.cn, tann5@mail.sysu.edu.cn, zhynong@mail.sysu.edu.cn

[2] School of Electronics and Information Technology, Sun Yat-sen University, Guangzhou 510006, People's Republic of China
yangmin1221@foxmail.com, huhaif@mail.sysu.edu.cn

Abstract. An equation can be transformed into an equivalent equation at a different level, which is termed equation equivalence or even generalized to be equation equivalency. In recent years, Zhang equivalency, more specifically, Zhang equation equivalency, i.e., a new equation equivalency originated from Zhang neurodynamics, has been proposed and investigated. Referring to Zhang equivalency and doing a careful investigation, we similarly find that an inequation can also be transformed into an equivalent inequation at a different level. The novel inequation equivalency named Zhang inequation equivalency (ZIE) is investigated in this paper. Then, ZIE is applied to acceleration-level redundant manipulator motion control. The configuration adjustment and cyclic motion generation of two types of redundant manipulators are investigated and simulated. Comparative experimental results verify the validity of the proposed ZIE. In fact, ZIE can also be applied in different actual projects according to practical requirements.

Keywords: Neurodynamics · Zhang inequation equivalency · Redundant manipulator motion control · Configuration adjustment · Cyclic motion generation

1 Introduction

In real life and world, the problem that can be easily solved at a deep level is universal [1]. For example, one can control the displacement (position) of the car by controlling its velocity during the task time interval. Similarly, one can control the velocity of the car by controlling its acceleration or even its jerk. Therefore, one can control the position of the car by controlling its acceleration or jerk. That is, the problem can be solved at different (usually, deeper) levels, e.g., the first time-derivative level and the second time-derivative level. As an example, during task interval $[0, 10]$ s, $\dot{x}(t) = \cos(t)$ with $x(0) = 0$ is equivalent

B. Luo et al. (Eds.): ICONIP 2023, LNCS 14447, pp. 238–252, 2024.
https://doi.org/10.1007/978-981-99-8079-6_19

to $x(t) = \sin(t)$. Besides, $\ddot{x}(t) = -\sin(t)$ with $x(0) = 0$ and $\dot{x}(0) = 1$ is also equivalent to $x(t) = \sin(t)$, conventionally termed mathematical equivalence. Note that this kind of equivalent problem is related to the initial value, e.g., $x(0)$. If the initial value is not correct (or, not appropriate), the deeper-level problem is not equivalent to the original problem. Hence, developing an equivalent problem without considering the initial value is more important, especially when the initial value cannot be controlled. Sometimes, the original problem is not convenient to be handled, while the equivalent problem at a deeper level is easy to be handled [2]. Therefore, transforming the problem into the equivalent problem at a deeper level is important, interesting and inspiring [3–5].

After eight years of investigation, in [1], concepts about mathematical equivalence were presented firstly, and then concepts about physical equivalency (PE, or termed physical mathematical equivalency, PME) were proposed. Meanwhile, concepts about Zhang equivalency (ZE, specifically, Zhang equation equivalency) were further proposed. The simplest equivalent design formula is $\dot{e}(t) = -\lambda e(t)$ with $\lambda \gg 0 \in \mathbb{R}$ being the sufficiently large design parameter, where $e(t)$ is the involved user-function or the defined error function (also termed Zhang function). The error function can be matrix-, vector-, or scalar-valued. The value of the design parameter λ should be set as large as possible or according to the actual situation. Evidently, as an ordinary derivative equation (ODE) [6], the solution of the design formula is $e(t) = e(0)\exp(-\lambda t) \to 0$ with $t \gg 0$. If one needs to get the equivalent equation of $x(t) = \sin(t)$, one can define error function $e(t) = x(t) - \sin(t)$ and apply $\dot{e}(t) = -\lambda e(t)$. Then, one obtains the equivalent equation $\dot{x}(t) - \cos(t) = -\lambda(x(t) - \sin(t))$. Further, if the equivalent problem at the second time-derivative level is desired to be obtained, one just needs to define the corresponding error function (i.e., another Zhang function) and apply the design formula again. Note that the above equivalency is a kind of physical equivalency, which is acceptable in practical mathematics and engineering. For instance, if $t = 1$ s and $\lambda = 10$, $e(1) = e(0)\exp(-10) \approx 0$, and this result is almost unaffected by any initial value $e(0)$.

In the previous work [1], ZE was proposed, investigated, and verified, introducing a question. That is, "Does inequation equivalency exist?" The inequation equivalency is also important and useful. In this case, by referring to ZE, a recently new inequation equivalency, i.e., Zhang inequation equivalency (ZIE), is presented and investigated in this paper. By utilizing ZIE, an inequation can be transformed into its equivalent inequation at a different level. Specifically, the detailed discussion and investigation are presented in the next section.

As a whole, the key of equivalency is the design formula. The equivalent design formula of ZE is $\dot{e}(t) = -\lambda e(t)$, which originates from the design formula of Zhang neurodynamics (ZN) [7,8]. Analogously, the equivalent design formula of ZIE is $\dot{e}(t) \leq -\lambda e(t)$. As a special class of recurrent neurodynamics (i.e., recurrent neural network) [9], ZN is an extremely powerful tool for solving various of time-varying problems [10–12]. For example, in [10], the time-varying system of nonlinear equations was solved with finite-time convergence by utilizing ZN method. In [11], the time-varying nonlinear equations were solved by using discrete-time ZN model. In [12], Xiao *et al.* solved the time-varying quadratic optimization with finite-time convergence and noise tolerance in a unified ZN framework. By developing ZN, the equivalency theory, being a kind of physical equivalency, is complete now. By applying the equivalency theory, the equivalent problem can be easily obtained, which is gradually physically equivalent to the original problem as time goes by, without relation to the initial value.

The remaining parts of this paper are organized as follows. In Sect. 2, ZIE is presented and investigated. In Sect. 3, ZIE is applied to redundant manipulator control. Experimental verification is conducted in Sect. 4, which illustrates the efficacy and validity of ZIE. The last section concludes the paper with some views of future research directions.

Particularly, the main contributions of the paper are listed as follows.

- The recently new inequation equivalency named ZIE is investigated.
- The ZIE is successfully applied to the redundant manipulator control theoretically, and can also be applied to other practical projects.
- Simulative experiments about the redundant manipulator control substantiate the efficacy and validity of ZIE.

2 Zhang Inequation Equivalency of Neurodynamics

In the scientific research, the time-varying inequation is often encountered:

$$\mathbf{x}(t) \leq \mathbf{0},$$

where $\mathbf{x}(t) \in \mathbb{R}^n$ is time-varying and $\mathbf{0} \in \mathbb{R}^n$ denotes the zero vector. The symbol "$\leq$" means that each element in the left part is less than or equal to the corresponding element in the right part. The above inequation may be difficult to be handled, while the inequation at the first time-derivative (velocity) level or the second time-derivative (acceleration) level may be easy to be handled. In other words, obtaining the inequation at a deeper level may be expected. By referring to the neurodynamics idea in ZN, the ZIE is thus proposed in this section. Note that, in this paper, we mainly focus on the experimental validity of ZIE. Therefore, in this section, we only give some theoretical results of ZIE directly and omit the rigorous proofs (to be shown in detail in our future research).

2.1 Velocity-Level ZIE of Neurodynamics

By referring to the neurodynamics idea in ZN, ZIE at the velocity level (i.e., the first time-derivative level) [13] is presented as follows.

Theorem 1. *Suppose that vector* $\mathbf{x}(t)$ *is continuously differentiable. With design parameter* $\lambda \gg 0$ *and time* $t \gg 0$,

$$\dot{\mathbf{x}}(t) \leq -\lambda \mathbf{x}(t) \tag{1}$$

is physically (mathematically) equivalent to $\mathbf{x}(t) \leq \mathbf{0}$.

Furthermore, the following corollary about velocity-level ZIE is acquired.

Corollary 1. *Suppose that vector* $\mathbf{x}(t)$*, its upper bound* $\mathbf{x}^+(t)$*, and its lower bound* $\mathbf{x}^-(t)$ *are continuously differentiable. With* $\lambda \gg 0$ *and* $t \gg 0$*,*

$$\dot{\mathbf{x}}^-(t) - \lambda(\mathbf{x}(t) - \mathbf{x}^-(t)) \leq \dot{\mathbf{x}}(t) \leq \dot{\mathbf{x}}^+(t) - \lambda(\mathbf{x}(t) - \mathbf{x}^+(t)) \tag{2}$$

is physically (mathematically) equivalent to $\mathbf{x}^-(t) \leq \mathbf{x}(t) \leq \mathbf{x}^+(t)$*.*

2.2 Acceleration-Level ZIE of Neurodynamics

By referring to the neurodynamics idea in ZN, ZIE at the acceleration level (i.e., the second time-derivative level) is presented as follows.

Theorem 2. *Suppose that vector* $\mathbf{x}(t)$ *is twice continuously differentiable. With design parameter* $\lambda \gg 0$ *and time* $t \gg 0$*,*

$$\ddot{\mathbf{x}}(t) \leq -2\lambda\dot{\mathbf{x}}(t) - \lambda^2\mathbf{x}(t) \tag{3}$$

is physically (mathematically) equivalent to $\mathbf{x}(t) \leq \mathbf{0}$*.*

Further, the following corollary about acceleration-level ZIE is acquired.

Corollary 2. *Suppose that vector* $\mathbf{x}(t)$*, its upper bound* $\mathbf{x}^+(t)$*, and its lower bound* $\mathbf{x}^-(t)$ *are twice continuously differentiable. With* $\lambda \gg 0$ *and* $t \gg 0$*,*

$$\begin{aligned} &\ddot{\mathbf{x}}^-(t) - 2\lambda(\dot{\mathbf{x}}(t) - \dot{\mathbf{x}}^-(t)) - \lambda^2(\mathbf{x}(t) - \mathbf{x}^-(t)) \leq \ddot{\mathbf{x}}(t) \\ &\leq \ddot{\mathbf{x}}^+(t) - 2\lambda(\dot{\mathbf{x}}(t) - \dot{\mathbf{x}}^+(t)) - \lambda^2(\mathbf{x}(t) - \mathbf{x}^+(t)) \end{aligned} \tag{4}$$

is physically (mathematically) equivalent to $\mathbf{x}^-(t) \leq \mathbf{x}(t) \leq \mathbf{x}^+(t)$*.*

Remarks. Using the design formula $\dot{\mathbf{x}}(t) \leq -\lambda\mathbf{x}(t)$, one can obtain the inequation at any level, physically equivalent to $\mathbf{x}(t) \leq \mathbf{0}$. Besides, as a Zhang function, $\mathbf{x}(t)$ can be matrix-, vector-, or scalar-valued. Note that the international unit of design parameter λ is Hz. The value of λ should be chosen on the basis of experimental requirements. Hence, originated from ZN, the complete theory about ZIE is established. Actually, the simplest design formula of ZIE is presented in this paper, and some complicated design formulas are desired to be found, which may be more effective in some specific situations.

3 Application to Redundant Manipulators

In the previous section, ZIE is presented and investigated. In this section, a specific application is considered to show its practicability. Redundant manipulator motion control can be a typical application of ZIE. In the actual operation, the physical limits of the redundant manipulator should be considered, which usually include joint angle limits, joint velocity limits, and joint acceleration limits [9,15]. The manipulator control scheme is convenient to be solved at a unified level, and thus the different-level limits should be transformed into the limits at a same level [16]. Fortunately, ZIE has the ability.

In consideration of the primary task, optimization index and physical limits of the redundant manipulator, the designed control scheme at joint acceleration level can be formulated as [17]:

$$\text{minimize} \quad \frac{1}{2}\ddot{\theta}^{\mathrm{T}}Q\ddot{\theta} + \mathbf{p}^{\mathrm{T}}\ddot{\theta}, \tag{5}$$

$$\text{subject to} \quad J(\theta)\ddot{\theta} = \mathbf{r}_{\mathrm{b}}, \tag{6}$$

$$\theta^{-} \leq \theta \leq \theta^{+}, \tag{7}$$

$$\bar{\theta}^{-} \leq \dot{\theta} \leq \bar{\theta}^{+}, \tag{8}$$

$$\acute{\theta}^{-} \leq \ddot{\theta} \leq \acute{\theta}^{+}, \tag{9}$$

$$\text{with} \quad \mathbf{r}_{\mathrm{b}} = \ddot{\mathbf{r}}_{\mathrm{d}} - \dot{J}(\theta)\dot{\theta} + 2\gamma(\dot{\mathbf{r}}_{\mathrm{d}} - J(\theta)\dot{\theta}) + \gamma^2(\mathbf{r}_{\mathrm{d}} - \mathbf{f}(\theta)),$$

where $\theta \in \mathbb{R}^n$, $\dot{\theta} \in \mathbb{R}^n$, and $\ddot{\theta} \in \mathbb{R}^n$ denote the joint angle vector, joint velocity vector, and joint acceleration vector, respectively. Matrix $Q \in \mathbb{R}^{n \times n}$ and vector $\mathbf{p} \in \mathbb{R}^n$ are defined according to actual needs, as exemplified in the next section. For a given manipulator, $\mathbf{f}(\cdot)$ is a smooth nonlinear function with known parameters and structure, J is the Jacobian matrix, and $\dot{J}$ denotes the time-derivative of J. Meanwhile, $\theta^{\pm}$, $\bar{\theta}^{\pm}$, and $\acute{\theta}^{\pm}$ represent the joint angle bounds (upper and lower ones, or traditionally termed joint angle limits), joint velocity bounds, and joint acceleration bounds, respectively, which are time-invariant. Additionally, $\mathbf{r}_{\mathrm{d}}$ denotes the desired path of the end-effector. Note that the variable "t" is usually omitted for simplicity, and $\gamma > 0$ is the ZN design parameter.

Control scheme (5)–(9) is relatively clear and easy to understand. By solving the control scheme in real time, the control information (e.g., θ) can be obtained, and the control task can be completed. However, the control scheme cannot be easily solved because there exist different-level limits (7), (8), and (9). The decision vector of the control scheme is $\ddot{\theta}$, which is at the acceleration level. However, (7) is at the angle level, and (8) is at the velocity level. That is, the control scheme is not a standard quadratic programming (QP). Therefore, applying ZIE to transform the different-level limits into a unified acceleration limit is a good choice. Specifically, the standard QP formulation is presented in the next subsection.

3.1 Standard QP Formulation

For conveniently handling control scheme (5)–(9), it should be reformulated as a standard QP at the same level (i.e., acceleration level). Specifically, the standard QP is formulated as follows:

$$\text{minimize} \quad \frac{1}{2}\mathbf{y}^{\mathrm{T}}Q\mathbf{y} + \mathbf{p}^{\mathrm{T}}\mathbf{y}, \tag{10}$$

$$\text{subject to} \quad A\mathbf{y} = \mathbf{b}, \tag{11}$$

$$\mathbf{y}^- \leq \mathbf{y} \leq \mathbf{y}^+, \tag{12}$$

where $\mathbf{y} = \ddot{\theta}$, $A = J(\theta)$, and $\mathbf{b} = \mathbf{r}_\mathrm{b}$. Besides,

$$\mathbf{y}^- = \max\{-2\lambda\dot{\theta} + \lambda^2(\theta^- - \theta), \lambda(\bar{\theta}^- - \dot{\theta}), \acute{\theta}^-\}, \tag{13}$$

$$\mathbf{y}^+ = \min\{-2\lambda\dot{\theta} + \lambda^2(\theta^+ - \theta), \lambda(\bar{\theta}^+ - \dot{\theta}), \acute{\theta}^+\}. \tag{14}$$

Note that the above functions "max" and "min" are element-wise operations. For example, with $i = 1, 2, \cdots, n$, the ith element of $\mathbf{y}^-$ is defined as

$$y_i^- = \max\{-2\lambda\dot{\theta}_i + \lambda^2(\theta_i^- - \theta_i), \lambda(\bar{\theta}_i^- - \dot{\theta}_i), \acute{\theta}_i^-\}.$$

The new limit $\mathbf{y}^- \leq \mathbf{y} \leq \mathbf{y}^+$ as well as new bounds $\mathbf{y}^-$ and $\mathbf{y}^+$ are obtained by applying ZIE. According to Corollaries 1 and 2, with new bounds defined as (13) and (14) as well as positive parameter λ, (12) is equivalent to (7)–(9).

3.2 Projection Neurodynamics Solver

By solving the standard QP (10)–(12) in real time t, the solution $\mathbf{y}$ (i.e., $\ddot{\theta}$) can be obtained for manipulator control. Therefore, for solving QP (10)–(12), the projection neurodynamics (PN) solver [17] is developed as follows.

With $\delta \in \mathbb{R}^+$ adjusting the convergence rate and with $\varsigma \in \mathbb{R}^+$ being large enough to numerically replace ∞, the PN solver [17] for QP (10)–(12) is developed as

$$\dot{\mathbf{z}} = \delta(I + M^\mathrm{T})(P_\Omega(\mathbf{z} - (M\mathbf{z} + \mathbf{c})) - \mathbf{z}), \tag{15}$$

where $\mathbf{z} = [\mathbf{y}; \mathbf{w}] \in \mathbb{R}^{n+m}$ and $\mathbf{c} = [\mathbf{p}; -\mathbf{b}] \in \mathbb{R}^{n+m}$ in MATLAB manner [18]. Meanwhile, $I \in \mathbb{R}^{(n+m)\times(n+m)}$ denotes an identity matrix, and $\mathbf{w} \in \mathbb{R}^m$ is the dual decision vector corresponding to (11). Besides,

$$M = \begin{bmatrix} Q & -A^\mathrm{T} \\ A & 0 \end{bmatrix} \in \mathbb{R}^{(n+m)\times(n+m)},$$

$$\mathbf{z}^- = \begin{bmatrix} \mathbf{y}^- \\ -\varsigma\mathbf{1}_\mathrm{v} \end{bmatrix} \in \mathbb{R}^{n+m}, \mathbf{z}^+ = \begin{bmatrix} \mathbf{y}^+ \\ \varsigma\mathbf{1}_\mathrm{v} \end{bmatrix} \in \mathbb{R}^{n+m},$$

in which $\mathbf{1}_\mathrm{v} = [1, \cdots, 1]^\mathrm{T} \in \mathbb{R}^m$. Moreover, with $i = 1, 2, \cdots, n+m$, the ith element of the Ω-projection operation is defined as

$$[P_\Omega(\mathbf{u})]_i = \begin{cases} z_i^-, & \text{if } u_i \leq z_i^-, \\ u_i, & \text{if } z_i^- < u_i < z_i^+, \\ z_i^+, & \text{if } u_i \geq z_i^+. \end{cases}$$

4 Experiments Verification

In this section, experiments on UR3 and UR10 manipulators are conducted to illustrate the applicability and efficacy of ZIE. Two kinds of manipulator tasks and two desired paths are considered.

Table 1. D-H (Denavit-Hartenberg) parameters of UR3 manipulator.

Joint	θ_i (rad)	d_i (m)	a_i (m)	α_i (rad)
1	θ_1	0.1519	0	$\pi/2$
2	θ_2	0	−0.2437	0
3	θ_3	0	−0.2133	0
4	θ_4	0.1124	0	$\pi/2$
5	θ_5	0.0854	0	$-\pi/2$
6	θ_6	0.0819	0	0

Table 2. Physical bounds of UR3 manipulator.

Joint	θ^- (rad)	θ^+ (rad)	$\bar{\theta}^-$ (rad/s)	$\bar{\theta}^+$ (rad/s)	$\acute{\theta}^-$ (rad/s^2)	$\acute{\theta}^+$ (rad/s^2)
1	$-\pi/2$	$\pi/2$	−0.5	0.5	−0.2	0.2
2	$-\pi$	0	−0.5	0.5	−0.2	0.2
3	$-\pi$	0	−0.5	0.5	−0.2	0.2
4	$-\pi/2$	$\pi/2$	−0.5	0.5	−0.2	0.2
5	0	π	−0.5	0.5	−0.2	0.2
6	$-\pi/2$	$\pi/2$	−0.5	0.5	−0.2	0.2

4.1 Configuration Adjustment

As a typical redundant manipulator, UR3 manipulator is usually used in experimental verification, which has six degrees of freedom and operates in the three-dimensional physical space. The Denavit-Hartenberg (D-H) parameters of the UR3 manipulator are presented in Table 1. The physical bounds of the UR3 manipulator are presented in Table 2. In this subsection, the UR3 manipulator is simulated to complete the configuration adjustment. That is, we control the manipulator from its initial configuration to the desired configuration. For example, the configuration may be adjusted to get an optimal manipulability [19]. Meanwhile, the physical limits should be satisfied when the manipulator configuration is being adjusted. The configuration adjustment task can be formulated as a quadratic index. The physical limits can be formulated as inequations. Therefore, solved at the acceleration level, control scheme (5)–(9) is applied to realizing the configuration adjustment. Because the tracking task of the end-effector is not considered when doing the configuration adjustment, (6) is removed from the control scheme.

Specifically, to realize the configuration adjustment (i.e., $\theta \to \theta_{\mathrm{d}}$, with θ_{d} denoting the desired joint angle configuration), the first error function is defined as $\mathbf{e}_1 = \theta - \theta_{\mathrm{d}}$ being the Zhang function. Then, applying ZN design formula [7] $\dot{\mathbf{e}}_1 = -\beta \mathbf{e}_1$ with $\beta > 0$ being the design parameter, one obtains $\dot{\theta} = -\beta(\theta - \theta_{\mathrm{d}})$. Then, defining $\mathbf{e}_2 = \dot{\theta} + \beta(\theta - \theta_{\mathrm{d}})$ as the second Zhang function, and applying $\dot{\mathbf{e}}_2 = -\beta \mathbf{e}_2$, one gets $\ddot{\theta} = -2\beta\dot{\theta} - \beta^2(\theta - \theta_{\mathrm{d}})$. That is, $\ddot{\theta} + 2\beta\dot{\theta} - \beta^2(\theta - \theta_{\mathrm{d}}) = \mathbf{0}$. Hence, the

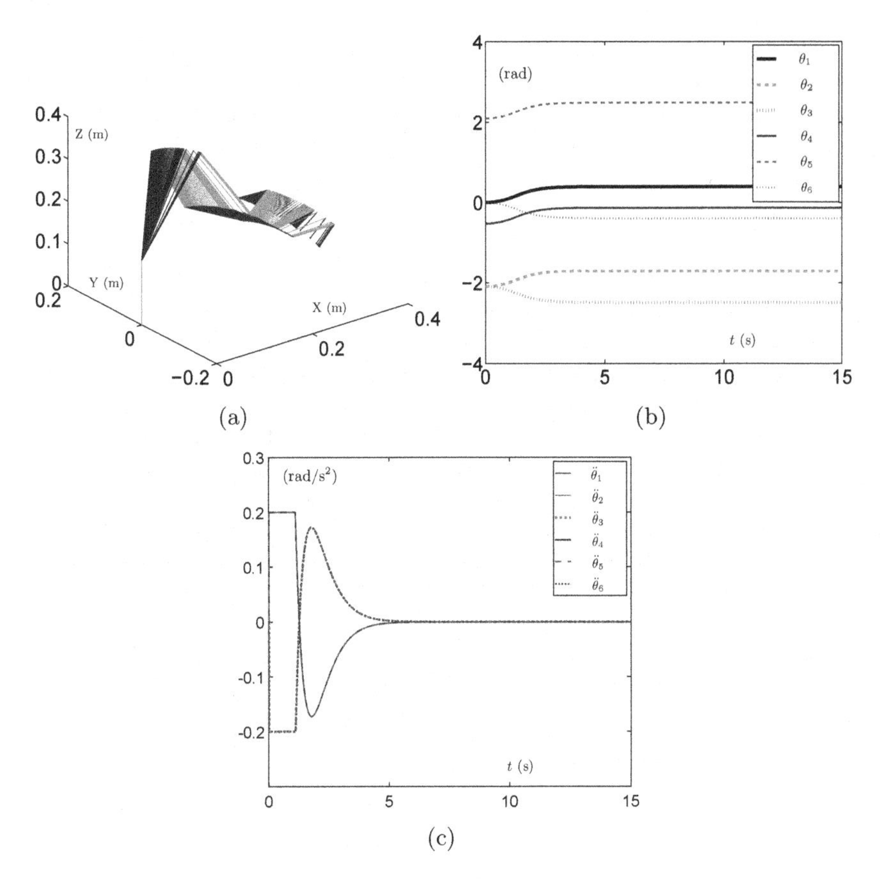

Fig. 1. Synthesized results of UR3 manipulator using bounds (13)–(14) and PN solver (15) for configuration adjustment. (a) Trajectories of manipulator. (b) Trajectories of joint angles. (c) Trajectories of joint accelerations.

quadratic index can be formulated as $(\ddot{\theta}+2\beta\dot{\theta}-\beta^2(\theta-\theta_{\rm d}))^{\rm T}(\ddot{\theta}+2\beta\dot{\theta}-\beta^2(\theta-\theta_{\rm d}))$. Hence, with decision vector $\ddot{\theta}$, matrix Q and vector $\mathbf{p}$ in (5) are defined as

$$Q = I, \quad \mathbf{p} = 2\beta\dot{\theta} + \beta^2(\theta - \theta_{\rm d}).$$

The initial joint angle vector (configuration) of the UR3 manipulator is $\theta(0) = [0; -2\pi/3; -2\pi/3; -\pi/6; 2\pi/3; 0]$ rad. Meanwhile, the desired joint angle vector is set as $\theta_{\rm d} = [\pi/8; -13\pi/24; -19\pi/24; -\pi/24; 19\pi/24; -\pi/8]$ rad. The corresponding parameters are set as $\delta = 10^4$, $\varsigma = 10^6$, $\gamma = 20$, $\lambda = 1$, and $\beta = 2$. The adjustment time interval is set as $[0, 15]$ s. Using PN solver (15) with transformed bounds (13)–(14), the simulated results of the UR3 manipulator are presented in Fig. 1. As seen from Fig. 1(a), the configuration of the manipulator is adjusted from its initial state to the desired state. Each joint angle is adjusted to the corresponding desired angle, as shown in Fig. 1(b). Besides, the physical limits are satisfied during the process of configuration adjustment. Particularly, joint accelerations are constrained by the acceleration limits, as presented in Fig. 1(c). From these results, one sees that the UR3 manipulator completes the configuration adjustment under the physical limits after a short time (about 5 s). In other words, ZIE is verified to be effective and useful.

4.2 Cyclic Motion Generation

In this subsection, UR10 manipulator is applied to performing the cyclic motion [17]. The UR10 manipulator has six degrees of freedom and operates in the three-dimensional physical space. Table 3 presents the D-H parameters of the UR10 manipulator. The physical bounds of the UR10 manipulator are presented in Table 4. Non-cyclic motion is that joint angles may not return to their initial values when the end-effector traces a closed path in its workspace. This may result in a joint-angle drift phenomenon and may induce a problem that the manipulator behavior is hard to predict. Cyclic motion means that the joint angles all return to the initial values when the tracking task is finished, and one does not need to adjust the configuration of the manipulator first when starting the next task. Hence, cyclic motion is meaningful and important.

Table 3. D-H (Denavit-Hartenberg) parameters of UR10 manipulator.

Joint	θ_i (rad)	d_i (m)	a_i (m)	α_i (rad)
1	θ_1	0.1273	0	$\pi/2$
2	θ_2	0	−0.6120	0
3	θ_3	0	−0.5723	0
4	θ_4	0.1639	0	$\pi/2$
5	θ_5	0.1157	0	$-\pi/2$
6	θ_6	0.0922	0	0

Table 4. Physical bounds of UR10 manipulator.

Joint	θ^- (rad)	θ^+ (rad)	$\bar{\theta}^-$ (rad/s)	$\bar{\theta}^+$ (rad/s)	$\acute{\theta}^-$ (rad/s^2)	$\acute{\theta}^+$ (rad/s^2)
1	$-\pi/2$	$\pi/2$	−1	1	−0.5	0.5
2	$-\pi$	0	−1	1	−0.5	0.5
3	−4.6	−1.6	−1	1	−0.5	0.5
4	$-\pi/2$	$\pi/2$	−1	1	−0.5	0.5
5	0	π	−1	1	−0.5	0.5
6	$-\pi/2$	$\pi/2$	−1	1	−0.5	0.5

Based on the previous work [17,20], in control scheme (5)–(9), cyclic motion can be realized by defining

$$Q = I, \quad \mathbf{p} = 2\nu\dot{\theta} + \nu^2(\theta - \theta(0)),$$

where $\nu > 0$ is the ZN design parameter, and $\theta(0)$ denotes the initial joint angle vector. Therefore, by solving at the joint acceleration level, control scheme (5)–(9) is applied to realizing the cyclic motion generation.

The initial joint angle vector of the UR10 manipulator is set as $\theta(0) = [0; -2\pi/3; -2\pi/3; -\pi/6; 2\pi/3; 0]$ rad. Corresponding parameters are set as $\delta = 10^4$, $\varsigma = 10^6$, $\gamma = 40$, $\lambda = 3$, and $\nu = 5$. At first, the end-effector is desired to track a circle path whose radius is 0.2 m. The tracking time interval is set as $[0, 20]$ s. Simulation results synthesized by the PN solver with transformed bounds (13)–(14) are presented in Fig. 2. The robot trajectories during the tracking process are shown in Fig. 2(a). The tracking task is completed successfully, as seen from Fig. 2(b) and Fig. 2(c). The tracking error is at the level of 10^{-4} m. Then, the profiles of joint angles, joint velocities and joint accelerations are presented in Fig. 2(d), Fig. 2(e), and Fig. 2(f), respectively. In addition, from Fig. 2(d), one can find that joint angles return to the initial values when the tracking task is finished. Joint angles and joint accelerations are bounded as shown in Fig. 2(d) and Fig. 2(f). That is, the physical limits are satisfied. In words, the tracking task is accomplished under the physical limits and with cyclic motion generation, which illustrates the efficacy of ZIE.

In the previous work [13,20], the conventional transformation of inequation limits (7)–(9) was presented for manipulator control. In the conventional way [13,20], limits (7)–(9) are transformed into $\mathbf{y}^- \leq \mathbf{y} \leq \mathbf{y}^+$, with defined bounds

$$\mathbf{y}^- = \max\{\kappa_\mathrm{p}(\mu\theta^- - \theta), \kappa_\mathrm{v}(\bar{\theta}^- - \dot{\theta}), \acute{\theta}^-\}, \tag{16}$$

$$\mathbf{y}^+ = \min\{\kappa_\mathrm{p}(\mu\theta^+ - \theta), \kappa_\mathrm{v}(\bar{\theta}^+ - \dot{\theta}), \acute{\theta}^+\}, \tag{17}$$

where $\mu \in (0, 1)$, $\kappa_\mathrm{p} > 0$, and $\kappa_\mathrm{v} > 0$. Note that there are three parameters in the conventional transformation about the inequation limits, which should be set appropriately. Besides, the conventional transformation of the inequation limits is not proved theoretically and in detail, and it may be not effective enough. For comparison, that conventional transformation is also used. Corresponding parameters are set as $\delta = 10^4$, $\varsigma = 10^6$, $\gamma = 40$, $\lambda = 3$, $\nu = 5$, $\kappa_\mathrm{p} = 5$, $\kappa_\mathrm{v} = 5$, and $\mu = 0.9$. Some simulation results synthesized by the PN solver with transformed bounds (16)–(17) are presented in Fig. 3. Joint velocities and joint accelerations satisfy the physical limits, and the values are less than the ones shown in Fig. 2(e) and Fig. 2(f), maybe an advantage of the conventional inequation transformation. However, as shown in Fig. 3(a), joint angles are not limited by physical bounds, illustrating that the conventional inequation transformation is ineffective. By contrast, adopting ZIE is feasible, as shown in Fig. 2.

To further verify the validity of ZIE in manipulator control, as a more complicated tracking task, the Lissajous path is desired to be tracked by the end-effector of the UR10 manipulator under the physical limits. Likewise, the transformed

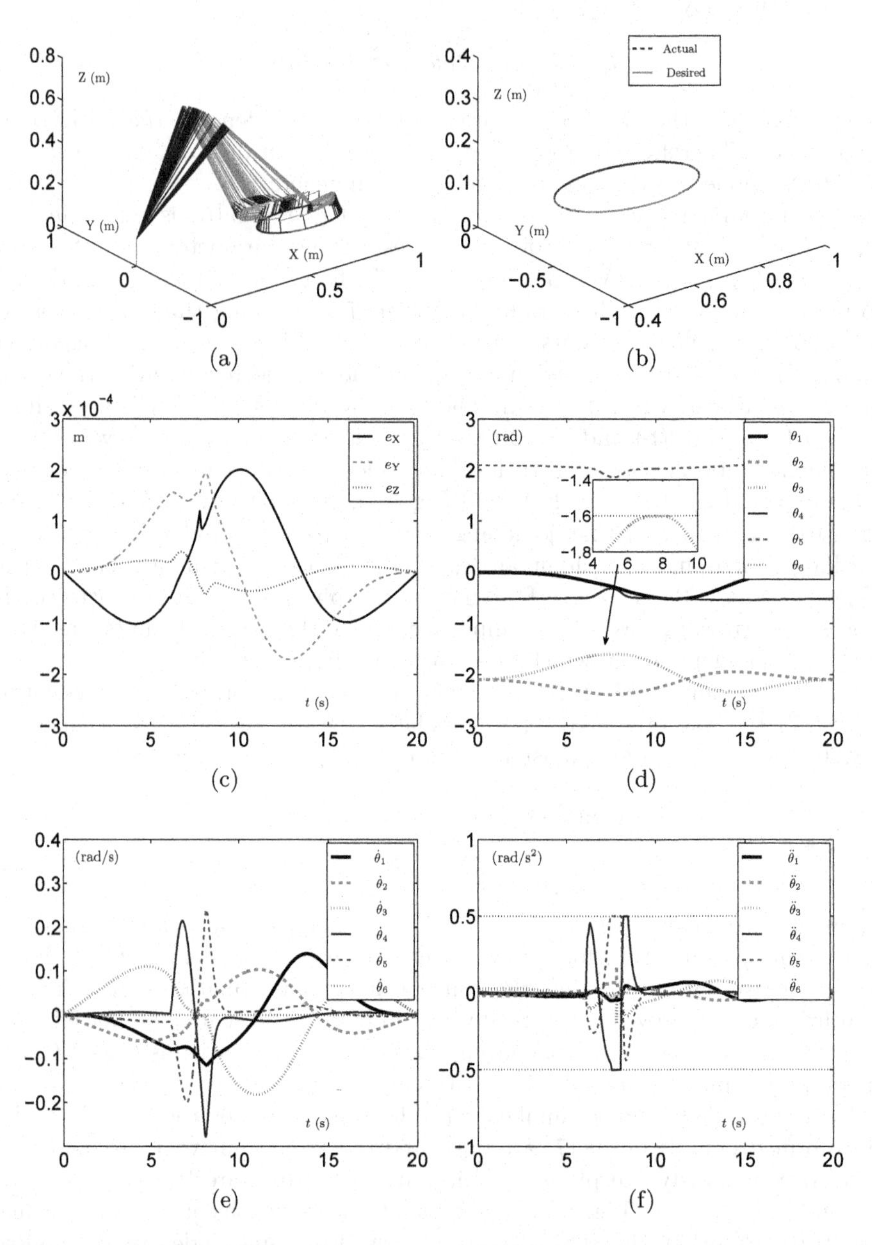

Fig. 2. Synthesized results of UR10 manipulator using bounds (13)–(14) for cyclic motion generation with desired circle path. (a) Trajectories of manipulator. (b) Actual trajectory of end-effector and desired path. (c) Tracking errors. (d) Trajectories of joint angles. (e) Trajectories of joint velocities. (f) Trajectories of joint accelerations.

bounds (13)–(14) are used, and the PN solver is applied. The tracking time interval is set as $[0, 30]$ s. Meanwhile, the corresponding parameters are set as $\delta = 10^4$, $\varsigma = 10^6$, $\gamma = 40$, $\lambda = 3$, and $\nu = 5$. The initial joint angle vector is also $\theta(0) = [0; -2\pi/3; -2\pi/3; -\pi/6; 2\pi/3; 0]$ rad. The simulated results are displayed in Fig. 4. The Lissajous path is tracked successfully, and the trajectories of UR10 manipulator are shown in Fig. 4(a). Besides, the profiles of joint angles and joint accelerations are presented in Fig. 4(b) and Fig. 4(c), respectively, which are both limited by the physical bounds. These results further verify the efficacy of ZIE. Note that some results of this experiment are omitted in this paper due to the page limitation.

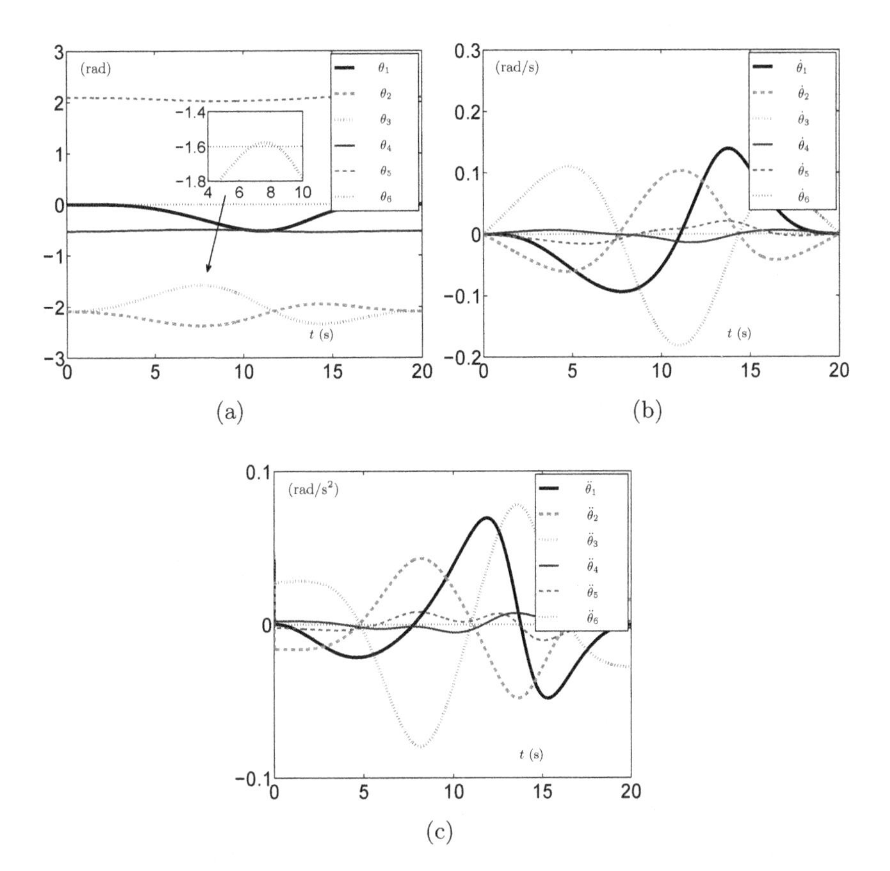

Fig. 3. Synthesized results of UR10 manipulator using bounds (16)–(17) for cyclic motion generation with desired circle path. (a) Trajectories of joint angles. (b) Trajectories of joint velocities. (c) Trajectories of joint accelerations.

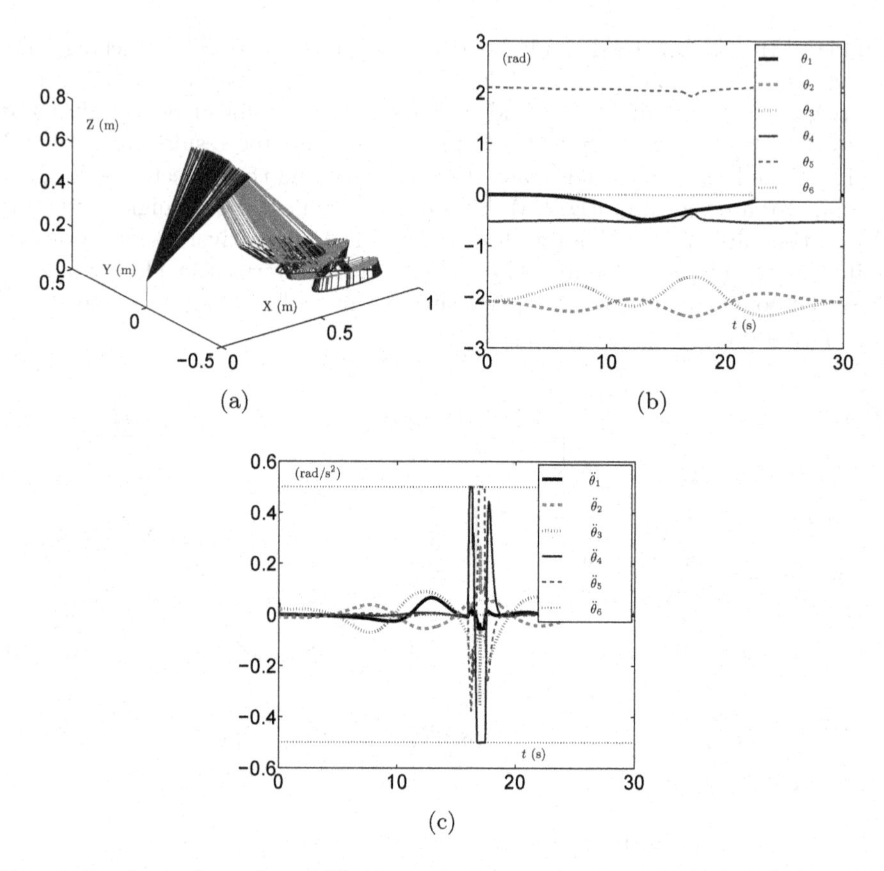

Fig. 4. Synthesized results of UR10 manipulator using bounds (13)–(14) for cyclic motion generation with desired Lissajous path. (a) Trajectories of manipulator. (b) Trajectories of joint angles. (c) Trajectories of joint accelerations.

5 Conclusion

In this paper, ZE has been introduced first. Then, the concept about the inequation equivalency has been introduced, and a recently new inequation equivalency named ZIE has been presented and investigated. Next, ZIE has been applied to the redundant manipulator control. Simulative experiments about redundant manipulator control have been conducted, further verifying the validity of ZIE. Developing other kinds of equation equivalency and inequation equivalency is one of future research directions. Besides, conducting more experiments including physical experiments about redundant manipulators is desired. Investigating other applications of ZIE is also a research direction.

Acknowledgements. This work is aided by the National Natural Science Foundation of China under Grant 61976230, the Project Supported by Guangdong Province Universities and Colleges Pearl River Scholar Funded Scheme under Grant 2018, and also the Key-Area Research and Development Program of Guangzhou under Grant 202007030004.

References

1. Zhang, Y., Yang, M., Qiu, B., Li, J., Zhu, M.: From mathematical equivalence such as Ma equivalence to generalized Zhang equivalency including gradient equivalency. Theor. Comput. Sci. **817**, 44–54 (2020)
2. Zhang, Y., Yang, M., Huang, H., Xiao, M., Hu, H.: New discrete-solution model for solving future different-level linear inequality and equality with robot manipulator control. IEEE Trans. Ind. Inform. **15**(4), 1975–1984 (2019)
3. Li, J., Mao, M., Zhang, Y., Qiu, B.: Different-level algorithms for control of robotic systems. Appl. Math. Model. **77**, 922–933 (2020)
4. Zhang, Y., Li, Z., Yang, M., Ming, L., Guo, J.: Jerk-level Zhang Neurodynamics equivalency of bound constraints, equation constraints, and objective indices for cyclic motion of robot-arm systems. IEEE Trans. Neural Netw. Learn. Syst. **34**(6), 3005–3018 (2021)
5. Tang, Z., Tan, N., Zhang, Y.: Velocity-layer Zhang equivalency for time-varying joint limits avoidance of redundant robot manipulator. IET Control Theor. Appl. **16**(18), 1909–1921 (2022)
6. Zhang, Y., Chou, Y., Chen, J., Zhang, Z., Xiao, L.: Presentation, error analysis and numerical experiments on a group of 1-step-ahead numerical differentiation formulas. J. Comput. Appl. Math. **239**(1), 406–414 (2013)
7. Jin, L., Li, S., Luo, X., Li, Y., Qin, B.: Neural dynamics for cooperative control of redundant robot manipulators. IEEE Trans. Ind. Inform. **14**(9), 3812–3821 (2018)
8. Zhang, Y., He, L., Hu, C., Guo, J.: General four-step discrete-time zeroing and derivative dynamics applied to time-varying nonlinear optimization. J. Comput. Appl. Math. **347**, 314–329 (2019)
9. Zhang, Y., Li, S., Kadry, S., Liao, B.: Recurrent neural network for kinematic control of redundant manipulators with periodic input disturbance and physical constraints. IEEE Trans. Cybern. **49**(12), 4194–4205 (2019)
10. Xiao, L., Zhang, Z., Li, S.: Solving time-varying system of nonlinear equations by finite-time recurrent neural networks with application to motion tracking of robot manipulators. IEEE Trans. Syst. Man Cybern. Syst. **49**(11), 2210–2220 (2019)
11. Guo, D., Xu, F., Li, Z., Nie, Z., Shao, H.: Design, verification, and application of new discrete-time recurrent neural network for dynamic nonlinear equations solving. IEEE Trans. Ind. Inform. **14**(9), 3936–3945 (2018)
12. Xiao, L., Li, K., Duan, M.: Computing time-varying quadratic optimization with finite-time convergence and noise tolerance: a unified framework for zeroing neural network. IEEE Trans. Neural Netw. Learn. Syst. **30**(11), 3360–3369 (2019)
13. Zhang, Y., Ge, S.S., Lee, T.H.: A unified quadratic-programming-based dynamical system approach to joint torque optimization of physically constrained redundant manipulators. IEEE Trans. Syst. Man Cybern. B, Cybern. **34**(5), 2126–2132 (2004)
14. Oppenheim, A.V., Willsky, A.S., Nawab, S.H.: Signals and Systems. Prentice Hall, New Jersey (1998)
15. Zhang, Y., Li, S., Gui, J., Luo, X.: Velocity-level control with compliance to acceleration-level constraints: a novel scheme for manipulator redundancy resolution. IEEE Trans. Ind. Inform. **14**(3), 921–930 (2018)
16. Jin, L., Li, S.: Distributed task allocation of multiple robots: a control perspective. IEEE Trans. Syst. Man Cybern. Syst. **48**(5), 693–701 (2018)

17. Zhang, Y., Zhang, Z.: Repetitive motion planning and control of redundant robot manipulators. Spinger-Verlag, Berlin (2013). https://doi.org/10.1007/978-3-642-37518-7
18. Mathews, J.H., Fink, K.D., Nawab, S.H.: Numerical Methods Using MATLAB, 4th edn. Prentice Hall, New Jersey (2004)
19. Jin, L., Li, S., La, H.M., Luo, X.: Manipulability optimization of redundant manipulators using dynamic neural networks. IEEE Trans. Ind. Electron. **64**(6), 4710–4720 (2017)
20. Zhang, Z., Zhang, Y.: Acceleration-level cyclic-motion generation of constrained redundant robots tracking different paths. IEEE Trans. Syst. Man Cybern. B, Cybern. **42**(4), 1257–1269 (2012)

High-Order Control Barrier Function Based Robust Collision Avoidance Formation Tracking of Constrained Multi-agent Systems

Dan Liu[1] and Junjie Fu[1,2(✉)]

[1] The School of Mathematics, Southeast University, Nanjing 210096, China
[2] Purple Mountain Laboratories, Nanjing 211111, China
fujunjie@seu.edu.cn

Abstract. In this work, we propose a high-order control barrier functions (HOCBFs) based safe formation tracking controller for second-order multi-agent systems subject to input uncertainties and both velocity and input constraints (VICs). First, a nominal velocity and input constrained formation tracking controller is proposed which using sliding mode control theory to eliminate the effects of the uncertain dynamics. Then, the HOCBFs-based collision avoidance conditions are derived for the followers where both collision among the agents and between the agents and the obstacles are considered. Finally, the collision avoidance formation tracking controller for the constrained uncertain second-order multi-agent systems is constructed by formulating a local quadratic programming (QP) problem for each follower. It is shown that under proper initial conditions, there always exist feasible control inputs such that collision avoidance can be guaranteed under both VICs of the agents. Simulation examples illustrate the effectiveness of the proposed control strategy.

Keywords: Multi-agent systems · velocity and input constraints · collision avoidance · high-order control barrier function

1 Introduction

Recent years have seen rapid development in formation control of multi-agent systems [1,2]. During the process of formation control, the agents need to coordinate with each other in order to maintain some specific pattern during the movement. In practice, agents are subject to the input saturation due to the limited actuation power of practical systems. Besides, velocity constraints also exist since a maximal velocity is usually imposed to ensure safe operation of the agents in the working environment, e.g., the requirement of limited velocity of vehicles for safety.

Supported by the National Natural Science Foundation of China under Grant No. 62173085.

B. Luo et al. (Eds.): ICONIP 2023, LNCS 14447, pp. 253–264, 2024.
https://doi.org/10.1007/978-981-99-8079-6_20

However, only few results have considered the VICs or either during the formation control process. In [3], the input bounded formation tracking of multiple Euler-Lagrange systems was achieved under the model-independent distributed controller. In [4], distributed leader-following consensus and formation navigation were investigated for double integrator multi-agent systems under VICs with general directed communication graphs.

Furthermore, to guarantee safety of the system, collision avoidance with other surrounding objects should also be considered for formation control of practical multi-agent systems. The common used methods include the potential function based strategy [5,6], the model predictive control (MPC) based control strategy [7,8] and the control barrier function (CBF) based collision avoidance method [9,10]. Note that the potential function based method generally cannot handle the VICs of the agents explicitly. The VICs can be incorporated easily with MPC, while the computational cost is generally high. Compared with the above two methods, the CBF based method has the advantages of handling input and/or velocity constraints explicitly and low computational cost. In [11], the safe formation tracking for second-order multi-agent systems was achieved based on the CBF method subject to the VICs. Recently, HOCBFs have been proposed in [12] which could handle the high-order dynamic model. However, the research of formation control with VICs is still lacking. Besides, most of the aforementioned results were developed only for precise model. It is obvious that practical control systems will be inevitably subject to external disturbances which is particularly dangerous for collision avoidance between agents since the collision avoidance conditions derived under the perfect model may become invalid with the uncertain dynamics.

In this paper, we develop a novel robust HOCBFs-based formation tracking controller for second-order multi-agent systems with input uncertainties and both VICs. In particular, we introduce HOCBFs to construct the collision avoidance conditions which can handle both VICs and are always feasible which is in contrast to existing results in [10,13]. The HOCBFs-based collision avoidance conditions are utilized to formulate a local QP problem for each follower which modify the nominal formation tracking controller in a minimal way.

Notations. $\|x\|_1$, $\|x\|$ and $\|x\|_\infty$ denote the 1-norm, Euclidean norm and the infinity norm of a vector $x \in \mathbb{R}^n$, respectively. $\text{sgn}(x) = [\text{sgn}(x_1), \ldots, \text{sgn}(x_n)]^T$ is the signum function. $\text{diag}\{x_1, \ldots, x_n\}$ represents a diagonal matrix with the diagonal elements $x_1, \ldots, x_n$.

2 Preliminaries and Problem Formulation

2.1 Graph Theory

A undirected graph $\mathcal{G} = (\mathcal{V}, \mathcal{E})$ consists of a vertex set $\mathcal{V} = \{p_1, \ldots, p_N\}$ and an edge set $\mathcal{E} = \{(p_i, p_j) | p_i, p_j \in \mathcal{V}\}$ where $(p_i, p_j) \in \mathcal{E}$ implies $(p_j, p_i) \in \mathcal{E}$. The neighboring set $\mathcal{N}_i$ of vertex p_i is $\{p_j | (p_j, p_i) \in \mathcal{E}\}$. A path from p_{i_1} to p_{i_k} is a sequence of distinct vertices $p_{i_1}, p_{i_2}, \ldots, p_{i_k}$ where $(p_{i_j}, p_{i_{j+1}}) \in \mathcal{E}$, $j = 1, \ldots, k-1$. Graph $\mathcal{G}$ contains a spanning tree if there exists a vertex, named as root, which has a directed path to all other vertices. The adjacency matrix A associated with graph $\mathcal{G}$ is defined as $a_{ij} > 0$ if $j \in \mathcal{N}_i, j \neq i$ and $a_{ij} = 0$, otherwise. The Laplacian matrix L is defined as $l_{ii} = \sum_{j=0}^{N} a_{ij}$ and $l_{ij} = -a_{ij}$. Without loss of generality, the leader is represented by a vertex p_0 and the followers are represented by a vertices $p_1, \cdots, p_N$. $\bar{\mathcal{G}}$ is a directed graph with vertex set $\mathcal{V}(\bar{\mathcal{G}}) = \mathcal{V}(\mathcal{G}) \bigcup \{p_0\}$. Let L denote the Laplacian matrix associated with $\mathcal{G}$, $A_0 = \text{diag}\{a_{10}, \cdots, a_{N0}\}$, we have the following lemma about the properties of $H = L + A_0$.

Lemma 1 ([14]). *If the commninication graph $\mathcal{G}$ of the followers is undirected and the graph $\bar{\mathcal{G}}$ contains a spanning tree with the leader as the root, then H is symmetric and positive definite.*

2.2 High-Order Control Barrier Function

Consider an affine control system:

$$\dot{x} = f(x) + g(x)u, \tag{1}$$

where $x \in \mathbb{R}^n$ and $u \in U \subset \mathbb{R}^q$, f and g are globally Lipschitz continuous.

Suppose a constraint $b(x,t) \geq 0$ is defined by a continuously differentiable function b, then the relative degree of b is also said as the relative degree of the constraint. For a constraint $b(x,t) \geq 0$ with relative degree m, $b : \mathbb{R}^n \times [0, \infty) \rightarrow \mathbb{R}$ and $\psi_0(x,t) := b(x,t)$, we define a sequence of functions $\psi_i : \mathbb{R}^n \times [0, \infty) \rightarrow \mathbb{R}, i \in \{1, \cdots, m\}$:

$$\psi_i(x,t) = \dot{\psi}_{i-1}(x,t) + \alpha_i(\psi_{i-1}(x,t)), i \in \{1, \cdots, m\}, \tag{2}$$

where $\alpha_i(\cdot), i \in \{1, \cdots, m\}$ denote class $\mathcal{K}$ functions of their argument.

We further define a sequence of sets $C_i(t), i \in \{1, \cdots, m\}$ associate with (2) as:

$$C_i(t) = \{x \in \mathbb{R}^n : \psi_{i-1}(x,t) \geq 0\}, i \in \{1, \cdots, m\}. \tag{3}$$

Definition 1 (HOCBF [12]**).** *Let $C_i(t), i \in \{1, \cdots, m\}$ be defined by (3) and $\psi_i(x,t), i \in \{0, 1, \cdots, m\}$ be defined by (2). A function $b : \mathbb{R}^n \times [0, \infty) \rightarrow \mathbb{R}$ is a HOCBF of relative degree m for system (1) if there exist differentiable class $\mathcal{K}$ functions $\alpha_i, i \in \{1, \cdots, m\}$ such that*

$$\sup_{u \in U} [\mathcal{L}_f^m b(x,t) + \mathcal{L}_g \mathcal{L}_f^{m-1} b(x,t) u + S(b(x,t)) + \alpha_m(\psi_{m-1}(x,t))] \geq 0, \tag{4}$$

for all $(x,t) \in C_1(t) \bigcap, \cdots, \bigcap C_m(t) \times [0, \infty)$. $\mathcal{L}_f$ and $\mathcal{L}_g$ denote the Lie derivatives along f and g, respectively. $S(\cdot)$ denotes the remaining Lie derivatives along f with degree at most $m-1$.

Theorem 1 ([12]). *Given a HOCBF $b(x,t)$ from Defition 1 with the associated sets $C_i(t), i \in \{1, \cdots, m\}$ defined by (3), if $x(0) \in C_1(0) \bigcap \cdots \bigcap C_m(0)$, then any Lipshitz continuous controller $u(t)$ that satisfies (4), $\forall t \geq 0$ renders $C_1(t) \bigcap \cdots \bigcap C_m(t)$ forward invariant for system (1).*

2.3 Problem Formulation

Consider a multi-agent system modeled as the following uncertain second-order systems:

$$\dot{p}_i = v_i, \quad \dot{v}_i = u_i + d_i, \quad i = 1, 2, \cdots, N, \tag{5}$$

where $p_i \in \mathbb{R}^n$, $v_i \in \mathbb{R}^n$ and $u_i \in \mathbb{R}^n$ represent the positions, velocities, and inputs of agent i, respectively. d_i represents the bounded uncertain dynamics and external disturbance which satisfies $\|d_i\|_\infty \leq \delta$ with $\delta > 0$ being a positive constant. A leader agent is defined with the dynamics:

$$\dot{p}_0 = v_0, \qquad \dot{v}_0 = u_0, \tag{6}$$

where $p_0 \in \mathbb{R}^n$, $v_0 \in \mathbb{R}^n$ and $u_0 \in \mathbb{R}^n$ are the positions, velocities and control inputs of the leader.

In the desired formation, each follower agent i has desired relative positions with respect to each other. Assign a formation vector r_i to agent $i, i = 0, 1, \cdots, N$ and let $r_{ij} = r_i - r_j$ represents the desired relative position between agent i and agent j.Then, the robust formation tracking problem for constrained system (5) and (6) is to design a control input u_i such that the VICs

$$\|v_i(t)\|_\infty \leq v^m, \quad \|u_i(t)\|_\infty \leq u^m, \tag{7}$$

with $v^m, u^m > 0$ are satisfied and it holds $p_i(t) - p_j(t) \to r_{ij}$ and $v_i(t) - v_0(t) \to 0$ as $t \to \infty$ where $i, j = 1, 2, \cdots, N$.

To guarantee safety of practical multi-agent systems, collision avoidance with respect to other agents and static obstacles should be considered. To solve this problem, we propose a robust HOCBFs-based control strategy and thereby the safe formation tracking subject to the VICs can be achieved. First, a nominal formation tracking controller is designed without considering the collision avoidance requirement under the VICs. Then, HOCBFs-based collision avoidance conditions are derived and then modify the nominal controller in a minimally invasive way.

3 Nominal Formation Tracking Controller

In this section, we design the nominal formation tracking controller without considering the collision-avoidance requirement. For the formation tracking system, the communication graph is assumed to satisfy the following assumption.

Assumption 1. *The commninication graph $\mathcal{G}$ of the followers is undirected and the graph $\bar{\mathcal{G}}$ contains a spanning tree with the leader as the root.*

Due to the existence of VICs of the follower agents, the states and inputs of the leader should also be bounded. Therefore, the following control structure of the leader is assumed:

$$u_0 = -k_0 v_0 + \bar{u}_0, \tag{8}$$

where $k_0 = \frac{u_0^m}{2v_0^m}$, $\|\bar{u}_0(t)\|_\infty \leq u_0^m/2$ and u_0^m, v_0^m are positive constants. Following the similar analysis as in [11], it can be shown that under the condition $\|v_0(0)\|_\infty \leq v_0^m$, we have $\|v_0(t)\|_\infty \leq v_0^m$ and $\|u_0(t)\|_\infty \leq u_0^m$. Moreover, it is assumed that $v_0^m \leq v^m$ and $u_0^m \leq u^m$.

First, for the multi-agent system (5) without regard to the disturbances, i.e.,

$$\dot{p}_i = v_i, \quad \dot{v}_i = u_i, \quad i = 1, \cdots, N, \tag{9}$$

the following controller is developed:

$$u_{i,nom} = -k_0 v_i - k_1 \mathrm{sgn}[k_0 \sum_{j=0}^{N} a_{ij}(p_i - p_j - r_i + r_j) + \sum_{j=0}^{N} a_{ij}(v_i - v_j)], \tag{10}$$

where $i = 1, 2, \cdots, N$, k_0 is the same as in (8), $k_1 > 0$ is the controller parameter.

Then the following nominal formation tracking controller for (5) and (6) subject to VICs (7) can be designed:

$$u_i = u_{i,nom} + u_{i,dis}, \quad u_{i,dis} = -\gamma \mathrm{sgn}(s_i), \tag{11}$$

where $s_i = v_i - v_i(0) - \int_0^t u_{i,nom} d\tau$, $\gamma > \delta$ and $u_{i,nom}$ is given by (10).

Theorem 2. *Under the initial condition that $\|v_i(0)\|_\infty \leq 2v_0^m(k_1 + \gamma + \delta)/u_0^m$, the velocity and input constrained formation tracking control for the multi-agent system (5) and (6) is achieved with the distributed controller (11) if*

$$2v_0^m(k_1 + \gamma + \delta)/u_0^m \leq v^m, \quad 2(k_1 + \gamma + \delta) \leq u^m, \quad k_1 > u_0^m/2. \tag{12}$$

Proof. With the control input (11), following similar steps as in [11], it can be shown that under the condition $\|v_i(0)\|_\infty \leq 2v_0^m(k_1 + \gamma + \delta)/u_0^m$ and (12) the VICs (7) will be satisfied. Then, we only need to show that the formation tracking will be achieved.

Consider the Lyapunov function $V = \frac{1}{2}\sum_{i=1}^{N} s_i^T s_i$, then we have $\dot{V} \leq -\sqrt{2}(\gamma - \delta)V^{1/2}$. Therefore, $s_i(t) = 0$ and $\dot{s}_i(t) = 0, i = 1, 2, \cdots, N$ for $t \geq t_1$ where $t_1 = \frac{\sqrt{2}V^{1/2}}{\gamma - \delta}$. Thus $\dot{v}_i = u_{i,nom}, i = 1, 2, \cdots, N$ for $t \geq t_1$. From the agent dynamics (5) and the facts $\|u_i(t)\|_\infty \leq u^m$, $\|d_i(t)\|_\infty \leq \delta$, it follows that all the signals in the closed-loop system will be bounded on the intervel $[0, t_1]$. When $t \geq t_1$, let $\tilde{p}_i = p_i - p_0 - r_i$, $\tilde{v}_i = v_i - v_0$, then we have

$$\dot{\tilde{p}}_i = \tilde{v}_i, \quad \dot{\tilde{v}}_i = -k_0\tilde{v}_i - k_1 \mathrm{sgn}[k_0 \sum_{j=0}^{N} a_{ij}(\tilde{p}_i - \tilde{p}_j) + \sum_{j=0}^{N} a_{ij}(\tilde{v}_i - \tilde{v}_j)] - \bar{u}_0. \tag{13}$$

Since (13) is decoupled in each dimension, we consider the qth dimension which satisfies

$$\dot{\tilde{p}}_i^q = \tilde{v}_i^q, \quad \dot{\tilde{v}}_i^q = -k_0\tilde{v}_i^q - k_1 \mathrm{sgn}[k_0 \sum_{j=0}^{N} a_{ij}(\tilde{p}_i^q - \tilde{p}_j^q) + \sum_{j=0}^{N} a_{ij}(\tilde{v}_i^q - \tilde{v}_j)^q] - \bar{u}_0^q.$$

Let $\eta_i^q = k_0\tilde{p}_i^q + \tilde{v}_i^q$ and $\eta^q = [\eta_1^q, \cdots, \eta_N^q]^T$ then $\dot{\eta}^q = -k_1\text{sgn}(H\eta^q) - U_0^q$ where $U_0^q = [\bar{u}_0^q, \cdots, \bar{u}_0^q]^T$. Consider the Lyapunov function $V^q = \frac{1}{2}\eta^{qT}H\eta^q$, then the derivative of V^q satisfies

$$\dot{V}^q = -k_1\eta^{qT}H\text{sgn}(H\eta^q) - U_0^{qT}H\eta^q \leq -(k_1 - \frac{u_0^m}{2})\|H\eta^q\|_1 \leq 0.$$

Note that as long as $\eta^q \neq 0$, we have $\dot{V}^q < 0$. Therefore, there exists a finite time after which it holds $\eta^q = 0$. Consequently, we have $k_0\tilde{p}_i + \tilde{v}_i = 0$ in finite time which leads to $\tilde{p}_i \to 0$ and $\tilde{v}_i \to 0$ as $t \to 0$. Hence, formation tracking control is achieved.

4 HOCBFs-Based Collision Avoidance Controller

In this section, we will develop a HOCBFs-based collision avoidance strategy for constrained system (5) and (6). The main idea is to constraint the admissible control inputs with proper HOCBFs and then the final collision avoidance formation tracking controller will be obtained by formulating a local QP problem which modifies the nonimial formation tracking controller in a minimal manner. The adopted nominal formation tracking controller for the follower agents is designed as $u_i = -k_0v_i + \bar{u}_i$ where $\bar{u}_i = \hat{u}_i = -k_1\text{sgn}[k_0\sum_{j=0}^{N} a_{ij}(p_i - p_j - r_i + r_j) - \gamma\text{sgn}(s_i)$.

4.1 HOCBFs-Based Collision Avoidance Condition

For a pair of follower agents i and j, denote the relative position and velocity as $\Delta p_{ij} = p_i - p_j$ and $\Delta v_{ij} = v_i - v_j$, respectively. We first consider the collision-avoidance requirements among the agents which can be expressed as $\|\Delta p_{ij}\| > D_s$ where D_s represents a minimal seperation distance which should be respected for all time. Inpired by ISSf [15], the safety set for agent i can be expressed as $S_{ij} = \{(p_i, v_i) \in \mathbb{R}^4 | h_{ij} \geq 0\}$ where $h_{ij} = \|\Delta p_{ij}\| - D_s' + 2\delta$ and $D_s' = D_s + 2\delta$. Noting that the relative degree of h_{ij} is 2, then we use h_{ij} to define a series of functions $\psi_{ik}^j, k = 0, 1, 2$ in the following form:

$$\psi_{i0}^j(t) = h_{ij}(t), \quad \psi_{i1}^j(t) = \dot{\psi}_{i0}^j(t) + q_1\psi_{i0}^j(t), \quad \psi_{i2}^j(t) = \dot{\psi}_{i1}^j(t) + q_2\psi_{i1}^j(t), \tag{14}$$

where q_1 and q_2 are positive constants.

The functions in (14) are associated with the following sequence of sets:

$$\begin{aligned} C_{i1}^j(t) &= \{(p_i, v_i) \in \mathbb{R}^{2n} : \psi_{i0}^j(t) \geq 0\}, \\ C_{i2}^j(t) &= \{(p_i, v_i) \in \mathbb{R}^{2n} : \psi_{i1}^j(t) \geq 0\}. \end{aligned} \tag{15}$$

Combining (5), (6) and (14), we have $\psi_{i1}^j(t) = \frac{\Delta p_{ij}^T \Delta v_{ij}}{\|\Delta p_{ij}\|} + q_1(\|\Delta p_{ij}\| - D_s' + 2\delta)$,

$$\psi_{i2}^j(t) = c_{ij} - (k_0 - q_2)\frac{\Delta p_{ij}^T \Delta v_{ij}}{\|\Delta p_{ij}\|} + \frac{\Delta p_{ij}^T}{\|\Delta p_{ij}\|}(\bar{u}_i - \bar{u}_j + d_i - d_j),$$

where $c_{ij} = \frac{\|\Delta v_{ij}\|^2}{\|\Delta p_{ij}\|} - \frac{(\Delta v_{ij}^T \Delta p_{ij})^2}{\|\Delta p_{ij}\|^3} + q_1 q_2(\|\Delta p_{ij}\| - D_s^{'} + 2\delta)$. Then the HOCBF-based constraint can be written as

$$-\frac{\Delta p_{ij}^T}{\|\Delta p_{ij}\|}(\bar{u}_i - \bar{u}_j + d_i - d_j) + (k_0 - q_2)\frac{\Delta p_{ij}^T \Delta v_{ij}}{\|\Delta p_{ij}\|} \leq c_{ij}. \tag{16}$$

Under the condition $q_1 q_2 \geq 1$, a sufficient condition for (16) to hold is

$$-\frac{\Delta p_{ij}^T}{\|\Delta p_{ij}\|}(\bar{u}_i - \bar{u}_j) + (k_0 - q_2)\frac{\Delta p_{ij}^T \Delta v_{ij}}{\|\Delta p_{ij}\|} \leq c_{ij}^{'}, \tag{17}$$

where $c_{ij}^{'} = \frac{\|\Delta v_{ij}\|^2}{\|\Delta p_{ij}\|} - \frac{(\Delta v_{ij}^T \Delta p_{ij})^2}{\|\Delta p_{ij}\|^3} + q_1 q_2(\|\Delta p_{ij}\| - D_s^{'})$. Note that $\Delta p_{ij} = -\Delta p_{ji}$ and $c_{ij}^{'} = c_{ji}^{'}$. Thus, the collision avoidance constraint (17) can be distributed to agents i and j as

$$\begin{aligned} -\frac{\Delta p_{ij}^T}{\|\Delta p_{ij}\|}\bar{u}_i + (k_0 - q_2)\frac{\Delta p_{ij}^T v_i}{\|\Delta p_{ij}\|} &\leq \frac{1}{2}c_{ij}^{'}, \\ -\frac{\Delta p_{ji}^T}{\|\Delta p_{ji}\|}\bar{u}_j + (k_0 - q_2)\frac{\Delta p_{ji}^T v_j}{\|\Delta p_{ji}\|} &\leq \frac{1}{2}c_{ij}^{'}. \end{aligned} \tag{18}$$

Next, we consider the collision avoidance requirements between the agents and the static obstacles O_l which can be expressed as $\|\Delta p_{iO_l}\| > D_{s_l}$, where $\Delta p_{iO_l} = p_i - p_{O_l}, l = 1, \cdots, N_O$ (N_O denotes the number of obstacles) and D_{s_l} represents the safe distance. Similarly, the collision avoidance set for agent i respect to obstacle O_l can be expressed as $S_{iO_l} = \{(p_i, v_i) \in \mathbb{R}^4 | h_{iO_l} \geq 0\}$ where $h_{iO_l} = \|\Delta p_{iO_l}\| - D_{s_l}^{'} + \delta$ and $D_{s_l}^{'} = D_{s_l} + \delta$ and its relative degree is 2. Following a similar derivation as in the collision avoidance among agents case, a series of functions $\psi_{ik}^{O_l}, k = 0, 1, 2$ are defined with h_{iO_l} in the following form:

$$\psi_{i0}^{O_l}(t) = h_{iO_l}(t), \quad \psi_{i1}^{O_l}(t) = \dot{\psi}_0^{O_l}(t) + q_1\psi_0^{O_l}(t), \quad \psi_{i2}^{O_l}(t) = \dot{\psi}_1^{O_l}(t) + q_2\psi_1^{O_l}(t), \tag{19}$$

along with a sequence of sets $C_{ik}^{O_l}, k = 1, 2$ associated (19)

$$\begin{aligned} C_{i1}^{O_l}(t) &= \{(p_i, v_i) \in \mathbb{R}^{2n} : \psi_{i0}^{O_l}(t) \geq 0\}, \\ C_{i2}^{O_l}(t) &= \{(p_i, v_i) \in \mathbb{R}^{2n} : \psi_{i1}^{O_l}(t) \geq 0\}. \end{aligned} \tag{20}$$

Combining (5), (6) and (19), we have $\psi_{i1}^{O_l}(t) = \frac{\Delta p_{iO_l}^T v_i}{\|\Delta p_{iO_l}\|} + q_1(\|\Delta p_{iO_l}\| - D_{s_l}^{'} + \delta)$,

$$\psi_{i2}^{O_l}(t) = c_{iO_l} + \frac{\Delta p_{iO_l}^T}{\|\Delta p_{iO_l}\|}(\bar{u}_i + d_i) - (k_0 - q_2)\frac{\Delta p_{iO_l}^T v_i}{\|\Delta p_{iO_l}\|},$$

where $c_{iO_l} = \frac{\|v_i\|^2}{\|\Delta p_{iO_l}\|} - \frac{(v_i^T \Delta p_{iO_l})^2}{\|\Delta p_{iO_l}\|^3} + q_1 q_2(\|\Delta p_{iO_l}\| - D_{s_l}^{'} + \delta)$. Then the HOCBF-based constraint can be written as

$$-\frac{\Delta p_{iO_l}^T}{\|\Delta p_{iO_l}\|}(\bar{u}_i + d_i) + (k_0 - q_2)\frac{\Delta p_{iO_l}^T v_i}{\|\Delta p_{iO_l}\|} \leq c_{iO_l}. \tag{21}$$

Under the condition $q_1 q_2 \geq 1$, a sufficient condition for (21) to hold is

$$-\frac{\Delta p_{iO_l}^T}{\|\Delta p_{iO_l}\|}\bar{u}_i + (k_O - q_2)\frac{\Delta p_{iO_l}^T v_i}{\|\Delta p_{iO_l}\|} \leq c_{iO_l}^{'}, \tag{22}$$

where $c_{iO_l}^{'} = \frac{\|v_i\|^2}{\|\Delta p_{iO_l}\|} - \frac{(v_i^T \Delta p_{iO_l})^2}{\|\Delta p_{iO_l}\|^3} + q_2 q_1(\|\Delta p_{iO_l}\| - D_{s_l}^{'})$.

4.2 HOCBFs-Based Collision Avoidance Formation Tracking

In order to ensure that the performance of formation tracking is respected to the highest degree possible, the collision avoidance strategy should modify the nominal formation tracking controller in a minimally invasive manner. Noting that the safety constraints (18) and (22) are linear in the control signal, we can add a quadratic cost that penalizes deviations from the nominal controller (in the least-squares sense), resulting in the following QP problem:

$$\begin{aligned}
\bar{u}_i^* = \arg \min_{\bar{u}_i \in \mathbb{R}^{2n}} \quad & J(\bar{u}_i) = \|\bar{u}_i - \hat{\bar{u}}_i\|_2 \\
s.t. - \frac{\Delta p_{ij}^T}{\|\Delta p_{ij}\|}\bar{u}_i + (k_O - q_2)\frac{\Delta p_{ij}^T v_i}{\|\Delta p_{ij}\|} & \leq \frac{1}{2}c_{ij}^{'}, j \in \{1, \cdots, N\} \setminus \{i\}, \\
-\frac{\Delta p_{iO_l}^T}{\|\Delta p_{iO_l}\|}\bar{u}_i + (k_O - q_2)\frac{\Delta p_{iO_l}^T v_i}{\|\Delta p_{iO_l}\|} & \leq c_{iO_l}^{'}, O_l \in \{1, \cdots, N_O\}, \\
\|\bar{u}_i\|_\infty & \leq k_1 + \gamma
\end{aligned}$$

for $i = 1, \cdots, N$. Then, the final control inputs of the follower agents are designed as

$$u_i = -k_O v_i + \bar{u}_i^*, \quad i = 1, \cdots, N. \tag{23}$$

The following theorem shows that under proper conditions, the QP problem (23) is always feasible.

Theorem 3. *Under the condition $|k_O - q_2|v^m \leq k_1 + \gamma$, the QP problem (18) is always feasible.*

For brevity, denote $C_i = (\bigcap_{j \in \mathcal{N}_c^i} C_{ij}) \bigcap (\bigcap_{l \in \mathcal{N}_O^i} C_{iO_l})$, where $C_{ij} = C_{i1}^j \bigcap C_{i2}^2$ and $C_{iO_l} = C_{i1}^{O_l} \bigcap C_{i2}^{O_l}$. Then, we have the following result which shows that collision avoidance of the multi-agent system is guaranteed with the QP based controller (23).

Theorem 4. *Consider the multi-agent system (5) and (6), suppose that for all agent $i = 1, \cdots, N$, the initial states satisfy $z_i(0) \in C_i$, $\|v_i(0)\|_\infty \leq 2v_0^m(k_1 + \gamma + \delta)/u_0^m$ and it holds $|k_O - q_2|v^m \leq k_1 + \gamma$. Then, with the control inputs given by (23), the multi-agent system is guaranteed to be collision free while the VICs (7) are satisfied.*

5 Simulation

In this section, several simulation examples are given to illustrate the performance of the proposed control strategy. We consider a network consists of one leader and five followers modeled by (5) and (6). The designated formation vectors r_i, $i = 0, 1, \cdots, 5$ are specified with $r_1 = [6, -6]^T$, $r_2 = [6, 6]^T$, $r_3 = [12, 12]^T$, $r_4 = [12, 0]^T$, $r_5 = [12, -12]^T$ and $r_0 = [0, 0]^T$. The communication graph with $A = [0, 1, 1, 0, 0; 1, 0, 0, 1, 0; 1, 0, 0, 1, 1; 0, 1, 1, 0, 1; 0, 0, 1, 1, 0]$ and $A_0 = \text{diag}\{1, 1, 0, 0, 0\}$ is chosen. The initial velocities of the agents are set to zero and the initial positions in the set $[-25, 25] \times [-25, 25]$.

The input of the leader is given as in (8) with $v_0^m = 0.5$, $u_0^m = 1$ and $\bar{u}_0 = [u_0^m/2, 0]^T$. It follows that $k_0 = 1$. The follower agents' velocity and input constraints correspond to $v^m = 2$ and $u^m = 4$, respectively. For the ith follower, the external disturbance $d_i = 0.2\sin(0.5t + \frac{i\pi}{2})$. Then, the upperbound of the disturbance can be set as $\delta = 0.2$. The radius of the agents and the static obstacles are set as $R_1 = 2$ and $R_2 = 3$, respectively. Then, the minimal safe distance $D_s = 2R_1$ and $D_{sl} = R_1 + R_2$. The controller parameters $k_1 = 1.5$ and $\gamma = 0.3$ are chosen according to Theorem 2.

A comparision with the widely studied potential function based collision avoidance method is performed. Take two agents i and j for example. Let $D_{ij} = \|\Delta p_{ij}\| - D_s$, then the repelling force acting on agent i is given by

$$f_{ij} = \begin{cases} 2\frac{\eta}{D_{ij}^2}\left(\frac{1}{D_{ij}} - \frac{1}{d}\right)\frac{p_i - p_j}{\|\Delta p_{ij}\|}, & if \quad D_{ij} \le d, \\ 0, & if \quad D_{ij} > d, \end{cases}$$

where $\eta > 0$ and $d > 0$ are design parameters. Thus the total force from all the other robots and obstcles can be obtained as $F_i = \sum_{j=1, j\neq i}^{N} f_{ij} + \sum_{l=1}^{N_O} f_{iO_l}$. Then, the following control input is considered to achieve collision avoidance formation tracking of the agents:

$$u_i = -k_0 v_i + \hat{\bar{u}}_i + F_i. \tag{24}$$

For the safe formation tracking task, the potential function parameters are taken as $\eta = 2.5$ and $d = 4$. To construct the HOCBFs-based constraints, the parameters $q_1 = q_2 = 1.8$ are chosed which satisfy $q_1 q_2 \ge 1$ and the condition in Theorem 3. The simulation results are given in Figs. 1, 2, 3. From Fig. 1 and Fig. 3, it can be seen that all the follower agents move into the desired formation and travel along with the leader, while the collision avoidance is also achieved under both the proposed HOCBFs-based collision avoidance controller and the potential function based collision avoidance controller (24). However, from Fig. 2, it is clear that both the velocity and inputs of the agents lie in the required bounded intervals with the proposed HOCBFs-based collision avoidance controller, while large repelling forces are generated to prevent collision which violates the input constraints of the agents with the potential function based collision avoidance controller (24).

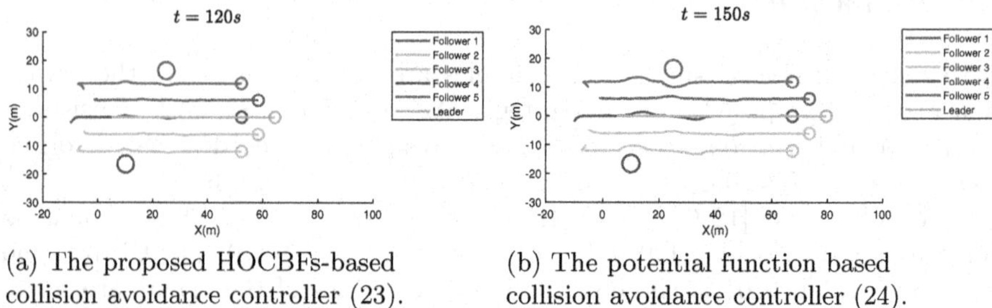

(a) The proposed HOCBFs-based collision avoidance controller (23).

(b) The potential function based collision avoidance controller (24).

Fig. 1. Trajectories of all agents under different controllers.

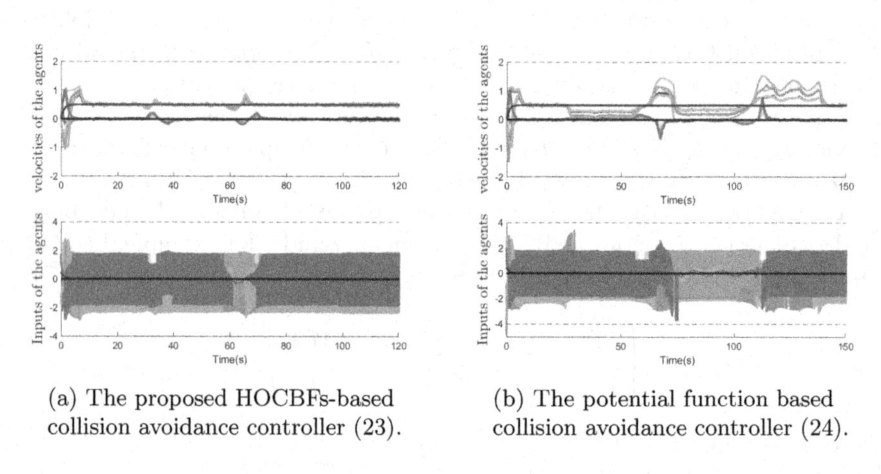

(a) The proposed HOCBFs-based collision avoidance controller (23).

(b) The potential function based collision avoidance controller (24).

Fig. 2. Velocities and inputs of the follower agents with different controllers.

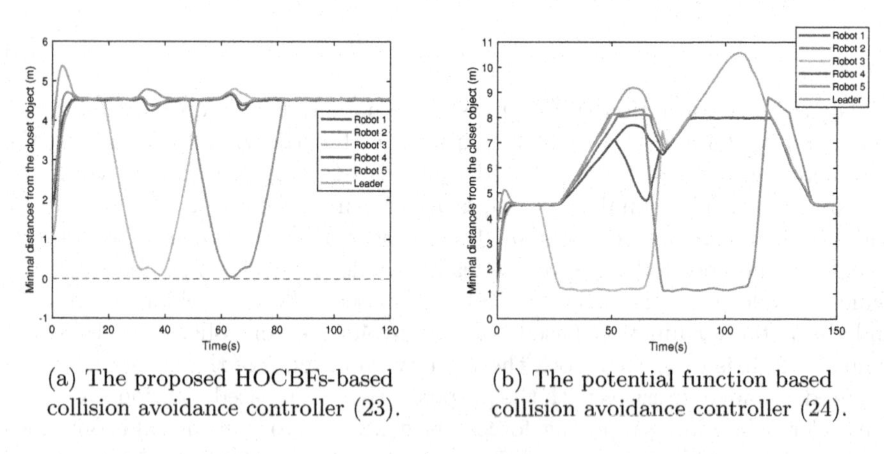

(a) The proposed HOCBFs-based collision avoidance controller (23).

(b) The potential function based collision avoidance controller (24).

Fig. 3. Minimum separation between each follower agent and its nearest object under different controllers.

6 Conclusion

In this work, we have considered the collision avoidance formation tracking problem for uncertain second-order multi-agent systems subject to both VICs. A novel HOCBFs-based safe formation tracking control strategy is proposed which guarantees collision avoidance under both VICs. A silding mode based nominal controller is firstly designed. Then, a local QP based formation controller is developed based on the HOCBFs collision avoidance conditions. The feasibility of the QP problems and the guaranteed collision avoidance are rigorously proven. Simulation results have been provided to demonstrat the effectiveness of the proposed controller.

References

1. Cao, Y., Yu, W., Ren, W., Chen, G.: An overview of recent progress in the study of distributed multi-agent coordination. IEEE Trans. Industr. Inf. **9**(1), 427–438 (2013)
2. Oh, K.K., Park, M.C., Ahn, H.S.: A survey of multi-agent formation control. Automatica **53**, 424–440 (2015)
3. Wang, L., He, H., Zeng, Z., Ge, M.: Model-independent formation tracking of multiple Euler-Lagrange systems via bounded inputs. IEEE Trans. Cybern. **51**(5), 2813–2823 (2021)
4. Fu, J., Wen, G., Yu, W., Huang, T.: Consensus of second-order multi-agent systems with both velocity and input constraints. IEEE Trans. Industr. Electron. **66**(10), 7946–7955 (2019)
5. Mastellone, S., Stipanović, D.M., Graunke, C.R., Intlekofer, K.A., Spong, M.W.: Formation control and collision avoidance for multi-agent non-holonomic systems: theory and experiments. Int. J. Robot. Res. **27**(1), 107–126 (2008)
6. Peng, Z., Wang, D., Li, T., Han, M.: Output-feedback cooperative formation maneuvering of autonomous surface vehicles with connectivity perservation and collision avoidance. IEEE Trans. Cybern. **50**(6), 2527–2535 (2020)
7. Murry, R.M., Dunbar, W.B.: Distributed receding horizon control for multi-vehicle formation stabilization. Automatica **42**(4), 549–558 (2016)
8. Wang, P., Ding, B.: A synthesis approach of distributed model predictive control of for homogeneous multi-agent system with collision avoidance. Int. J. Control **87**(1), 52–63 (2014)
9. Borrmann, U., Wang, L., Ames, A.D., Egerstedt, M.: Control barrier certificates for safe swarm behavior. IFAC-PapersOnLine **48**(27), 68–73 (2015)
10. Wang, L., Ames, A.D., Egerstedt, M.: Safety barrier certificates for collisions-free multirobot systems. IEEE Trans. Rob. **33**(3), 661–674 (2017)
11. Fu, J., Lv, Y., Wen, G., Yu, X., Huang, T.: Velocity and input constrained coordination of second-order multi-agent systems with relative output information. IEEE Trans. Network Sci. Eng. **7**(3), 1925–1938 (2020)
12. Wei, X., Belta, C.: High order control barrier functions. IEEE Trans. Autom. Control **67**(7), 3655–3662 (2022)
13. Fu, J., Wen, G., Yu, X., Huang, T.: Robust collision-avoidance formation navigation of velocity and input-constrained multirobot systems. IEEE Trans. Cybern., 1–13 (2023)

14. Zhang, H., Li, Z., Qu, Z., Lewis, F.L.: On constructing Lyapunov functions for multi-agent systems. Automatica **58**, 39–42 (2015)
15. Alan, A., Taylor, A.J., He, C.R., Orosz, G., Ames, A.D.: Safe controller synthesis with tunable input-to-state safe control barrier functions. IEEE Control Syst. Lett. **6**, 908–913 (2022)

Decision Support System Based on MLP: Formula One (F1) Grand Prix Study Case

Jakub Więckowski[1], Bartosz Paradowski[2], Bartłomiej Kizielewicz[1], Andrii Shekhovtsov[1], and Wojciech Sałabun[1(✉)]

[1] West Pomeranian University of Technology in Szczecin, ul. Żołnierska 49, 71-210 Szczecin, Poland
{jakub-wieckowski,bartlomiej-kizielewicz,andrii-shekhovtsov, wojciech.salabun}@zut.edu.pl

[2] National Telecommunications Institute, ul. Szachowa 1, 04-894 Warsaw, Poland
b.paradowski@il-pib.pl

Abstract. Neural networks are widely used due to the adaptability of models to many problems and high efficiency. These solutions are also gaining popularity in the design of Decision Support Systems. It leads to increased use of such techniques to support the decision-maker in practical problems.

In this paper, we propose an Artificial Neural Network Decision Support System (ANN-DSS) based on Multilayer Perceptron. The model structure was determined by searching the optimal hyperparameters with Tree-structured Parzen Estimator. Based on the qualification results, the proposed system was directed to evaluate the Formula 1 divers' best lap time during the race. Obtained rankings were compared with reference rankings using the WS rank similarity. Model performance proves to be highly consistent in rankings predictions, which makes it a reliable tool for the given problem.

Keywords: Decision Support Systems · Neural Network · Formula One · Decision Making

1 Introduction

In times of technological development and basing our actions on information systems, efficient and reliable techniques must support the proposed solutions. Moreover, in many aspects of life, we use systems whose aim is to help us choose a rational variant or make a ranking of available decision options [24]. Such systems are called Decision Support Systems (DSS) [38]. Depending on their purpose, they are built on various techniques, such as Multi-Criteria Decision Analysis (MCDA), Neural Network (NN), or Stochastic Optimization (SO) [18]. DSSs result in proposed rankings shown to the decision-maker, who makes the final decision. Thus, they provide preference values calculated based on the methods used, supported by solid arguments in favor of one alternative over another [25].

B. Luo et al. (Eds.): ICONIP 2023, LNCS 14447, pp. 265–276, 2024.
https://doi.org/10.1007/978-981-99-8079-6_21

The use of Neural Networks in Decision Support Systems is a popular approach [37]. Neural Networks are a core component of Artificial Intelligence (AI), aimed at solving complex problems [40]. In the literature, works devoted to developing DSS systems using different types of Neural Networks can be found [12]. The use of Artificial Neural Network (ANN), Fuzzy Neural Network (FNN), or Deep Neural Network (DNN) has proven to be effective [6,14]. It guarantees robust results with the high efficiency of these DSS. The combination of these two techniques has a high practical potential in many fields, which can be used to design other dedicated Decision Support Systems based on Neural Networks.

The literature review shows that dedicated DSSs based on ANN are used in many areas where they prove to be efficient tools [11]. Systems for decision support in dentistry [15], finance [28], stock market [7], or production management [5] are being developed. Sport is also becoming a popular area for the use of such systems [35]. The results are often influenced by the use of Information Systems (IS) supporting the training process [42]. They also make it possible to create predictions based on historical data. Therefore, the system users have additional knowledge, which translates into an advantage over their opponents. It makes dedicated Decision Support Systems desirable due to the possibility of their practical use.

Decision Support Systems with high efficiency and short processing times are crucial in Formula 1 (F1). With the development of technology, racing teams have paid more and more attention to IS [1]. The vast amount of sensor measurement data coming in during races must be processed in real-time to provide engineers with the most recent data [41]. Effective data analysis helps to gain an advantage over rivals, which is extremely valuable due to the minimal differences between drivers [21]. Race predictions are fundamental there [17]. The qualifying round before the race determines the starting order for the race. However, the performance in the race is decisive for the final position.

The main contribution of this paper is the proposition of an Artificial Neural Network Decision Support System (ANN-DSS) based on Multilayer Perceptron (MLP) to predict the ranking of the fastest lap times during a race based on recorded times by drivers in qualifying. Race performance is extremely important as it is a significant factor in the final result. Hence, a tool that predicts the fastest lap time at a race pace, featuring fast and effective performance, can be invaluable. Such predictions are made on historical data and can provide interesting insights for Formula One teams and their investors. Moreover, this approach can prove itself usable in other disciplines.

The rest of the paper is organized as follows. In Sect. 2, the literature review regarding the research approaches to the Formula 1 field is included. Section 3 presents the main assumptions of core elements of Artificial Neural Networks. Section 4 shows the study case of developing the ANN-DSS dedicated to ranking drivers' race pace based on qualification results. Finally, in Sect. 5, the summary and the conclusions drawn from the research are presented.

2 Literature Review

With the prestige and popularity of the sport Formula One, it is attracting a growing audience and a broad academic community. The former benefits from the spectacle and duels between drivers on the track, while the latter has the opportunity to study the complex processes involved in the performance of racing cars. In addition, due to the integration of the many areas that make up the overall F1 sport, attention is also directed toward investigating marketing mechanisms, the quality of materials, the impact of aerodynamic elements, or assessing the quality of drivers' performances throughout a season. It makes the sport a complex and challenging research area constantly being explored to improve performance and find more effective applications.

There is much research in different sub-areas on this topic. The researches all over the world devote their studies to marketing [20,43], technical solutions for the race cars [8,39], performance of the drivers during the races [9,23], and many other interesting topics, such as investigating car collisions [30], looking for the best driver [13], and many other [19,26,31].

There are also works in which authors tried to use different computational methods, such as simulations, Artificial Neural Networks, or multilevel modeling, to evaluate different cars and drivers. Bekker et al. tried to plan a formula one race strategy using discrete simulations [2]. Heilmeier et al. used ANN to make race strategy decisions in the context of motorsport [16]. A similar study was also conducted by O'Hanlon, who tries to use Neural Networks to predict the final rankings of the Formula One championship in 2021 [29]. Earlier, Stoppels also tried to use an ANN-based model to predict the results of the race [36]. Rockerbie and Eatson use a regression model to evaluate the performance of a car and a driver during the F1 race [33].

3 Preliminaries

3.1 Multilayer Perceptron

Multilayer perceptron (MLP) is a type of feed-froward artificial neural network that was first presented by Frank Rosenblatt. This type of ANN consists of one input layer, one or more hidden layers, and one output layer. It is one of the most popular and widely used types of ANN [27,32].

In order to train MLP, different optimizers are used. In this paper, we will use the Stochastic Gradient Descent (SGD) optimizer, which can provide great accuracy in most cases. The other important thing is tuning hyperparameters, which might greatly influence results. In our study, we have used Tree-structured Parzen Estimator [4] for this purpose. As an activation function, the ReLU function is used, as it is very popular in recent applications and research. This activation function, because of its simple equation (1), provides a computationally efficient way to calculate the value of hidden neurons.

$$\phi(z) = \max(0, z) \tag{1}$$

3.2 WS Rank Similarity Coefficient

WS rank similarity coefficient is based on the distance between positions in rankings, regarding the greater importance of differences in the top of the ranking than in the bottom [34]. It can be calculated with the following formula (2):

$$WS = 1 - \sum_{i=1}^{n} \left(2^{-x_i} \frac{|x_i - y_i|}{\max\{|x_i - 1|, |x_i - N|\}} \right) \tag{2}$$

In beforehand Equation, N corresponds to the size of the ranking, and values x_i and y_i are rank values from two rankings.

4 Study Case

One of the major problems in machine learning is the architecture of the neural network and how to define it. For this task, numerous different methods and approaches were created. In our study, we have used Tree-structured Parzen Estimator [4] to provide the optimal architecture for evaluating F1 rankings. In searching for optimal architecture, the loss function needs to be determined. For this case, as the problem is similar to a regression, the Mean Squared Error (MSE) was used as a loss function. The search was minimized in terms of MSE for prediction and test set. The searching process is depicted in Fig. 1.

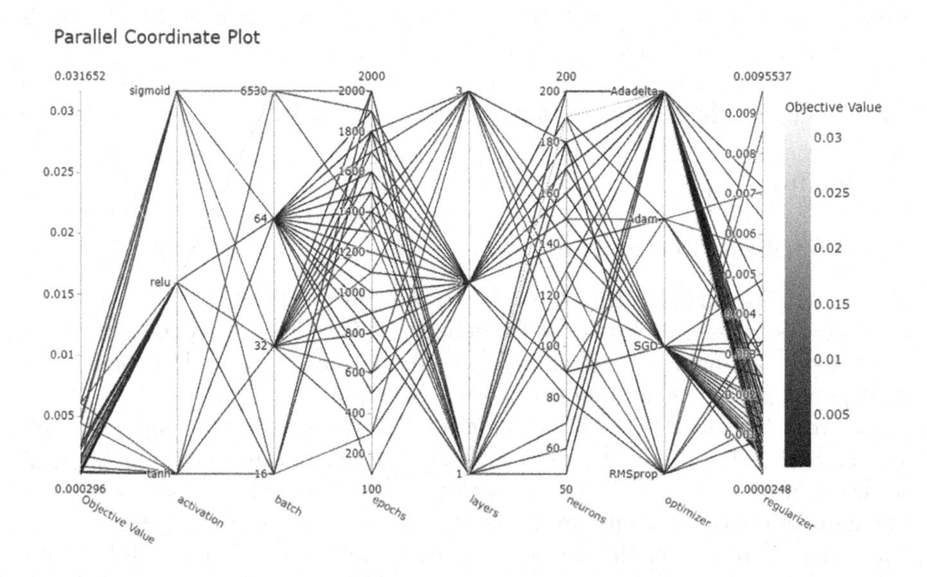

Fig. 1. Results obtained in process of searching the optimal model parameters.

With mentioned search, the optimal architecture that was found presents as shown in Fig. 2. Additionally, for the batch size were checked sizes of 16, 32, 64, and all possible records, and it turned out that the most optimal size is 64. The rest of the tuned hyperparameters are as follows: regularizer l2 type of value 0.0000446, two hidden layers, and 1600 epochs.

The specific nature of the sport, Formula 1, means that even the most minor differences can impact the final result [3]. It is confirmed by the fact that differences between drivers in the qualifying session are measured with an accuracy of 0.001 of a second. Additionally, the difference between a group of top drivers at certain tracks may be so slight that 0.1 of a second may decide the 1st or 5th place on the grid. It translates into the need to pay attention to every detail in order to optimize team performance. The difference in races between drivers' times is often more noticeable than in qualifying times. It is influenced by many factors such as tyre type and its condition, fuel level, race stage, and track position [10].

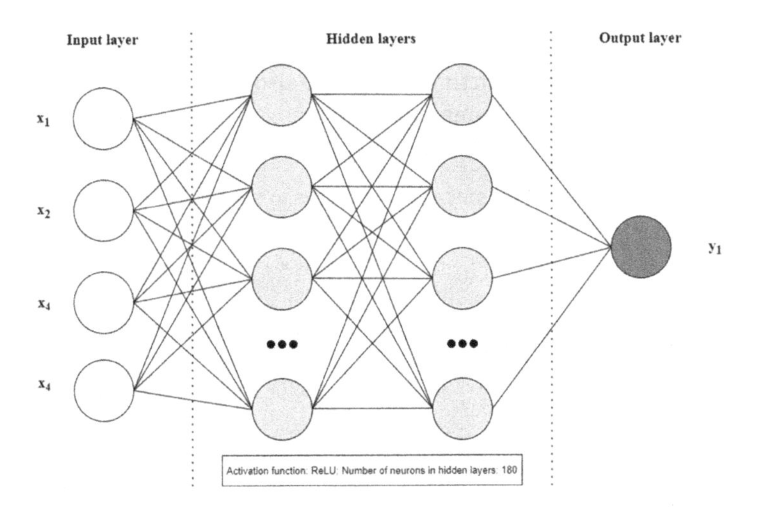

Fig. 2. MLP architecture determined in process of searching the optimal parameters.

On the other hand, consistently underperforming on a race lap-by-lap can result in a loss of position. Therefore, race time prediction systems can help forecast a driver's chances to maintain position, overtake the competitor in front, or plan strategy. Because of the need for real-time data processing, these systems need to be fast and efficient. It, in turn, prompts the planning of such models so that they use relatively small amounts of data while ensuring the reliability of the results.

In the proposed Decision Support System based on the Artificial Neural Network, we decided to use 4 parameters as MLP input: qualifying position, the best time in the 1st qualifying session (Q1), where all drivers participate, the best time in the 2nd qualifying session (Q2), where the fastest 15 drivers from Q1 start, and the best time in the 3rd qualifying session (Q3), where the 10 fastest drivers from Q2 drive. From this data, the model's task is to predict the fastest time on a single lap in the race. The driver ranking is then determined based on these results.

That information can be crucial for planning the race strategy of a given driver or evaluating the chances of getting a given position in the race. It is also worth mentioning that the driver who achieves the fastest single lap time during the race and finishes a minimum of 10th in the race receives 1 point towards the general classification. This could be vital in the battle for the championship title, as demonstrated by the 2021 season, where the difference between Max Verstappen (1st) and Lewis Hamilton (2nd) was only 8 points (395.5, 387.5).

The Neural Network structure presented above was used in the process of learning the prediction model. Data from a publicly available database containing information on Formula 1 races over the years 1950–2022 was used for learning [22]. The data was preprocessed and then used as input to the Neural Network. The data was divided into a training set of size (6530, 4) and a test set of size (459, 4). Data captured between 1950 and 2020 were used to train the model, while data from the completed 2021 season were used as a test set to check the effectiveness of the fitting. It should also be mentioned that the data were normalized using the min-max method.

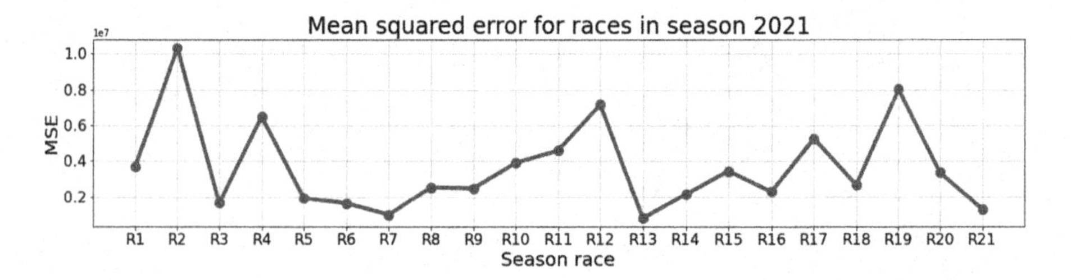

Fig. 3. Mean Squared Error (MSE) determined for test data for F1 Season 2021.

The loss function in the model was set to Mean Squared Error (MSE), as its one of the metrics dedicated to regression problems. By applying a search process for optimal model parameters, the obtained MSE values were 1.5471e−03 for the training set and 3.1732e−04 for the test set. Then, verification of the effectiveness of the learned MLP model was performed on measurement data obtained during the 2021 season. MSE and Mean Absolute Error (MAE) metrics were used for each race analyzed. Figures 3 and 4 show plots representing the values of these metrics for each of the races considered. It is worth noting that MSE gave higher values at the beginning and end of the season while maintaining relatively low values in the middle. In contrast, the MAE run showed that the values were more diversified. However, it is also possible to note significant similarities in the determined metrics trends, particularly evident in the second half of the analyzed season.

Afterward, based on the predicted fastest race lap times, a ranking of the drivers was made, and their order was calculated. These were used to calculate

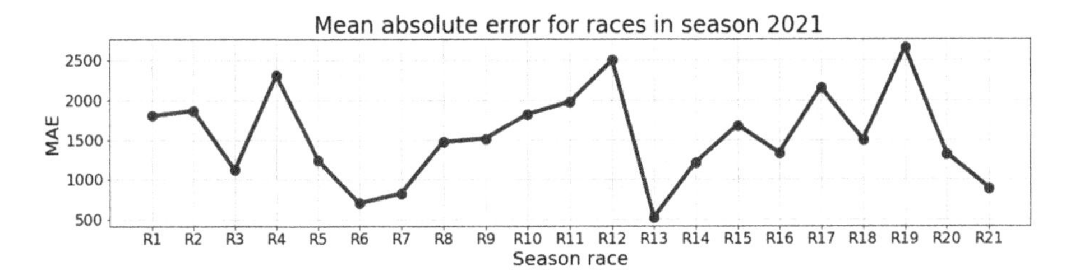

Fig. 4. Mean Absolute Error (MAE) determined for test data for F1 Season 2021.

the similarity between the hierarchy determined using the MLP model and the actual ranking of the fastest laps in the race recorded by the drivers. In addition, the similarity of the rankings was also calculated for the order of drivers from Q1, Q2, and Q3. These rankings were calculated based on the best times for all the drivers in the 1st, 2nd and 3rd qualifying sessions.

Table 1. Statistics for race fastest lap ranking similarity in F1 Season 2021.

Metric	MLP	Q1	Q2	Q3
Mean	0.861	0.761	0.814	0.845
STD	0.088	0.162	0.125	0.103

The averaged data is shown in Table 1, where, besides the mean, the standard deviation (STD) is also presented. It can be seen that the MLP model used guarantees the highest correlation value of rankings with the lowest standard deviation. Therefore, it makes it a more reliable indicator in predicting the order of the drivers' fastest laps in the race than the order of the times recorded by the drivers in the qualifying session.

To visualize the distribution of rankings similarity, the data was presented using boxplots in Fig. 5. It can be noticed that the MLP guarantees the most coherent ranking similarity with the highest mean similarity value. On the other hand, comparing results to the similarities for Q1, the calculated spread was significantly greater. Furthermore, similarities for the ranking of Q2 and Q3 were close to each other, slightly in advance of the Q3 results. However, the determined spread of yielded WS ranking similarities was more remarkable than in the case of MLP comparison. It proves that the MLP provides more reliable and regular results prediction than others.

Figure 6 compares MLP, Q1, Q2, and Q3 rankings for subsequent races in F1 Season 2021. Position in the ranking was calculated based on the preference obtained from the WS coefficient. It is worth mentioning the number of first places in compared rankings. The determined MLP model was ranked first 4 times. Times from the Q1 session achieved the highest similarity with reference ranking only 1 time. Times from Q2 and Q3 sessions were ranked best 7 and

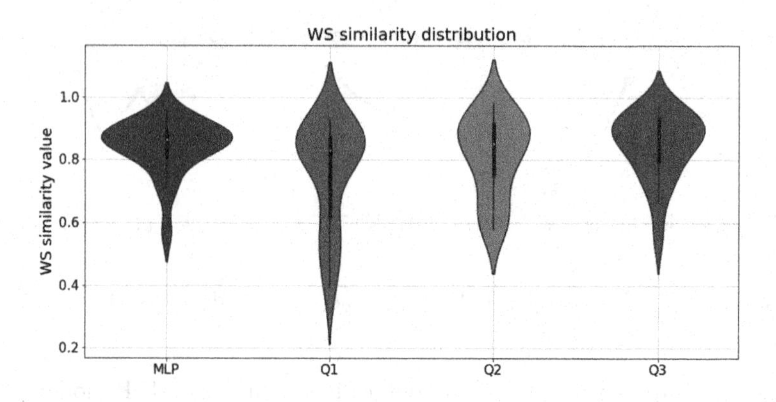

Fig. 5. WS ranking similarity distribution for races in F1 Season 2021.

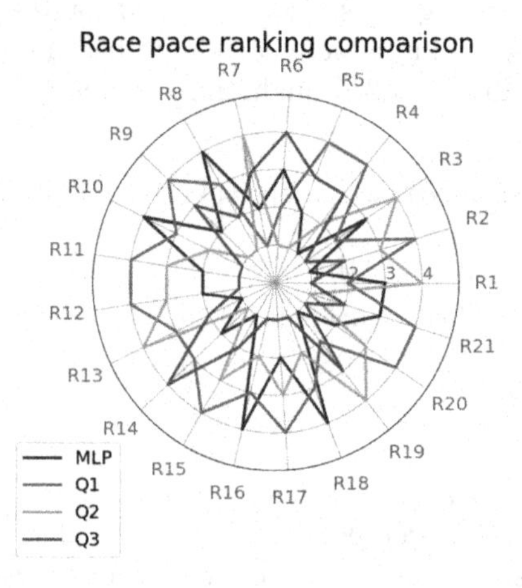

Fig. 6. Race fastest lap ranking comparison for each race in F1 Season 2021.

9 times, respectively. However, MLP was placed on 2nd (9 times) and 3rd (3 times) place more frequently than Q2 (2nd 4 times, 3rd 6 times) and Q3 (2nd 5 times, 3rd 5 times). Despite the smaller number of wins, MLP provided more consistent predictions of drivers' fastest lap ranking than others.

Moreover, the determined Artificial Neural Network model was used to predict rankings of race fastest lap for the first 3 races in F1 Season 2022. Obtained classifications of the MLP model and reference ranking were presented in Fig. 7. Drivers' problems, accidents, and not finished races result in different amounts of drivers compared for each race. It can be seen that a significant number of positions overlap each other by comparing reference and predicted ranking.

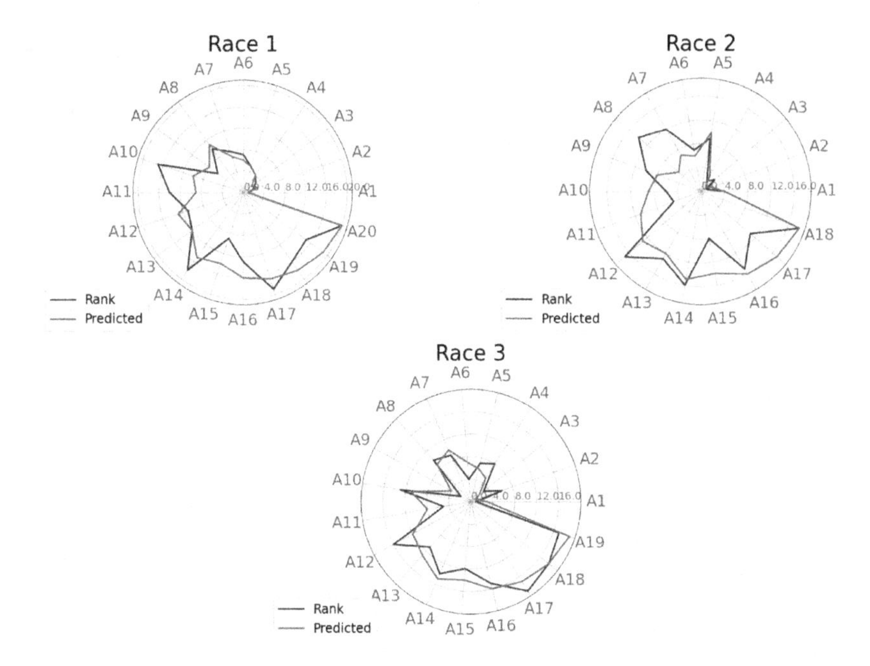

Fig. 7. Race fastest lap ranking comparison for first 3 races in F1 Season 2022.

Table 2 presents the comparison of WS ranking similarity between reference ranking and ranking determined based on results predicted by the MLP model and those defined based on Q1, Q2, and Q3 times. The best mean similarity was achieved by the MLP and equaled 0.941. Second place belongs to Q3 rankings with a 0.015 loss. This example also shows that the MLP proved to be more consistent in drivers' fastest lap rankings prediction.

Table 2. WS similarity for fastest lap ranking for first 3 races in F1 Season 2022.

Race	MLP	Q1	Q2	Q3
1	0.974	0.943	0.941	0.996
2	0.951	0.884	0.877	0.869
3	0.898	0.817	0.887	0.913
Mean	0.941	0.881	0.901	0.926

5 Conclusion

The use of Neural Networks in complex problems is an effective solution, and the results obtained using these techniques give high reliability. ANN methods are widely used in many areas. One of them is the prediction of results based on the use of regression. The high performance and adaptability of the Neural Network to complex problems allow the use of these techniques in Decision Support Systems, whose purpose is to provide the expert with suggested preference values.

A comprehensive search of the optimal model parameters allowed the MLP network structure to be defined, which was then used to predict the fastest lap times in the race based on the times achieved by Formula 1 drivers in qualifying. The predictions determined by the MLP model showed that they had high efficiency in predicting the drivers' ranking times. It allowed the development of a reliable ANN-DSS, a tool that can be used for practical problems in this area. However, the proposed model has some limitations. The ANN used in the decision support process is based on MLP and, therefore cannot be suitable for all types of problems and may not be generalized to other domains. Application of such DSS should be preceded by the verification if this approach is suitable for the interested domain or not.

For further directions, it is worth considering performing a more complex search for the model's optimal parameters, as all possibilities were not included in the search. Moreover, it would be meaningful to test the proposed approach to develop ANN-DSS for other practical fields and verify its reliability.

References

1. Aversa, P., Furnari, S., Haefliger, S.: Business model configurations and performance: a qualitative comparative analysis in formula one racing, 2005–2013. Ind. Corp. Chang. **24**(3), 655–676 (2015)
2. Bekker, J., Lotz, W.: Planning formula one race strategies using discrete-event simulation. J. Oper. Res. Soc. **60**(7), 952–961 (2009)
3. Bell, A., Smith, J., Sabel, C.E., Jones, K.: Formula for success: multilevel modelling of formula one driver and constructor performance, 1950–2014. J. Quantitative Anal. Sports **12**(2), 99–112 (2016)
4. Bergstra, J., Bardenet, R., Bengio, Y., Kégl, B.: Algorithms for hyper-parameter optimization. Advances in neural information processing systems 24 (2011)
5. Beşikçi, E.B., Arslan, O., Turan, O., Ölçer, A.I.: An artificial neural network based decision support system for energy efficient ship operations. Comput. Oper. Res. **66**, 393–401 (2016)
6. Bisoi, R., Parhi, P., Dash, P.: Hybrid modified weighted water cycle algorithm and deep analytic network for forecasting and trend detection of forex market indices. Int. J. Knowl.-Based Intell. Eng. Syst. (Preprint), 1–21 (2022)
7. Boonpeng, S., Jeatrakul, P.: Decision support system for investing in stock market by using oaa-neural network. In: 2016 Eighth International Conference on Advanced Computational Intelligence (ICACI). pp. 1–6. IEEE (2016)
8. Bopaiah, K., Samuel, S.: Analysis of energy recovery system of formula one cars. Tech. rep. (2021)
9. Celik, O.B.: Survival of formula one drivers. Soc. Sci. Q. **101**(4), 1271–1281 (2020)
10. Chandra, S., Lee, A., Gorrell, S., Jensen, C.G.: Cfd analysis of pace formula-1 car. Comput.-Aided Des. Appl., PACE **1**, 1–14 (2011)

11. Delen, D., Sharda, R.: Artificial neural networks in decision support systems. In: Handbook on Decision Support Systems 1, pp. 557–580. Springer, Cham (2008). https://doi.org/10.1007/978-3-540-48713-5_26
12. Efendigil, T., Önüt, S., Kahraman, C.: A decision support system for demand forecasting with artificial neural networks and neuro-fuzzy models: a comparative analysis. Expert Syst. Appl. **36**(3), 6697–6707 (2009)
13. Eichenberger, R., Stadelmann, D., et al.: Who is the best formula 1 driver? an economic approach to evaluating talent. Economic Analysis and Policy **39**(3), 389 (2009)
14. Fong, S.J., Li, G., Dey, N., Crespo, R.G., Herrera-Viedma, E.: Composite monte carlo decision making under high uncertainty of novel coronavirus epidemic using hybridized deep learning and fuzzy rule induction. Appl. Soft Comput. **93**, 106282 (2020)
15. Goh, W.P., Tao, X., Zhang, J., Yong, J.: Decision support systems for adoption in dental clinics: a survey. Knowl.-Based Syst. **104**, 195–206 (2016)
16. Heilmeier, A., Thomaser, A., Graf, M., Betz, J.: Virtual strategy engineer: Using artificial neural networks for making race strategy decisions in circuit motorsport. Appl. Sci. **10**(21), 7805 (2020)
17. Henderson, D.A., Kirrane, L.J.: A comparison of truncated and time-weighted plackett-luce models for probabilistic forecasting of formula one results. Bayesian Anal. **13**(2), 335–358 (2018)
18. Jahani, A., Feghhi, J., Makhdoum, M.F., Omid, M.: Optimized forest degradation model (ofdm): an environmental decision support system for environmental impact assessment using an artificial neural network. J. Environ. Planning Manage. **59**(2), 222–244 (2016)
19. Javed, H., Samuel, S.: Energy optimal control for formula one race car. Tech. rep, SAE Technical Paper (2022)
20. Jensen, J.A., Cobbs, J.B., Albano, B., Tyler, B.D.: Analyzing price premiums in international sponsorship exchange: What drives marketing costs in formula one racing? J. Advert. Res. **61**(1), 44–57 (2021)
21. Judde, C., Booth, R., Brooks, R.: Second place is first of the losers: An analysis of competitive balance in formula one. J. Sports Econ. **14**(4), 411–439 (2013)
22. Kaggle: Formula 1 world championship (1950–2022) (2021). https://www.kaggle.com/datasets/rohanrao/formula-1-world-championship-1950-2020. Accessed 29 May 2021
23. van Kesteren, E.J., Bergkamp, T.: Bayesian analysis of formula one race results: Disentangling driver skill and constructor advantage. arXiv preprint arXiv:2203.08489 (2022)
24. Kizielewicz, B., Shekhovtsov, A., Sałabun, W.: pymcdm-the universal library for solving multi-criteria decision-making problems. SoftwareX **22**, 101368 (2023)
25. Krishankumar, R., Pamucar, D.: Solving barrier ranking in clean energy adoption: an mcdm approach with q-rung orthopair fuzzy preferences. Int. J. Knowl.-Based Intell. Eng. Syst **27**, 55–72 (2023)
26. Lee, A., Leng, H.K.: The marketing of the 2014 formula one Singapore grand prix on Facebook. Pamukkale J. Sport Sci. **7**(3), 14–22 (2016)
27. Martínez Comesaña, M., Febrero-Garrido, L., Troncoso-Pastoriza, F., Martínez-Torres, J.: Prediction of building's thermal performance using lstm and mlp neural networks. Appl. Sci. **10**(21), 7439 (2020)
28. Mettenheim, H.J.v., Breitner, M.H.: Robust decision support systems with matrix forecasts and shared layer perceptrons for finance and other applications (2010)

29. O'Hanlon, E.: Using Supervised Machine Learning to Predict the Final Rankings of the 2021 Formula One Championship. Ph.D. thesis, Dublin, National College of Ireland (2022)
30. Piezunka, H., Lee, W., Haynes, R., Bothner, M.S.: Escalation of competition into conflict in competitive networks of formula one drivers. Proc. Natl. Acad. Sci. **115**(15), E3361–E3367 (2018)
31. Piquero, A.R., Piquero, N.L., Han, S.: Identifying the most successful formula 1 drivers in the turbo era. Open Sports Sci. J. **14**(1) (2021)
32. Ramkumar, M., Babu, C.G., Kumar, K.V., Hepsiba, D., Manjunathan, A., Kumar, R.S.: Ecg cardiac arrhythmias classification using dwt, ica and mlp neural networks. J. Phys. Conf. Series **1831**, 012015. IOP Publishing (2021)
33. Rockerbie, D.W., Easton, S.T.: Race to the podium: Separating and conjoining the car and driver in f1 racing. Applied Economics, pp. 1–14 (2022)
34. Sałabun, W., Urbaniak, K.: A new coefficient of rankings similarity in decision-making problems. In: International Conference on Computational Science, pp. 632–645. Springer (2020)
35. Schelling, X., Robertson, S.: A development framework for decision support systems in high-performance sport. Int. J. Comput. Sci. Sport **19**(1), 1–23 (2020)
36. Stoppels, E.: Predicting race results using artificial neural networks. Master's thesis, University of Twente (2017)
37. Sutton, R.T., Pincock, D., Baumgart, D.C., Sadowski, D.C., Fedorak, R.N., Kroeker, K.I.: An overview of clinical decision support systems: benefits, risks, and strategies for success. NPJ Digital Med. **3**(1), 1–10 (2020)
38. Toslak, M., Ulutaş, A., Üreа, S., Stević, Ž.: Selection of peanut butter machine by the integrated psi-sv-marcos method. Int. J. Knowl.-Based Intell. Eng. Syst. **27**, 73–86 (2023)
39. Vadgama, T.N., Patel, M.A., Thakkar, D.D.: Design of formula one racing car. Int. J. Eng. Res. Technol. **4**(04), 702–712 (2015). ISSN: 2278–0181
40. Vishnukumar, H.J., Butting, B., Müller, C., Sax, E.: Machine learning and deep neural network-artificial intelligence core for lab and real-world test and validation for adas and autonomous vehicles: Ai for efficient and quality test and validation. In: 2017 Intelligent Systems Conference (IntelliSys). pp. 714–721. IEEE (2017)
41. Waldo, J.: Embedded computing and formula one racing. IEEE Pervasive Comput. **4**(3), 18–21 (2005)
42. Wang, T.: Sports training auxiliary decision support system based on neural network algorithm. Neural Computing and Applications, pp. 1–14 (2022)
43. Watanabe, Y., Gilbert, C., Aman, M.S., Zhang, J.J.: Attracting international spectators to a sport event held in asia: The case of formula one petronas malaysia grand prix. International Journal of Sports Marketing and Sponsorship (2018)

Theoretical Analysis of Gradient-Zhang Neural Network for Time-Varying Equations and Improved Method for Linear Equations

Changyuan Wang and Yunong Zhang(✉)

School of Computer Science and Engineering, Sun Yat-sen University, Guangzhou 510006, People's Republic of China
zhynong@mail.sysu.edu.cn

Abstract. Solving time-varying equations is fundamental in science and engineering. This paper aims to find a fast-converging and high-precision method for solving time-varying equations. We combine two classes of feedback neural networks, i.e., gradient neural network (GNN) and Zhang neural network (ZNN), to construct a continuous gradient-Zhang neural network (GZNN) model. Our research shows that GZNN has the advantages of high convergence precision of ZNN and fast convergence speed of GNN in certain cases, i.e., all the eigenvalues of Jacobian matrix of the time-varying equations multiplied by its transpose are larger than 1. Furthermore, we conduct the different detailed mathematical proof and theoretical analysis to establish the stability and convergence of the GZNN model. Additionally, we discretize the GZNN model by utilizing time discretization formulas (i.e., Euler and Taylor-Zhang discretization formulas), to construct corresponding discrete GZNN algorithms for solving discrete time-varying problems. Different discretization formulas can construct discrete algorithms with varying precision. As the number of time sampling instants increases, the precision of discrete algorithms can be further improved. Furthermore, we improve the matrix inverse operation in the GZNN model and develop inverse-free GZNN algorithms to solve linear problems, effectively reducing their time complexity. Finally, numerical experiments are conducted to validate the feasibility of GZNN model and the corresponding discrete algorithms in solving time-varying equations, as well as the efficiency of the inverse-free method in solving linear equations.

Keywords: Gradient-Zhang Neural Network · Time-Varying Equations · Discrete Algorithms

1 Introduction

Time-varying problems are fundamental in the fields of science and engineering and are widely applied in various domains such as redundant manipulator control, optimization, and circuit systems [1,13,16–18,24]. With the advancement

B. Luo et al. (Eds.): ICONIP 2023, LNCS 14447, pp. 277–291, 2024.
https://doi.org/10.1007/978-981-99-8079-6_22

of society, there is an increasing demand for real-time computations in engineering applications, which has led to the attention of many scholars and researchers towards time-varying equations [6,8,9,15,28]. For example, in redundant manipulator control, it is essential to develop time-varying dynamic systems based on the constraints of the manipulator arm to facilitate motion planning and control [6]. Over the past two decades, with the development of artificial intelligence, many scholars and researchers have proposed neural network-based methods to solve time-varying problems. This paper focuses on two classes of feedback neural networks: gradient neural network (GNN) and Zhang neural network (ZNN) [19,27,28]. Traditional GNN models overlook the time derivative information of time-varying problems and assume that the problems remain static within the sampling interval [27,28]. Consequently, they are less effective in solving time-varying problems. To enhance accuracy, Zhang et al. introduced the ZNN model, which utilizes a vector-valued error function (as a Zhang function) to monitor the solution process [18,19]. Recently, some researchers have combined these two classes of neural networks to create a new neural network called gradient-Zhang neural network (GZNN) [7,14,23,25]. The GZNN model harnesses the fast convergence of GNN model and the high precision of ZNN model, particularly for solving some specific time-varying problems (e.g., when all of the eigenvalues of $J(\boldsymbol{x},t)J^{\mathrm{T}}(\boldsymbol{x},t)$ are larger than 1, where $J(\boldsymbol{x},t)$ represents the Jacobian matrix of the time-varying equations). The GZNN model has been applied to problems such as time-varying matrix inversion [14] and time-varying linear equation systems solving [23]. This paper focuses on GZNN model for solving time-varying vector equations and provides the initial theoretical proof, analysis, and one inverse-free improvement method. The main contributions of this paper are as follows.

- By combining GNN and ZNN, we utilize continuous GZNN model and the corresponding discrete algorithms to solve time-varying equations.
- We provide the initial theoretical proof and analysis on the GZNN model.
- We propose an inverse-free method for GZNN to reduce the time complexity of the model when solving time-varying linear equations.
- Through simulation experiments, we verify the feasibility of the continuous GZNN model, discrete GZNN algorithms, and the inverse-free algorithms.

Before completing this section, let us provide an overview of the work presented throughout the entire paper. In Sect. 2, a description of the time-varying equations is provided. Section 3 introduces the continuous GZNN model and presents the initial theoretical proof and analysis of the model. Section 4 presents the discrete GZNN algorithms, and Sect. 5 introduces an inverse-free improvement method for the continuous GZNN model and its discrete algorithms in solving time-varying linear equations. Section 6 presents simulation experiments to validate the methods presented in this paper. Finally, in Sect. 7, a comprehensive conclusion is drawn for the paper.

2 Problem Description

Generally, the time-varying equation can be described as below [27]:

$$\boldsymbol{f}(\boldsymbol{x}, t) = \mathbf{0}, \tag{1}$$

where $\boldsymbol{x} = [x_1(t), x_2(t), \cdots, x_n(t)]^{\mathrm{T}}$ is the vector variable that varies with time t and needs to be determined in real time, and $\boldsymbol{f}(\boldsymbol{x}, t)$ is a n-dimension function.

Furthermore, we expand the time-varying Eq. (1) as the following group of detailed equations:

$$\begin{cases} f_1(x_1(t), x_2(t), \cdots, x_n(t), t) = 0, \\ f_2(x_1(t), x_2(t), \cdots, x_n(t), t) = 0, \\ \quad \vdots \qquad\qquad \vdots \\ f_n(x_1(t), x_2(t), \ldots, x_n(t), t) = 0. \end{cases} \tag{2}$$

In this paper, we have two basic assumptions about time-varying equations.

Assumption 1. *The time-varying function $f_k(x_1(t), x_2(t), \cdots, x_n(t), t)$ in (2) is smooth with respect to all the variables including x_k $(k = 1, 2, \cdots, n)$ and t, i.e., the time-varying function is infinitely differentiable, and the partial derivatives of arbitrary order with respect to all the variables are bounded.*

Assumption 2. *The Jacobian matrix $J(\boldsymbol{x}, t)$ of the time-varying equation $\boldsymbol{f}(\boldsymbol{x}, t)$ is nonsingular. More specifically, we describe the Jacobian matrix $J(\boldsymbol{x}, t)$ as*

$$J(\boldsymbol{x}, t) = \frac{\partial \boldsymbol{f}(\boldsymbol{x}, t)}{\partial \boldsymbol{x}} = \begin{bmatrix} \frac{\partial f_1}{\partial x_1} & \frac{\partial f_1}{\partial x_2} & \cdots & \frac{\partial f_1}{\partial x_n} \\ \frac{\partial f_2}{\partial x_1} & \frac{\partial f_2}{\partial x_2} & \cdots & \frac{\partial f_2}{\partial x_n} \\ \vdots & \vdots & \ddots & \vdots \\ \frac{\partial f_n}{\partial x_1} & \frac{\partial f_n}{\partial x_2} & \cdots & \frac{\partial f_n}{\partial x_n} \end{bmatrix}.$$

3 Continuous Solution Models

In this section, we first present two classical feedback neural networks, i.e., GNN and ZNN [28], and then introduce the GZNN by combining these two networks.

3.1 Gradient Neural Network

GNN model views the time-varying problems from a static perspective and ignores the problems' time derivative information. So one cannot obtain high precision solutions through the GNN model.

The continuous GNN model defines a scalar error function (i.e., so-called energy function) described as below [28]:

$$\epsilon(t) = \frac{1}{2}\|\boldsymbol{f}(\boldsymbol{x}, t)\|_2^2 \in \mathrm{R}, \tag{3}$$

where $\|\cdot\|_2$ denotes the two-norm of a vector.

By performing the design formula of GNN, i.e., $\dot{\boldsymbol{x}}(t) = -\lambda \partial \epsilon(t)/\partial \boldsymbol{x}$, where $\dot{\boldsymbol{x}}(t)$ denotes $\mathrm{d}\boldsymbol{x}(t)/\mathrm{d}t$ and λ is the hyperparameter which influences the model convergence speed, we obtain the following continuous GNN model:

$$\begin{aligned} \dot{\boldsymbol{x}}(t) &= -\lambda \frac{\partial \epsilon(t)}{\partial \boldsymbol{x}} \\ &= -\lambda \left(\frac{\partial \boldsymbol{f}(\boldsymbol{x}, t)}{\partial \boldsymbol{x}} \right)^{\mathrm{T}} \boldsymbol{f}(\boldsymbol{x}, t) \\ &= -\lambda J^{\mathrm{T}}(\boldsymbol{x}, t) \boldsymbol{f}(\boldsymbol{x}, t). \end{aligned} \tag{4}$$

3.2 Zhang Neural Network

In order to improve the solution precision, Zhang et al. developed the ZNN model. The ZNN model defines a vector error function (i.e., so-called Zhang function) [27]:

$$\boldsymbol{e}(t) = \boldsymbol{f}(\boldsymbol{x}, t) \in \mathrm{R}^n. \tag{5}$$

By performing the design formula of ZNN, i.e., $\dot{\boldsymbol{e}}(t) = -\lambda \boldsymbol{e}(t)$, we obtain the following implicit continuous ZNN model:

$$\frac{\partial \boldsymbol{f}(\boldsymbol{x}, t)}{\partial \boldsymbol{x}} \dot{\boldsymbol{x}}(t) + \frac{\partial \boldsymbol{f}(\boldsymbol{x}, t)}{\partial t} + \lambda \boldsymbol{f}(\boldsymbol{x}, t) = \mathbf{0}. \tag{6}$$

The implicit continuous ZNN model (6) can be further converted into the explicit ZNN model:

$$\dot{\boldsymbol{x}}(t) = -J^{-1}(\boldsymbol{x}, t) \frac{\partial \boldsymbol{f}(\boldsymbol{x}, t)}{\partial t} - \lambda J^{-1}(\boldsymbol{x}, t) \boldsymbol{f}(\boldsymbol{x}, t). \tag{7}$$

In general, we view the first item in the right side of Eq. (7), i.e., $-J^{-1}(\boldsymbol{x}, t)(\partial \boldsymbol{f}(\boldsymbol{x}, t)/\partial t)$, as the feedforward item, while the second item which is related to the error function, i.e., $-\lambda J^{-1}(\boldsymbol{x}, t) \boldsymbol{f}(\boldsymbol{x}, t)$, is viewed as the feedback item.

The ZNN model makes full use of the time derivative information of the time-varying equations and it can achieve higher convergence precision compared with GNN model [28].

3.3 Gradient-Zhang Neural Network

In order to develop a method with fast convergence speed and high convergence precision, some scholars combined the two neural networks and construct the gradient-Zhang neural networks [7,14,23,25].

More specifically, by substituting the feedback item of ZNN model with the item in the right side of GNN model (4), we obtain the continuous GZNN model:

$$\dot{\boldsymbol{x}}(t) = -J^{-1}(\boldsymbol{x}, t) \frac{\partial \boldsymbol{f}(\boldsymbol{x}, t)}{\partial t} - \lambda J^{\mathrm{T}}(\boldsymbol{x}, t) \boldsymbol{f}(\boldsymbol{x}, t), \tag{8}$$

which corresponds to an implicit GZNN model:

$$J(\boldsymbol{x},t)\dot{\boldsymbol{x}}(t) + \frac{\partial \boldsymbol{f}(\boldsymbol{x},t)}{\partial t} + \lambda J(\boldsymbol{x},t)J^{\mathrm{T}}(\boldsymbol{x},t)\boldsymbol{f}(\boldsymbol{x},t) = \mathbf{0}. \tag{9}$$

The feasibility of this network is validated in some time-varying problems solving, such as time-varying linear equations solving, and time-varying matrix inversion, through experiments in the past work. In this paper, we give a different and detailed theoretical proof of GZNN model in solving time-varying equations.

Theorem 1. *When utilizing the continuous GZNN model (8) to solve time-varying Eq. (1), the two-norm of the equation, i.e.,* $\|\boldsymbol{f}(\boldsymbol{x},t)\|_2$*, converges exponentially and stably to zero.*

Proof. Firstly, we make a transformation on the Eq. (8):

$$J(\boldsymbol{x},t)\dot{\boldsymbol{x}}(t) + \frac{\partial \boldsymbol{f}(\boldsymbol{x},t)}{\partial t} = -\lambda J(\boldsymbol{x},t)J^{\mathrm{T}}(\boldsymbol{x},t)\boldsymbol{f}(\boldsymbol{x},t), \tag{10}$$

i.e.,

$$\frac{\mathrm{d}\boldsymbol{f}(\boldsymbol{x},t)}{\mathrm{d}t} = -\lambda J(\boldsymbol{x},t)J^{\mathrm{T}}(\boldsymbol{x},t)\boldsymbol{f}(\boldsymbol{x},t). \tag{11}$$

In other words, similar to ZNN, the GZNN model defines a vector error function (i.e., so-called Zhang function) $\boldsymbol{\beta}(t) = \boldsymbol{f}(\boldsymbol{x},t)$, but its design principle is as follows:

$$\dot{\boldsymbol{\beta}}(t) = -\lambda J(\boldsymbol{x},t)J^{\mathrm{T}}(\boldsymbol{x},t)\boldsymbol{\beta}(t). \tag{12}$$

Since $J(\boldsymbol{x},t)$ is nonsingular, $J(\boldsymbol{x},t)J^{\mathrm{T}}(\boldsymbol{x},t)$ is a positive definite matrix.

We now prove its stability by constructing a Lyapunov function candidate [3,21]:

$$l(t) = \frac{1}{2}\boldsymbol{\beta}^{\mathrm{T}}(t)\boldsymbol{\beta}(t) \geq 0. \tag{13}$$

Therefore, we have

$$\frac{\mathrm{d}l(t)}{\mathrm{d}t} = \dot{\boldsymbol{\beta}}^{\mathrm{T}}(t)\boldsymbol{\beta}(t) = -\lambda\boldsymbol{\beta}^{\mathrm{T}}(t)J(\boldsymbol{x},t)J^{\mathrm{T}}(\boldsymbol{x},t)\boldsymbol{\beta}(t) \leq 0. \tag{14}$$

Hence, the GZNN model is stable and converges to the solution to the time-varying vector equation.

Next, we research the convergence speed of the continuous GZNN model.

Let the minimum eigenvalue of $J(\boldsymbol{x},t)J^{\mathrm{T}}(\boldsymbol{x},t)$ be $v_1(t)$ and the maximum eigenvalue be $v_2(t)$, where $v_2(t) \geq v_1(t) > 0$. From Eq. (14), we see that

$$\dot{l}(t) = -\lambda\boldsymbol{\beta}^{\mathrm{T}}(t)J(\boldsymbol{x},t)J^{\mathrm{T}}(\boldsymbol{x},t)\boldsymbol{\beta}(t) \leq -\lambda v_1(t)\boldsymbol{\beta}^{\mathrm{T}}(t)\boldsymbol{\beta}(t) = -2\lambda v_1(t)l(t), \tag{15}$$

$$\dot{l}(t) = -\lambda\boldsymbol{\beta}^{\mathrm{T}}(t)J(\boldsymbol{x},t)J^{\mathrm{T}}(\boldsymbol{x},t)\boldsymbol{\beta}(t) \geq -\lambda v_2(t)\boldsymbol{\beta}^{\mathrm{T}}(t)\boldsymbol{\beta}(t) = -2\lambda v_2(t)l(t). \tag{16}$$

In other words, $-2\lambda v_2(t)l(t) \leq \dot{l}(t) \leq -2\lambda v_1(t)l(t)$.

Note that, according to Assumption 1, the time-varying functions in equation group (2) are infinitely differentiable, and the partial derivatives of arbitrary order with respect to all the variables are bounded. So every element in the $J(\boldsymbol{x}, t)$ is smooth and bounded, and the eigenvalues of $J(\boldsymbol{x}, t)J^{\mathrm{T}}(\boldsymbol{x}, t)$ are bounded.

Let $a/2$ be the lower bound of $v_1(t)$ ($a > 0$), and $b/2$ be the upper bound of $v_2(t)$, i.e., $0 < a/2 \leq v_1(t) \leq v_2(t) \leq b/2$. Then we have

$$-\lambda bl(t) \leq \dot{l}(t) \leq -\lambda al(t). \tag{17}$$

For $\dot{l}(t) \leq -\lambda al(t)$, we construct the following differential equation:

$$\dot{l}(t) = -\lambda al(t) - k(t), \tag{18}$$

where $k(t) \geq 0$.

$\dot{l}(t) = -\lambda \boldsymbol{\beta}^{\mathrm{T}}(t)J(\boldsymbol{x}, t)J^{\mathrm{T}}(\boldsymbol{x}, t)\boldsymbol{\beta}(t)$ is obtained by performing a finite number of additions, subtractions, and multiplications on a finite number of functions. That is, $\dot{l}(t)$ is also a smooth and bounded function, which means $k(t)$ is smooth and bounded.

By taking the Laplace transform [11], we obtain

$$sL(s) - L(0) = -\lambda aL(s) - K(s), \tag{19}$$

where $L(s) = \int_0^{+\infty} l(t)\exp(-st)\mathrm{d}t$, $K(s) = \int_0^{+\infty} k(t)\exp(-st)\mathrm{d}t$. Thus,

$$L(s) = \frac{1}{s + \lambda a}L(0) - \frac{1}{s + \lambda a}K(s). \tag{20}$$

By taking the inverse Laplace transform, we obtain

$$l(t) = \exp(-\lambda at)l(0) - \exp(-\lambda at)\int_0^t \exp(\lambda a\tau)k(\tau)\mathrm{d}\tau. \tag{21}$$

Because $\exp(-\lambda at)\int_0^t \exp(\lambda a\tau)k(\tau)\mathrm{d}\tau \geq 0$, we get $l(t) \leq \exp(-\lambda at)l(0)$. So $l(t)$ is exponential or hyperexponential.

Similarly, from $\dot{l}(t) \geq -\lambda bl(t)$, we deduce that $l(t)$ is exponential or subexponential. So $l(t) = \boldsymbol{\beta}(t)^{\mathrm{T}}\boldsymbol{\beta}(t)/2$ exhibits exponential convergence, which deduces $\|\boldsymbol{f}(\boldsymbol{x}, t)\|_2$ exponentially converges to zero. ■

More specifically, we give a detailed description on the speed of continuous GZNN model:

$$\exp(-\lambda bt)l(0) \leq l(t) \leq \exp(-\lambda at)l(0). \tag{22}$$

Note that, in regard to continuous ZNN model, we can also define a Lyapunov function candidate, i.e., $h(t) = (1/2)\boldsymbol{e}^{\mathrm{T}}(t)\boldsymbol{e}(t)$ [5]. Performing the design formula of ZNN model, we obtain

$$\dot{h}(t) = \dot{\boldsymbol{e}}^{\mathrm{T}}(t)\boldsymbol{e}(t) = -\lambda\boldsymbol{e}^{\mathrm{T}}(t)\boldsymbol{e}(t) = -2h(t). \tag{23}$$

Solving the differential equation $\dot{h}(t) = -2h(t)$, we observe the convergence speed of the ZNN model:

$$h(t) = \exp(-2\lambda t)h(0). \tag{24}$$

Comparing with (22), we summarize the following:

- If $a > 2$, i.e., the minimum eigenvalue of $J(\boldsymbol{x},t)J^{\mathrm{T}}(\boldsymbol{x},t)$ is larger than 1, the continuous GZNN model converges faster than the ZNN model.
- If $b < 2$, i.e., the maximum eigenvalue of $J(\boldsymbol{x},t)J^{\mathrm{T}}(\boldsymbol{x},t)$ is smaller than 1, the continuous GZNN model converges slower than the ZNN model.

4 Discrete Gradient-Zhang Neural Network Algorithms

Based on the analysis in the previous section, we have developed a continuous model for solving time-varying vector equations. In order to adapt to the development of digital circuits and numerical algorithms, Zhang et al. developed some time discretization formulas, to discretize the continuous model. In this paper, we present two time discretization formulas, i.e., Euler forward formula and Taylor-Zhang discretization formula, and utilize them to discretize the continuous GZNN model [4,8,12,20].

For ease of description, we use t_k to represent the k-th time sampling point, i.e., $t_k = k\tau$, where τ is the sampling gap. Correspondingly, $\boldsymbol{x}_k$ represents $\boldsymbol{x}(t_k)$.

According to [4,20], we get the following time discretization formulas:

- Euler forward formula:

$$\dot{\boldsymbol{x}}_k = \frac{1}{\tau}(\boldsymbol{x}_{k+1} - \boldsymbol{x}_k) + O(\tau); \tag{25}$$

- Taylor-Zhang discretization formula:

$$\dot{\boldsymbol{x}}_k = \frac{1}{2\tau}(2\boldsymbol{x}_{k+1} - 3\boldsymbol{x}_k + 2\boldsymbol{x}_{k-1} - \boldsymbol{x}_{k-2}) + O(\tau^2). \tag{26}$$

Combined with continuous GZNN model (8), we get the following two discrete GZNN algorithms, which are called as Euler discrete GZNN (E-DGZNN) and Taylor-Zhang discrete GZNN (TZ-DGZNN) algorithms:

- E-DGZNN algorithm:

$$\boldsymbol{x}_{k+1} = -\lambda\tau J^{\mathrm{T}}(\boldsymbol{x}_k,t_k)\boldsymbol{f}(\boldsymbol{x}_k,t_k) - \tau J^{-1}(\boldsymbol{x}_k,t_k)\frac{\partial \boldsymbol{f}(\boldsymbol{x}_k,t_k)}{\partial t} + \boldsymbol{x}_k + O(\tau^2); \tag{27}$$

- TZ-DGZNN algorithm:

$$\begin{aligned}\boldsymbol{x}_{k+1} =& -\lambda\tau J^{\mathrm{T}}(\boldsymbol{x}_k,t_k)\boldsymbol{f}(\boldsymbol{x}_k,t_k) - \tau J^{-1}(\boldsymbol{x}_k,t_k)\frac{\partial \boldsymbol{f}(\boldsymbol{x}_k,t_k)}{\partial t}\\ &+ \frac{3}{2}\boldsymbol{x}_k - \boldsymbol{x}_{k-1} + \frac{1}{2}\boldsymbol{x}_{k-2} + O(\tau^3).\end{aligned} \tag{28}$$

In many scenarios, obtaining the Jacobian matrix and the time derivative ($J(\boldsymbol{x}_k,t)$ and $\partial \boldsymbol{f}(\boldsymbol{x}_k,t)/\partial t$) of a time-varying equation is challenging. In such cases, we can use the following approximation rule to estimate them, as described in [4]:

$$\begin{aligned}\frac{\partial \boldsymbol{g}(x,y)}{\partial x} =& \frac{1}{60\tau}(147\boldsymbol{g}(x,y) - 360\boldsymbol{g}(x-\tau,y) + 450\boldsymbol{g}(x-2\tau,y) - 400\boldsymbol{g}(x-3\tau,y) \\ &+ 225\boldsymbol{g}(x-4\tau,y) - 72\boldsymbol{g}(x-5\tau,y) + 10\boldsymbol{g}(x-6\tau,y)) + O(\tau^6).\end{aligned} \tag{29}$$

Note that the equation above represents an approximation method for computing the partial derivative of function $\boldsymbol{g}(x,y)$ with respect to x. The approximation is based on a finite difference scheme and has an error term denoted by $O(\tau^6)$. Compared with the discrete GZNN algorithms we present, the approximation method with an $O(\tau^6)$ error term does not damage the algorithm precision.

5 Inverse-Free GZNN Algorithms on Linear Equations

Although the GZNN model improves the convergence speed and accuracy compared with GNN and ZNN under specific conditions, observing from Eqs. (8), (27), and (28), we see that GZNN requires the inversion of relevant matrices, which incurs significant time complexity. Therefore, in this section, we explore the inverse-free rule of the GZNN model for time-varying linear equations, aiming to find low time complexity methods, through the replacement of matrix-inverse serial processing with matrix-update parallel processing.

According to [2,15,26], we can use ZNN to estimate the inverse of the Jacobian matrix for time-varying linear problems. Let us define the following error function (i.e., another Zhang function) [10,22,26]:

$$E(t) = P^{-1}(t) - J(\boldsymbol{x},t), \tag{30}$$

where $P(t)$ represents the inverse of the Jacobian matrix. According to the design principle of ZNN, i.e., $\mathrm{d}E(t)/\mathrm{d}t = -\lambda E(t)$, we can construct the following continuous model for inversion [2,10,15,22,26]:

$$-P^{-1}(t)\dot{P}(t)P^{-1}(t) - \frac{\mathrm{d}J(\boldsymbol{x},t)}{\mathrm{d}t} = -\lambda(P^{-1}(t) - J(\boldsymbol{x},t)), \tag{31}$$

which can be further simplified as

$$\dot{P}(t) = -P(t)\frac{\mathrm{d}J(\boldsymbol{x},t)}{\mathrm{d}t}P(t) + \lambda P(t) - \lambda P(t)J(\boldsymbol{x},t)P(t). \tag{32}$$

In time-varying linear systems, the Jacobian matrix $J(\boldsymbol{x},t)$ is independent of $\boldsymbol{x}$. Therefore, we have $\mathrm{d}J(\boldsymbol{x},t)/\mathrm{d}t = \partial J(\boldsymbol{x},t)/\partial t$. Let us define $M(t) = J(\boldsymbol{x},t)$. Combining Eqs. (32) and (8), we obtain the following inverse-free continuous model called as ZNN-GZNN model [2,10,15,22,26]:

$$\begin{cases}\dot{P}(t) = -P(t)\dot{M}(t)P(t) + \lambda P(t) - \lambda P(t)M(t)P(t), \\ \dot{\boldsymbol{x}}(t) = -P(t)\dfrac{\partial \boldsymbol{f}(\boldsymbol{x},t)}{\partial t} - \lambda M^{\mathrm{T}}(t)\boldsymbol{f}(\boldsymbol{x},t).\end{cases} \tag{33}$$

Similarly, by combining different time discretization formulas, we can derive discrete rules with different precision. In this paper, in order to avoid the loss of precision of the second equation in (33) after discretization, we use the following five-instant discretization formula [8] to estimate the inverse of the Jacobian matrix:

$$\dot{P}_k = \frac{1}{18\tau}(8P_{k+1} + P_k - 6P_{k-1} - 5P_{k-2} + 2P_{k-3}) + O(\tau^3). \tag{34}$$

Then, we present two discrete ZNN-GZNN algorithms with different precision: Euler discrete ZNN-GZNN algorithm (E-DZNN-GZNN) and Taylor-Zhang discrete ZNN-GZNN (TZ-DZNN-GZNN) algorithm. E-DZNN-GZNN algorithm is given by

$$\begin{aligned} P_k = &-\frac{9}{4}\tau P_{k-1}\dot{M}_{k-1}P_{k-1} + \frac{9}{4}\lambda\tau P_{k-1} - \frac{9}{4}\lambda\tau P_{k-1}M_{k-1}P_{k-1} \\ &-\frac{1}{8}P_{k-1} + \frac{3}{4}P_{k-2} + \frac{5}{8}P_{k-3} - \frac{1}{4}P_{k-4}, \end{aligned} \tag{35}$$

$$\boldsymbol{x}_{k+1} = -\tau P_k \frac{\partial \boldsymbol{f}(\boldsymbol{x}_k, t_k)}{\partial t} - \lambda\tau M_k^{\mathrm{T}} \boldsymbol{f}(\boldsymbol{x}_k, t_k) + \boldsymbol{x}_k. \tag{36}$$

TZ-DZNN-GZNN algorithm is given by

$$\begin{aligned} P_k = &-\frac{9}{4}\tau P_{k-1}\dot{M}_{k-1}P_{k-1} + \frac{9}{4}\lambda\tau P_{k-1} - \frac{9}{4}\lambda\tau P_{k-1}M_{k-1}P_{k-1} \\ &-\frac{1}{8}P_{k-1} + \frac{3}{4}P_{k-2} + \frac{5}{8}P_{k-3} - \frac{1}{4}P_{k-4}, \end{aligned} \tag{37}$$

$$\boldsymbol{x}_{k+1} = -\tau P_k \frac{\partial \boldsymbol{f}(\boldsymbol{x}_k, t_k)}{\partial t} - \lambda\tau M_k^{\mathrm{T}} \boldsymbol{f}(\boldsymbol{x}_k, t_k) + \frac{3}{2}\boldsymbol{x}_k - \boldsymbol{x}_{k-1} + \frac{1}{2}\boldsymbol{x}_{k-2}. \tag{38}$$

6 Simulation Experiments

In this section, we perform three groups of simulation experiments to validate the feasibility of the GZNN and its inverse-free methods.

6.1 Example 1

We firstly construct a four dimension time-varying equation shown as below:

$$\boldsymbol{f}(\boldsymbol{x}, t) = \begin{pmatrix} \ln(x_1) - \frac{1}{1+t} \\ x_1 x_2 - \exp\left(\frac{1}{1+t}\right)\sin(t) \\ x_1^2 - \sin(t)x_2 + x_3 - 2 \\ x_1^2 - x_2^2 + x_3 + x_4 - t \end{pmatrix} = \mathbf{0}. \tag{39}$$

Note that the Jacobian matrix of the above equation is nonsingular:

$$J(\boldsymbol{x},t) = \begin{bmatrix} \frac{1}{x_1} & 0 & 0 & 0 \\ x_2 & x_1 & 0 & 0 \\ 2x_1 & -\sin(t) & 1 & 0 \\ 2x_1 & -2x_2 & 1 & 1 \end{bmatrix}. \tag{40}$$

For comparison, the theoretical solution of Eq. (39) is given as

$$\boldsymbol{x} = \begin{pmatrix} x_1 \\ x_2 \\ x_3 \\ x_4 \end{pmatrix} = \begin{pmatrix} \exp\left(\frac{1}{1+t}\right) \\ \sin(t) \\ 2 - \exp\left(\frac{2}{1+t}\right) + \sin^2(t) \\ t-2 \end{pmatrix} = \mathbf{0}. \tag{41}$$

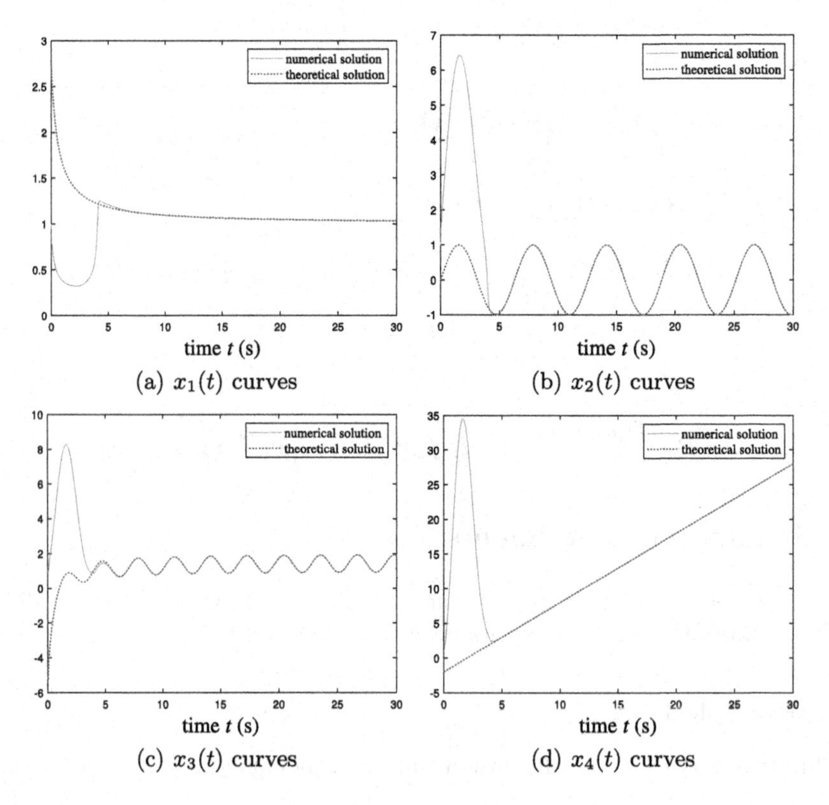

Fig. 1. Curves of theoretical solution and numerical solution synthesized by implicit continuous GZNN model (9)

In Fig. 1, we present the curves of theoretical solution and the numerical solution synthesized by the continuous GZNN model (9). We set $\lambda = 10$. One can find out that the curves of theoretical solution and the numerical solution coincides after about 5 s. In Fig. 2, we show the curves of residual errors denoted by $\|\boldsymbol{f}(\boldsymbol{x},t)\|_2$ synthesized by the implicit continuous GZNN model, E-DGZNN algorithm, and TZ-DGZNN algorithm. The hyperparameters of the two discrete algorithms are both set as $\lambda = 0.05$, and we utilize the estimation rule (29) to estimate the Jacobian matrix and the time derivative, i.e., $J(\boldsymbol{x}_k,t)$ and $\partial\boldsymbol{f}(\boldsymbol{x}_k,t)/\partial t$. For every discrete GZNN algorithm, we set the sampling gaps as 0.1 s, 0.01 s, and 0.001 s, respectively. At 1000 s, the residual errors of E-DGZNN are 0.9273, 0.0625, 0.0058, which corresponds to the residual error pattern $O(\tau)$. At 2000 s, the residual errors of E-DGZNN are $1.756 \times 10^{-2}, 1.591 \times 10^{-4}, 1.58 \times 10^{-6}$, which corresponds to the residual error pattern $O(\tau^2)$. So, if we reduce the sampling gap by ten times, the residual error reduces one hundred times. In this experiment group, we find that the discrete GZNN algorithms may be too sensitive about the hyperparameter λ, so we set λ to be small, leading to the slow convergence speeds in Fig. 2 (a) and (b).

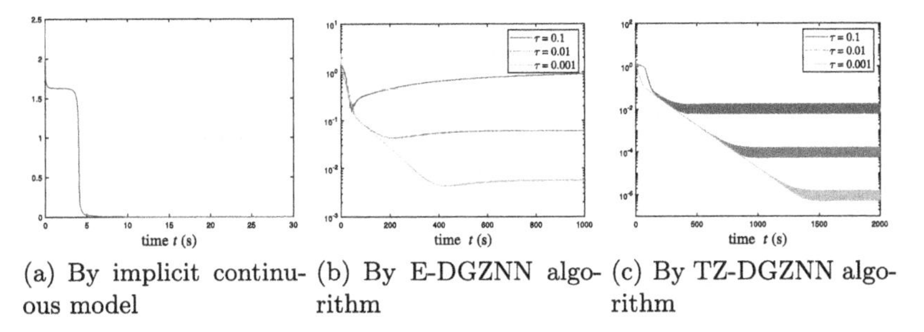

(a) By implicit continuous model (b) By E-DGZNN algorithm (c) By TZ-DGZNN algorithm

Fig. 2. Residual error $\|\boldsymbol{f}(\boldsymbol{x},t)\|_2$ curves synthesized by implicit continuous GZNN model (9), E-DGZNN algorithm (27), and TZ-DGZNN algorithm (28), with Y-axis corresponding to residual errors.

6.2 Example 2

In this experiment group, we construct two time-varying linear equations to compare the convergence speed of the GZNN and ZNN methods. The two time-varying linear equations are shown as follows.

$$\boldsymbol{f}_1(\boldsymbol{x},t) = \begin{pmatrix} \left(\frac{1}{3}+\frac{1}{2}\sin(t)\right)x_1 + \frac{1}{10}\cos(t)x_2 - \sin(t) \\ -\frac{1}{10}\cos(t)x_1 + \left(\frac{1}{3}+\frac{1}{2}\sin(t)\right)x_2 - \cos(t) \end{pmatrix} = \boldsymbol{0}, \tag{42}$$

$$\boldsymbol{f}_2(\boldsymbol{x},t) = \begin{pmatrix} \left(3+\frac{1}{2}\sin(t)\right)x_1 + \frac{1}{10}\cos(t)x_2 - \sin(t) \\ -\frac{1}{10}\cos(t)x_1 + \left(3+\frac{1}{2}\sin(t)\right)x_2 - \cos(t) \end{pmatrix} = \boldsymbol{0}. \tag{43}$$

The Jacobian matrices of the two equations are shown as follows.

$$J_1(\boldsymbol{x},t)=\begin{bmatrix}\frac{1}{3}+\frac{1}{2}\sin(t) & \frac{1}{10}\cos(t)\\ -\frac{1}{10}\cos(t) & \frac{1}{3}+\frac{1}{2}\sin(t)\end{bmatrix},\ J_2(\boldsymbol{x},t)=\begin{bmatrix}3+\frac{1}{2}\sin(t) & \frac{1}{10}\cos(t)\\ -\frac{1}{10}\cos(t) & 3+\frac{1}{2}\sin(t)\end{bmatrix}.$$

Besides, we have

$$J_1(\boldsymbol{x},t)J_1^{\mathrm{T}}(\boldsymbol{x},t)=\begin{bmatrix}\left(\frac{1}{3}+\frac{1}{2}\sin(t)\right)^2-\frac{1}{100}\cos^2(t) & 0\\ 0 & \left(\frac{1}{3}+\frac{1}{2}\sin(t)\right)^2-\frac{1}{100}\cos^2(t)\end{bmatrix},$$

$$J_2(\boldsymbol{x},t)J_2^{\mathrm{T}}(\boldsymbol{x},t)=\begin{bmatrix}\left(3+\frac{1}{2}\sin(t)\right)^2-\frac{1}{100}\cos^2(t) & 0\\ 0 & \left(3+\frac{1}{2}\sin(t)\right)^2-\frac{1}{100}\cos^2(t)\end{bmatrix}.$$

This means that the eigenvalues of $J_1(\boldsymbol{x},t)J_1^{\mathrm{T}}(\boldsymbol{x},t)$ are all smaller than 1, while the eigenvalues of $J_2(\boldsymbol{x},t)J_2^{\mathrm{T}}(\boldsymbol{x},t)$ are all larger than 1. From the analysis

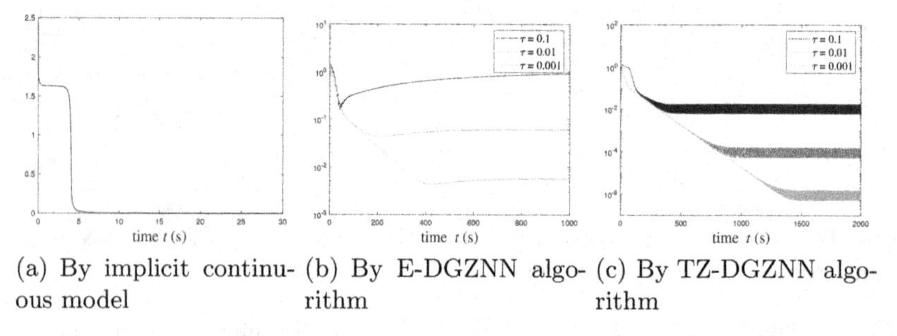

(a) By implicit continuous model (b) By E-DGZNN algorithm (c) By TZ-DGZNN algorithm

Fig. 3. Residual error $\|\boldsymbol{f}(\boldsymbol{x},t)\|_2$ curves synthesized by implicit continuous ZNN (6) and GZNN (9) models, as well as E-DZNN discretizing (7) by Euler forward formula (25)] and E-DGZNN (27) algorithms, with Y-axis corresponding to residual errors.

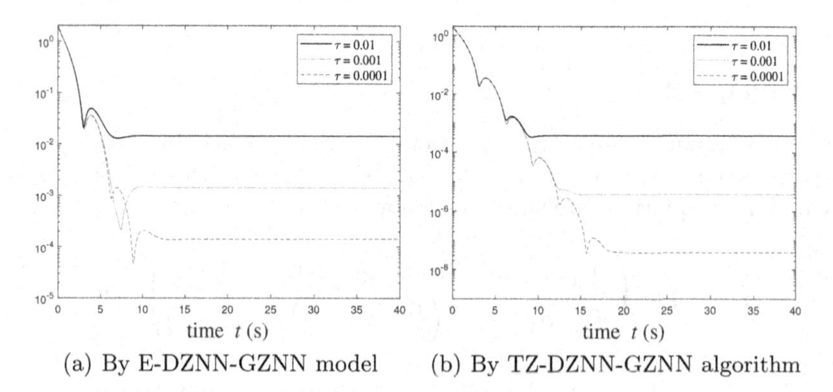

(a) By E-DZNN-GZNN model (b) By TZ-DZNN-GZNN algorithm

Fig. 4. Residual error $\|\boldsymbol{f}(\boldsymbol{x},t)\|_2$ curves synthesized by E-DZNN-GZNN algorithm (35)-(36), and TZ-DZNN-GZNN algorithm (37)-(38), with Y-axis corresponding to residual errors.

in Sect. 3, the convergence speed of GZNN method solving Eq. (42) is slower than ZNN model, and it is just opposite when solving Eq. (43), which coincides well with our experiment results displayed in Fig. 3. In this experiment group, we set the hyperparameter $\lambda = 10$ when using the implicit continuous ZNN (6) and GZNN (9) model, while $\lambda = 0.1$ when using the corresponding discrete algorithms, i.e., E-DZNN which is obtained by utilizing Euler forward formula (25) to discretize the continuous ZNN model (7), and E-DGZNN (27). Similar to Example 1, we utilize the estimation rule (29) to estimate the Jacobian matrix and the time derivative, i.e., $J(\boldsymbol{x}_k, t)$ and $\partial \boldsymbol{f}(\boldsymbol{x}_k, t)/\partial t$. More intuitively, Fig. 3 (a) and (b) show the residual errors of the two equations synthesized by the implicit continuous GZNN and ZNN models, while Fig. 3 (c) and (d) show the residual errors synthesized by Euler discrete GZNN and ZNN algorithms.

6.3 Example 3

In this example, the E-DZNN-GZNN and TZ-DZNN-GZNN algorithms are employed to solve a time-varying linear equation system, i.e., $A(t)\boldsymbol{x}(t) = \boldsymbol{b}(t)$, which is shown as below.

$$A(t) = \begin{bmatrix} \sin t & \cos t \\ -\cos t & \sin t \end{bmatrix}, \quad \boldsymbol{b}(t) = \begin{bmatrix} \sin t \\ \cos t \end{bmatrix}. \tag{44}$$

This experiment group assumes that the time derivative of Jacobian matrix $\mathrm{d}J(\boldsymbol{x}, t)/\mathrm{d}t$, i.e., $\dot{A}(t)$, is known. We set the time sampling gaps as 0.01 s, 0.001 s, and 0.0001 s, respectively and the hyperparameter λ as 1. The feasibility of the E-DZNN-GZNN and TZ-DZNN-GZNN algorithms for solving time-varying linear vector equation is illustrated in Fig. 4, with convergence precision of $O(\tau)$ and $O(\tau^2)$, respectively.

7 Conclusion

In this paper, we have combined two classic neural network approaches, namely, gradient neural network (GNN) and Zhang neural network (ZNN), and presented gradient-Zhang neural network (GZNN) for solving time-varying equations. Through mathematical analysis, we have provided the theoretical proof about the correctness of GZNN in solving time-varying problems. Furthermore, we have introduced an inverse-free method for GZNN to reduce the time complexity when solving time-varying linear problems. Finally, through experiments, we have validated the feasibility of GZNN and its inverse-free method.

However, there is still significant room for further development of GZNN. When solving nonlinear time-varying vector equations, GZNN still involves matrix inversion operations, requiring further improvement. Through experiments, we have observed that GZNN is sensitive to the hyperparameter λ. Larger values of λ can lead to divergence of the discrete algorithms, while smaller values can reduce the convergence rate and accuracy of the discrete algorithms. Developing adaptive strategies for determining the hyperparameter is crucial for improving the model. Lastly, applying GZNN to a wider range of time-varying problems, such as time-varying optimization, robotic arm control, and signal processing, would be an interesting direction for future research.

Overall, although GZNN has presented promising results in this study, there are still challenges and opportunities for further improvement and application in various domains.

Acknowledgements. This work is aided by the National Natural Science Foundation of China (with number 61976230), the Project Supported by Guangdong Province Universities and Colleges Pearl River Scholar Funded Scheme (with number 2018), and the Key-Area Research and Development Program of Guangzhou (with number 202007030004), with corresponding author Yunong Zhang.

References

1. Chan, P.K., Chen, D.Y.: A CMOS ISFET interface circuit with dynamic current temperature compensation technique. IEEE Trans. Circuits Syst. I Regul. Pap. **54**(1), 119–129 (2007)
2. Guo, D., Zhang, Y.: Zhang neural network, Getz-Marsden dynamic system, and discrete-time algorithms for time-varying matrix inversion with application to robots' kinematic control. Neurocomputing **97**, 22–32 (2012)
3. Hassoun, M.H.: Fundamentals of Artificial Neural Networks. MIT Press, Cambridge (1995)
4. Hildebrand, F.B.: Introduction to Numerical Analysis. Courier Corporation, North Chelmsford (1987)
5. Jin, L., Li, S., Liao, B., Zhang, Z.: Zeroing neural networks: a survey. Neurocomputing **267**, 597–604 (2017)
6. Jin, L., Zhang, Y., Li, S., Zhang, Y.: Modified ZNN for time-varying quadratic programming with inherent tolerance to noises and its application to kinematic redundancy resolution of robot manipulators. IEEE Trans. Ind. Electron. **63**(11), 6978–6988 (2016)
7. Li, J., Wu, G., Li, C., Xiao, M., Zhang, Y.: GMDS-ZNN variants having errors proportional to sampling gap as compared with models 1 and 2 having higher precision. In: Proceedings of International Conference on Systems and Informatics, pp. 728–733 (2018)
8. Li, J., Zhang, Y., Mao, M.: Five-instant type discrete-time ZND solving discrete time-varying linear system, division and quadratic programming. Neurocomputing **331**, 323–335 (2019)
9. Li, S., Li, Y.: Nonlinearly activated neural network for solving time-varying complex Sylvester equation. IEEE Trans. Cybern. **44**(8), 1397–1407 (2014)
10. Liao, B., Zhang, Y.: Different complex ZFs leading to different complex ZNN models for time-varying complex generalized inverse matrices. IEEE Trans. Neural Netw. Learn. Syst. **25**(9), 1621–1631 (2014)
11. Oppenheim, A.V., Willsky, A.S., Nawab, S.H., Ding, J.J.: Signals and Systems. Prentice Hall, New Jersey (1997)
12. Qiu, B., Guo, J., Li, X., Zhang, Z., Zhang, Y.: Discrete-time advanced zeroing neurodynamic algorithm applied to future equality-constrained nonlinear optimization with various noises. IEEE Trans. Cybern. **52**(5), 3539–3552 (2022)

13. Sun, C., Ye, M., Hu, G.: Distributed time-varying quadratic optimization for multiple agents under undirected graphs. IEEE Trans. Autom. Control **62**(7), 3687–3694 (2017)
14. Wu, D., Zhang, Y., Guo, J., Li, Z., Ming, L.: GMDS-ZNN model 3 and its ten-instant discrete algorithm for time-variant matrix inversion compared with other multiple-instant ones. IEEE Access **8**, 228188–228198 (2020)
15. Yang, M., Zhang, Y., Hu, H.: Inverse-free DZNN models for solving time-dependent linear system via high-precision linear six-step method. IEEE Trans. Neural Netw. Learn. Syst. 1–12 (2022)
16. Yang, M., Zhang, Y., Zhang, Z., Hu, H.: Adaptive discrete ZND models for tracking control of redundant manipulator. IEEE Trans. Ind. Inf. **16**(12), 7360–7368 (2020)
17. Yu, H., Sung, Y.: Least squares approach to joint beam design for interference alignment in multiuser multi-input multi-output interference channels. IEEE Trans. Signal Process. **58**(9), 4960–4966 (2010)
18. Zhang, Y.: Analysis and design of recurrent neural networks and their applications to control and robotic systems. The Chinese University of Hong Kong (2003)
19. Zhang, Y., Ge, S.: Design and analysis of a general recurrent neural network model for time-varying matrix inversion. IEEE Trans. Neural Netw. **16**(6), 1477–1490 (2005)
20. Zhang, Y., Gong, H., Yang, M., Li, J., Yang, X.: Stepsize range and optimal value for Taylor-Zhang discretization formula applied to zeroing neurodynamics illustrated via future equality-constrained quadratic programming. IEEE Trans. Neural Netw. Learn. Syst. **30**(3), 959–966 (2019)
21. Zhang, Y., Li, Z.: Zhang neural network for online solution of time-varying convex quadratic program subject to time-varying linear-equality constraints. Phys. Lett. A **373**(18), 1639–1643 (2009)
22. Zhang, Y., Ling, Y., Yang, M., Yang, S., Zhang, Z.: Inverse-free discrete ZNN models solving for future matrix pseudoinverse via combination of extrapolation and ZeaD formulas. IEEE Trans. Neural Netw. Learn. Syst. **32**(6), 2663–2675 (2021)
23. Zhang, Y., Wang, C.: Gradient-Zhang neural network solving linear time-varying equations. In: Proceedings of IEEE Conference on Industrial Electronics and Applications, pp. 396–403 (2022)
24. Zhang, Y., Wang, J., Xia, Y.: A dual neural network for redundancy resolution of kinematically redundant manipulators subject to joint limits and joint velocity limits. IEEE Trans. Neural Netw. **14**(3), 658–667 (2003)
25. Zhang, Y., Wu, G., Qiu, B., Li, W., He, P.: Euler-discretized GZ-type complex neuronet computing real-time varying complex matrix inverse. In: Proceedings of Chinese Control Conference, pp. 3914–3919 (2017)
26. Zhang, Y., Xie, Y., Tan, H.: Time-varying Moore-Penrose inverse solving shows different Zhang functions leading to different ZNN models. In: Proceedings of Advances in Neural Networks - ISNN 2012, pp. 98–105 (2012)
27. Zhang, Y., Yi, C., Guo, D., Zheng, J.: Comparison on Zhang neural dynamics and gradient-based neural dynamics for online solution of nonlinear time-varying equation. Neural Comput. Appl. **20**, 1–7 (2011)
28. Zhang, Y., Yi, C., Ma, W.: Comparison on gradient-based neural dynamics and Zhang neural dynamics for online solution of nonlinear equations. In: Proceedings of International Symposium on Advances in Computation and Intelligence, pp. 269–279 (2008)

EdgeMA: Model Adaptation System for Real-Time Video Analytics on Edge Devices

Liang Wang[1], Nan Zhang[2], Xiaoyang Qu[2(✉)], Jianzong Wang[2], Jiguang Wan[1], Guokuan Li[1], Kaiyu Hu[3], Guilin Jiang[4], and Jing Xiao[2]

[1] Huazhong University of Science and Technology, Wuhan, China
[2] Ping An Technology (Shenzhen) Co., Ltd., Shenzhen, China
quxiaoy@gmail.com
[3] Stony Brook University, Stony Brook, NY, USA
[4] Hunan Chasing Financial Holdings Co., Ltd., Changsha, China

Abstract. Real-time video analytics on edge devices for changing scenes remains a difficult task. As edge devices are usually resource-constrained, edge deep neural networks (DNNs) have fewer weights and shallower architectures than general DNNs. As a result, they only perform well in limited scenarios and are sensitive to data drift. In this paper, we introduce EdgeMA, a practical and efficient video analytics system designed to adapt models to shifts in real-world video streams over time, addressing the data drift problem. EdgeMA extracts the gray level co-occurrence matrix based statistical texture feature and uses the Random Forest classifier to detect the domain shift. Moreover, we have incorporated a method of model adaptation based on importance weighting, specifically designed to update models to cope with the label distribution shift. Through rigorous evaluation of EdgeMA on a real-world dataset, our results illustrate that EdgeMA significantly improves inference accuracy.

Keywords: Edge Computing · Deep Neural Network · Video Analytics · Data Drift · Model Adaptation

1 Introduction

Real-time video analytics has become a promising application in the field of computer vision, which is powered by deep neural network (DNN) models, e.g., ResNet [7] and EfficientNet [28]. Video analytics applications such as traffic monitoring [17] use local cameras that continuously generate high-quality video streams to understand the environment. Most of these applications have to be carried out with real-time feedback. Therefore, edge computing is favored in video analytics because it eliminates the need for costly network overhead associated with uploading videos to the cloud and also reduces latency.

Due to limited computing resources, edge devices typically use lightweight DNN models [8,34] for video analytics. Lightweight models have fewer weights

L. Wang and N. Zhang—These two authors have contributed to this work equally.

B. Luo et al. (Eds.): ICONIP 2023, LNCS 14447, pp. 292–304, 2024.
https://doi.org/10.1007/978-981-99-8079-6_23

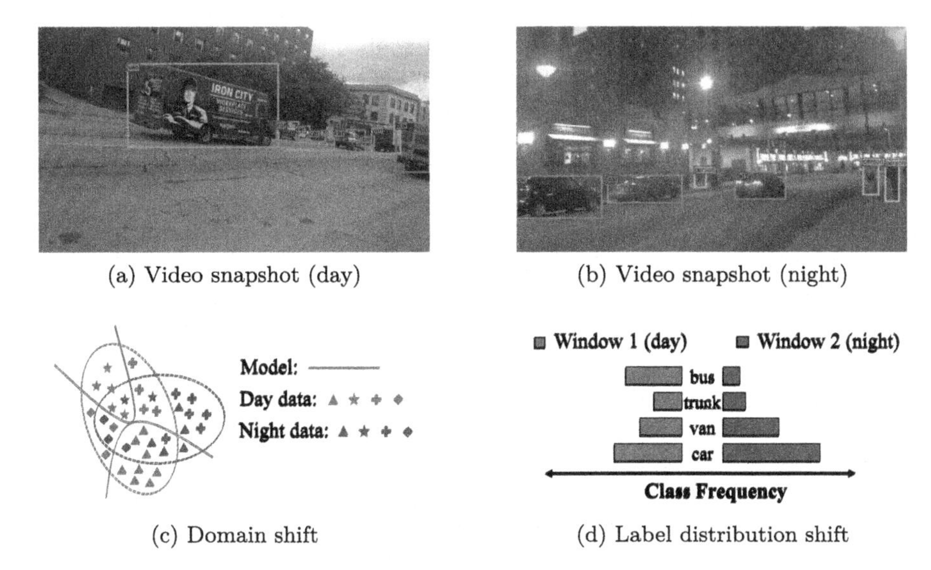

(a) Video snapshot (day) (b) Video snapshot (night)

(c) Domain shift (d) Label distribution shift

Fig. 1. (a) and (b) show the crossroad traffic in the video during the day and night, respectively. (c) shows the domains from day to night and the results of inference labels for data. We count the frequency of objects within each time window, and (d) shows the statistics results of class label distribution.

and shallower architectures, often making them unsuitable for delivering high accuracy. In addition, when employed in the actual field, they are vulnerable to the data drift [21] problem: a mismatch between initial training data and live video data. Data drift results in substantial declines in accuracy as it violates the underlying assumption of DNN models that the training data from the past should resemble the test data in the future. This further aggravates the challenge of real-time video analytics.

In video analytics serving, domain and label distribution shifts are typical data drift phenomena [1,12,15,16,19]. As shown in Fig. 1, street cameras encounter varying scenes over time. The label distribution of video varies over time, reducing the edge model's accuracy. In addition, the domain goes from day to night. While the model trained on the data collected based on daytime conditions fails to work well when deployed in the dark, degrading the accuracy.

Having a static model throughout the entire life-long inference process often negatively impacts performance, particularly with edge models. Due to their constrained capacity to learn object variations, edge DNNs require regular updates to accommodate changing object distributions and ensure optimal accuracy. We must consider the migration from domain to domain (e.g., weather, light).

In this work, we propose EdgeMA (**Edge** **M**odel **A**dapter), a video analytics system that resolves data drift in real-time video analytics on edge devices. EdgeMA copes with changes in the image domain and the label distribution over time, providing the model adaptation for edge computing. In summary, the main contributions of our work are the following:

- We propose a novel system EdgeMA for real-time video analytics on edge devices to cope with both domain shift and label distribution shift.
- We design the lightweight domain detector to effectively identify and address domain shift.
- We detect label distribution shift and allow for real-time model retraining while ensuring the overall system remains resource-efficient.
- In our experiment on a real-world dataset, EdgeMA substantially outperforms scenarios without automatic model adaptation.

2 Related Work

Video Analytics Systems. Existing work has contributed to the creation of efficient video analytics systems. VideoStorm [33] investigates quality-lag requirements in video queries. Focus [9] uses low-cost models to index videos. Chameleon [13] exploits correlations in camera content to amortize profiling costs. NoScope [14] enhances inference speed through cascading models and filtering mechanisms. Clownfish [23] amalgamates inference outcomes from optimized models on edge devices with delayed results from comprehensive models in the cloud. However, these systems predominantly employ offline-trained models, overlooking the potential impact of model adaptation and data drift on accuracy.

Edge Computing. The proliferation of the Internet of Things (IoT) has driven the production and usage of diverse hardware devices/sensors across the globe. These devices are able to collect data and then send it to the server for storage or processing, allowing end-users to access and extract the information as per their requirements [27]. Nevertheless, cloud computing has begun to reveal some issues. The centralized nature of cloud computing, handling data generated by global end devices, leads to numerous challenges, such as reduced throughput, increased latency, bandwidth constraints, data privacy concerns, and augmented costs. These challenges become particularly pressing in IoT applications, which demand rapid and low-latency data processing, analytics, and result delivery [24, 25]. To combat the aforementioned challenges associated with cloud computing, a novel computing paradigm known as edge computing, has attracted widespread attention [10]. In essence, edge computing offloads data processing, storage, and computing tasks traditionally assigned to the cloud to the network's edge, in close proximity to the terminal devices. This transition helps minimize data transmission and device response times, alleviate network bandwidth pressure, decrease data transmission costs, and facilitate a decentralized system [5].

Data Drift. Edge DNNs can only memorize a limited number of object appearances and scenarios. Hence, they are particularly vulnerable to data drift [12,21], which arises when real-time video data significantly diverges across domains. Variations of scene density (for instance, during rush hour) and lighting conditions (such as daytime versus nighttime) over time pose challenges for surveillance cameras attempting precise object detection. Additionally, the distribution of object classes changes over time, leading to a decrease in the precision of the edge model [29]. Because of their restricted ability to memorize changes, edge models necessitate continuous retraining with the most recent data and shifting object distributions to maintain high accuracy. Video-analytics systems are beginning to adopt continuous learning to adapt to changing video scenes and improve inference accuracy [1,15,16]. Ekya [1] provides sophisticated resource-sharing mechanisms for efficient model retraining. AMS [16] dynamically adjusts the frame sampling rate on edge devices depending on scene changes, mitigating the need for frequent retraining. RECL [15] combines Ekya and AMS, and offers more rapid responses by choosing an appropriate model from a repository of historical models during analysis. EdgeMA further considers lightweight retraining processing at the edge.

3 System Design

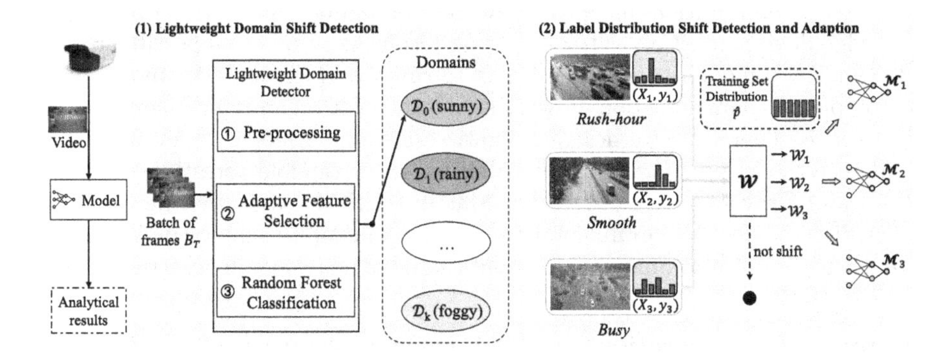

Fig. 2. EdgeMA system overview.

Figure 2 presents an overview of EdgeMA. Each edge device continuously performs inference for video analytics and buffers sampled video frames over time. The EdgeMA at the edge device runs iteratively on each new buffered batch of frames B_T, deciding whether to adapt the lightweight specialized model. The fundamental information upon which our system depends, when analyzing video feeds from static cameras or any other videos where the camera remains stationary, is that their domains and label distributions exhibit spatio-temporal locality. The model adaptation algorithm consists of two phases: (1) lightweight domain shift detection and (2) label distribution shift detection and adaptation.

In the first phase, EdgeMA detects in which domain the current batch lies. In our system, each domain corresponds to a training set containing a few data samples (e.g., 2000 images) from the domain scenario and an existing model pre-trained on a general dataset. After detecting, if the domain has changed, the model is fine-tuned using the corresponding training set to adapt to the specialized domain.

In the second phase, the edge device detects if the label distribution shifts. If the label distribution changes, the model also requires retraining with the dataset of the current domain. We assume that it makes sense to give more importance to the classes of training data whose frequency of appearance is nearest to the features of the live data to improve accuracy for different label distributions.

When our training set is small, retraining can perform well with low computing costs [6]. And for concreteness, we describe our design for object detection, but the system is general and can be adapted to other tasks. Next, we illustrate the two phases of model adaptation in detail.

3.1 Lightweight Domain Shift Detection

The variations of illumination, weather, or style in different domains can seriously affect the model performance. Cross-domain detection for video is the challenge that model adaptation must overcome. Previous works [11,32] can use heavily computation-intensive networks for domain shift detection. However, edge devices must allocate computing resources as much as possible to the video analytics model. It is not desirable to run the video analytics model with the complex domain detection network simultaneously on a resource-constrained edge device. Therefore, we design a lightweight domain detector for EdgeMA. We detect the video's environment condition by a random forest classifier [3] (RFC) and Gray Level Co-occurrence Matrix (GLCM)-based texture features, to detect the video's environmental conditions. This approach not only ensures high accuracy but also minimizes computational expenses and conserves energy.

Pre-processing Operation. EdgeMA requires conversion of the frame to grayscale, wherein each pixel value represents spectral intensity ranging between 0 and 255. The grayscale image, represented by g, is derived by computing the luminance of the color image using equation (1):

$$g = 0.299 \times R + 0.587 \times G + 0.114 \times B \tag{1}$$

Adaptive Feature Selection. For image classification, texture features, dependent on the repetitive patterns across an image region, are predominantly utilized. The GLCM of an image I of dimensions $n * m$ characterizes the texture by quantifying occurrences of pixels with specific absolute values based on a spatial offset, is defined as:

$$C(g_i,g_j)=\sum_{p=1}^{n}\sum_{q=1}^{m}\begin{cases}1, \text{if } I(p,q)=g_i \text{ and } I(p+\Delta_x,q+\Delta_y)=g_j\\0, \text{otherwise}\end{cases} \tag{2}$$

where g_i and g_j represent the gray level values of an image. For spatial positions p and q within the image I, the offset (Δ_x, Δ_y) is determined by both the direction and distance for which the matrix is computed. GLCM features are computed with directions $\theta \in \{0°, 45°, 90°, 135°\}$, distance $d \in [1, 30]$, and the 6 prevailing texture properties [20], including contrast, correlation, homogeneity, angular second moment, dissimilarity, and energy. In total, 4*30*6 = 720 features are derived from GLCM.

Moreover, we employ the AdaBoost method [26] for feature selection to ascertain the relative importance of features. AdaBoost not only selects the most significant features but also assigns weights to weak classifiers to enhance classification performance. The algorithm based on AdaBoost starts by choosing a feature as an initial weak learner. During each iteration t, the primary steps of the selection process are shown as follows. (1) Normalize the weights of the images as $w^{[t]} = \left\{w_I^{[t]}\right\}_I$, where I is from the training image set. (2) Choose the feature F_j that yields the least classification error summed over all images, weighted by $w^{[t]}$. (3) Increase the value S_j, denoting the importance score of the feature F_j. (4) Adjust the weights for each image sample $I(w^{[t]})$. The weights should correlate with the error rate E for that image. Through this method, images correctly classified by the chosen feature have diminished influence, whereas the weights of misclassified images increase. The final output of the algorithm is the vector S, signifying the relative importance of the original features.

Random Forest Classification. RFC [3] is a prominent method in machine learning, particularly for high-dimensional classification. RFC comprises a collection of decision trees, with each tree generated using a random vector drawn independently from the feature vector. During training, each tree's training set is formed from a bootstrap sample of the data, selecting frames with replacement. Each tree then casts a vote for the most probable class to predict the output. To develop and evaluate our proposed model, the classifier was trained multiple times, recording the best accuracy achieved. The frames were classified into diverse domain classes.

When EdgeMA is executed, it, by default, retrieves the last ten frames from the batch, converts them to grayscale, and extracts features to ascertain the domain. To evaluate our proposed detector, we train it multiple times and record the highest accuracy, as discussed in Sect. 4.1.

3.2 Label Distribution Shift Detection and Model Adaptation

After identifying the specific domain, EdgeMA fine-tunes the model using the corresponding domain training set. Fine-tuning employs importance weighting [4] (IW). This powerful technique applies a weight of importance to each class in the training set based on the distributional feature, which captures similarity to the live data distribution.

The label distribution of the training set X_S for the current domain is $P_S(X_S)$ (default IID). And the label distribution of the dataset X_T consisting of all N

frames in the batch B_T is $P_T(X_T)$. The distribution $P_T(X_T)$ of different windows is non-IID and dynamically varying. Figure 3 shows the streaming schema of main operations.

Estimate P_T. The model M identifies K classes in total, and the predicted value of these is $f(*)$, then the distribution $P_T(X_T)$ is as follows:

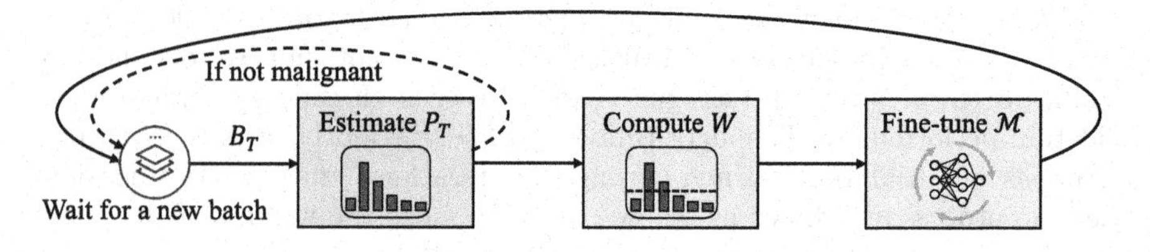

Fig. 3. Streaming schema of the main operations.

$$P_T(X_T) = \left\{ \sum_{j=1}^{N} \frac{\{f(x_j) == i\}}{N} \right\}_{i=1}^{K} \tag{3}$$

where $\{f(x_j) == i\}$ is a Boolean operation (1 if equal, 0 if not). The distribution outcome is a vector.

Compute W. Similarly to [18], the importance weight $W(X)$ is calculated as follows. It explains that the class is given a higher weight with a higher probability of occurrence:

$$W(X) = \frac{P_T(X_T)}{P_S(X_S)} = C_{h_0}^{-1} q_{h_0} \tag{4}$$

where h_0 is the initial neural network trained on X_S, C_{h_0} is the confusion matrix of h_0 in source label distribution P_S, and q_{h_0} is a K-dimension vector $q_{h_0}[i] = P_T(h_0(X) = y_i)$.

Fine-tune M. Finally, model M is fine-tuned with W to cope with the ever-changing distribution. To reduce time cost, we utilize *coordinate descent* [22, 31]. In this approach, EdgeMA retrains a limited subset of parameters (e.g., 20%) during each fine-tuning phase, with the edge device refining them over k iterations. The final optimization target for fine-tuning the model is shown below:

$$\frac{1}{n} \sum_{i=1}^{n} \frac{P_T(y_i)}{P_S(y_i)} f(x_i, y_i) \rightarrow \mathbb{E}_{x,y \sim P_S} \left(\frac{P_T(y)}{P_S(y)} f(x, y) \right) \tag{5}$$

Notably, fine-tuning can take up computing resources on the edge device. EdgeMA needs to determine the malignancy of a shift and reduces model update frequency. EdgeMA allows the model lags *if not malignant.* In practice, distributions shift constantly, and often these changes are benign. We employ the Kullback-Leibler (KL) divergence to quantify the discrepancy between the distribution P_M at the final stage of fine-tuning and the current distribution P_T. The KL divergence is defined as:

$$d = D_{KL}(P_T \| P_M) = \mathbb{E}_{x \sim P_T}\left[\log \frac{P_T(X_T)}{P_M(X_M)}\right] \tag{6}$$

By configuring a distance threshold D to determine whether adaptive learning is required, if $d < D$, it means that the model still matches the current label distribution.

4 Evaluation

In this section, we assess the performance of EdgeMA. Specifically, our evaluation seeks to answer two primary questions: (1) How effective is the lightweight detector, and what constitutes its peak performance? (2) How does EdgeMA cope with domain and label distribution shifts, and how does the effect of model adaptation perform in comparison to the baseline model?

4.1 Effectiveness of Lightweight Domain Detector

We evaluate EdgeMA on the task of object detection with the UA-DETRAC [30] dataset, which contains various environments, including sunny, rainy, cloudy, and night. We first trained our lightweight detector multiple times to develop and assess it for detecting the four domain environments.

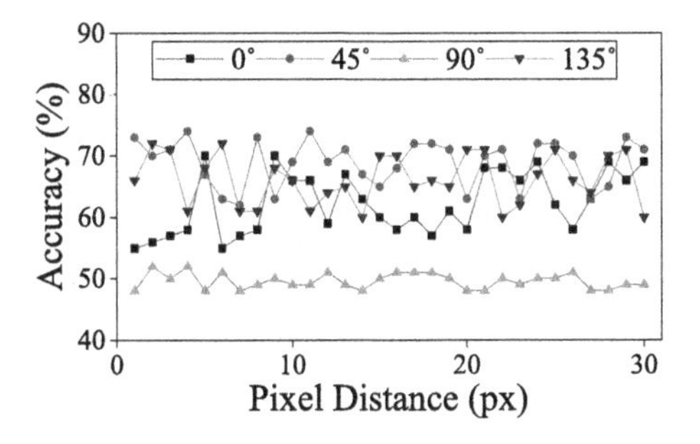

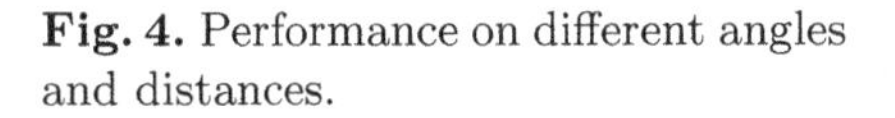
Fig. 4. Performance on different angles and distances.

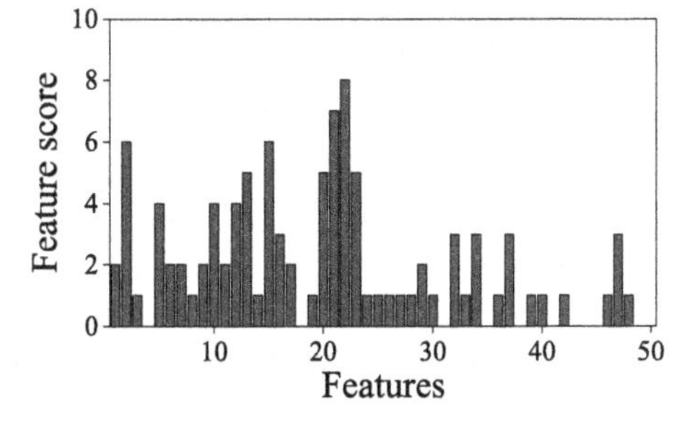

Fig. 5. Feature Importance.

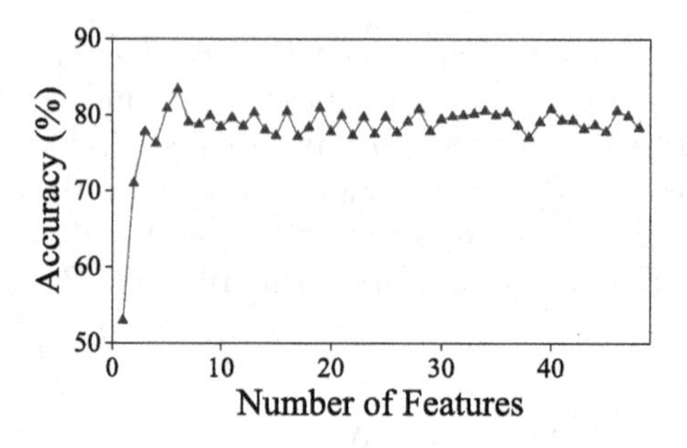

Fig. 6. Performance on number of best features.

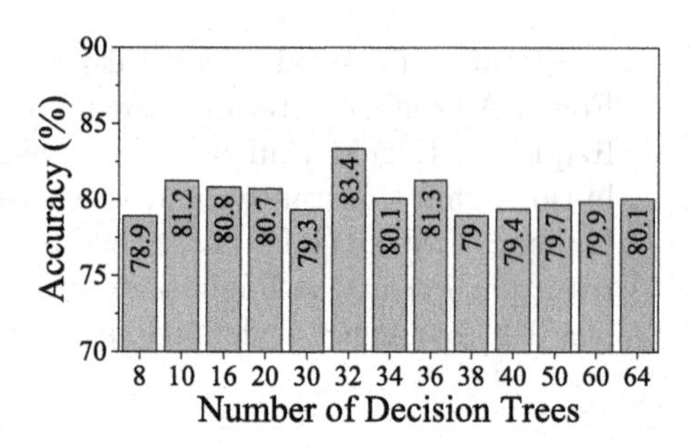

Fig. 7. Classifier accuracy.

Figure 4 shows the accuracy for a collection of 6 features from GLCM using different pixel distances $d \in [1, 30]$ and different angles $0°, 45°, 90°, 135°$. Then we select the 2 best neighboring pixel distances for each of the 4 angles (d = {5,9} for 0°, d = {4,11} for 45°, d = {2,4} for 90°, d = {2,6} for 135°). Consequently, the total number of features is calculated as 6*2*4 = 48 features.

The AdaBoost selects the top-performing features from the pool of 48 based on their feature importance. We set the iteration parameter for AdaBoost to 100. We number these 48 features. Among these, F_{22} has the highest score, which denotes the correlation of the GLCM matrix with d = 4 and angle 45°. Figure 5 shows the importance of all features.

Then we analyze the accuracy by varying the number of top-ranked features based on their scores, as illustrated in Fig. 6. The highest accuracy, 83.38%, is achieved with six selected features.

We set 32 as the number of decision trees in the above evaluation. In addition, we evaluate other numbers, and the results are shown in Fig. 7. Assigning 32 trees for the random forest still shows the best performance.

4.2 Overall Improvements of Model Adaptation

The UA-DETRAC dataset comprises over 140 thousand frames with timestamps, and we utilize these frames, concatenating part of them to form a one-hour-long video stream, 25 frames per second. We use Nvidia Jetson AGX Xavier boards as edge devices. For these devices, we use the YOLOv4 [2] with the Resnet18 [7] model backbone. We select an additional 2000 images as the training set, and we find that choosing the fine-tuning iteration k = 8 is the ideal trade-off between accuracy and training time.

To validate the robustness and efficacy of our model adaptation, we compare the detection accuracy (mAP metric) of the static lightweight YOLOv4 and the models fine-tuned in different domain conditions. The results are shown in Fig. 8. The adaptative solution delivers higher detection accuracy since the

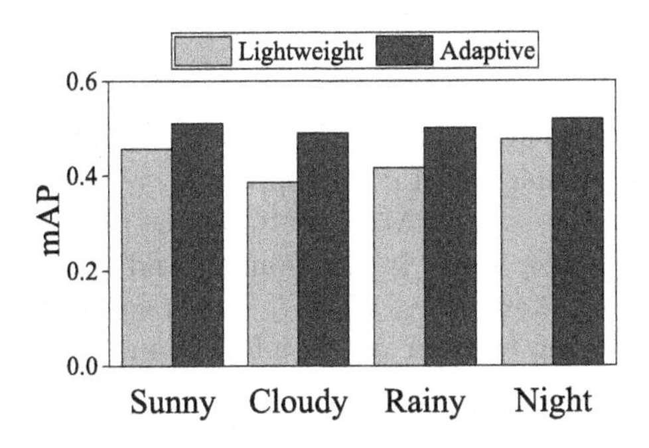

Fig. 8. Overall mAP.

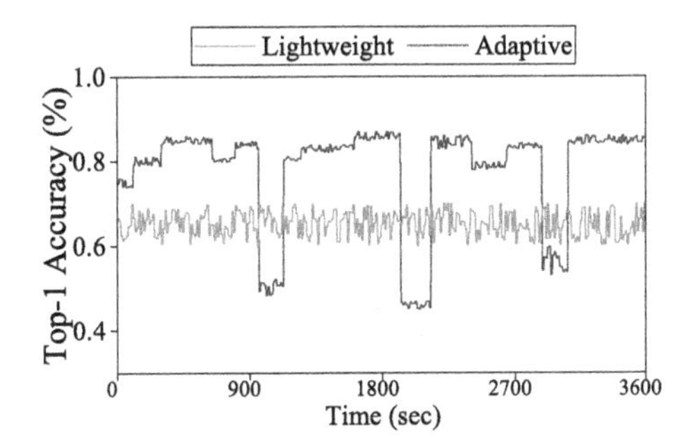

Fig. 9. Top-1 Acc over time.

model is specifically adapted to the revealed shift. Figure 9 shows the result of the lightweight model and model adaptation on the inference top-1 accuracy of the video over time. There are several intervals when the accuracy significantly drops (about 3–5 min per interval) as the model lags during scene changes. Nevertheless, the average performance of model adaptation is much higher than using only a lightweight model for inference. The adaptive solution has a top-1 accuracy above 0.8 most of the time, while the fixed lightweight model solution is stable in the range of 0.6–0.7.

Table 1. Comparison of two different schemes.

Scheme	Fraction of Parameters	mAP@0.5	FPS_{avg}	Fine-tuning Time Cost
Lightweight	–	0.406	59.6	–
Model Adaptation	5%	0.470	56.3	95 s
	10%	0.493	54.9	123 s
	20%	0.515	52.2	192 s
	100%	0.523	37.5	887 s

Table 1 summarizes the performance of two different schemes. Fractions for model adaptation represent the percentage of how many model parameters for each fine-tuning. Selecting only 20% of model parameters performs very effectively, achieving >0.1 better mAP score than the lightweight model. It results in only a 0.008 loss of mAP on average, but it reduces the fine-tuning time cost from 887 s for full-model fine-tuning to 192 s. In addition, due to the usage of computing resources for model fine-tuning, it shows a slight speed advantage over the model adaptation scheme.

5 Conclusion

In this paper, we propose EdgeMA, an innovative framework designed for video analytics on edge devices, which utilizes model adaptation supported by data drift detection and adaptive retraining methods. EdgeMA fine-tunes the model based on importance weighting after detecting shifts in the domain and label distribution during inference. Our evaluation shows that EdgeMA delivers overall higher accuracy compared to the static setting without model adaptation.

Acknowledgements. This work is supported by the Key Research and Development Program of Guangdong Province under Grant No.2021B0101400003, the National Natural Science Foundation of China under Grant No.62072196, and the Creative Research Group Project of NSFC No.61821003.

References

1. Bhardwaj, R., et al.: Ekya: continuous learning of video analytics models on edge compute servers. In: 19th USENIX Symposium on Networked Systems Design and Implementation (NSDI 2022), pp. 119–135 (2022)
2. Bochkovskiy, A., Wang, C.Y., Liao, H.Y.M.: Yolov4: optimal speed and accuracy of object detection. arXiv preprint arXiv:2004.10934 (2020)
3. Cutler, A., Cutler, D.R., Stevens, J.R.: Random forests. In: Ensemble Machine Learning, pp. 157–175. Springer, Heidelberg (2012). https://doi.org/10.1007/978-1-4419-9326-7_5
4. Fang, T., Lu, N., Niu, G., Sugiyama, M.: Rethinking importance weighting for deep learning under distribution shift. Adv. Neural. Inf. Process. Syst. **33**, 11996–12007 (2020)
5. Ghosh, A.M., Grolinger, K.: Edge-cloud computing for internet of things data analytics: embedding intelligence in the edge with deep learning. IEEE Trans. Ind. Inf. **17**(3), 2191–2200 (2020)
6. Gu, Y., Du, Z., Zhang, H., Zhang, X.: An efficient joint training framework for robust small-footprint keyword spotting. In: Yang, H., Pasupa, K., Leung, A.C.-S., Kwok, J.T., Chan, J.H., King, I. (eds.) ICONIP 2020. LNCS, vol. 12532, pp. 12–23. Springer, Cham (2020). https://doi.org/10.1007/978-3-030-63830-6_2
7. He, K., Zhang, X., Ren, S., Sun, J.: Deep residual learning for image recognition. In: Proceedings of the IEEE Conference on Computer Vision and Pattern Recognition, pp. 770–778 (2016)
8. Howard, A.G., et al.: Mobilenets: efficient convolutional neural networks for mobile vision applications. arXiv preprint arXiv:1704.04861 (2017)
9. Hsieh, K., et al.: Focus: querying large video datasets with low latency and low cost. In: 13th {USENIX} Symposium on Operating Systems Design and Implementation ({OSDI} 2018), pp. 269–286 (2018)
10. Hua, H., Li, Y., Wang, T., Dong, N., Li, W., Cao, J.: Edge computing with artificial intelligence: a machine learning perspective. ACM Comput. Surv. **55**(9), 1–35 (2023)
11. Ibrahim, M.R., Haworth, J., Cheng, T.: Weathernet: recognising weather and visual conditions from street-level images using deep residual learning. ISPRS Int. J. Geo Inf. **8**(12), 549 (2019)

12. Jia, Y., Zhang, X., Lan, L., Luo, Z.: Counterfactual causal adversarial networks for domain adaptation. In: Neural Information Processing: 29th International Conference, ICONIP 2022, Virtual Event, 22–26 November 2022, Proceedings, Part VI, pp. 698–709. Springer, Heidelberg (2023). https://doi.org/10.1007/978-981-99-1645-0_58
13. Jiang, J., Ananthanarayanan, G., Bodik, P., Sen, S., Stoica, I.: Chameleon: scalable adaptation of video analytics. In: Proceedings of the 2018 Conference of the ACM Special Interest Group on Data Communication, pp. 253–266 (2018)
14. Kang, D., Emmons, J., Abuzaid, F., Bailis, P., Zaharia, M.: Noscope: optimizing neural network queries over video at scale. Proc. VLDB Endow. **10**(11) (2017)
15. Khani, M., et al.: RECL: responsive resource-efficient continuous learning for video analytics. In: 20th USENIX Symposium on Networked Systems Design and Implementation (NSDI 2023), pp. 917–932 (2023)
16. Khani, M., Hamadanian, P., Nasr-Esfahany, A., Alizadeh, M.: Real-time video inference on edge devices via adaptive model streaming. In: Proceedings of the IEEE/CVF International Conference on Computer Vision, pp. 4572–4582 (2021)
17. Liao, X.C., Qiu, W.J., Wei, F.F., Chen, W.N.: Combining traffic assignment and traffic signal control for online traffic flow optimization. In: Neural Information Processing: 29th International Conference, ICONIP 2022, Virtual Event, 22–26 November 2022, Proceedings, Part VI, pp. 150–163. Springer, Heidelberg (2023). https://doi.org/10.1007/978-981-99-1645-0_13
18. Lipton, Z., Wang, Y.X., Smola, A.: Detecting and correcting for label shift with black box predictors. In: International Conference on Machine Learning, pp. 3122–3130. PMLR (2018)
19. Liu, C., Qu, X., Wang, J., Xiao, J.: Fedet: a communication-efficient federated class-incremental learning framework based on enhanced transformer. arXiv preprint arXiv:2306.15347 (2023)
20. Maji, K., Sharma, R., Verma, S., Goel, T.: RVFL classifier based ensemble deep learning for early diagnosis of alzheimer's disease. In: Neural Information Processing: 29th International Conference, ICONIP 2022, Virtual Event, 22–26 November 2022, Proceedings, Part III, pp. 616–626. Springer, Heidelberg (2023). https://doi.org/10.1007/978-3-031-30111-7_52
21. Moreno-Torres, J.G., Raeder, T., Alaiz-Rodríguez, R., Chawla, N.V., Herrera, F.: A unifying view on dataset shift in classification. Pattern Recogn. **45**(1), 521–530 (2012)
22. Nesterov, Y.: Efficiency of coordinate descent methods on huge-scale optimization problems. SIAM J. Optim. **22**(2), 341–362 (2012)
23. Nigade, V., Wang, L., Bal, H.: Clownfish: edge and cloud symbiosis for video stream analytics. In: 2020 IEEE/ACM Symposium on Edge Computing (SEC), pp. 55–69. IEEE (2020)
24. Qin, M., Chen, L., Zhao, N., Chen, Y., Yu, F.R., Wei, G.: Power-constrained edge computing with maximum processing capacity for iot networks. IEEE Internet Things J. **6**(3), 4330–4343 (2018)
25. Qu, X., Wang, J., Xiao, J.: Quantization and knowledge distillation for efficient federated learning on edge devices. In: 2020 IEEE 22nd International Conference on High Performance Computing and Communications; IEEE 18th International Conference on Smart City; IEEE 6th International Conference on Data Science and Systems (HPCC/SmartCity/DSS), pp. 967–972. IEEE (2020)
26. Rojas, R., et al.: Adaboost and the super bowl of classifiers a tutorial introduction to adaptive boosting. Freie University, Berlin, Technical Report (2009)

27. Shi, W., Dustdar, S.: The promise of edge computing. Computer **49**(5), 78–81 (2016)
28. Tan, M., Le, Q.: Efficientnet: rethinking model scaling for convolutional neural networks. In: International Conference on Machine Learning, pp. 6105–6114. PMLR (2019)
29. Wang, L., et al.: Shoggoth: towards efficient edge-cloud collaborative real-time video inference via adaptive online learning. arXiv preprint arXiv:2306.15333 (2023)
30. Wen, L., et al.: UA-DETRAC: a new benchmark and protocol for multi-object detection and tracking. Comput. Vision Image Underst. **193**, 102907 (2020)
31. Wright, S.J.: Coordinate descent algorithms. Math. Program. **151**(1), 3–34 (2015)
32. Xie, R., Yu, F., Wang, J., Wang, Y., Zhang, L.: Multi-level domain adaptive learning for cross-domain detection. In: Proceedings of the IEEE/CVF International Conference on Computer Vision Workshops, pp. 0–0 (2019)
33. Zhang, H., Ananthanarayanan, G., Bodik, P., Philipose, M., Bahl, P., Freedman, M.J.: Live video analytics at scale with approximation and delay-tolerance. In: 14th USENIX Symposium on Networked Systems Design and Implementation (2017)
34. Zhang, X., Zhou, X., Lin, M., Sun, J.: Shufflenet: an extremely efficient convolutional neural network for mobile devices. In: Proceedings of the IEEE Conference on Computer Vision and Pattern Recognition, pp. 6848–6856 (2018)

Mastering Complex Coordination Through Attention-Based Dynamic Graph

Guangchong Zhou[1,2], Zhiwei Xu[1,2], Zeren Zhang[1,2], and Guoliang Fan[1(✉)]

[1] Institute of Automation, Chinese Academy of Sciences, Beijing, China
{zhouguangchong2021,xuzhiwei2019,zhangzeren2021,guoliang.fan}@ia.ac.cn
[2] School of Artificial Intelligence, University of Chinese Academy of Sciences, Beijing, China

Abstract. The coordination between agents in multi-agent systems has become a popular topic in many fields. To catch the inner relationship between agents, the graph structure is combined with existing methods and improves the results. But in large-scale tasks with numerous agents, an overly complex graph would lead to a boost in computational cost and a decline in performance. Here we present DAGMIX, a novel graph-based value factorization method. Instead of a complete graph, DAGMIX generates a dynamic graph at each time step during training, on which it realizes a more interpretable and effective combining process through the attention mechanism. Experiments show that DAGMIX significantly outperforms previous SOTA methods in large-scale scenarios, as well as achieving promising results on other tasks.

Keywords: Multi-Agent Reinforcement Learning · Coordination · Value Factorization · Dynamic Graph · Attention

1 Introduction

Multi-agent systems (MAS) have become a popular research topic in the last few years, due to their rich application scenarios like auto-driving [16], cluster control [26], and game AI [1,22,29]. Communication constraints exist in many MAS settings, meaning only local information is available for agents' decision-making. A lot of research discussed multi-agent reinforcement learning (MARL), which combines MAS and reinforcement learning techniques. The simplest way is to employ single-agent reinforcement learning on each agent directly, as the *independent Q-learning* (IQL) [20] does. But this approach may not converge as each agent faces an unstable environment caused by other agents' learning and exploration. Alternatively, we can learn decentralized policies in a centralized fashion, known as the *centralized training and decentralized execution* (CTDE) [9] paradigm, which allows sharing of individual observations and global information during training but limits the agents to receive only local information while executing.

Focusing on the cooperative tasks, the agents suffer from the credit assignment problem. If all agents share a joint value function, it would be hard to

B. Luo et al. (Eds.): ICONIP 2023, LNCS 14447, pp. 305–318, 2024.
https://doi.org/10.1007/978-981-99-8079-6_24

measure each agent's contribution to the global value, and one agent may mistakenly consider others' credits as its own. Value factorization methods decompose the global value function as a mixture of each agent's individual value function, thus solving this problem from the principle. Some popular algorithms such as VDN [18], QMIX [14], and QTRAN [17] have demonstrated impressive performance in a range of cooperative tasks.

The above-mentioned algorithms focus on constructing the mixing function mathematically but ignore the physical topology of the agents. For example, players in a football team could form different formations like '4-3-3' or '4-3-1-2' (shown in Fig. 1), and a player has stronger links with the nearer teammates. An intuitive idea is to use graph neural networks (GNN) to extract the information contained in the structure. *Multi-agent graph network* (MAGNet) [11] generates a real-time graph to guide the agents' decisions based on historical and current global information. In QGNN [6], the model encodes every agent's trajectory by a gated recurrent unit (GRU), which is followed by a GNN to produce a local Q-value of each agent. These methods do make improvements on original algorithms, but they violate the communication constraints, making it hard to extend to more general scenarios. Xu et al. modify QMIX and propose *multi-graph attention network* (MGAN) [27], which constructs all agents as a graph and utilizes multiple graph convolutional networks (GCN [5]) to jointly approximate the global value. Similar but not the same, *Deep Implicit Coordination Graph* (DICG) [7] processes individual Q-values through a sequence of GCNs instead of parallel ones. These algorithms fulfill CTDE and significantly outperform the benchmarks. However, these methods prefer fully connected to sparse when building the graph. As the number of agents grows, the complexity of a complete graph increases by square, leading to huge computational costs and declines in the results.

Fig. 1. Topology of football players.

In this paper, we proposed a cooperative multi-agent reinforcement learning algorithm called *Dynamic Attention-based Graph MIX* (DAGMIX). DAGMIX presents a more reasonable and intuitive way of value mixing to estimate Q_{tot} more accurately, thus providing better guidance for the learning of agents. Specifically, DAGMIX generates a partially connected graph rather than fully connected according to the agents' attention on each other at every time step during training. Later we perform graph attention on this dynamic graph to help integrate individual Q-values. Like other value factorization methods, DAGMIX perfectly fulfills the CTDE paradigm, and it also meets the monotonicity assumption, which ensures the consistency of global optimal and local optimal policy. Experiments on StarCraft multi-agent challenge (SMAC) [15] show that our work is comparable to the baseline algorithms. DAGMIX outperforms previous SOTA methods significantly in some tasks, especially those with numerous agents and asymmetric environmental settings which are considered as super-hard scenarios.

2 Background

2.1 Dec-POMDP

A fully cooperative multi-agent environment is usually modeled as a decentralized partially observable Markov decision process (Dec-POMDP) [12], consisting of a tuple $G = \langle \mathcal{S}, \boldsymbol{\mathcal{U}}, \mathcal{P}, \mathcal{Z}, r, \mathcal{O}, n, \gamma \rangle$. At each time-step, $s \in \mathcal{S}$ is the current global state of the environment, but each agent $a \in \mathcal{A} := \{1, ..., n\}$ receives only a unique local observation $z_a \in \mathcal{Z}$ produced by the observation function $\mathcal{O}(s, a) : \mathcal{S} \times \mathcal{A} \rightarrow \mathcal{Z}$. Then every agent a will choose an action $u_a \in \mathcal{U}$, and all individual actions form the joint action $\boldsymbol{u} = [u_1, ..., .u_n] \in \boldsymbol{\mathcal{U}} \equiv \mathcal{U}^n$. After the interaction between the joint action $\boldsymbol{u}$ and current state s, the environment changes to state s' according to state transition function $\mathcal{P}(s' \| s, \boldsymbol{u}) : \mathcal{S} \times \mathcal{U} \times \mathcal{S} \rightarrow [0, 1]$. All the agents in Dec-POMDP share the same global reward function $r(s, \boldsymbol{u}) : \mathcal{S} \times \boldsymbol{\mathcal{U}} \rightarrow \mathbb{R}$. $\gamma \in [0, 1)$ is the discount factor.

In Dec-POMDP, each agent a chooses action based on its own action-observation history $\tau_a \in T \equiv (\mathcal{Z} \times \mathcal{U})$, thus the policy of each agent a can be written as $\pi_a(u_a | \tau_a) : T \times \mathcal{U} \rightarrow [0, 1]$. The joint action-value function can be computed by the following equation: $Q^{\pi}(s_t, \boldsymbol{u}_t) = \mathbb{E}_{s_{t+1:\infty}, \boldsymbol{u}_{t+1:\infty}} [R_t | s_t, \boldsymbol{u}_t]$, where $\boldsymbol{\pi}$ is the joint policy of all agents. The goal is to maximize the discounted return $R^t = \sum_{l=0}^{\infty} \gamma^l r_{t+l}$.

2.2 Value Factorization Methods

Credit assignment is a key problem in cooperative MARL problems. If all agents share a joint value function, it would be hard for a single agent to tell how much it contributes to global utilization. Without such feedback, learning is easy to fail.

In value factorization methods, every agent has its own value function for decision-making. The joint value function is regarded as an integration of individual ones. To ensure that the optimal action of each agent is consistent with the global optimal joint action, all value decomposition methods comply with the *Individual Global Max* (IGM) [14] conditions described below:

$$\arg\max_{\boldsymbol{u}} Q_{\text{tot}}(\boldsymbol{\tau}, \boldsymbol{u}) = \begin{pmatrix} \arg\max_{u_1} Q_1(\tau_1, u_1) \\ \vdots \\ \arg\max_{u_n} Q_n(\tau_n, u_n) \end{pmatrix},$$

where $Q_{tot} = f(Q_1, ..., Q_n)$, $Q_1, ..., Q_n$ denote the individual Q-values, and f is the mixing function.

For example, VDN assumes the function f is a plus operation, while QMIX makes monotonicity constraints to meet the IGM conditions as below:

$$\text{(VDN)} \qquad Q_{tot}(s, u_a) = \sum_{a=1}^{n} Q_a(s, u_a).$$

$$\text{(QMIX)} \qquad \frac{\partial Q_{tot}(\boldsymbol{\tau}, \boldsymbol{u})}{\partial Q_a(\tau_a, u_a)} \geq 0, \quad \forall a \in \{1, \ldots, n\}.$$

2.3 Self Attention

Consider a sequence of vectors $\boldsymbol{\alpha} = (\mathbf{a}_1, ..., \mathbf{a}_n)$, the motivation of self-attention is to catch the relevance between each pair of vectors [21]. An attention function can be described as mapping a query $(\mathbf{q})$ and a set of key-value $(\mathbf{k}, \mathbf{v})$ pairs to an output $(\mathbf{o})$, where the query, keys, and values are all translated from the original vectors. The output is derived as below:

$$w_{ij} = \text{softmax}\left(s\left(\mathbf{q}^i, \mathbf{k}^j\right)\right) = \frac{\exp\left(s\left(\mathbf{q}^i, \mathbf{k}^j\right)\right)}{\sum_t \exp\left(s\left(\mathbf{q}^i, \mathbf{k}^t\right)\right)},$$

$$\mathbf{o}_i = \sum_j w_{ij} \mathbf{v}_j,$$

where $s(\cdot, \cdot)$ is a user-defined function to measure similarity, usually dot-product. In practice, a multi-head implementation is usually employed to enable the model to calculate attention from multiple perspectives.

2.4 Graph Neural Network

Graph Neural Networks (GNNs) are a class of deep learning methods designed to directly perform inference on data described by graphs and easily deal with node-level, edge-level, and graph-level tasks.

The main idea of GNN is to aggregate the node's own information and its neighbors' information together using a neural network. Given a graph denoted as $\boldsymbol{G} = (\boldsymbol{V}, \boldsymbol{E})$ and consider a K-layer GNN structure, the operation of the k-th layer can be formalized as below [3]:

$$\mathbf{h}_v^k = \sigma\left(\left[\mathbf{W}_k \cdot \text{AGG}\left(\left\{\mathbf{h}_u^{k-1}, \forall u \in \mathcal{N}(v)\right\}\right), \mathbf{B}_k \mathbf{h}_v^{k-1}\right]\right),$$

where $\mathbf{h}_\mathbf{v}^\mathbf{k}$ refers to the value of node v at k-th layer, $\mathcal{N}$ is the neighborhood function to get the neighbor nodes, and AGG (...) is the generalized aggregator which can be a function such as Mean, Pooling, LSTM and so on.

Unlike other deep networks, GNN could not improve its performance by simply deepening the model, since too many layers in GNN usually lead to severe over-smoothing or over-squashing. Besides, the computational complexity of GNN is closely related to the amount of nodes and edges. Fortunately, DAGMIX tries to generate an optimized graph structure, making it suitable for applying GNN.

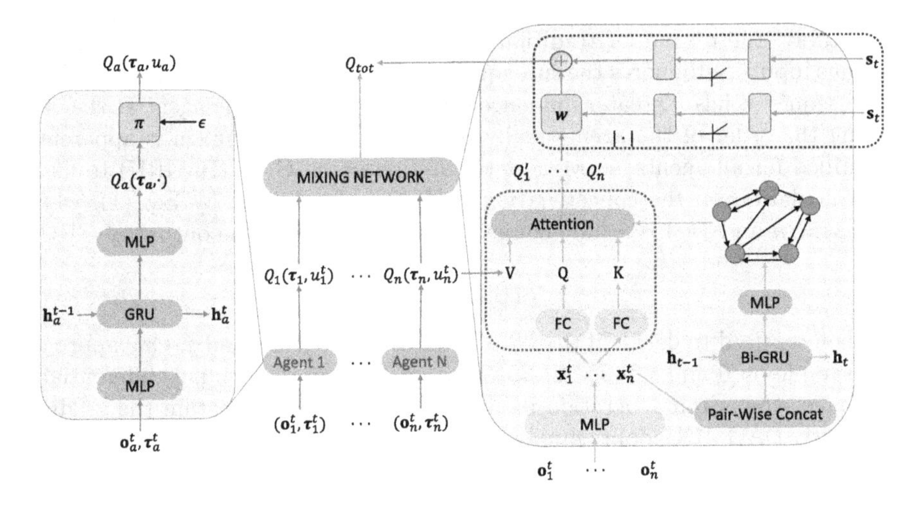

Fig. 2. The overall framework of DAGMIX. The individual Q network is shown in the blue box, and the mixing network is shown in the green box. (Color figure online)

3 DAGMIX

In this section, we'd like to introduce a new method called DAGMIX. Compared to other related studies, DAGMIX enables a dynamic graph in the training process of cooperative MARL. It fulfills the CTDE paradigm and IGM conditions perfectly and provides a more precise and explainable estimation of all agents' joint action, thus reaching a better performance than previous methods. The overall architecture is shown in Fig. 2.

3.1 Dynamic Graph Generation

Each agent corresponds to a node in the graph. Instead of relying on a fully connected graph, DAGMIX employs a dynamic graph that generates a real-time structure at each time step. Inspired by G2ANet [8], we adopt the hard-attention mechanism. It computes the attention weights between all the nodes and then sets the weights to 0 or 1 representing if there are connections between each pair of nodes in the dynamic graph. Nonetheless, DAGMIX removes the communication between agents in G2ANet and enables fully decentralized decision-making.

Assume that we take out an episode from the replay buffer. First, we encode the observation o_a^t of each agent a at each time-step t as the embedding $\mathbf{x}_a^t$. Then we perform a pair-wise concatenation on all the embedding vectors, which yields a matrix $\mathbf{x}_{n\times n}^t$ where $\mathbf{x}_{ij}^t = \left(\mathbf{x}_i^t, \mathbf{x}_j^t\right), \quad i, j = 1, ..., n$. This matrix $\mathbf{x}^t$ can be regarded as a batch of sequential data since it has a batch size of n corresponding to n agents, and every agent a has a sequence from $(\mathbf{x}_a^t, \mathbf{x}_1^t)$ to $(\mathbf{x}_a^t, \mathbf{x}_n^t)$. We could naturally think of using a sequential model like GRU to process every sequence and calculate the attention between agents.

Notably, the output of traditional GRU only depends on the current and previous inputs but ignores the subsequent ones. Therefore, the agent at the front of the sequence has a greater influence on the output than the agent at the end, making the order of the agents really crucial. Such a mechanism is apparently unjustified for all agents, so we adopt a bi-directional GRU (Bi-GRU) to fix it. The calculation on the concatenation $\left(\mathbf{x}_i^t, \mathbf{x}_j^t\right)$ is formalized by Eq. (1), where $i, j = 1, ..., n$ and $f(\cdot)$ is a fully connected layer to embed the output of GRU.

$$\mathbf{a}_{i,j} = f\left(\text{Bi-GRU}\left(\mathbf{x}_i, \mathbf{x}_j\right)\right). \tag{1}$$

Then we need to decide if there's a link between agent i and j. One approach is to sample on 0 and 1 based on $\mathbf{a}_{i,j}$, but simply sampling is not differentiable and makes the graph generation module untrainable. To maintain the gradient continuity, DAGMIX adopts gumbel-softmax [4] as follows:

$$\text{gum}(\mathbf{x}) = \text{softmax}\left(\left(\mathbf{x} + \log\left(\lambda e^{-\lambda \mathbf{x}}\right)\right)/\tau\right), \tag{2}$$

where $\text{gum}(\cdot)$ means the gumbel-softmax function, λ is the hyperparameter of the exponential distribution, and τ is the hyperparameter referring to the temperature. As the temperature decreases, the output of Gumbel-softmax gets closer to one-hot. In DAGMIX, Gumbel-softmax returns a vector of length 2.

$$\left(\,\cdot\,, \mathbf{A}_{i,j}\right)^{\mathrm{T}} = \text{gum}\left(f\left(\text{Bi-GRU}\left(\mathbf{x}_i, \mathbf{x}_j\right)\right)\right). \tag{3}$$

Combining Eq. (2) and Eq. (1) we get Eq. (3). $\mathbf{A}$ is the adjacency matrix whose element $\mathbf{A}_{i,j}$ is the second value in the output of Gumbel-softmax, indicating whether there's an edge from agent j to agent i. Finally, we get a graph structure similar to the one illustrated on the right side of Fig. 2. It is characterized by a sparse connectivity pattern that exhibits an expansion in sparsity as the graph scales up.

3.2 Value Mixing Network

As we've already got a graph structure (shown in the right of Fig. 2), we design an attention-based value integration on the dynamic graph, which can be viewed as performing self-attention on the original observations.

More specifically, there are two independent channels to process the observation. On the one hand, the Q network receives current observation as input and outputs an individual Q value estimation, which serves as the value in self-attention. On the other hand, the observation is encoded with an MLP in the mixing network, which is followed by two fully connected layers representing $\mathbf{W}_q$ and $\mathbf{W}_k$. The query and key of agent a in self attention are derived through $\boldsymbol{q}_a = \mathbf{W}_q \mathbf{x}_a$ and $\boldsymbol{k}_a = \mathbf{W}_k \mathbf{x}_a$ respectively.

Besides, restricted by the dynamic graph, we only calculate the attention scores in agent a's neighborhood $\mathcal{N}(a) = \{i \in \mathcal{A} | \mathbf{A}_{a,i} = 1\}$, which can be regarded as a mask operation based on the adjacency matrix $\mathbf{A}$. The calculation process in matrix form is shown below:

$$\begin{aligned} \boldsymbol{Q} = \boldsymbol{X}\mathbf{W}_q, \quad \boldsymbol{K} = \boldsymbol{X}\mathbf{W}_k, \quad \boldsymbol{V} = (Q_1, ..., Q_n)^{\mathrm{T}}, \\ \left(Q_1^{'}, ..., Q_n^{'}\right)^{\mathrm{T}} = \mathrm{softmax}\left(\mathrm{Mask}\left(\frac{\boldsymbol{Q}\boldsymbol{K}^{\mathrm{T}}}{\sqrt{d_k}}\right)\right)\boldsymbol{V}, \end{aligned} \tag{4}$$

where $\boldsymbol{X} = (\mathbf{x}_1, ..., \mathbf{x}_n)^{\mathrm{T}}$, and d_k is the hidden dimension of $\mathbf{W}_k$. According to Eq. (4), DAGMIX blends the original Q-values of all agents and transforms them into $\left(Q_1^{'}, ..., Q_n^{'}\right)$.

In the last step, we need to combine $\left(Q_1^{'}, ..., Q_n^{'}\right)$ into a global Q_{tot}. As the dotted black box in Fig. 2 shows, we adopt a QMIX-style mixing module, making DAGMIX benefit from the global information. The hypernetworks [2] take in global state $\mathbf{s}_t$ and output the weights $\mathbf{w}$ and bias b, thus the Q_{tot} is calculated as:

$$Q_{tot} = \left(Q_1^{'}, ..., Q_n^{'}\right)\mathbf{w} + b \tag{5}$$

The operation of taking absolute values guarantees weights in $\mathbf{w}$ non-negative. And in Eq. 4, the coefficients of any Q_a are also non-negative thanks to the softmax function. Equation 6 ensures DAGMIX fits in the IGM assumption perfectly.

$$\frac{\partial Q_{tot}}{\partial Q_a} = \sum_{i=1}^{n} \frac{\partial Q_{tot}}{\partial Q_i^{'}} \frac{\partial Q_i^{'}}{\partial Q_a} \geq 0, \quad \forall a \in \{1, ..., n\} \tag{6}$$

It is noteworthy that DAGMIX is not a specific algorithm but a framework. As depicted by the green box in Fig. 2, the part in the red dotted box can be replaced by any kind of GNN (e.g., GCN, GraphSAGE [3], etc.), while the mixing module in the black dotted box can be substituted with any value factorization method. DAGMIX enhances the performance of the original algorithm, especially on large-scale problems.

3.3 Loss Function

Like other value factorization methods, DAGMIX is trained end-to-end. The loss function is set to TD-error, which is similar to value-based SARL algorithms [19]. We denote the parameters of all neural networks as θ which is optimized by minimizing the following loss function:

$$\mathcal{L}(\theta) = \left(y_{tot} - Q_{tot}(\boldsymbol{\tau}, \boldsymbol{u}|\theta)\right)^2, \tag{7}$$

where $y_{tot} = r + \gamma \max_{\boldsymbol{u}'} Q_{tot}\left(\boldsymbol{\tau}', \boldsymbol{u}'|\theta^-\right)$ is the target joint action-value function. θ^- denotes the parameters of the target network. The training algorithm is displayed in Algorithm 1.

Algorithm 1: DAGMIX

Initialize replay buffer $\mathbf{D}$
Initialize $[Q_a], Q_{tot}$ with random parameters θ, initialize target parameters $\theta^- = \theta$
while *training* **do**
  for *episode* $\leftarrow 1$ **to** M **do**
    Start with initial state $\mathbf{s}^0$ and each agent's observation $\mathbf{o}_a^0 = \mathcal{O}\left(\mathbf{s}^0, a\right)$
    Initialize an empty episode recorder E **for** $t \leftarrow 0$ **to** T **do**
      For every agent a, with probability ϵ select action u_a^t randomly
      Otherwise select $u_a^t = \arg\max_{u_a^t} Q_a\left(\tau_a^t, u_a^t\right)$
      Take joint action $\boldsymbol{u}^t$, and retrieve next state $\mathbf{s}^{t+1}$, next observations $\boldsymbol{o}^{t+1}$ and reward r^t
      Store transition $\left(\mathbf{s}^t, \boldsymbol{o}^t, \boldsymbol{u}^t, r^t, \mathbf{s}^{t+1}, \boldsymbol{o}^{t+1}\right)$ in E
    end
    Store episode data E in $\mathbf{D}$
  end
  Sample a random mini-batch data $\mathbf{B}$ with batch size N from $\mathbf{D}$
  for $t \leftarrow 0$ **to** $T-1$ **do**
    Extract transition $\left(\mathbf{s}^t, \boldsymbol{o}^t, \boldsymbol{u}^t, r^t, \mathbf{s}^{t+1}, \boldsymbol{o}^{t+1}\right)$ from $\mathbf{B}$
    For every agent a, calculate $Q_a\left(\tau_a^t, u_a^t|\theta\right)$
    Generate the dynamic graph $\mathcal{G}^t$ using $\boldsymbol{o}^t$ based on Equation (3)
    According to the structure of $\mathcal{G}^t$, calculate $Q_{tot}\left(\boldsymbol{\tau}^t, \boldsymbol{u}^t|\theta\right)$ based on Equation (4) and (5)
    With target network, calculate
    $Q_a\left(\tau_a^{t+1}, u_a^{t+1}|\theta^-\right) = \max Q_a\left(\tau_a^{t+1}, \cdot\, |\theta^-\right)$
    Calculate $Q_{tot}\left(\boldsymbol{\tau}^{t+1}, \boldsymbol{u}^{t+1}|\theta^-\right)$
  end
  Update θ by minimizing the total loss in Equation (7)
  Update target network parameters $\theta^- = \theta$ periodically
end

4 Experiments

4.1 Settings

SMAC is an environment for research in the field of collaborative multi-agent reinforcement learning (MARL) based on Blizzard's StarCraft II RTS game. It consists of a set of StarCraft II micro scenarios that aim to evaluate how well independent agents are able to learn coordination to solve complex tasks. The version of StarCraft II is 4.6.2(B69232) in our experiments, and it should be noted that results from different client versions are not always comparable. The difficulty of the game AI is set to *very hard* (7). To conquer a wealth of challenges with varying levels of difficulty in SMAC, algorithms should adapt to different scenarios and perform well both in single-unit control and group coordination. SMAC has become increasingly popular recently for its ability to comprehensively evaluate MARL algorithms.

The detailed information of the challenges used in our experiments is shown in Table 1. All the selected challenges have a large number of agents to control, and some of them are even asymmetric or heterogeneous, making it extremely hard for MARL algorithms to handle these tasks.

Our experiment is based on Pymarl [15]. To judge the performance of DAGMIX objectively, we adopt several most popular value factorization methods (VDN, QMIX, and QTRAN) as well as more recent methods including Qatten [28], QPLEX [23], W-QMIX [13], MAVEN [10], ROMA [24] and RODE [25] as baselines. The hyperparameters of these baseline algorithms are set to the default in Pymarl.

Table 1. Information of selected challenges.

Challenge	Ally Units	Enemy Units	Level of Difficulty
2s3z	2 Stalkers 3 Zealots	2 Stalkers 3 Zealots	Easy
3s5z	3 Stalkers 5 Zealots	3 Stalkers 5 Zealots	Easy
1c3s5z	1 Colossus 3 Stalkers 5 Zealots	1 Colossus 3 Stalkers 5 Zealots	Easy
8m_vs_9m	8 Marines	9 Marines	Hard
MMM2	1 Medivac 2 Marauders 7 Marines	1 Medivac 3 Marauders 8 Marines	Super Hard
bane_vs_bane	4 Banelings 20 Zerglings	4 Banelings 20 Zerglings	Hard
25m	25 Marines	25 Marines	Hard
27m_vs_30m	27 Marines	30 Marines	Super Hard

4.2 Validation

On every challenge, we have run each algorithm 2 million steps for 5 times with different random seeds and recorded the changes in win ratio. To reduce the contingency in validation, we take the average win ratio of the recent 15 tests as the current results. The performances of DAGMIX and the baselines are shown in Fig. 3, where the solid line represents the median win ratio of the five experiments using the corresponding algorithm, and the 25–75% percentiles of the win ratios are shaded. Detailed win ratio data is displayed in Table 2.

Among these baselines, QMIX is recognized as the SOTA method and the main competitor of DAGMIX due to its stability and effectiveness on most tasks. It can be clearly observed that the performance of DAGMIX is very close to the

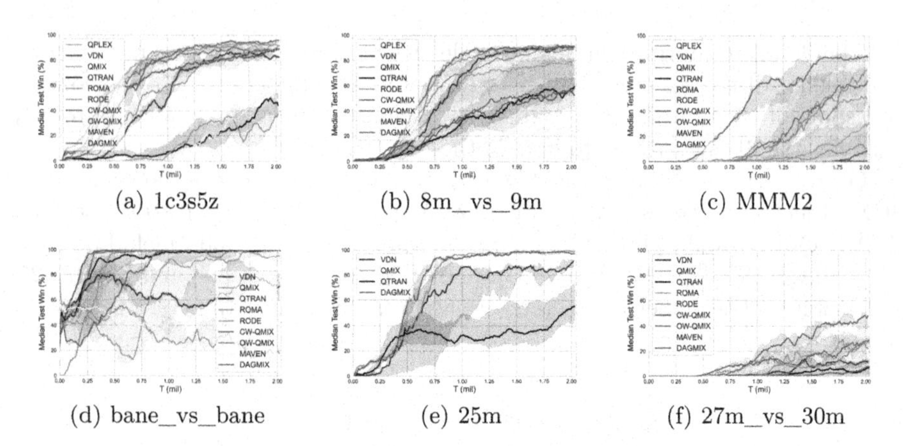

(a) 1c3s5z (b) 8m_vs_9m (c) MMM2

(d) bane_vs_bane (e) 25m (f) 27m_vs_30m

Fig. 3. Overall results in different challenges.

Table 2. Median performance and of the test win ratio in different scenarios.

	DAGMIX	VDN	QMIX	QTRAN
1c3s5z	93.75	**96.86**	94.17	11.67
8m_vs_9m	92.71	**94.17**	**94.17**	70.63
MMM2	**86.04**	4.79	45.42	0.63
bane_vs_bane	**100**	94.79	99.79	99.58
25m	**99.79**	87.08	**99.79**	35.20
27m_vs_30m	**49.58**	9.58	33.33	18.33

best baseline in relatively easy challenges such as *1c3s5z* and *8m_vs_9m*. As the scenarios become more asymmetric and complex, DAGMIX begins to show its strengths and exceed the performances of the baselines. In *bane_vs_bane* and *25m*, the win ratio curve of DAGMIX lies on the left of QMIX, indicating a faster convergence rate. *MMM2* and *27m_vs_30m* are classified as super hard scenarios on which most algorithms perform very poorly, due to the huge amount of agents and significant disparity between the opponent's troops and ours. However, Fig. 3(c) and 3(f) show that our method makes tremendous progress on the performance, as DAGMIX achieves unprecedented win ratios on these two tasks. Notably, the variance of DAGMIX's training results is relatively small, indicating its robustness against random perturbations.

4.3 Ablation

We conducted ablation experiments to demonstrate that the dynamic graph generated by the hard-attention mechanism is key to the success of DAGMIX. First, while many studies have improved original models by replacing the fully connected layer with the attention layer, we doubt if DAGMIX also benefits from

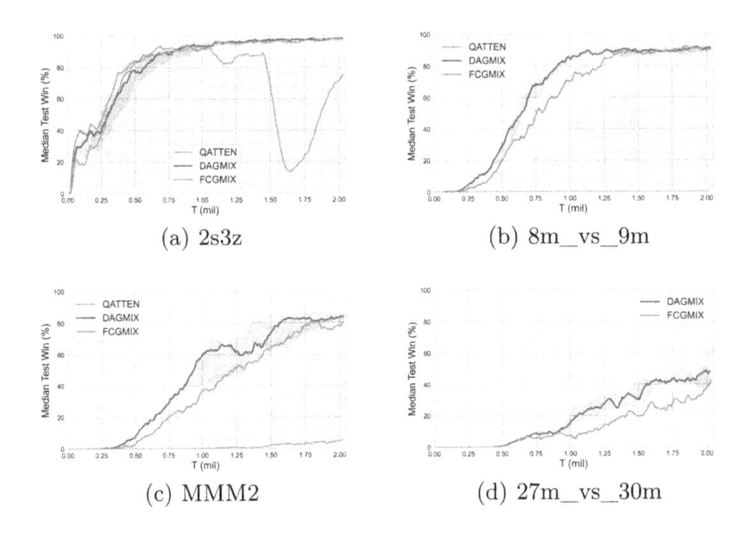

(a) 2s3z (b) 8m_vs_9m

(c) MMM2 (d) 27m_vs_30m

Fig. 4. Ablation experiments results.

such tricks, so we select Qatten [28] as the control group which also aggregates individual Q-values through attention mechanism but lacks a graph structure. Second, to confirm the effectiveness of the sparse dynamic graph over a fully connected one, we set all values in the adjacency matrix **A** to 1 which is referred to as *Fully Connected Graph MIX* (FCGMIX) for comparison.

From Fig. 4 we find Qatten gets a disastrous win ratio with high variance, indicating that simply adding an attention layer does not necessarily improve performance. When there are fewer agents, the fully connected graph is more comprehensive and does not have much higher complexity than the dynamic graph, so in *2s3z* FCGMIX performs even slightly better than DAGMIX. However, as the number of agents increases and the complete graph becomes more complex, DAGMIX is gaining a clear advantage thanks to the optimized graph structure.

It was noted in Sect. 3.2 that DAGMIX is a framework that enhances the base algorithm on large-scale problems. Here we combine VDN with our dynamic graph, denoted as DAGVDN, to investigate the improvements upon VDN, as well as DAGMIX upon QMIX. The results are presented in Fig. 5. Similar to DAGMIX, DAGVDN outperforms VDN slightly in relatively easy tasks, while it shows significant superiority in more complex scenarios that VDN couldn't handle.

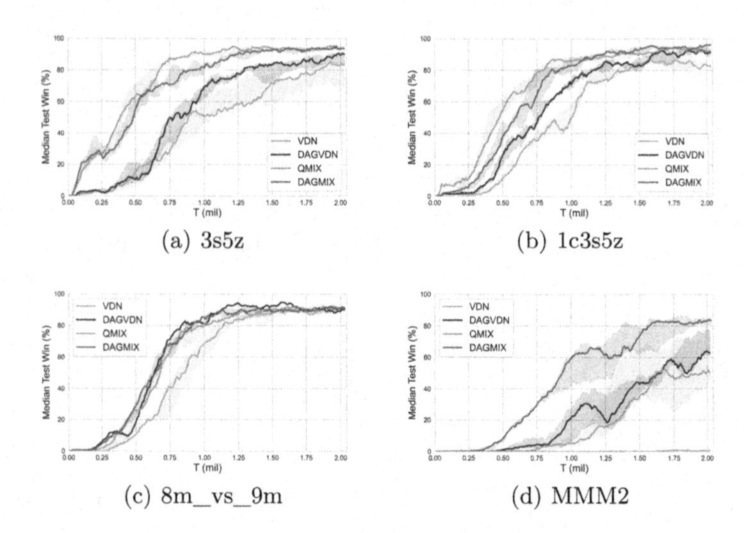

(a) 3s5z (b) 1c3s5z

(c) 8m_vs_9m (d) MMM2

Fig. 5. The results of DAGMIX and DAGVDN compared to original algorithms.

5 Conclusion

In this paper, we propose DAGMIX, a cooperative MARL algorithm based on value factorization. It combines the individual Q-values of all agents through operations on a real-time generated dynamic graph and provides a more interpretable and precise estimation of the global value to guide the training process. DAGMIX catches the intrinsic relationship between agents and allows end-to-end learning of decentralized policies in a centralized manner.

Experiments on SMAC show the prominent superiority of DAGMIX when dealing with large-scale and asymmetric problems. In other scenarios, DAGMIX still demonstrates very stable performance which is comparable to the best baselines. We believe DAGMIX provides a reliable solution to multi-agent coordination tasks.

There are still points of improvement for DAGMIX. First, it seems practical to apply multi-head attention in later implementation. Besides, the dynamic graph in DAGMIX is not always superior to the complete graph when dealing with problems with fewer agents. On the basis of the promising findings presented in this paper, work on the remaining issues is continuing and will be presented in future papers.

Acknowledgements. This work was supported by the Strategic Priority Research Program of the Chinese Academy of Science, Grant No. XDA27050100.

References

1. Berner, C., et al.: Dota 2 with large scale deep reinforcement learning. arXiv preprint arXiv:1912.06680 (2019)
2. Ha, D., Dai, A., Le, Q.V.: Hypernetworks. arXiv preprint arXiv:1609.09106 (2016)
3. Hamilton, W., Ying, Z., Leskovec, J.: Inductive representation learning on large graphs. In: Advances in Neural Information Processing Systems, vol. 30 (2017)
4. Jang, E., Gu, S., Poole, B.: Categorical reparameterization with gumbel-softmax. arXiv preprint arXiv:1611.01144 (2016)
5. Kipf, T.N., Welling, M.: Semi-supervised classification with graph convolutional networks. arXiv preprint arXiv:1609.02907 (2016)
6. Kortvelesy, R., Prorok, A.: QGNN: value function factorisation with graph neural networks. arXiv preprint arXiv:2205.13005 (2022)
7. Li, S., Gupta, J.K., Morales, P., Allen, R., Kochenderfer, M.J.: Deep implicit coordination graphs for multi-agent reinforcement learning. arXiv preprint arXiv:2006.11438 (2020)
8. Liu, Y., Wang, W., Hu, Y., Hao, J., Chen, X., Gao, Y.: Multi-agent game abstraction via graph attention neural network. In: Proceedings of the AAAI Conference on Artificial Intelligence, vol. 34, pp. 7211–7218 (2020)
9. Lowe, R., Wu, Y.I., Tamar, A., Harb, J., Pieter Abbeel, O., Mordatch, I.: Multi-agent actor-critic for mixed cooperative-competitive environments. In: Advances in Neural Information Processing Systems, vol. 30 (2017)
10. Mahajan, A., Rashid, T., Samvelyan, M., Whiteson, S.: MAVEN: multi-agent variational exploration. In: Advances in Neural Information Processing Systems, vol. 32 (2019)
11. Malysheva, A., Kudenko, D., Shpilman, A.: MAGNet: multi-agent graph network for deep multi-agent reinforcement learning. In: 2019 XVI International Symposium "Problems of Redundancy in Information and Control Systems" (REDUNDANCY), pp. 171–176. IEEE (2019)
12. Oliehoek, F.A., Amato, C.: A Concise Introduction to Decentralized POMDPs. Springer, Cham (2016). https://doi.org/10.1007/978-3-319-28929-8
13. Rashid, T., Farquhar, G., Peng, B., Whiteson, S.: Weighted QMIX: expanding monotonic value function factorisation for deep multi-agent reinforcement learning. In: Advances in Neural Information Processing Systems, vol. 33, pp. 10199–10210 (2020)
14. Rashid, T., Samvelyan, M., Schroeder, C., Farquhar, G., Foerster, J., Whiteson, S.: QMIX: monotonic value function factorisation for deep multi-agent reinforcement learning. In: International Conference on Machine Learning, pp. 4295–4304. PMLR (2018)
15. Samvelyan, M., et al.: The StarCraft multi-agent challenge. arXiv preprint arXiv:1902.04043 (2019)
16. Shamsoshoara, A., Khaledi, M., Afghah, F., Razi, A., Ashdown, J.: Distributed cooperative spectrum sharing in UAV networks using multi-agent reinforcement learning. In: 2019 16th IEEE Annual Consumer Communications & Networking Conference (CCNC), pp. 1–6. IEEE (2019)
17. Son, K., Kim, D., Kang, W.J., Hostallero, D., Yi, Y.: QTRAN: learning to factorize with transformation for cooperative multi-agent reinforcement learning. CoRR abs/1905.05408 (2019). https://arxiv.org/abs/1905.05408
18. Sunehag, P., et al.: Value-decomposition networks for cooperative multi-agent learning. arXiv preprint arXiv:1706.05296 (2017)

19. Sutton, R.S., Barto, A.G.: Reinforcement Learning: An Introduction. MIT Press, Cambridge (2018)
20. Tampuu, A., et al.: Multiagent cooperation and competition with deep reinforcement learning. PLoS ONE **12**(4), e0172395 (2017)
21. Vaswani, A., et al.: Attention is all you need. In: Advances in Neural Information Processing Systems, vol. 30 (2017)
22. Vinyals, O., et al.: Grandmaster level in StarCraft II using multi-agent reinforcement learning. Nature **575**(7782), 350–354 (2019)
23. Wang, J., Ren, Z., Liu, T., Yu, Y., Zhang, C.: QPLEX: duplex dueling multi-agent Q-learning. arXiv preprint arXiv:2008.01062 (2020)
24. Wang, T., Dong, H., Lesser, V., Zhang, C.: ROMA: multi-agent reinforcement learning with emergent roles. arXiv preprint arXiv:2003.08039 (2020)
25. Wang, T., Gupta, T., Mahajan, A., Peng, B., Whiteson, S., Zhang, C.: RODE: learning roles to decompose multi-agent tasks. arXiv preprint arXiv:2010.01523 (2020)
26. Xu, D., Chen, G.: Autonomous and cooperative control of UAV cluster with multi-agent reinforcement learning. Aeronaut. J. **126**(1300), 932–951 (2022)
27. Xu, Z., Zhang, B., Bai, Y., Li, D., Fan, G.: Learning to coordinate via multiple graph neural networks. In: Mantoro, T., Lee, M., Ayu, M.A., Wong, K.W., Hidayanto, A.N. (eds.) ICONIP 2021. LNCS, vol. 13110, pp. 52–63. Springer, Cham (2021). https://doi.org/10.1007/978-3-030-92238-2_5
28. Yang, Y., et al.: Qatten: a general framework for cooperative multiagent reinforcement learning. arXiv preprint arXiv:2002.03939 (2020)
29. Ye, D., et al.: Towards playing full MOBA games with deep reinforcement learning. In: Advances in Neural Information Processing Systems, vol. 33, pp. 621–632 (2020)

SORA: Improving Multi-agent Cooperation with a Soft Role Assignment Mechanism

Guangchong Zhou[1,2], Zhiwei Xu[1,2], Zeren Zhang[1,2], and Guoliang Fan[1(✉)]

[1] Institute of Automation, Chinese Academy of Sciences, Beijing, China
{zhouguangchong2021,xuzhiwei2019,zhangzeren2021,guoliang.fan}@ia.ac.cn
[2] School of Artificial Intelligence, University of Chinese Academy of Sciences, Beijing, China

Abstract. Role-based multi-agent reinforcement learning (MARL) holds the promise of achieving scalable multi-agent cooperation by decomposing complex tasks through the concept of roles and has enjoyed great success in various tasks. However, conventional role-based MARL methods typically assign a single role to each agent, limiting the agent's behavior in certain scenarios. In real life, an individual usually performs multiple responsibilities in a given task. To meet such situations, we propose a novel soft role assignment (SORA) process that enables an agent to play multiple roles simultaneously. Concretely, SORA first generates a role distribution via the attention mechanism to interpret the agent's identity as a combination of different roles. To ensure consistent behavior with an agent's assigned role, we also introduce role-specific Q networks for decision-making. By virtue of these advances, our proposed method makes a prominent improvement over the prior state-of-the-art approaches on StarCraft multi-agent challenges and Google Research Football.

Keywords: Role-based Multi-agent Reinforcement Learning · Cooperation · Soft Role Assignment · Role Distribution

1 Introduction

Multi-agent systems (MAS) provide a powerful framework for understanding and solving complex problems that involve multiple agents, such as MOBA games [1,19,25], autonomous driving [14], robotic control [2], etc. Traditional approaches to MAS (e.g., rule-based or game-theoretic methods) have limitations in complex and dynamic environments, where agents must adapt to changing circumstances and learn from experience, and this is where multi-agent reinforcement learning (MARL) comes into play. It learns effective strategies and behaviors in MAS through the use of reinforcement signals, with each agent's goal being to maximize its own cumulative reward over time. MARL has shown promising results in various applications, while it also presents several challenges that are not present in single-agent reinforcement learning, such as the presence of non-stationarity, partial observability, and coordination issues. For

B. Luo et al. (Eds.): ICONIP 2023, LNCS 14447, pp. 319–331, 2024.
https://doi.org/10.1007/978-981-99-8079-6_25

example, *independent Q-learning* (IQL) [17] directly applies Q-learning on every single agent and treats other agents as part of the environment, which leads to a lack of cooperation and an unstable training process. To address the above issues, most MARL algorithms adopt the *centralized training and decentralized execution* (CTDE) [8] paradigm, which allows the sharing of all individual and global information during training but restricts the agents to solely receiving local information during execution.

To alleviate the complexity of the model and improve the training efficiency, the parameter sharing technique is commonly employed in MARL, which enables agents to utilize the same neural network for decision-making, rather than each agent having its own independent network. However, parameter sharing also leads to homogeneous policies, and it is widely observed that agents often behave similarly, indicating that the joint policy is stuck in a local optimum [6,9].

In a basketball game, players carry out different responsibilities in accordance with their assigned roles, such as guards or forwards, to achieve teamwork. To align with this concept, a group of hierarchical learning methods called role-based MARL has been proposed. These methods facilitate the coordination between agents by assigning specialized roles to different entities according to their capabilities and objectives, resulting in varied behaviors of agents conditioned on their respective roles. Under the CTDE paradigm, the key problem is the inference of agents' roles during execution without global information. ROMA [21] generates policies for agents according to the individual roles represented by stochastic latent variables, which are obtained from their local observations. In RODE [22], each role assigned to the agents is associated with a specific sub-task and policy. By searching for the optimal solution within the designated sub-space, agents are able to reduce the complexity of the overall problem-solving process. The LDSA algorithm [23] involves an encoder to generate role embeddings and a decoder to derive the parameters of the Q-networks from these embeddings. At runtime, each agent selects the Q-network that is tailored to its sampled role and produces optimized outcomes.

Previous research has typically imposed singular roles on agents, but we argue that an agent can take multiple roles concurrently in many real-life MAS. For example, in the attacking phase of a football match, the midfielder is responsible for assisting offensive plays as well as keeping an eye on the opponent's forward to prevent a counterattack, which means the midfielder can be regarded as a combination of the attacking and defending roles. Following this idea, we propose a *SOft Role Assignment* (SORA) workflow for multi-agent cooperation. SORA utilizes a role encoder to transform one-hot vectors into role representations, which are later employed to create role-specific Q networks that contain extensive role information. Moreover, SORA designs a novel attention-based soft role assignment process and computes the agent's Q value estimation based on its soft role. The overall architecture of SORA is well-crafted and can be combined with any value factorization method. The experiment results have demonstrated the superiority of our proposed method over other role-based MARL algorithms. In the majority of tested scenarios, SORA is comparable to or even outperforms the previous state-of-the-art (SOTA) methods in terms of win ratio and variance.

2 Preliminaries

2.1 Problem Formulation

A fully cooperative MAS is typically represented by a decentralized partially observable Markov decision process (Dec-POMDP) [10], which is composed of a tuple $G = \langle \mathcal{S}, \boldsymbol{\mathcal{U}}, \mathcal{P}, \mathcal{Z}, r, \mathcal{O}, n, \gamma \rangle$. At each time-step, the current global state of the environment is denoted by $s \in \mathcal{S}$, while each agent $a \in \mathcal{A} := 1, \ldots, n$ only receives a unique local observation $z_a \in \mathcal{Z}$ generated by the observation function $\mathcal{O}(s, a) : \mathcal{S} \times \mathcal{A} \rightarrow \mathcal{Z}$. Subsequently, every agent a selects an action $u_a \in \mathcal{U}$, and all individual actions are combined to form the joint action $\boldsymbol{u} = [u_1, \ldots, u_n] \in \boldsymbol{\mathcal{U}} \equiv \mathcal{U}^n$. The interaction between the joint action $\boldsymbol{u}$ and the current state s leads to a change in the environment to state s' as dictated by the state transition function $\mathcal{P}(s'|s, \boldsymbol{u}) : \mathcal{S} \times \mathcal{U} \times \mathcal{S} \rightarrow [0, 1]$. All agents in the Dec-POMDP share the same global reward function $r(s, \boldsymbol{u}) : \mathcal{S} \times \boldsymbol{\mathcal{U}} \rightarrow \mathbb{R}$, and $\gamma \in [0, 1)$ represents the discount factor.

Definition 1 (Roles). *The cooperative multi-agent task* $G = \langle \mathcal{S}, \mathcal{U}, \mathcal{P}, \mathcal{Z}, r, \mathcal{O}, n, \gamma \rangle$ *invokes a set of* K *roles* $\{\phi_1, ..., \phi_K\} \equiv \Phi$, *where* K *is set manually. Each role* $\phi_k \in \Phi$ *holds a tuple* $\langle x_{\phi_k}, \mathcal{A}_{\phi_k}, \pi_{\phi_k} \rangle$, *where* $x_{\phi_k} \in \mathbb{R}^m$ *is the embedding vector of* ϕ_k, *and* $\mathcal{A}_{\phi_k}$ *is a set of agents assigned with role* ϕ_k, *satisfying* $\cup_{k=1}^{K} \mathcal{A}_{\phi_k} = \mathcal{A}$. *Each* $a_i \in \mathcal{A}_{\phi_k}$ *shares the policy network* π_{ϕ_k}.

On the basis of introducing roles into Dec-POMDP, each agent a selects a role ϕ_k based on its own action-observation history $\tau_a \in T \equiv (\mathcal{Z} \times \mathcal{U})$, thus the policy of each agent a can be written as $\pi_{\phi_k}(u_a|\tau_a) : T \times \mathcal{U} \rightarrow [0, 1]$. The joint action-value function can be computed by the following equation: $Q^{\pi}(s_t, \boldsymbol{u}_t) = \mathbb{E}_{s_{t+1:\infty}, \boldsymbol{u}_{t+1:\infty}} [R_t | s_t, \boldsymbol{u}_t]$, where $\boldsymbol{\pi}$ is the joint policy of all agents. The goal is to maximize the discounted return $R^t = \sum_{l=0}^{\infty} \gamma^l r_{t+l}$.

2.2 Value Factorization Methods

Credit assignment is a key problem in cooperative MARL problems. When agents share a joint value function, it is challenging for an individual agent to discern its impact on the collective performance. Insufficient feedback raises the likelihood of learning failure.

Value factorization methods assume that each agent has a specific value function for decision-making. The integration of these individual functions creates the joint value function. To guarantee that the optimal action for each agent aligns with the global optimal joint action, all value decomposition methods satisfy the *Individual Global Max* (IGM) [12] conditions, which are described below:

$$\arg\max_{\boldsymbol{u}} Q_{\text{tot}}(\boldsymbol{\tau}, \boldsymbol{u}) = \begin{pmatrix} \arg\max_{u_1} Q_1(\tau_1, u_1) \\ \vdots \\ \arg\max_{u_n} Q_n(\tau_n, u_n) \end{pmatrix},$$

where $Q_{tot} = f(Q_1, ..., Q_n)$, $Q_1, ..., Q_n$ denote the individual Q-values, and f is the mixing function.

QMIX, the most well-known value factorization algorithm, is highly regarded for its effectiveness across a range of scenarios. To fulfill the IGM conditions, QMIX confines the parameters of the value mixing network to non-negative values. This is achieved by satisfying the inequality as listed below:

$$\frac{\partial Q_{tot}(\boldsymbol{\tau}, \boldsymbol{u})}{\partial Q_a\left(\tau_a, u_a\right)} \geq 0, \quad \forall a \in \{1, \ldots, n\}.$$

2.3 Attention Mechanism

Humans selectively focus on a part of all information while ignoring other visible information in cognition. Inspired by this, researchers have proposed the attention mechanism [18]. Consider a set of information stored in key-value pairs $(\boldsymbol{K}, \boldsymbol{V}) = [(\boldsymbol{k}_1, \boldsymbol{v}_1), ..., (\boldsymbol{k}_N, \boldsymbol{v}_N)]$ and a query vector $\boldsymbol{q}$, the attention mechanism outputs a weighted sum of all values, where the weights are obtained by calculating the similarity between the query and each key. The formalized process is shown below:

$$w_{ij} = \mathrm{softmax}\left(s\left(\mathbf{q}^i, \mathbf{k}^j\right)\right) = \frac{\exp\left(s\left(\mathbf{q}^i, \mathbf{k}^j\right)\right)}{\sum_t \exp\left(s\left(\mathbf{q}^i, \mathbf{k}^t\right)\right)}, \tag{1}$$

$$\mathbf{o}_i = \sum_j w_{ij} \mathbf{v}_j,$$

where $s(\cdot, \cdot)$ measures similarity between vectors, typically via dot product. In practice, a multi-head approach is usually employed to enable the model to calculate attention from multiple perspectives.

3 Method

This section introduces the specific structure and implementation of *SOft Role Assignment* (SORA), a hierarchical multi-agent reinforcement learning framework that guides agents' decision-making by assigning roles in a soft manner. SORA fulfills the CTDE paradigm and achieves better performance than previous methods.

3.1 Role Representation

Different individuals play different roles in both virtual and real multi-agent tasks. Previous works commonly rely on prior knowledge and assign roles according to manually designed rules (e.g., dividing basketball players into guards, forwards, and centers based on their body size). Nevertheless, these approaches are impractical when dealing with an unknown environment and lacking the required expertise. In order to eliminate the reliance on human knowledge and

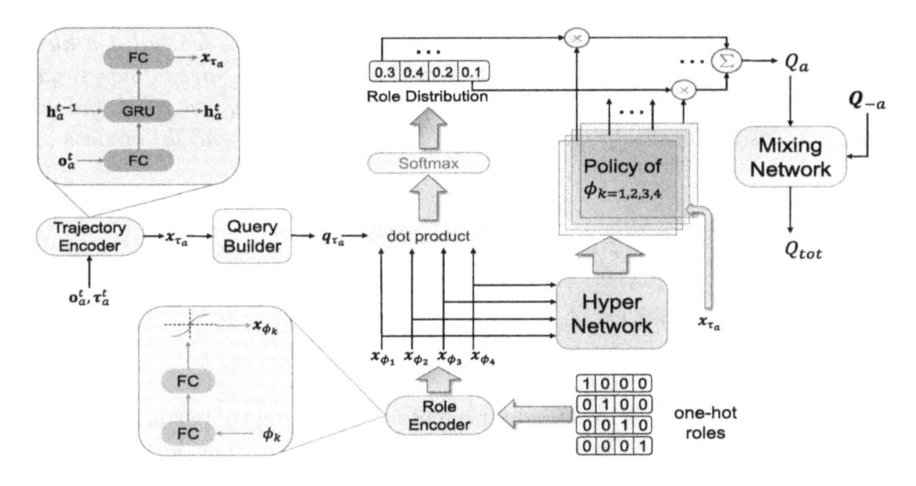

Fig. 1. The overall framework of SORA.

enable autonomous learning of role allocation, it is necessary to first design a proper method for embedding role information.

Given K distinct roles, each role k could be represented using a one-hot vector $\boldsymbol{\phi}_k$ of length K. However, such a vector is too sparse to effectively express the role's features. As illustrated in the green box in Fig. 1), we build a role encoder $f_r(\cdot;\theta_r)$ with two fully connected layers and a tanh activation function to transform $\boldsymbol{\phi}_k$ into a more sophisticated representation $\boldsymbol{x}_{\phi_k}$:

$$\boldsymbol{x}_{\phi_k} = f_r(\boldsymbol{\phi}_k;\theta_r), \quad \forall i \in \{1, ..., K\}$$

To promote the variety of behaviors among agents with specific roles, it is necessary to maintain the distinctiveness of each role representation. We preserve the identity of each role by maximizing the distance between role representations, which is equivalent to minimizing the identity loss below:

$$\mathcal{L}_i = \mathbb{E}_{\mathcal{D}}\left[-\sum_{1\le i<j\le K} d\left(\boldsymbol{x}_{\phi_i}, \boldsymbol{x}_{\phi_j}\right)\right], \tag{2}$$

where D is the sampled batch, $d(\cdot,\cdot)$ is a distance metric, and K is the number of roles.

3.2 Role Distribution

Under CTDE paradigm, each agent a can only make decisions based on its own observation $\boldsymbol{o}_a$ and history $\boldsymbol{\tau}_a$. We referred to [5] to build the trajectory encoder $f_t(\cdot;\theta_t)$, which consists of a Fully Connected layer (FC) taking $\boldsymbol{o}_a$ as input, a Gated Recurrent Unit (GRU) [3] taking $\boldsymbol{\tau}_a$ as input, and another FC that outputs the trajectory embedding $\boldsymbol{x}_{\tau_a}$ (depicted in the blue box in Fig. 1).

Definition 2 (Hard and Soft Role). *In role-based MARL, we call it a* **hard role** *assignment if the agent is classified into one of the K given roles specifically. In contrast, a* **soft role** *is an integration of all the given roles and can be described as a probability distribution. Their correlations and differences are formalized as:*

$$\begin{aligned} &\sum_{k=1}^{K} \mathcal{P}\left(a \in \mathcal{A}_{\phi_k}\right) = 1 \\ &\mathcal{P}\left(a \in \mathcal{A}_{\phi_k}\right) \in \{0,1\} \quad \textit{(Hard Role)} \\ &\mathcal{P}\left(a \in \mathcal{A}_{\phi_k}\right) \in [0,1] \quad \textit{(Soft Role)} \end{aligned}$$

As argued previously, it is practical for agents to concurrently perform multiple roles. Therefore, we assign soft roles to the agents to depict their identities. The attention mechanism is introduced to determine the proportion of different roles by calculating a role distribution.

In the process of role assignment, an agent's role can be inferred according to its trajectory of behavior [21,23,26], which implies that the trajectory embedding $\boldsymbol{x}_{\tau_a}$ can be transformed into a query $\boldsymbol{q}_{\tau_a} = f_q\left(\boldsymbol{x}_{\tau_a}\right)$ for choosing the roles, and the K role representations are the corresponding keys $\boldsymbol{K} = \left[\boldsymbol{x}_{\phi_1}, ..., \boldsymbol{x}_{\phi_K}\right]$ to be matched with the query. According to Eq. 1, we compute the dot product of an agent's query and each role's representation to get the similarity scores, which are then transformed into a distribution of the agent's assigned roles through a softmax function.

It is proved that introducing a changing latent variable into the network would make the training process more unstable. To alleviate this problem, the variation of an agent's role distribution between two consecutive time steps is restricted. For example, RODE [22] updates the role of the agent periodically to maintain its stationary. SORA stabilizes the training by facilitating a smooth transition of an agent's role within an episode, which is practically formalized as follows:

$$\mathcal{L}_s = \mathbb{E}_{\mathcal{D}}\left[\sum_{t=1}^{T} D_{\mathrm{KL}}\left(\Pr\left(\Phi^t | \boldsymbol{x}_{\tau}^t\right) \| \Pr\left(\Phi^{t-1} | \boldsymbol{x}_{\tau}^{t-1}\right)\right)\right] \tag{3}$$

Here, $\Phi^t \in [0,1]^K$ is the agent's role distribution at time step t, and D_{KL} refers to Kullback-Leibler Divergence which is used as a similarity metric for two distributions.

3.3 Role-Specific Q Networks

In role-based MARL, each agent considers both its current observation and assigned role to determine its action u_a. For efficiency, we reuse the trajectory embedding $\boldsymbol{x}_{\tau_a}$ introduced in Sect. 3.2 instead of building a new observation encoder, while it is still crucial to build another network which models the policy $p\left(u_a | \Phi_a, \boldsymbol{x}_{\tau_a}\right)$.

It is impractical to simply concatenate the role distribution Φ_a and the trajectory embedding $\boldsymbol{x}_{\tau_a}$ together, as the high-dimension input would result in more parameters and reduced performance. Instead, we employ hypernetworks [4] which uses the role representation $\boldsymbol{x}_{\phi_k}$ to generate the parameters $\theta_{\phi_k} = f_h(\boldsymbol{x}_{\phi_k})$ of a role-specific Q network Q_{ϕ_k}. Next, we feed the trajectory embedding of agent a into all role-specific Q networks to derive role-specific Q values, which are then computed as a dot product with the role distribution Φ_a as follows:

$$\boldsymbol{q}_k = Q_{\phi_k}(\boldsymbol{x}_{\tau_a}, \cdot; \theta_{\phi_k}) \tag{4}$$

$$\begin{aligned} Q_a(\boldsymbol{x}_{\tau_a}, \cdot) &= \Phi_a \cdot [\boldsymbol{q}_1, ..., \boldsymbol{q}_K] \\ &= \sum_{k=1}^{K} w_k \cdot \boldsymbol{q}_k, \end{aligned} \tag{5}$$

where $\boldsymbol{q}_k \in \mathbb{R}^{|\mathcal{U}|}$ is a vector containing the outputs of Q_{ϕ_k} with $\boldsymbol{x}_{\tau_a}$ and each action $u \in \mathcal{U}$ as input, and w_k is the weight of role k in Φ_a. The agent a selects its action greedily based on Q_a:

$$u_a = \arg\max_{u \in \mathcal{U}} Q_a(\boldsymbol{x}_{\tau_a}, u) \tag{6}$$

3.4 Overall Optimization Objective

We can refer to any value decomposition method [11,12,15,16,20,24] to integrate the derived individual Q values into a global Q value Q_{tot}. In this paper, we adopt QMIX [12] for value mixing, which updates the network by minimizing the TD loss below:

$$\mathcal{L}_{\mathrm{TD}}(\theta) = \mathbb{E}_D\left[\left(y^{tot} - Q_{tot}(\boldsymbol{\tau}, \boldsymbol{u}, s; \theta)\right)^2\right], \tag{7}$$

where $y^{tot} = r + \gamma \max_{\boldsymbol{u}'} Q_{tot}(\boldsymbol{\tau}', \boldsymbol{u}', s'; \theta^-)$ and θ^- represents the parameters of the target network. In SORA, the TD loss is combined with two additional losses introduced above to form the overall loss function:

$$\mathcal{L}_{\mathrm{TD}}(\theta) = \mathcal{L}_{\mathrm{TD}}(\theta) + \lambda_i \mathcal{L}_i(\theta_r) + \lambda_s \mathcal{L}_s(\theta_r, \theta_t), \tag{8}$$

where λ_i and λ_s are hyperparameters that adjust the importance of their corresponding losses.

4 Experiments

4.1 Environments and Settings

To objectively evaluate the effectiveness of SORA, we adopt two popular cooperative multi-agent environments, StarCraft Multi-Agent Challenge (SMAC) [13] and Google Research Football (GRF) [7], as testbeds. SMAC offers a diversity of micro-war scenarios in StarCraft II, which requires algorithms to perform well both in single-unit control and coordination. Every challenge in SMAC can be classified into one of three groups: *easy*, *hard*, and *super hard*, according to the difficulty level of winning this challenge. Notably, the difficulty of the enemy AI is set to 7 (the highest, referring to *very hard*), and the version of StarCraft II used in our experiments is 4.6.2, from which the results are not always comparable with those from other versions. GRF is a simulation environment developed by Google Research and has been well received by the reinforcement learning community. It provides a user-friendly platform for training and evaluating agents in various areas, such as control, planning, and multi-agent cooperation.

We compare SORA with several state-of-the-art algorithms, including the most popular baseline QMIX [12] and three recent role-based methods ROMA [21], RODE [22], and LDSA [23]. Our implementation is based on the framework of Pymarl [13] with default hyperparameters. For the sake of fairness, the number of roles in SORA is set to 4 in both SMAC and GRF scenarios to maintain consistency with RODE and LDSA.

In our experiments, we test each algorithm on 5 different random seeds in order to reduce the randomness and enhance the reliability of the results. The number of iterations is set to 2 million or 5 million in SMAC, and 4 million in GRF, based on the characteristics of distinct scenarios.

4.2 Performances

As the SOTA methods already reach extremely high win ratios on the *easy* challenges in SMAC, we only select *hard* and *super hard* challenges to evaluate the performance of SORA. The win ratio curves of the chosen SMAC and GRF scenarios are depicted in Fig. 2, where the solid lines represent the median of win rates and the shaded areas are the 25–75% win percentage range. Detailed results of the mean and standard deviation are displayed in Table 1. The highest win rate and lowest deviation among all methods in every scenario are highlighted in italic, and the best among the role-based methods are marked in bold.

It can be clearly seen that our proposed method SORA shows the best overall performance. Compared to the baselines, SORA achieves nearly SOTA win rates in all the selected scenarios. The superiority of SORA is particularly evident in *2c_vs_64zg*, *5m_vs_6m*, *27m_vs_30m* and *corner* (Fig. 2(b), (d), (f), and (h)), with both prominent higher win rates and earlier starts. We believe the reason is that the optimized role assignment process accelerates the learning of cooperative behavior in agents. It is also noteworthy that the results of SORA have lower variances, which demonstrate the more stable performance of SORA than those of the baselines.

4.3 Ablations

We conducted ablation experiments to address the following questions: 1. Does the superior performance of SORA come from the added parameters in the Q networks? 2. Is the role assignment process necessary in SORA? 3. How does

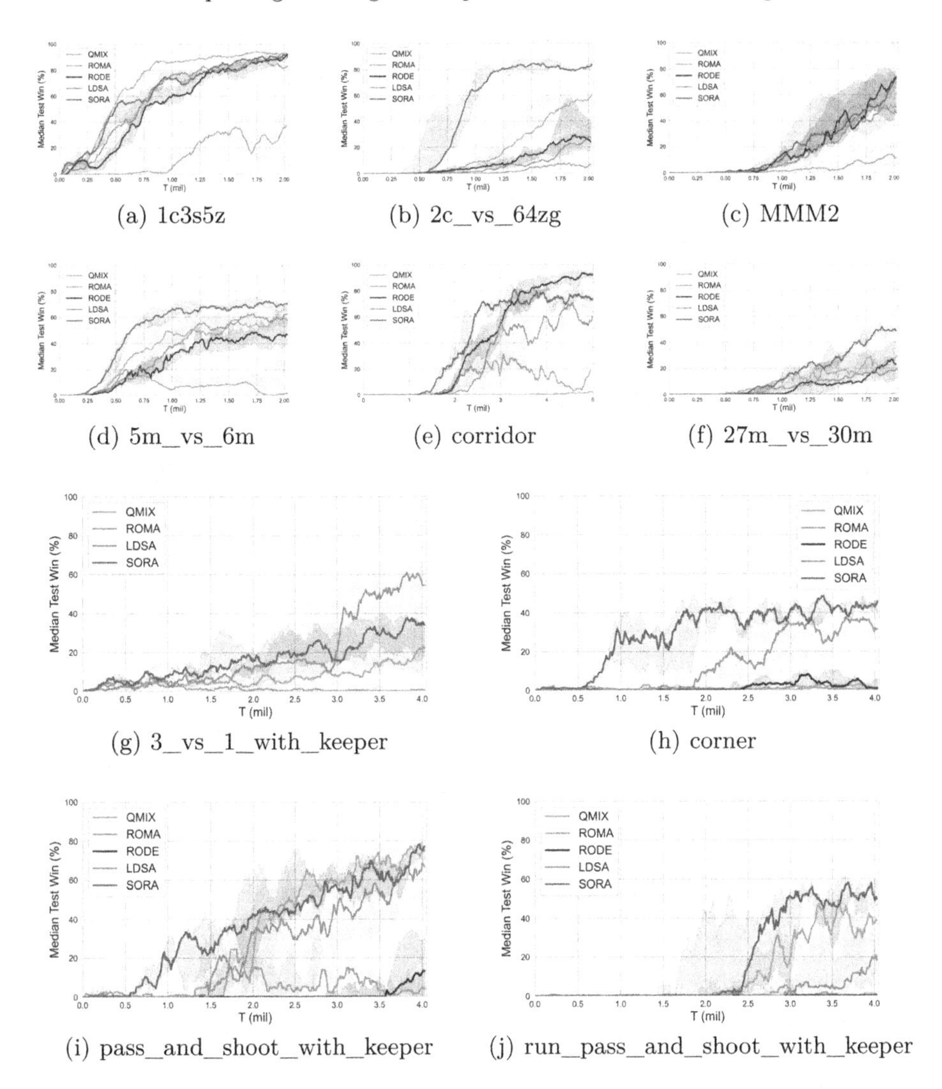

(a) 1c3s5z (b) 2c_vs_64zg (c) MMM2

(d) 5m_vs_6m (e) corridor (f) 27m_vs_30m

(g) 3_vs_1_with_keeper (h) corner

(i) pass_and_shoot_with_keeper (j) run_pass_and_shoot_with_keeper

Fig. 2. Overall results in SMAC and GRF.

the number of roles affect SORA's performance? 4. Whether the two additional regularizers help improve the performance of SORA?

To answer the first question, we manually expand the scale of the Q network in QMIX to ensure it has similar parameter counts with SORA, and we call this version of QMIX as QMIX_Large. It can be clearly seen from Fig. 3 that SORA achieves significantly better results than QMIX_Large, which strongly demonstrates the conclusion that the superiority of SORA is not due to the added parameters but rather to the well-designed role assignment mechanism.

Table 1. Mean and standard deviation (represented as a percentage with respect to the mean value) of the test win rates in different scenarios.

	QMIX	ROMA	RODE	LDSA	SORA (Ours)
MMM2	62.625 ± 31.56%	14.875 ± 77.15%	63.28 ± 28.01%	49.92 ± 24.34%	*71.29 ± 12.78%*
2c_vs_64zg	53.58 ± 34.28%	5.875 ± 66.96%	32.33 ± 61.49%	32.99 ± 37.34%	*84.29 ± 5.80%*
27m_vs_30m	25.42 ± 48.43%	0 ± NaN	27.75 ± 73.62%	20.42 ± 51.05%	*48.5 ± 9.94%*
1c3s5z	*93.13 ± 2.29%*	40.25 ± 57.98%	**91.46 ± 2.56%**	87.08 ± 7.53%	90.08 ± 5.52%
5m_vs_6m	60.71 ± 14.28%	16.92 ± 119.53%	49.17 ± 31.37%	58.71 ± 18.52%	*68.87 ± 11.16%*
3s5z_vs_3s6z	*36.42* ± 74.92%	7.96 ± 119.78%	15.10 ± 122.19%	16.58 ± 121.25%	**21.65** ± *73.63%*
corridor	19.0 ± 180.88%	21.25 ± 74.00%	*92.14 ± 3.82%*	56.33 ± 54.46%	73.42 ± 6.97%
academy_3_vs_1 _with_keeper	26.71 ± 72.02%	2.17 ± 158.65%	/	*45.08* ± 65.64%	34.50 ± *51.76%*
academy _corner	3.33 ± 109.90%	1.04 ± 59.33%	3.21 ± 104.19%	33.54 ± 49.05%	*44.33 ± 9.57%*
academy_pass _and_shoot _with_keeper	*75.21 ± 9.65%*	8.375 ± 112.42%	21.33 ± 111.06%	52.79 ± 53.13%	**71.75 ± 21.57%**
academy_run_ pass_and_shoot _with_keeper	29.0 ± 87.01%	0.33 ± 63.74%	14.96 ± 194.48%	23.79 ± 72.91%	*51.625 ± 50.03%*

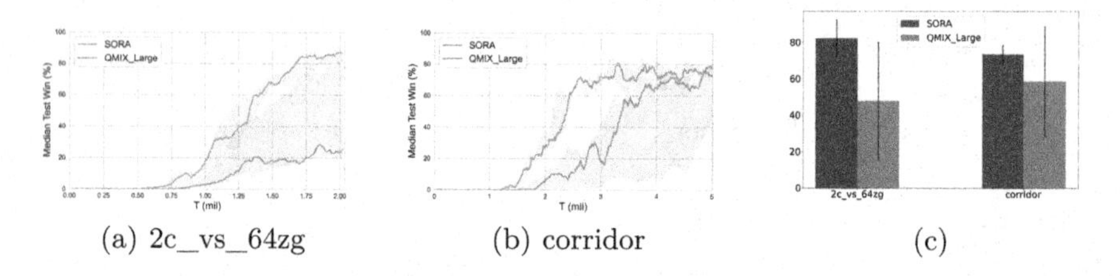

(a) 2c_vs_64zg (b) corridor (c)

Fig. 3. Ablation study of larger agent network.

For the second question, we replace the attention-based role distribution of each agent in SORA with a randomly generated distribution (represented as SORA_RandR), in order to demonstrate the significance of the role assignment process. From Fig. 4 we can clearly see that the random role leads to a decrease in winning percentage and an increase in variance compared to the original implementation.

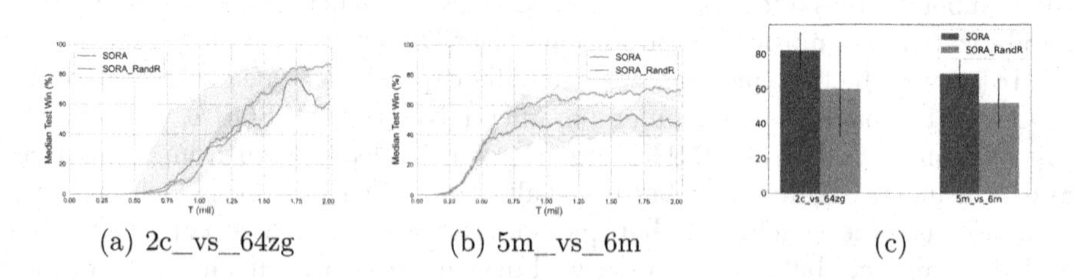

(a) 2c_vs_64zg (b) 5m_vs_6m (c)

Fig. 4. Ablation study of randomly choosing roles.

An empirical viewpoint on the third question is that the performance of SORA boosts as the number of roles increases due to the additional parameters. However, our experiments yield different conclusions. A range of settings was tested, and the results are presented in Fig. 5. Different settings are distinguished by the suffix (e.g., the number of roles is set to 6 in SORA_6R). No significant correlation between the performance and the set role numbers is observed, and the key is to choose appropriate settings for specific tasks. To some extent, this ablation experiment addresses the first question as well that more parameters do not necessarily lead to improved performance.

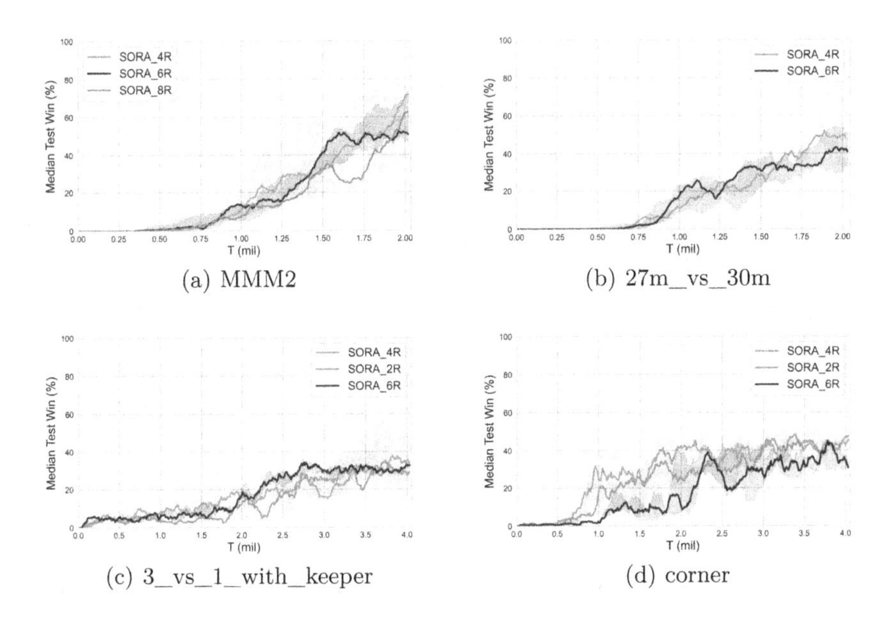

Fig. 5. Ablation study of randomly choosing roles.

Finally, we examine the contribution of the two regularizers in SORA to answer the fourth question. We first remove the regularizer $\mathcal{L}_i$ which facilitates the distinctiveness of each role representation x_{ϕ_k} and denote this version as SORA_NoI. Second, the regularizer $\mathcal{L}_s$ is omitted, and we represent this version as SORA_NoS. As shown in Fig. 6, SORA_NoI has slightly lower win rates than SORA, which demonstrates the benefit of distinct role representations. SORA_NoS performs significantly worse than SORA, and the results are more unstable on different random seeds and exhibit higher variances, indicating the importance of controlling the smoothness of an agent's role transition between two consecutive time steps.

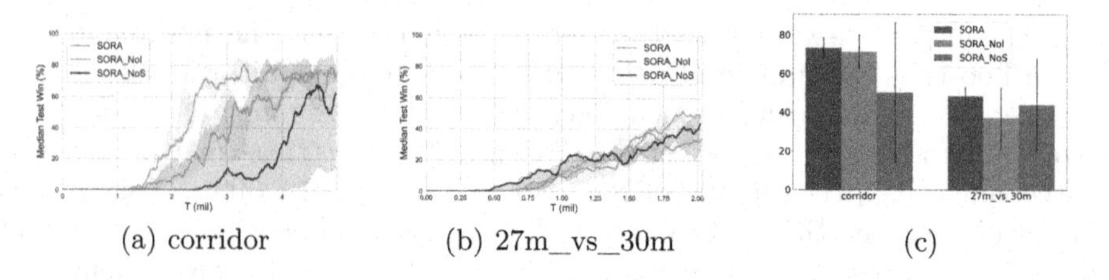

(a) corridor (b) 27m_vs_30m (c)

Fig. 6. Ablation study of the regularizers.

5 Conclusion

This paper proposes an innovative SOft Role Assignment (SORA) process, which improves conventional role-based multi-agent reinforcement learning approaches at the mechanism level. Specifically, SORA allows an agent to perform multiple responsibilities rather than choosing only one from the given roles, enabling agents to attain flexible strategies and handle different situations. We compared SORA with several state-of-the-art methods on Starcraft multi-agent challenges and Google Research Football to evaluate its performance, as well as we conducted ablation experiments for further investigation. The results have promisingly demonstrated the superiority of our proposed method.

Acknowledgment. This work was supported by the Strategic Priority Research Program of the Chinese Academy of Science, Grant No. XDA27050100.

References

1. Berner, C., et al.: Dota 2 with large scale deep reinforcement learning. arXiv preprint arXiv:1912.06680 (2019)
2. Capitan, J., Spaan, M.T., Merino, L., Ollero, A.: Decentralized multi-robot cooperation with auctioned POMDPs. Int. J. Robot. Res. **32**(6), 650–671 (2013)
3. Chung, J., Gulcehre, C., Cho, K., Bengio, Y.: Empirical evaluation of gated recurrent neural networks on sequence modeling. arXiv preprint arXiv:1412.3555 (2014)
4. Ha, D., Dai, A., Le, Q.V.: Hypernetworks. arXiv preprint arXiv:1609.09106 (2016)
5. Hausknecht, M., Stone, P.: Deep recurrent Q-learning for partially observable MDPs. In: 2015 AAAI Fall Symposium Series (2015)
6. Hu, S., Xie, C., Liang, X., Chang, X.: Policy diagnosis via measuring role diversity in cooperative multi-agent RL. In: International Conference on Machine Learning, pp. 9041–9071. PMLR (2022)
7. Kurach, K., et al.: Google research football: a novel reinforcement learning environment (2020)
8. Lowe, R., et al.: Multi-agent actor-critic for mixed cooperative-competitive environments. In: Advances in Neural Information Processing Systems, vol. 30 (2017)
9. Mahajan, A., Rashid, T., Samvelyan, M., Whiteson, S.: MAVEN: multi-agent variational exploration. In: Advances in Neural Information Processing Systems, vol. 32 (2019)

10. Oliehoek, F.A., Amato, C.: A Concise Introduction to Decentralized POMDPs. Springer, Cham (2016). https://doi.org/10.1007/978-3-319-28929-8
11. Rashid, T., Farquhar, G., Peng, B., Whiteson, S.: Weighted QMIX: expanding monotonic value function factorisation for deep multi-agent reinforcement learning. In: Advances in Neural Information Processing Systems, vol. 33, pp. 10199–10210 (2020)
12. Rashid, T., Samvelyan, M., Schroeder, C., Farquhar, G., Foerster, J., Whiteson, S.: QMIX: monotonic value function factorisation for deep multi-agent reinforcement learning. In: International Conference on Machine Learning, pp. 4295–4304. PMLR (2018)
13. Samvelyan, M., et al.: The StarCraft multi-agent challenge. arXiv preprint arXiv: 1902.04043 (2019)
14. Shamsoshoara, A., Khaledi, M., Afghah, F., Razi, A., Ashdown, J.: Distributed cooperative spectrum sharing in uav networks using multi-agent reinforcement learning. In: 2019 16th IEEE Annual Consumer Communications & Networking Conference (CCNC), pp. 1–6. IEEE (2019)
15. Son, K., Kim, D., Kang, W.J., Hostallero, D., Yi, Y.: QTRAN: learning to factorize with transformation for cooperative multi-agent reinforcement learning. CoRR abs/1905.05408 (2019). http://arxiv.org/abs/1905.05408
16. Sunehag, P., et al.: Value-decomposition networks for cooperative multi-agent learning. arXiv preprint arXiv:1706.05296 (2017)
17. Tampuu, A., et al.: Multiagent cooperation and competition with deep reinforcement learning. PLOS ONE **12**(4), e0172395 (2017)
18. Vaswani, A., et al.: Attention is all you need. In: Advances in Neural Information Processing Systems, vol. 30 (2017)
19. Vinyals, O., et al.: Grandmaster level in StarCraft II using multi-agent reinforcement learning. Nature **575**(7782), 350–354 (2019)
20. Wang, J., Ren, Z., Liu, T., Yu, Y., Zhang, C.: QPLEX: duplex dueling multi-agent Q-learning. arXiv preprint arXiv:2008.01062 (2020)
21. Wang, T., Dong, H., Lesser, V., Zhang, C.: ROMA: multi-agent reinforcement learning with emergent roles. arXiv preprint arXiv:2003.08039 (2020)
22. Wang, T., Gupta, T., Mahajan, A., Peng, B., Whiteson, S., Zhang, C.: RODE: learning roles to decompose multi-agent tasks. arXiv preprint arXiv:2010.01523 (2020)
23. Yang, M., Zhao, J., Hu, X., Zhou, W., Zhu, J., Li, H.: LDSA: learning dynamic subtask assignment in cooperative multi-agent reinforcement learning. In: Advances in Neural Information Processing Systems, vol. 35, pp. 1698–1710 (2022)
24. Yang, Y., et al.: Qatten: a general framework for cooperative multiagent reinforcement learning. arXiv preprint arXiv:2002.03939 (2020)
25. Ye, D., et al.: Towards playing full MOBA games with deep reinforcement learning. In: Advances in Neural Information Processing Systems, vol. 33, pp. 621–632 (2020)
26. Zeng, X., Peng, H., Li, A.: Effective and stable role-based multi-agent collaboration by structural information principles. arXiv preprint arXiv:2304.00755 (2023)

Outer Synchronization for Multi-derivative Coupled Complex Networks with and without External Disturbance

Han-Yu Wu[1] and Qingshan Liu[2(✉)]

[1] School of Cyber Science and Engineering, Southeast University, Nanjing 210096, China
wuhanyu@seu.edu.cn

[2] School of Mathematics, Southeast University, Nanjing 210096, China
qsliu@seu.edu.cn

Abstract. This paper investigates the outer synchronization of multi-derivative coupled complex networks (MDCCNs), and further studies the outer H_∞ synchronization between two MDCCNs with external disturbance. For the outer synchronization, a synchronization criterion is proposed by using adaptive control strategy, which is proved based on Lyapunov functional and the Barbalat's lemma. For the outer H_∞ synchronization, an adaptive state controller and parameter updating scheme are devised for MDCCNs with external disturbance. Finally, the validity of the presented criteria is demonstrated by providing two simulation examples.

Keywords: External disturbance · Multi-derivative coupled complex networks · Outer synchronization · Outer H_∞ synchronization

1 Introduction

In recent years, the synchronization of complex networks (CNs) has aroused increasing attention, and many synchronization criteria have been derived for multiple kinds of CNs. [1–4]. In [1], the authors introduced a type of stochastic delayed CNs, and derived a synchronization criterion by exploiting stochastic analysis approach. In [2], the authors considered synchronization for CNs with time-varying delay, and proposed some sufficient conditions by leveraging on linear matrix inequalities and Lyapunov theory. In [4], the authors investigated exponential synchronization for CNs by employing periodically intermittent noise and feedback control.

Nevertheless, the great majority of existing researches concentrate on the synchronization for CNs with state coupling [1,2,4–6]. Actually, the derivative

This work was partially supported by the National Natural Science Foundation of China under Grants 62276062.

B. Luo et al. (Eds.): ICONIP 2023, LNCS 14447, pp. 332–343, 2024.
https://doi.org/10.1007/978-981-99-8079-6_26

coupling is another crucial coupling form in CNs [7–9], which reflects the influence of state change of other nodes. In [9], the authors dealt with the exponential synchronization of derivative coupled CNs with proportional delay, and put forward some sufficient criteria by means of pinning control, impulsive control and comparison principle. Recently, considering the diversity of different influencing factors, synchronization for multi-derivative coupled CNs (MDCCNs) was explored [10,11]. In [11], the authors gave two classes of MDCCNs with positive definite or positive semidefinite matrix, and acquired several lag output synchronization criteria by designing output feedback and adaptive output feedback controllers.

Moreover, the external disturbance commonly exists in CNs, which may destroy or even break the synchronization. Therefore, many researchers have tackled the H_∞ synchronization problem of CNs [12–15]. Wang et al. utilized adaptive control and integral inequality techniques to investigate the H_∞ synchronization of directed CNs with time delays [12]. The H_∞ synchronization for a type of switched CNs was studied by utilizing switching impulsive controller consisting of impulsive control and state-feedbacks parts [13]. Fan and Ma focused on the H_∞ synchronization of Takagi-Sugeno CNs, and analyzed the influence of hybrid delay via pinning control strategy [14].

However, the synchronization among different nodes in a single CNs is considered in many existing papers. Noted that the synchronization between two CNs universally exists in real world, which is called outer synchronization [16,17]. In fact, various phenomena show the performance of outer synchronization such as infectious disease spreading between two communities, and the balance of different species. Consequently, it is significant and interesting to study the outer synchronization between two CNs. In [16], the authors considered drive-response coupled CNs with different nodes, and proposed several criteria for insuring asymptotic outer synchronization and exponential synchronization between two CNs. Particularly, the outer synchronization between two derivative coupled CNs has also gained growing attention in recent years [8,18]. By [18], the CNs without and with derivative coupling were introduced, and several outer synchronization conditions between two given CNs were established by combining impulsive and pinning control. However, the outer synchronization between MDCCNs with and without external disturbance has not yet been studied.

Inspired by the above analysis, this paper investigates the outer synchronization and H_∞ synchronization for MDCCNs without and with external disturbance. The contributions of this work are summarized as follows. First, two types of MDCCNs without and with external disturbance are presented, which generalize the traditional CNs models to some extent. Second, the multi-derivative couplings considered in this paper describes the dynamics of CNs more accurate than the derivative coupling in [8,18]. Third, the convergence rate of network synchronization error is compared under the cases of the derivative coupling and the multi-derivative couplings. Finally, we further study the outer H_∞ synchronization since the presence of external disturbance may destroy the network synchronization.

2 Outer Synchronization for MDCCNs

In this section, the outer synchronization for MDCCNs is studied by employing Lyapunov functional and the Barbalat's lemma. An adaptive controller and a parameter updating strategy are employed to ensure the outer synchronization between two MDCCNs.

2.1 Network Model

The MDCCNs discussed in this paper is represented by:

$$\dot{m}_k(t) = h(m_k(t)) + \sum_{l=1}^{\mu}\sum_{z=1}^{\xi} c_l a_{kz}^l \Gamma^l \dot{m}_z(t), \tag{1}$$

where $k = 1, 2, \ldots, \xi$; $m_k(t) = (m_{k1}(t), m_{k2}(t), \ldots, m_{kn}(t))^T \in \mathbb{R}^n$ represents the kth node state; $h(m_k(t)) = (h_1(m_{k1}(t)), h_2(m_{k2}(t)), \ldots, h_n(m_{kn}(t)))^T \in \mathbb{R}^n$; $0 < c_l \in \mathbb{R}$, $A^l = (a_{kz}^l)_{\xi\times\xi} \in \mathbb{R}^{\xi\times\xi}$ and $0 < \Gamma^l \in \mathbb{R}^{n\times n}$ respectively signify coupling strength, outer coupling matrix and inner coupling matrix of the lth coupling form. Moreover, $a_{kz}^l \in \mathbb{R}$ is defined as:

$$a_{kz}^l = \begin{cases} a_{zk}^l > 0, & \text{if } (k, z) \in \varsigma, \\ -\sum\limits_{\substack{i=1\\ i\neq k}}^{N} a_{ki}^l, & \text{if } z = k, \\ 0, & \text{otherwise}, \end{cases}$$

where $\varsigma \subset \{1, 2, \ldots, \xi\} \times \{1, 2, \ldots, \xi\}$ is the undirected edge set corresponding the lth coupling form in the network (1).

The function $h_r(\cdot)(r = 1, 2, \ldots, n)$ fulfills

$$|h_r(\varrho_1) - h_r(\varrho_2)| \leqslant \phi_r|\varrho_1 - \varrho_2|, \ \forall \varrho_1, \varrho_2 \in \mathbb{R},$$

where $\phi_r > 0$, $|\cdot|$ represents absolute value. We set $\Phi = \text{diag}(\phi_1^2, \phi_2^2, \ldots, \phi_n^2)$.

Remark 1. The state coupling and derivative coupling are two kinds of important forms of coupling in CNs. For the former, many authors have carried out extensive research [1,2,4]. On one hand, owing to various derivatives of node state may lead to different changes in other nodes, some investigators have preliminarily explored the synchronization for derivative coupled CNs [8,18]. On the other hand, considering the diversity of different influencing factors in CNs, the MDCCNs model was considered [10,11]. However, the synchronization among nodes rather than networks was discussed in [10,11]. Therefore, this paper further consider the outer synchronization between two MDCCNs.

In this paper, we take (1) as the drive network, the response MDCNNs is described as follows:

$$\dot{w}_k(t) = h(w_k(t)) + \sum_{l=1}^{\mu}\sum_{z=1}^{\xi} c_l b_{kz}^l(t)\Gamma^l \dot{w}_z(t) + v_k(t), \tag{2}$$

where $k = 1, 2, \ldots, \xi$; $w_k(t) = (w_{k1}(t), w_{k2}(t), \ldots, w_{kn}(t))^T \in \mathbb{R}^n$; $v_k(t) \in \mathbb{R}^n$ represents the control input; $B^l(t) = (b_{kz}^l(t))_{\xi\times\xi} \in \mathbb{R}^{\xi\times\xi}$ means outer coupling matrix in the lth coupling form.

Noted that the network (2) has the same topology structure as the network (1) under different coupling forms. Furthermore, $b_{kz}^l(0)$ fulfills the same requirement as a_{kz}^l. But, $b_{kz}^l(t)$ does not need to satisfy the similar definition of a_{kz}^l for $t > 0$.

Definition 1. *The outer synchronization can be realized between the drive network (1) and response network (2) if*

$$\lim_{t\to+\infty} \|w_k(t) - m_k(t)\| = 0, k = 1, 2, \ldots, \xi,$$

where $\|\cdot\|$ *is the Euclidean norm.*

Lemma 1 *(Barbalat's lemma [19]). If a differentiable function $\zeta(t)$ has a finite limit as $t \to +\infty$ and $\dot{\zeta}(t)$ is uniformly continuous, then $\dot{\zeta}(t) \to 0$ as $t \to +\infty$.*

2.2 Synchronization Criterion for MDCCNs

Denote $d_k(t) = w_k(t) - m_k(t) \in \mathbb{R}^n$. By (1) and (2), one gets

$$\begin{aligned}\dot{d}_k(t) =& h(w_k(t)) - h(m_k(t)) + \sum_{l=1}^{\mu}\sum_{z=1}^{\xi} c_l b_{kz}^l(t)\Gamma^l \dot{w}_z(t) \\ &- \sum_{l=1}^{\mu}\sum_{z=1}^{\xi} c_l a_{kz}^l \Gamma^l \dot{m}_z(t) + v_k(t), k = 1, 2, \ldots, \xi.\end{aligned} \tag{3}$$

For ensuring the outer synchronization between networks (1) and (2), we propose the following adaptive controller

$$v_k(t) = -q_k(t)d_k(t), \quad \dot{q}_k(t) = s_k d_k^T(t)d_k(t), \tag{4}$$

and parameter updating strategy

$$\dot{b}_{kz}^l(t) = -\theta d_k^T(t)\Gamma^l \dot{w}_z(t), \tag{5}$$

where $k, z = 1, 2, \ldots, \xi$, $l = 1, 2, \ldots, \mu$, $\theta > 0$, $s_k > 0$ and $q_k(0) > 0$.

Theorem 1. *Combining the adaptive controller (4) and parameter updating strategy (5), the networks (1) and (2) can realize outer synchronization.*

Proof: The Lyapunov function for the network (3) is defined by:

$$\begin{aligned} V(t) =& \sum_{k=1}^{\xi} d_k^T(t)d_k(t) + \sum_{k=1}^{\xi} \frac{1}{s_k}(q_k(t) - q^*)^2 \\ &+ \frac{1}{\theta} \sum_{l=1}^{\mu} \sum_{k=1}^{\xi} \sum_{z=1}^{\xi} c_l (b_{kz}^l(t) - a_{kz}^l)^2 \\ &- \sum_{l=1}^{\mu} c_l d^T(t)(A^l \otimes \Gamma^l)d(t), \end{aligned}$$

where $d(t) = (d_1^T(t), d_2^T(t), \ldots, d_\xi^T(t))^T$, q^* is a constant, $\otimes$ represents Kronecker product and $A^l \otimes \Gamma^l \leqslant 0$.

Then, one derives

$$\begin{aligned} \dot{V}(t) =& 2\sum_{k=1}^{\xi} d_k^T(t)\Big(h(w_k(t)) - h(m_k(t)) + \sum_{l=1}^{\mu}\sum_{z=1}^{\xi} c_l b_{kz}^l(t)\Gamma^l \dot{w}_z(t) \\ &- \sum_{l=1}^{\mu}\sum_{z=1}^{\xi} c_l a_{kz}^l \Gamma^l \dot{m}_z(t) - q_k(t)d_k(t)\Big) + 2\sum_{k=1}^{\xi}(q_k(t) - q^*)d_k^T(t)d_k(t) \\ &- 2\sum_{l=1}^{\mu}\sum_{k=1}^{\xi}\sum_{z=1}^{\xi} c_l(b_{kz}^l(t) - a_{kz}^l)d_k^T(t)\Gamma^l \dot{w}_z(t) - 2\sum_{l=1}^{\mu} c_l d^T(t)(A^l \otimes \Gamma^l)\dot{d}(t) \\ =& 2\sum_{k=1}^{\xi} d_k^T(t)\Big(h(w_k(t)) - h(m_k(t)\Big) \\ &+ 2\sum_{l=1}^{\mu}\sum_{k=1}^{\xi}\sum_{z=1}^{\xi} c_l a_{kz}^l d_k^T(t)\Gamma^l(\dot{w}_z(t) - \dot{m}_z(t)) \\ &- 2\sum_{l=1}^{\mu} c_l d^T(t)(A^l \otimes \Gamma^l)\dot{d}(t) - 2q^*\sum_{k=1}^{\xi} d_k^T(t)d_k(t) \\ =& 2\sum_{k=1}^{\xi} d_k^T(t)\Big(h(w_k(t)) - h(m_k(t))\Big) - 2q^*\sum_{k=1}^{\xi} d_k^T(t)d_k(t). \end{aligned} \tag{6}$$

In addition, one has

$$2\sum_{k=1}^{\xi} d_k^T(t)(h(w_k(t)) - h(m_k(t))) \leqslant \sum_{k=1}^{\xi} d_k^T(t)\Phi d_k(t). \tag{7}$$

From (6) and (7), one obtains

$$\dot{V}(t) \leqslant (\phi_r^2 - 2q^*)d^T(t)d(t) \leqslant (\phi_{max}^2 - 2q^*)d^T(t)d(t),$$

where $\phi_{max} = \max\{\phi_1, \phi_2, \ldots, \phi_n\}$.

Taking $q^* \geqslant (\phi_{max}^2 + \eta)/2, \eta > 0$, we have

$$\dot{V}(t) \leqslant -\eta d^T(t)d(t). \tag{8}$$

From $V(t)$ and (8), one derives that $V(t)$ is bounded and has a finite limit as $t \rightarrow +\infty$. Then, one concludes that $d(t), q_k(t)$ and $b_{kz}^l(t)$ are all bounded.

Furthermore, from (8) we have

$$\eta \lim_{t\rightarrow+\infty} \int_0^t d^T(\beta)d(\beta)d\beta \leqslant -\lim_{t\rightarrow+\infty} \int_0^t \dot{V}(\beta)d\beta = V(0) - \lim_{t\rightarrow+\infty} V(t).$$

As a consequence, $\int_0^t d^T(\beta)d(\beta)d\beta$ has a finite limit as $t \rightarrow +\infty$. In the light of (3)–(5) and the requirement of $h_k(\cdot)$, $\dot{d}(t)$ is bounded. In other words, $d^T(t)d(t)$ is also bounded. Hence, $d^T(t)d(t)$ is uniformly continuous. Based on Lemma 1, one has $\lim_{t\rightarrow+\infty} d^T(t)d(t) = 0$. Namely, $\lim_{t\rightarrow+\infty} \|d_k(t)\| = \lim_{t\rightarrow+\infty} \|w_k(t) - m_k(t)\| = 0$.

Therefore, the outer synchronization between networks (1) and (2) can be realized. □

3 Outer H_∞ Synchronization for MDCCNs with External Disturbance

In this section, the outer H_∞ synchronization for MDCCNs with external disturbance is further discussed. A sufficient condition is acquired based on the definition of H_∞ synchronization, inequality technique and adaptive control.

3.1 Network Model

The MDCCNs with external disturbance is given by:

$$\dot{m}_k(t) = h(m_k(t)) + \sum_{l=1}^{\mu}\sum_{z=1}^{\xi} c_l a_{kz}^l \Gamma^l \dot{m}_z(t) + \varepsilon_k(t), \tag{9}$$

where $k = 1, 2, \ldots, \xi, m_k(t), h(m_k(t)), c_l, a_{kz}^l$ and Γ^l have the same definitions as those in the network (1); $\varepsilon_k(t) = (\varepsilon_{k1}(t), \varepsilon_{k2}(t), \ldots, \varepsilon_{kn}(t))^T \in \mathbb{R}^n$ represents external disturbance, which satisfies

$$\int_0^t \varepsilon_r^T(t)\varepsilon_r(t)dt < +\infty, \ \forall t \geqslant 0.$$

Remark 2. In [8,18], the authors have studied the outer synchronization between two derivative coupled CNs. As a matter of fact, external disturbance [13–15] generally exists in CNs, which may result in the fail of network synchronization. Consequently, it is necessary and interesting to consider the outer H_∞ synchronization between two CNs with external disturbance.

Similar to Sect. 2, we take (9) as the drive network and consider the following response network:

$$\dot{w}_k(t) = h(w_k(t)) + \sum_{l=1}^{\mu}\sum_{z=1}^{\xi} c_l b_{kz}^l(t)\Gamma^l \dot{w}_z(t) + v_k(t), \tag{10}$$

where $k = 1, 2, \ldots, \xi, w_k(t), b_{kz}^l$ and $v_k(t)$ take the same definitions as those in network (2).

Definition 2. *We say that the drive network (9) and response network (10) reach the H_∞ synchronization if*

$$\int_0^{t_j} d^T(t)d(t)dt \leqslant V(0) + \delta^2 \int_0^{t_j} \varepsilon^T(t)\varepsilon(t)dt$$

holds for $t_j \geqslant 0$ and some nonegative function $V(\cdot)$ with $\delta > 0$.

3.2 H_∞ Synchronization Criterion for MDCCNs with External Disturbance

Denote $d_k(t) = w_k(t) - m_k(t) \in \mathbb{R}^n$. By (9) and (10), one has

$$\begin{aligned}\dot{d}_k(t) =& h(w_k(t)) - h(m_k(t)) + \sum_{l=1}^{\mu}\sum_{z=1}^{\xi} c_l b_{kz}^l(t)\Gamma^l \dot{w}_z(t) \\ &- \sum_{l=1}^{\mu}\sum_{z=1}^{\xi} c_l a_{kz}^l \Gamma^l \dot{m}_z(t) + v_k(t) - \varepsilon_k(t).\end{aligned} \tag{11}$$

Theorem 2. *Combining the adaptive controller (4) and parameter updating strategy (5), the networks (9) and (10) can reach outer H_∞ synchronization.*

Proof: Taking the same Lyapunov function $V(t)$ as that in (7) for the network (11), one gets

$$\begin{aligned}\dot{V}(t) =& 2\sum_{k=1}^{\xi} d_k^T(t)\Big(h(w_k(t)) - h(m_k(t)) + \sum_{l=1}^{\mu}\sum_{z=1}^{\xi} c_l b_{kz}^l(t)\Gamma^l \dot{w}_z(t) - \varepsilon_k(t) \\ &- \sum_{l=1}^{\mu}\sum_{z=1}^{\xi} c_l a_{kz}^l \Gamma^l \dot{m}_z(t) - q_k(t)d_k(t)\Big) + 2\sum_{k=1}^{\xi}(q_k(t) - q^*)d_k^T(t)d_k(t) \\ &- 2\sum_{l=1}^{\mu}\sum_{k=1}^{\xi}\sum_{z=1}^{\xi} c_l(b_{kz}^l(t) - a_{kz}^l)d_k^T(t)\Gamma^l \dot{w}_z(t) \\ &- 2\sum_{l=1}^{\mu} c_l d^T(t)(A^l \otimes \Gamma^l)\dot{d}(t) - 2d^T(t)\varepsilon(t) \\ \leqslant& d^T(t)(I_\xi \otimes \Phi - 2q^* I_{\xi n})d(t) - 2d^T(t)\varepsilon(t) \\ \leqslant& (\phi_{max}^2 - 2q^*)d^T(t)d(t) - 2d^T(t)\varepsilon(t).\end{aligned} \tag{12}$$

According to (12), we have

$$\begin{aligned}
&\dot{V}(t) + d^T(t)d(t) - \delta^2\varepsilon^T(t)\varepsilon(t) \\
&\leqslant(\phi_{max}^2 - 2q^*)d^T(t)d(t) - 2d^T(t)\varepsilon(t) + d^T(t)d(t) - \delta^2\varepsilon^T(t)\varepsilon(t) \\
&= [\phi_{max}^2 - 2(q^* - 1)]d^T(t)d(t) - (d(t) + \delta\varepsilon(t))^T(d(t) + \delta\varepsilon(t)).
\end{aligned}$$

Taking $q^* \geqslant (\phi_{max}^2 + 2)/2$, one has

$$d^T(t)d(t) \leqslant -\dot{V}(t) - \delta^2\varepsilon^T(t)\varepsilon(t). \tag{13}$$

From (13), one obtains

$$\begin{aligned}
\int_0^{t_j} d^T(t)d(t)dt \leqslant& V(0) - V(t_j) + \delta^2\int_0^{t_j}\varepsilon^T(t)\varepsilon(t)dt \\
\leqslant& V(0) + \delta^2\int_0^{t_j}\varepsilon^T(t)\varepsilon(t)dt
\end{aligned} \tag{14}$$

for any $t_j \geqslant 0$.

Therefore, the outer H_∞ synchronization between two MDCCNs with external disturbance can be achieved. □

4 Numerical Examples

In this section, two examples are provided to substantiate the validity of the presented outer synchronization and H_∞ synchronization criteria.

Example 1. Consider the following MDCCNs:

$$\begin{aligned}
\dot{m}_k(t) =& h(m_k(t)) + 0.1\sum_{z=1}^{5} a_{kz}^1\Gamma^1\dot{m}_z(t) \\
&+ 0.1\sum_{z=1}^{5} a_{kz}^2\Gamma^2\dot{m}_z(t) + 0.2\sum_{z=1}^{5} a_{kz}^3\Gamma^3\dot{m}_z(t),
\end{aligned} \tag{15}$$

where $k = 1, 2, \ldots, 5$, $h_f(s) = 0.125(|s+1|-|s-1|)$, $f = 1, 2, 3$, $\Gamma^1 = \text{diag}(0.2, 0.3, 0.4)$, $\Gamma^2 = \text{diag}(0.1, 0.4, 0.5)$,

$$\Gamma^3 = \begin{pmatrix} 0.2 & 0 & 0.1 \\ 0 & 0.3 & 0 \\ 0 & 0 & 0.1 \end{pmatrix},\ A^1 = \begin{pmatrix} -0.3 & 0 & 0.2 & 0.1 & 0 \\ 0 & -0.4 & 0.4 & 0 & 0 \\ 0.2 & 0.4 & -0.6 & 0 & 0 \\ 0.1 & 0 & 0 & -0.4 & 0.3 \\ 0 & 0 & 0 & 0.3 & -0.3 \end{pmatrix},$$

$$A^2 = \begin{pmatrix} -0.5 & 0 & 0.2 & 0.3 & 0 \\ 0 & -0.7 & 0.7 & 0 & 0 \\ 0.2 & 0.7 & -0.9 & 0 & 0 \\ 0.3 & 0 & 0 & -0.5 & 0.2 \\ 0 & 0 & 0 & 0.2 & -0.2 \end{pmatrix}, A^3 = \begin{pmatrix} -0.5 & 0 & 0.4 & 0.1 & 0 \\ 0 & -0.2 & 0.2 & 0 & 0 \\ 0.4 & 0.2 & -0.6 & 0 & 0 \\ 0.1 & 0 & 0 & -0.2 & 0.1 \\ 0 & 0 & 0 & 0.1 & -0.1 \end{pmatrix}.$$

Obviously, $h_r(\cdot)$ satisfies $|h_r(s_1) - h_r(s_2)| \leqslant 0.25|s_1 - s_2|$ for $r = 1, 2, 3$. The corresponding response network is defined by:

$$\begin{aligned}\dot{w}_k(t) &= h(w_k(t)) + 0.1\sum_{z=1}^{5} b_{kz}^1(t)\Gamma^1\dot{w}_z(t) + 0.1\sum_{z=1}^{5} b_{kz}^2(t)\Gamma^2\dot{w}_z(t)\\ &+ 0.2\sum_{z=1}^{5} b_{kz}^3(t)\Gamma^3\dot{w}_z(t) + v_k(t),\end{aligned} \tag{16}$$

where $k = 1, 2, \cdots, 5$.

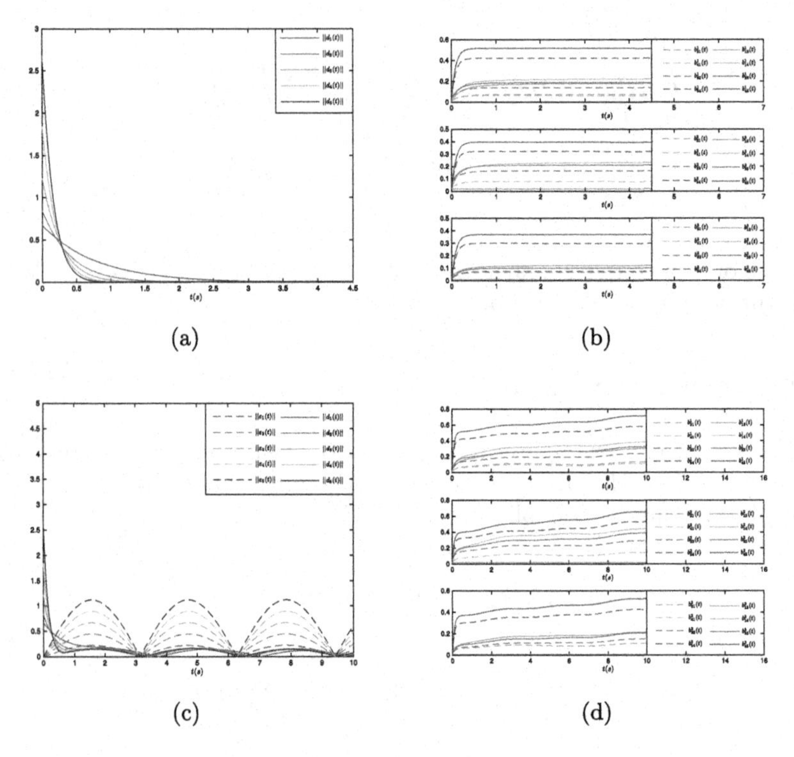

Fig. 1. (a) Trajectories of $\|d_k(t)\|$ $(k = 1, 2, \cdots, 5)$ in Case 1; (b) The curve of $b_{kz}^l(t)$ $(l = 1, 2, 3;\ k, z = 1, 2, \cdots, 5)$ in Case 1; (c) Trajectories of $\|\varepsilon_k(t)\|$ and $\|d_k(t)\|$ $(k = 1, 2, \cdots, 5)$ in Case 2; (d) The curve of $b_{kz}^l(t)$ $(l = 1, 2, 3;\ k, z = 1, 2, \cdots, 5)$ in Case 2.

The adaptive controller (17) and parameter updating strategy (18) are chosen:

$$v_k(t) = -q_k(t)d_k(t), \quad \dot{q}_k(t) = s_k d_k^T(t)d_k(t), \tag{17}$$

$$\dot{b}_{kz}^l(t) = -0.7d_k^T(t)\Gamma^l\dot{w}_z(t), \tag{18}$$

in which $k = 1, 2, \cdots, 5, q_k(0) = 1.2k, s_k = 0.1k$.

Case 1: Under adaptive controller (17) and parameter updating scheme (18), the networks (15) and (16) with above listed parameters can realize outer synchronization. The simulation results are shown in Fig. 1(a) and (b).

Case 2: We take the external disturbance $\varepsilon_k(t) = (0.01k\sin(2t), 0.2k\sin(t), 0.1k\sin(t))^T$ for $k = 1, 2, \ldots, 5$. Under adaptive feedback controller (17) and parameter updating scheme (18), the networks (15) and (16) with above given parameters can realize the outer H_∞ synchronization. The simulation results are portrayed in Fig. 1(c) and (d).

To compare the convergence rate of network synchronization error under the cases of the derivative coupling and the multi-derivative couplings, the following numerical example is provided.

Example 2. Consider the MDCCNs as follows:

$$\dot{m}_k(t) = h(m_k(t)) + \sum_{l=1}^{\mu}\sum_{z=1}^{\xi} c_l a_{kz}^l \Gamma^l \dot{m}_z(t), \tag{19}$$

where $k = 1, 2, \ldots, 5, h_f(s) = 0.25(|s+1| - |s-1|), f = 1, 2, 3, \mu = 1, 3, 5, c_l = 0.1l$,

$$\Gamma^l = \frac{9+l}{10}\begin{pmatrix} 0.5 & 0 & 0.1 \\ 0 & 0.4 & 0 \\ 0 & 0 & 0.3 \end{pmatrix}, A^l = (0.4 + 0.1l)\begin{pmatrix} -0.4 & 0 & 0.2 & 0.2 & 0 \\ 0 & -0.5 & 0.5 & 0 & 0 \\ 0.2 & 0.5 & -0.7 & 0 & 0 \\ 0.2 & 0 & 0 & -0.5 & 0.3 \\ 0 & 0 & 0 & 0.3 & -0.3 \end{pmatrix}.$$

Obviously, $h_r(\cdot)$ satisfies $|h_r(s_1) - h_r(s_2)| \leqslant 0.5|s_1 - s_2|$ for $r = 1, 2, 3$.

$$\dot{w}_k(t) = h(w_k(t)) + \sum_{l=1}^{\mu}\sum_{z=1}^{\xi} c_l b_{kz}^l(t) \Gamma^l \dot{w}_z(t), \tag{20}$$

where $k, z = 1, 2, \ldots, 5$. The values of c_l and Γ^l are the same as those in network (19).

The adaptive controller (21) and parameter updating strategy (22) are designed:

$$v_k(t) = -q_k^{\mu}(t) d_k(t), \quad \dot{q}_k(t) = s_k d_k^T(t) d_k(t), \tag{21}$$

$$\dot{b}_{kz}^l(t) = -0.005 d_k^T(t) \Gamma^l \dot{w}_z(t), \tag{22}$$

where $k = 1, 2, \cdots, 5, \mu = 1, 3, 5$ represent the numbers of coupling form. $q_k^1(0) = 0.2k, q_k^3(0) = 0.6k, q_k^5(0) = k, s_k = 0.1k$.

Case 1: Under adaptive controller (21) and parameter updating scheme (22), the networks (19) and (20) with above listed parameters can realize outer synchronization. Figure 2 (a) depicts the trajectories of $\sum_{k=1}^{\xi} \|d_k^{\mu}(t)\|$ ($\xi = 5, \mu = 1, 3, 5$), which compares the convergence rate of synchronization error of norm sum between different numbers of derivative couplings. From the results, we can see that the convergence rate of $\sum_{k=1}^{\xi} \|d_k^{\mu}(t)\|$ improves as the number of coupling form μ increases.

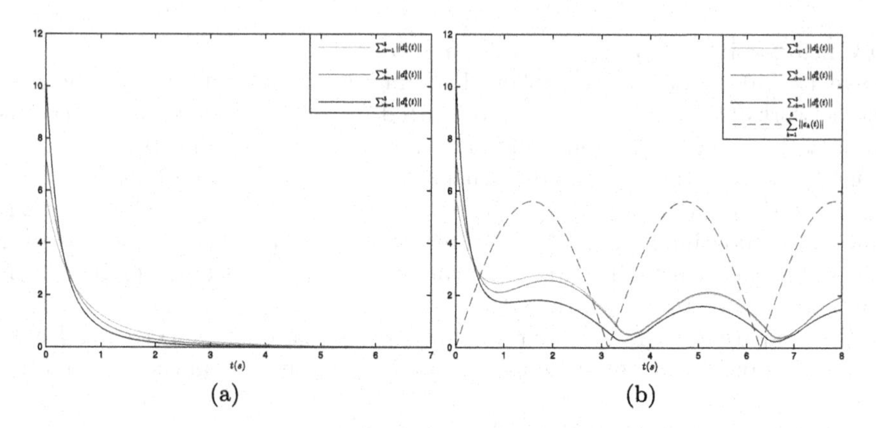

Fig. 2. (a) Trajectories of $\sum_{k=1}^{\xi} \|d_k^{\mu}(t)\|$ ($\xi = 5, \mu = 1, 3, 5$); (b) Trajectories of $\sum_{k=1}^{\xi} \|d_k^{\mu}(t)\| (\xi = 5, \mu = 1, 3, 5)$ and $\sum_{k=1}^{\xi} \|\varepsilon_k(t)\|$ ($\xi = 5$).

Case 2: We take the external disturbance $\varepsilon_k(t) = (0.2k\sin(t), 0.3k\sin(t), 0.1k \sin(t))^T$ for $k = 1, 2, \ldots, 5$. Under adaptive feedback controller (21) and parameter updating scheme (22), the networks (19) and (20) with above given parameters can realize the outer H_∞ synchronization. Figure 2(b) shows the trajectories of $\sum_{k=1}^{\xi} \|d_k^{\mu}(t)\|$ ($\xi = 5, \mu = 1, 3, 5$) and $\sum_{k=1}^{\xi} \|\varepsilon_k(t)\|$, which reflects the convergence rate of synchronization error in the case of external disturbance.

5 Conclusion

In this paper, the outer synchronization and H_∞ synchronization for two types of MDCCNs have been studied. The adaptive control strategy has been devised to guarantee the synchronization of MDCCNs. Based on the Barbalat's lemma and Lyapunov functional approach, the outer synchronization condition for MDCCNs without external disturbance has been derived. Moreover, an adaptive controller and parameter updating scheme have been designed to assure the outer H_∞ synchronization of MDCCNs with external disturbance. Finally, simulation results have been proposed to illustrate the validity of the derived synchronization criteria.

References

1. Song, B., Park, J.H., Wu, Z.-G., Zhang, Y.: Global synchronization of stochastic delayed complex networks. Nonlinear Dyn. **70**, 2389–2399 (2012)
2. Tang, J., Zou, C., Zhao, L.: A general complex dynamical network with time-varying delays and its novel controlled synchronization criteria. IEEE Syst. J. **10**(1), 46–52 (2016)
3. Chen, X., Cao, J., Park, J.H., Zong, G., Qiu, J.: Finite-time complex function synchronization of multiple complex-variable chaotic systems with network transmission and combination mode. J. Vib. Control **24**(22), 5461–5471 (2018)

4. He, X., Zhang, H.: Exponential synchronization of complex networks via feedback control and periodically intermittent noise. J. Franklin Inst. **359**, 3614–3630 (2022)
5. Gao, Z., Li, Y., Wang, Y., Liu, Q.: Distributed tracking control of structural balance for complex dynamical networks based on the coupling targets of nodes and links. Complex Intell. Syst. **9**(1), 881–889 (2023)
6. Gao, Z., Xiong, J., Zhong, J., Liu, F., Liu, Q.: Adaptive state observer design for dynamic links in complex dynamical networks. Comput. Intell. Neurosci. **2020**, 1–8 (2020)
7. Xu, Y., Zhou, W., Fang, J., Sun, W., Pan, L.: Topology identification and adaptive synchronization of uncertain complex networks with non-derivative and derivative coupling. J. Franklin Inst. **347**, 1566–1576 (2010)
8. Yuan, X., Li, J., Li, J.: Adaptive synchronization of unknown complex dynamical networks with derivative and distributed time-varying delay couplings. Int. J. Fuzzy Syst. **20**(4), 1088–1097 (2018). https://doi.org/10.1007/s40815-017-0424-9
9. Tang, Z., Park, J.H., Wang, Y., Feng, J.: Adaptively synchronize the derivative coupled complex networks with proportional delay. IEEE Trans. Syst. Man Cybern. Syst. **51**(8), 4969–6979 (2019)
10. Zhao, L.-H., Wang, J.-L., Zhang, Y.: Lag output synchronization for multiple output coupled complex networks with positive semidefinite or positive definite output matrix. J. Franklin Inst. **357**, 414–436 (2020)
11. Wang, J.-L., Wang, D.-Y., Wu, H.-N., Huang, T.: Output synchronization of complex dynamical networks with multiple output or output derivative couplings. IEEE Trans. Cybern. **51**(2), 927–937 (2021)
12. Wang, G., Yin, Q., Shen, Y., Jiang, F.: H_∞ synchronization of directed complex dynamical networks with mixed time-delays and switching structures. Circ. Syst. Sig. Process. **32**, 1575–1593 (2013)
13. Zong, G., Yang, D.: H_∞ synchronization of switched complex networks: a switching impulsive control method. Commun. Nonlinear Sci. Numer. Simul. **77**, 338–348 (2019)
14. Fan, G., Ma, Y.: Non-fragile delay-dependent pinning H_∞ synchronization of T-S fuzzy complex networks with hybrid coupling delays. Inf. Sci. **608**, 1317–1333 (2022)
15. Li, J., Jiang, H., Wang, J., Hu, C., Zhang, G.: H_∞ exponential synchronization of complex networks: aperiodic sampled-data-based event-triggered control. IEEE Trans. Cybern. **52**(8), 7968–7980 (2022)
16. Zhang, L., Lei, Y., Wang, Y., Chen, H.: Generalized outer synchronization between non-dissipatively coupled complex networks with different-dimensional nodes. Appl. Math. Modell. **55**, 248–261 (2018)
17. Sun, Y., Wu, H., Chen, Z., Zheng, X., Chen, Y.: Outer synchronization of two different multi-links complex networks by chattering-free control. Phys. A **584**, 126354 (2021)
18. Zheng, S.: Pinning and impulsive synchronization control of complex dynamical networks with non-derivative and derivative coupling. J. Franklin Inst. **354**, 6341–6363 (2017)
19. Slotine, J.J., Li, W.: Applied Nonlinear Control. Prentice-Hall, Englewood Cliffs (1991)

A Distributed Projection-Based Algorithm with Local Estimators for Optimal Formation of Multi-robot System

Yuanyuan Yue[1], Qingshan Liu[1(✉)], and Ziming Zhang[2]

[1] School of Mathematics, Southeast University, Nanjing 210096, China
{yyyue,qsliu}@seu.edu.cn

[2] Aviation Equipment Monitoring and Control and Reverse Engineering Laboratory, State Wuhu Machinery Factory, Wuhu 241007, China
zimingzhang@nuaa.edu.cn

Abstract. In general, the optimal formation problem can be modeled as a standard constrained optimization problem according to the shape theory. By adding local supplementary estimators, it can be further modeled as a distributed constrained optimization problem. Then a distributed projection-based algorithm is designed for solving this problem. The aim of the algorithm is to drive a group of robots to move to the desired geometric pattern by minimizing the total travel distance of robots from the initial positions. It is worth noticing that, as long as the graph of the communication network among the robots is undirected and connected, the global convergence of the algorithm can be guaranteed. Moreover, all of the robots finally form an ideal formation in the limited space. Finally, simulation results are provided to verify the effectiveness of the proposed distributed algorithm.

Keywords: distributed optimal formation · projection operator · shape theory · multi-robot system

1 Introduction

Multi-robot systems have been intensively researched in recent years due to their wide applications, such as unmanned vehicle control, environmental monitoring and automated container transportation [1–4]. Among the various tasks of multi-robot systems, the formation problem is of critical importance because proper robot organization can dramatically improve coordination efficiency.

In addition, the formation of multi-robot systems is widely used in load transportation [5], satellite formation flying [3], and unmanned aerial vehicle formation [6], among other applications. The goal of multi-robot formation control is to direct robots to realize and maintain a predetermined geometric shape.

This work was partially supported by the National Natural Science Foundation of China under Grant 62276062, and the Third Batch of Projects of Anhui Key Research and Development Plan in 2021 (2021008).

B. Luo et al. (Eds.): ICONIP 2023, LNCS 14447, pp. 344–355, 2024.
https://doi.org/10.1007/978-981-99-8079-6_27

Based on coordination schemes, the multi-robot formation control problem can be solved using leader-following control [7,8], behavior-based control (for example, obstacle avoidance and formation keeping) [9,10], and virtual structure control [11], where the formation is regarded as a single structure. According to the sensing capabilities of robots and interaction topology [12], formation control schemes can be categorized into the following groups: The first is the position-based approach [10], which assumes a global coordinate system and has a tight communication topology and low sensing capabilities. The second is the displacement-based approach [13], which communication topology and sensing capability are moderate and each robot can sense the positions of its neighbors. The third is the distance-based approach [13], which has a loose communication topology and strong sensing capabilities. In terms of analytical method, the formation of multi-robot system includes consensus-based [7], optimization-based [14] and model predictive control (MPC)-based methods [15].

The preceding paragraph describes formation control strategies for multi-robot systems. However, identifying the desired formation has received little attention. Usually, a concise way to describe formation is the shape theory [16], which focuses solely on the fundamental geometric connection between robots instead of considering factors like distance, size, and orientation. As described in [16], two formations have the same shape if they are in the same equivalence class. How to choose one formation from an equivalence class is the crux of the matter when trying to determine a desired formation. In the groundbreaking work [17], finding the optimal formation is modeled as a standard constrained optimization problem, which can be solved using various numerical algorithms. Similarly, the optimal formation problem is reformulated as a nonsmooth convex optimization in [14], and it is resolved using a recurrent neural network. In [18], a finite-time convergence algorithm is proposed for solving the formation problem with limitations on convex regions. In [19], the optimal formation problem, which is converted to a mixed norm constrained optimization problem, is solved by a projection-based iteration algorithm. However, they are all not modeled in a distributed way, which has the advantage of rapid convergence and low communication consumption.

So, we consider the distributed optimal formation of multi-robot systems in this paper. First, by introducing some auxiliary variables to estimate the states of the first two robots, the problem of determining the optimal formation is modeled as a distributed optimization problem, while many existing works for the optimal formation are centralized [14,19]. Then, we propose the projection-based iteration formulas for resolving the distributed optimization problem based on the distributed method, which greatly reduces communication consumption between robots. Finally, an illustrative example is supplied to verify the effectiveness of the proposed distributed method.

2 Problem Formulation for Optimal Formation

We are concerned with the optimal formation problem of a multi-robot system with m robots on a two-dimensional plane to ensure that robots constitute a formation as same as a given geometric shape, where the icon representing the shape (formation) consists of m points $\{s_1, s_2, \ldots, s_m\}$, denoted by S. Without loss of generality, let s_1 always lie at the origin and s_2 always be on the x axis. The multi-robot system starts with the formation $(\xi_1, \xi_2, \ldots, \xi_m) \in \mathbb{R}^{2m}$, where ξ_i is the initial position of robot i. To minimize the overall moving distance, the optimal formation problem is typically formulated as a constrained optimization problem as follows [19]:

$$\begin{aligned} \min_{P} \ & \tfrac{1}{2}\textstyle\sum_{i=1}^{m} \|p_i - \xi_i\|_2^2, \\ \text{s.t.} \ & P \in [S], \\ & p_i \in \Omega_i, \quad i = 1, 2, \ldots, m, \end{aligned} \tag{1}$$

in which p_i is the i-th robot's final location, $P = (p_1, p_2, \ldots, p_m)$ is the final formation and $\|\cdot\|_2$ denotes the 2-norm of vector. We denote $[S] = \{rRS + \mathbf{1}_m d^T : r \in \mathbb{R}_+, R \in SO(2), d \in \mathbb{R}^2\}$ as equivalence class of formations [16], which means that the elements of $[S]$ are generated under rotation, scale, and translation transformations from S. $\Omega_i \in \mathbb{R}^2$ is a non-empty and closed convex set that limits the moving range of robot i.

To efficiently deal with the constraint $P \in [S]$ in (1), we convert it into a series of linear equality constraints [16]:

$$M(p_i - p_1) = M_i(p_2 - p_1) \quad (i = 3, 4, \ldots, m), \tag{2}$$

where $M = \begin{pmatrix} \|s_2\|_2 & 0 \\ 0 & \|s_2\|_2 \end{pmatrix}$, $M_i = \begin{pmatrix} s_i^x & -s_i^y \\ s_i^y & s_i^x \end{pmatrix}$, $s_i = (s_i^x, s_i^y)^T$. From (2), we can observe that robot i need know the following information: p_i, p_1, p_2, M_i, and M. In other words, robots 1 and 2 have to communicate with every other robot, thus incurring significant communication consumption, as shown in Fig. (1a).

In order to counteract this drawback, the local estimators are introduced to allow each robot i $(i = 3, 4, ..., m)$ estimate the values of p_1 and p_2. In this way, robots 1 and 2 do not have to communicate with every other robot, as shown in Fig. (1b).

Assume that the i-th robot's estimates to the first and second robots' positions are p_{i1} and p_{i2} respectively. Then, the condition (2) can be transformed into

$$M(p_i - p_{i1}) = M_i(p_{i2} - p_{i1}) \quad (i = 3, 4, \ldots, m), \tag{3}$$

that is,

$$\begin{pmatrix} M & M_i - M & -M_i \end{pmatrix} \begin{pmatrix} p_i \\ p_{i1} \\ p_{i2} \end{pmatrix} = 0 \quad (i = 3, 4, \ldots, m). \tag{4}$$

Let $B_i = \begin{pmatrix} M & M_i - M & -M_i \end{pmatrix}$, $\tilde{p}_i = (p_i^T, p_{i1}^T, p_{i2}^T)^T$, then (4) can be expressed as

$$B_i \tilde{p}_i = 0 \quad (i = 3, 4, \ldots, m). \tag{5}$$

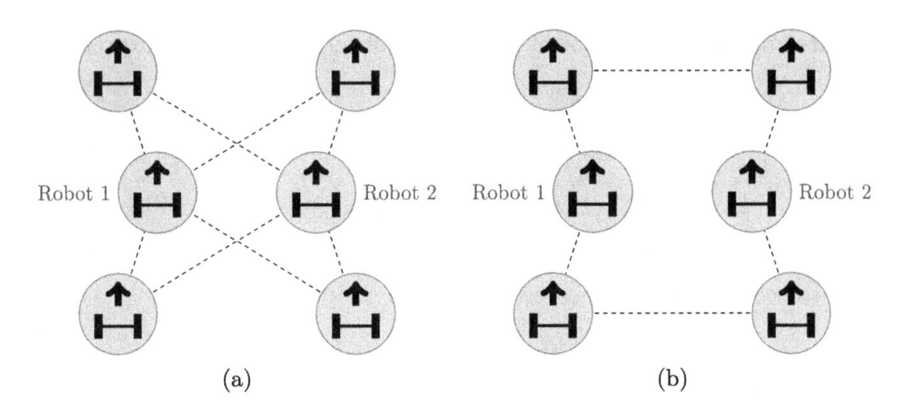

Fig. 1. (a): Robots 1 and 2 must communicate with every other robot in (2). (b): With local estimators, robots 1 and 2 only need to interact with at least one robot respectively, greatly reducing the communication load of multi-robot system.

Due to the fact that all robots have the same p_1 and p_2, the following conditions for consistence are added:

$$\begin{cases} p_{31} = p_{41} = \cdots = p_{m1} = p_1, \\ p_{32} = p_{42} = \cdots = p_{m2} = p_2. \end{cases} \tag{6}$$

We plug in the first robot's estimated value for p_2 and the second robot's estimated value for p_1 in order to design the distributed method while considering the network topology as a whole. Consequently, (6) is further expressed as follows:

$$\begin{cases} p_1 = p_{21} = p_{31} = p_{41} = \cdots = p_{m1}, \\ p_{12} = p_2 = p_{32} = p_{42} = \cdots = p_{m2}. \end{cases} \tag{7}$$

The compact form of (7) is written as

$$(L \otimes I_2)p^1 = 0, \quad (L \otimes I_2)p^2 = 0, \tag{8}$$

where $p^1 = (p_1^T, p_{21}^T, \ldots, p_{m1}^T)^T$, $p^2 = (p_{12}^T, p_2^T, \ldots, p_{m2}^T)^T$, $\otimes$ is the Kronecker product, I_2 denotes two-dimensional identity matrix. In this paper, we suppose that the interaction graph G among robots is undirected and connected, so the Laplace matrix $L = \{l_{ij}\} \in \mathbb{R}^{m \times m}$ corresponding to G has the following properties: $0 = \lambda_1 < \lambda_2 \leq \cdots \leq \lambda_m$, where λ_i is the i-th smallest eigenvalue.

To keep robots from getting too close to each other during formation switching, it is necessary to constrain the relative position between robots 1 and 2. Hence, we investigate the case for adding a free anchor constraint: $p_1 - p_2 = b$. Based on the notation introduced above, we can apply constraints $p_1 - p_{12} = b$ and $p_2 - p_{21} = -b$ to robots 1 and 2, respectively.

Combining the objective function in (1), formation constraint (5), consistency constraint (8), anchor constraint and boundary constraint $p_i \in \Omega_i$, the distributed optimization problem of multi-robot formation can be expressed as

$$\begin{array}{ll} \min\limits_{p_i} & \frac{1}{2}\sum_{i=1}^{m} \|p_i - \xi_i\|_2^2, \\ \text{s.t.} & p_1 - p_{12} = b, \quad p_2 - p_{21} = -b, \\ & B_i \tilde{p}_i = 0 \quad (i = 3, 4, \ldots, m), \\ & (L \otimes I_2)p^1 = 0, \quad (L \otimes I_2)p^2 = 0, \\ & p_i \in \Omega_i \quad (i = 1, 2, \ldots, m). \end{array} \tag{9}$$

For the convenience of theoretical analysis, we denote

$$\tilde{p} = (p_1^T, p_{12}^T, p_2^T, p_{21}^T, p_3^T, p_{31}^T, p_{32}^T, \ldots, p_m^T, p_{m1}^T, p_{m2}^T)^T,$$

$$\tilde{c} = \text{diag}\{1, 0, 1, 0, 1, 0, 0, \ldots, 1, 0, 0\} \otimes I_2,$$

$$\tilde{B} = \begin{pmatrix} I_2 & -I_2 & O_2 & O_2 & O_{2\times 6(m-2)} \\ O_2 & O_2 & I_2 & -I_2 & O_{2\times 6(m-2)} \\ 0_{2(m-2)\times 2} & 0_{2(m-2)\times 2} & 0_{2(m-2)\times 2} & 0_{2(m-2)\times 2} & \text{blkdiag}\{B_3, \ldots, B_m\} \end{pmatrix},$$

$$\tilde{b} = (b^T, -b^T, 0_{2(m-2)}^T)^T,$$

$$\tilde{\xi} = (\xi_1^T, 0_2^T, \xi_2^T, 0_2^T, \xi_3^T, 0_2^T, 0_2^T, \ldots, \xi_m^T, 0_2^T, 0_2^T)^T,$$

$$\tilde{L}_1 = \left\{ \begin{array}{cccccccccccc} l_{11} & 0 & 0 & l_{12} & 0 & l_{13} & 0 & \cdots & 0 & l_{1m} & 0 \\ 0 & 0 & 0 & 0 & 0 & 0 & 0 & \cdots & 0 & 0 & 0 \\ 0 & 0 & 0 & 0 & 0 & 0 & 0 & \cdots & 0 & 0 & 0 \\ l_{21} & 0 & 0 & l_{22} & 0 & l_{23} & 0 & \cdots & 0 & l_{2m} & 0 \\ 0 & 0 & 0 & 0 & 0 & 0 & 0 & \cdots & 0 & 0 & 0 \\ l_{31} & 0 & 0 & l_{32} & 0 & l_{33} & 0 & \cdots & 0 & l_{3m} & 0 \\ 0 & 0 & 0 & 0 & 0 & 0 & 0 & \cdots & 0 & 0 & 0 \\ \vdots & \vdots & \vdots & \vdots & \vdots & \vdots & \vdots & \ddots & \vdots & \vdots & \vdots \\ 0 & 0 & 0 & 0 & 0 & 0 & 0 & \cdots & 0 & 0 & 0 \\ l_{m1} & 0 & 0 & l_{m2} & 0 & l_{m3} & 0 & \cdots & 0 & l_{mm} & 0 \\ 0 & 0 & 0 & 0 & 0 & 0 & 0 & \cdots & 0 & 0 & 0 \end{array} \right\},$$

$$\tilde{L}_2 = \left\{ \begin{array}{cccccccccccc} 0 & 0 & 0 & 0 & 0 & 0 & 0 & \cdots & 0 & 0 & 0 \\ 0 & l_{11} & l_{12} & 0 & 0 & l_{13} & 0 & \cdots & 0 & 0 & l_{1m} \\ 0 & l_{21} & l_{22} & 0 & 0 & l_{23} & 0 & \cdots & 0 & 0 & l_{2m} \\ 0 & 0 & 0 & 0 & 0 & 0 & 0 & \cdots & 0 & 0 & 0 \\ 0 & 0 & 0 & 0 & 0 & 0 & 0 & \cdots & 0 & 0 & 0 \\ 0 & 0 & 0 & 0 & 0 & 0 & 0 & \cdots & 0 & 0 & 0 \\ 0 & l_{31} & l_{32} & 0 & 0 & l_{33} & 0 & \cdots & 0 & 0 & l_{3m} \\ \vdots & \vdots & \vdots & \vdots & \vdots & \vdots & \vdots & \ddots & \vdots & \vdots & \vdots \\ 0 & 0 & 0 & 0 & 0 & 0 & 0 & \cdots & 0 & 0 & 0 \\ 0 & 0 & 0 & 0 & 0 & 0 & 0 & \cdots & 0 & 0 & 0 \\ 0 & l_{m1} & l_{m2} & 0 & 0 & l_{m3} & 0 & \cdots & 0 & 0 & l_{mm} \end{array} \right\},$$

$$\tilde{\Omega} = \Omega_1 \times \Omega_2 \times \Omega_2 \times \Omega_1 \times \Omega_3 \times \Omega_1 \times \Omega_2 \times \cdots \times \Omega_m \times \Omega_1 \times \Omega_2,$$

where diag$\{\cdot\}$ denotes the diagonal matrix, blkdiag$\{\cdot\}$ denotes the block diagonal matrix and $\times$ is the Cartesian product.

Therefore, (9) can be described as follows:

$$\begin{aligned} \min_{\tilde{p}} \quad & \tfrac{1}{2}\|\tilde{c}\tilde{p} - \tilde{\xi}\|_2^2, \\ \text{s.t.} \quad & \tilde{B}\tilde{p} = \tilde{b}, \quad \tilde{p} \in \tilde{\Omega}, \\ & (\tilde{L}_1 \otimes I_2)\tilde{p} = 0, \quad (\tilde{L}_2 \otimes I_2)\tilde{p} = 0. \end{aligned} \tag{10}$$

3 Algorithm Description and Convergence Analysis

First of all, the following lemma illustrates the necessary and sufficient conditions for the optimal solutions of (10).

Lemma 1 *[20]. $\tilde{p}^* \in \mathbb{R}^{6m-4}$ is an optimal solution to problem (10) if and only if there exist $\tilde{y}^* \in \mathbb{R}^{2m}$, $\tilde{\theta}^* \in \mathbb{R}^{6m-4}$ and $\tilde{\eta}^* \in \mathbb{R}^{6m-4}$ such that $(\tilde{p}^*, \tilde{y}^*, \tilde{\theta}^*, \tilde{\eta}^*)$ satisfies*

$$\begin{cases} \tilde{p}^* = \phi(\tilde{p}^* - \sigma(\tilde{c}^T(\tilde{c}\tilde{p}^* - \tilde{\xi}) + \tilde{B}^T\tilde{y}^* + \tilde{L}_1 \otimes I_2\tilde{\theta}^* + \tilde{L}_2 \otimes I_2\tilde{\eta}^*)), \\ \tilde{B}\tilde{p}^* = \tilde{b}, \\ \tilde{L}_1 \otimes I_2\tilde{p}^* = 0, \\ \tilde{L}_2 \otimes I_2\tilde{p}^* = 0, \end{cases} \tag{11}$$

where $\sigma > 0$ is a constant, $\phi : \mathbb{R}^{6m-4} \to \tilde{\Omega}$ is a projection operator.

3.1 Algorithm Description

We define $\tilde{p}(k) = \tilde{p}^k$, $\tilde{y}(k) = \tilde{y}^k$, $\tilde{z}(k) = \tilde{z}^k$ and $\tilde{w}(k) = \tilde{w}^k$. From (11), a distributed algorithm is designed to solve problem (10)

$$\begin{cases} \tilde{p}^{k+1} = \phi(\tilde{p}^k - \sigma(\tilde{c}^T(\tilde{c}\tilde{p}^k - \tilde{\xi}) + \tilde{B}^T(\tilde{y}^k + \tilde{B}\tilde{p}^k - \tilde{b}) + \tilde{z}^k + \tilde{L}_1 \otimes I_2\tilde{p}^k \\ \qquad\quad + \tilde{w}^k + \tilde{L}_2 \otimes I_2\tilde{p}^k)), \\ \tilde{y}^{k+1} = \tilde{y}^k + \tilde{B}\tilde{p}^{k+1} - \tilde{b}, \\ \tilde{z}^{k+1} = \tilde{z}^k + \tilde{L}_1 \otimes I_2\tilde{p}^{k+1}, \\ \tilde{w}^{k+1} = \tilde{w}^k + \tilde{L}_2 \otimes I_2\tilde{p}^{k+1}, \end{cases} \tag{12}$$

where $\tilde{z} = \left(z_1^T, z_{12}^T, z_2^T, z_{21}^T, z_3^T, z_{31}^T, z_{32}^T, \ldots, z_m^T, z_{m1}^T, z_{m2}^T\right)^T$, $\tilde{y} = \left(y_1^T, y_2^T, y_3^T, \ldots, y_m^T\right)^T$, $\tilde{w} = \left(w_1^T, w_{12}^T, w_2^T, w_{21}^T, w_3^T, w_{31}^T, w_{32}^T, \ldots, w_m^T, w_{m1}^T, w_{m2}^T\right)^T$.

It stands to reason that $(\tilde{p}^*, \tilde{y}^*, \tilde{z}^*, \tilde{w}^*)$ satisfies the equations in (11) if and only if it is also an equilibrium point of system (12) where $\tilde{z}^* = \tilde{L}_1 \otimes I_2\tilde{\theta}^*$ and $\tilde{w}^* = \tilde{L}_2 \otimes I_2\tilde{\eta}^*$.

Next, we present the Algorithm (12) in component form in order to fully understand the algorithm's distributed processing mechanism.

For robot 1, the iterative formula is

$$\begin{cases} p_1^{k+1} = \phi_1(p_1^k - \sigma(p_1^k - \xi_1 + y_1^k + p_1^k - b - p_{12}^k + z_1^k \\ \qquad + \sum_{j\in N_1} a_{1j}(p_1^k - p_{j1}^k))), \\ p_{12}^{k+1} = \phi_2(p_{12}^k - \sigma(-y_1^k - p_1^k + p_{12}^k + b_3 + w_{12}^k \\ \qquad + \sum_{j\in N_1} a_{1j}(p_{12}^k - p_{j2}^k))), \\ y_1^{k+1} = y_1^k + p_1^{k+1} - p_{12}^{k+1} - b, \\ z_1^{k+1} = z_1^k + \sum_{j\in N_1} a_{1j}(p_1^{k+1} - p_{j1}^{k+1}), \\ w_{12}^{k+1} = w_{12}^k + \sum_{j\in N_1} a_{1j}(p_{12}^{k+1} - p_{j2}^{k+1}), \end{cases} \tag{13}$$

where $p_{22} = p_2$ if $a_{12} \neq 0$, N_1 denotes the set of robots that can communicate with robot 1.

For robot 2, the iterative formula is

$$\begin{cases} p_2^{k+1} = \phi_2(p_2^k - \sigma(p_2^k - \xi_2 + y_2^k + p_2^k + b - p_{21}^k + w_2^k + \sum_{j\in N_2} a_{2j}(p_2^k - p_{j2}^k))), \\ p_{21}^{k+1} = \phi_1(p_{21}^k - \sigma(-y_2^k - p_2^k + p_{21}^k - b + z_{21}^k + \sum_{j\in N_2} a_{2j}(p_{21}^k - p_{j1}^k))), \\ y_2^{k+1} = y_2^k + p_2^{k+1} - p_{21}^{k+1} + b, \\ z_{21}^{k+1} = z_{21}^k + \sum_{j\in N_2} a_{2j}(p_{21}^{k+1} - p_{j1}^{k+1}), \\ w_2^{k+1} = w_2^k + \sum_{j\in N_2} a_{2j}(p_2^{k+1} - p_{j2}^{k+1}), \end{cases} \tag{14}$$

where $p_{11} = p_1$ if $a_{21} \neq 0$, N_2 denotes all neighbor robots of robot 2.

For robot i $(i = 3, 4, \ldots, m)$, the iterative formula is

$$\begin{cases} p_i(k+1) = \phi_i(p_i^k - \sigma(p_i^k - \xi_i + M(y_i^k + B_i\tilde{p}_i^k))), \\ p_{i1}(k+1) = \phi_1(p_{i1}^k - \sigma((M_i^T - M)(y_i^k + B_i\tilde{p}_i^k) + z_{i1}^k + \sum_{j\in N_i} a_{ij}(p_{i1}^k - p_{j1}^k))), \\ p_{i2}(k+1) = \phi_2(p_{i2}^k - \sigma(-M_i^T(y_i^k + B_i\tilde{p}_i^k) + w_{i2}^k + \sum_{j\in N_i} a_{ij}(p_{i2}^k - p_{j2}^k))), \\ y_i(k+1) = y_i^k + B_i\tilde{p}_i^{k+1}, \\ z_{i1}(k+1) = z_{i1}^k + \sum_{j\in N_i} a_{ij}(p_{i1}^k - p_{j1}^k), \\ w_{i2}(k+1) = w_{i2}^k + \sum_{j\in N_i} a_{ij}(p_{i2}^k - p_{j2}^k), \end{cases} \tag{15}$$

where $p_{11} = p_1$ if $a_{i1} \neq 0$, $p_{22} = p_2$ if $a_{i2} \neq 0$, $\tilde{p}_i = (p_i^T, p_{i1}^T, p_{i2}^T)^T$. N_i denotes all neighbor robots of robot i.

3.2 Convergence Analysis

For further analysis, another equivalent algorithm is introduced as follows:

$$\begin{cases} \tilde{p}^{k+1} = \phi(\tilde{p}^k - \sigma(\tilde{c}^T(\tilde{c}\tilde{p}^k - \tilde{\xi}) + \tilde{B}^T(\tilde{y}^k + \tilde{B}\tilde{p}^k - \tilde{b}) + \tilde{L}_1 \otimes I_2(\tilde{\theta}^k + \tilde{p}^k) \\ \qquad + \tilde{L}_2 \otimes I_2(\tilde{\eta}^k + \tilde{p}^k))), \\ \tilde{y}^{k+1} = \tilde{y}^k + \tilde{B}\tilde{p}^{k+1} - \tilde{b}, \\ \tilde{\theta}^{k+1} = \tilde{\theta}^k + \tilde{p}^{k+1}, \\ \tilde{\eta}^{k+1} = \tilde{\eta}^k + \tilde{p}^{k+1}. \end{cases} \tag{16}$$

Theorem 1. *The algorithms in* (12) *and* (16) *are equivalent if the initial conditions meet* $\tilde{z}_0 = (\tilde{L}_1 \times I_2)\tilde{\theta}_0$ *and* $\tilde{w}_0 = (\tilde{L}_2 \times I_2)\tilde{\eta}_0$ *for any* $\tilde{\theta}_0 \in \mathbb{R}^{6m-4}$ *and* $\tilde{\eta}_0 \in \mathbb{R}^{6m-4}$.

Proof. The proof is similar to that of Lemma 3 in [20] and is omitted here. □

Next, we prove the convergence of Algorithm (16), which also holds for Algorithm (12) as per Theorem 1.

Lemma 2 *[21]. Let A and B are square matrices of m and n dimensions respectively, $\lambda_1, \lambda_2, \ldots, \lambda_m$ and $\mu_1, \mu_2, \ldots, \mu_n$ are eigenvalues of A and B respectively. Then $\lambda_i \mu_j$ $(i = 1, 2, \ldots, m, j = 1, 2, \ldots, n)$ are eigenvalues of $A \otimes B$.*

Assume that $\tilde{p}^*$ is an optimal solution to (10). From Lemma 1, there exist $\tilde{y}^*$, $\tilde{\theta}^*$and $\tilde{\eta}^*$ such that the equations in (11) hold, thus we can define and draw some conclusions about the following functions:

$$V_1(\tilde{p}^k) = \|\tilde{p}^k - \tilde{p}^*\|_2^2, \; V_2(\tilde{y}^k) = \|\tilde{y}^k - \tilde{y}^*\|_2^2,$$
$$V_3(\tilde{\theta}^k) = (\tilde{\theta}^k - \tilde{\theta}^*)^T \tilde{L}_1 \otimes I_2(\tilde{\theta}^k - \tilde{\theta}^*), V_4(\tilde{\eta}^k) = (\tilde{\eta}^k - \tilde{\eta}^*)^T \tilde{L}_2 \otimes I_2(\tilde{\eta}^k - \tilde{\eta}^*).$$

Lemma 3 *[20]. The following inequalities are satisfied:*

(1) $V_1(\tilde{p}^{k+1}) - V_1(\tilde{p}^k) \leq -\|\tilde{p}^{k+1} - \tilde{p}^k\|_2^2 + 2\sigma\tilde{c}^T\tilde{c}\|\tilde{p}^{k+1} - \tilde{p}^k\|_2^2 - 2\sigma(\tilde{B}\tilde{p}^{k+1} - \tilde{b})^T(\tilde{y}^k - \tilde{y}^* + \tilde{B}\tilde{p}^k - \tilde{b}) - 2\sigma(\tilde{L}_1 \otimes I_2\tilde{p}^{k+1})^T(\tilde{\theta}^k - \tilde{\theta}^* + \tilde{p}^k) - \sigma(\tilde{L}_2 \otimes I_2\tilde{p}^{k+1})^T(\tilde{\eta}^k - \tilde{\eta}^* + \tilde{p}^k)$;
(2) $V_2(\tilde{y}^{k+1}) - V_2(\tilde{y}^k) \leq 2(\tilde{B}\tilde{p}^{k+1} - \tilde{b})^T(\tilde{y}^k - \tilde{y}^* + \tilde{B}\tilde{p}^k - \tilde{b}) + \|\tilde{B}(\tilde{p}^{k+1} - \tilde{p}^k)\|^2 - \|\tilde{B}\tilde{p}^k - \tilde{b}\|^2$;
(3) $V_3(\tilde{\theta}^{k+1}) - V_3(\tilde{\theta}^k) = 2(\tilde{L}_1 \otimes I_2\tilde{p}^{k+1})^T(\tilde{\theta}^k - \tilde{\theta}^* + \tilde{p}^k) + (\tilde{p}^{k+1} - \tilde{p}^k)^T\tilde{L}_1 \otimes I_2(\tilde{p}^{k+1} - \tilde{p}^k) - (\tilde{p}^k)^T\tilde{L}_1 \otimes I_2\tilde{p}^k$;
(4) $V_4(\tilde{\eta}^{k+1}) - V_4(\tilde{\eta}^k) = 2(\tilde{L}_2 \otimes I_2\tilde{p}^{k+1})^T(\tilde{\eta}^k - \tilde{\eta}^* + \tilde{p}^k) + (\tilde{p}^{k+1} - \tilde{p}^k)^T\tilde{L}_2 \otimes I_2(\tilde{p}^{k+1} - \tilde{p}^k) - (\tilde{p}^k)^T\tilde{L}_2 \otimes I_2\tilde{p}^k$.

Theorem 2. *For any initial value* $(\tilde{p}_0, \tilde{y}_0, \tilde{\theta}_0, \tilde{\eta}_0)$*, the state* $\tilde{p}_k$ *in* (16) *is globally convergent to the optimal solution of problem* (10) *if*

$$\sigma \leq \frac{1}{\lambda_{\max}(2\tilde{c}^T\tilde{c} + \tilde{B}^T\tilde{B} + \tilde{L}_1 \otimes I_2 + \tilde{L}_2 \otimes I_2))},$$

where $\lambda_{\max}$ *is the maximum eigenvalue of corresponding matrix.*

Proof. Suppose that $\tilde{p}^*$ to be an optimal solution of (10). By Lemma 1, there exist $\tilde{y}^*$, $\tilde{\theta}^*$ and $\tilde{\eta}^*$ such that the equalities in (11) are fulfilled.

Define the candidate Lyapunov function as:

$$V(\tilde{p}^k,\tilde{y}^k,\tilde{\theta}^k,\tilde{\eta}^k)=V_1(\tilde{p}^k)+\sigma\big(V_2(\tilde{y}^k)+V_3(\tilde{\theta}^k)+V_4(\tilde{\eta}^k)\big). \tag{17}$$

Then according to Lemma 3, we have

$$\begin{aligned}
&V(\tilde{p}^{k+1},\tilde{y}^{k+1},\tilde{\theta}^{k+1},\tilde{\eta}^{k+1}) - V(\tilde{p}^k,\tilde{y}^k,\tilde{\theta}^k,\tilde{\eta}_k)\\
=&V_1(\tilde{p}^{k+1}) - V_1(\tilde{p}^k) + \sigma\big(V_2(\tilde{y}^{k+1}) - V_2(\tilde{y}^k)\big) + \sigma\big(V_3(\tilde{\theta}^{k+1}) - V_3(\tilde{\theta}^k)\big)\\
&+ \sigma\big(V_4(\tilde{\eta}^{k+1}) - V_4(\tilde{\eta}^k)\big)\\
\le& - \|\tilde{p}^{k+1} - \tilde{p}^k\|_2^2 + 2\sigma\tilde{c}^T\tilde{c}\|\tilde{p}^{k+1} - \tilde{p}^k\|_2^2 - 2\sigma(\tilde{B}\tilde{p}^{k+1} - \tilde{b})^T(\tilde{y}^k - \tilde{y}^* + \tilde{B}\tilde{p}^k - \tilde{b})\\
&- 2\sigma(\tilde{L}_1 \otimes I_2\tilde{p}^{k+1})^T(\tilde{\theta}^k - \tilde{\theta}^* + \tilde{p}^k) - 2\sigma(\tilde{L}_2 \otimes I_2\tilde{p}^{k+1})^T(\tilde{\eta}^k - \tilde{\eta}^* + \tilde{p}^k)\\
&+ 2\sigma(\tilde{B}\tilde{p}^{k+1} - \tilde{b})^T(\tilde{y}^k - \tilde{y}^* + \tilde{B}\tilde{p}^k - \tilde{b}) + \sigma\|\tilde{B}(\tilde{p}^{k+1} - \tilde{p}^k)\|^2 - \sigma\|\tilde{B}\tilde{p}^k - \tilde{b}\|^2\\
&+ 2\sigma(\tilde{L}_1 \otimes I_2\tilde{p}^{k+1})^T(\tilde{\theta}^k - \tilde{\theta}^* + \tilde{p}^k) + \sigma(\tilde{p}^{k+1} - \tilde{p}_k)^T\tilde{L}_1 \otimes I_2(\tilde{p}^{k+1} - \tilde{p}^k)\\
&- \sigma(\tilde{p}^k)^T\tilde{L}_1 \otimes I_2\tilde{p}^k + 2\sigma(\tilde{L}_2 \otimes I_2\tilde{p}^{k+1})^T(\tilde{\eta}^k - \tilde{\eta}^* + \tilde{p}^k)\\
&+ \sigma(\tilde{p}^{k+1} - \tilde{p}^k)^T\tilde{L}_2 \otimes I_2(\tilde{p}^{k+1} - \tilde{p}^k) - \sigma(\tilde{p}^k)^T\tilde{L}_2 \otimes I_2\tilde{p}^k\\
\le& - \|\tilde{p}^{k+1} - \tilde{p}^k\|_2^2 + 2\sigma\tilde{c}^T\tilde{c}\|\tilde{p}^{k+1} - \tilde{p}^k\|_2^2 + \sigma\|\tilde{B}(\tilde{p}^{k+1} - \tilde{p}^k)\|^2 - \sigma\|\tilde{B}\tilde{p}^k - \tilde{b}\|^2\\
&+ \sigma(\tilde{p}^{k+1} - \tilde{p}^k)^T\tilde{L}_1 \otimes I_2(\tilde{p}^{k+1} - \tilde{p}^k) - \sigma(\tilde{p}^k)^T\tilde{L}_1 \otimes I_2\tilde{p}^k\\
&+ \sigma(\tilde{p}^{k+1} - \tilde{p}^k)^T\tilde{L}_2 \otimes I_2(\tilde{p}^{k+1} - \tilde{p}^k) - \sigma(\tilde{p}^k)^T\tilde{L}_2 \otimes I_2\tilde{p}^k\\
=& - (\tilde{p}^{k+1} - \tilde{p}^k)^T\big(I - \sigma(2\tilde{c}^T\tilde{c} + \tilde{B}^T\tilde{B} + \tilde{L}_1 \otimes I_2 + \tilde{L}_2 \otimes I_2)\big)(\tilde{p}^{k+1} - \tilde{p}^k)\\
&- \sigma\|\tilde{B}\tilde{p}^k - \tilde{b}\|^2 - \sigma(\tilde{p}^k)^T\tilde{L}_1 \otimes I_2\tilde{p}^k - \sigma(\tilde{p}^k)^T\tilde{L}_2 \otimes I_2\tilde{p}^k.
\end{aligned}$$

If $\sigma \le 1/\lambda_{max}(2\tilde{c}^T\tilde{c} + \tilde{B}^T\tilde{B} + \tilde{L}_1 \otimes I_2 + \tilde{L}_2 \otimes I_2)$, the matrix $I - \sigma(2\tilde{c}^T\tilde{c} + \tilde{B}^T\tilde{B} + \tilde{L}_1 \otimes I_2 + \tilde{L}_2 \otimes I_2)$ is positive definite. And according to the definition of $\tilde{L}_1$ and $\tilde{L}_2$ as well as Lemma 2, we have $\tilde{L}_1 \otimes I_2$ and $\tilde{L}_2 \otimes I_2$ are positive semi-definite. Consequently, $V(\tilde{p}_{k+1},\tilde{y}_{k+1},\tilde{\theta}_{k+1},\tilde{\eta}_{k+1}) - V(\tilde{p}_k,\tilde{y}_k,\tilde{\theta}_k,\tilde{\eta}_k) \le 0$.

The rest of proof is similar to that of Theorem 2 in [20], so we leave it out here due to page limit. □

4 Illustrative Example

To verify the proposed distributed algorithm, we show the simulation results of the distributed optimal formation control schemes applied to 33 robots working in a two-dimensional environment. The robot's position is represented by its coordinates. The task is to find a new formation like Fig. 2 based on the principle of minimum total travel distance. Figure 3 shows the preliminary assembly of the multi-robot system. The optimal formation problem in this paper is formulated as (10) in a distributed way. And the communication graph among robots has a ring structure.

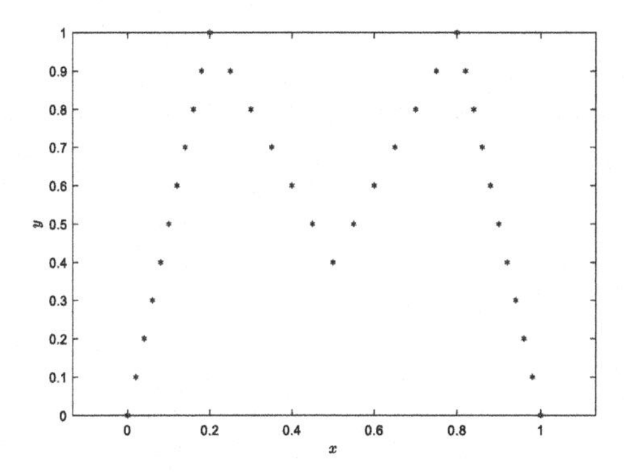

Fig. 2. Reference shape icon M.

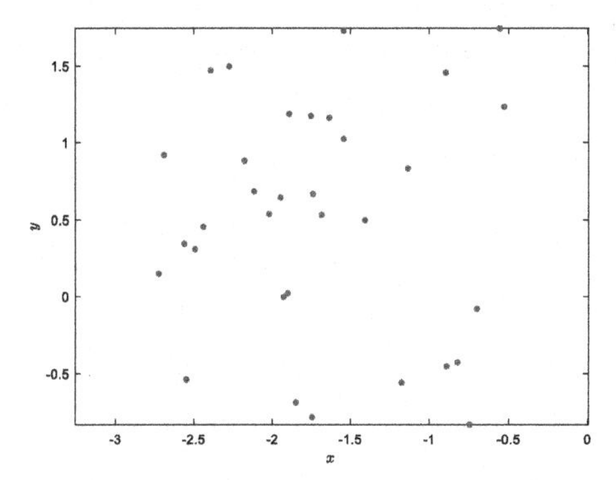

Fig. 3. The preliminary assembly of the multi-robot system.

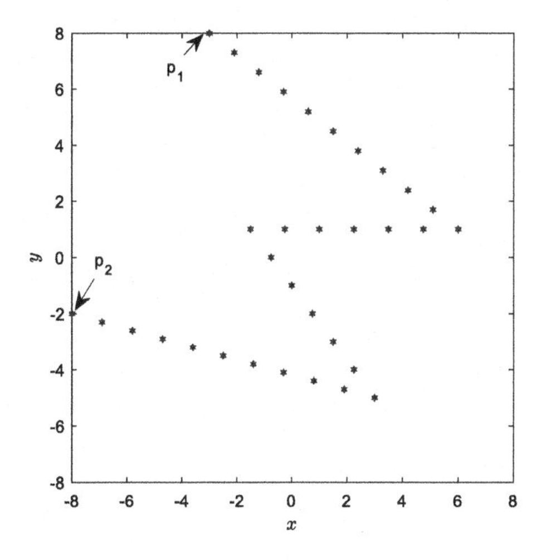

Fig. 4. Optimal formation obtained by using the Algorithm (16) with the limited region $\Omega = [-8, 8]^{2m}$.

The limited area Ω_i of robot i ($i = 1, 2, ..., 33$) is set as $[-8, 8]^2$. The optimal formation obtained by using the Algorithm (16) is shown in Fig. 4. In addition, Fig. 4 depicts the simulation formation on an example with two anchors p_1 and p_2 satisfying $p_1 - p_2 = (5, 10)^T$, indicating that the shape rotates along the direction $(5, 10)^T$.

5 Conclusion

According to the shape theory and utilizing auxiliary estimators, the optimal formation of multi-robot system has been formulated as a distributed optimization problem in this paper. Then a distributed projection-based algorithm has been presented to deal with this problem. The states of all robots obtained by the proposed distributed algorithm are convergent globally under the assumption that the communication graph is undirect and connected. Moreover, all robots have formed an ideal formation within the limited area. At the end, simulation results are given to show the effectiveness of the designed distributed technique.

References

1. Yue, L., Fan, H.: Dynamic scheduling and path planning of automated guided vehicles in automatic container terminal. IEEE/CAA J. Automatica Sinica **9**(11), 2005–2019 (2022)
2. Zuo, Z., Liu, C., Han, Q.L., Song, J.: Unmanned aerial vehicles: control methods and future challenges. IEEE/CAA J. Automatica Sinica **9**(4), 601–614 (2022)
3. Li, J., Chen, L., Cao, M., Li, C.: Satellite formation flying control by using only angle measurements. IEEE Trans. Aerosp. Electron. Syst. **59**(2), 1439–1451 (2023)
4. Timothy, H.C., Geoffrey, A.H., Volkan, I.: Search and pursuit-evasion in mobile robotics. Auton. Robot. **31**(4), 299–316 (2011)
5. Chen, X., He, S., Zhang, Y., Tong, L.C., Shang, P., Zhou, X.: Yard crane and AGV scheduling in automated container terminal: a multi-robot task allocation framework. Transp. Res. Part C Emerg. Technol. **114**, 241–271 (2020)
6. Ouyang, Q., Wu, Z., Cong, Y., Wang, Z.: Formation control of unmanned aerial vehicle swarms: a comprehensive review. Asian J. Control **25**(1), 570–593 (2023)
7. Xiao, H., Chen, C.L.P.: Leader-follower consensus multi-robot formation control using neurodynamic-optimization-based nonlinear model predictive control. IEEE Access **7**, 43581–43590 (2019)
8. Yoo, S.J., Park, B.: Connectivity preservation and collision avoidance in networked nonholonomic multi-robot formation systems: unified error transformation strategy. Automatica **103**, 274–281 (2019)
9. Chen, Z., Jiang, C., Guo, Y.: Distance-based formation control of a three-robot system. In: 2019 Chinese Control And Decision Conference (CCDC), pp. 5501–5507 (2019)
10. Pang, Z.H., Zheng, C.B., Sun, J., Han, Q.L., Liu, G.P.: Distance- and velocity-based collision avoidance for time-varying formation control of second-order multi-agent systems. IEEE Trans. Circ. Syst. II **68**(4), 1253–1257 (2021)
11. Huang, J., Zhou, S., Tu, H., Yao, Y., Liu, Q.: Distributed optimization algorithm for multi-robot formation with virtual reference center. IEEE/CAA J. Automatica Sinica **9**(4), 732–734 (2022)

12. Oh, K.K., Park, M.C., Ahn, H.S.: A survey of multi-agent formation control. Automatica **53**, 424–440 (2015)
13. Cao, K., Qiu, Z., Xie, L.: Relative docking and formation control via range and odometry measurements. IEEE Trans. Control Netw. Syst. **7**(2), 912–922 (2020)
14. Wang, Y., Cheng, L., Hou, Z., Yu, J., Tan, M.: Optimal formation of multirobot systems based on a recurrent neural network. IEEE Trans. Neural Netw. Learn. Syst. **27**(2), 322–333 (2016)
15. Fukushima, H., Kon, K., Matsuno, F.: Model predictive formation control using branch-and-bound compatible with collision avoidance problems. IEEE Trans. Rob. **29**(5), 1308–1317 (2013)
16. Derenick, J.C., Spletzer, J.R.: Convex optimization strategies for coordinating large-scale robot formations. IEEE Trans. Rob. **23**(6), 1252–1259 (2007)
17. Spletzer, J., Fierro, R.: Optimal positioning strategies for shape changes in robot teams. In: Proceedings of the 2005 IEEE International Conference on Robotics and Automation, pp. 742–747 (2005)
18. Yang, Z., Pan, X., Zhang, Q., Chen, Z.: Finite-time formation control for first-order multi-agent systems with region constraints. Front. Inf. Technol. Electron. Eng. **22**(1), 134–140 (2021)
19. Liu, Q., Wang, M.: A projection-based algorithm for optimal formation and optimal matching of multi-robot system. Nonlinear Dyn. **104**(1), 439–450 (2021)
20. Zhao, Y., Liu, Q.: A consensus algorithm based on collective neurodynamic system for distributed optimization with linear and bound constraints. Neural Netw. **122**, 144–151 (2019)
21. Lancaster, P., Tismenetsky, M.: The Theory of Matrices: With Applications. Academic Press (1985)

A Stochastic Gradient-Based Projection Algorithm for Distributed Constrained Optimization

Keke Zhang[1,2], Shanfu Gao[1], Yingjue Chen[1], Zuqing Zheng[3], and Qingguo Lü[1,2(✉)]

[1] College of Computer Science, Chongqing University, Chongqing 400044, People's Republic of China
qglv@cqu.edu.cn

[2] Key Laboratory of Smart Grid, Chengdu 610065, Sichuan Province, People's Republic of China

[3] School of Automation, Central South University, Changsha 410000, People's Republic of China

Abstract. This paper investigates a category of constrained convex optimization problems, where the collective objective function is represented as the sum of all local objective functions subjected to local bounds and equality constraints. This kind of problems is important and can be formulated form a variety of applications, such as power control, sensor networks and source localization. To solve this problem more reliable and effective, we propose a novel distributed stochastic gradient-based projection algorithm under the presence of noisy gradients, where the gradients are infiltrated by arbitrary but uniformly bounded noise sample through local gradient observation. The proposed algorithm allows the adoption of constant step-size, which guarantees it can possess faster convergence rate compared with existing distributed algorithms with diminishing step-size. The effectiveness of the proposed algorithm is verified and testified by simulation experiments.

Keywords: Stochastic gradient algorithm · Distributed constrained optimization · Multi-agent systems · Constant step-size · Linear convergence

This work was supported in part by the project funded by Sichuan Key Laboratory of Smart Grid (Sichuan University) under Grant 2023-IEPGKLSP-KFYB01; in part by the project funded by Hubei Province Key Laboratory of Intelligent Information Processing and Real-time Industrial System (Wuhan University of Science and Technology) under Grant ZNXX2022004; in part by the Natural Science Foundation of Chongqing under Grant CSTB2022NSCQ-MSX1627; in part by the Chongqing Postdoctoral Science Foundation under Grant 2021XM1006; in part by the China Postdoctoral Science Foundation under Grant 2021M700588; in part by the Fundamental Research Funds for the Central Universities under grant 2023CDJXY-039; in part by the Natural Science Foundation of China under Grant 62302068.

B. Luo et al. (Eds.): ICONIP 2023, LNCS 14447, pp. 356–367, 2024.
https://doi.org/10.1007/978-981-99-8079-6_28

1 Introduction

Distributed optimization strives to achieve optimal control or decision making by leveraging the computational nodes to locally manipulate their private data and propagate relevant local information throughout a designated network. Over the years, the advancement in control theory related to multi-agent networks has sparked a growing interest in distributed optimization, which holds great promise for various engineering fields, including but not limited to sensor networks [1, 2], smart grids [3,4], resource allocation [5,6], and machine learning [7,8]. The majority of challenges encountered in these fields can be modeled as distributed optimization problems, characterized by the collaborative minimization of the sum of all local objective functions subject to local variable constraints. This process is achieved through communication with neighboring nodes within the network, enabling a distributed optimization approach that enhances overall system performance while preserving data privacy and communication efficiency.

In most of existing literature involved so far, the solution methodologies of distributed optimization problems can be divided into two main methods: the primal domain methods and the dual domain methods. In the primal domain methods, typical distributed optimization algorithms are mainly based on subgradient methods, such as subgradient descent [9] and dual averaging subgradient [10]. To meet the demand for dealing with more practical situations, various extensions of [9,10] are studied e.g., random networks [11], time-varying directed networks [12], and communication delays [13]. However, this class of gradient-based algorithms requires that their step-sizes are diminished to guarantee exact convergence, because of which the convergence rate is limited to slow. To accelerate the convergence rate, a distributed algorithm with constant step-size is proposed in [14] through utilizing Nesterov gradient-based method, which guarantees not only exact convergence but also faster rate. Considering that the primal domain method based algorithms have intuitive form and low computation burden, similar way has been adopted by subgradient-based tracking algorithm [15,16] recently. Furthermore, some related research along this line has been extended to directed networks [17–19], time-varying or stochastic networks [20,21].

Unlike primal domain methods, a given optimization problem is usually formulated into constrained model by the dual domain methods firstly, then the primal variables are solved through minimizing some augmented Lagrangian related functions for a fixed dual variable at each iteration, and after that the values of dual variables are renewed accordingly [22]. In particular, the performance of distributed augmented Lagrange (AL) methods usually indicates that they have a linear convergence rate, which is faster than some distributed gradient-based methods in the primal domain. Specially, the decentralized alternating direction method of multipliers (ADMM) in [23] is one of a typical category in the AL methods, and many effective extensions of distributed algorithms [24,25] are developed based on this method. Although the distributed ADMM methods achieve linear convergence rate, its high computation burden is inevitably incurred since the local problem assigned for each node needs to be optimized at each iteration. Hence, to reduce the complexity of computation, the linearized ADMM algorithms [26,27] are proposed through utilizing the saddle point methods, and the update of primal variables is executed through a gradient-like step at each iteration. Moreover, an exact first-order algorithm (EXTRA) is developed in [28] based on the saddle point methods, which provides a base for some interesting algorithms [8,15,16,20,29,30]. Especially, the results in [26] and [28] indicate that both the linearized ADMM algorithms and EXTRA achieve fast convergence while enjoy low computation burden. The algorithms [9,10,12,14–16,20,22–28] mentioned above all assume that the communications are perfect without any disturbance under the ideal conditions. However, noises always exist in the communication channels, which is often inevitable. In particular, the gradient estimation of local objective function is often corrupted by noise in the gradient-based optimization, and there are various sources responsible for this phenomenon, such as quantization errors and sampling.

Considering the practical situation, the distributed implementation of stochastic gradient algorithms has gained significant interest in recent years. In [31], to accommodate the presence of subgradient errors and noisy communication links, a distributed asynchronous algorithm for constrained optimization is proposed with two diminishing step-sizes, where the estimates are proven to converge to a random point of the optimal solution almost surely. For the time-varying directed networks, a stochastic gradient-push algorithm is proposed in [32] to accommodate the presence of noisy gradient. If the objective functions are Lipschitz continuous and strongly convex, its convergence rate is proved to be asymptotically faster than the one in [12]. When the distributed convex optimization subjects to inequality constraints, a stochastic gradient algorithm is introduced [33] to deal with the presence of noisy gradient, and two convergence rates are established for both strongly and non-strongly convex optimization. It is worth noting that the aforementioned methods [31–33] all utilize the diminishing step-sizes to guarantee the convergence, and slow convergence rate is inevitably caused. Recent literature on this similar topic can be found in [34–39], to name a few.

The literature from recent years, including [39,40], and [41], are particularly relevant to the work presented in this paper. Ram *et al.* introduced a distributed stochastic subgradient projection algorithm in [39], which study the influences of stochastic subgradient (known as subgradient with stochastic error) on the convergence of the algorithm over deterministic switching networks. If the expected error is diminished with uniformly bound, the algorithm in [39] can converge to the optimal solution with diminishing step-sizes in probability one. However, when the step-size is constant, the performance of the algorithm in [39] only attains a bound to the optimal solution after a finite number of iterations. Liu *et al.* [40] and Lei *et al.* [41] developed distributed algorithms for constrained convex optimization over time-invariant and undirected network, and they all utilizes constant step-size at the gradient descent step which guarantees a fast convergence rate, but both of them did not take the noisy gradient into account. In addition, Although the analysis method in [41] is more intuitive than that in [40], the algorithm in [41] can only deal with variable constraints instead of equality constraints in [40]. From the above analysis, it is worth to develop a stochastic gradient-based algorithm to against noisy gradient for distributed constrained optimization, while improve its convergence speed and simplify its analysis process.

Contributions: (i)We propose a novel distributed stochastic gradient-based projection algorithm through adopting the primal-dual method over time-invariant undirected network. Unlike the existing methods [9,10,14–16,20,26,28] which only consider unconstrained optimization problem under perfect communications, the proposed algorithm can effectively solve a class of constrained optimization problems with the existence of noisy gradient. (ii)The proposed algorithm allows the utilization of constant step-size for the local optimization, and explicit upper bounds for the selection of constant step-size are also given for cases with or without constraints. Compared with related work [40], a linear convergence rate for the proposed algorithm for the case without constraints is shown, which can be used as a reference to estimate the convergence rate of the case with constraints.

Outlines: Sect. 2 describes the concerned problem. Section 3 provides the proposed algorithm. Section 4 offers some numerical examples. Section 4 concludes this article.

2 Problem Statement

In this section, the formal formulation of the distributed optimization problem with constraints is first presented, and the reformulation of the problem is introduced afterwards.

2.1 Problem Formulation

This paper focuses on a network with m nodes collaboratively solve the following constrained convex optimization problem:

$$\begin{aligned} &\min_{\bar{x}\in\mathbb{R}^n} \bar{f}(\bar{x}) = \sum_{i=1}^{m} f_i(\bar{x}) \\ &\text{s.t. } B_i\bar{x} = b_i,\ \bar{x}\in\mathcal{X},\ i=\{1,\cdots,m\} \end{aligned} \tag{1}$$

where $f_i : \mathbb{R}^n \to \mathbb{R}$ is the local objective function, $B_i \in \mathbb{R}^{q_i\times n}$, $b_i \in \mathbb{R}^{q_i}$ $(0 \leq q_i < n)$, $\mathcal{X}_i \subseteq \mathbb{R}^n$ is a nonempty closed convex set, and the set $\mathcal{X} = \prod_{i=1}^m \mathcal{X}_i$ is the Cartesian product. Specifically, all the local nodes have the capacity of communication and calculation, and the f_i, B_i, b_i and $\mathcal{X}_i$ are the local private information for node i which only can be accessed by itself.

2.2 Problem Reformulation

According to properties of Laplacian matrix, the problem (1) can be reformulated as follow:

$$\begin{aligned} &\min_{x\in\mathbb{R}^{mn}} f(x) = \sum_{i=1}^{m} f_i(x_i) \\ &\text{s.t. } Bx = b\,,\ \mathcal{L}x = 0,\ x\in\mathcal{X} \end{aligned} \tag{2}$$

where B =blkdiag$\{B_1, \cdots, B_m\} \in \mathbb{R}^{q \times mn}(q = \sum_{i=1}^{m} q_i)$ is the block diagonal matrix, $b = ((b_1)^{\mathrm{T}}, \cdots, (b_m)^{\mathrm{T}})^{\mathrm{T}} \in \mathbb{R}^q$ is column vector, and $\mathcal{L} = \mathcal{L}_m \otimes \mathrm{I}_n \in \mathbb{R}^{mn \times mn}$ is the Laplacian matrix.

For dealing with the equality constraints in problem (2), an augmented Lagrangian function is constructed as follows:

$$\mathscr{L}(x, y, z) = f(x) + y^{\mathrm{T}}(Bx - b) + z^{\mathrm{T}}\mathcal{L}x, \tag{3}$$

where $y \in \mathbb{R}^q, z \in \mathbb{R}^{mn}$ are the vectors form of augmented Lagrangian multipliers corresponding to the coupling constraints $Bx = b$ and $\mathcal{L}x = 0$, respectively.

Then, the problem (2) can be reformulated as

$$\min\nolimits_{x \in \mathcal{X}} \max\nolimits_{y \in \mathbb{R}^q, z \in \mathbb{R}^{mn}} \mathscr{L}(x, y, z) \tag{4}$$

Assumption 1: *Each objective function $f_i, i = 1, \cdots, m$, is σ_i-smoothness and μ_i-strong convexity, where $0 < \mu_i \leq \sigma_i$. In the following analysis, we adopt $\hat{\mu} \triangleq \max_{i=1}^{m}\{\mu_i\}$ and $\check{\sigma} \triangleq \min_{i=1}^{m}\{\sigma_i\}$.*

Assumption 2: *The underlying undirected graph $\mathcal{G}$ is connected.*

3 Optimization Algorithm

To solve the problem (1), we propose a stochastic gradient-based projection algorithm (Algorithm 1) through adopting the primal-dual method [42].

Through stacking all the local ones $x_{i,k}$, $\widehat{\nabla f}_i(x_{i,k})$, $y_{i,k}$ and $z_{i,k}$, for $i = \{1, \cdots, m\}$ on top of one another at iteration k, we rewrite the Algorithm 1 in a compact vector form as follow:

$$\begin{aligned} x_{k+1} = \mathcal{P}_{\mathcal{X}}[&x_k - \alpha(\widehat{\nabla f}(x_k) - \alpha B^{\mathrm{T}}(y_k + Bx_k - b) \\ &+ \mathcal{L}x_k + \mathcal{L}z_k)] \end{aligned} \tag{5a}$$

$$y_{k+1} = y_k + Bx_{k+1} - b \tag{5b}$$

$$z_{k+1} = z_k + \alpha\mathcal{L}x_{k+1} \tag{5c}$$

where $x_k = [x_{1,k}^{\mathrm{T}}, \cdots, x_{m,k}^{\mathrm{T}}]^{\mathrm{T}} \in \mathbb{R}^{mn}$, $\widehat{\nabla f}(x_k) = [\widehat{\nabla f}_1^{\mathrm{T}}(x_{1,k}), \cdots, \widehat{\nabla f}_m^{\mathrm{T}}(x_{m,k})]^{\mathrm{T}} \in \mathbb{R}^{mn}$, $y_k = [y_{1,k}^{\mathrm{T}}, \cdots, y_{m,k}^{\mathrm{T}}]^{\mathrm{T}} \in \mathbb{R}^q$, $z_k = [z_{1,k}^{\mathrm{T}}, \cdots, z_{m,k}^{\mathrm{T}}]^{\mathrm{T}} \in \mathbb{R}^{mn}$.

Remark 1 *: Since communication noise is often inevitable in practical networks, the situation may be occurred that node $i \in \mathcal{V}$ can not access its exact gradient but substitution of gradient embroiled with noise sample due to the existence of observation noises. Hence, we consider the case that the noise sample for node $i \in \mathcal{V}$ only generates from the access to its local gradient observation, while each node i has access to the unbiased estimates of its local gradient. That is, for a arbitrarily given point $x_{i,k} \in \mathbb{R}^n$, the node i can not observe the exact gradient $\nabla f_i(x_{i,k})$ but the one with noise $\widehat{\nabla f}_i(x_{i,k}) = \nabla f_i(x_{i,k}) + \vartheta_{i,k}$ at the k-th iteration for $\forall k \geq 1$, and $\mathbb{E}[\widehat{\nabla f}_i(x_{i,k})] = \nabla f_i(x_{i,k})$, which has more practical significance.*

Algorithm 1. A Stochastic Gradient-based Projection Algorithm for Distributed Constrained Optimization

1: **Initialization:** Let $x_{i,0}, z_{i,0} \in \mathbb{R}^n, y_{i,0} \in \mathbb{R}^{q_i}$ for $i \in \mathcal{V}$.
2: **Repeat**
3: **Set** $k = 0$
4: **For** $i = 1$ **to** m **do**
5: **Local gradients estimation**

$$\widehat{\nabla f}_i(x_{i,k}) = \nabla f_i(x_{i,k}) + \vartheta_{i,k}$$

6: **Local variables updates**
7: Update variable $x_{i,k}$ according to:

$$\begin{aligned} x_{i,k+1} = \mathcal{P}_{\mathcal{X}_i}[x_{i,k} - \alpha\widehat{\nabla f}_i(x_{i,k}) - \alpha(B_i)^{\mathrm{T}}(y_{i,k} + \\ B_i x_{i,k} - b_i) - \alpha\textstyle\sum_{j=1,j\neq i}^{m} a_{ij}(x_{i,k} - x_{j,k}) \\ -\alpha\textstyle\sum_{j=1,j\neq i}^{m} a_{ij}(z_{i,k} - z_{j,k}))] \end{aligned}$$

8: Update variable $y_{i,k}$ according to:

$$y_{i,k+1} = y_{i,k} + B_i x_{i,k+1} - b_i$$

9: Update variable $w_{i,k}$ according to:

$$z_{i,k+1} = z_{i,k} + \alpha\sum_{j=1,j\neq i}^{m} a_{ij}(x_{i,k+1} - x_{j,k+1})$$

10: **End for**
11: **Set** $k = k + 1$ and **repeat**.
12: **Until** a predefined stopping rule is satisfied.

Assumption 3: *The noise samples for gradients are drawn i.i.d. from some probability space $\mathcal{F} = (\Omega, \mathcal{B}, \mathcal{P})$ such that the mean of independent random vector $\vartheta_{i,k}$ is zero, and its mean square is uniformly bounded with probability one. That is, for $\forall k \geq 1$,*

$$\mathbb{E}[\vartheta_{i,k}|\mathcal{F}_{k-1}] = 0, \ \mathbb{E}[\|\vartheta_{i,k}\|^2|\mathcal{F}_{k-1}] \leq \hat{\nu}^2 \tag{6}$$

where $\hat{\nu}$ is some deterministic constant. In addition, we adopt $\mathcal{F}_0 \subset \mathcal{F}_1 \subset \cdots$ to indicate the σ-algebra of $\mathcal{F}_k$ which is generated by the entire history of the algorithm up to iteration k, i.e., for $\forall k \geq 1$, $\mathcal{F}_k = \{(x_{i,0}, i \in \mathcal{V}); \vartheta_{i,l}, 1 \leq l \leq k-1\}$.

4 Numerical Simulation Result

In this section, a traditional minimization problem is used to test the performance of the proposed algorithm to verify its effectiveness. In particular, we consider nodes of a network cooperatively settling the following distributed optimization problem:

$$\min f(\bar{x}) = \frac{1}{2}\sum_{i=1}^{m} \|D_i\bar{x} - d_i\|^2$$
$$\text{s.t. } B_i^{\mathrm{T}}\bar{x} = b_i,\ 0 \leq \bar{x} \leq 5,\ \forall i \in \{1, \cdots, m\}$$

where $f_i(\bar{x}) = \|D_i\bar{x} - d_i\|^2$ is the local objective function, $\bar{x} \in \mathbb{R}^n$, $D_i \in \mathbb{R}^{q_i \times n}$, $d_i \in \mathbb{R}^{q_i}$, $B_i \in \mathbb{R}^n$, $b_i \in \mathbb{R}^m$, respectively. Let $n = 5$, $m = 10$, $q_i = 5$ for $\forall i \in \{1, 2, \cdots, m\}$, and and $B_1, B_6 = [1,1,0,0,0]$, $B_2, B_7 = [0,0,1,1,0]$, $B_3, B_8 = [1,0,0,1,1]$, $B_4, B_9 = [1,0,1,0,1]$, $B_5, B_{10} = [1,1,0,0,1]$. The components of D_i are equal to 1, d_i and b_i are randomly selected in $[0,5]$ and $[0,10]$, respectively. The estimation error is defined as $E(k) = \log_{10}((1/m)\sum_{i=1}^{m} \|x_{i,k} - x^*\|)$.

4.1 Algorithm 1 Under Different Scenarios

This case study involves an investigation of the performance of the proposed algorithm under various scenarios. From Fig. 1, it is shown that the proposed algorithm has a fast convergence rate with or without equality and bound constraints, and the one with constant step-size has faster convergence rate than that one with diminishing step-size ($\alpha_k = 1/k$) under the case with or without noisy gradients, but the fluctuation of the former is larger than the latter.

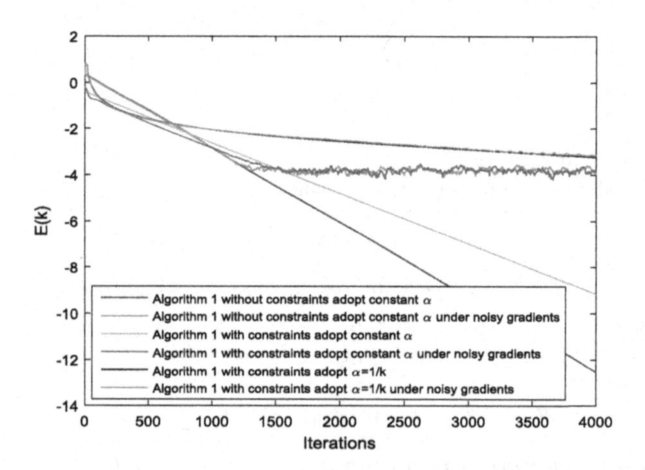

Fig. 1. Evolution of $E(k)$ for Algorithm 1 under different scenarios.

4.2 Effects of Network Sparsity

This case study examines the impact of network sparsity on the performance of the proposed algorithm. The results in Fig. 2 indicates that the convergence rate of the proposed algorithm becomes faster when the sparsity of network becomes dense.

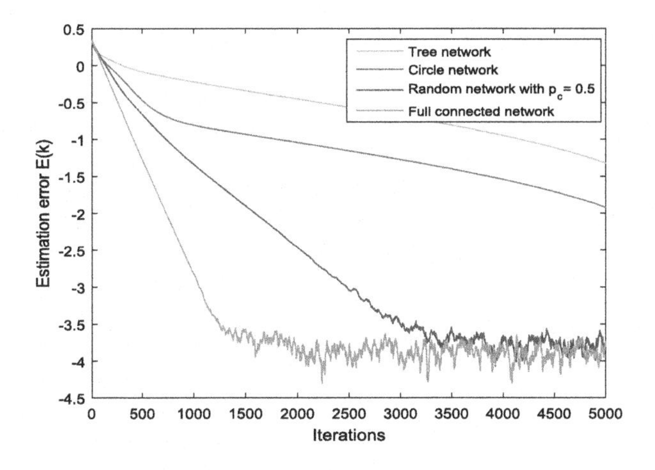

Fig. 2. Evolution of $E(k)$ for Algorithm 1 with constraints under different networks.

4.3 Effects of Adjacency Edge Weights

This section of the case study investigates the impact of various adjacency edge weights on the convergence of the proposed algorithm. The weights a_{ij}, $\forall i \in \{1, \cdots, m\}$ are randomly chosen in $[0.1, 0.4]$, $[0.4, 0.7]$ and $[0.7, 1.0]$, respectively. The results in Fig. 3 indicate that increasing the adjacency edge weights can accelerate the convergence rate of the proposed algorithm.

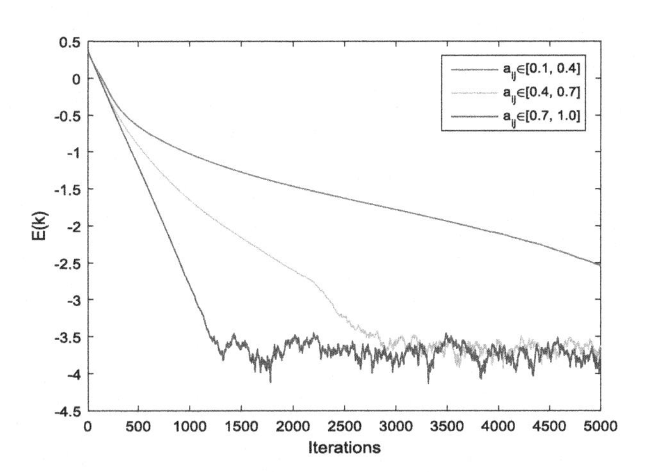

Fig. 3. Time evolution of $E(k)$ for Algorithm 1 with constraints under different adjacency edge weights.

4.4 Effects of Constant Step-Size

It is worth noting that the selection of a constant step-size also has an impact on the convergence rate of the proposed algorithm. Therefore, this case study aims to compare the performance of the proposed algorithm under different step-size. The simulation results in Fig. 4 indicate that the convergence rate of the proposed algorithm is fluctuating not always increasing when $0.02 < \alpha < 0.1$. The main cause of fluctuation is the proposed algorithm has already converged to the optimal solution at the given point under the noisy gradients, but the accuracy is wandering in a certain range when a certain accuracy is reached, and the range depends on the size of the noise. Therefore, when the constant step-size does not exceed a certain upper bound (in this case the upper bound for constant step-size is 0.1), the convergence rate of the proposed algorithm can be accelerated through increasing the value of the constant step-size.

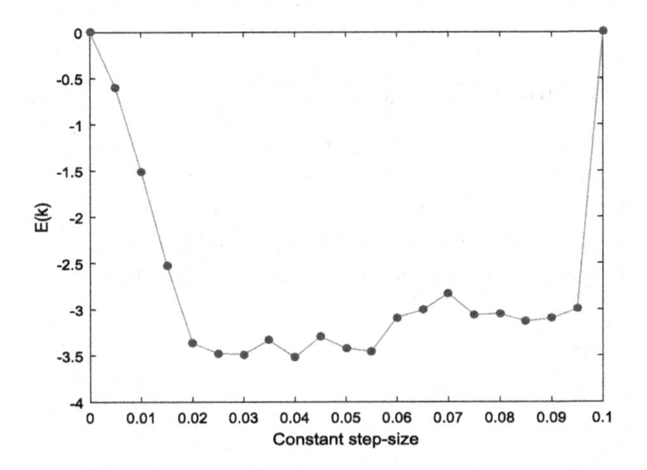

Fig. 4. Evolution of $E(k)$ for Algorithm 1 with constraints at the 300-th iteration with different constant step-size.

5 Conclusion

In this paper, we developed a distributed stochastic gradient-based projection algorithm with constant step-size for a class of constrained optimization problems based on primal-dual methods. The proposed algorithm allowed the adoption of constant step-size, which guaranteed it could possess faster convergence rate compared with existing distributed algorithms with diminishing step-size. The simulation results effectively demonstrate the efficacy and superiority of the proposed algorithm. As a result, future work could investigate the extension of the algorithm to more general directed and time-varying networks.

References

1. Lü, Q., Liao, X., Deng, S., Li, H.: A decentralized stochastic algorithm for coupled composite optimization with linear convergence. IEEE Trans. Sig. Inf. Process. Netw. **8**, 627–640 (2022)
2. Zhu, S., Chen, C., Li, W., Yang, B., Guan, X.: Distributed optimal consensus filter for target tracking in heterogeneous sensor networks. IEEE Trans. Cybern. **43**(6), 1963–1976 (2013)
3. Lü, Q., Liao, X., Li, H., Deng, S., Gao, S.: Distributed Optimization in Networked Systems: Algorithms and Applications. Springer, Singapore (2023). https://doi.org/10.1007/978-981-19-8559-1
4. Zhang, H., Li, H., Zhu, Y., Wang, Z., Xia, D.: A distributed stochastic gradient algorithm for economic dispatch over directed network with communication delays. Int. J. Electr. Power Energy Syst. **110**, 759–771 (2019)
5. Lakshmanan, H., Defarias, D.: Decentralized resource allocation in dynamic networks of agents. SIAM J. Optim. **19**(2), 911–940 (2008)
6. Li, H., et al.: Decentralized dual proximal gradient algorithms for non-smooth constrained composite optimization problems. IEEE Trans. Parallel Distrib. Syst. **32**(10), 2594–2605 (2021)
7. Cevher, V., Becker, S., Schmidt, M.: Convex optimization for big data: scalable, randomized, and parallel algorithms for big data analytics. IEEE Sig. Process. Mag. **31**(5), 32–34 (2014)
8. Zhang, K., Liao, X., Lü, Q.: Privacy-protected decentralized dual averaging push with edge-based correlated perturbations over time-varying directed networks. IEEE Trans. Netw. Sci. Eng. **9**(6), 4145–4158 (2022)
9. Nedic, A., Ozdaglar, A.: Distributed subgradient methods for multi-agent optimization. IEEE Trans. Autom. Control **54**(1), 48–61 (2009)
10. Lü, Q., et al.: Privacy-preserving decentralized dual averaging for online optimization over directed networks. IEEE Trans. Ind. Cyber-Phys. Syst. **1**, 79–91 (2023). https://doi.org/10.1109/TICPS.2023.3291667
11. Lobel, I., Ozdaglar, A.: Convergence analysis of distributed subgradient methods over random networks. In: 46th Annual Allerton Conference on Communication, Control, and Computing, pp. 353–360 (2008)
12. Nedic, A., Olshevsky, A.: Distributed optimization over time-varying directed graphs. IEEE Trans. Autom. Control **60**(3), 601–615 (2015)
13. Nedic, A., Ozdaglar, A.: Convergence rate for consensus with delays. J. Global Optim. **47**(3), 437–456 (2010)
14. Jakovetic, D., Xavier, J., Moura, J.: Fast distributed gradient methods. IEEE Trans. Autom. Control **59**(5), 1131–1146 (2014)
15. Qu, G., Li, N.: Harnessing smoothness to accelerate distributed optimization. IEEE Trans. Control Netw. Syst. **5**(3), 1245–1260 (2018)
16. Nedic, A., Olshevsky, A., Shi, W., Uribe, C.: Geometrically convergent distributed optimization with uncoordinated step-sizes. In: Proceedings of the American Control Conference, pp. 3950–3955 (2017)
17. Xi, C., Xin, R., Khan, U.: ADD-OPT: accelerated distributed directed optimization. IEEE Trans. Autom. Control **63**(5), 1329–1339 (2018)
18. Xin, R., Khan, U.: A linear algorithm for optimization over directed graphs with geometric convergence. IEEE Control Syst. Lett. **2**(3), 325–330 (2018)
19. Li, H., Huang, C., Chen, G., Liao, X., Huang, T.: Distributed consensus optimization in multiagent networks with time-varying directed topologies and quantized communication. IEEE Trans. Cybern. **47**(8), 2044–2057 (2017)

20. Nedic, A., Olshevsky, A., Shi, W.: Achieving geometric convergence for distributed optimization over time-varying graphs. SIAM J. Optim. **27**(4), 2597–2633 (2017)
21. Xu, J., Zhu, S., Soh, Y., Xie, L.: Convergence of asynchronous distributed gradient methods over stochastic networks. IEEE Trans. Autom. Control **63**(2), 434–448 (2018)
22. Terelius, H., Topcu, U., Murray, R.: Decentralized multi-agent optimization via dual decomposition. In: Proceedings of the 18th World Congress, pp. 7391–7397 (2011)
23. Boyd, S., Parikh, N., Chu, E., Peleato, B., Eckstein, J.: Distributed optimization and statistical learning via the alternating direction method of multipliers. Found. Trends Mach. Learn. **3**(1), 122 (2011)
24. Iutzeler, F., Bianchi, P., Ciblat, P., Hachem, W.: Explicit convergence rate of a distributed alternating direction method of multipliers. IEEE Trans. Autom. Control **61**(4), 892–904 (2016)
25. Shi, W., Ling, Q., Yuan, K., Wu, G., Yin, W.: On the linear convergence of the ADMM in decentralized consensus optimization. IEEE Trans. Sig. Process. **62**(7), 1750–1761 (2014)
26. Ling, Q., Shi, W., Wu, G., Ribeiro, A.: DLM: decentralized linearized alternating direction method of multipliers. IEEE Trans. Sig. Process. **63**(15), 4051–4064 (2015)
27. Aybat, N., Wang, Z., Lin, T., Ma, S.: Distributed linearized alternating direction method of multipliers for composite convex consensus optimization. IEEE Trans. Autom. Control **63**(1), 5–20 (2018)
28. Shi, W., Ling, Q., Wu, G., Yin, W.: EXTRA: an exact first-order algorithm for decentralized consensus optimization. SIAM J. Optim. **25**(2), 944–966 (2015)
29. Wang, Z., Li, H.: Edge-based stochastic gradient algorithm for distributed optimization. IEEE Trans. Netw. Sci. Eng. **7**, 1421–1430 (2020). https://doi.org/10.1109/TNSE.2019.2933177
30. Li, H., Lü, Q., Huang, T.: Convergence analysis of a distributed optimization algorithm with a general unbalanced directed communication network. IEEE Trans. Netw. Sci. Eng. **6**(3), 237–248 (2019)
31. Srivastava, K., Nedic, A.: Distributed asynchronous constrained stochastic optimization. IEEE J. Sel. Top. Sign. Proces. **5**(4), 772–790 (2011)
32. Nedic, A., Olshevsky, A.: Stochastic gradient-push for strongly convex functions on time-varying directed graphs. IEEE Trans. Autom. Control **61**(12), 3936–3947 (2016)
33. Yuan, D., Ho, D., Hong, Y.: On convergence rate of distributed stochastic gradient algorithm for convex optimization with inequality constraints. SIAM J. Control. Optim. **54**(5), 2872–2892 (2016)
34. Touri, B., Nedic, A., Ram, S.: Asynchronous stochastic convex optimization over random networks: error bounds. In: Information Theory and Applications Workshop, pp. 1–10 (2010)
35. Lei, J., Chen, H., Fang, H.: Asymptotic properties of primal-dual algorithm for distributed stochastic optimization over random networks with imperfect communications. SIAM J. Control. Optim. **56**(3), 2159–2188 (2018)
36. Pu, S., Nedic, A.: A distributed stochastic gradient tracking method. In: Proceedings of the IEEE Conference on Decision and Control, pp. 963–968 (2019)
37. Sayin, M., Vanli, N.D., Kozat, S., Basar, T.: Stochastic subgradient algorithms for strongly convex optimization over distributed networks. IEEE Trans. Netw. Sci. Eng. **4**(4), 248–260 (2017)

38. Sirb, B., Ye, X.: Decentralized consensus algorithm with delayed and stochastic gradients. SIAM J. Optim. **28**(2), 1232–1254 (2018)
39. Ram, S., Nedic, A., Veeravalli, V.: Distributed stochastic subgradient projection algorithms for convex optimization. J. Optim. Theor. Appl. **147**(3), 516–545 (2010)
40. Liu, Q., Yang, S., Hong, Y.: Constrained consensus algorithms with fixed step size for distributed convex optimization over multiagent networks. IEEE Trans. Autom. Control **62**(8), 4259–4265 (2017)
41. Lei, J., Chen, H., Fang, H.: Primal-dual algorithm for distributed constrained optimization. Syst. Control Lett. **96**, 110–117 (2016)
42. Bertsekas, D., Tsitsiklis, J.: Parallel and Distributed Computation: Numerical Methods. Prentice-Hall (1989)

FalconNet: Factorization for the Light-Weight ConvNets

Zhicheng Cai and Qiu Shen(✉)

School of Electronic Science and Engineering, Nanjing University, Nanjing, China
caizc@smail.nju.edu.cn, shenqiu@nju.edu.cn

Abstract. Designing light-weight CNN models with little parameters and Flops is a prominent research concern. However, three significant issues persist in the current light-weight CNNs: i) the lack of architectural consistency leads to redundancy and hindered capacity comparison, as well as the ambiguity in causation between architectural choices and performance enhancement; ii) the utilization of a single-branch depth-wise convolution compromises the model representational capacity; iii) the depth-wise convolutions account for large proportions of parameters and Flops, while lacking efficient method to make them light-weight. To address these issues, we factorize the four vital components of light-weight CNNs from coarse to fine and redesign them: i) we design a light-weight overall architecture termed LightNet, which obtains better performance by simply implementing the basic blocks of other light-weight CNNs; ii) we abstract a Meta Light Block, which consists of spatial operator and channel operator and uniformly describes current basic blocks; iii) we raise RepSO which constructs multiple spatial operator branches to enhance the representational ability; iv) we raise the concept of receptive range, guided by which we raise RefCO to sparsely factorize the channel operator. Based on above four vital components, we raise a novel light-weight CNN model termed as FalconNet. Experimental results validate that FalconNet can achieve higher accuracy with lower number of parameters and Flops compared to existing light-weight CNNs.

Keywords: Light-weight CNN · Neural Architecture Design · Computer Vision

1 Introduction

Convolutional Neural Networks (CNNs) possess superior representational capability and have dominated in various computer vision tasks. Consider the implementation on mobile devices for real-world applications, the computational and storage resources are always limited, requiring CNNs to be light-weight while maintaining competitive performance. Depth-wise separable convolution (DS-Conv) [12] factorizes the regular convolution into depth-wise convolution (DW-Conv) and point-wise convolution (PW-Conv), which decreases a large amount

This work was supported in part by the National Natural Science Foundation of China under Grant 62071216, 62231002 and U1936202.

B. Luo et al. (Eds.): ICONIP 2023, LNCS 14447, pp. 368–380, 2024.
https://doi.org/10.1007/978-981-99-8079-6_29

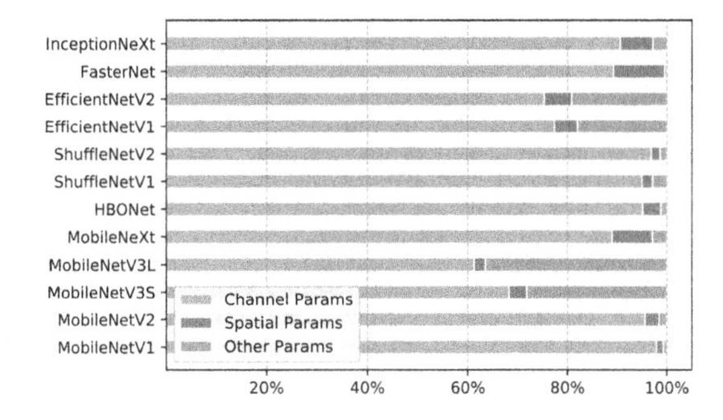

Fig. 1. The parameter proportions of channel parameters, spatial parameters and other parameters (ignore the classifier head).

of computation and parameters and has been a fundamental design component for subsequent light-weight CNNs [11,14,19,22–24,26,28]. Subsequently, many light-weight models with different overall architectures [11,22–24] and basic blocks [1,14,26,28] are raised. However, the architectural inconsistencies cause redundant structures and obscure the relationship between architectural choices and performance enhancements.

Moreover, while some works [21,25] factorize the DW Conv into parallel low-rank branches to save computational cost, the main stem for lighter models lies in the PW Conv. As Fig. 1 exhibited, the PW Conv (broadly, the densely-connected linear layers processing the channel information) accounts for the majority of the light-weight model parameters, while the DW Conv (broadly, conv layers processing the spatial information) only makes up a small proportion. While some works tends to split the channels [3,18,20,27], the information communication among channels is insufficient, thus damaging the performance.

This paper aims to design a light-weight CNN model with small amount of parameters while maintaining competitive performance. In light of the aforementioned issues, we factorize the four vital components of constructing light-weight CNNs, namely, overall architecture, meta basic block, spatial operator and channel operator. To be specific:

- We first design a light-weight overall architecture termed as ***LightNet***, which refers to the four-stage structure of modern CNNs, each stage is stacked with basic blocks. Better performance can be obtained by simply implementing the basic blocks of other light-weight CNNs on LightNet.
- We abstract and analyze a ***Meta Light Block***, consisting of spatial operator and channel operator (specifically, PW-Conv), for light-weight CNN model design. The paradigm of meta basic block uniformly describes the current basic blocks, inferring that the framework of the meta block provides the basic ability to the model, while the differences of model performances essentially come from different structural instantiations [26].
- We introduce ***RepSO*** as the spatial operator. Under the guidance of weight magnitude, RepSO constructs multiple extra branches to compensate for the

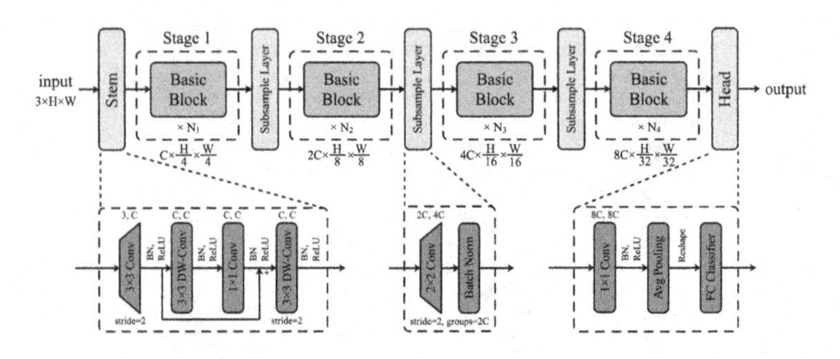

Fig. 2. LightNet overall architecture.

reduced parameters and enhance representational capacity. In inference, these diverse branches are equivalently converted into a single branch.

- We introduce the concept of ***Receptive Range*** for channel dimension. Receptive range elaborates the way of connection between the output and input neurons in the PW-Conv, which claims that one output neuron should attend to all of the input neurons directly or indirectly to obtain the full receptive range. Based on the concept of receptive range we further raise ***RefCO*** which factorizes the PW-Conv through introducing sparsity to the dense channel connection. Moreover, RefCO utilizes structural reparameterization which constructs multiple ***sparsely factorized PW-Convs*** to compensate for the reduction of channel connections.
- Finally, we raise a novel light-weight CNN model termed as ***FalconNet*** based on above four vital components. Experimental results show that FalconNet can achieve higher accuracy with less parameters and Flops compared to existing light-weight CNNs.

2 Related Works

2.1 Light-Weight CNN Models

Many light-weight CNN models with reduced parameters and Flops are proposed. IGC [20] raises interleaved group convolutions to decompose the regular convolution into multi-layer group convolutions. ShuffleNets [18,27] utilize 1×1 group convolutions and shuffles the grouped channels. MobileNetV1 [12] and Xception [2] propose the depth-wise separable convolution to decouple the regular convolution into depth-wise convolution and point-wise convolution. MobileNetV2 [19] introduces the inverted residual block. MobileNetV3 [11] enhances MobileNetV2 with SE module [13] and neural architecture search. MobileNeXt [28] introduces sandglass block to alleviate information loss by flipping the structure of inverted residual block. HBONet [14] raises harmonious bottleneck on two orthogonal dimensions to improve representation. EfficientNet [22–24] proposes a compound scaling method to scale depth, width and resolution uniformly. FasterNet [1] raises partial convolution to conduct convolution on part of the channels to reduce Flops while increasing FLOPS.

2.2 Structural Reparameterization

Structural Reparameterization [4–6,8] is a reparameterization methodology to parameterize a structure with the parameters transformed from another structure. ACNet [4] constructs vertical and horizontal convolution branches in training and converts them into the original branch in inference. RepVGG [8] constructs identity shortcuts parallel to the 3×3 convolution and converts the shortcuts into the 3×3 branches. DBB [6] constructs inception-like diverse branches of different scales and complexities to enrich the feature space.

3 Method

3.1 Design the Overall Architecture

Although certain light-weight CNNs utilize similar basic building blocks, their overall architectures differ, resulting in unnecessary redundancy that could be consolidated through unification. Moreover, certain light-weight CNNs incorporate distinct basic building blocks, yet they still exhibit varying overall architectures. This leads to an unfair comparison of capacity between the basic blocks and obscures the causal relationship between architectural choices and performance improvements. Hence we firstly intend to construct an overall light-weight architecture. Referring to the modern architectures of powerful CNN [1,17,25] and ViT [9,15] models, we raise ***LightNet*** sketched in Fig. 2.

Stem. Stem refers to the beginning part of the model. Instead of using single conv layer with large stride, we desire to capture more details by several conv layers. After the first 3×3 regular conv layer with stride 2, we add one $3\times$ DW-Conv layer, followed by a 1×1 PW-Conv layer and another 3×3 DW-Conv layer for subsampling [7]. There also exists a shortcut as shown in Fig. 2.

Stages. Stages 1–4 comprises several repeated *basic blocks.* According to the stage compute ratio of 1:1:3:1 [16,17], the block number in each stage is set as [3, 3, 9, 3]. Following the pyramid principle [10] and considering reducing the parameters, the channel dimension in each stage is set as [32, 64, 128, 256].

Subsampling Layers. We add separate subsampling layers between stages. We use 2×2 conv layer with C groups and stride 2 for halving the spatial resolution. The 2×2 conv layer also doubles the channel dimension. A batch normalization layer is arranged subsequently to stablize training.

Head. Head is the last part of model. A 1×1 conv layer is used to further mix the information, then globally average-pooled to obtain the feature vectors, which is subsequently input into the classifier and obtain the final output.

Experiments show that simply implementing the basic blocks of other light-weight CNNs on LightNet achieves better performance than the original models.

3.2 Explore the Meta Basic Block

Basic blocks are the pivotal component for light-weight CNNs. As exhibited in Fig. 3, different basic blocks can be generally abstracted into the ***Meta Basic***

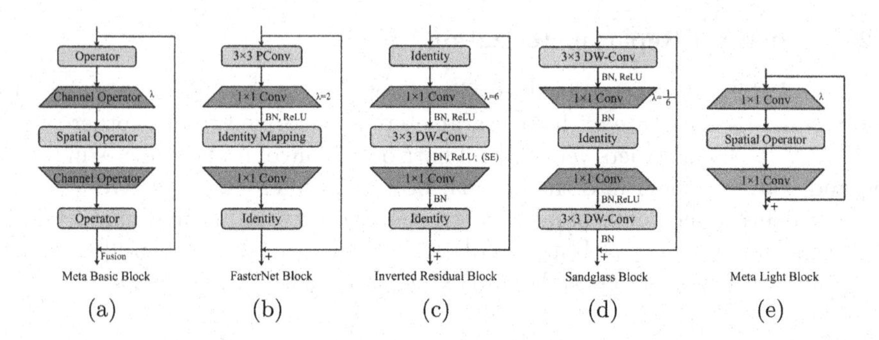

(a) (b) (c) (d) (e)

Fig. 3. Abstracted unified Meta Basic Block and Meta Light Block for light-weight models. Some correspondingly instantiated basic blocks are also exhibit.

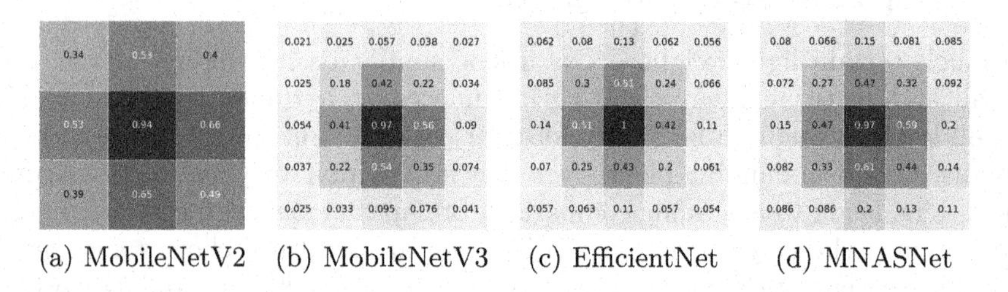

(a) MobileNetV2 (b) MobileNetV3 (c) EfficientNet (d) MNASNet

Fig. 4. The average kernel magnitude matrices of MobileNetV2, MobileNetV3, EfficientNet and MNASNet trained on ImageNet.

Block (Fig. 3(a)), which is alternately composed of spatial and channel operators (i.e., PW-Conv). The framework of the meta basic block provides the basic ability to the model, which means instantiating these spatial operators as non-learnable identity mappings can still achieve effective performance, while the differences of model performances essentially come from different structural instantiations of the meta basic block, e.g., FasterNet block instantiates the first spatial operator as PConv and instantiates the other two spatial operators as identity mapping, as shown in Fig. 3(b). Through conducting extensive experiments (Sect. 4), it is observed that the first and the last operator layers make no benefit to the enhancement of model performance, while the second spatial operator layer between two channel operator layers is significant. Thus we further simplify meta basic block into ***Meta Light Block*** (Fig. 3(d)), which consists of two PW-Conv layers (with an expansion ratio λ) and a single spatial operator layer in between.

3.3 Strengthen the Spatial Operator

Though the Meta Light Block guarantees the fundamental performance, powerful spatial operator can significantly enhance the representational capacity. Thus we construct multiple branches of versatile spatial operators to enrich the representation space. It is assumed that a position of the kernel tends to be more

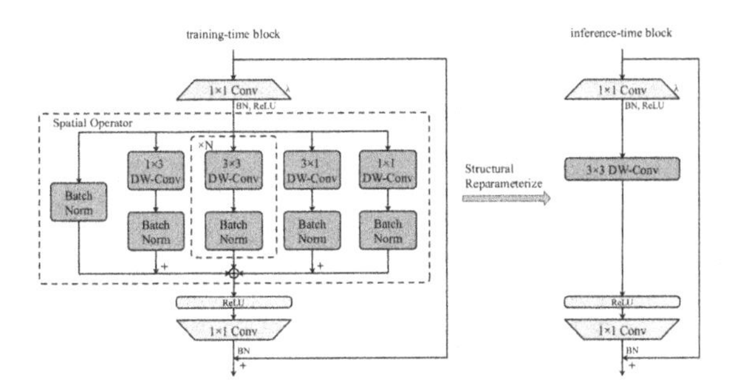

Fig. 5. Meta Light Block with RepSO

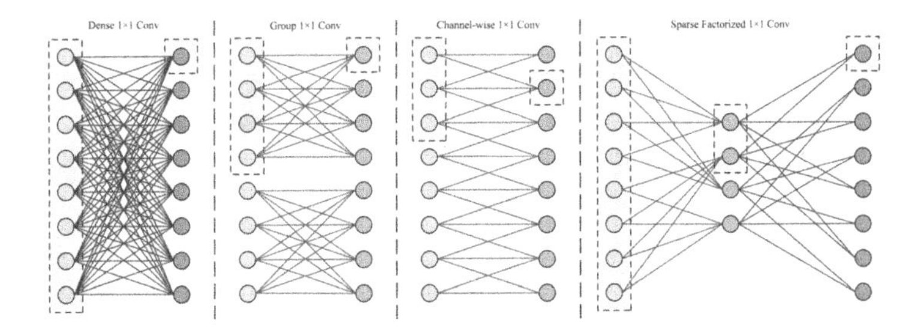

Fig. 6. Connections and corresponding receptive ranges of different 1 × 1 Convs

significant if it has a larger magnitude [4]. We first calculate and visualize the average kernel magnitude matrices of four popular light-weight CNNs as shown in Fig. 4. It is observed that the outermost circle of positions in 5 × 5 kernel have negligible importance compared to the central 3 × 3 positions, thus we use 3 × 3 conv kernels which also have fewer parameters and Flops. To compensate for the reduced feature channels, we construct N (N = 3 by default) parallel 3 × 3 DW-Conv branches. Moreover, Fig. 4 shows that the positions in the skeleton pattern of the 3 × 3 kernel account for much importance compared to these corner positions, thus we additionally construct 1 × 3 and 3 × 1 DW-Conv branches. In addition, it is observed that the central position always possesses the highest importance score, thus we construct an extra 1 × 1 DW-Conv branch for further enhancement. Last but not least, as the meta block provides the fundamental capacity, we also add a shortcut to the spatial operator layer. The obtained multi-branch operator is termed as ***RepSO*** (***Re****parameterized* ***S****patial* ***O****perator*), as sketched in Fig. 5 left, all these seven branches are individually equipped with a batch normalization layer, then the normalized outputs of each branch are element-wise added. According to the additivity and homogeneity of convolution and the methodology of structural reparameterization [4], these seven branches can be equivalently converted to a single 3 × 3 DW-Conv branch in inference as sketched in Fig. 5 right, which produces no extra inference cost.

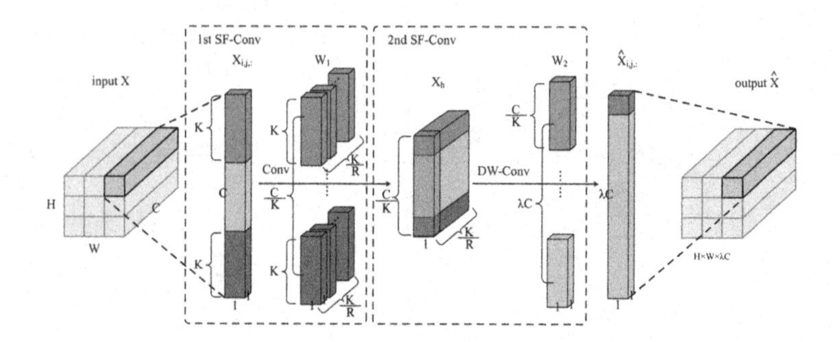

Fig. 7. Sparse Factorized 1 × 1 Convolution.

3.4 Factorize the Channel Operator

Receptive Range. For that 1 × 1 Conv accounts for large proportion of model parameters and Flops, we want to make it light-weight. Essentially, we can change the connections between input and output neurons to change the amount of parameters. Hence we propose the concept of ***Receptive Range*** as the guideline for establishing these connections. The concept of receptive range is introduced specifically for channel dimension and analogous to the receptive field for spatial dimension. The value of the receptive range of a certain output neuron is the number of input neurons that it attend to directly (by a weight) or indirectly (attend to a set of hidden neurons which attend to the input neurons). Figure 6 exhibits the connections and corresponding receptive ranges of different 1 × 1 Convs. Suppose C is the number of input neurons (channel numbers), as observed, each output neuron of dense 1×1 Conv can attend to all the input neurons directly, thus the receptive range is C. For group 1 × 1 Conv with g groups, the receptive range is $\frac{C}{G}$, thus different groups can establish information connections, causing significant reduce of representation capacity. For channel-wise 1 × 1 Conv [3] with window size of k, the receptive range is k, thus each output neuron can only attend to small amount of input neurons, leading to insufficient channel information aggregating. Consequently, to make the channel information aggregated sufficiently thus guarantee the model representation capacity, each output neuron should have a full receptive range, namely, each output neuron should be connected to all of the input neurons directly or indirectly.

Sparsely Factorized 1 × 1 Conv. With the guidance of full receptive range, we introduce sparsity to the densely connected 1 × 1 conv emulating the spatial convolution. We raise the ***Sparsely Factorized* 1 × 1 *Conv*** (***SF-Conv*** for short), which is proposed to be a new paradigm of channel sparse connectivity. As Fig. 7 exhibits, SF-Conv factorizes a densely connected 1 × 1 conv into two sparsely connected 1 × 1 convs, namely, 1st and 2nd SF-Convs, to guarantee the full receptive range. Given input feature map of tensor $\mathrm{X} \in \mathbb{R}^{H \times W \times C_{in}}$, where $H, W$is the spatial size, and C_{in} is the input channel numbers. The output feature map is $\hat{\mathrm{X}} \in \mathbb{R}^{H \times W \times C_{out}}$, where $C_{out} = \lambda C_{in} = \lambda C$. For certain operation, we only consider the number of parameters since $Flops = H \times W \times Params$

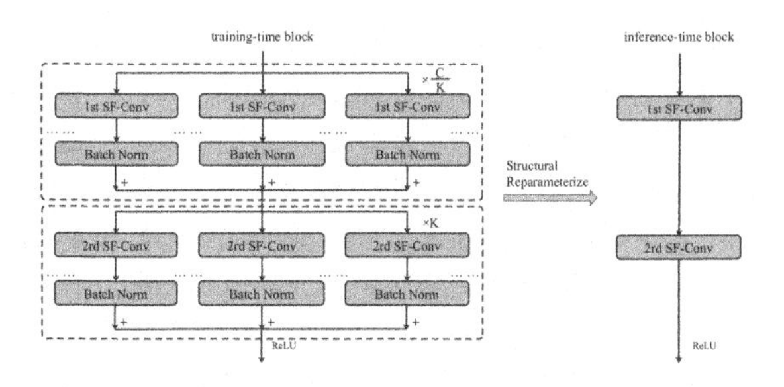

Fig. 8. RefCO Channel Operator

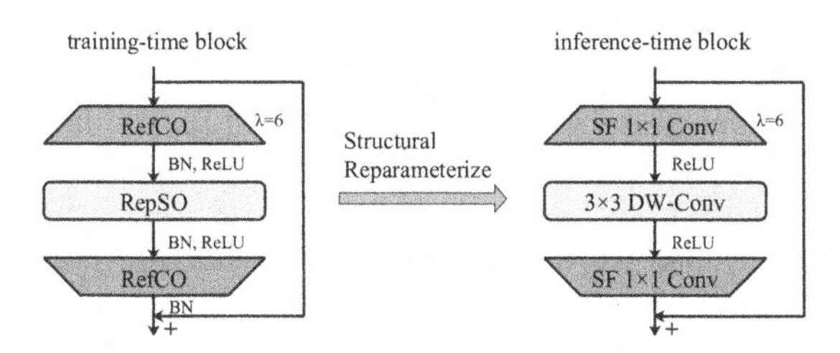

Fig. 9. Meta Light Block with RepSO and RefCO for FalconNet

where H, W are fixed. For a SF-Conv, we can consider $\mathrm{X}_{i,j,:} \in \mathbb{R}^{C \times 1 \times 1}$ for each pixel (i, j) in X as an actual input feature map with a spatial size of $C_{in} \times 1$ and 1 channel, thus we can conduct standard convolution on it, which introduces sparsity connection. The 1st SF-Conv has a channel reduction coefficient R (hyper-parameter, $R = 2$ by default) to control the neuron number of the hidden feature map X_h thus control the parameters. Thus there are $\frac{C}{R}$ neurons in X_h. Suppose a convolution kernel $\hat{\mathrm{W}}_1$ with spatial size of $K \times 1$. Consider the case of single channel, sliding $K \times 1$ on the input $C \times 1$ with stride S generates $\frac{C-K+S}{S}$ output neurons. Thus to generate $\frac{C}{R}$ hidden neurons, the convolution kernel should have $\frac{C}{R}/\frac{C-K+S}{S}$ channels, that is $\mathrm{W}_1 \in \mathbb{R}^{(\frac{C}{R}/\frac{C-K+S}{S}) \times 1 \times K \times 1}$, and the feature map of the hidden neurons is $\mathrm{X}_h \in \mathbb{R}^{\frac{C-K+S}{S} \times 1 \times (\frac{C}{R}/\frac{C-K+S}{S})}$. To achieve the minimal connections while obtain the maximum receptive range, the stride S it to be $S = K$. Moreover, to enhance the representational capacity and increase the degree of freedom, we make the kernel weights unshared in the spatial dimension, that is, each set of input neurons is operated by a individual set of weights. Cconsequently, the weights of 1st SF-Conv is $\mathrm{W}_1 \in \mathbb{R}^{\frac{K}{R} \times \frac{C}{K} \times K \times 1}$, and the output of 1st SF-Conv is $\mathrm{X}_h \in \mathbb{R}^{\frac{C}{K} \times 1 \times \frac{K}{R}}$. Thus 1st SF-Conv has $\frac{CK}{R}$ parameters.

For each output neuron of 1st SF-Conv has a receptive range of K, while for the set of neurons in a certain channel of X_h has a total receptive range of C. Thus through attending to the set of neurons in a certain channel of X_h, the output neuron of SF-Conv can have a full receptive range with minimal number of parameters. Thus the 2nd SF-Conv takes $\mathrm{X}_h \in \mathbb{R}^{\frac{C}{K} \times 1 \times \frac{K}{R}}$ as the input and conducts a DW-Conv on it, thus the weights of 2nd SF-Conv W_2 has a spatial kernel size of $\frac{C}{K} \times 1$. To obtain the λC output channels, W_2 should have a width multiplier of $\lambda C / \frac{K}{R}$, thus $W_2 \in \mathbb{R}^{\lambda C \times 1 \times \frac{C}{K} \times 1}$ with $\frac{\lambda C^2}{K}$ parameters. The output of the 2nd SF-Conv is $\hat{X}_{i,j,:} \in \mathbb{R}^{\lambda C \times 1 \times 1}$ for the pixel (i, j) in $\hat{X}$. Through operating the shared SF-Conv in all spatial positions of X, we can obtain the $\hat{X}$. SF-Conv has a total parameters of $\frac{CK}{R} + \frac{\lambda C^2}{K}$. Given a certain channel reduction ratio R, to achieve the least parameters, the kernel size K is set as $K = \sqrt{\lambda C R}$, which is dynamically adjusted according to C, λ and R. The total number of parameters of SF-Conv becomes $2C\sqrt{\frac{\lambda C}{R}}$. For a dense 1×1 conv with weights $W_d \in \mathbb{R}^{\lambda C \times C \times 1 \times 1}$ has λC^2 parameters. Therefore, SF-Conv exhibits a parameter count that is only $\frac{2}{\sqrt{\lambda C R}}$ of PW-Conv. In this way, SF-Conv can introduce sparsity to the channel connections, as well as maintain the full receptive range (as shown in Fig. 6), thus reducing the number of parameters and Flops while obtain a competitive representation capacity.

RefCO. We further raise ***RefCO*** as the ***Re***parameterized ***f***actorized ***C***hannel ***O***perator. RefCO also employs the structural reparameterization methodology to compensate for the reduced parameter count and enhance representations. As Fig. 8 exhibits, in training, RefCO firstly constructs $\frac{C}{K}$ parallel 1st SF-Conv branches and added the outputs, then constructs K parallel 2nd SF-Conv branches and added the K output feature maps to obtain the final output. In inference, these parallel 1st/2nd SF-Conv branches can be equivalently converted into a single SF-Conv branch.

3.5 FalconNet

Based on above four vital components, namely, LightNet overall architecture, Meta Light Block, RepSO spatial operator and RefCO channel operator, we obtain a novel light-weight CNN model termed as ***FalconNet*** (***Fa***ctorization for the ***l***ight-weight ***con***v***Net***). Figure 9 exhibits the Meta Light Block with RepSO and RefCO, which is utilized as the basic block for FalconNet. Later experimental results validate that FalconNet can achieve higher accuracy with lower number of parameters and Flops compared to existing light-weight CNNs.

4 Experiments

4.1 Configurations

We conduct abundant experiments on three challenging benchmark datasets, CIFAR-10, CIFAR-100 and Tiny-ImageNet-200, to validate the effectiveness and superiority of the four vital components illustrated above. For the training configuration, we adopt an SGD optimizer with momentum of 0.9, batch size of 256, and weight decay of 4×10^{-5}, as the common practice [7,19]. We use a learning rate schedule with a 5-epoch warmup, initial value of 0.1, and cosine

Table 1. Results of various light-weight CNN models and LightNets with corresponding basic blocks on CIFAR-10, CIFAR-100 and Tiny-ImageNet-200. Width multiplier is used for some LightNets to compensate for the largely reduced channels of certain basic blocks, such as Sandglass and ShuffleNet blocks. The numbers of parameters and Flops in inference are also exhibited.

Basic Block	Model	CIFAR-10	CIFAR-100	Tiny-ImageNet-200	Params	Flops
DSC Block	MobileNetV1	93.18%	72.47%	64.36%	4.23M	588.91M
	LightNet×3.5	93.22%	74.26%	64.38%	3.59M	536.06M
Residual DSC Block	ResNet-18	93.89%	74.52%	65.02%	11.72M	1844.08M
	ResNet-34	93.95%	74.82%	65.21%	21.84M	3698.78M
	LightNet×3.5	93.70%	75.01%	64.66%	3.59M	536.06M
ShuffleNetV1 Block	ShuffleNetV1	91.35%	68.51%	56.86%	1.81M	138.75M
	LightNet×4.0	91.61%	69.42%	58.94%	1.05M	132.86M
ShuffleNetV2 Block	ShuffleNetV2	92.31%	70.08%	60.38%	2.28M	154.87M
	LightNet×3.5	93.50%	73.56%	64.10%	2.01M	219.65M
Inverted Residual Block	MobileNetV2	93.43%	74.03%	66.30%	3.56M	353.01M
	MobileNetV3S	91.89%	70.50%	60.18%	2.94M	66.89M
	MobileNetV3L	94.07%	73.56%	65.54%	5.48M	238.85M
	EfficientNetV1-B0	94.21%	75.68%	66.62%	4.98M	404.42M
	EfficientNetV2-S	93.85%	75.42%	67.18%	21.14M	2915.26M
	LightNet	94.84%	76.92%	68.18%	3.32M	526.51M
Sandglass Block	MobileNeXt	93.01%	67.42%	58.12%	3.31M	310.04M
	LightNet×6.0	94.09%	75.43%	65.76%	3.77M	607.85M
HBO Block	HBONet	92.32%	72.39%	64.20%	4.56M	326.99M
	LightNet	92.44%	72.85%	65.92%	3.83M	209.23M
FasterNet Block	FasterNet-T0	92.94%	68.02%	58.32%	3.64M	310.98M
	LightNet	93.45%	74.14%	64.78%	3.36M	509.10M
RepSO Block	**LightNet**	94.96%	78.24%	69.72%	3.32M	526.51M
RepSO+RefCO Block	**FalconNet**	94.85%	78.04%	69.46%	2.39M	333.14M

annealing for 300 epochs to guarantee the compete convergence. The data augmentation uses random cropping and horizontal flipping. The input resolution is uniformly resized to 224×224. All the models are random initialized with LeCun initialization and trained with the same training configuration from scratch.

4.2 Performance Evaluation

We evaluate the performance of various existing light-weight CNN models, including MobileNetV1/V2/V3 [11,12,19], MobileNeXt [28], EfficientNetV1/V2 [23,24], ShuffleNetV1/V2 [18,27], HBONet [14], and FasterNet [1], as well as two heavy-weight CNNs, i.e., ResNet-18 and ResNet-34 [10]. Then we implement the basic blocks of these light-weight CNN models to our LightNet overall architecture and compare the performance with existing light-weight CNNs. Table 1 shows the experimental results. As can be observed, LightNet can achieve better performance by simply implementing the basic blocks. For example, LightNet with Inverted Residual Block surpasses EfficientNetV1 by 0.63%, 1.24% and 1.56% on CIFAR-10/100 and Tiny-ImageNet-200 respectively with only 66% of the parameters (there exists a trade-off that the Flops is enhanced by 30%). Moreover, LightNet with ShuffleNetV2 Block significantly surpasses ShuffleNetV2 by 3.48% and 3.72% on CIFAR-100 and Tiny-ImageNet-200 with 88% parameters (trade-off of 40% more Flops). Besides, LightNet with Sandglass Block significantly surpasses MobileNeXt by 8.01% and 9.94% on CIFAR-100 and Tiny-ImageNet-200 with 14% more parameters, and LightNet with FasterNet Block surpasses FasterNet by 6.08% and 6.46% on CIFAR-100 and Tiny-ImageNet-200 with 8% less parameters. In addition, when compared to the heavy-weight CNN models ResNet-18/34, LightNet with Residual DSC Block (obtained by replacing the standard convolution in the bottleneck block of ResNet with DS-Conv) still achieves competitive performance, while the numbers of parameters are reduced by 70% and 84% respectively. The experimental results validate the effectiveness and efficiency of LightNet overall architecture compared to the overall architectures of existing light-weight CNN models.

Then we evaluate the performance of LightNet with RepSO Block (Meta Light Block with RepSO), it is observed in Table 1 that RepSO can enhance the performance of LightNet significantly and achieves the highest accuracy on all of the three datasets, for example, it surpasses LightNet with Inverted Residual Block by 1.32% and 1.54% on CIFAR-100 and Tiny-ImageNet-200 respectively, while maintaining the inference parameters and Flops unchanged. Thus validates the effectiveness of RepSO which boosts the morel performance significantly without incurring any extra inference costs.

We then evaluate the performance of FalconNet, which equips LightNet with Meta Light Block of RepSO and RefCO. As Table 1 exhibits, FalconNet can achieve higher accuracy than other existing light-weight CNNs while possessing less parameters and Flops. Moreover, compared to FalconNet without RefCO (namely, LightNet with RepSO Block), FalconNet significantly reduces the number of parameters and Flops by 28% and 37% while still achieving competitive accuracy with negligible decline of 0.11%, 0.20% and 0.26% on CIFAR-10, CIFAR-100 and Tiny-ImageNet-200 respectively. This validates the effectiveness of RefCO which significantly reduce the number of parameters and Flops while maintaining good representation capacity and competitive performance.

5 Conclusion

This paper factorizes the light-weight CNNs from coarse to fine and obtain four vital components, namely, overall structure, basic block, spatial operator and channel operator. In light of the existing issues of these components, this paper respectively raise LightNet, Meta Light Block, RepSO, Receptive Range and RefCO. Furthermore, this paper raises FalconNet, which achieves higher accuracy with fewer parameters and Flops compared to existing light-weight CNNs.

References

1. Chen, J., et al.: Run, don't walk: chasing higher flops for faster neural networks. arXiv preprint arXiv:2303.03667 (2023)
2. Chollet, F.: Xception: deep learning with depthwise separable convolutions. In: Proceedings of the IEEE Conference on Computer Vision and Pattern Recognition, pp. 1251–1258 (2017)
3. Gao, H., Wang, Z., Cai, L., Ji, S.: ChannelNets: compact and efficient convolutional neural networks via channel-wise convolutions. IEEE Trans. Pattern Anal. Mach. Intell. **43**(8), 2570–2581 (2021)
4. Ding, X., Guo, Y., Ding, G., Han, J.: ACNet: strengthening the kernel skeletons for powerful CNN via asymmetric convolution blocks. In: Proceedings of the IEEE/CVF International Conference on Computer Vision, pp. 1911–1920 (2019)
5. Ding, X., et al.: ResRep: lossless CNN pruning via decoupling remembering and forgetting. In: Proceedings of the IEEE/CVF International Conference on Computer Vision, pp. 4510–4520 (2021)
6. Ding, X., Zhang, X., Han, J., Ding, G.: Diverse branch block: building a convolution as an inception-like unit. In: Proceedings of the IEEE/CVF Conference on Computer Vision and Pattern Recognition, pp. 10886–10895 (2021)
7. Ding, X., Zhang, X., Han, J., Ding, G.: Scaling up your kernels to 31 × 31: revisiting large kernel design in CNNs. In: Proceedings of the IEEE/CVF Conference on Computer Vision and Pattern Recognition, pp. 11963–11975 (2022)
8. Ding, X., Zhang, X., Ma, N., Han, J., Ding, G., Sun, J.: RepVGG: making VGG-style convnets great again. In: Proceedings of the IEEE/CVF Conference on Computer Vision and Pattern Recognition, pp. 13733–13742 (2021)
9. Dong, X., et al.: CSWin transformer: a general vision transformer backbone with cross-shaped windows. In: Proceedings of the IEEE/CVF Conference on Computer Vision and Pattern Recognition, pp. 12124–12134 (2022)
10. He, K., Zhang, X., Ren, S., Sun, J.: Deep residual learning for image recognition. In: Proceedings of the IEEE Conference on Computer Vision and Pattern Recognition, pp. 770–778 (2016)
11. Howard, A., et al.: Searching for MobileNetV3. In: Proceedings of the IEEE International Conference on Computer Vision, pp. 1314–1324 (2019)
12. Howard, A.G., et al.: MobileNets: efficient convolutional neural networks for mobile vision applications. arXiv preprint arXiv:1704.04861 (2017)
13. Hu, J., Shen, L., Sun, G.: Squeeze-and-excitation networks. In: Proceedings of the IEEE Conference on Computer Vision and Pattern Recognition, pp. 7132–7141 (2018)
14. Li, D., Zhou, A., Yao, A.: HBONet: harmonious bottleneck on two orthogonal dimensions. In: Proceedings of the IEEE/CVF International Conference on Computer Vision, pp. 3316–3325 (2019)
15. Liu, Z., et al.: Swin Transformer V2: scaling up capacity and resolution. In: Proceedings of the IEEE/CVF Conference on Computer Vision and Pattern Recognition, pp. 12009–12019 (2022)
16. Liu, Z., et al.: Swin transformer: hierarchical vision transformer using shifted windows. In: Proceedings of the IEEE/CVF International Conference on Computer Vision, pp. 10012–10022 (2021)
17. Liu, Z., Mao, H., Wu, C.Y., Feichtenhofer, C., Darrell, T., Xie, S.: A convnet for the 2020s. In: Proceedings of the IEEE/CVF Conference on Computer Vision and Pattern Recognition, pp. 11976–11986 (2022)

18. Ma, N., Zhang, X., Zheng, H.-T., Sun, J.: ShuffleNet V2: practical guidelines for efficient CNN architecture design. In: Ferrari, V., Hebert, M., Sminchisescu, C., Weiss, Y. (eds.) Computer Vision – ECCV 2018. LNCS, vol. 11218, pp. 122–138. Springer, Cham (2018). https://doi.org/10.1007/978-3-030-01264-9_8
19. Sandler, M., Howard, A., Zhu, M., Zhmoginov, A., Chen, L.C.: MobileNetV2: inverted residuals and linear bottlenecks. In: Proceedings of the IEEE Conference on Computer Vision and Pattern Recognition, pp. 4510–4520 (2018)
20. Sun, K., Li, M., Liu, D., Wang, J.: IGCV3: interleaved low-rank group convolutions for efficient deep neural networks. arXiv preprint arXiv:1806.00178 (2018)
21. Szegedy, C., Vanhoucke, V., Ioffe, S., Shlens, J., Wojna, Z.: Rethinking the inception architecture for computer vision. In: Proceedings of the IEEE Conference on Computer Vision and Pattern Recognition, pp. 2818–2826 (2016)
22. Tan, M., et al.: MnasNet: platform-aware neural architecture search for mobile. In: Proceedings of the IEEE/CVF Conference on Computer Vision and Pattern Recognition, pp. 2820–2828 (2019)
23. Tan, M., Le, Q.: EfficientNet: rethinking model scaling for convolutional neural networks. In: International Conference on Machine Learning, pp. 6105–6114. PMLR (2019)
24. Tan, M., Le, Q.: EfficientNetV2: smaller models and faster training. In: International Conference on Machine Learning, pp. 10096–10106. PMLR (2021)
25. Yu, W., Zhou, P., Yan, S., Wang, X.: InceptionNeXt: when inception meets ConvNeXt. arXiv preprint arXiv:2303.16900 (2023)
26. Zhang, J., et al.: Rethinking mobile block for efficient neural models. arXiv preprint arXiv:2301.01146 (2023)
27. Zhang, X., Zhou, X., Lin, M., Sun, J.: ShuffleNet: an extremely efficient convolutional neural network for mobile devices. In: Proceedings of the IEEE Conference on Computer Vision and Pattern Recognition, pp. 6848–6856 (2018)
28. Zhou, D., Hou, Q., Chen, Y., Feng, J., Yan, S.: Rethinking bottleneck structure for efficient mobile network design. In: Vedaldi, A., Bischof, H., Brox, T., Frahm, J.-M. (eds.) ECCV 2020, Part III. LNCS, vol. 12348, pp. 680–697. Springer, Cham (2020). https://doi.org/10.1007/978-3-030-58580-8_40

An Interactive Evolutionary Algorithm for Ceramic Formula Design

Wen-Xiang Song, Wei-Neng Chen(✉), and Ya-Hui Jia

South China University of Technology, Guangzhou, China
cschenwn@scut.edu.cn

Abstract. The ceramic industry is a representative traditional industry in Guangdong Province, where its degree of informatization is low, and the design of ceramic formula mainly depends on human experience. To intelligently generate ceramic formulas, two main challenges are raised, i.e., the evaluation of a ceramic formula by actual firing is expensive, and the historical accumulated actual data are limited. To solve this problem, this paper models the ceramic formula design process as an expensive constrained multi-objective optimization problem. Based on the mathematical model, we propose an interactive hybrid metaheuristic evolutionary algorithm, cEDA to optimize the production cost and meanwhile satisfy the category constraints, chemical component constraints and material constraints. It consists of three key components, nondominated sorting, materials selection and proportion allocation to search for qualified ceramic formulas. To incorporate domain expertise, a classification-based interactive optimization method is introduced in cEDA. After two rounds of interaction, the acceptance rate of the generated formulas by the algorithm has increased from 18% to 87.5%, which demonstrates the effectiveness of the proposed algorithm.

Keywords: ceramic formula design · evolutionary algorithm · interactive optimization

1 Introduction

Ceramic is a kind of industrial product which is made of various materials and fired at high temperature [7]. With the development of science and technology, ceramics have infiltrated into various aspects of our life, becoming one of the essential materials. Ceramics with different functions are fired based on different formulas. Ceramic formula design including type selection and proportion determination of raw materials, has great influence on the physical and chemical properties of the final ceramic product [10]. In this traditional industry, an obvious feature is the low degree of information and automation. In order to obtain a satisfactory ceramic formula, workers usually switch the choice of raw materials and fine-tune the corresponding proportions based on their experience. It is a complicated process requiring a large number of repeated fire experiments for evaluation, and greatly depends on a few of historical classical formulas created

B. Luo et al. (Eds.): ICONIP 2023, LNCS 14447, pp. 381–394, 2024.
https://doi.org/10.1007/978-981-99-8079-6_30

by the experienced experts. This old-fashioned ceramic formula design process consumes a great deal of manpower and material resources.

The current difficulties of ceramic formula design exhibit in the three following aspects. 1) The raw materials coming from the same mine change over time. Ceramic formulas are redesigned frequently to ensure consistent quality of ceramic product. 2) The selection of raw materials and the determination of proportion together form a NP-hard combinatorial optimization problem, that cannot be solved by traditional mathematical optimization methods. Workers need to go through a large number of orthogonal experiments to obtain an appropriate formula. 3) The quality measurement of the final ceramic product usually depends on physical and chemical performance testing and experts' judgments instead of a mathematical formula.

So far, a few studies have tried computational intelligent algorithms on ceramic formula design. Luo et al. [9] applied improved evolutionary algorithms(EAs) and Liu et al. [6] combined genetic algorithms with a complex method to obtain formulas whose chemical component percentage is close to the target formula's. Yang et al. [8] train a back propagation(BP) neural network to predict the chemical composition of a ceramic formula. These studies truly make some progress and alleviate the hardship in ceramic formula design.

In a real-world ceramic formula design algorithm, the following three challenging issues must be considered. 1) Multiple constraints, including the proportion limitation of each category, each material, and each chemical component should be taken into account. 2) Evaluation of a ceramic formula is expensive due to the high cost of firing experiment for the final product. Meanwhile, there are limited data of historical ceramic formulas. Thus, it is difficult to train an effective model for evaluation. 3) The ceramic formula design method needs to be integrated with the long-term accumulated human experience, so that the obtained ceramic formula conforms to the intention of human experts.

Evolutionary computation(EC) is an effective technique to solve optimization problems [3], especially combinatorial optimization problems [1]. Some popular algorithms, such as ant colony optimization (ACO) [4], estimation of distribution algorithm (EDA) [5] have been successfully applied to a number of real-word optimization problems. These gradient-free optimizers can meet the requirement for ceramic formula design in the aspects of raw material selection and proportion allocation.

Facing above-mentioned challenges, this paper aims to integrate multi-modal, data-driven and interactive evolutionary algorithms to tackle the intelligent ceramic formula design problem. The contributions of the proposed algorithm are shown as follows.

1. The ceramic formula design is modeled as a constrained multi-modal multi-objective optimization problem. A customized EDA denoted as cEDA, is proposed to select raw materials and determine the proportion. The process of cEDA solving a formula is similar to human formula design, where a matrix is used to assist in selecting raw materials and EDA is adopted to determine the proportion of raw materials according to the Gaussian distribution.

2. An interactive data-driven mechanism is designed to introduce expert experience into the algorithm to improve the quality of results. Since some expert experiences cannot be modeled, we authorize the experts to judge whether some generated formulas are qualified. According to the small set of formulas evaluated by experts, a classification model is trained to guide the optimization procedure to generate more acceptable formulas.

The rest of this paper is organized as follows. Section 2 defines the ceramic formula optimization problem in mathematical expression. Section 3 illustrates the architecture of the proposed interactive hybrid metaheuristic algorithm, cEDA. Section 4 shows the experimental results and Sect. 5 draws the final conclusion.

2 Problem Description and Formulation

Given a set $\mathcal{N}$ containing n kinds of raw materials, the optimization of ceramic formula is to obtain a set of diverse formulas with minimum production cost. First, we need to select a subset $\mathcal{M}$ of $\mathcal{N}$ to determine which raw materials will be used. The size of $\mathcal{M}$ is denoted as m. The corresponding proportion and cost per unit of the ith raw material are denoted as X_i and q_i respectively. A feasible formula should satisfy the following four constrains:

1. Category Constraint. Each raw material belongs to a category. Eight categories are taken into account in this paper. Taking the raw materials in the kth category as a set $\mathcal{C}_k$, the total proportion of materials in $\mathcal{C}_k$ in a formula should be limited in an interval $[L_k, U_k]$.
2. Material Constraint. The proportion of the ith raw material is also limited in an interval $[l_i, u_i]$.
3. Chemical Constraint. Different raw materials have different chemical components. In total, eight chemical components are considered in this paper including SiO_2, Al_2O_3, Fe_2O_3, TiO_2, CaO, MgO, K_2O and Na_2O. I.L which indicates the ignition loss can be approximated as a chemical component for processing as well. The content of the eth chemical component in a formula is denoted as g_e. The content of the eth chemical component in the ith raw material is denoted as p_{ei}. The eth chemical component has a limited interval $[lc_e, uc_e]$.
4. Cost Constraint. The ceramic formulas under a cost limit C can be considered acceptable.

The optimization of ceramic formula design is formulated mathematically as a constrained multi-modal cost minimization problem as follows.

$$min \quad Q = \sum_{i=1}^{n} q_i X_i \tag{1}$$

s.t.

$$L_k \leq \sum X_j \leq U_k, j \in \mathcal{C}_k \tag{2}$$

$$l_i \leq X_i \leq u_i,\ if\ \phi(X_i) = 1 \tag{3}$$

$$lc_e \leq \sum_{i=1}^{n} p_{ei} X_i \leq uc_e \tag{4}$$

$$\sum_{i=1}^{n} X_i = 1 \tag{5}$$

$$\sum_{i=1}^{n} \phi(X_i) = m$$
$$\phi(X_i) = \begin{cases} 1 & ,if\ X_i > 0 \\ 0 & ,if\ X_i = 0 \end{cases} \tag{6}$$

$$Q \leq C \tag{7}$$

The objective (1) of the problem is to minimize the production cost Q. Constraint (2) indicates a limitation of each considered category' proportion. Constraint (3) stipulates a limitation of each selected raw material' proportion. Constraint (4) guarantees that the content of each chemical components is in a reasonable interval. Constraint (5) makes sure that the total percentage of raw materials is equal to 1. Constraint (6) is a cardinality constraint which ensures that the number of selected raw materials is m. Constraints(7) guarantees that the obtained solution is acceptable.

In a real production environment, the chemical constraint is usually a soft constraint where some chemical components are allowed to exceed the feasible range slightly. Thus, to handle this constraint effectively, we adopt a common constraint handling method that transforms this constraint into another objective. Denoting g_e as the proportion of the eth kind of chemical component in a ceramic formula, calculated by (8), we define f_e as the penalty function of the eth chemical component in (9). By adding up the penalty function values of all chemical components, another objective F is formed as (10). After transforming the chemical component constraints into an optimization objective, the original single-objective problem becomes a bi-objective multimodal optimization problem whose constraints are (2)(3)(5)(6)(7).

$$g_e = \sum_{i=1}^{n} p_{ei} X_i \tag{8}$$

$$f_e = \begin{cases} \left(\dfrac{g_e - lc_e}{uc_e - lc_e} \right)^2 & ,g_e < lc_e \\ \left(\dfrac{g_e - uc_e}{uc_e - lc_e} \right)^2 & ,g_e > uc_e \\ 0 & ,lc_e \leq g_e \leq uc_e \end{cases} \tag{9}$$

$$min \quad F = \sum f_e, e \in \{1, 2, \ldots, 9\} \tag{10}$$

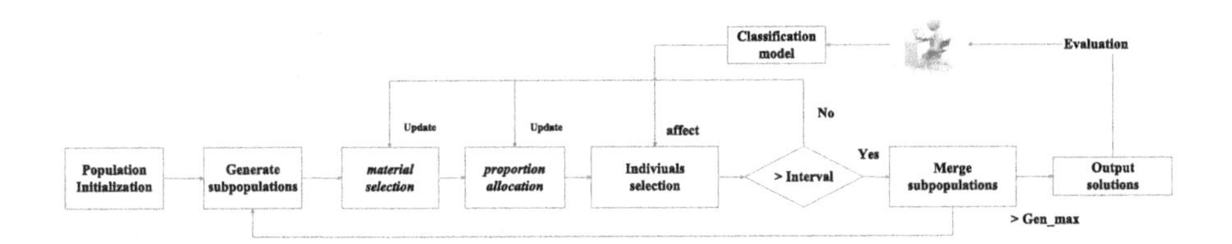

Fig. 1. Flow chart of cEDA

3 Method

In the problem description, we treat each ceramic formula as an n-dimensional solution, where the values of $n - m$ dimensions are all 0. To facilitate solution representation and generation, we simplify the solution to $2m$ dimensions in the optimization procedure, where the previous m dimensions are the serial numbers of the selected materials, and the latter m dimensions are the corresponding proportions. Estimation of distribution algorithm(EDA) is stochastic optimization technique that explores the space of potential solutions by building and sampling explicit probabilistic models of promising candidate solutions. [5] Aiming at obtaining ceramic formulas with lower cost, we propose a customized EDA for ceramic formula design denoted as cEDA. This algorithm consists of nondominated sorting (to handle the two objectives), material selection, and proportion allocation. Figure 1 illustrates the entire process of cEDA. It has five main steps.

1. Initialization. Parameters such as population size, maximum generation, subpopulation size, stopping criteria, etc. are set. At the first run of the algorithm, the m materials of an individual are randomly chosen and the ith chosen raw material's proportion is sampled according to a preset Gaussian distribution $N(\overline{\mu}_i, \overline{\sigma}_i)$. When the algorithm performs again, the initialization of the population will be conducted based on the sampling information obtained by the last application of cEDA.
2. Division. This step is to aggregate individuals with similar costs in a subpopulation and divide the whole population into multiple subpopulations. After certain generations, all subpopulations will be merged and partitioned again by applying the same method.
3. Individual generation. In each subpopulation, material selection and proportion allocation (Algorithm 1) are executed sequentially to generate new individuals.
4. Selection. Newly generated individuals are evaluated in terms of the two objectives, cost function (1) and penalty function (10). After combining the parent and child generations, we use the selection method of NSGA-II [2] to choose individuals with better fitness values based on a rank given by fast nondominated sorting and a crowd distance given by crowding calculation.
5. Update. The selection matrix related to material selection and the distribution of each material related to proportion allocation are updated by the chosen individuals obtained from the previous selection step.

Material Selection. When ceramic formulas are designed manually, the materials are usually determined empirically before the proportions are determined. For individual generation in cEDA, we follow the same process to select raw materials in the first place in order to shrink the search space.

Since individuals in the same subpopulation have similar material selection cost, all the selected materials of individuals in the same subpopulation compose a raw material pool. The raw materials of a newborn individual are chosen from the material pool of the corresponding subpopulation. The establishment of raw material pools narrows the search space and results in a smaller cost gap for individuals in the same subpopulation. cEDA maintains a selection matrix S, which is shared by all subpopulations that is shown below as (11).

$$S = \begin{bmatrix} \tau_{11} & \tau_{12} & \cdots & \tau_{1n} \\ \tau_{21} & \tau_{22} & \cdots & \tau_{2n} \\ \vdots & \vdots & \ddots & \vdots \\ \tau_{n1} & \tau_{n2} & \cdots & \tau_{nn} \end{bmatrix} \tag{11}$$

where τ_{ij} indicates the probability of choosing the jth material after choosing the ith one. τ_{ii} indicates the probability of choosing the ith raw material as the first chosen raw material. Based on the matrix, a binary selection method is proposed for material selection. Firstly, two numbers i, j are randomly generated from the material pool. τ_{ii} and τ_{jj} are compared that the one with bigger τ value is picked as the first material. Assuming the ith material is selected as the first one, we generate another two numbers j and k randomly, and compare τ_{ij} and τ_{ik}. Still, the one with larger τ value is chosen as the next material. This process is repeated until m materials are selected.

$$\begin{cases} \Delta_{ij} = \sum \frac{\beta \times p_i \times p_j \times Th}{N_j \times rank} \ (rank < Th) \\ \tau_{ij} = (1-\alpha)\tau_{ij} + \Delta_{ij} \\ \tau_{ii} = \frac{\sum \tau_t}{L} (t \in A_i) \\ \tau = \tau_{min} (\tau < \tau_{min}) \\ \tau = \tau_{max} (\tau > \tau_{max}) \end{cases} \tag{12}$$

After each iteration in a subpopulation, the maintained selection matrix will be updated according to (12). In this equation, i and j represent two materials sequentially selected in a solution. p_i and p_j are the corresponding proportions. α and β are coefficients controlling the decay and the increase of τ_{ij}, respectively. N_j is the number of raw material pools that contain the jth raw material. τ_{min} and τ_{max} set the lower bound and the upper bound of the matrix element. $rank$ represents the rank of an individual calculated by nondominated sorting. Th is a threshold for $rank$. When updating τ_{ii}, the average value of the top L largest matrix elements in the ith row is assigned to τ_{ii}. L is an adjustable parameter and the top L largest matrix elements in the i row form a set A_i.

After sampling m serial numbers for one individual by binary selection method, we validate that the sum of selected materials' lower and upper bound in each category will not conflict with the corresponding category constraint. Otherwise, we resample the serial numbers. During the solution construction process, we maintain two variables DL_i and DU_i to keep track the dynamic lower and upper bounds of the ith categories for an individual. In the process of material selection, we constantly update DL_i and DU_i by the sum of selected materials' upper and lower bound in the ith category respectively.

Algorithm 1. proportion allocation

Input: $[DL_1, DL_2, \ldots, DL_k]$:dynamic lower bound of each category; $[DU_1, DU_2, \ldots, DU_k]$:dynamic upper bound of each category; $[l_1, l_2, \ldots, l_m]$: lower bound of each selected materials; $[u_1, u_2, \ldots, u_m]$: upper bound of each selected materials; k : category number;

Output: $[p_1, p_2, \ldots, p_m]$: allocated proportion of each selected raw materials;

1: $available\_proportion = 1$
2: **for** $i = 1$ to m **do**
3: $\quad L = U = available\_proportion$
4: $\quad [ll_1, ll_2, \ldots, ll_k] \longleftarrow [0, 0, ..., 0]$
5: $\quad [lu_1, lu_2, \ldots, lu_k] \longleftarrow [0, 0, ..., 0]$
6: $\quad g \longleftarrow$ the serial number of a category which the ith raw material is in
7: $\quad$ **for** $j = i + 1$ to m **do**
8: $\quad\quad t \longleftarrow$ the serial number of a category which the jth raw material is in
9: $\quad\quad ll_t = ll_t + l_j, lu_t = lu_t + u_j$
10: $\quad$ **end for**
11: $\quad$ **for** $j = 1$ to k **do**
12: $\quad\quad$ **if** $lu_j \neq 0$ and $j \neq g$ **then**
13: $\quad\quad\quad L = L - min(lu_j, DU_j), U = U - max(ll_j, DL_j)$
14: $\quad\quad$ **end if**
15: $\quad$ **end for**
16: $\quad L = L - lu_g, U = U - ll_g$
17: $\quad L = max(L, DL_g - lu_g, l_i),\ U = min(U, DU_g - ll_g, u_i)$
18: $\quad pi \longleftarrow$ random sampling according to Gaussian distribution $N(\overline{\mu}_i, \overline{\sigma}_i)$
19: $\quad p_i = min(p_i, U), p_i = max(p_i, L)$
20: $\quad available\_proportion = available\_proportion - p_i$
21: $\quad DL_g = max(0, DL_g - p_i), DU_g = max(0, DU_g - p_i)$
22: **end for**
23: **return** $[p_1, p_2, \ldots, p_m]$

Proportion Allocation. The next step after material selection is to determine the proportion of each material. In cEDA, each material constructs a Gaussian distribution. cEDA maintains the mean and deviation values of all distributions and updates them after each iteration in the subpopulations. The updating rules are shown in (13) and (14), where $\overline{\mu_i}$ and $\overline{\sigma_i}$ are the mean and deviation of the distribution of the ith raw material. H is the number of individuals that have selected the ith raw material and p_h is the corresponding proportion of each individual. θ is a coefficient that controls the speed of updating and ε is a coefficient to prevent the increment of deviation being zero.

$$\overline{\mu_i} = \begin{cases} (1-\theta)\overline{\mu_i} + \frac{\theta \times \sum_{h=1}^{H} p_h}{H} & ,H > 0 \\ (1-\theta)\overline{\mu_i} + \theta \times l_i & ,H = 0 \end{cases} \tag{13}$$

$$\overline{\sigma_i} = \begin{cases} (1-\theta)\overline{\sigma_i} + \theta\left(\sqrt{\frac{\sum_{h=1}^{H}(p_h - \overline{\mu_i})^2}{H}} + \varepsilon\right), H > 0 \\ (1-\theta)\overline{\sigma_i} + \theta(u_i - l_i)\ , H = 0 \end{cases} \tag{14}$$

Algorithm 1 gives the details of the proportion allocation. A specific boundary calculation method is designed to handle category constraint and material constraint in Algorithm 1(line 3-17), which determines a feasible region for proportion allocation of the ith selected material. Assuming that the ith material is in the gth category, the required proportion of each category are determined by the $(i+1)$th to the last unallocated raw materials' bounds(line 7-10). After that, the influence of the required proportion of each category on feasible region is taken into account(line 11-15). Then the influence of the required proportion of the other selected materials in the gth category to the feasible region is considered(line 16). Finally, the influence of the gth category constraint and the ith material constraint to the feasible region is considered(line 17).

For the ith selected material, we draw a random sample from the corresponding Gaussian distribution as the proportion and limit it in the feasible region. The available proportion that can be assigned decreases gradually from 1 to 0 as the proportion of each material is determined. Dynamic lower and upper bounds of each category for a individual, DL and DU,which is influenced by the assigned proportion, will decrease.

Classification-Based Optimization. In the process of problem modeling, there are extensive domain knowledge and experience can be utilized to make the generated solutions closer to real expectations. If users input evaluated ceramic formulas, a classification model can be trained to guide the selection procedure in the later optimization procedure.

The features of a formula are the material serial numbers, the proportion of each category and the proportion of chemical components which can both be judgement basis for users and input features for classification model. Users classify the formulas into cl categories $\{1, 2, ..., cl\}$, based on the acceptability. 1 means the most unacceptable category and cl represents the most acceptable category. After users evaluate the results obtained by cEDA, a classification model can be trained.

The objective function we tend to modify is the penalty function, and the modification method is shown as (15), where F is the penalty function in (10) and cl is the classification result for an individual. Individuals with high evaluation results had higher cl values. The greater the classification result is, the smaller the modified penalty function value, F', is. k is a coefficient affecting the gap between F and F' , which should be determined suitably.

$$F' = \frac{F}{cl^k} \quad ,k \in N \tag{15}$$

Table 1. Chemical Component Constraints

	SiO_2	Al_2O_3	Fe_2O_3	TiO_2	CaO	MgO	K_2O	Na_2O	I.L
Lower bound	68.20	18.30	0	0	0	0.85	3.30	1.80	0
Upper bound	69.20	18.70	1.35	0.40	0.70	1.15	3.90	2.20	4.50

Table 2. Category Constraints

	calcined clay	clay	high alumina sand	feldspar	recycled feldspar	recycled material	cosolvent	bentonite
Lower bound	0	10	0	2	5	0	0.5	5
Upper bound	2	20	10	60	10	3	3	8

Table 3. An Acceptable Ceramic Formula

Serial number	74	58	45	32	50	76	29	10	22	78	57	36	20	70	85	65
proportion	11.8	0.0	9.3	12.7	0.4	1.8	1.7	0.0	16.0	0.0	7.0	7.2	1.0	19.9	7.0	4.2

By using classification results of individuals to modify the objective function value, individuals who accept higher degree of classification results will have a greater chance of surviving through the selection mechanism. As a result, the optimization procedure is directed to generate more solutions with higher level of acceptance.

4 Experimental Results

In this part, a number of ceramic formulas are generated according to users' requirement. The process of interaction with users and the adjustment of the model is investigated. The experiments are conducted on the Windows operating system with i5-8250U CPU at 1.6 GHz and 8 GB of main memory. The calculation is carried out using the python programming language.

Eight chemical components closely related to ceramics' physical and chemical properties are taken into consideration in this paper, including SiO_2, Al_2O_3, Fe_2O_3, TiO_2, CaO, MgO, K_2O and Na_2O. Besides, I.L which stands for ignition loss is concerned as well. Each material can be classified into one of the eight categories which are calcined clay, clay, high alumina sand, feldspar, recycled feldspar, recycled material, cosolvent, and bentonite. The chemical component constraint and category constraint given by users are illustrated in Tables 1 and 2 respectively. In a ceramic formula, 16 raw materials ought to be selected from 78 given materials, and meanwhile the corresponding proportions ought to be determined.

4.1 Visualization of Optimization Process

In the application of cEDA, the population size is set as 700. The number of subpopulations is set as 70. Therefore, there are 10 individuals in each subpopulation. As for parameter initialization, we set τ_{min} and τ_{max} as 0 and 100 in (12) for the selection matrix. α and β are set to 0.05 and 0.0001 to control the updating pace. Th and L are set to 5 and 10 to implement the elite strategy. In (13) and (14), the mean $\overline{\mu_i}$ and the deviation $\overline{\sigma_i}$ of the ith raw material's proportion are initialized as 20 and 15. θ and ε are set to 0.02 and 0.1, respectively.

Acceptable ceramic formulas that meet users' requirements should have a chemical penalty function value less than 0.1 and a cost less than 150, which means a slight chemical violation is acceptable. The optimization process ends if more than 200 acceptable ceramic formulas are generated.

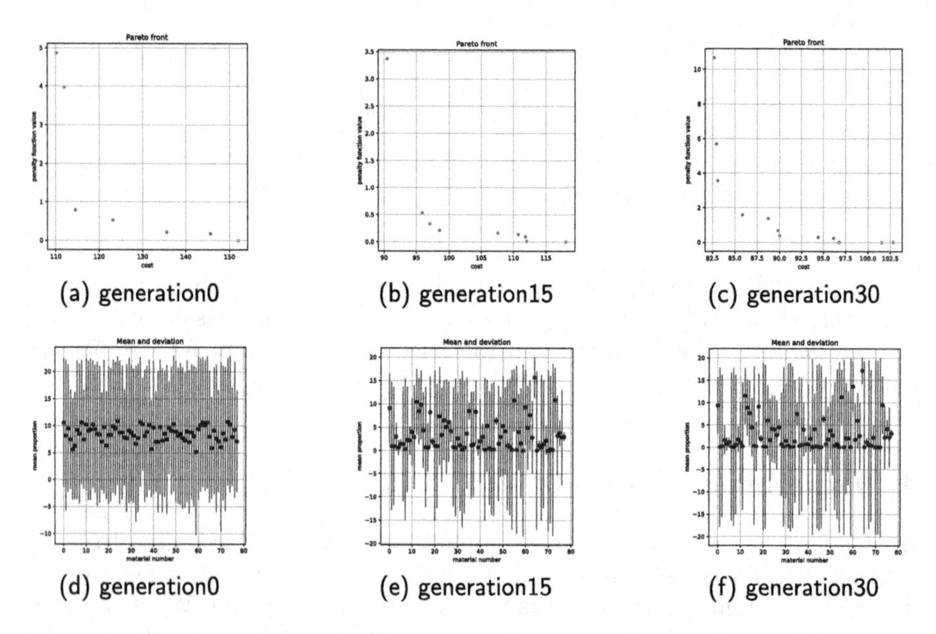

Fig. 2. Visualization Of Optimization Process. The graphs from left to right are for generation 0, 15, and 30 respectively. The top and bottom rows show Pareto frontier and the mean and deviation of each raw material respectively. In the scatter diagrams of the Pareto frontier, the horizontal and vertical coordinate are the cost function value and the penalty function value respectively. In the error bar diagrams, the horizontal and vertical coordinate represent the serial number of raw materials and the corresponding proportion.

After 30 generations in this experiment, 208 ceramic formulas are generated and recorded. An example of the generated formulas is shown in Table 3. Important indicators of optimization process are shown in Fig. 2, in which these indicators are visually displayed in the form of graphs every fifteen generations of the optimization process. It can be observed that with the increase of iteration, the Pareto frontier moves toward a better direction and the number of solutions located at the Pareto frontier also tend to increase. In the 30th generation of the Pareto frontier, the optimal cost per ton is close to 82.5, and the minimum

Table 4. Results Of Evaluation In The First Round

unacceptable	Slightly acceptable	Mostly acceptable	acceptable
135	16	19	38

penalty function value is close to 0. In the error bar diagrams, the general trend can be seen that the mean values of the raw materials' proportions tend to be stable. The deviations of some raw materials with high selection frequency increase, and the mean values of the raw material that are not often selected gradually approach the given lower bound.

4.2 The Investigation of Interactive Optimization Process

After obtaining 208 ceramic formulas through cEDA, interactive experiments are conducted with users, hoping to introduce some experience of domain experts into the optimization procedure. The interactive process in the experiment has been practiced for two rounds.

In the first round, the users evaluated 208 ceramic formulas, and the results are shown in Table 4. The users divided the formulas into four classes according to the acceptability degree. 38 out of 208 formulas are acceptable, accounting for 18%. We randomly divided 208 formulas into training set and test set in a ratio of 6:4, and trained a decision tree model. The trained model has reached 89% classification accuracy on the test set. Then, in the following selection step, we have used the decision tree to evaluate the individuals of the population. The evaluation results are used to affect the individual selection through (15).

Table 5. Results Of Classification By Decision Tree

	unacceptable	Slightly acceptable	Mostly acceptable	acceptable
k = 0	138	19	13	34
k = 1	69	23	20	71
k = 2	51	18	19	103

Table 6. Results Of Classification By MLP

	unacceptable	acceptable
k = 0	20	21
k = 4	19	40

For k in (15), we have taken 0, 1 and 2, respectively. Table 5 shows the classification results of the trained decision tree model on the generated ceramic formulas. It illustrates that when k is equal to 0, the distribution of each category is similar to the one shown in Table 4, since (15) will not influence the individual selection when k = 0. When k is equal to 2, the number of formulas in the acceptable category increases sharply, indicating that the individual selection is strongly guided by human experience. Then, we have given the 103 acceptable formulas generated when k equals 2 to the users for re-evaluation. The result is that 32 formulas are considered acceptable, accounting for 31%.

Although this acceptable percentage is higher than 18%, it is still not good enough. There are multiple reasons contribute to this result. 1) the users' evaluation contains subjective factors, and the 208 ceramic formulas cannot fully reflect the expertise of the users. 2) Some constraints given by users are not clear enough, which leads to inconformity between the users' idea and generated formulas.

In the second round of interaction, the users modified the limit of chemical penalty function to 0.0001, which indicates a stricter adherence to the chemical constraint. The stopping criterion for cEDA is also changed to 40 feasible formulas or 60 iterations.

Based on these new rules, cEDA generated 40 ceramic formulas that meet the constraints within 35 generations. They were sent to the users for evaluation again. At this time, users only classify the formulas as acceptable and unacceptable, and the number of formulas in both categories was 20, indicating that the percentage of qualified formulas has increased to 50%. Then, same as in the first round, we trained a classifier of multi-layer perceptron(MLP) with 85% accuracy on the test set. It is used to classify each individual in the population. Through (15), the classification results are utilized to influence individual selection and guide the optimization process.

For k values in (15), we take 0 and 4 to make a comparison and the results are shown in Table 6. It can be observed that the number of acceptable formulas recognized by the classifier increases remarkably, when k equals to 4. The 40 acceptable formulas generated when k equals 4 were passed to users for evaluation, and 35 of them were recognized as qualified formulas, accounting for 87.5%. After two rounds of interaction, the ceramic formula design method is adjusted to a state, where most generated ceramic formulas can be satisfactory to users.

Fig. 3. The fired experimental ceramic tiles

The experts selected two of the available formulas and sent them to the laboratory for firing experiment. The fired experimental ceramic tiles are shown in the Fig. 3. The left one and the middle one are generated by cEDA. The right one is designed by the expert. The firing experiment shows that the whiteness and strength of the two ceramic tiles whose formulas are obtained by cEDA indeed can meet the production requirements of industry. According to the feedback from the laboratory, these obtained formulas are excellent alternatives to existing formulas.

5 Conclusion

The goal of this paper is to propose an efficient alternative to replace the traditional time-consuming and laborious ceramic formula design method. This goal has been successfully achieved by proposing an interactive data-driven algorithm that was verified by real fire experiment. A customized cEDA algorithm that can handle multiple constraints nicely was designed with multi-objective optimization technique. Experimental results showed that the algorithm could intelligently obtain a set of diverse feasible formulas satisfying both hard and soft constraints in the search space.

Furthermore, the interaction mechanism have successfully endowed the algorithm ability to incorporate expertise. Experiments showed that after two rounds of interaction with users, the generated ceramic formulas has reached acceptance rate of 87.5%, indicating that cEDA can be used in real production environment.

Acknowledgment. This work was supported in part by the National Natural Science Foundation of China under Grants 61976093 and in part by Guangdong Regional Joint Foundation Key Project 2022B1515120076.

References

1. Chen, C.H., Chou, F.I., Chou, J.H.: Optimization of robotic task sequencing problems by crowding evolutionary algorithms. IEEE Trans. Syst. Man Cybern. Syst. **52**, 6870–6885 (2021)
2. Deb, K., Pratap, A., Agarwal, S., Meyarivan, T.: A fast and elitist multiobjective genetic algorithm: NSGA-II. IEEE Trans. Evol. Comput. **6**(2), 182–197 (2002)
3. Hong, Y.D., Lee, B.: Evolutionary optimization for optimal hopping of humanoid robots. IEEE Trans. Industr. Electron. **64**(2), 1279–1283 (2016)
4. Li, T.G., Fu, C.L., Guan, P., Yu, T.B., Wang, W.S.: Machine tool selection based on AHP and ACO. In: Applied Mechanics and Materials, vol. 44, pp. 874–878. Trans Tech Publications (2011)
5. Li, Z., Liu, P., Deng, C., Guo, S., He, P., Wang, C.: Evaluation of estimation of distribution algorithm to calibrate computationally intensive hydrologic model. J. Hydrol. Eng. **21**(6), 04016012 (2016)
6. Liu, B., Cheng, X., Xiao, X.: Green ceramic body and glaze formulation system. In: Proceedings of 2011 International Conference on Electronic & Mechanical Engineering and Information Technology, vol. 6, pp. 3105–3108. IEEE (2011)

7. Raupp-Pereira, F., Hotza, D., Segadães, A., Labrincha, J.: Ceramic formulations prepared with industrial wastes and natural sub-products. Ceram. Int. **32**(2), 173–179 (2006)
8. Yang, H., Yang, Y.: Application of improved BP algorithm to the optimized formulation of ceramics glaze. In: 2011 3rd International Conference on Computer Research and Development, vol. 1, pp. 402–405. IEEE (2011)
9. Luo, Y., Yang Yun, C.Y.: Design of ceramic formula based on genetic algorithms. J. YanBian University **45**(2), 166–170 (2019)
10. YongPing, Z.L. (ed.): Ceramic Technology. 1, HuNan University Press, 1 edn. (2005)

Using Less but Important Information for Feature Distillation

Xiang Wen[1(✉)], Yanming Chen[2], Li Liu[1], Choonghyun Lee[1,3], Yi Zhao[3,4], and Yong Gong[1(✉)]

[1] Nanhu Academy of Electronics and Information Technology, JiaXing, China
1374454341@qq.com, gongyong@cnaeit.com

[2] School of Compute Science and Technology, Anhui University, Hefei, China

[3] College of Information Science and Electronic Engineering, Zhejiang University, Hangzhou, China

[4] School of Integrated Circuits, East China Normal University, Shanghai, China

Abstract. The purpose of feature distillation is that using the teacher network to supervise student network so that the student network can mimic the intermediate layer representation of the teacher network. The most intuitive way of feature distillation is to use the Mean-Square Error (MSE) to optimize the distance of feature representation at the same level for both networks. However, one problem in feature distillation is that the dimension of the intermediate layer feature maps of the student network may be different from that of the teacher network. Previous work mostly elaborated a projector to transform feature maps to the same dimension. In this paper, we proposed a simple and straightforward feature distillation method without additional projector to adapt the feature dimension inconsistency between the teacher and the student networks. We consider the redundancy of the data and show that it is not necessary to use all the information when performing feature distillation. In detail, we propose a cut-off operation for channel alignment and use singular value decomposition (SVD) for knowledge alignment so that only important information is transferred to the student network to solve the dimension inconsistency problem. Extensive experiments on several different models show that our method can improve the performance of student networks.

Keywords: Neural network · Knowledge distillation · Model compression

1 Introduction

Currently, deep convolutional neural networks(CNNs) are playing a very important role in a wide variety of vision tasks and have achieved amazing success. CNNs can run very smoothly on some devices with powerful computational resources and give computational results in real time. However, in the real world, the computing resources of the devices used in many applications are very limited, such as mobile devices, vehicle systems, embedded devices, etc. It is very

B. Luo et al. (Eds.): ICONIP 2023, LNCS 14447, pp. 395–407, 2024.
https://doi.org/10.1007/978-981-99-8079-6_31

difficult for CNN models to run smoothly on these devices. A good solution is to compress the model and use smaller models with less resource requirements. Therefore, many methods for model compression and improving the performance of small models have been proposed, including network pruning [10], and knowledge distillation [8,14], etc.

The goal of knowledge distillation is to distill the knowledge from the teacher network (large network) into the student network (small network) to improve the performance of the student network. The teacher network will play the role of a supervisor to supervise the training process of the student network, so that the student network will not only learn the knowledge from the ground truth labels, but also learn the class predictions probabilities (as known as soft targets) or intermediate layer representation of the teacher network.

The concept of knowledge distillation (KD) is proposed in [8], where the student network additionally learns the class prediction probabilities of the teacher network. However, KD [8] is only applicable to classification tasks. Therefore, many efforts [1,13,14,19] considering feature distillation, which impels the student network to mimic the intermediate layer representation of the teacher network, to further improve the performance of the student network and expand the the applicable tasks of knowledge distillation. The most intuitive way of feature distillation is that we can optimize the Mean-Square Error (MSE) loss of the intermediate layer feature maps between the teacher and student networks. However, there are two questions when using MSE: One is that the number of extracted feature maps is mismatched when the output channels of the intermediate layer of the teacher network and the student network are different. Another problem is that the size of each pair of feature maps from the teacher and student networks, respectively, may not be the same. For these two questions, many previous methods elaborate a projector to adjust the number of output channels and use pooling operation to scale the feature map size to fit the MSE. The feature gap between the teacher and the student networks was then measured to force the student network to learn the representation of the teacher network through feedback propagation mechanism. As shown in Fig. 1(a), for simplicity, we merged the pooling operation into the projector, this is also true for Fig. 2. It is a highly complex task that elaborating projectors. Pooling operation also is prone to lose important information.

So, in this paper, we consider the redundancy of parameters and how to use less but more important information to more effectively supervise the training process of the student network, instead of additionally designing projectors to align the dimension of the feature maps and supervise each position of the student network as in previous work [1,19] or computing probability distribution information in the feature space as in [13]. We think there are some features of teacher networks that are not important and not worth learning. We also believe that we do not need to supervise every detail of the student network during the training process. Just like human learning. Students do not need to be supervised all the time and teachers will also filter the important knowledge to teach students. Based on the above, we propose a simple and effective framework, which

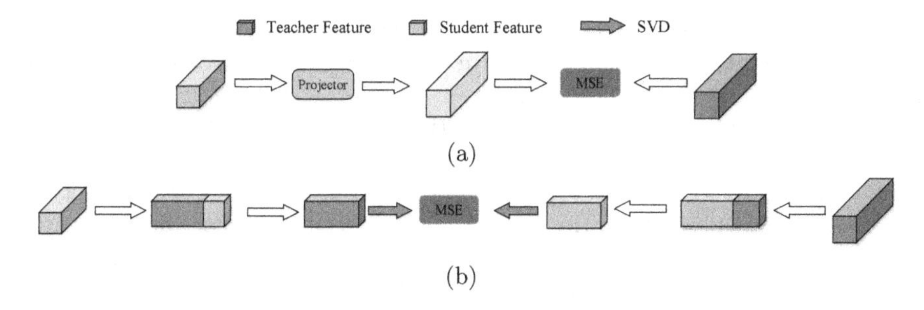

Fig. 1. (a) Previous feature knowledge framework. (b) We propose to use less but more important information for feature distillation by singular value decomposition (SVD).

aims to achieve more efficient feature distillation using less but more important information, as shown in Fig. 1(b). First, when the number of intermediate layer output channels is different between the teacher network and the student network. We use cut-off operation to enable channel alignment. Second, for a given pair of teacher-student feature map, we decompose the two feature maps using the singular value decomposition (SVD) and measure the gap between the singular value vectors instead of computing the error of the original matrices. This will bring two benefits. One is the alignment of important information when the shape (height and width) of a pair of feature maps is the same. Another case is when the shapes of a pair of feature maps are different. We only need to choose the larger singular values to evaluate the gap of important information, and the rest of the very small singular values can be completely ignored.

Our contribution can be summarized as follows:

- We propose a novel feature distillation method without additional projector to adapt the feature dimension inconsistency between the teacher and the student networks and show that it is not necessary to use all the feature information when performing feature distillation.
- When the number of intermediate feature maps is different between the teacher and the student networks, we show that we can greatly improve the performance of the student network by computing only the gap of a part of feature information.
- For a given pair of teacher-student feature map, we compare the gap of the singular values of the two feature maps instead of transferring the feature matrices to the same shape and measuring the distance at each position.
- Extensive experiments on several different models have demonstrated the effectiveness of our method.

2 Related Work

The concept of knowledge distillation is proposed in KD [8], which considers a pair of teacher-student networks and constrains that the student network to learn not only ground truth labels but also soft targets from the teacher network. Soft targets express more similar feature between data and can provide more information.

FitNet [14] extends knowledge distillation to feature distillation. FitNet [14] additionally computes the MSE of the feature maps from the teacher and student networks to force the intermediate representation of the student network to be as close as possible to the teacher network. Subsequently, various feature distillation algorithms [1,13,19] were proposed. AT [19] proposed spatial attention as mechanism of transfering knowledge from one network to another. VID [1] proposed a feature distillation algorithm based on a Gaussian distribution that maximizes the variational lower bound of mutual information between two networks. These methods all require aligning the feature dimension.

PKT [13] tries to match the probability distribution of the data between the feature space of the teacher and the student networks to avoid designing a projector. But it is very cumbersome and time-consuming to estimate the distribution and design the probability density function. In this paper, we proposed a novel and simple feature distillation framework, which considers the redundancy of the data without the need to design an additional projector and also using MSE as the loss function for feature distillation.

3 Method

In this section, we normalize the neural network training and distillation process, and then describe how our method uses less feature information to distill the student network.

3.1 Knowledge Distillation

Generally, the training process of CNNs can be shown as in Fig. 2. Given a image for a network (such as the student network of Fig. 2), it will undergo multiple convolution modules to produce multiple intermediate layer feature maps $A_s^l \in \mathbb{R}^{c_s^l \times h_s^l \times g_s^l}$, l denotes the l-th convolutional layer, where $1 \leq l \leq L$, L is the number of convolutional layers, c_s^l is the number of output channels, h_s^l and g_s^l are height and width of feature maps. The feature maps generated by the penultimate layer will then be passed through a classifier consisting of a fully connected layer and a softmax activation function to output a class probability vector $y_s \in \mathbb{R}^K$ for a K classification task. The loss function between the predicted class probability vector y_s and the one-hot ground truth labels y_{gt} will be computed to optimize the performance of the network using the feedback propagation mechanism.

$$y_s^i = \frac{\exp\left(z_s^i/T\right)}{\sum_i^K \exp\left(z_s^i/T\right)} \tag{1}$$

Vanilla KD [8] uses the softmax activation function with the temperature parameter T to soften the logits z_s and z_t of the student and teacher networks,

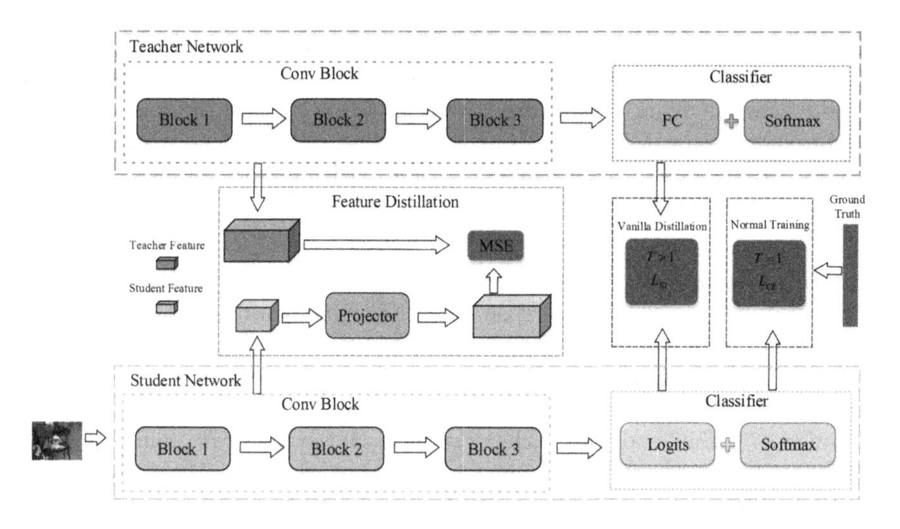

Fig. 2. Normal training, vanilla knowledge distillation and feature distillation for deep convolutional neural networks.

as shown in Eq. 1. Then using soft targets from teacher network to additionally supervise the student network enables student network to learn more information and improve the performance of the student network, as the "Vanilla Distillation" module in Fig. 2. Therefore, as shown in Eq. 2, the KD [8] loss function consists of two components: L_{CE} is the conventional cross entropy loss and L_{KL} is the Kullback-Leibler divergence of class probabilities y_t and y_s from the teacher and student networks, respectively.

$$L_{KD} = L_{CE}(y_{gt}, y_s) + L_{KL}(y_t, y_s) \tag{2}$$

Feature distillation [1,14,19] want to transfer the middle layer feature information from the teacher network to the student network. These works usually elaborate a projector that converts the feature maps of student network to the same dimension of the teacher network and then to compute MSE at each position, as shown in Fig. 2.

We believe that the mechanism for measuring difference at each location is inefficient, pooling operations can also cause significant information loss. We consider that designing a projector in feature distillation to align the feature dimension of the teacher and the student networks is an avoidable operation. Based on this, we give a new perspective on feature distillation, where we think about and want to discriminate the redundancy and importance of feature information, to improve the performance of the student network by using less supervisory information.

3.2 Our Method

As considered in previous work [5,10], there are very many redundant parameters and many unimportant channels in the CNNs. So, when performing feature distillation, we think that we do not need to use all the feature information of the teacher network, and the student network also does not need to be supervised by the teacher network all the time. We can only compare the gap of less but more important information for a pair of teacher and student network. As shown in Fig. 3, we propose a novel feature distillation framework to more effectively distill student network.

Channel Alignment. Feature distillation aims to minimize the difference in intermediate representation between the student network and the teacher network. An intuition is that we can use the MSE as a loss function during feature distillation to provide additional gradient information for the student network and supervise the student network to output intermediate layer representation as similar as possible to the teacher network. However, MSE requires that the dimensions of the two variables must be identical, which is not always true for feature distillation. So, we need to adjust the feature dimensions when the number of output channels of convolution layer or the shape (height and width) of the feature maps are different.

First, when the number of output channels (which indicates the number of two-dimensional feature maps) of the selected intermediate layer is different for a pair of teacher-student networks. Previous works [1,14,19] usually elaborate a projector with convolutional layers to switch the number of feature maps of the student network to the same as the teacher network, as shown in the feature distillation module in Fig. 2.

Different from the previous work, in this paper, we aim to extract the important information in the original feature matrices, rather than transforming the feature information additionally. We show that using only a part of the feature information can also provide a good supervision effect. As shown in Fig. 1(b) and Fig. 3(a), we propose a simple cut-off operation for the output channel alignment. Specifically, after we extract the output feature maps of the penultimate layer(or the last block, e.g., residual block [6]). If the number of output feature maps of teacher and student networks are unequal. We propose a cut-off operation to select the same number of feature maps for teacher and student networks to align the output channels for feature distillation, as in Eq. 3.

$$c_{\min} = min\left(c_p^t, c_p^s\right), \begin{cases} A_t^p = A_t^p\left[0 : c_{\min}\right] \\ A_s^p = A_s^p\left[0 : c_{\min}\right] \end{cases} \tag{3}$$

where $A_t^p \in \mathbb{R}^{c_t^p \times h_t^p \times g_t^p}$ and $A_s^p \in \mathbb{R}^{c_s^p \times h_s^p \times g_s^p}$ denote the feature maps of teacher and student networks, respectively. The syntax in the Eq. 3 is presented using the syntax of python.

This method is effective because two feature maps located within the same layer are not completely orthogonal and irrelevant to each other, and they may contain similar feature information. Previous work on structured pruning [10] also pointed out that the filters in the CNNs are redundant and that the activations generated by redundant filters are unimportant. These unimportant activations can be safely removed. So, selecting all feature maps during feature

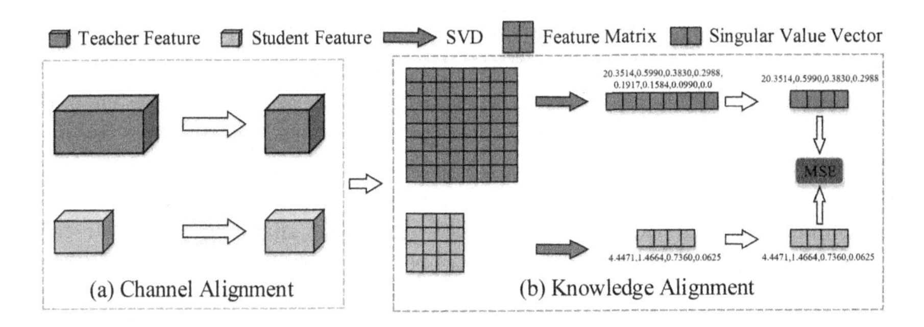

Fig. 3. This figure corresponds to the feature distillation module of Fig. 2. (a) Channel alignment. (b) SVD and Knowledge alignment. We show the detailed singular value vectors. The teacher feature map is an 8×8 matrix from ResNet32 $\times$ 4. The student feature map is an 4×4 matrix from ShuffleNetV1.

distillation is a redundant operation that may ultimately provide bad gradient information for the student network. Furthermore, even if only the i-th feature map $A_s^{p,i}$ is supervised. It will only compute the gradient of the filter $W_s^{p,i}$ in this layer, but will provide additional gradient information for all parameters in the previous layers by the chain rule. So, it is not at all necessary to use all the features in the feature distillation.

The experiments in Sect. 4 also prove our explanation perfectly. In addition, as shown in Subsect. 4.3, we also show that even using fewer feature maps can still provide enough gradient information and greatly improve the performance of the student network.

Knowledge Alignment. Generally, given two two-dimensional matrices, the MSE loss function will always evaluate the difference in each corresponding position. When a pair of feature maps has different feature shapes (height and width). We can use the pooling operation to transfer the feature maps to the same shape as the previous methods [1,14,19]. But we think this is an inefficient way and may also lead to the losing of important feature information. Even if two feature matrices have the same shape, we think it is not necessary to evaluate the gap at each position.

In this section, we take a novel and simple technique to solve this problem. We want to transfer important knowledge to the student network while ignoring other irrelevant and unimportant noise information. As shown in Fig. 3(b), we use SVD to decompose the feature matrices and to transfer important feature information to the student network. SVD is a very effective technique for extracting important information from matrices and is also used in low-rank decomposition [4] to speed up the inference of CNNs.

For a given two-dimensional matrix $A \in \mathbb{R}^{m \times k}$. The SVD is defined as $A = USV^T$, where $U \in \mathbb{R}^{m \times m}$ is the left singular value matrix, $V \in \mathbb{R}^{k \times k}$ is the right singular value matrix. Both U and V are orthogonal matrices. $S \in \mathbb{R}^{m \times k}$ is a diagonal matrix with the singular values on the diagonal. For convenience, we simplify S to a vector consisting of singular values in descending order.

As shown in Fig. 3. After aligning the channels, we decompose each two-dimensional feature map using SVD and extract the singular value vector. In contrast to computing the difference between the original matrices, we compute the difference between the singular value vectors so that the student network produces similar singular values to the teacher network. Using larger singular values to supervise larger singular value can realize important information alignment. This is because the large singular values usually express more important information contained in the original matrix. Finally, Our loss is as shown in Eq. 4 (S_t and S_s are the singular value vectors of the teacher and student networks, respectively), where λ is the hyperparameter, λ is set to 1 in this paper.

$$L_{KD} = L_{CE}(y_{gt}, y_s) + \lambda L_{MSE}(S_t, S_s) \tag{4}$$

When the shape of the feature maps is same, comparing the singular value vectors of two feature maps can automatically compare the feature information gap at different levels of importance for two feature maps. Our proposed singular value-based feature distillation method works well even if when the feature shapes are not same. This is because when the shape of the feature maps of the student and teacher networks are different, we can approximate the gap between two feature maps by computing only the gap between the top-k singular values in the two singular value vectors. This gap represents the difference in the primary information of the feature maps, while the remaining secondary or even redundant information does not provide guidance for the student network.

As shown in Fig. 3(b). We show the extracted singular value vectors. After the feature map is decomposed by SVD. The sum of the top 10% or even the top 1% of singular values in the singular value vector may reach 90% of the total sum of all singular values. Other very small singular values can be ignored.

4 Experiments

4.1 Experiment Setting

We choose the datasets of CIFAR-100 [9] and ImageNet [3] respectively for our experiments. CIFAR-100 is a 100 classification dataset, it contains 50K training images and 10K test images. The ImageNet dataset is the most convincing dataset for the classification task, which consists of 1.2 million and 50K of images of 224×224 pixel size as the training and validation sets.

The experiments were conducted on several well-known networks, including VGG [16], ResNet [6], WideResNet [18], MobileNet [15], ShuffleNet [11,20].

Our experiment setting follows previous works [2,17]. We use SGD optimizer with 0.9 Nesterov momentum train all network. On CIFAR-100 dataset, the initial learning rate is set to 0.01 for MobileNet/ShuffleNet and 0.05 for VGG/ResNet/WideResNet. We train all network for 240 epoch and the learning rate is divide by 10 at 150th,180th and 210th epochs. The weight decay is set to 5×10^{-4} and the mini-batch size is set to 64. On ImageNet dataset, the initial learning rate is set to 0.1 and divide by 10 at 30th, 60th, 90th of the total 120 training epochs. The weight decay is set to 1×10^{-4} and the mini-batch size is set to 256.

4.2 Comparison of Test Accuracy

In this section, we will show the effectiveness of our method. We show in detail the experimental results with the same or different combinations of architectures.

Results on CIFAR-100. The experiment results on CIFAR-100 are shown in Tables 1 and 2. Our method can significantly improve the performance of student networks and has better performance in most of the comparison data.

Table 1 shows ehe results of the experiments where the teacher and the student networks are same architecture. In Table 1, we extract the output feature maps of the penultimate layer for distillation, and the obtained feature maps all have the same dimension. So we don't need to do alignment operation and can distill directly student network using the singular value vector. Furthermore, our method and KD are complementary and it can be seen that combining KD can further improve the performance of student networks in some combinations.

Table 2 shows the results of the experiments where the teacher and the student networks are different architecture (in addition to the combination of WRN40-2 and WRN40-1). In these experiments, we extract the output feature map from the last convolution block (MobileNetV2 and ShuffleNetV2 also add a convolution layer after the last convolution block). The dimension of the output feature maps of these teacher-student combinations is not the same. As we have seen, our method can still greatly improve the performance of the student network. This completely validates our idea that we do not need to consider all the

Table 1. Results on CIFAR-100 with the teacher and student have same feature dimension.

Mechanism	Teacher	ResNet32 × 4	ResNet56	ResNet110	WRN40-2	VGG13
	acc	79.42	72.34	74.31	75.61	74.64
	Student	resnet8 × 4	Resnet20	Resnet32	WRN16-2	VGG8
	acc	72.50	69.06	71.14	73.26	70.36
Logits	KD [8]	73.33	70.66	73.08	74.92	72.98
Feature Distillation	FitNet [14]	73.50	69.21	71.06	73.58	71.02
	PKT [13]	73.64	70.34	71.82	74.54	72.88
	RKD [12]	71.90	69.61	71.82	73.35	71.48
	CRD [17]	**75.51**	71.16	73.48	75.48	**73.94**
	AT [19]	73.44	70.55	72.31	74.08	71.43
	VID [1]	73.09	70.38	72.61	74.11	71.23
	Ours	74.3	71.21	73.44	**76.16**	73.13
	Ours+KD	73.95	**71.46**	**73.75**	75.54	73.43

Table 2. Results on CIFAR-100 with the teacher and student have different feature dimension.

Mechanism	Teacher	ResNet32 × 4	ResNet32 × 4	VGG13	WRN40-2	WRN40-2
	acc	79.42	79.42	74.64	75.61	75.61
	Student	ShuffleNetV1	ShuffleNetV2	MobileNetV2	WRN40-1	ShuffleNetV1
	acc	70.50	71.82	64.6	71.98	70.50
Logits	KD [8]	74.07	74.45	67.37	73.54	74.83
Feature Distillation	FitNet [14]	73.59	73.54	64.14	72.24	73.73
	PKT [13]	74.10	74.69	67.13	73.54	73.89
	RKD [12]	72.28	73.21	64.52	72.22	72.21
	CRD [17]	75.11	75.65	**69.73**	**74.14**	76.05
	AT [19]	71.73	72.73	59.40	72.77	73.32
	VID [1]	73.38	73.40	65.56	73.30	73.61
	Ours	**76.15**	76.65	67.32	74.02	77.03
	Ours+KD	75.78	**76.73**	69.00	74.06	**77.53**

Table 3. Results on ImageNet. ResNet18 as student, ResNet34 as teacher.

acc(1,5)	Teacher	Student	KD[8]	AT[19]	OFD[7]	CRD[17]	Ours
top-1	73.31	69.75	70.66	70.69	70.81	71.17	**71.56**
top-5	91.42	89.07	89.88	90.01	89.98	90.13	**90.47**

information in feature distillation. Our method can transfer knowledge from the teacher network to the student network more effectively with less supervisory information.

It is worth emphasizing that although our method did not achieve the best results in all comparative experiments. But our method saves a lot of overhead in computing gradient information, especially as shown in Table 2. For example, for the combination of ResNet32×4 and ShuffleNetV1, the number of extracted feature maps are 256 of 8×8 and 960 of 4×4, respectively. So, after performing channel alignment and knowledge alignment, we only need to compute the MSE of 256 pairs of one-dimensional vectors of length 4, as shown in Fig. 3. For the student network, we only evaluate the gradient information of 256 feature maps compared to other methods that evaluate the gradient information of all 960 feature maps. And as shown in Table 5, we can further use fewer feature maps without reducing the distillation effect. In addition, our approach and some other feature distillation mechanisms are complementary. In the future, we will further explore the effectiveness of our approach to replace the projector.

It is also worth noting that the performance improvement of the ShuffleNet series are significant. We suppose that the shuffle operation can improve the efficiency of our method and enhance the generalization ability of the student network when distillation is performed with less information.

Results on ImageNet. We use the ImageNet dataset to further demonstrate the efficiency and effectiveness of our method. The experiment results on ImageNet are shown in Table 3. We set ResNet34 as the teacher network and ResNet18 as the student network. The accuracy of top-1 and top-5 is reported. Our method improves the performance of the student network by 1.81 in top-1 accuracy.

4.3 Ablation Study

In this section, we verify the effectiveness of our proposed knowledge alignment and channel alignment.

Table 4. The effect of singular values on CIFAR-100.

Teacher	ResNet32 × 4	ResNet32 × 4	VGG13	WRN40-2	WRN40-2
acc	79.42	79.42	74.64	75.61	75.61
Student	ShuffleNetV1	ShuffleNetV2	MobileNetV2	WRN40-1	ShuffleNetV1
acc	70.50	71.82	64.6	71.98	70.50
M	73.23	73.98	66.07	74.32	74.68
Ours	76.15	76.65	67.32	74.02	77.03

Table 5. The effect of the number of channels on the CIFAR-100.

Teacher	ResNet32 × 4	ResNet32 × 4	VGG13	WRN40-2	WRN40-2
acc	79.42	79.42	74.64	75.61	75.61
Student	ShuffleNetV1	ShuffleNetV2	MobileNetV2	WRN40-1	ShuffleNetV1
acc	70.50	71.82	64.6	71.98	70.50
1/4	75.44	75.99	66.5	71.95	75.84
1/2	76.05	76.48	66.78	73.13	76.48
3/4	76.14	76.6	67.23	73.97	77.04
1	76.15	76.65	67.32	74.02	77.03

First, we verify the effectiveness of larger singular values. We compare the distillation effect of using singular values and directly using the original matrix (marked as "M"). When two two-dimensional matrices do not have the same shape, we cut off the height and width of the matrix in the same way as the channel alignment, as shown in Eq. 5. Our proposed knowledge alignment can greatly improve distillation efficiency, as shown in Table 4.

$$\begin{cases} A_p^t = A_p^t\left[0 : c_{\min}, 0 : h_{\min}, 0 : g_{\min}\right] \\ A_p^s = A_p^s\left[0 : c_{\min}, 0 : h_{\min}, 0 : g_{\min}\right] \end{cases} \tag{5}$$

We also use fewer feature maps for ablation study. We chose 1/2, 1/4 and 3/4 of C_{min} for the ablation study. As shown in Table 5, we can see that using 1/4 of the number of channels can greatly improve the performance of the student network. When set to 1/2, the performance of the student network is very close to the results of our standard feature distillation. When set to 3/4, the student network can outperform even our standard feature distillation.

5 Conclusion

In this paper, we propose a novel and simple feature distillation method that uses less but more important information. We propose a channel alignment operation and a knowledge alignment operation to improve the performance of student networks more effectively. We also provide a new perspective, where we do not need to observe all feature information during feature distillation. Extensive experiments can prove the effectiveness and efficiency of our method.

Acknowledgment. This work is supported by the National Key Research and Development Program of China under Grant NO.2020AAA0109001, the "Ling Yan" Program for Tackling Key Problems in Zhejiang Province under Grant No.2022C01098.

References

1. Ahn, S., Hu, S.X., Damianou, A.C., Lawrence, N.D., Dai, Z.: Variational information distillation for knowledge transfer. In: CVPR, pp. 9163–9171 (2019)
2. Chen, D., Mei, J., Zhang, H., Wang, C., Feng, Y., Chen, C.: Knowledge distillation with the reused teacher classifier. In: CVPR, pp. 11923–11932 (2022)
3. Deng, J., Dong, W., Socher, R., Li, L., Li, K., Fei-Fei, L.: ImageNet: a large-scale hierarchical image database. In: CVPR, pp. 248–255 (2009)
4. Denton, E.L., Zaremba, W., Bruna, J., LeCun, Y., Fergus, R.: Exploiting linear structure within convolutional networks for efficient evaluation. In: NIPS, pp. 1269–1277 (2014)
5. Han, S., Pool, J., Tran, J., Dally, W.J.: Learning both weights and connections for efficient neural network. In: NIPS, pp. 1135–1143 (2015)
6. He, K., Zhang, X., Ren, S., Sun, J.: Deep residual learning for image recognition. In: CVPR, pp. 770–778 (2016)
7. Heo, B., Kim, J., Yun, S., Park, H., Kwak, N., Choi, J.Y.: A comprehensive overhaul of feature distillation. In: ICCV, pp. 1921–1930 (2019)
8. Hinton, G.E., Vinyals, O., Dean, J.: Distilling the knowledge in a neural network. CoRR abs/1503.02531 (2015)
9. Krizhevsky, A., Hinton, G., et al.: Learning multiple layers of features from tiny images (2009)
10. Liu, Z., Li, J., Shen, Z., Huang, G., Yan, S., Zhang, C.: Learning efficient convolutional networks through network slimming. In: ICCV, pp. 2755–2763 (2017)
11. Ma, N., Zhang, X., Zheng, H.-T., Sun, J.: ShuffleNet V2: practical guidelines for efficient CNN architecture design. In: Ferrari, V., Hebert, M., Sminchisescu, C., Weiss, Y. (eds.) Computer Vision – ECCV 2018. LNCS, vol. 11218, pp. 122–138. Springer, Cham (2018). https://doi.org/10.1007/978-3-030-01264-9_8

12. Park, W., Kim, D., Lu, Y., Cho, M.: Relational knowledge distillation. In: CVPR, pp. 3967–3976 (2019)
13. Passalis, N., Tefas, A.: Learning deep representations with probabilistic knowledge transfer. In: Ferrari, V., Hebert, M., Sminchisescu, C., Weiss, Y. (eds.) ECCV 2018. LNCS, vol. 11215, pp. 283–299. Springer, Cham (2018). https://doi.org/10.1007/978-3-030-01252-6_17
14. Romero, A., Ballas, N., Kahou, S.E., Chassang, A., Gatta, C., Bengio, Y.: FitNets: hints for thin deep nets. In: ICLR (Poster) (2015)
15. Sandler, M., Howard, A.G., Zhu, M., Zhmoginov, A., Chen, L.: MobileNetV2: inverted residuals and linear bottlenecks. In: CVPR, pp. 4510–4520 (2018)
16. Simonyan, K., Zisserman, A.: Very deep convolutional networks for large-scale image recognition. In: ICLR (2015)
17. Tian, Y., Krishnan, D., Isola, P.: Contrastive representation distillation. In: ICLR (2020)
18. Zagoruyko, S., Komodakis, N.: Wide residual networks. In: BMVC (2016)
19. Zagoruyko, S., Komodakis, N.: Paying more attention to attention: improving the performance of convolutional neural networks via attention transfer. In: ICLR (Poster) (2017)
20. Zhang, X., Zhou, X., Lin, M., Sun, J.: ShuffleNet: an extremely efficient convolutional neural network for mobile devices. In: CVPR, pp. 6848–6856 (2018)

Efficient Mobile Robot Navigation Based on Federated Learning and Three-Way Decisions

Chao Zhang[1], Haonan Hou[1], Arun Kumar Sangaiah[2,3](✉), Deyu Li[1](✉), Feng Cao[1], and Baoli Wang[4]

[1] School of Computer and Information Technology, Shanxi University, Taiyuan, China
lidysxu@163.com

[2] International Graduate Institute of Artificial Intelligence, National Yunlin University of Science and Technology, Douliu, Taiwan
aksangaiah@ieee.org

[3] Department of Electrical and Computer Engineering, Lebanese American University, Byblos, Lebanon

[4] School of Mathematics and Information Technology, Yuncheng University, Yuncheng, China

Abstract. In the context of Industry 5.0, the significance of MRN (Mobile Robot Navigation) cannot be overstated, as it is crucial for facilitating the collaboration between machines and humans. To augment MRN capabilities, emerging technologies such as federated learning (FL) are being utilized. FL enables the consolidation of knowledge from numerous robots located in diverse areas, enabling them to collectively learn and enhance their navigation skills. By integrating FL into MRN systems, Industry 5.0 can effectively utilize collaborative intelligence for efficient and high-quality production processes. When considering information representation in MRN, the adoption of picture fuzzy sets (PFSs), which expand upon the concept of intuitionistic fuzzy sets, offers significant advantages in effectively handling information inconsistencies in practical situations. Specifically, by leveraging the benefits of multi-granularity (MG) probabilistic rough sets (PRSs) and three-way decisions (3WD) within the FL framework, an efficient MRN approach based on FL and 3WD is thoroughly investigated. Initially, the adjustable MG picture fuzzy (PF) PRS model is developed by incorporating MG PRSs into the PF framework. Subsequently, the PF maximum deviation method is utilized to calculate various weights. In order to determine the optimal granularity of MG PF membership degrees, the CODAS (Combinative Distance based ASsesment) method is employed, known for its flexibility in handling both quantitative and qualitative attributes whereas effectively managing incomplete or inconsistent data with transparency and efficiency. After determining the optimal granularity, the MRN method grounded in FL and 3WD is established. Finally, a realistic case study utilizing MRN data from the Kaggle database is performed to validate the feasibility of our method.

B. Luo et al. (Eds.): ICONIP 2023, LNCS 14447, pp. 408–422, 2024.
https://doi.org/10.1007/978-981-99-8079-6_32

Keywords: Mobile Robot Navigation · Federated Learning · Multi-Granularity · Three-Way Decision · CODAS

1 Introduction

The term "Industry 5.0" refers to the upcoming stage of industrial development that builds upon earlier phases, including Industry 4.0 [1]. In the setting of Industry 5.0, where the optimization of industrial processes is focused on through the fusion of advanced techniques such as computational intelligence, hyperautomation and cloud of things [2], MRN has emerged as a vital technology. In this context, mobile robots play a vital role in improving the efficiency and flexibility of manufacturing and logistics operations by autonomously navigating in dynamic environments. By leveraging advanced sensors, mapping, and localization technologies, mobile robots can perform tasks such as material handling, inventory management, and quality inspection with high accuracy and speed. Moreover, mobile robots can adapt to changing production requirements and collaborate with human workers to achieve higher productivity and safety. Therefore, MRN stands as a pivotal catalyst for Industry 5.0, holding the potential to revolutionize the operational landscape of industries in the forthcoming era. FL functions as a machine learning structure which facilitates collaboration among multiple devices or nodes to collectively train a model without the need to share their data with a central server [3]. FL aims to maintain the confidentiality of data while enabling the development of a model that encompasses the expertise of all the devices engaged in the procedure [4]. This paper aims to explore the enhancement of overall explainability in industrial processes through the utilization of an efficient MRN technique customized for FL. The primary focus is on assessing the method's efficacy and its ability to shed light on the intricate aspects present within the industrial environment. To accomplish this goal, the authors utilize the form of PFSs, 3WD and CODAS to construct a comprehensive model. The rationale for employing these models is outlined below.

With its clear advantages in conveying decision information compared to other approaches [5,6], PF decision-making has gradually gained prominence as a significant research area within the field of intelligent decision-making [7,8]. Granular computing [9,10] is an instrumental tool employed to tackle diverse decision-making problems. Yao [11], in his notable contributions, enhanced the decision-theoretic rough set to establish a significant granular computing technique. Additionally, he introduced the concept of 3WD, which has proven to be a highly proficient approach for analyzing uncertain information. This makes 3WD an essential component of granular computing methodologies. By leveraging 3WD models and incorporating the Bayesian theory, decision risks can be significantly mitigated. In decision-theoretic rough sets, traditional loss functions are commonly used to establish threshold values for decision-makers. However, they often fall short in accurately discerning personalized threshold values for individual decision objects. To overcome these limitations, this paper explores an enhanced model that expands decision-driven 3WD approaches into the PF context. It achieves this by incorporating relative loss functions.

Considering the need to incorporate multiple thresholds in decision-driven 3WD approaches under the PF context, this procedure can be constructed from a typical issue of multi-attribute decision-making. To address this task, the CODAS method is adopted in this paper. The CODAS method is a versatile technique in the realm of decision analysis, which aims to evaluate and rank options in light of a set of predefined attributes. At its core, the CODAS method relies on measuring the distance between options and attributes, which serves as an indicator of the level of satisfaction or preference for each object in terms of each attribute. The CODAS method [12] offers flexibility by accommodating both qualitative and quantitative attributes, and it stands out for its transparency, efficiency, and ability to handle incomplete or contradictory data. As a result, the CODAS method finds applicability across diverse domains such as engineering, finance and healthcare, where it assists in realistic decision analysis.

Based on the above analysis, it is imperative to examine the adjustable MG PF PRS that is founded on 3WD. The first step involves introducing the model of adjustable MG PF PRSs. Subsequently, the relative loss function is computed via employing the 3WD approach, leading to the establishment of the PF 3WD method. In conclusion, an MRN data set in the Kaggle database is used as the data source, and an MRN information system is suggested. This system combines the PF 3WD method with FL to provide the ultimate classification and ranking outcomes.

In light of the above statement, the paper makes the following key contributions:

1. An efficient MRN method based on FL and 3WD is investigated;
2. The framework of 3WD is used to expand the PF concept, and the adjustable MG PF PRS model is established;
3. The feasibility and effectiveness of the PF 3WD method, utilizing adjustable MG PF PRSs and CODAS, are verified through MRN cases in the Kaggle database.

2 The Proposed Approach

2.1 PFSs

PFSs provide benefits in managing uncertain problem modeling in practical situations, making them valuable in real-world scenarios. Their application for representing decision information has generated considerable attention in the intelligent decision-making area, resulting in the emergence of PF decision-making as a prominent and actively researched area. In what follows, a detailed introduction to the fundamental concept of PFSs will be presented.

Definition 1. [13] Suppose U is a universe set. Then a PFS P on U can be provided by

$$P=\{(x, < \mu_P(x), \eta_P(x), \nu_P(x) >)|x \in U\}, \tag{1}$$

where $\mu_P \in [0,1]$ indicates the agreement degree, $\eta_P \in [0.1]$ is called the neutrality degree, and $\nu_P \in [0,1]$ is called the opposition degree of x in P, μ_P, η_P and ν_P fulfill the restrictive condition: $0 \leq \mu_P(x) + \eta_P(x) + \nu_P(x) \leq 1$. Then, $\rho_P = 1 - (\mu_P(x) + \eta_P(x) + \nu_P(x))$ is called the refusal membership degree of x in P. Additionally, $< \mu_P(x), \eta_P(x), \nu_P(x) >$ is named a PF number (PFN).

2.2 MG PF PRSs Based on FL

FL can effectively protect data privacy and make data have better security. We apply MG PF PRSs to the framework of FL, so that the data set is segmented according to the attribute dimension, and then classified and trained according to attribute characteristics. Suppose a decision system has f decision-makers, U is an object set, V is an attribute set, $\boldsymbol{u}$ is an attribute weight set, $\boldsymbol{w} = \{\boldsymbol{w}_1, \boldsymbol{w}_2, \cdots, \boldsymbol{w}_f\}^T$ is a decision-maker weight set, $\boldsymbol{w}_i \in [0,1]$ and $\sum_{i=1}^{f} \boldsymbol{w}_i = 1$, R_i is the PF relationship of $U \times V$, D is a standard evaluation set. In conclusion, a PF information system (PFIS) can be eventually built.

Definition 2. Suppose (U, V, R_i, D) is a PFIS. For any $x_j \in U\,(j = 1, 2, \cdots, m)$ and $y_k \in V\,(k = 1, 2, \cdots, n)$, $\psi_D^{R_i}(x_j)$ represents the MG PF membership degree of x_j in terms of R_i, which can be expressed by

$$\psi_D^{R_i}(x_j) = \left(\frac{\sum_{y_k \in V} \boldsymbol{u}_k R(x_j, y_k)\,(D_{y_k})^c}{\sum_{y_k \in V} \boldsymbol{u}_k R(x_j, y_k)}\right)^c. \tag{2}$$

Based on the MG PF membership degree $\psi_D^{R_i}(x_j)$, an MG PF PRS model can be set up.

Definition 3. Suppose (U, V, R_i, D) is a PFIS, $\psi_D^{R_i}(x_j)$ is an MG PF membership degree. Suppose numerical values in $\psi_D^{R_i}(x_j)$ are arranged in a growing sequence. $\psi_D^{R_{\tau(i)}}(x_j)$ is the $i-th$ in $\psi_D^{R_i}(x_j)$, $\psi_D^{R_{\tau(i)}}(x_j)$ is the MG PF membership of x_j. The two thresholds represented by PFNs are modeled by α and β with $\beta < \alpha$. Then the MG PF upper and lower approximations in the matter of a PFIS can be provided by

$$\underline{\sum_{i=1}^{f} R_i}^{\psi,\alpha,\eta}(D) = \{\psi_D^{R_{\tau(i)}}(x_j) \geq \alpha \,|\, x_j \in U\}; \tag{3}$$

$$\overline{\sum_{i=1}^{f} R_i}^{\psi,\beta,\eta}(D) = \{\psi_D^{R_{\tau(i)}}(x_j) > \beta \,|\, x_j \in U\}, \tag{4}$$

where $(\underline{\sum_{i=1}^{f} R_i}^{\psi,\alpha,\eta}(D), \overline{\sum_{i=1}^{f} R_i}^{\psi,\beta,\eta}(D))$ is an adjustable MG PF PRS. Then η represents the risk coefficient of decision-makers, that is, when $\eta = 1$ and $\eta = \frac{1}{f}$, the decision-maker tends to completely avoid risks or completely pursue risks. When $\eta \in [\frac{1}{f}, 1]$, the decision-maker's risk preference follows a changing pattern, gradually shifting from full risk-taking to complete risk-aversion. The positive, negative, and boundary domains of MG PF PRSs can be denoted by

$$POS^{\psi}_{\alpha,\beta}(D) = \underline{\sum_{i=1}^{f} R_i}^{\psi,\alpha,\eta}(D);$$

$$NEG^{\psi}_{\alpha.\beta}(D) = U - \overline{\sum_{i=1}^{f} R_i}^{\psi,\beta,\eta}(D);$$

$$BND^{\psi}_{\alpha.\beta}(D) = \overline{\sum_{i=1}^{f} R_i}^{\psi,\beta,\eta}(D) - \underline{\sum_{i=1}^{f} R_i}^{\psi,\alpha,\eta}(D).$$

2.3 Problem Backgrounds

Assuming that there are f experts who evaluate the robot performance in an MRN process. Let U denote the mobile robot data, V be the collection of evaluation indices, $\boldsymbol{u}$ be the collection of evaluation index weights, and $\boldsymbol{w}$ be the group weight set. Based on these assumptions, the relationship of PFSs can be determined, and the standard evaluation set of PFSs is computed. Finally, according to the above U, V, R_i and D, the PFIS (U, V, R_i, D) can be established. In this paper, by combining with the system architecture of FL, the horizontal FL is added in the experiment, which can make the data not leaked.

2.4 Model Applications

This section presents the development of the PF MRN approach for studying the accuracy of mobile data in MRN. To accurately compute relative loss functions, it is essential to create a new rating matrix $P = (\psi_{ji})_{m \times n}$ in PFISs via utilizing the single PF membership degree $\psi^{R_i}_{\mathrm{D}}(x_j)$. To be specific, within the rated matrix P, every $\psi_{ji} \in [\psi^i_{\min}, \psi^i_{\max}]$ is represented by the lowest and highest values of R_i denoted by $\psi^i_{\min}$ and $\psi^i_{\max}$ respectively. Table 1 below displays the PF relative loss functions of x_j after incorporating the coefficient $\sigma \in [0, 0.5]$ which is utilized to indicate the risk preferences of a group of decision-makers [14].

Table 1. PF relative loss functions x_j

	$R(P)$	$\neg R(P)$
a_P	0	$\neg\psi_j$
a_B	$\sigma\psi_j$	$\sigma(\neg\psi_j)$
a_N	ψ_j	0

At last, ξ_j and ζ_j of x_j can be provided by

$$\xi_j = \frac{(1-\sigma)(\neg\psi_j)}{(1-\sigma)(\neg\psi_j)+\sigma\psi_j}; \tag{5}$$

$$\zeta_j = \frac{\sigma(\neg\psi_j)}{\sigma(\neg\psi_j)+(1-\sigma)\psi_j}. \tag{6}$$

The PF membership degrees, which are computed based on the aforementioned threshold values, are utilized to classify different outcomes. The initial ranking of all objects can be determined based on $POS^{\psi}_{\xi,\zeta}(D) \succ BND^{\psi}_{\xi,\zeta}(D) \succ NEG^{\psi}_{\xi,\zeta}(D)$, and the final ranking is established by employing PF score functions.

The weight set of evaluation indicators corresponding to each object is calculated by

$$\boldsymbol{u}^{R_i}_{ik} = \frac{\sum_{j=1}^{m}\sum_{l=1}^{m}\left(d\left(R_i(x_j,y_k),R_i(x_l,y_k)\right)\right)}{\sum_{k=1}^{n}\sum_{j=1}^{m}\sum_{l=1}^{m}\left(d\left(R_i(x_j,y_k),R_i(x_l,y_k)\right)\right)}. \tag{7}$$

The weight of decision-makers is further determined by utilizing the deviation maximization tool.

$$\boldsymbol{w}_i = \frac{\sum_{j=1}^{n}\sum_{o=1}^{n}(d(\boldsymbol{u}^{R_i}_{ij},\boldsymbol{u}^{R_i}_{io}))}{\sum_{i=1}^{f}\sum_{j=1}^{n}\sum_{o=1}^{n}(d(\boldsymbol{u}^{R_i}_{ij},\boldsymbol{u}^{R_i}_{io}))}. \tag{8}$$

Finally, the parameter σ is determined via consulting to the decision-maker's risk preferences.

2.5 CODAS

CODAS is a decision-making approach employed to assess and prioritize alternatives according to various criteria. It encompasses both quantitative and qualitative factors and has the ability to handle incomplete or inconsistent data. The CODAS method calculates the distances between alternatives and reference points, which are then aggregated with appropriate weights to determine the entire manifestation of each alternative. By taking into account both the positive and negative aspects of the criteria, CODAS offers a comprehensive evaluation that assists in decision-making processes. Assuming that we have n alternatives and m attributes, the proposed method can be classified into the following steps:

Step 1. Formulate the specific decision matrix, which is illustrated below:

$$X = [x_{ij}]_{n\times m} = \begin{bmatrix} x_{11} & x_{12} & \cdots & x_{1m} \\ x_{21} & x_{22} & \cdots & x_{2m} \\ \vdots & \vdots & \vdots & \vdots \\ x_{n1} & x_{n2} & \cdots & x_{nm} \end{bmatrix}. \tag{9}$$

In this context, x_{ij} $(x_{ij} \geq 0)$ represents the general measure of the $i-th$ option on the $j-th$ criterion.

Step 2. Calculate the normalized decision matrix via employing the linear normalization of performance measures in the following manner:

$$n_{ij} = \begin{cases} \frac{x_{ij}}{\max\limits_{i} x_{ij}} & if \ \ j \in N_b \\ \frac{\min\limits_{i} x_{ij}}{x_{ij}} & if \ \ j \in N_c \end{cases}. \tag{10}$$

In this context, N_b denotes the collection of benefit attributes, whereas N_c pertains to the collection of cost attributes.

Step 3. Calculate the weighted normalized decision matrix via determining the normalized performance measures that are weighted as follows:

$$r_{ij} = \boldsymbol{w}_j n_{ij}. \tag{11}$$

In this context, $\boldsymbol{w}_j$ denotes the weight assigned to the $j-th$ attribute, while $\sum_{j=1}^{m} \boldsymbol{w}_j = 1$ denotes another variable or set of attributes.

Step 4. Find the negative-ideal point via the following steps:

$$ns = [ns_j]_{1\times m}; \tag{12}$$

$$ns_j = \min_i r_{ij}. \tag{13}$$

Step 5. Calculate the Euclidean and Taxicab distances between the negative-ideal point and all alternatives, which are presented below:

$$E_i = \sqrt{\sum\nolimits_{j=1}^{m} (r_{ij} - ns_j)^2}; \tag{14}$$

$$T_i = \sum\nolimits_{j=1}^{m} |r_{ij} - ns_j|. \tag{15}$$

Step 6. Create the relative rated matrix, which is presented below:

$$Ra = [h_{jk}]_{n\times n}; \tag{16}$$

$$h_{ik} = (E_i - E_k) + (\psi(E_i - E_k) \times (T_i - T_k)). \tag{17}$$

Here, $k \in \{1, 2, \cdots, n\}$ and ψ denote a threshold function that is used to confirm if the Euclidean distances of two alternatives are equal. The function is expressed below:

$$\psi(x) = \begin{cases} 1 \ \ if & |\mathrm{x}| \geq \tau \\ 0 \ \ if & |\mathrm{x}| < \tau \end{cases}. \tag{18}$$

The function includes a threshold parameter τ, which can be set by experts. It is recommended to set the parameter between 0.01 and 0.05. When the difference between the Euclidean distances of two alternatives is below τ, the Taxicab distance is also utilized to compare them.

Step 7. Find the evaluation score of every alternative, which is presented below:

$$H_i = \sum_{k=1}^{n} h_{ik}. \tag{19}$$

Step 8. Arrange the alternatives in a descending sequence based on their evaluation score (H_i). The alternative with the largest H_i value is considered the most favorable selection for all alternatives.

According to the above calculation process of CODAS, its core steps are constructed and summarized in Algorithm 1.

3 Experimental Preparation

According to the weight calculation method described above, MG PF 3WD and the CODAS method, the following decision steps in the matter of PF MRN are built and summed up in Algorithm 2.

The current section shows a case analysis in the field of robotic navigation utilizing the Kaggle data set. The first step involves the demonstration of the presented PF MRN method in a real-world scenario. Furthermore, comparative and sensitivity analyses are conducted to reveal the applicability and stability of the presented PF MRN method. Lastly, experimental analysis is made to acknowledge the validity of our method.

Algorithm 1: The algorithm of CODAS

Input: The intermediate decision matrix
Output: The sorting result of objects

1 **for** $i = 1$ *to* n, $j = 1$ *to* m **do**
2 | Obtain a decision-making matrix (X) by Eq. (9)
3 **end**
4 **for** $i = 1$ *to* n, $j = 1$ *to* m **do**
5 | Calculate the normalized decision matrix n_{ij} and r_{ij} by Eqs. (10)-(11)
6 **end**
7 **for** $i = 1$ *to* n, $j = 1$ *to* m **do**
8 | Calculate the negative-ideal solution ns_j by Eqs. (12)-(13)
9 **end**
10 **for** $j = 1$ *to* m, $k = 1$ *to* n **do**
11 | Calculate the relative assessment matrix h_{jk} according to the Euclidean and Taxicab distances of alternatives by Eqs. (14)-(18)
12 **end**
13 **for** $j = 1$ *to* m, $k = 1$ *to* n **do**
14 | Calculate the assessment score of each alternative H_i by Eq. (19)
15 **end**
16 **return** The sorting result of m objects

Algorithm 2: The algorithm of PF MRN via MG 3WD

Input: A PFIS
Output: The decision result of m objects
1 **for** $i = 1$ *to* f, $j = 1$ *to* m, *and* $k = 1$ *to* n **do**
2 | Obtain an MG PFIS
3 **end**
4 **for** $i = 1$ *to* f, $j = 1$ *to* m *and* $k = 1$ *to* n **do**
5 | Calculate each team's weight and confirm the parameter σ and η by Eqs.(3)-(4) and Eqs.(7)-(8)
6 **end**
7 **for** $i = 1$ *to* f, $j = 1$ *to* m **do**
8 | Calculate PF relative loss functions of each object by Table 1
9 **end**
10 **for** $i = 1$ *to* f, $j = 1$ *to* m **do**
11 | Calculate two thresholds in terms of each decision-maker by Eqs. (5)-(6)
12 **end**
13 **for** $i = 1$ *to* f *and* $j = 1$ *to* m **do**
14 | Calculate E_i, T_i and H_i by incorporating experimental data into the CODAS method
15 **end**
16 **for** $j = 1$ *to* m **do**
17 | Determine the positive, boundary and negative domains via adjustable MG PRSs and the optimal threshold
18 **end**
19 **return** Three decision domains and the decision result of m objects

4 Experimental Analysis

4.1 The Case Study Based on the Proposed PF MRN Approach

This section focuses on analyzing the accuracy of SCITOS G5 robot navigation mobile data using the Kaggle data set (source: www.kaggle.com/datasets/jimschacko/wall-following-robot-navigation-dataset). The following paragraphs provide a comprehensive overview of the data set.

The MRN data set includes 24 attributes and 4 actions. The 4 actions refer to the SCITOS G5 mobile robot moving clockwise for 4 rounds while following the room's wall, represented as $R = \{R_1, R_2, R_3, R_4\}$. Out of the 24 available attributes, we choose the 1st, 5th, 9th, 13th, 17th, and 21st attributes represented by $V = \{y_1, y_2, y_3, y_4, y_5, y_6\}$. We choose 12 options, which we consider as 12 robots represented by $U = \{x_1, x_2, \cdots, x_{12}\}$. This paper considers the average value as the optimal route data for MRN. If the MRN data set $X = \{x_2, x_4, x_6, x_8, x_{10}, x_{12}\}$ is used, it is likely to yield better results.

To meet the requirements of a PFIS, we follow several steps to process the rated data set as follows.

The first step is to transform the raw data into a fuzzy type, which is completed by using the following formula: $v_{jk} = 1 - \frac{u_{jk} - u_{average}}{\max|u_{jk} - u_{average}|}$. Next, the obtained fuzzy data are changed into the PF types, and the resulting normalized PF component is denoted by $p'_{jk} = \{v_{jk}, (1 - v_{jk}), \min(v_{jk}, (1 - v_{jk}))\}$.

The formula employed to obtain the standard rated set D is shown as the specific set $D = \{< y_1, (\frac{\sum_{j=1}^{r}(v_{jk})}{r}, \frac{\sum_{j=1}^{r}(1-v_{jk})}{r}, \frac{\sum_{j=1}^{r}\min\{v_{jk},1-v_{jk}\}}{r}) > \cdot < y_6, (\frac{\sum_{j=1}^{r}(v_{jk})}{r}, \frac{\sum_{j=1}^{r}(1-v_{jk})}{r}, \frac{\sum_{j=1}^{r}\min\{v_{jk},1-v_{jk}\}}{r}) >\}$.
Moreover, we set the coefficient σ as $\sigma = \{0.25, 0.25, 0.25\}^T$.

Determining the weight is crucial in attribute measurement, and in this paper, the deviation maximization method is utilized for this purpose. The objective weight in accordance to R_i are $u_1 = 0.287699$, $u_2 = 0.459551$, $u_3 = 0.397982$, $u_4 = 0.254766$.

According to Definition 2, it is calculated by the PF membership degree $\psi_D^R(x_j)$ of each patient. After that, the result of an adjustable PF membership degree $\psi_D^{R_{\tau(i)}}(x_j)$ can be obtained when $\eta = \frac{2}{3}$.

Subsequently, the PF relative loss function and two thresholds for three actions are derived. The CODAS method can determine the ultimate threshold. Figure 1 displays the final sorting outcome of the experiment. According to Figure 1, we can obtain the sorting result: $x_6 \prec x_1 \prec x_3 \prec x_{10} \prec x_2 \prec x_5 \prec x_{12} \prec x_{11} \prec x_9 \prec x_4 \prec x_7 \prec x_8$. Thus, it can be seen of these 12 robots, the 8-th robot is more likely to precise. Additionally, the sorting results of all objects by comparing the magnitude of $\psi_D^{R_{\tau(i)}}(x_j)$ and the thresholds are displayed in Table 2.

Table 2. The classification results of all objects

Domains	Objects	Numbers
$POS_{\xi,\varsigma}^{\psi}(P)$	$x_2, x_4, x_5, x_7, x_8, x_9, x_{11}, x_{12}$	8
$BND_{\xi,\varsigma}^{\psi}(P)$	x_1, x_3, x_6, x_{10}	4
$NEG_{\xi,\varsigma}^{\psi}(P)$	$\emptyset$	0

4.2 Comparative Analysis

The validity of our method is evaluated by analyzing its comparison with different types of PF 3WD. Furthermore, it is important to standardize the decision matrix to ensure comparability between different methods and eliminate the influence of variable dimensions on the final outcomes, as the dimensions of various decision variables may vary.

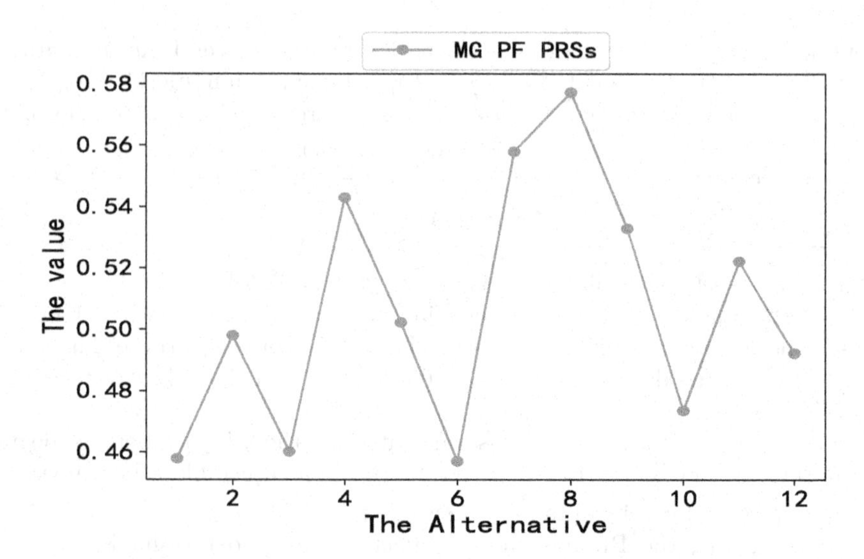

Fig. 1. Sorting results of all alternatives

1. The sorting outcome of the final approach is listed in Fig. 2a, in comparison to the PF VIKOR method [15]. Similarly, the ranking result of the final approach is demonstrated in Fig. 2b, in contrast to the PF TOPSIS method [15]. The sorting results exhibit slight variations because the proposed method combines the concepts of PFSs and 3WD. As a result, the decision loss is reduced, leading to more reasonable decision-making outcomes.
2. Our method is compared to the group decision-making techniques in light of PF weighted arithmetic (WA) operator and PF weighted geometric (WG) operator suggested in Literature [16]. The ranking outcome of the final approach is displayed in Fig. 3b. The comparison between our method and the aggregation operator method via 3WD [17] yields the calculation results as presented in Fig. 3a. The ranking results of the three decision-making methods exhibit some variations. This is due to the susceptibility of the mentioned PF weighted integration operator method to outliers. In contrast, the method proposed in this paper effectively mitigates the influence of outliers and provides a certain level of stability in decision-making outcomes.
3. The comparison between our method and the decision indicator set method discussed in Literature [18] yields the following results in Fig. 4.

In conclusion, the method developed in this paper exhibits distinct advantages in terms of information representation, fusion and analysis. A modifiable version of the PF information fusion and analysis method has been established, which can be considered as an efficient PF group decision-making approach.

4.3 Sensitivity Analysis

This section aims to confirm and examine the stability of our method while analyzing the influence of η and σ on decision outcomes. Figure 5a and Fig. 5b display the ranking outcomes for various η and σ values. Additionally, Table 3 and Table 4 present the categorization outcomes for different η and σ values, respectively.

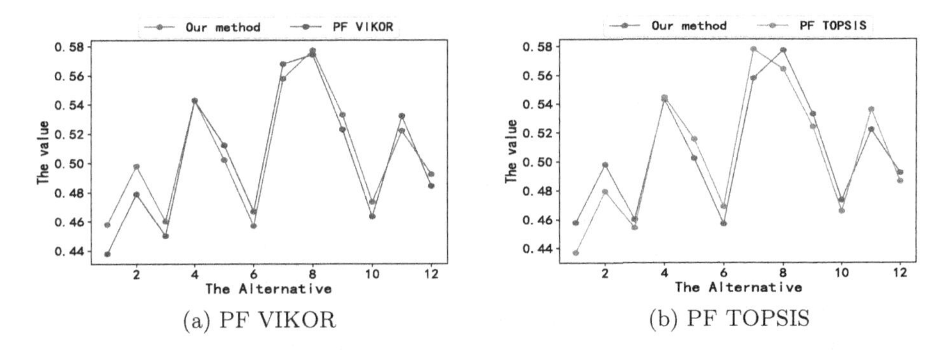

(a) PF VIKOR (b) PF TOPSIS

Fig. 2. Comparative analysis with PF VIKOR and PF TOPSIS

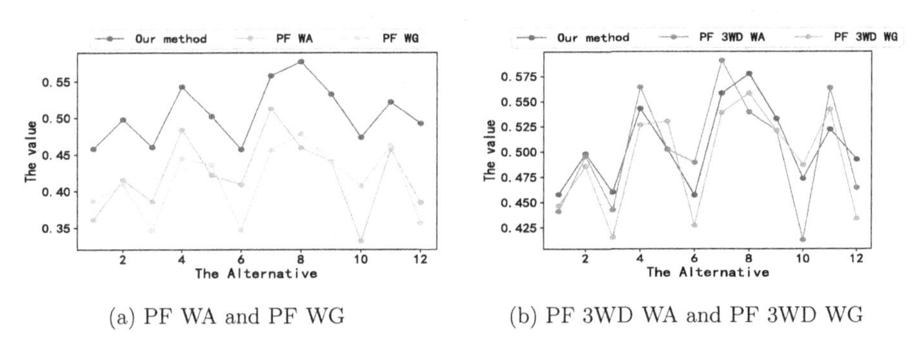

(a) PF WA and PF WG (b) PF 3WD WA and PF 3WD WG

Fig. 3. Comparative analysis with aggregation operators

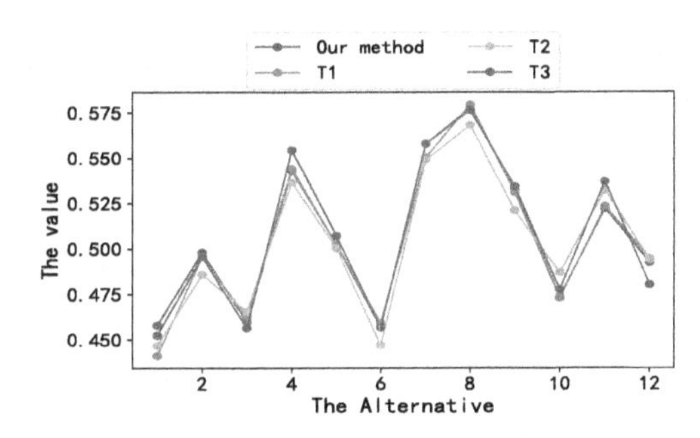

Fig. 4. Comparative analysis with the decision indicator set method

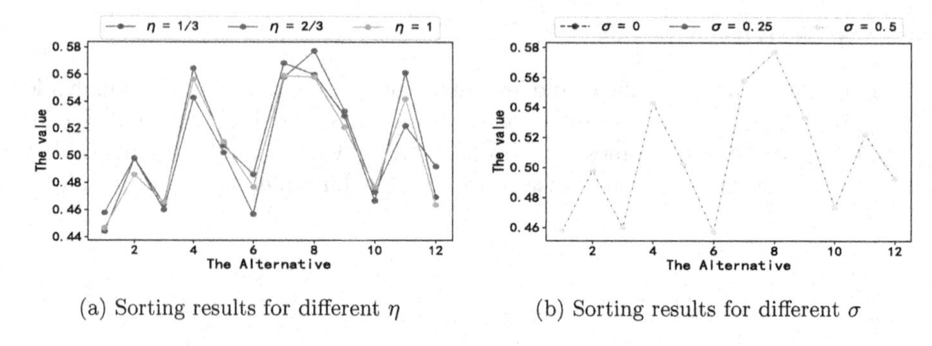

(a) Sorting results for different η (b) Sorting results for different σ

Fig. 5. Sorting results for different η and σ values

Table 3 and Fig. 5a indicate that as parameter η increases, the classification results vary slightly across different η. Specifically, As η grows from $\frac{1}{3}$ to 1, the best object changes, and the ranking results are slightly different. These observations suggest that differences in decision-makers' risk attitudes may impact certain decision outcomes. Conversely, Table 4 and Fig. 5b demonstrate that boosting σ from 0 to 0.5 does not affect the final ranking outcomes. Therefore, considering the explored sensitivity analysis, it can be inferred that the proposed method for group decision-making utilizing PFSs demonstrates the robust stability.

Table 3. Classification results for different η

η	$POS^{\psi}_{\xi,\varsigma}(D)$	$BND^{\psi}_{\xi,\varsigma}(D)$	$NEG^{\psi}_{\xi,\varsigma}(D)$
$\eta = \frac{1}{3}$	$x_2, x_4, x_5, x_6, x_7, x_8, x_9, x_{11}$	x_1, x_3, x_{10}, x_{12}	$\emptyset$
$\eta = \frac{2}{3}$	$x_2, x_4, x_5, x_7, x_8, x_9, x_{11}, x_{12}$	x_1, x_3, x_6, x_{10}	$\emptyset$
$\eta = 1$	$x_2, x_4, x_5, x_6, x_7, x_8, x_9, x_{11}$	x_1, x_3, x_{10}, x_{12}	$\emptyset$

Table 4. Classification results for different σ

σ	$POS^{\psi}_{\xi,\varsigma}(D)$	$BND^{\psi}_{\xi,\varsigma}(D)$	$NEG^{\psi}_{\xi,\varsigma}(D)$
$\sigma = 0$	$x_2, x_4, x_5, x_7, x_8, x_9, x_{11}, x_{12}$	x_1, x_3, x_6, x_{10}	$\emptyset$
$\sigma = 0.25$	$x_2, x_4, x_5, x_7, x_8, x_9, x_{11}, x_{12}$	x_1, x_3, x_6, x_{10}	$\emptyset$
$\sigma = 0.5$	$x_2, x_4, x_5, x_6, x_7, x_8, x_9, x_{11}$	x_1, x_3, x_6, x_{10}	$\emptyset$

5 Conclusions

As robots hold a significant position in the realm of artificial intelligence, this paper investigates an MRN model based on MG PF 3WD. In particular, the

PFIS concept is initially developed to capture data related to the robot's navigation process. Furthermore, a new PF MRN approach is formulated by utilizing the adjustable MG PF PRSs and CODAS techniques. Additionally, the FL structure is employed to integrate and execute these methods. Ultimately, a case study is conducted utilizing the Kaggle data set for MRN, and several experiments showcase the efficacy of the proposed approach.

Subsequent research can delve into various significant areas to enhance the model introduced in this paper. Below are some of the essential topics that can be considered:

1. Since the utilization rate of MRN is not high so far, in order to improve its utilization rate, this paper suggests exploring additional artificial intelligence applications aided by publicly available datasets.
2. The theoretical knowledge framework of MG 3WD has been established, however it still needs to be further improved, it is crucial to examine generalized MG tools and sequential 3WD.
3. The application scope of the method in this paper has certain limitations. In order to expand it, it may be worthwhile to explore additional aspects, such as clustering analysis, the evidence theory, and decision-making tools in large groups.

Acknowledgment. This work is supported by the National Natural Science Foundation of China (62272284; 62072291; 62072294; 61972238; 61703363), the Special Fund for Science and Technology Innovation Teams of Shanxi (202204051001015), the Cultivate Scientific Research Excellence Programs of Higher Education Institutions in Shanxi (CSREP) (2019SK036), the Graduate Education Innovation Programs of Shanxi University (SXU2022Y256), and the Training Program for Young Scientific Researchers of Higher Education Institutions in Shanxi.

References

1. Thakur, P., Sehgal, V.K.: Emerging architecture for heterogeneous smart cyber-physical systems for industry 5.0. Comput. Ind. Eng. **162**, 107750 (2021)
2. Sangaiah, A.K., Rostami, A.S., Hosseinabadi, A.A.R., Shareh, M.B., Javadpour, A., Bargh, S.H., Hassan, M.M.: Energy-aware geographic routing for real-time workforce monitoring in industrial informatics. IEEE Internet Things J. **8**(12), 9753–9762 (2021)
3. Li, Q.B., et al.: A survey on federated learning systems: vision, hype and reality for data privacy and protection. IEEE Trans. Knowl. Data Eng. **35**(4), 3347–3366 (2019)
4. Zhang, C., Xie, Y., Bai, H., Yu, B., Li, W.H., Gao, Y.: A survey on federated learning. Knowl.-Based Syst. **216**, 106775 (2021)
5. Lin, M.W., Huang, C., Chen, R.Q., Fujita, H., Wang, X.: Directional correlation coefficient measures for Pythagorean fuzzy sets: Their applications to medical diagnosis and cluster analysis. Complex Intell. Syst. **7**(2), 1025–1043 (2021)
6. Lin, M.W., Li, X.N., Chen, R.Q., Fujita, H., Lin, J.: Picture fuzzy interactional partitioned Heronian mean aggregation operators: an application to MADM process. Artif. Intell. Rev. **55**(2), 1171–1208 (2022)

7. Singh, A., Kumar, S.: Picture fuzzy set and quality function deployment approach based novel framework for multi-criteria group decision making method. Eng. Appl. Artif. Intell. **104**, 104–395 (2021)
8. Amanathulla, S., Bera, B., Pal, M.: Balanced picture fuzzy graph with application. Artif. Intell. Rev. **54**(7), 5255–5281 (2021)
9. Yao, J.T., Vasilakos, A.V., Pedrycz, W.: Granular computing: perspectives and challenges. IEEE Trans. Cyb. **43**(6), 1977–1989 (2013)
10. Yao, Y.Y.: Three-way decision and granular computing. Int. J. Approx. Reason. **103**, 107–123 (2018)
11. Yao, Y.Y.: Three-way decisions with probabilistic rough sets. Inform. Sci. **180**(3), 341–353 (2010)
12. Ghorabaee, M.K., Zavadskas, E.K., Turskis, Z., Antucheviciene, J.: A new combinative distance-based assessment (CODAS) method for multi-criteria decision-making. Econ. Comput. Econ. Cyb. **50**(3), 25–44 (2016)
13. Cuong, B.C., Kreinovich, V.: Picture fuzzy sets-A new concept for computational intelligence problems. In: Proceedings of the Third World Congress on Information and Communication Technologies (WICT 2013), Vietnam, Hanoi, pp. 1–6 (2013)
14. Jia, F., Liu, P.D.: A novel three-way decision model under multiple-criteria environment. Inform. Sci. **471**, 29–51 (2019)
15. Qiyas, M., Abdullah, S., Al-Otaibi, Y.D., Aslam, M.: Generalized interval-valued picture fuzzy linguistic induced hybrid operator and TOPSIS method for linguistic group decision-making. Soft. Comput. **25**(7), 5037–5054 (2021)
16. Opricovic, S., Tzeng, G.H.: Compromise solution by MCDM methods: a comparative analysis of VIKOR and TOPSIS. Eur. J. Oper. Res. **156**(2), 445–455 (2004)
17. Zhang, C., Ding, J.J., Zhan, J.M., Li, D.Y.: Incomplete three-way multi-attribute group decision making based on adjustable multigranulation Pythagorean fuzzy probabilistic rough sets. Int. J. Approx. Reason. **147**, 40–59 (2022)
18. Zhang, C., Ding, J.J., Li, D.Y., Zhan, J.M.: A novel multi-granularity three-way decision making approach in q-rung orthopair fuzzy information systems. Int. J. Approx. Reason. **138**, 161–187 (2021)

GCM-FL: A Novel Granular Computing Model in Federated Learning for Fault Diagnosis

Xueqing Fan[1], Chao Zhang[1](✉), Arun Kumar Sangaiah[2,3](✉), Yuting Cheng[1], Anna Wang[1], and Liyin Wang[1]

[1] School of Computer and Information Technology, Shanxi University, Taiyuan, China
czhang@sxu.edu.cn
[2] International Graduate Institute of Artificial Intelligence, National Yunlin University of Science and Technology, Douliu, Taiwan
aksangaiah@ieee.org
[3] Department of Electrical and Computer Engineering, Lebanese American University, Byblos, Lebanon

Abstract. In the realm of industrial production, maintaining continuous monitoring and implementing precise diagnostics of mine ventilators (MVs) holds a critical role in minimizing faults and accidents. Hence, it becomes imperative to devise an efficient and precise fault diagnosis (FD) technique for MVs. This paper endeavors to address the FD challenge of MVs by integrating federated learning (FL) with granular computing models. FL offers a partial solution to the issues of security and privacy associated with data sharing, which ensures data security while establishing an FD system for MVs. This system harnesses the adjustable multi-granularity (MG) triangular fuzzy (TF) probabilistic rough set (PRS) model to enhance the model's interpretability. In this study, the TF concept is introduced into the structure of three-way decisions (TWD) to tackle uncertainty and multi-performance attributes. We introduce the notion of an MG TF information system (IS) and propose an adjustable MG TF PRS model. The ELECTRE (Elimination Et Choice Translating Reality) method is employed to determine the optimal threshold. Furthermore, to validate the efficiency of the proposed model, we establish a TF multi-attribute group decision-making (MAGDM) approach using MV data within the MG and TWD frameworks. Finally, we verify the method's applicability through comparative analysis experiments. The experimental outcomes demonstrate the method's effectiveness and practicality in diagnosing faults for MVs.

Keywords: Federated Learning · Fault Diagnosis · Granular Computing · ELECTRE · Three-Way Decision

1 Introduction

FL is a newly emerging machine learning structure pioneered by Google in 2016 [3], which offers an innovative approach to user data sharing. In recent

B. Luo et al. (Eds.): ICONIP 2023, LNCS 14447, pp. 423–435, 2024.
https://doi.org/10.1007/978-981-99-8079-6_33

years, FL has found applications in data-sensitive domains such as intelligent manufacturing, healthcare, finance and education, yielding promising results [8]. Moreover, the onset of the fifth industrial revolution [4] has provided an opportunity for FL to address, to a certain extent, the security issues associated with data sharing processes.

Moreover, due to the inherent nature of humans to perceive and understand the world through granular computing, the principles associated with granular computing are extensively applied in practical settings. Since Zadeh [16] proposed the granular computing theory in 1997, the research on granular computing has been continuously developing. There are two major categories of granular computing models. One is based on information granulation, such as the fuzzy set [14], and the other is based on MG computing, such as the rough set [5] and TWD [9].

In order to address the limitations of traditional fuzzy sets, various generalized fuzzy sets have been proposed, such as triangular fuzzy sets (TFSs). TFSs are favored for their simplicity, ease of interpretation and versatility, making them well-suited for addressing a wide range of practical problems. Furthermore, MAGDM [17] often requires multiple decision-makers (DMs) to make choices based on different sets of attribute information. These individual decisions are then fused together to reach a final consensus decision. It is noteworthy that MAGDM and MG TWD share significant similarities in terms of data representation and the process of aggregating data from multiple DMs.

This paper presents a two-fold approach to address critical issues in data sharing and security. First, it incorporates FL methods to enhance data protection and security measures. Second, it employs granular computing techniques to tackle MAGDM, mitigating the problem of limited explanatory power stemming from localized model training. Consequently, this research integrates the field of granular computing to address real-world challenges in industrial production, with the ultimate goal of enhancing diagnostic accuracy and overall effectiveness across various outcomes.

1.1 Motivations

In the actual production process of industry, to handle the FD issue of MVs, many MAGDM models are used to model this issue, which will face the following challenges:

The first challenge is to propose a fuzzy IS to record the uncertainty and complexity of FD.

The second challenge is that the traditional MG rough set model tends to be either too optimistic or pessimistic in terms of information fusion. Therefore, it is necessary to use an improved MG rough set model to enhance the accuracy and rationality of decision-making problems.

The third challenge is to find an optimal granularity selection method that can assist in determining a more appropriate threshold when employing the MG TWD approach to solve actual MAGDM.

1.2 Contributions

In light of the above challenges, the current paper makes the following contributions:

Contribution 1: The introduction of triangular fuzzy numbers (TFNs) as a foundational element in constructing a fuzzy IS for the purpose of recording FD.

Contribution 2: The proposal of an adjustable MG TF PRS model to enhance the interpretability and rationality of the system.

Contribution 3: The development of an FD method based on MG TWD using real-world data from MVs as a specific example.

2 Related Works

FL, a distributed machine learning technique, has garnered significant attention and adoption across diverse domains. This popularity is primarily attributable to FL's capacity to effectively address concerns related to data security and privacy [20]. Granular computing represents a novel computing paradigm that centers around the concept of granularity. This paradigm aims to formulate, simplify, merge and divide complex real-world problems and data for more efficient processing and analysis [12]. The synergy of FL and granular computing presents an opportunity to construct models that are not only secure but also highly efficient and accurate. It is worth noting that within the framework of granular computing, three notable methods and algorithms have gained widespread recognition and acclaim for their applicability in tackling complex problems, i.e., fuzzy sets, rough sets and TWD.

2.1 TFSs

The notion of fuzzy sets, introduced by Zadeh [15], offers a means to represent the imprecise and uncertain nature of real-world problems. Unlike traditional sets in set theory, fuzzy sets exhibit a fuzzy characteristic whereby the membership of elements is not absolute or definite, but rather indeterminate or vague to some extent. Among the various forms of generalized fuzzy sets, the TFS stands out as notable, owing to its simplicity in calculation, ease of comprehension, and wide applicability. In this paper, we will review the definitions of TFSs, TFNs and triangular fuzzy relations (TFRs), and build our model in the context of TFSs.

2.2 PRSs and MG TWD

The PRS theory [6] inherits the fundamental characteristics of rough sets and offers a viable solution for dealing with problems characterized by incomplete and imprecise information. Moreover, probabilistic models enable the modeling and analysis of uncertainties, enhancing their applicability to practical problems. The MG TWD approach serves as a valuable method for addressing decision problems that entail multiple levels and perspectives. It facilitates a comprehensive evaluation and balance at each granularity, ultimately yielding a relatively optimal decision outcome [18]. Yao [11] has demonstrated that three decisions can effectively interpret the three domains of rough sets, which correspond to acceptance, non-commitment, and rejection, respectively.

In the analysis stage of MG TWD [10], the problem can be separated into sub-parts of different granularity, and the PRS is used to extract information and model uncertainty for each sub-problem, so as to provide support for the subsequent decision. In the decision stage of PRSs, the results of PRSs under different granularity can be comprehensively considered by evaluating and weighing MG, and the final decision result can be obtained.

PRSs and MG TWD-based decision-making [13] integrate concepts from the probability theory, the rough set theory and multi-attribute decision making. This approach excels at handling problems characterized by uncertainty and complexity, making it a powerful support tool for decision-making processes.

2.3 ELECTRE

To improve the effectiveness of decision-making, Roy [7] introduced the ELECTRE method, specifically designed to handle potential conflicts among decision criteria. The ELECTRE method evaluates the relative importance of alternative solutions by considering consistency and inconsistency indices. In this study, the ELECTRE method is employed to score and select the decision matrix generated by multiple DMs, ultimately leading to the identification of the best decision outcome.

3 The Proposed Granular Computing Model in FL for FD

Definition 1. [1,2] If three real numbers fulfill $x^L \leq x^M \leq x^U$, and x^L, x^M, x^U are all within the range of $[0,1]$, then $x = (x^L, x^M, x^U)$ is called a TFN with a membership degree of

$$f_x(a) = \begin{cases} 0, a < x^L \\ \frac{a-x^L}{x^M-x^L}, x^L \leq a \leq x^M \\ \frac{a-x^U}{x^M-x^U}, x^M \leq a \leq x^U \\ 0, x^U \leq a \end{cases}. \tag{1}$$

In practical problem-solving scenarios, it is often imperative to represent the set associated with a fuzzy concept in a more descriptive manner, beyond just a numerical value. Hence, to more comprehensively depict the set linked to fuzzy concepts, the notion of TFSs is presented below.

Definition 2. [2] Let U be a universe, a TFS A on U is expressed by $A = \{\langle x, h_A(x)\rangle \,|x \in U\}$.

To gain a deeper understanding of decision scenarios in complex cases, we introduce the concept of TFRs within the structure of two universes.

Definition 3. Let U and V be two universes, a TFR can be expressed by $R = \{\langle (x,y), h_R(x,y)\rangle \,|\, (x,y) \in U \times V\}$, where $h_R(x,y)$ denotes the membership degree of the relation between x and y.

Suppose there are m alternatives, respectively by $x_j (j = 1, 2, \cdots, m)$, denoted by U with n attributes, respectively by $y_k (k = 1, 2, \cdots, n)$, denoted by V, and the weight u corresponds to the attribute form. The decision matrix $D = (d_{jk})_{m\times n}$ can be constructed, and d_{jk} represents the evaluated value of x_j relative to y_k. There are f DMs, each DM provides a decision matrix, w is the DM weight set. In sum, $TFIS = (U, V, R_i, A)$ can be built, where U is the alternative set, V is the attribute set, R_i is the TFR on $U \times V$, where $R_i = \cup_{y_k \in V} R_{i_{y_k}}$, $R_{i_{y_k}}$ is the value range of y_k relative to the relationship R_i, and A is the standard evaluation set.

Definition 4. Given a TFIS (U, V, R_i, A), for any $x_j \in U$ and $y_k \in V$, the MG TF membership degree of x_j relative to the relation R_i is listed below:

$$\varphi_A^{R_i}(x_j) = \left(\frac{\sum\limits_{y_k\in V} u_k R_i(x_j, y_k)(A(y_k))^c}{\sum\limits_{y_k\in V} u_k R_i(x_j, y_k)}\right)^c. \tag{2}$$

Definition 5. Given the MG TFIS (U, V, R_i, A), $\varphi_A^{R_i}(x_j)$ is the MG TF membership. Let all the values of $\varphi_A^{R_i}(x_j)$ be arranged in a growing sequence, $\varphi_A^{R_{\sigma(i)}}(x_j)$ is the value ranked at the position i in $\varphi_A^{R_i}(x_j)$, and $\varphi_A^{R_{\sigma(i)}}(x_j)$ is called $\varphi_A^{R_{\sigma(i)}}(x_j)$ of adjustable MG TF membership degrees of (U, V, R_i, A).

Definition 6. Given a TFIS (U, V, R_i, A), for any $x_j \in U$, the upper and lower approximations of adjustable MG TF PRSs with respect to (U, V, R_i, A) with $\alpha, \beta \in [0, 1]$ with $\beta < \alpha$ can be given by

$$\underline{\sum_{i=1}^{f} R_i}^{\varphi,\alpha,\eta}(A) = \left\{\varphi_A^{R_{\sigma(i)}}(x_j) \geq \alpha | x_j \in U\right\}; \tag{3}$$

$$\overline{\sum_{i=1}^{f} R_i}^{\varphi,\beta,\eta}(A) = \left\{\varphi_A^{R_{\sigma(i)}}(x_j) > \beta | x_j \in U\right\}, \tag{4}$$

where $(\underline{\sum_{i=1}^{f} R_i}^{\varphi,\alpha,\eta}(A), \overline{\sum_{i=1}^{f} R_i}^{\varphi,\beta,\eta}(A))$ is titled as an adjustable MG TF PRS. The parameter $\eta = \frac{i}{f}$ represents the risk coefficient of the DM, and $\eta \in [\frac{1}{f}, 1]$ means that the DM goes from preferring to completely avoid risk to completely pursue risk. The following formulas can be obtained:

$$POS^{\varphi}_{\alpha,\beta}(A) = \underline{\sum_{i=1}^{f} R_i}^{\varphi,\alpha,\eta}(A); \tag{5}$$

$$NEG^{\varphi}_{\alpha,\eta}(A) = U - \overline{\sum_{i=1}^{f} R_i}^{\varphi,\beta,\eta}(A); \tag{6}$$

$$BND^{\varphi}_{\alpha,\beta}(A) = \overline{\sum_{i=1}^{f} R_i}^{\varphi,\beta,\eta}(A) - \underline{\sum_{i=1}^{f} R_i}^{\varphi,\alpha,\eta}(A). \tag{7}$$

The concrete steps of ELECTRE are presented below.

Step 1: Normalize the decision matrix to form a new matrix $P = (p_{jk})_{m \times n}$.

Minimization objective: $p_{jk} = \frac{\frac{1}{d_{jk}}}{\sqrt{\sum_{j=1}^{m} \frac{1}{d_{jk}^2}}}$;

Maximization objective: $p_{jk} = \frac{d_{jk}}{\sqrt{\sum_{j=1}^{m} d_{jk}^2}}$.

Step 2: The weighted decision matrix $Q = (q_{jk})_{m \times n}$ is obtained by weighting the decision matrix at the basis. Q is obtained through the formula $q_{jk} = p_{jk} \cdot u_k$.

Step 3: Obtain the concordance and discordance sets through comparison. For each pair of decision alternatives, x_r and x_s $(r, s = 1, 2, \cdots, m, r \neq s)$, the attribute set is divided into two subsets. If all the attributes of alternative x_r are superior to those of alternative x_s, a concordance set is formed, denoted as $C(r, s) = \{k|q_{rk} > q_{sk}\}$. The discordance set, denoted by $H(r, s) = \{k|q_{rk} < q_{sk}\}$, contains the attributes for which alternative x_r is inferior to alternative x_s.

Step 4: Compute the concordance and discordance indices. The concordance index for $C(r, s)$ is acquired using the following formula: $C_{rs} = \sum_{k*} u_{k*}$. Here, $k*$ is the set of attributes in the concordance set. The discordance index represents the degree of $x_r \to x_s$ inconsistency and is acquired using the following formula:

$$H_{rs} = \frac{\sum_{k+} |q_{rk+} - q_{sk+}|}{\sum_{k} |q_{rk} - q_{sk}|}. \tag{8}$$

Step 5: Determine the preference relationship between different decision alternatives. If $C_{rs} \geq \overline{C}$ and $H_{rs} \leq \overline{H}$, then decision alternative x_r is preferred to x_s. Here, $\overline{C}$ and $\overline{H}$ is the average of C_{rs} and H_{rs}. By referring to the above steps, we develop a TF ELECTRE method.

First, in light of Definition 4, we acquire the membership degree of each element x_j with respect to the relation R_i to obtain a new matrix D, where d_{jk} denotes the rated value of x_j relative to y_k. Next, in an effort to objectively calculate the relative loss function, the relative loss function of x_j in the classical fuzzy IS is further generalized to the context of the TF system. The TF relative loss function of x_j is listed in the following table. We use a parameter σ to represent the DM's risk appetite, which increases from 0 to 0.5, representing the expert's preference for risks from complete pursuit to complete avoidance. Consider a set of states $\Omega = \{V, \neg V\}$, which includes two complementary states representing membership or non-membership in a state. The action set is denoted by $O = \{o_p, o_b, o_n\}$. These actions indicate whether x_j belongs to the positive domain $POS(V)$, the boundary domain $BND(V)$, or the negative domain $NEG(V)$, respectively. We acquire the relative loss function of x_j below, as shown in Table 1.

Table 1. Relative loss functions

	V	$\neg V$
o_p	0	$\sum_{k=1}^{n} 1 - d_{jk}$
o_b	$\sigma \sum_{k=1}^{n} d_{jk}$	$\sigma(\sum_{k=1}^{n} 1 - d_{jk})$
o_n	$\sum_{k=1}^{n} d_{jk}$	0

By consulting the relative loss functions of x_j in the table above and the MG TWD structure, we can obtain two thresholds for x_j, denoted by α and β.

$$\alpha = \frac{(1-\sigma)(1-\sum_{k=1}^{n} d_{jk})}{(1-\sigma)(1-\sum_{k=1}^{n} d_{jk}) + \sigma \sum_{k=1}^{n} d_{jk}}; \tag{9}$$

$$\beta = \frac{\sigma(1-\sum_{k=1}^{n} d_{jk})}{\sigma(1-\sum_{k=1}^{n} d_{jk}) + (1-\sigma) \sum_{k=1}^{n} d_{jk}}. \tag{10}$$

Based on the formula provided above, we can compute the experts and their associated decision thresholds, creating the threshold matrix. Subsequently, we can consolidate these thresholds using the ELECTRE method based on the threshold matrix. Following this, the positive region, negative region, and boundary region are defined as per Definition 6. Afterward, the adjustable TF membership degree is employed to assign each decision scheme to its respective domain. Lastly, the score function is utilized to rank all alternatives.

Moreover, the whole decision steps can be summarized as Algorithm 1.

Algorithm 1: The algorithm of TF MAGDM via MG TWD and ELECTRE

Input : TFIS (U, V, R_i, A)

Output: Ranking results of FD

1. The attribute weights corresponding to experts are calculated according to the evaluation matrix.
2. The expert weights are calculated based on the attribute weights obtained above.
3. The MG TF membership of each type of fault is calculated.
4. The MG TF membership, which is adjustable for each type of fault, is calculated.
5. The membership results with moderate risk are used to calculate the relative loss function of the experts.
6. The threshold corresponding to each expert is calculated.
7. The above threshold is integrated based on ELECTRE.
8. The positive, negative and boundary regions are determined.
9. The final diagnosis ranking is determined according to the score function.

In order to better show the sequence of the above algorithm, we use a flow chart to show the logical dependencies between the steps, as shown in Fig. 1.

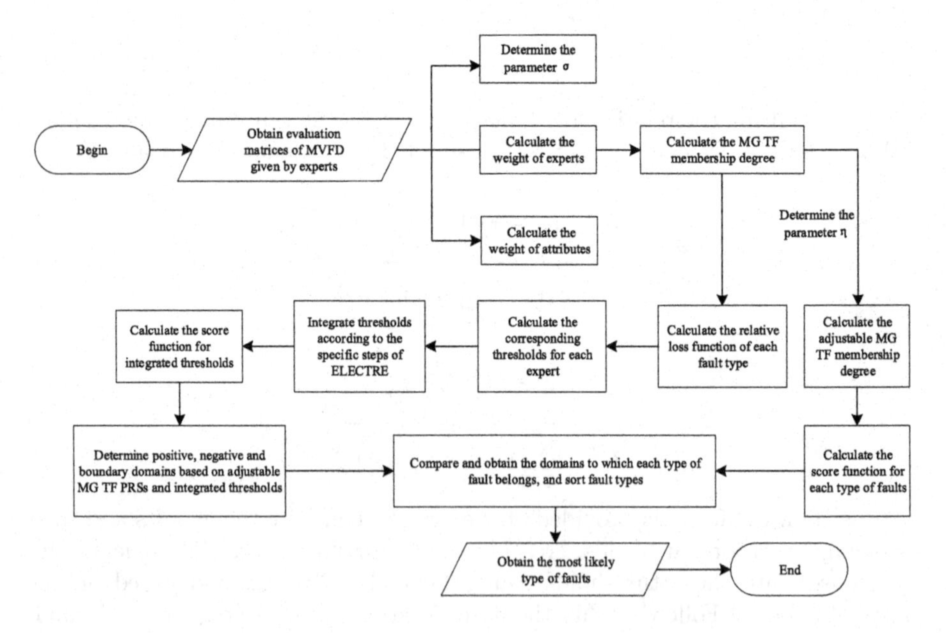

Fig. 1. The flow chart for Algorithm 1

4 Experimental Analysis

In this section, we will explore a case analysis focusing on TFSs in the context of MVFD to showcase the practicality of the previously proposed MG TF PRS model. To exemplify the feasibility of the previously introduced MG TF PRS model, a realistic case analysis involving TFSs in the domain of MVFD is thoroughly examined.

Let U be the set comprising eight faults: x_1 represents the unbalance fault, x_2 represents the misalignment fault, x_3 represents the mechanical looseness fault, x_4 represents the blade fault, x_5 represents the outer ring of bearing fault, x_6 represents the inner ring of bearing fault, x_7 represents the rolling element fault, and x_8 represents the holder fault. Furthermore, let y_1 denote the vibration frequency caused by the unbalance fault, y_2 denote the vibration frequency caused by the unbalance fault, y_3 denote the vibration frequency caused by the blade fault, y_4 denote the vibration frequency multiplied by the vibration frequency caused by the unbalance fault, y_5 denote the characteristic frequency of outer rings, y_6 denote the characteristic frequency of inner rings, y_7 denote the characteristic frequency of rolling elements, and y_8 denote the characteristic frequency of holders. The set V consists of eight vibration frequencies. R_i represents the relationship between fault types and characteristics of MV, as provided by the $i-th$ expert. We have a total of three experts.

$$A = \{\langle y_1, 0.5\rangle, \langle y_2, 0.6\rangle, \langle y_3, 0.3\rangle, \langle y_4, 0.7\rangle, \langle y_5, 0.1\rangle, \langle y_6, 0.2\rangle, \langle y_7, 0.2\rangle, \langle y_8, 0.1\rangle\}.$$

Since there is only one membership degree available in the data set [19] cited in this article, we need to construct the other two required membership degrees. One membership degree is acquired by subtracting the non-membership degree from 1, and the other membership degree is obtained by taking the average of the two already computed membership degrees.

Next, we employ the proposed MVFD algorithm to address the practical problem described above.

We first calculate the weights of the experts and the attributes. The results of the attribute weights of Expert 1 are

$$u = \{0.2002, 0.1278, 0.0653, 0.2308, 0.1159, 0.1013, 0.0866, 0.0720\},$$

and the results of the expert weights are $w = \{0.3373, 0.3496, 0.3131\}$.

We select $\sigma = 0.25$ to calculate the MG TF membership degree, and select $\eta = \frac{2}{3}$ to calculate the adjustable MG TF membership. The results are as follows:

$$\varphi_A^{R_{\sigma(2)}}(x_1) = \{0.5000, 0.4909\ 0.4931\}.$$

The calculated relative loss function corresponding to Expert 1 turns out to be shown in Table 2.

The calculated threshold result for Expert 1 is shown in Table 3.

In what follows, the result of integrated thresholds by ELECTRE is listed in Table 4.

Table 2. The calculated relative loss function

	V	$\neg V$
o_p	[0.0000, 0.0000, 0.0000]	[0.5000, 0.5091, 0.5069]
o_b	[0.1250, 0.1227, 0.1233]	[0.1267, 0.1273, 0.1250]
o_n	[0.5000, 0.4909, 0.4931]	[0.0000, 0.0000, 0.0000]

Table 3. The calculated threshold result

	$\alpha_j^{R_i}$	$\beta_j^{R_i}$
x_1	[0.7500, 0.7568, 0.7551]	[0.2500, 0.2569, 0.2552]
x_2	[0.7500, 0.8251, 0.8256]	[0.2500, 0.3439, 0.3447]
x_3	[0.7500, 0.8351, 0.8360]	[0.2500, 0.3602, 0.3617]
x_4	[0.7500, 0.7999, 0.8001]	[0.2500, 0.3076, 0.3078]
x_5	[0.7500, 0.5828, 0.5710]	[0.2500, 0.1343, 0.1288]
x_6	[0.7500, 0.6446, 0.6363]	[0.2500, 0.1677, 0.1627]
x_7	[0.7500, 0.6562, 0.6480]	[0.2500, 0.1749, 0.1699]
x_8	[0.7500, 0.6343, 0.6238]	[0.2500, 0.1616, 0.1556]

The threshold values are compared with $\varphi_A^{R_{\sigma(2)}}(x)$. First, the threshold value and the score function of element x in $\varphi_A^{R_{\sigma(2)}}(x)$ are calculated, respectively. Then according to Definition 6, because $s(\beta_j(x_1)) < s(\varphi_A^{R_{\sigma(2)}}(x_1)) < s(\alpha_j(x_1))$, $x_1 \in BND_{\alpha,\beta}^{\varphi}(A)$. By the same steps, we acquire $POS_{\alpha,\beta}^{\varphi}(A) = \{\emptyset\}$, $BND_{\alpha,\beta}^{\varphi}(A) = \{x_1, x_2, x_3, x_4, x_5, x_6, x_7, x_8\}$, $NEG_{\alpha,\beta}^{\varphi}(A) = \{\emptyset\}$. We acquire the final sorting result is $x_4 \succ x_2 \succ x_1 \succ x_3 \succ x_8 \succ x_6 \succ x_5 \succ x_7$. Thus, the blade fault is the most likely type of failures.

Next, we compare our method with several methods, including TOPSIS, TODIM, TWD+MABAC, TWD+TOPSIS, TWD+VIKOR and TWD+ TODIM, and the conclusions of the experiment are presented in Fig. 2 and Fig. 3.

According to Fig. 2, we can see that our method and TODIM method have the same optimal ranking, indicating that our method is effective.

According to Fig. 3, we can see our method and the TWD + MABAC, TWD + TOPSIS, TWD + VIKOR, TWD + TODIM method of the experimental results are $x_4 \succ x_2 \succ x_1 \succ x_3 \succ x_8 \succ x_6 \succ x_5 \succ x_7$. Therefore, the effectiveness of our experiment can be verified.

Table 4. The result of integrated thresholds by ELECTRE

	α_j	β_j
x_1	$[0.7500, 0.7568, 0.7551]$	$[0.2500, 0.2693, 0.2653]$
x_2	$[0.7500, 0.8251, 0.8256]$	$[0.2500, 0.3407, 0.3414]$
x_3	$[0.7500, 0.8351, 0.8360]$	$[0.2500, 0.3499, 0.3526]$
x_4	$[0.7500, 0.7999, 0.8001]$	$[0.2500, 0.3083, 0.3078]$
x_5	$[0.7500, 0.5828, 0.5710]$	$[0.2500, 0.1598, 0.1519]$
x_6	$[0.7500, 0.6446, 0.6363]$	$[0.2500, 0.1913, 0.1840]$
x_7	$[0.7500, 0.6562, 0.6480]$	$[0.2500, 0.1991, 0.1917]$
x_8	$[0.7500, 0.6343, 0.6238]$	$[0.2500, 0.1886, 0.1804]$

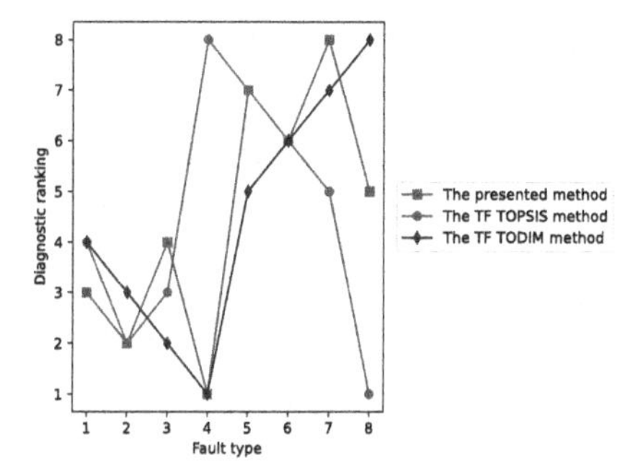

Fig. 2. The comparison of presented method with the classic MAGDM methods

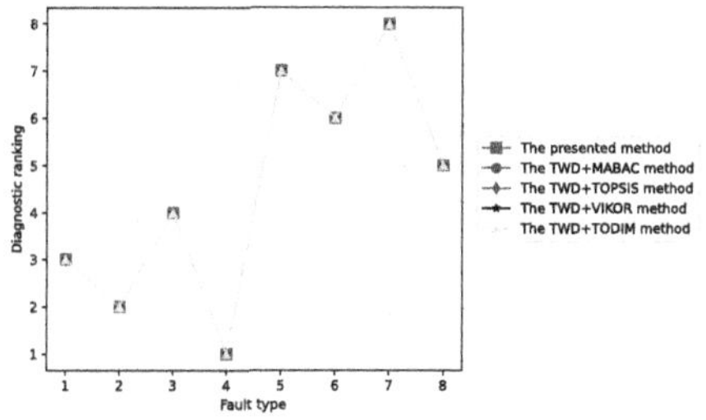

Fig. 3. The comparison of presented method with the classic MAGDM methods under TWD

In conclusion, the strength of our method is evident in its capability to overcome subjective thresholds and limitations associated with extreme information fusion. The method establishes an information fusion and analysis approach via adjustable MG TF PRSs, which can be considered as a viable TF group decision-making scheme.

5 Conclusions

This paper primarily focuses on the FD problem of MVs utilizing a MAGDM method. To develop an intelligent FD system for MVs, we integrate the concepts of FL, particle size calculation, and Industry 5.0. FL ensures data safety, real-time sensor data collection, and equipment status monitoring. MG TWD is applied to predict potential fault types, and an adjustable MG TF PRS model is established. This approach enables the development of an efficient, accurate, and interpretable FD method for MVs. Additionally, the FD process of MVs directly impacts the safety of mine workers, aligning with the people-oriented philosophy of Industry 5.0 and satisfying its requirements. Future research can further enhance the efficiency and accuracy of FD by integrating the FD of MVs with emerging technologies and methods. This will contribute to ensuring human safety and fostering industry development and innovation.

Acknowledgment. This work is supported by the National Natural Science Foundation of China (62272284; 62072294; 61972238), the National Undergraduate Innovation and Entrepreneurship Training Program (20230014), the Special Fund for Science and Technology Innovation Teams of Shanxi (202204051001015), the 21st Undergraduate Innovation and Entrepreneurship Training Program of Shanxi University (202210108012), the Cultivate Scientific Research Excellence Programs of Higher Education Institutions in Shanxi (CSREP) (2019SK036), and the Training Program for Young Scientific Researchers of Higher Education Institutions in Shanxi.

References

1. Gani, A., Assarudeen, S.: A new operation on triangular fuzzy number for solving fuzzy linear programming problem. Appl. Math. Sci. **6**(11), 525–532 (2012)
2. van Laarhoven, P.J.M., Pedrycz, W.: A fuzzy extension of Saaty's priority theory. Fuzzy Sets Syst. **11**(1–3), 229–241 (1983)
3. McMahan, H., Moore, E., Ramage, D., Hampson, S., Arcas, B.: Communication-efficient learning of deep networks from decentralized data. ArXiv (2016). https://doi.org/10.48550/arXiv.1602.05629
4. Nahavandi, S.: Industry 5.0-a human-centric solution. Sustain. **11**(16), 4371 (2019)
5. Pawlak, Z.: Rough sets. Int. J. Comput. Inform. Sci. **11**(5), 341–356 (1982)
6. Qian, Y., Zhang, H., Sang, Y., Liang, J.: Multigranulation decision-theoretic rough sets. Int. J. Approx. Reason. **55**(1), 225–237 (2014)
7. Roy, B.: The outranking approach and the foundations of electre methods. Theor. Decis. **31**(1), 49–73 (1991). https://doi.org/10.1007/BF00134132
8. Yang, Q., Liu, Y., Chen, T., Tong, Y.: Federated machine learning: concept and applications. ACM Trans. Intell. Syst. Technol. **10**(2), 1–19 (2019)
9. Yao, Y.: Three-way decisions with probabilistic rough sets. Inform. Sci. **180**(3), 341–353 (2010)
10. Yao, Y.: Three-way decisions and cognitive computing. Cogn. Comput. **8**(4), 543–554 (2016). https://doi.org/10.1007/s12559-016-9397-5
11. Yao, Y.: Three-way decision and granular computing. Int. J. Approx. Reason. **103**, 107–123 (2018)
12. Yao, Y.: Three-way granular computing, rough sets, and formal concept analysis. Int. J. Approx. Reason. **116**, 106–125 (2020)
13. Yao, Y.: The geometry of three-way decision. Appl. Intel. **51**(9), 6298–6325 (2021). https://doi.org/10.1007/s10489-020-02142-z
14. Zadeh, L.: Fuzzy sets. Inform Control **8**(3), 338–353 (1965)
15. Zadeh, L.: The concept of a linguistic variable and its application to approximate reasoning-ii. Inform. Sci. **8**(4), 301–357 (1975)
16. Zadeh, L.: Toward a theory of fuzzy information granulation and its centrality in human reasoning and fuzzy logic. Fuzzy Sets Syst. **90**(2), 111–127 (1997)
17. Zhan, J., Ye, J., Ding, W., Liu, P.: A novel three-way decision model based on utility theory in incomplete fuzzy decision systems. IEEE Trans. Fuzzy Syst. **30**(7), 2210–2226 (2022)

18. Zhang, C., Ding, J., Zhan, J., Sangaiah, A., Li, D.: Fuzzy intelligence learning based on bounded rationality in IoMT systems: a case study in Parkinson's disease. IEEE Trans. Comput. Soc. Syst. **10**(4), 1607–1621 (2023)
19. Zhang, C., Li, D., Mu, Y., Song, D.: A pythagorean fuzzy multigranulation probabilistic model for mine ventilator fault diagnosis. CompLex **2018**, 7125931 (2018)
20. Zhang, C., Xie, Y., Bai, H., Yu, B., Li, W., Gao, Y.: A survey on federated learning. Knowl.-Based Syst. **216**, 106775 (2021)

Adaptive Load Frequency Control and Optimization Based on TD3 Algorithm and Linear Active Disturbance Rejection Control

Yuemin Zheng[1], Jin Tao[2(✉)], Qinglin Sun[1(✉)], Hao Sun[1], Mingwei Sun[1], and Zengqiang Chen[1]

[1] Nankai University, Tianjin 300350, China
{sunql,sunh,chenzq}@nankai.edu.cn
[2] Silo AI, Helsinki 00100, Finland
taoj@nankai.edu.cn

Abstract. This paper presents the Twin Delayed Deep Deterministic Policy Gradient (TD3) algorithm optimized Linear Active Disturbance Rejection Control (LADRC) approach to tackle the problem of frequency deviation resulting from load disturbance and Renewable Energy Sources (RESs) in interconnected power systems. The LADRC approach employs a Linear Extended State Observer (LESO) to estimate the disturbance information in each area and utilizes a Proportional-Derivative (PD) controller to eliminate the disturbance. Simultaneously, the TD3 algorithm is trained in to acquire the adaptive controller parameters. In order to improve the convergence of the TD3 algorithm, a Lyapunov-reward shaping function is adopted. Finally, the proposed method is applied to two-area interconnected power system, comprising thermal, hydro, and gas power plants in each area, as well as RESs such as a noise-based wind turbine and photovoltaic (PV) system. The simulation results indicate that the proposed method is a highly effective approach for load frequency control.

Keywords: Load frequency control · TD3 · LADRC

1 Introduction

The problem of load frequency control (LFC) in power systems primarily ensures the normal operation of the system by maintaining frequency stability. The conventional sources of power generation mainly comprise thermal power and hydropower [1,2]. In recent years, with the growing awareness of environmental protection, renewable energy sources (RESs) such as wind power [3] and solar power [4] have gained traction and are gradually being developed. However, renewable energy systems can also induce system frequency oscillations, which can result in severe equipment damage in some cases [5]. Thus, designing a control method to regulate the generator's output power and maintain the system frequency within an acceptable range is of significant importance.

Supported by College of Artificial Intelligence, Nankai University.

B. Luo et al. (Eds.): ICONIP 2023, LNCS 14447, pp. 436–447, 2024.
https://doi.org/10.1007/978-981-99-8079-6_34

Currently, load frequency control methods primarily rely on PID controllers. For example, Shabani et al. [6] applied the PID controller with low pass filter to a traditional three-area interconnected power system, and the controller parameters were optimized by imperialist competitive algorithm (ICA). Can et al. [7] proposed a PI-(1+DD) LFC controller for a two-area interconnected thermal power system integrated with RESs such as photovoltaic (PV) system and wind energy, and the optimal controller parameters were determined by the Grey Wolf Optimization (GWO) algorithm. Moreover, some published works are presented in Table 1. And it can be observed that the majority of current load frequency control approaches are implemented by incorporating intelligent optimization algorithms on the basis of conventional controllers.

Table 1. Some published works about LFC with RESs.

Ref.	Power sources	Controller	Optimization algorithm	Year
[8]	Thermal, hydro and gas power plant	FOPID	PSO	2018
[9]	Thermal, wind, solar and fuel cells	IPD	GA	2019
[10]	Thermal	SMC	BA	2021
[11]	Thermal	Fuzzy control	LMA	2022
[12]	Thermal, hydro, gas, wind and solar power	ID-T	AOA	2022
[13]	Thermal, hydro, wind and solar power	LADRC	SAC	2022
[14]	Thermal, hydro, diesel, gas, wind farms	PID	GOA	2023
[15]	Thermal, hydro, gas and wind	3DOF-PID	GOA	2023

Linear active disturbance rejection control (LADRC) is a control strategy that builds upon PID methodology, which is mainly composed of a linear extended state observer (LESO) and a PD controller [16,17]. LADRC can efficiently estimate and eliminate unknown disturbance signals while disregarding the model's intricacies, making it a successful application across various fields, such as servo system [18], micro-hydro plant [19], and motion control of complex dynamic systems [20]. Given that power systems may encounter unknown load disturbances, this paper employs LADRC for control purposes.

Nevertheless, there is no standardized optimization approach for tuning controller parameters. Currently, optimization methods can be classified into two types: offline parameter tuning and adaptive parameter tuning. The first one uses heuristic algorithms such as population algorithms [21,22] and evolutionary algorithms. And the adaptive optimization methods are usually implemented using fuzzy control [23] and neural networks [24]. Compared with offline parameters, adaptive parameters are obviously more adaptable to changes in the control system. In recent years, deep reinforcement learning (DRL) algorithms [25–27] have garnered significant interest from researchers due to their potent computing and decision-making abilities. Consequently, employing DRL to optimize controller parameters has emerged as a promising research avenue [28,29]. The Twin Delayed Deep Deterministic Policy Gradient (TD3) algorithm [30] is an enhanced version of Deep Deterministic Policy Gradient (DDPG) algorithm [31], which can deal with the problem of continuous action space. However, the design of reward functions in DRL algorithms needs to be set for specific environments. Inspired by Lyapunov reward function [32], this paper designs a Lyapunov reward shaping based TD3 algorithm for LFC systems with renewable energy.

Overall, the main contributions of this paper can be summarized as follows:

- An Lyapunov reward shaping based TD3 algorithm is designed for LFC system with renewable energies.
- The LADRC controller is designed for a two-area interconnected power system integrated with noise-based wind turbine and PV system.
- The improved TD3 algorithm is utilized to optimize controller parameters, and the simulation results are compared with published works [12,33].

2 A Two-Area Interconnected Power System with RESs

The model investigated in this paper is a two-area interconnected power system, with each area contains thermal power plant, hydro power plant, and gas turbine plant, as shown in Fig. 1. For a power system under a deregulated environment, the objective of LFC is to maintain the frequency deviation Δf_1, Δf_2 and tie-line exchanged power ΔP_{tie} at 0. Generally, the tie-line bias control (TBC) mode is used to realize load frequency control, that is, area control area (ACE) is used as controller input, and the expression of ACE is presented in Eq. 1, where i stands for area number. In addition, for the explanation of the symbols illustrated in Fig. 1, please refer to Table 2.

$$ACE_i = B_i \Delta f_i + \Delta P_{tie} \tag{1}$$

Table 2. Expression of symbols of Fig. 1.

Terminology	Symbols and Values
Frequency bias constant	$B_1 = B_2 = 0.4312$ MW/Hz
Speed regulation constant	$R_1 = R_2 = 2.4$ Hz/MW
Time constant of Thermal plant	$T_{sg} = 0.06$s, T_t=0.3 s, T_r=10.2 s
Gain of Thermal plant	$K_r = 0.3$, KT=0.5747
Time constant of Hydro plant	T_{gh}=0.2 s, $T_{rs} = 4.9$s, $T_{rh} = 28.749$s, T_w=1.1 s
Gain of Hydro plant	$KH = 0.2873$
Time constant of Gas plant	B_{gs}=0.049 s, $X_g = 0.6$s, $Y_g = 1.1$s, $T_{cr} = 0.01$s, $T_f = 0.239$s, $T_{cd} = 0.2$s
Gain of Gas plant	$C_g = 1$, $KG = 0.138$
Time constant and gain of generator	$T_p = 11.49$ s, $K_p = 68.9655$

In Fig. 1, ΔP_{d1} and ΔP_{d2} represent the load disturbances added to the two areas, respectively; ΔP_{res1} and ΔP_{res2} are power output deviations of renewable energies; and the power output deviations of thermal power plant, hydro power plant and gas turbine plant are indicated by the symbols of ΔP_{tpi}, ΔP_{hpi} and

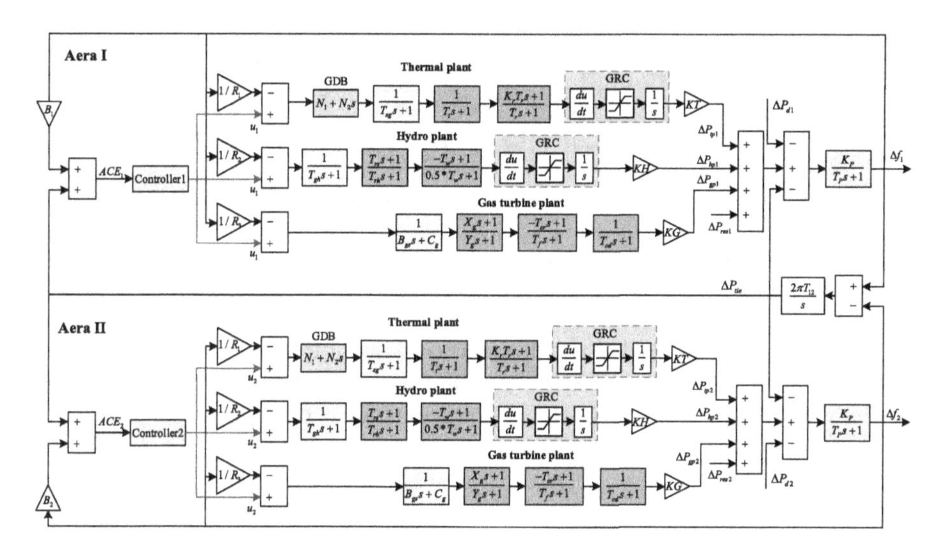

Fig. 1. Structure diagram of the studied power system [12].

ΔP_{gpi}, respectively. Besides, non-linear links such as governor deadband (GDB) and generation rate constraint (GRC) are also considered in this paper, and the expression of GDB is

$$GDB = 0.8 - \frac{0.2}{\pi}s \tag{2}$$

According to Fig. 1, the following expression can be obtained

$$\begin{cases} \Delta f_i = (\Delta P_{tpi} + \Delta P_{hpi} + \Delta P_{gpi} + \Delta P_{rei} - \Delta P_{di} - \Delta P_{tie})\, G_{pi} \\ \Delta \dot{P}_{tie} = 2\pi T_{12}(\Delta f_1 - \Delta f_2) \end{cases} \tag{3}$$

with

$$\begin{cases} \Delta P_{tpi} = \dfrac{KT\,(0.2 - 0.8/\pi)\,(K_r T_r s + 1)}{(T_{sg}s + 1)\,(T_t s + 1)\,(T_r s + 1)} \left(u_i - \dfrac{1}{R_1}\Delta f_i\right) \\ \Delta P_{hpi} = \dfrac{KH\,(T_{rs}s + 1)\,(-0.5T_w s + 1)}{(T_{gh}s + 1)\,(T_{rh}s + 1)\,(0.5T_w s + 1)} \left(u_i - \dfrac{1}{R_2}\Delta f_i\right) \\ \Delta P_{gpi} = \dfrac{KG\,(X_g s + 1)\,(-T_{cr}s + 1)}{(B_{gs}s + C_g)\,(Y_g s + 1)\,(T_f s + 1)\,(T_{cd}s + 1)} \left(u_i - \dfrac{1}{R_3}\Delta f_i\right) \end{cases} \tag{4}$$

and $G_{pi} = \frac{K_p}{T_p s+1}$.

2.1 Noise-Based PV System and Wind Turbine

This paper considers two common renewable energy sources: solar energy and wind energy. And the schematic diagram of the studied noise-based PV system and wind turbine is presented in Fig. 2.

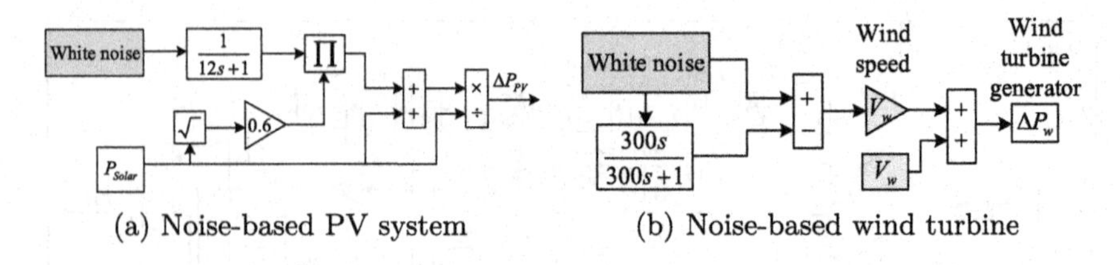

Fig. 2. Schematic diagram of noise-based PV system and wind turbine

In this PV system model, the power variations of PV system are determined by considering the variance from the uniform and nonuniform insolation. And for the wind turbine, the mechanical power output is determined by Eq. 5,

$$P_w = \frac{1}{2}\rho A_r C_p V_w^3 \tag{5}$$

where $\rho(\mathrm{kg/m^3})$ is air density; $A_r(\mathrm{m^2})$ is blade swept area; V_w is the wind speed. And C_p is the rotor power coefficient, presented in Eq. 6. Moreover, for the parameters of wind turbine, please refer to Ref. [12].

$$\begin{cases} C_p = c_1\left(\dfrac{c_2}{\lambda_I} - c_3\beta - c_4\right)e^{-\frac{c_5}{\lambda_I}} + c_6\lambda \\ \dfrac{1}{\lambda_I} = \dfrac{1}{\lambda + 0.08\beta} - \dfrac{0.035}{\beta^3 + 1} \end{cases} \tag{6}$$

2.2 Transfer Function Processing

Based on the above introduction, it is possible to derive the transfer function of each area for the power system. For convenience, define $G_{tp} = \frac{KT(0.2-0.8/\pi)(K_r T_r s+1)}{(T_{sg}s+1)(T_t s+1)(T_r s+1)}$, $G_{hp} = \frac{KH(T_{rs}s+1)(-0.5T_w s+1)}{(T_{gh}s+1)(T_{rh}s+1)(0.5T_w s+1)}$ and $G_{gp} = \frac{KG(X_g s+1)(-T_{cr}s+1)}{(B_{gs}s+C_g)(Y_g s+1)(T_f s+1)(T_{cd}s+1)}$ in Eq. 4. Firstly, by combining Eqs. 3 and 4, we can obtain the expressions for Δf_i and the controller output u_i as follows:

$$\Delta f_i = \frac{G_{tp} + G_{hp} + G_{gp}}{\frac{1}{G_{pi}} + \frac{G_{tp}}{R_1} + \frac{G_{hp}}{R_2} + \frac{G_{gp}}{R_3}}u_i + \frac{\Delta P_{rei} - \Delta P_{di} + \Delta P_{tie}}{\frac{1}{G_{pi}} + \frac{G_{tp}}{R_1} + \frac{G_{hp}}{R_2} + \frac{G_{gp}}{R_3}} \tag{7}$$

Then according to Eqs. 1 and 7, ACE_i yields to

$$ACE_i = \frac{B_i\left(G_{tp} + G_{hp} + G_{gp}\right)}{\frac{1}{G_{pi}} + \frac{G_{tp}}{R_1} + \frac{G_{hp}}{R_2} + \frac{G_{gp}}{R_3}}u_i + \frac{B_i\left(\Delta P_{rei} - \Delta P_{di} + \Delta P_{tie}\right)}{\frac{1}{G_{pi}} + \frac{G_{tp}}{R_1} + \frac{G_{hp}}{R_2} + \frac{G_{gp}}{R_3}} + \Delta P_{tie} \tag{8}$$

3 Controller Design Process for LFC

In this paper, Δf, ΔP_{tie} and ACE are all required to be stabilized to 0. Based on the concept of decentralized control [34], it is possible to design a controller for each area under the premise of ignoring the tie-line exchanged power. Besides, we consider the power output deviation of renewable energy as disturbances since its integration into the system can affect the stability of the frequency deviation. Then, taking Area I as an example, the following describes the design process of LADRC load frequency controller.

LADRC is mainly comprised of a LESO and a PD controller, where LESO is employed to estimate the unknown disturbance. It is worth noting that LADRC does not depend on the specific information of the model but only requires knowledge of the system order. According to Eq. 8, since $\frac{ACE_1}{u_i} = B_i \frac{G_{tp}+G_{hp}+G_{gp}}{\frac{1}{G_{pi}}+\frac{G_{tp}}{R_1}+\frac{G_{hp}}{R_2}+\frac{G_{gp}}{R_3}}$, the system order can be derived as 2. Then a second-order system originated from Eq. 8 can be obtained as

$$A\ddot{C}E_1 = f_a\left(ACE_1, \varDelta P_{d1}, \varDelta P_{re1}\right) + bu_1 \tag{9}$$

where f_a denotes the total disturbance function containing internal modeling information as well as load disturbance and renewable energy output power deviation disturbance; b is the constant coefficient. Usually, we will replace b with an adjustable parameter b_0,

$$\begin{aligned} A\ddot{C}E_1 &= f_a + (b - b_0)u_1 + b_0 u_1 \\ &= f + b_0 u_1 \end{aligned} \tag{10}$$

with f being the final total disturbance.

Define the state as $x_1 = A\dot{C}E_1$, $x_2 = ACE_1$ and $x_3 = f$. Then, a third-order LESO is organized

$$\begin{cases} \dot{z}_1 = z_2 - l_1\left(z_1 - x_1\right) \\ \dot{z}_2 = z_3 + b_0 u_i - l_2\left(z_1 - x_1\right) \\ \dot{z}_3 = -l_3\left(z_1 - x_1\right) \end{cases} \tag{11}$$

where z_1, z_2 and z_3 are observed states of x_1, x_2 and x_3, respectively; l_1, l_2 and l_3 represent the observer gains. When the observer gains take appropriate values, the observed values are nearly equal to the actual state values, that is, $z_3 \approx f$. Next, a PD controller is used to eliminate the disturbance, as shown in Eq. 12, where y_r is the target value of ACE_1, i.e. $y_r = 0$. And k_p and k_d are controller parameters.

$$u_1 = \frac{k_p\left(y_r - z_1\right) - k_d z_2 - z_3}{b_0} \tag{12}$$

By substituting Eq. 12 into Eq. 11, it can be observed that the disturbance in the system has been effectively canceled out. The above is the design process of LADRC. And it can be found that the parameters that need to be adjusted in this controller are b_0, l_1, l_2, l_3, k_p and k_d. Using the pole configuration method, the observer gains and PD controller parameters can be configured to the poles $-\omega_o(\omega_o > 0)$ and $-\omega_c(\omega_c > 0)$ [35], respectively. Then we have $l_1 = 3\omega_o$, $l_2 = 3{\omega_o}^2$, $l_3 = {\omega_o}^3$, $k_p = \omega_c^2$ and $k_d = 2\omega_c$. As a result, the parameters to be tuned are b_0, ω_o and ω_c for one area.

4 Lyapunov-Reward Shaping Based TD3 Optimized LADRC

The TD3 algorithm is a kind of DRL, and its pseudo code is shown in Algorithm 1. In simple terms, the TD3 algorithm aims to train an actor network to determine the optimal action in the current state. In the training process, the critic network is employed to evaluate the value function $Q(s, a)$ corresponding to the current state and action. The detailed principle will not be explained here, and readers can refer to Ref. [36].

Algorithm 1. TD3 Algorithm

Initialize critic networks $Q_{\theta_1}(s,a)$, $Q_{\theta_2}(s,a)$ and actor network $\pi_\phi(s)$ with random parameters θ_1, θ_2 and ϕ.
Initialize target networks $Q_{\theta'_1}(s,a)$, $Q_{\theta'_2}(s,a)$ and $\pi_{\phi'}(s)$ with weights $\theta'_1 \leftarrow \theta_1$, $\theta'_2 \leftarrow \theta_2$ and $\phi' \leftarrow \phi$.
Initialize replay buffer $\mathcal{D}$.
if $t \leq T$ **then**
  Select action with exploration noise $a \sim \pi_\phi(s) + \epsilon$, $\epsilon \sim \mathcal{N}(0, \sigma)$,observe reward r and new states s'.
  Store transition tuple (s, a, r, s') in $\mathcal{D}$.
  Sample mini-batch of m transitions (s, a, r, s') from $\mathcal{D}$.
  $a' \leftarrow \pi_{\phi'}(s') + \epsilon$, $\epsilon \sim$clip$(\mathcal{N}(0,\sigma), -c, c)$.
  $y \leftarrow r + \gamma \text{min}_{i=1,2} Q_{\theta'_i}(s', a')$.
  Update critics by $\theta_i \leftarrow \arg\min_{\theta_i} \frac{1}{m} \sum (y - Q_{\theta_i}(s,a))^2$.
  if t mod κ **then**
    Update ϕ by the deterministic policy gradient: $\nabla_\phi J(\phi) = \frac{1}{m} \sum \nabla_a Q_{\theta_1}(s,a) \Big|_{a=\pi_\phi(s)} \nabla_\phi \pi_\phi(s)$.
    Update target networks by moving average method: $\theta'_i \leftarrow \tau\theta_i + (1-\tau)\theta'_i$, $\phi' \leftarrow \tau\phi + (1-\tau)\phi'$.
  end if
end if

In this paper, only the tuning of the parameters ω_o and ω_c is considered, then the action space is

$$\{a_{11}, a_{12}, a_{21}, a_{22} \in \mathcal{A} \,|a_{11} = \omega_{o1}, a_{12} = \omega_{c1}, a_{21} = \omega_{o2}, a_{22} = \omega_{c2}\} \quad (13)$$

And the state space is chosen as

$$\left\{s_{11}, s_{12}, s_{21}, s_{22} \in \mathcal{S} \,\middle|\, s_{11} = ACE_1, s_{12} = A\dot{C}E_1, s_{21} = ACE_2, s_{22} = A\dot{C}E_2\right\} \quad (14)$$

As for the reward function, this paper utilized the following Lyapunov-reward shaping function [32]:

$$R^{lyap} = R(s,a) + \lambda(\gamma R(s',a') - R(s,a)) \quad (15)$$

with

$$R_t = R(s,a) = -2\sum_{i=1}^{n}\left[(100ACE_i(t))^2 + \left(10A\dot{C}E_i(t)\right)^2\right] \tag{16}$$

5 Simulation Comparison Results Analysis

During the training process, for each episode, the simulation time interval is set to 0.01 s, and the simulation time is 15 s. The hyperparameters used in the algorithm are selected as $\gamma = 0.99$, $\lambda = 0.3$ and $\tau = 0.005$. To showcase the efficiency of the proposed reward function shown in Eqs. 15 and 16, we compared the performance of it with TD3 algorithm using the reward function in Eq. 16 on the LADRC load frequency control system. For the control system, the parameters of b_0 for both two areas were set to be 20, and the action space for DRL was limited as

$$\begin{cases} \omega_{oi} \in [5,15], i = 1,2 \\ \omega_{ci} \in [1,6], i = 1,2 \end{cases} \tag{17}$$

The results of episode rewards are displayed in Fig. 3, indicating that the proposed method can achieve faster stabilization.

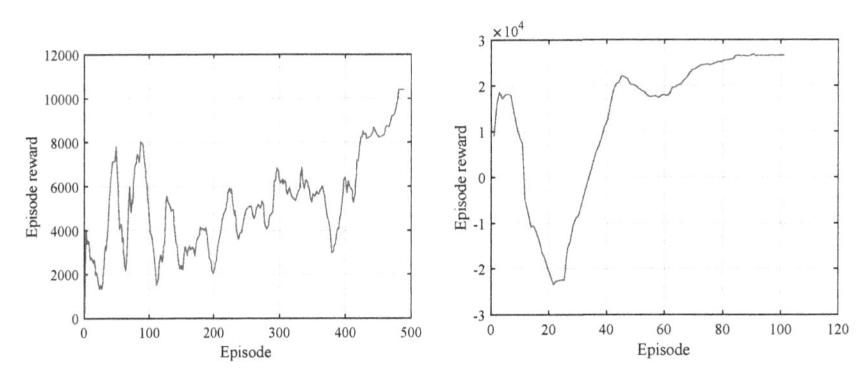

(a) Episode reward for TD3 using normal reward (b) Episode reward for TD3 using Lyapunov-reward shaping

Fig. 3. Episode reward

To validate the effectiveness of the actor network based on the Lyapunov-reward shaping, it is applied to the following scenario: The step load disturbance $\Delta P_{d1} = 0.01$ pu and $\Delta P_{d2} = 0.05$ pu are added to the system at 10 s and 150 s, respectively. The PV system is connected from 80 s, and the wind turbine is connected from 220 s. And the simulation results are shown in Figs. 4 and 5, where 'LADRC-TD3-nr' denotes that the controller parameters are obtained by TD3 algorithm with normal reward function as Eq. 16, while 'LADRC-TD3-Lr' represents that the controller parameters are obtained by TD3 algorithm with

Lyapunov-reward shaping function. And the proposed method obviously outperforms the existing methods I-TD [33] and ID-T [10], as well as LADRC-TD3 with normal reward function in terms of overshoot and settling time. Figure 5 shows the changing process the controller parameters with different reward functions.

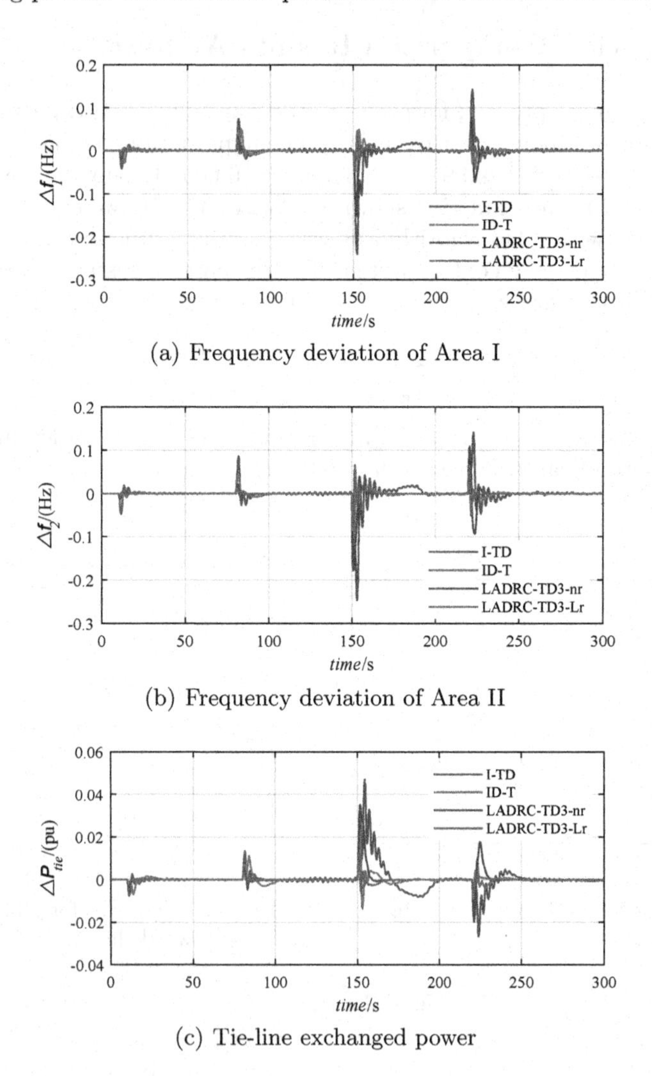

(a) Frequency deviation of Area I

(b) Frequency deviation of Area II

(c) Tie-line exchanged power

Fig. 4. Response curves of frequency deviation and tie-line exchanged power

6 Conclusion

This paper proposed a LADRC load frequency control approach based on TD3 algorithm optimization to overcome the frequency instability caused by load

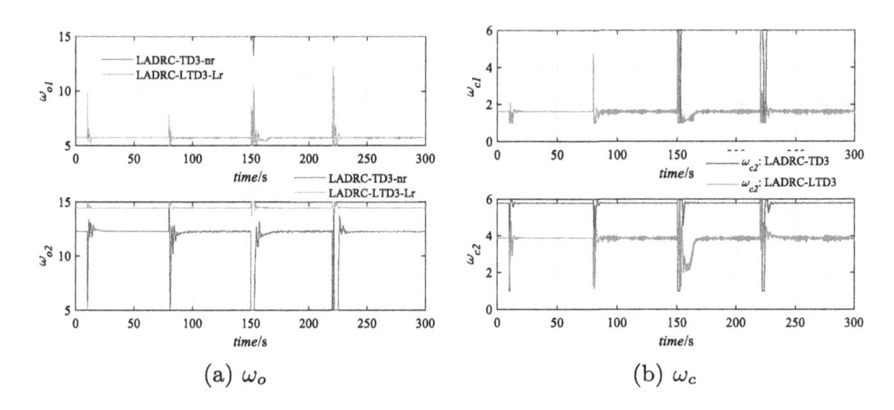

Fig. 5. Adaptive parameters of LADRC

disturbance and RESs. And a Lyapunov-reward shaping function is adopted to improve the convergence of TD3 algorithm. The proposed method is applied to a two-area interconnected power system, including thermal, hydro, and gas power plants in each area. Additionally, a noise-based wind turbine and PV system are also taken into account. Finally, the simulation results display comparison curves on frequency deviation and tie-line exchanged power, by the I-TD, ID-T, LADRC controllers optimized using TD3 with and without Lyapunov-reward shaping function, which implied that the proposed LADRC-TD3 with Lyapunov-reward is an effective method to deal with load frequency control.

Acknowledgment. This work was supported by the National Natural Science Foundation of China (Grant Nos. 61973172, 61973175, 62003175, 62003177 and 62073177) and Key Technologies Research and Development Program of Tianjin (Grant No. 19JCZDJC32800).

References

1. Zheng, Y., Chen, Z., Huang, Z., et al.: Active disturbance rejection controller for multi-area interconnected power system based on reinforcement learning. Neurocomputing **425**, 149–159 (2021)
2. Guha, D., Roy, P.-K., Banerjee, S.: Quasi-oppositional JAYA optimized 2-degree-of-freedom PID controller for load-frequency control of interconnected power systems. Int. J. Model. Simul. **42**(1), 63–85 (2022)
3. Guha, D., Roy, P.-K., Banerjee, S.: Equilibrium optimizer-tuned cascade fractional-order 3DOF-PID controller in load frequency control of power system having renewable energy resource integrated. Int. Trans. Electr. Energy Syst. **31**(1), e12702 (2021)
4. Abd-Elazim, S.-M., Ali, E.-S.: Load frequency controller design of a two-area system composing of PV grid and thermal generator via firefly algorithm. Neural Comput. Appl. **30**, 607–616 (2018)

5. Irudayaraj, A.-X.-R., Wahab, N.-I.-A., Premkumar, M., et al.: Renewable sources-based automatic load frequency control of interconnected systems using chaotic atom search optimization. Appl. Soft Comput. **119**, 108574 (2022)
6. Shabani, H., Vahidi, B., Ebrahimpour, M.: A robust PID controller based on imperialist competitive algorithm for load-frequency control of power systems. ISA Trans. **52**(1), 88–95 (2013)
7. Can, O., Ozturk, A., Eroğlu, H., et al.: A novel grey wolf optimizer based load frequency controller for renewable energy sources integrated thermal power systems. Electr. Power Compon. Syst. **49**(15), 1248–1259 (2022)
8. Morsali, J., Zare, K., Hagh, M.-T.: Comparative performance evaluation of fractional order controllers in LFC of two-area diverse-unit power system with considering GDB and GRC effects. J. Electr. Syst. Inf. Technol. **5**(3), 708–722 (2018)
9. Kler, D., Kumar, V., Rana, K.-P.-S.: Optimal integral minus proportional derivative controller design by evolutionary algorithm for thermal-renewable energy-hybrid power systems. IET Renew. Power Gener. **13**(11), 2000–2012 (2019)
10. Shouran, M., Anayi, F., Packianather, M.: The bees algorithm tuned sliding mode control for load frequency control in two-area power system. Energies **14**(18), 5701 (2021)
11. Shakibjoo, A.-D., Moradzadeh, M., Moussavi, S.-Z., et al.: Load frequency control for multi-area power systems: a new type-2 fuzzy approach based on Levenberg-Marquardt algorithm. ISA Trans. **121**, 40–52 (2022)
12. Ahmed, M., Magdy, G., Khamies, M., et al.: Modified TID controller for load frequency control of a two-area interconnected diverse-unit power system. Int. J. Electr. Power Energy Syst. **135**, 107528 (2022)
13. Zheng, Y., Tao, J., Sun, Q., et al.: Deep reinforcement learning based active disturbance rejection load frequency control of multi-area interconnected power systems with renewable energy. J. Franklin Inst. (2022, in press). https://doi.org/10.1016/j.jfranklin.2022.10.007
14. Gouran-Orimi, S., Ghasemi-Marzbali, A.: Load frequency control of multi-area multi-source system with nonlinear structures using modified grasshopper optimization algorithm. Appl. Soft Comput. **137**, 110135 (2023)
15. Biswas, S., Roy, P.K., Chatterjee, K.: FACTS-based 3DOF-PID controller for LFC of renewable power system under deregulation using GOA. IETE J. Res. **69**(3), 1486–1499 (2023)
16. Jin, H., Song, J., Lan, W., et al.: On the characteristics of ADRC: a PID interpretation. Sci. China Inf. Sci. **63**, 1–3 (2020)
17. Gao, Z.: Active disturbance rejection control: a paradigm shift in feedback control system design. In: 2006 American Control Conference. IEEE. Minneapolis, USA (2006). https://doi.org/10.1109/ACC.2006.1656579
18. Liu, C., Luo, G., Duan, X., et al.: Adaptive LADRC-based disturbance rejection method for electromechanical servo system. IEEE Trans. Ind. Appl. **56**(1), 876–889 (2019)
19. Guo, B., Bacha, S., Alamir, M., et al.: LADRC applied to variable speed micro-hydro plants: experimental validation. Control. Eng. Pract. **85**, 290–298 (2019)
20. Zheng, Y., Tao, J., Sun, Q., et al.: An intelligent course keeping active disturbance rejection controller based on double deep Q-network for towing system of unpowered cylindrical drilling platform. Int. J. Robust Nonlinear Control **31**(17), 8463–8480 (2021)
21. Feng, H., Ma, W., Yin, C., et al.: Trajectory control of electro-hydraulic position servo system using improved PSO-PID controller. Autom. Constr. **127**, 103722 (2021)

22. Xu, B., Cheng, Z., Zhang, R., et al.: PSO optimization of LADRC for the stabilization of a quad-rotor. In: Proceedings of 2020 12th International Conference on Measuring Technology and Mechatronics Automation (ICMTMA), pp. 437–441. IEEE, Thailand (2020). https://doi.org/10.1109/ICMTMA50254.2020.00100
23. Wang, Z., Yu, C., Li, M., et al.: Vertical profile diving and floating motion control of the underwater glider based on fuzzy adaptive LADRC algorithm. J. Mar. Sci. Eng. **9**(7), 698 (2021)
24. Liu, W., Zhao, T., Wu, Z., et al.: Linear active disturbance rejection control for hysteresis compensation based on backpropagation neural networks adaptive control. Trans. Inst. Meas. Control. **43**(4), 915–924 (2021)
25. Osband, I., Blundell, C., Pritzel, A., et al.: Deep exploration via bootstrapped DQN. In: Advances in Neural Information Processing Systems, vol. 29 (2016)
26. Qiu, C., Hu, Y., Chen, Y., et al.: Deep deterministic policy gradient (DDPG)-based energy harvesting wireless communications. IEEE Internet Things J. **6**(5), 8577–8588 (2019)
27. Bhatnagar, S., Sutton, R.-S., Ghavamzadeh, M., et al.: Natural actor-critic algorithms. Automatica **45**(11), 2471–2482 (2009)
28. Zheng, Y., Tao, J., Sun, Q., et al.: DDPG-based active disturbance rejection 3D path-following control for powered parafoil under wind disturbances. Nonlinear Dyn. **111**, 11205–11221 (2023)
29. Okafor, E., Udekwe, D., Ibrahim, Y., et al.: Heuristic and deep reinforcement learning-based PID control of trajectory tracking in a ball-and-plate system. J. Inf. Telecommun. **5**(2), 179–196 (2021)
30. Dankwa, S., Zheng, W.: Twin-delayed DDPG: a deep reinforcement learning technique to model a continuous movement of an intelligent robot agent. In: Proceedings of the 3rd International Conference on Vision, Image and Signal Processing, pp. 1–5. Association for Computing Machinery, New York (2019). https://doi.org/10.1145/3387168.3387199
31. Qiu, C., Hu, Y., Chen, Y., et al.: Deep deterministic policy gradient (DDPG)-based energy harvesting wireless communications. IEEE Internet Things J. **6**(5), 8577–8588 (2019)
32. Dong, Y., Tang, X., Yuan, Y.: Principled reward shaping for reinforcement learning via Lyapunov stability theory. Neurocomputing **393**, 83–90 (2020)
33. Morsali, J., Zare, K., Hagh, M.-T.: Comparative performance evaluation of fractional order controllers in LFC of two-area diverse-unit power system with considering GDB and GRC effects. J. Electr. Syst. Inf. Technol. **5**(3), 708–722 (2018)
34. Tan, W.: Decentralized load frequency controller analysis and tuning for multi-area power systems. Energy Convers. Manage. **52**(5), 2015–2023 (2021)
35. Gao, Z.: Scaling and bandwidth-parameterization based controller tuning. In: Proceedings of the 2003 American Control Conference, pp. 4989–4996. IEEE (2003). https://doi.org/10.1109/ACC.2003.1242516
36. Fujimoto, S., Hoof, H., Meger, D.: Addressing function approximation error in actor-critic methods. In: Proceedings of the 35th International Conference on Machine Learning, vol. 80, pp. 1587–1596. PMLR (2018). https://doi.org/10.48550/arXiv.1802.09477

Theory-Guided Convolutional Neural Network with an Enhanced Water Flow Optimizer

Xiaofeng Xue[1], Xiaoling Gong[2], Jacek Mańdziuk[3,4], Jun Yao[5], El-Sayed M. El-Alfy[6], and Jian Wang[1](✉)

[1] College of Science, China University of Petroleum (East China), Qingdao, China
wangjiannl@upc.edu.cn
[2] College of Control Science and Engineering, China University of Petroleum (East China), Qingdao, China
[3] Faculty of Mathematics and Information Science, Warsaw University of Technology, Warsaw, Poland
[4] Institute of Computer Science, AGH University of Science and Technology, Kraków, Poland
[5] College of Petroleum Engineering, China University of Petroleum (East China), Qingdao, China
[6] College of Computing and Mathematics, King Fahd University of Petroleum and Minerals, Dhahran 31261, Saudi Arabia

Abstract. Theory-guided neural network recently has been used to solve partial differential equations. This method has received widespread attention due to its low data requirements and adherence to physical laws during the training process. However, the selection of the punishment coefficient for including physical laws as a penalty term in the loss function undoubtedly affects the performance of the model. In this paper, we propose a comprehensive theory-guided framework using a bilevel programming model that can adaptively adjust the hyperparameters of the loss function to further enhance the performance of the model. An enhanced water flow optimizer (EWFO) algorithm is applied to optimize upper-level variables in the framework. In this algorithm, an opposition-based learning technic is used in the initialization phase to boost the initial group quality; a nonlinear convergence factor is added to the laminar flow operator to upgrade the diversity of the group and expand the search range. The experiments show that competitive performance of the method in solving stochastic partial differential equations.

This work was supported in part by the National Key R&D Program of China under Grant 2019YFA0708700; in part by the National Natural Science Foundation of China under Grant 62173345; and in part by the Fundamental Research Funds for the Central Universities under Grant 22CX03002A; and in part by the China-CEEC Higher Education Institutions Consortium Program under Grant 2022151; and in part by the Introduction Plan for High Talent Foreign Experts under Grant G2023152012L; and in part by the "The Belt and Road" Innovative Talents Exchange Foreign Experts Project under Grant DL2023152001L; and in part by SDAIA-KFUPM Joint Research Center for Artificial Intelligence under grant no. JRC-AI-RFP-04.

B. Luo et al. (Eds.): ICONIP 2023, LNCS 14447, pp. 448–461, 2024.
https://doi.org/10.1007/978-981-99-8079-6_35

Keywords: theory-guided neural network · stochastic partial differential equation · bilevel programming · water flow optimizer · penalty

1 Introduction

Stochastic partial differential equations (SPDEs) are a type of complex control system typically used to describe uncertain input random fields. A large number of numerical methods based on grid methods have been developed to solve these SPDEs [1,9]. Refining the anisotropic heterogeneous substrates using large-scale or high-resolution fine-grid blocks is a common method in modern geological modeling. As numerical simulators require significant computational resources, it is necessary to upscale the geological model from fine to coarse scale [7,14].

Recently, data-driven machine learning has revolutionized computer vision, reinforcement learning, and many other scientific and engineering fields [3,15]. However, several limitations exist, such as the need for large volumes of training data to ensure the precision of the model and lack of conformity to physical laws. Physics-driven neural networks, formed by combining data-driven neural networks with physical priors, are attracting increasing attention to solve various partial differential equation problems.

Physics-informed neural network (PINN) was presented by Raissi et al. [10] for the forward and backward problem of SPDEs by integrating physical knowledge into the neural network structure. Wang et al. [12,13] proposed a theory-guided convolutional neural network (TgCNN), and replaced the numerical solver of the coarse-grid block flow equation with the trained TgCNN to improve the scaling efficiency. In [11], two networks were constructed to deal with the inner observed points and satisfy the boundary conditions, so as to embed the physical laws.

In recent years, intelligent optimization algorithms have developed rapidly and been widely applied, achieving significant results in hyperparameter optimization [4]. Selecting appropriate hyperparameters within a certain range is critical to the performance of the model. However, manual tuning requires researchers to be extremely familiar with model hyperparameters. Therefore, automatic tuning is gaining more attention in adjusting model hyperparameters [6].

In this paper, a comprehensive theory-guided framework was proposed using a bilevel programming model, and an enhanced water flow optimizer (EWFO) algorithm. This method embeds prior knowledge into the network training process and adaptively adjusts hyperparameters in the loss function. In the second section, we discuss the stochastic partial differential equation and WFO algorithm. In the third section, the network structure and the EWFO algorithm are presented. In the fourth section, some experimental results are presented. At the end, the conclusion of this paper is given.

2 Preliminaries

2.1 Frame of Governing Equations

The incompressible single-phase steady-state subsurface flow problem is controlled by the following SPDE:

$$\begin{cases} \nabla \cdot (K(x) \nabla H(x)) = 0, & x \in \Gamma, \\ H(x) = p(x), & x \in \Gamma_D, \\ K(x) \nabla H(x) \cdot n(x) = q(x), x \in \Gamma_N, \end{cases} \tag{1}$$

where H is the hydraulic head $[L]$ ($[L]$ is any consistent unit of length); Γ is the N-dimensional domain $R^N(N > 1)$; K represents the uncertain hydraulic conductivity tensor $[LT^{-1}]$ ($[T]$ is any consistent unit of time); Γ_D and Γ_N denote the Dirichlet and Newman boundary, respectively; $p(x)$ is a stochastic variable on Dirichlet boundary $[L]$; $n(x)$ is the unit normal vector when x lies on the Newman boundary; $q(x)$ is random variable on the Newman boundary $[LT^{-1}]$.

To characterize and generate the inhomogeneously distributed hydraulic conductivity fields, we use the Karhunen-Loeve expansion (KLE) [2]. The log-transformed hydraulic conductivity $lnK(x,\omega)$, $\omega \in \Omega$ (Ω is probability space) can be considered as a stochastic field. In the following, $M(x,\omega)$ is used instead of $lnK(x,\omega)$:

$$M(x,\omega) = \sum_{i=1}^{\infty} \sqrt{\lambda_i} f_i(x)\xi_i(\omega) + \widehat{M(x,\omega)}, \tag{2}$$

where $\widehat{M(x,\omega)}$ indicates the mean of the stochastic field; $\xi_i(\omega)$ represents orthogonal Gaussian random variables with mean of zero and variance of unit; $f_i(x)$ and λ_i represent the covariance eigenfunction and eigenvalue, respectively. We can truncate the infinite series in Eq. (2) according to the decay rate $\xi_i(\omega)$ and thus approximate the random field. It is commonly understood that a greater number of retained terms in expansion yields increased accuracy, but with the drawback of increased dimensionality of the random field. The stochastic field can be represented as:

$$M(x,\omega) = \sum_{i=1}^{n} \sqrt{\lambda_i} f_i(x)\xi_i(\omega) + \widehat{M(x,\omega)}, \tag{3}$$

where n indicates the amount of terms retained, determined by the corresponding eigenvalues λ_i. More specifically, the truncation number n can be calculated by the energy percentage $\sum_{i=1}^{n} \lambda_i / \sum_{i=1}^{\infty} \lambda_i$. Each set of stochastic variables stands for a unique realization of the conductivity field. The retained energy is truncated when it reaches a predetermined energy, which significantly reduces the dimension of the stochastic field and thus saves computational resources.

Finite difference (FD) method can be utilized to solve the flow equation of different coarse grid blocks. The discretized form of Eq. (1) is illustrated as follows:

$$\frac{Kx_{i+1/2,j}(H_{i+1,j}-H_{i,j})-Kx_{i-1/2,j}(H_{i,j}-H_{i-1,j})}{(\triangle x)^2}+ \\ \frac{Ky_{i,j+1/2}(H_{i,j+1}-H_{i,j})-Ky_{i,j-1/2}(H_{i,j}-H_{i,j-1})}{(\triangle y)^2}=0, \tag{4}$$

where K_x and K_y represent the horizontal and vertical hydraulic conductivity component; i and j represent horizontal and vertical grid block indices. More specifically, $Kx_{i+1/2,j}$ represents the grid block (i, j) and $(i+1, j)$ at the junction of the conductivity, which is given as:

$$Kx_{i+1/2,j} = \frac{2}{1/Kx_{i,j} + 1/Kx_{i+1,j}}. \tag{5}$$

Tx and Ty are the transmissibility between two adjacent grid blocks:

$$Tx_{i+1/2,j} = \frac{2 \triangle y}{1/Kx_{i,j} + 1/Kx_{i+1,j}}, \tag{6}$$

$$Ty_{i,j+1/2} = \frac{2 \triangle x}{1/Kx_{i,j} + 1/Kx_{i,j+1}}. \tag{7}$$

By multiplying Eq. (4) by $\triangle x \triangle y$ and replacing some of its terms with Eqs. (6) and (7), we obtain:

$$\frac{Tx_{i+1/2,j}(H_{i+1,j}-H_{i,j})}{\triangle x} - \frac{Tx_{i-1/2,j}(H_{i,j}-H_{i-1,j})}{\triangle x}+ \\ \frac{Ty_{i,j+1/2}(H_{i,j+1}-H_{i,j})}{\triangle y} - \frac{Ty_{i,j-1/2}(H_{i,j}-H_{i,j-1})}{\triangle y}=0. \tag{8}$$

The following boundary conditions are imposed on Eq. (1):

$$H(x,y)\mid_{x=0} = H(x,y)\mid_{x=L_x} + \triangle H_1, \quad v_x(x,y)\mid_{x=0} = v_x(x,y)\mid_{x=L_x}; \\ H(x,y)\mid_{y=0} = H(x,y)\mid_{y=L_y} + \triangle H_2, \quad v_y(x,y)\mid_{y=0} = v_y(x,y)\mid_{y=L_y}; \tag{9}$$

where L_x and L_y are the coarse grid block length in horizontal and vertical directions, respectively; v_x and v_y are Darcy velocities; $\triangle H_1$ and $\triangle H_2$ are the steady hydraulic head differences between the boundaries.

2.2 Water Flow Optimizer

Inspired by two kinds of fluid flow in environment: laminar and turbulent flow, water flow optimizer (WFO) was presented by Luo [8]. Suppose the problem to be solved is minimizing the objective function. Analogously water always flows from high to low in nature, the water particles can be considered as solutions, the water particle's position is equivalent to the solution value, and water potential energy is considered to be the fitness value.

(1) *Laminar Operator:* Due to the viscosity of the water, the speed of the particles may vary. Therefore, the following formula is used for modeling:

$$y_i(t) = x_i(t) + s * \hat{\boldsymbol{d}}, \quad \forall i \in \{1, 2, \cdots, m\}, \tag{10}$$

where i indicates the index of water particles, m is the amount of water particles, and t represents the iteration number. $x_i(t)$ represents the i-th water particle position at the t-th iteration, $y_i(t)$ is the possible moving positions of the water particles, and s is randomly and uniformly taken in $[0, 1]$. $\hat{\boldsymbol{d}}$ indicates the direction of joint motion of all particles at t-th iteration, defined by two different water particles. The formula for $\hat{\boldsymbol{d}}$ is shown below:

$$\hat{\boldsymbol{d}} = x_h(t) - x_k(t), \quad (h \neq k, f(x_h(t)) \leq f(x_k(t))) \tag{11}$$

where $f(x_h(t))$ and $f(x_k(t))$ denote the h-th and k-th particles' potential energy, respectively. This means that all the other particles move in the direction of particle h. Therefore, h is designated as the guiding particle due to its observed minimum potential energy among those examined. While the value of $\hat{\boldsymbol{d}}$ remains constant during each iteration, the value of s in Eq. (10) fluctuates across particles.

(2) *Turbulent Operator:* The potential displacement of water particles are obtained by oscillations on randomly chosen dimensions, as shown below:

$$y_i(t) = (\cdots, x_i^{\alpha-1}(t), v, x_i^{\alpha+1}(t), \cdots), \tag{12}$$

where $\alpha \in \{1, 2, \cdots, n\}$ denotes a randomly chosen component in n dimensions. Mutation value v calculation method is as follows:

$$v = \begin{cases} \psi(x_i^{\alpha}(t), x_k^{\alpha}(t)), if \quad r < p_e, \\ \varphi(x_i^{\alpha}(t), x_k^{\gamma}(t)), otherwise, \end{cases} \tag{13}$$

where $k \in \{1, 2, \cdots, m\}$ denotes a randomly selected index of different water particles, $k \neq i$; r represents a stochastic number, $r \in [0, 1]$; α denotes a random selection of different dimensions, $\alpha \neq \gamma$; p_e denotes that a control parameter known as the eddying probability, $p_e \in (0, 1)$; ψ denotes eddying transformation; φ denotes general cross-layer shift transformation. Thus, the turbulence operator achieves the eddying transformation ψ by probability p_e, while probability $1-p_e$ achieves φ. The structure of the eddying function is as follows:

$$\psi(x_i^{\alpha}(t), x_k^{\alpha}(t)) = x_i^{\alpha}(t) + \rho * \theta * \cos(\theta), \tag{14}$$

where $\theta \in [-\pi, \pi]$ represents a stochastic number, and ρ indicates determined dynamically by the self-adaptive rule, which is defined as:

$$\rho =\mid x_i^{\alpha}(t) - x_k^{\alpha}(t) \mid, \tag{15}$$

where ρ represents the shear force applied by particle k to particle i.

The following transformation functions are used to approximate the common behavior of water particles on layers:

$$\varphi(x_i^{\alpha-1}(t), x_k^{\gamma}(t)) = (ub^{\alpha} - lb^{\alpha}) * \frac{x_k^{\gamma}(t) - lb^{\gamma}}{ub^{\gamma} - lb^{\gamma}} + lb^{\alpha}, \tag{16}$$

where lb and ub represent the lower and upper limits, respectively; $x_k^{\gamma}(t)$ is the position of the k-th particle in the γ-th dimension.

(3) *Evolutionary Rule:* The simulation probability of laminar flow is $p_l \in (0, 1)$ and the simulation probability of turbulence is $1 - p_l$. Laminar probability p_l is analogous to the Reynolds number threshold, as it determines whether to implement the laminar operator or the turbulent operator. During the iterative process, the objective is constantly being optimized. If a potential moving position offers lower energy, it is chosen as the next position. If not, the particle keeps at its present position. Namely,

$$x_i(t+1) = \begin{cases} y_i(t), \ if \quad f(y_i(t)) < f(x_i(t)), \\ x_i(t), \ otherwise. \end{cases} \tag{17}$$

3 Proposed Methodology

In this section, a comprehensive theory-guided framework based on a bilevel programming model is presented. Meanwhile, an enhanced water flow optimizer (EWFO) algorithm is designed to optimize the upper-level variables, resulting in more stable outcomes.

3.1 Construction of Neural Network

Firstly, the lower-level problem of the bilevel programming model is investigated. We apply a convolutional neural network embedded with physical laws to solve the flow equations of grid blocks. The neural network model is shown as:

$$\widehat{H} = nn(x, \theta), \tag{18}$$

where θ represents the parameters in the neural network, both the weights and the bias; $\widehat{H}$ indicates the predicted hydraulic head; nn represents the neural network.

If the predicted hydraulic is in accordance with the physical laws, specifically, Eq. (8), the residual below will asymptotically approach zero:

$$\begin{aligned} R(K;\theta) = & \frac{Tx_{i+1/2,j}(\widehat{H}_{i+1,j} - \widehat{H}_{i,j})}{\triangle x} \\ & - \frac{Tx_{i-1/2,j}(\widehat{H}_{i,j} - \widehat{H}_{i-1,j})}{\triangle x} \\ & + \frac{Ty_{i,j+1/2}(\widehat{H}_{i,j+1} - \widehat{H}_{i,j})}{\triangle y} \\ & - \frac{Ty_{i,j-1/2}(\widehat{H}_{i,j} - \widehat{H}_{i,j-1})}{\triangle y} \\ & \rightarrow 0. \end{aligned} \tag{19}$$

In order to minimize Eq. (19) during training and thus force the network's predictions to conform to Eq. (8), the residual term of the network's loss function is represented as below:

$$Loss_{GE}(\theta) = \frac{1}{N_{grid}} \sum_{i=1}^{N_{grid}} \| R(K_i; \theta) \|_2^2, \tag{20}$$

where N_{grid} indicates for the amount of grids in each realization. It is worth noting that the training of Eq. (19) does not require the use of any label values.

Similarly the predicted hydraulic $\widehat{H}$ can be substituted into the periodic boundary condition Eq. (9) to obtain the following equation:

$$\begin{aligned} &Loss_{BC-H}(\theta) \\ &= \tfrac{1}{N_{grid-b}} \textstyle\sum_{j=1}^{N_{grid-b}} \| (\widehat{H} \mid_{x=0} - \widehat{H} \mid_{x=L_x}) - \triangle H_1 \|_2^2 \\ &+ \tfrac{1}{N_{grid-b}} \textstyle\sum_{j=1}^{N_{grid-b}} \| (\widehat{H} \mid_{y=0} - \widehat{H} \mid_{y=L_y}) - \triangle H_2 \|_2^2, \\ &Loss_{BC-v}(\theta) \\ &= \tfrac{1}{N_{grid-b}} \textstyle\sum_{j=1}^{N_{grid-b}} \| (\widehat{v_x} \mid_{x=0} - \widehat{v_x} \mid_{x=L_x}) \|_2^2 \\ &+ \tfrac{1}{N_{grid-b}} \textstyle\sum_{j=1}^{N_{grid-b}} \| (\widehat{v_y} \mid_{y=0} - \widehat{v_y} \mid_{y=L_y}) \|_2^2, \end{aligned} \tag{21}$$

where N_{grid-b} represents the amount of grids at boundaries; $\widehat{v_x}$ and $\widehat{v_y}$ are the Darcy velocities obtained from the network predictions, which can be worked out from the Darcy's law as follows:

$$\widehat{v_x} = -K_x \cdot \nabla \widehat{H}, \tag{22}$$

$$\widehat{v_y} = -K_y \cdot \nabla \widehat{H}. \tag{23}$$

The loss function of our model is given as below:

$$Loss(\theta) = \lambda_1 Loss_{GE}(\theta) + \lambda_2 Loss_{BC-H}(\theta) + \lambda_3 Loss_{BC-v}(\theta), \tag{24}$$

where λ_1, λ_2 and λ_3 are hyperparameters that balance the importance between the different functions in Eq. (24). The method of adding penalty terms based on the laws of physics to the loss function necessitates the coefficients of the penalty terms to be sufficiently large to coerce the network into satisfying prior knowledge. If λ_1, λ_2 and λ_3 are constants, it is obvious that the requirement cannot be satisfied.

Using the convolutional neural network, the output of the network is embedded in physical conditions such as SPDE as the regularization term in the loss function, minimizing the loss function while forcing the network predictions to conform to the physical conditions.

3.2 Enhanced Water Flow Optimizer

Inspired by WFO, enhanced water flow optimizer (EWFO) has innovated in the initialization stage and laminar flow operator to obtain a stronger and more stable search capability than WFO.

(1) *Elite opposition-based Learning technic:* The optimal performance of global optimization algorithms is influenced by the quality of the initial solution. High-quality initial group can accelerate the algorithm convergence and help find the global optimal solution. Standard WFO algorithm [8] tends to use random initialization to initialize the group.

In this section, we apply elite opposition-based learning technic [16] to the initialization stage of EWFO, take advantage of the property that elite particles contain more effective information to construct its opposite group, and chose the better particles from the present group and the reverse group as the initial solution which enhances the quality and diversity of the initial group.

Suppose $x_i = (x_i^1, x_i^2, \cdots, x_i^n)$ is an elite particle in an n-dimensional search space, then its opposite solution $\breve{x}_i = (\breve{x}_i^1, \breve{x}_i^2, \cdots, \breve{x}_i^n)$ is defined by the following equation:

$$\breve{x}_i^j = \xi * (lb^j + ub^j) - x_i^j, \quad i \in \{1, 2, \cdots, m\}, j \in \{1, 2, \cdots, n\} \tag{25}$$

where $\xi \in [0, 1]$ is a random number. The addition of ξ avoids polarization of the distribution of particles in the two groups. The elite group and the opposite group are combined to obtain a new group, from which m particles with better fitness are selected to form the initial group.

(2) *Adaptive nonlinear convergence factor:* In the standard WFO algorithm, the value of s in the Eq. (10) describing the state of motion of the particle for the laminar flow operator is taken randomly. This approach leads to too random positions of water particles in the next step, which is detrimental to the fast convergence of the algorithm and reduces the diversity and flexibility of the group. To solve this problem, a nonlinear convergence factor is introduced, whose specific expression is as follows:

$$a = 1 - (\frac{t}{T_{max}})^{\lambda} * (e + \mu) + \eta * \tau, \tag{26}$$

$$s = 2 * \beta * a - a, \tag{27}$$

where λ, μ, τ are constant coefficients; $\eta \in [0, 1]$ is a random number; β is randomly and uniformly taken in [0, 1]; t refers to the number of current iterations; T_{max} represents maximum iteration number. The convergence factor a reduces nonlinearly as the number of iterations increases. In the initial phase a has a lower degree of decay and the water particles are able to move in larger steps to better find the global optimal solution. At a later stage, the degree of decay of a increases and the water particles move in smaller steps, allowing a more accurate search for the optimal solution. As a result water particles are more dispersed. This approach more effectively balances the exploitation capability during global search with the mining capability during local search.

Algorithm 1: EWFO

Input: number of particles m, dimensions n, the minimum value of particle initialization lb, the maximum value of particle initialization ub, the maximum number of iterations T_{max}

Output: the best solution found

1 **for** $i = 1$ *to* m **do**
2   **for** $j = 1$ *to* n **do**
3   Generate random elite position, $x_i^j = lb^j + rand() * (ub^j - lb^j)$.
      Generate the opposite position, $\check{x}_i^j = rand() * (lb^j + ub^j) - x_i^j$
4   Calculate the potential energy of elite and opposite groups, $f_i = f(x_i)$ and $\check{f}_i = f(\check{x}_i)$;
5 /* —— Choose the initial group of m particles with better fitness values from the elite group and the opposing group —— */
6 **for** $t = 1$ *to* T_{max} **do**
7   **if** $p_l > rand()$ **then**
8   The particle with the median potential energy in a population, x_k
      Determine a laminar direction, $\hat{\boldsymbol{d}} = x_h(t) - x_k(t)$;
9     $a = 1 - (\frac{t}{T_{max}})^{\lambda} * (e + \mu) + \eta * \tau$;
10    $s = 2 * rand() * a - a$.
11   **for** $i = 1$ *to* m **do**
12      $y_i(t) = x_i(t) + s * \hat{\boldsymbol{d}}$
13 **else**
14   **for** $i = 1$ *to* m **do**
15   Initialize a trial position, $y_i \leftarrow x_i$;
16   Choose a particle at random, $k \in \{1, 2, \cdots, m\}$ and $k \neq i$;
17   Choose a random dimension, $\alpha \in \{1, 2, \cdots, n\}$;
18   **if** $p_e > rand()$ **then**
19        $\rho = \mid x_i^\alpha - x_k^\alpha \mid$;
20        $\sigma = 2 * \pi * rand() - \pi$;
21        $y_i^\alpha = x_i^\alpha + \rho * \cos(\sigma)$;
22   **else**
23   Choose another dimension at random, $\gamma \in \{1, 2, \cdots, n\}$ and $\gamma \neq \alpha$. $y(x_i^{\alpha-1}(t), x_k^\gamma(t)) = (ub^\alpha - lb^\alpha) * \frac{x_k^\gamma(t) - lb^\gamma}{ub^\gamma - lb^\gamma} + lb^\alpha$
24 **for** $i = 1$ *to* m **do**
25   **if** $f(y_i) < f_i$ **then**
26      $f_i = f(y_i)$ and $x_i \leftarrow y_i$
27   **if** $f_i < f_g$ **then**
28        $f_g = f_i, x_g \leftarrow x_i$ and $g \leftarrow i$;
29 return the best solution found;

4 Experiments

We experimentally compare the proposed TgCNN optimized using EWFO with the standard TgCNN and the TgCNN optimized by the standard WFO algorithm on the same dataset. A square physical domain of $100[L]$ is used, with a correlation length of $20[L]$ in the horizontal and vertical directions. KLE was utilized to generate hydraulic conductivity fields, and the random log-transformed hydraulic conductivity used has a mean of 0 and a variance of 1. The experimental results of all models are the average of 10 runs. Each run of the training process is 500 iterations. Assuming isotropic hydraulic conductivity and $weight = 0.9$. All experiments are conducted on Python 3.8 software platform with 3.10 GHz CPU and 16 GB of memory.

The fine-scale geological model is divided into 100×100 grid blocks. The inputs to all three TgCNN models are 10×10 coarse grid block images which are composed of high-resolution grid blocks, and the outputs of the network models are all predicted values containing boundaries, so the outputs are 12×12 stress images. In this paper, KLE was utilized to generate five fine-scale hydraulic conductivity fields. In addition, the constant head difference of the periodic boundary condition is set to $\triangle H_1 = 1$ and $\triangle H_2 = 0$. We adopt the Adam optimizer [5] to train all the neural networks.

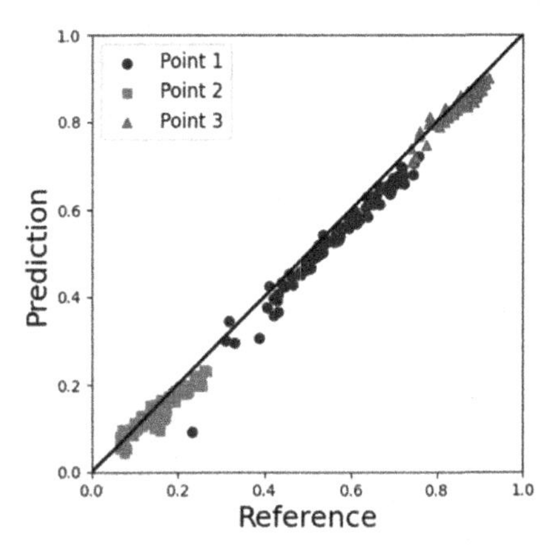

Fig. 1. Comparison of predicted solution and numerical solution (Point 1: x = 5, y = 5; Point 2: x = 9, y = 9; Point 3: x = 2, y = 2).

Figure 1 depicts the scatter plot illustrating the hydraulic heads measurements collected at three discrete points. It is worth noting that the horizontal coordinate in Fig. 1 is the labeled value and the vertical coordinate is the predicted value, not the location of the point taken. Figure 2 shows the effect of different amounts of training data on the accuracy of the model. As the amount of training data increases, the accuracy of the model improves, but the rate of improvement gradually slows down. Figure 3 demonstrates that the images of the labeled and predicted values in the two-dimensional case are generally consistent. The last column gives the error between the two, and it can be observed that there is a relatively large error at a few positions and a small error overall. It is evident that the EWFO-TgCNN model constructed herein accurately predicts hydraulic head solutions for various hydraulic head patches, even in the absence of labeled data. The relative L_2 error and R^2 scores of the three models are given in Table 1. The formula for calculating the two evaluation criteria is given below:

$$L_2 = \frac{\| P_{pred} - P_{ref} \|_2}{\| P_{ref} \|_2}, \tag{28}$$

$$R^2 = 1 - \frac{\sum_{i=1}^{N_{grid}} (P_{pred,i} - P_{ref,i})^2}{\sum_{i=1}^{N_{grid}} (P_{ref,i} - \bar{P}_{ref,i})^2}, \tag{29}$$

where $P_{pred,i}$ is the predictions from our model at the i-th block; $P_{ref,i}$ is the reference value from numerical solutions at the i-th block; $\| \cdot \|_2$ represents the standard Euclidean norm; $\bar{P}_{ref,i}$ is the mean of $P_{ref,i}$. Through ten experiments, the standard deviations of the R^2 scores of the three models were calculated to be 0.012005453 for TgCNN, 0.005361467 for TgCNN optimized using WFO, and 0.002886972 for TgCNN optimized using EWFO. Experimental results with standard deviations show our proposed EWFO-TgCNN is indeed more stable than the other two models.

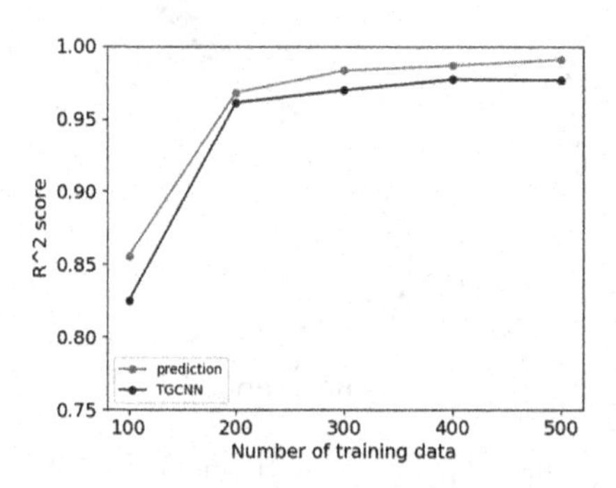

Fig. 2. Comparison of R^2 scores of the deep learning method with different amounts of training data

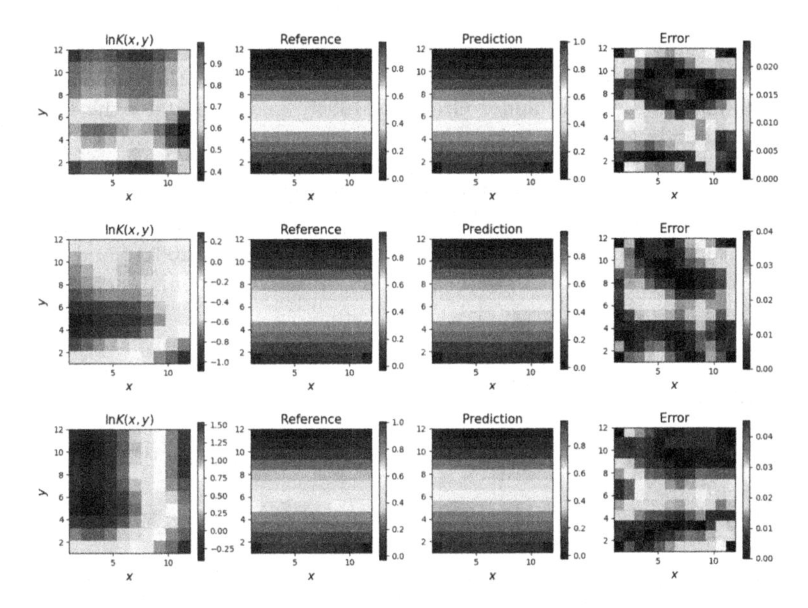

Fig. 3. Comparison of hydraulic conductivity fields, reference, predicted values and errors in the two-dimensional case

Table 1. The average of several model indicators.

Models	L_2 mean	L_2 var	R^2 mean	R^2 var
TgCNN	7.666e-02	7.012e-04	9.766e-01	4.153e-04
WFO-TgCNN	5.804e-02	6.060e-04	9.866e-01	1.235e-04
EWFO-TgCNN	**4.760e-02**	**5.454e-04**	**9.907e-01**	**8.725e-05**

5 Conclusion

At present, research on models for solving partial differential equations with embedded physical laws is flourishing. In this paper, a comprehensive theory-guided framework for solving stochastic partial differential equations is presented. With the help of the bilevel programming model, a convolutional neural network is applied to the lower-level problem with the objective of minimizing the loss function and forcing the network to conform to the physical prior. The EWFO algorithm optimizes the upper-level variables and adaptively assigns weights to penalty terms. The validity and stability of the model are verified by extensive comparative experiments with 2 related algorithms. Our model not only reduces computational resources without using labeled data, but also satisfies physical laws, thus making the model more robust. Despite the advancements demonstrated by the proposed approach in computational precision, potential for further amelioration remains.

Acknowledgements. This work was supported in part by the National Key R&D Program of China under Grant 2019YFA0708700; in part by the National Natural Science Foundation of China under Grant 62173345; and in part by the Fundamental Research Funds for the Central Universities under Grant 22CX03002A; and in part by the China-CEEC Higher Education Institutions Consortium Program under Grant 2022151; and in part by the Introduction Plan for High Talent Foreign Experts under Grant G2023152012L; and in part by the "The Belt and Road" Innovative Talents Exchange Foreign Experts Project under Grant DL2023152001L; and in part by SDAIA-KFUPM Joint Research Center for Artificial Intelligence under grant no. JRC-AI-RFP-04.

References

1. Pao-Liu, C.: Stochastic Partial Differential Equations. CRC Press, Boca Raton (2014)
2. Ghanem, R.G., Spanos, P.D.: Stochastic finite elements: a spectral approach. Courier Corporation (2003)
3. Gong, X., Yu, L., Wang, J., Zhang, K., Bai, X., Pal, N.R.: Unsupervised feature selection via adaptive autoencoder with redundancy control. Neural Netw. **150**, 87–101 (2022)
4. Jones, D.R., Schonlau, M., Welch, W.J.: Efficient global optimization of expensive black-box functions. J. Global Optim. **13**(4), 455 (1998)
5. Kingma, D.P., Ba, J.: Adam: a method for stochastic optimization. arXiv preprint arXiv:1412.6980 (2014)
6. Li, J., Wang, X., Xue, G., Zhang, H., Wang, J.: Sparse broad learning system via a novel competitive swarm optimizer. In: 2022 IEEE 6th Advanced Information Technology, Electronic and Automation Control Conference (IAEAC), pp. 1697–1701. IEEE (2022)
7. Liao, Q., Lei, G., Zhang, D., Patil, S.: Analytical solution for upscaling hydraulic conductivity in anisotropic heterogeneous formations. Adv. Water Resour. **128**, 97–116 (2019)
8. Luo, K.: Water flow optimizer: a nature-inspired evolutionary algorithm for global optimization. IEEE Trans. Cybern. **52**(8), 7753–7764 (2021)
9. Pardoux, É.: Stochastic Partial Differential Equations: An Introduction. Springer, Cham (2021). https://doi.org/10.1007/978-3-030-89003-2
10. Raissi, M., Perdikaris, P., Karniadakis, G.E.: Physics-informed neural networks: a deep learning framework for solving forward and inverse problems involving nonlinear partial differential equations. J. Comput. Phys. **378**, 686–707 (2019)
11. Wang, J., Pang, X., Yin, F., Yao, J.: A deep neural network method for solving partial differential equations with complex boundary in groundwater seepage. J. Petrol. Sci. Eng. **209**, 109880 (2022)
12. Wang, N., Chang, H., Zhang, D.: Efficient uncertainty quantification and data assimilation via theory-guided convolutional neural network. SPE J. **26**(06), 4128–4156 (2021)
13. Wang, N., Liao, Q., Chang, H., Zhang, D.: Deep-learning-based upscaling method for geologic models via theory-guided convolutional neural network. arXiv preprint arXiv:2201.00698 (2021)
14. Wen, X.-H., Gómez-Hernández, J.J.: Upscaling hydraulic conductivities in heterogeneous media: an overview. J. Hydrol. **183**(1–2), 9–27 (1996)

15. Zhang, B., Gong, X., Wang, J., Tang, F., Zhang, K., Wei, W.: Nonstationary fuzzy neural network based on FCMnet clustering and a modified CG method with Armijo-type rule. Inf. Sci. **608**, 313–338 (2022)
16. Zhou, Y., Wang, R., Luo, Q.: Elite opposition-based flower pollination algorithm. Neurocomputing **188**, 294–310 (2016)

An End-to-End Dense Connected Heterogeneous Graph Convolutional Neural Network

Ranhui Yan[1] and Jia Cai[2](✉)

[1] School of Statistics and Mathematics, Guangdong University of Finance and Economics, Guangzhou, China

[2] School of Digital Economics, Guangdong University of Finance and Economics, Guangzhou, China

jiacai1999@gdufe.edu.cn

Abstract. Graph convolutional networks (GCNs) are powerful models for graph-structured data learning task. However, most existing GCNs may confront with two major challenges when dealing with heterogeneous graph: (1) Predefined meta-paths are required to capture the semantic relations between nodes from different types, which may not exploit all the useful information in the graph; (2) Performance degradation and semantic confusion may happen with the growth of the network depth, which limits their ability to capture long-range dependencies. To meet these challenges, we propose Dense-HGCN, an end-to-end dense connected heterogeneous convolutional neural network to learn node representation. Dense-HGCN computes the attention weights between different nodes and incorporates the information of previous layers into each layer's aggregation process via a specific fuse function. Moreover, Dense-HGCN leverages multi-scale information for node classification or other downstream tasks. Experimental results on real-world datasets demonstrate the superior performance of Dense-HGCN in enhancing the representational power compared with several state-of-the-art methods.

Keywords: Graph neural network · Heterogeneous graph · Multi-scale

1 Introduction

Graph Neural Networks (GNNs) [2,3,7,11,17] are powerful neural network models for handling graph-structured data, which naturally represents real-world data such as social networks, citation networks, world wide web, drug reactions, and protein-protein interaction. There is a vast literature on homogeneous graph

The work described in this paper was supported partially by the National Natural Science Foundation of China (12271111), Special Support Plan for High Level Talents of Guangdong Province (2019TQ05X571), Foundation of Guangdong Educational Committee (2019KZDZX1023), Project of Guangdong Province Innovative Team (2020WCXTD011).

B. Luo et al. (Eds.): ICONIP 2023, LNCS 14447, pp. 462–475, 2024.
https://doi.org/10.1007/978-981-99-8079-6_36

learning, including, but not limited to, ChebNet [3], GCN [11], GraphSAGE [7], and graph attention neural networks (GAT) [17]. However, in real-world, a graph usually contains diverse types of nodes and edges. It can yield more comprehensive information and rich semantics, which is called heterogeneous graph. Node representation learning in graph is essential, which can boost a variety of tasks such as node classification and clustering, link prediction, and relation prediction etc. To efficiently learn node representation, we need to consider not only the node features, but also the topological structure of the graph, i.e., extracting useful features from nodes and their neighbors.

Meta-path [15] is a widely considered approach defined on a heterogeneous graph [14], which describes a composite relation between two node types (same or different). Taking the DBLP (an academic graph) dataset as an example, from Fig. 1, we can see that the meta-path: $Author - Paper - Author$ (abbreviated as APA) describes the relation of co-authorship of a paper. Similarly, a meta-path APC expresses the semantics of an author submitting his/her paper to a conference. Most GCNs [4,6,9,18,19] need predefined meta-paths to learn

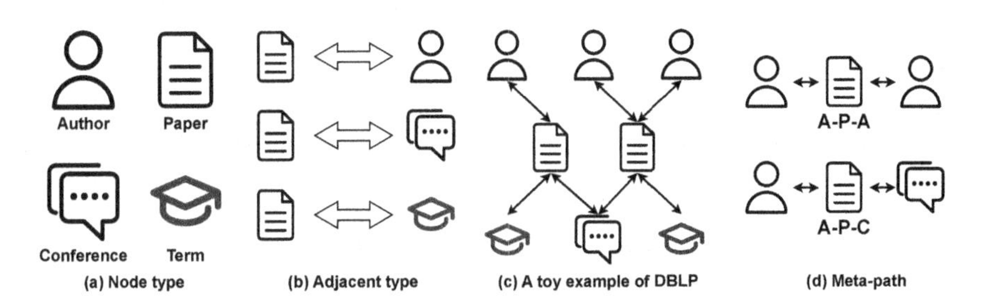

Fig. 1. An illustrative example of a heterogeneous graph (DBLP). (a) Four types of nodes (i.e., author, paper, conference, term). (b) Three types of links (i.e., Paper-Author, Paper-Conference, and Paper-Term). (c) A heterogeneous graph DBLP consists four types of nodes and two types of connections. (d) Two meta-paths involved in DBLP, i.e., $Author - Paper - Author$ and $Author - Paper - Conference$.

node representation. The predefined meta-path is a dominant factor in determining the model performance. However, exploring and exploiting useful predefined meta-paths needs professional knowledge for specific tasks. To address this issue, [21,22] conducted aggregation procedure automatically without using predefined meta-paths. On contrary, they utilized attention mechanism to all possible length limited meta-paths.

Nevertheless, they are suffering from semantic confusion problem. Similar to the over-smoothing problem in homogeneous graph, semantic confusion means that the node embeddings tend to be similar and become indistinguishable with the growth of the network depth. Semantic confusion phenomenon occurs when considering the semantic level fusion problem, which is highly related with meta-path while over-smoothing is not. Let us examine the sensitivity of ie-HGCN [21]

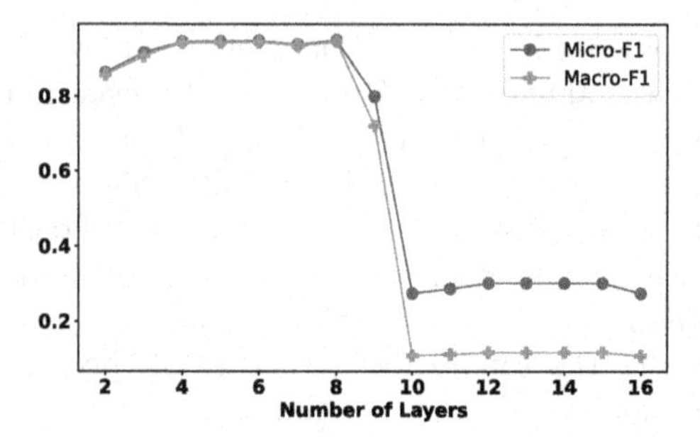

Fig. 2. Performance of ie-HGCN on DBLP dataset with distinct number of layers.

to the number of layers with a simple experiment, as Fig. 2 indicated, the performance of the model drops severely as the number of the layer increased to 10. It follows that the semantic confusion phenomenon greatly affect the performance of heterogeneous graph neural networks. To overcome the limitations of predefined meta-paths and semantic confusion simultaneously, we design an end-to-end **D**ense **H**eterogeneous **G**raph **C**onvolustional Neural **N**etwork (Dense-HGCN) in this paper. Dense-HGCN considers feature matrices and adjacency matrices as input directly, and learns task-specific node representation through multiple convolutional layers iteratively. There are four key steps in Dense-HGCN: projection, dense connected node-level aggregation, type-level aggregation, integration of multi-scale embeddings. Specifically, similar to [21], Dense-HGCN first applies type-specific linear transformation to project different types of nodes into the same semantic space. Then, to obtain informative embeddings in each hidden layer, Dense-HGCN combines the embeddings of the previous layers by a specific fuse function to build dense connection and perform node-level aggregation, which forms the dense connected architecture. Thereafter, it evaluates the inter-type importance by using attention mechanism [16], and performs type-level aggregation via these attentions. Finally, Dense-HGCN integrates multi-scale information with scale-attention for node classification task or other downstream tasks. Thus, Dense-HGCN can learn distinguishable comprehensive node representation based on range subgraphs.

2 Related Work

GNN models have shown great performance on homogeneous graph. GCN [11] simplifies the spectral graph convolution operation used in [2] and [3]. GAT [17] and GraphSAGE [7] aggregate neighbors based on attention mechanism and random sampling respectively. EIGNN [12] utilizes residual connection to maintain node distinguishability as the network grows deeper, while [20] and [13] design deep architecture with multi-scale information.

However, heterogeneous graph is a more realistic graph of the real world. Therefore, many research has been focused on heterogeneous graph recently [1,5]. Meta-path2vec [4] performed random walk based on meta-path to learn node embeddings. Heterogeneous graph attention network (HAN) [18] aggregated nodes of different types under the semantic space of some specific meta-paths using node-level attention and semantic-level attention simultaneously. MAGNN [6] employed relational rotation encoder to obtain the embeddings of meta-path, then performs aggregation similar to HAN. Graph transformer networks (GTN) [22] learned node embeddings by conducting convolution on multiple candidate graphs that are based on all possible meta-paths of arbitrary length, and ie-HGCN [21] considered all possible meta-paths within a limited length and evaluated the importance along the path, which further enhances the interpretability and scalability.

HPN [9] tried to alleviate the semantic confusion problem by adding the initial feature matrix, Xiong et al. [19] added similarity regularization loss and hierarchical fusion to avoid this problem. However, both of them still need predefined meta-paths, which highly depends on professional knowledge. Constructing efficient heterogeneous graph neural networks remain open.

3 Preliminary

In this section, we introduce some important concepts related to heterogeneous graph and give an example to illustrate them.

Definition 1 ***Heterogeneous Graph*** [14]. *A heterogeneous graph is denoted as $\mathcal{G} = (\mathcal{V}, \mathcal{E}, \phi, \psi)$, where $\mathcal{V}$ and $\mathcal{E}$ are the sets of nodes and edges, respectively, $\phi : \mathcal{V} \rightarrow \mathcal{A}$ and $\psi : \mathcal{E} \rightarrow \mathcal{R}$ are mapping functions that assign a node type and an edge type (relation) to each node and edge, respectively, $\mathcal{A}$ and $\mathcal{R}$ are the sets of node types and edge types, respectively, where $|\mathcal{A}| + |\mathcal{R}| > 2$.*

Definition 2 ***Adjacent Types.*** *Two node types Γ_i and Γ_j in a heterogeneous graph are adjacent types if there exists an edge type R_k such that $\psi^{-1}(R_k) \subseteq \mathcal{V}_{\Gamma_i} \times \mathcal{V}_{\Gamma_j}$, i.e., there is an edge type R_k connecting node type Γ_i and node type Γ_j. We denote the adjacent sets of type Γ_i by $\mathcal{N}_{\Gamma_i}$, i.e., $\mathcal{N}_{\Gamma_i} = \{\Gamma_j \in \mathcal{A} | \exists R_k \in \mathcal{R}, \psi^{-1}(R_k) \subseteq \mathcal{V}_{\Gamma_i} \times \mathcal{V}_{\Gamma_j}\}$.*

4 The Proposed Model

In this section, we present our model Dense-HGCN for node embedding learning on heterogeneous graph. Figure 3 demonstrates the workflow of the Dense-HGCN model.

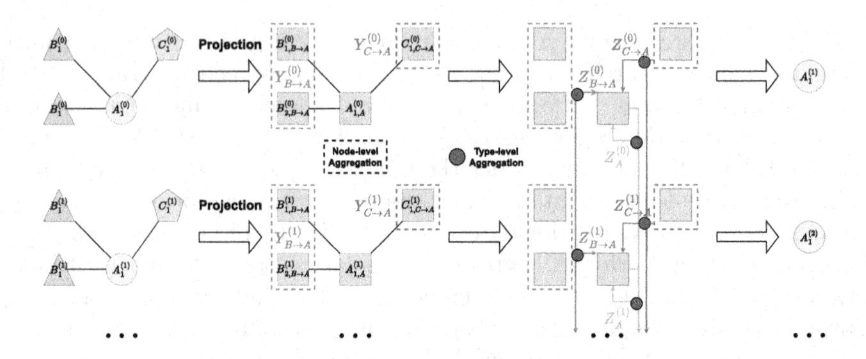

Fig. 3. Workflow of the Dense-HGCN model.

4.1 Projection

For a heterogeneous graph, different node types lie in different feature spaces or semantic spaces. Therefore, we need to apply transformations before node aggregation to make different types of node features comparable. To achieve an end-to-end model, we perform the same transformation as done in [21], which also facilitates message passing along paths of arbitrary length.

Specifically, for the node embeddings of type Γ_i in the ℓth layer $H_{\Gamma_i}^{(\ell)} \in \mathcal{R}^{n_{\Gamma_i} \times d_{\Gamma_i}^{(\ell)}}, \ell = 0, 1, \cdots, n-1$, where $H_{\Gamma_i}^{(0)}$ is the original feature matrix, and a set of its adjacent types embeddings $\{H_{\Gamma_j}^{(\ell)} \in \mathcal{R}^{n_{\Gamma_j} \times d_{\Gamma_j}^{(\ell)}}, \Gamma_j \in \mathcal{N}_{\Gamma_i}\}$, we need to project them and their adjacent nodes into the same feature space with $d_{\Gamma_i}^{\prime(\ell)}$ dimensions by multiplying transformation matrices $W_{\Gamma_i}^{(\ell)} \in \mathcal{R}^{d_{\Gamma_i}^{(\ell)} \times d_{\Gamma_i}^{\prime(\ell)}}$ and $\{W_{\Gamma_j \to \Gamma_i}^{(\ell)} \in \mathcal{R}^{d_{\Gamma_j}^{(\ell)} \times d_{\Gamma_i}^{\prime(\ell)}}, \Gamma_j \in \mathcal{N}_{\Gamma_i}\}$, respectively,

$$Y_{\Gamma_i}^{(\ell)} = H_{\Gamma_i}^{\ell} W_{\Gamma_i}^{(\ell)}, \tag{1}$$

$$Y_{\Gamma_j \to \Gamma_i}^{(\ell)} = H_{\Gamma_j}^{(\ell)} W_{\Gamma_j \to \Gamma_i}^{(\ell)}, \tag{2}$$

where $\Gamma_j \to \Gamma_i$ denotes the relation from Γ_j to Γ_i. Thus, nodes of type Γ_i and all of its adjacent nodes are located in the same feature space after transformation.

4.2 Dense Connected Node-Level Aggregation

After transforming different types of nodes into the same feature space, we can aggregate information from adjacent types through adjacency matrices of Γ_i to Γ_j, i.e., $\{A_{\Gamma_i - \Gamma_j} \in \mathcal{R}^{n_{\Gamma_i} \times n_{\Gamma_j}} | \Gamma_j \in \mathcal{N}_{\Gamma_i}\}$.

For learning more informative node embeddings, we design a densely connected aggregation rule inspired by Snowball [13] and DenseNet [8] and inherit three main merits from them: (1) reusing features of each layer, which can help to mitigate the semantic confusion phenomenon; (2) allowing the stacking of more layers to absorb long-range information of the nodes; (3) alleviating gradient vanishing problem.

In practice, we are using fusion function $f(\mathcal{M})$ to aggregate different types of nodes with multi-scale information. We will give a detailed discussion in the sequel. Define node-level aggregation as follows:

$$Z_{\Gamma_i}^{(\ell)} = f(\mathcal{M}_{\Gamma_i}^{(\ell)}) = f(Y_{\Gamma_i}^{(0)}, \ldots, Y_{\Gamma_i}^{(\ell)}) \tag{3}$$

$$Z_{\Gamma_j \to \Gamma_i}^{(\ell)} = \hat{A}_{\Gamma_i - \Gamma_j} f(\mathcal{M}_{\Gamma_j \to \Gamma_i}^{(\ell)}) = \hat{A}_{\Gamma_i - \Gamma_j} f(Y_{\Gamma_j \to \Gamma_i}^{(0)}, \ldots, Y_{\Gamma_j \to \Gamma_i}^{(\ell)}), \tag{4}$$

where $\hat{A}_{\Gamma_i - \Gamma_j} = D_{\Gamma_i - \Gamma_j}^{-1} A_{\Gamma_i - \Gamma_j}$ is a normalized adjacency matrix of type Γ_i to its adjacent type Γ_j, and $D_{\Gamma_i - \Gamma_j} \in \mathcal{R}^{n_{\Gamma_i} \times n_{\Gamma_i}}$ is the degree matrix of $A_{\Gamma_i - \Gamma_j}$, $M_{\Gamma_i}^{(\ell)}$ and $M_{\Gamma_j \to \Gamma_i}^{(\ell)}$ store the transformed multi-scale information of type Γ_i and its adjacent type Γ_j respectively. Each row of $Z_{\Gamma_j \to \Gamma_i}^{(\ell)}$ contains the information of all the node types Γ_j that are connecting with the specific node type Γ_i, i.e. the ith row of $Z_{\Gamma_j \to \Gamma_i}^{(\ell)}$ contains convolved message of node type Γ_j, which have edges with the ith node belongs to type Γ_i. Alternatively, we can view each $Z_{\Gamma_j \to \Gamma_i}^{(\ell)}$ as a part of contribution from type Γ_j to learn type Γ_i's embedding of the next layer. Before aggregating all of $\{Z_{\Gamma_j \to \Gamma_i}^{(\ell)} | \forall \Gamma_j \in \mathcal{N}_{\Gamma_i}\}$ and $Z_{\Gamma_i}^{(\ell)}$, i.e. type-level aggregation, it needs to calculate the inter-type attention which implies the importance of the adjacent type to the target type.

There are two different points between our approach and ie-HGCN. On one hand, in ie-HGCN, f takes the identity mapping function. On the other hand, we aggregate the information of ℓ layers $Y_{\Gamma_i}^{(s)}$ $(s = 1, \cdots, \ell)$, thus integrating multi-scale information.

4.3 Type-Level Aggregation

Before computing the inter-type attention, we evaluate the importance of adjacent types for each target node, since each node may have dissimilar tendency on attention with different node types. For instance, let us consider node classification task for DBLP dataset, some authors may prefer to submit their paper to some specific conference, and some of authors will keep a strong personal style on the title of the paper. Hence, they leave larger attention to nodes with type 'conference' or 'paper' respectively.

To learn the importance mentioned above, we firstly transform $Z_{\Gamma_i}^{(\ell)}$ and $Z_{\Gamma_j \to \Gamma_i}^{(\ell)}$, then compute the similarity between them and the learnable attention vector q. Moreover, since the transformed embeddings and the attention vector q will be optimized according to the labeled data. Therefore, we take the average of the attention vector (column-wise) as inter-type attention for generality.

$$e_{\Gamma_i}^{(\ell)} = mean\{tanh[(Z_{\Gamma_i}^{(\ell)} E_{\Gamma_i}^{(\ell)}) q_{\Gamma_i}^{(l)}]\}, \tag{5}$$

$$e_{\Gamma_j \to \Gamma_i}^{(\ell)} = mean\{tanh[(Z_{\Gamma_j \to \Gamma_i}^{(\ell)} E_{\Gamma_i}^{(\ell)}) q_{\Gamma_i}^{(\ell)}]\}, \tag{6}$$

where $E_{\Gamma_i}^{(\ell)} \in \mathcal{R}^{d_{\Gamma_i}^{\prime(\ell)} \times d_a^{(\ell)}}$ is a parametric transformation matrix, and $q_{\Gamma_i}^{(\ell)} \in \mathcal{R}^{d_a^{(\ell)} \times 1}$. Furthermore, we normalize the inter-type attention for numerical stability and explainability via softmax function:

$$\alpha_{\Gamma_i}^{(\ell)} = \frac{\exp(e_{\Gamma_i}^{(\ell)})}{\exp(e_{\Gamma_i}^{(\ell)}) + \sum_k \exp(e_{\Gamma_k \to \Gamma_i}^{(\ell)})}, \tag{7}$$

$$\alpha_{\Gamma_j \to \Gamma_i}^{(\ell)} = \frac{\exp(e_{\Gamma_j \to \Gamma_i}^{(\ell)})}{\exp(e_{\Gamma_i}^{(\ell)}) + \sum_k \exp(e_{\Gamma_k \to \Gamma_i}^{(\ell)})}. \tag{8}$$

Now we obtain the node embeddings of type Γ_i by type-level aggregation with inter-type attention:

$$H_{\Gamma_i}^{(\ell+1)} = \sigma\Big(\alpha_{\Gamma_i}^{(\ell)} Z_{\Gamma_i}^{(\ell)} + \sum_{\Gamma_j \in \mathcal{N}_{\Gamma_i}} \alpha_{\Gamma_j \to \Gamma_i}^{(\ell)} Z_{\Gamma_j \to \Gamma_i}^{(\ell)}\Big), \tag{9}$$

where σ is a non-linear activation function such as ReLU.

4.4 Integration of Multi-scale Embeddings

Notice that each embedding $H_{\Gamma_i}^{(\ell)}$ $(\ell = 0, \cdots, n-1)$ carries ℓ-order neighbors information for the nodes in type Γ_i, and each scale have different importance for specific learning task. Thus, to enhance the expressive power of heterogeneous graph neural networks, we consider n multi-scale embeddings $[H_{\Gamma_i}^{(0)}, \ldots, H_{\Gamma_i}^{(n-1)}]$ of a specific node type for prediction after feeding data through n layers (the input layer is the first layer). During the aggregation process, we aim to catch long-range information. The neighbors of the nodes might not be equally important. Thus, we perform a simple concatenation via attention to combine the outputs of the n layers $\{H^{(\ell)}, \ell = 0, \cdots, n-1\}$, i.e.,

$$\Big[(\beta^{(0)} H_{\Gamma_i}^{(0)}) || \ldots || (\beta^{(n-1)} H_{\Gamma_i}^{(n-1)})\Big], \tag{10}$$

where $\beta^{(i)}$ $(i = 0, \cdots, n-1)$ are learnable weights for the embeddings of each scale that learns the importance of each layer's embeddings, and $\beta^{(i)}$ are initialized uniformly with summation in 1.

For node classification task, we feed node embeddings into a linear layer and utilize softmax function subsequently:

$$softmax\Big(\Big[(\beta^{(0)} H_{\Gamma_i}^{(0)}) || \ldots || (\beta^{(n-1)} H_{\Gamma_i}^{(n-1)})\Big] W_c\Big), \tag{11}$$

where W_c is parametric matrix. Moreover, for semi-supervised node classification task, we use the cross entropy loss function to optimize the parameters through back propagation:

$$L = -\sum_{i \in \mathcal{L}} \sum_{j \in C} \mathcal{Y}_{i,j} \ln(output_{i,j}), \tag{12}$$

where $\mathcal{L}$ is a set of indices of the labeled node, C is a set of indices of the traget class that we aim to predict, and $\mathcal{Y}$ denotes the ground-truth label matrices.

4.5 Fuse Function

To fuse multi-scale information of adjacent types in node-level aggregation, we examine three candidate functions: Concatenate, Max-pooling and Mean-pooling.

$$f(\mathcal{M}) = Concat(\mathcal{M}) = (Y_{\bullet}^{(0)} || \dots || Y_{\bullet}^{(l)}), \tag{13}$$

where || denotes row-wise concatenation operator,

$$f(\mathcal{M}) = Max(\mathcal{M}) = Max(Y_{\bullet}^{(0)}, \dots, Y_{\bullet}^{(l)}), \tag{14}$$

and

$$f(\mathcal{M}) = Mean(\mathcal{M}) = Mean(Y_{\bullet}^{(0)}, \dots, Y_{\bullet}^{(l)}). \tag{15}$$

5 Experiments

In this section, we present extensive experiments to evaluate the performance of the proposed Dense-HGCN on two widely used real-world heterogeneous graph structural datasets that addressed in Table 1.

5.1 Datasets

- **DBLP**. A subset of the DBLP computer science bibliography website, it contains four types of nodes: Author (A), Paper (P), Conference (C) and Term (T). The authors are divided into four areas: database, data mining, artificial intelligence, and information retrieval. Each author is described by a bag-of-words representation of the keywords in their paper.
- **IMDB**. A subset of the Internet Movie Database (IMDB), it's a heterogeneous graph containing three types of nodes: Movie (M), Actor (A), and Director (D). The movies are divided into three classes (action, comedy, drama) according to their genre. Movie features correspond to elements of a bag-of-words representation of its plot keywords.

Table 1. Statistics of the DBLP and IMDB datasets.

Dataset	Types	Number	Features
DBLP	Author	4,057	128
	Paper	14,328	128
	Term	7,723	128
	Conference	20	128
IMDB	Actor	4,278	128
	Movie	5,257	128
	Director	2,081	128

5.2 Implementing Details

The proposed Dense-HGCN is compared with the following baselines: metapath2vec [4], HAN [18], MAGNN [6], HPN [9], and ie-HGCN [21]. We randomly divide each dataset into training set, validation set and test set where the ratio of the training set is $x\%$, and the rest of data are divided equally for validation set and test set, where $x \in \{20, 40, 60, 80\}$. Adam is considered as the default optimizer [10]. The details of the baselines and Dense-HGCN are as follows:

- **Dense-HGCN**: According to the three fuse functions described above, we consider three variants of Dense-HGCN: Dense-HGCN$_{max}$, Dense-HGCN$_{mean}$, Dense-HGCN$_{concat}$. Specifically, we set the depth of the model as 10. The dimensions of the hidden layer are the same for all the node types in a fixed layer, and we set them as 64. For the attention calculation of each layer, we set the dimension of type-level attention d_a to 64, the dimension of q to 128, the learning rate to 0.001, the dropout rate of each layer (except for the output layer) to 0.6, and ℓ_2 regularization of parameters to 0.0001, parameters are initialized by Xavier uniform distribution.
- **Baselines**: For metapath2vec, HAN, MAGNN, HPN, we employ the metapath used in [6]. For DBLP dataset, the mata-path set considered are stated as the following: APA, APTPA, APCPA, and MDM, MAM, DMD, while for IMDB dataset, DMAMD, AMA, AMDMA are selected. For ie-HGCN, we set the parameters the same as in [21].

5.3 Node Classification

To compare the performance of different methods for node classification task. Each method is run 10 times randomly, and we report the average Micro-F1 and Macro-F1 to evaluate the performance of all the models.

The experimental results are shown in Table 2. As per results in Table 2, we can see that the proposed Dense-HGCN has achieved the state-of-the-art performance on both DBLP and IMDB datasets. Based on the experimental results, three variants of Dense-HGCN perform better than all the baselines on DBLP dataset, and Dense-HGCN stay in comparable performance on IMDB dataset under the 20% training proportion. All of the three variants show no significant difference on the above mentioned two datasets. Compared with the baselines that need predefined meta-paths, e.g., metapath2vec, HAN, HPN, the end-to-end method ie-HGCN performs better, due to the reason that it considers all the length limited meta-paths and learns node embeddings along the metapath. Although long-range information may be trivial but still help to improve the classification results. We can see that Dense-HGCN performs slightly better than ie-HGCN. This maybe due to the introduction of multi-scale information, which can efficiently incorporates long-range information into the Dense-HGCN model.

Table 2. Experimental results for node classification task.

Dataset	Metric	Training	metapath2vec	HAN	MAGNN	HPN	ie-HGCN	DenseHGCN$_{concat}$	DenseHGCN$_{max}$	DenseHGCN$_{mean}$
DBLP	Micro-F1	20%	0.9353	0.9301	0.9357	0.9236	0.9400	**0.9465**	0.9436	0.9449
		40%	0.9277	0.9297	0.9374	0.9334	0.9414	**0.9491**	0.9418	**0.9488**
		60%	0.9359	0.9362	0.9429	0.9260	0.9464	**0.9531**	**0.9501**	0.9488
		80%	0.9360	0.9356	0.9447	0.9252	0.9539	**0.9567**	0.9544	**0.9567**
	Macro-F1	20%	0.9296	0.9169	0.9313	0.9174	0.9356	**0.9422**	0.9394	**0.9408**
		40%	0.9247	0.9196	0.9330	0.9282	0.9360	**0.9441**	0.9366	**0.9442**
		60%	0.9293	0.9214	0.9384	0.9192	0.9425	**0.9493**	**0.9467**	**0.9453**
		80%	0.9302	0.9250	0.9389	0.9193	0.9494	**0.9527**	**0.9500**	**0.9526**
IMDB	Micro-F1	20%	0.5272	0.5527	**0.5960**	0.5934	0.5891	0.5923	0.5921	**0.5961**
		40%	0.5588	0.6290	0.6050	0.6230	0.6320	0.6306	**0.6377**	**0.6383**
		60%	0.5432	0.6460	0.6088	0.6016	0.6704	0.6717	**0.6846**	0.6799
		80%	0.5701	0.6355	0.6153	0.6285	0.6939	0.6937	**0.7047**	0.6991
	Macro-F1	20%	0.5255	0.5547	**0.5935**	**0.5934**	0.5865	**0.5919**	**0.5920**	**0.5949**
		40%	0.5497	0.6261	0.6027	0.6224	0.6320	0.6306	**0.6379**	**0.6385**
		60%	0.5428	0.6458	0.6066	0.5893	0.6711	0.6715	**0.6845**	**0.6805**
		80%	0.5688	0.6314	0.6144	0.6175	0.6953	0.6949	**0.7054**	0.7003

5.4 Node Clustering

In addition, we conduct experiments for node clustering task. Node embeddings are obtained by each trained model, and we study the embeddings of the labeled nodes (author in DBLP and movie in IMDB) by employing K-means algorithm. Since the result of K-means is highly related to the initial centroid, we run K-Means 10 times for each model and report the average normalized mutual information (NMI) and adjusted rank index (ARI) to evaluate the performance of each model.

As the results in Table 3 demonstrated, the proposed Dense-HGCN achieves the best performance on all the datasets in terms of ARI and NMI. Especially, Dense-HGCN$_{mean}$ outperforms the best baseline by 3% on DBLP dataset, and outperform the best baseline – HAN slightly on IMDB dataset.

Table 3. Experimental results for node clustering task.

Dataset	Metric	metapath2vec	HAN	MAGNN	HPN	ie-HGCN	Dense-HGCN$_{concat}$	Dense-HGCN$_{max}$	Dense-HGCN$_{mean}$
DBLP	**ARI**	0.8060	0.8295	0.8384	0.8131	0.8256	0.8641	0.8450	**0.8656**
	NMI	0.7665	0.7749	0.7850	0.7694	0.7795	0.8084	0.7983	**0.8160**
IMDB	**ARI**	0.0635	**0.1917**	0.1703	0.1881	0.1587	0.1392	0.1858	**0.1926**
	NMI	0.0604	0.1831	0.1650	0.1807	0.1652	0.1822	**0.1964**	**0.1988**

5.5 Parameters Study

In this section, we investigate the effect of parameters on Dense-HGCN in terms of accuracy (Micro-F1) with training ratio of 60%.

Number of the Layers. We assume the dense connected architecture in the Dense-HGCN model can mitigate the semantic confusion problem. To verify this, we gradually increase the number of layers from 2 to 16 with the other parameters fixed. The experimental results are shown in Fig. 4. According to the result, we found that Dense-HGCN achieves stable performance even after stacking multiple layers. On DBLP dataset, Dense-HGCN performs well after stacking 4 layers or more and fluctuates slightly. On IMDB dataset, the performance fluctuates more, but still in a acceptable level. However, Dense-HGCN maintains high level performance when the model gets deeper, which means it can mitigate semantic confusion and be able to capture long-range information.

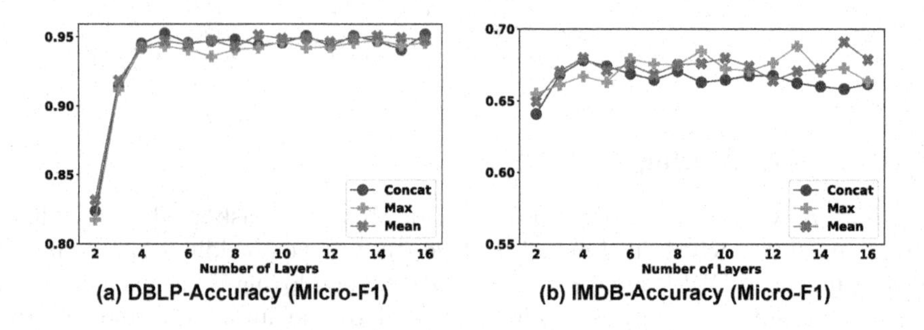

Fig. 4. Classification performance of Dense-HGCN w.r.t. the number of layers.

Dimension of the Embeddings and Attention Vector. To test the sensitivity of Dense-HGCN with respect to the dimension of the node embeddings and attention vector, we gradually increase the dimension of the node embeddings and the output size of each hidden layer from 8 to 256, and we fix the dimension of the attention vector d_a. Based on the experimental results in Fig. 5, we can see that Dense-HGCN$_{concat}$ is more sensitive to the dimension of the node embeddings than other variants. As the dimension increases, the performance of Dense-HGCN$_{concat}$ does not vary too much.

Now we consider the effect of the dimension of the attention vector q. Similarly, we gradually increase the dimension of attention vector d_a from 8 to 256 and fix the dimension of the node embeddings. As per results in Fig. 6, it's obvious that Dense-HGCN is not sensitive to the dimension of the attention vector, all the variants of Dense-HGCN perform stable w.r.t to different dimensions.

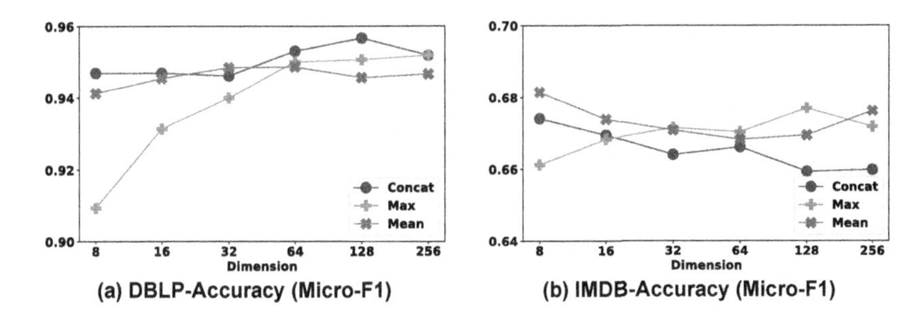

Fig. 5. Accuracy of Dense-HGCN w.r.t. the dimension of the node embeddings.

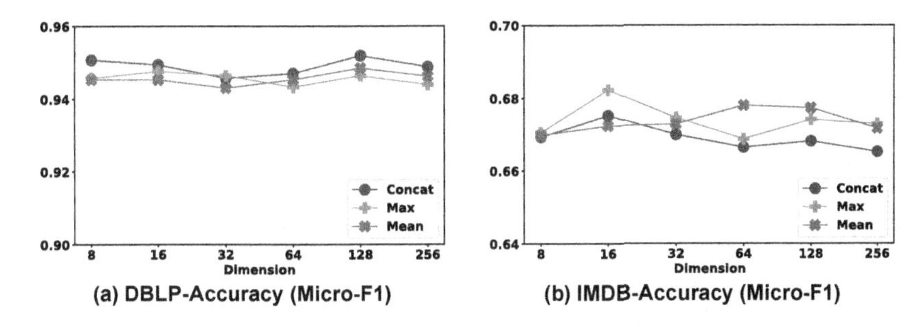

Fig. 6. Accuracy of Dense-HGCN w.r.t. the dimension of the attention vector.

6 Conclusion

In this paper, we propose Dense-HGCN to learn the node embeddings on heterogeneous graph, which can capture the information along a path of any length without predefined meta-paths, and alleviate the semantic confusion phenomenon. In each layer, it projects each type of nodes into the same semantic space that facilitates the aggregation of inter-type, and fuses the projected embeddings of previous layers by a specific fuse function, which constitutes the dense connection. Then, we perform node-level aggregation through adjacent matrix and type-level aggregation via attention mechanism. Finally, we assign a learnable vector on the embeddings of each layer for classification, which evaluates the importance of each scale information, and use them as attention for classification. Dense-HGCN achieves the state-of-the-art experimental results on both node classification and node clustering tasks.

References

1. Ahn, H., Yang, Y., Gan, Q., Moon, T., Wipf, D.P.: Descent steps of a relation-aware energy produce heterogeneous graph neural networks. In: Conference on Neural Information Processing Systems, pp. 38436–38448. NIPS, Curran Associates, Inc. (2022)

2. Bruna, J., Zaremba, W., Szlam, A., LeCun, Y.: Spectral networks and locally connected networks on graphs. In: International Conference on Learning Representation, ICLR (2014)
3. Defferrard, M., Bresson, X., Vandergheynst, P.: Convolutional neural networks on graphs with fast localized spectral filtering. In: Conference on Neural Information Processing Systems, pp. 3844–3852. NIPS, Curran Associates Inc., Red Hook (2016)
4. Dong, Y., Chawla, N.V., Swami, A.: Metapath2vec: scalable representation learning for heterogeneous networks. In: ACM International Conference on Knowledge Discovery and Data Mining, KDD, pp. 135–144. Association for Computing Machinery, New York (2017)
5. Feng, J., Wang, Z., Li, Y., Ding, B., Wei, Z., Xu, H.: MGMAE: molecular representation learning by reconstructing heterogeneous graphs with A high mask ratio. In: Hasan, M.A., Xiong, L. (eds.) ACM International Conference on Information & Knowledge Management, CIKM, pp. 509–519. ACM (2022)
6. Fu, X., Zhang, J., Meng, Z., King, I.: MAGNN: metapath aggregated graph neural network for heterogeneous graph embedding. In: International World Wide Web Conference, WWW, pp. 2331–2341. Association for Computing Machinery, New York (2020)
7. Hamilton, W.L., Ying, R., Leskovec, J.: Inductive representation learning on large graphs. In: Conference on Neural Information Processing Systems, pp. 1025–1035. NIPS, Curran Associates Inc., Red Hook (2017)
8. Huang, G., Liu, Z., Van Der Maaten, L., Weinberger, K.Q.: Densely connected convolutional networks. In: IEEE Conference on Computer Vision and Pattern Recognition, CVPR, pp. 2261–2269 (2017)
9. Ji, H., Wang, X., Shi, C., Wang, B., Yu, P.S.: Heterogeneous graph propagation network. IEEE Trans. Knowl. Data Eng. **35**(1), 521–532 (2023)
10. Kingma, D.P., Ba, J.: Adam: a method for stochastic optimization. In: International Conference on Learning Representations, ICLR (2015)
11. Kipf, T.N., Welling, M.: Semi-supervised classification with graph convolutional networks. In: International Conference on Learning Representations, ICLR (2017)
12. Liu, J., Kawaguchi, K., Hooi, B., Wang, Y., Xiao, X.: EIGNN: efficient infinite-depth graph neural networks. In: Conference on Neural Information Processing Systems, NIPS, pp. 18762–18773 (2021)
13. Luan, S., Zhao, M., Chang, X.W., Precup, D.: Break the ceiling: stronger multi-scale deep graph convolutional networks. In: Conference on Neural Information Processing SystemsNIPS, , pp. 10943–10953. Curran Associates, Inc. (2019)
14. Sun, Y., Han, J.: Mining heterogeneous information networks: a structural analysis approach. ACM SIGKDD Explor. Newsl. **14**(2), 20–28 (2013)
15. Sun, Y., Han, J., Yan, X., Yu, P.S., Wu, T.: Pathsim: meta path-based top-k similarity search in heterogeneous information networks. Proc. VLDB Endow. **4**(11), 992–1003 (2011)
16. Vaswani, A., et al.: Attention is all you need. In: Conference on Neural Information Processing Systems, NIPS, pp. 6000–6010. Curran Associates Inc., Red Hook (2017)
17. Veličković, P., Cucurull, G., Casanova, A., Romero, A., Lió, P., Bengio, Y.: Graph attention networks. In: International Conference on Learning Representations, ICLR (2018)
18. Wang, X., et al.: Heterogeneous graph attention network. In: World Wide Web Conference, WWW, pp. 2022–2032. Association for Computing Machinery, New York (2019)

19. Xiong, Z., Cai, J.: Deep heterogeneous graph neural networks via similarity regularization loss and hierarchical fusion. In: IEEE International Conference on Data Mining Workshops, ICDMW, pp. 759–768(2022)
20. Xu, K., Li, C., Tian, Y., Sonobe, T., Kawarabayashi, K., Jegelka, S.: Representation learning on graphs with jumping knowledge networks. In: International Conference on Machine Learning, ICML, vol. 80, pp. 5449–5458. PMLR (2018)
21. Yang, Y., Guan, Z., Li, J., Zhao, W., Cui, J., Wang, Q.: Interpretable and efficient heterogeneous graph convolutional network. IEEE Trans. Knowl. Data Eng. **35**(2), 1637–1650 (2023)
22. Yun, S., Jeong, M., Kim, R., Kang, J., Kim, H.J.: Graph transformer networks. In: Neural Information Processing Systems. Curran Associates Inc., Red Hook (2019)

Actor-Critic with Variable Time Discretization via Sustained Actions

Jakub Łyskawa[1(✉)] and Paweł Wawrzyński[2]

[1] Warsaw University of Technology, Pl. Politechniki 1, 00-661 Warsaw, Poland
jakub.lyskawa.dokt@pw.edu.pl

[2] Ideas NCBR, ul. Chmielna 69, 00-801 Warsaw, Poland

Abstract. Reinforcement learning (RL) methods work in discrete time. In order to apply RL to inherently continuous problems like robotic control, a specific time discretization needs to be defined. This is a choice between sparse time control, which may be easier to train, and finer time control, which may allow for better ultimate performance. In this work, we propose SusACER, an off-policy RL algorithm that combines the advantages of different time discretization settings. Initially, it operates with sparse time discretization and gradually switches to a fine one. We analyze the effects of the changing time discretization in robotic control environments: Ant, HalfCheetah, Hopper, and Walker2D. In all cases our proposed algorithm outperforms state of the art.

Keywords: reinforcement learning · frame skipping · robotic control

1 Introduction

Reinforcement Learning (RL) is an area of machine learning that focuses on maximizing expected rewards' sum in the Markov Decision Process [25]. Such approach may be applied to difficult problems such as robotic control, video games, and healthcare [7,24,31]. It may result in control policy that is robust to unpredicted events and is able to solve control problems that are difficult or impossible to be solved by human engineers [20].

The interaction with the environment is typically assumed to be the most expensive part of the RL process. Thus, when comparing RL methods, an important aspect to consider is the sample efficiency. It is defined as the speed of the learning process with respect to the number of training samples collected. An algorithm with higher sample efficiency will achieve a desired policy using less environment data and, given a specific amount of data, such algorithm will likely obtain a better policy [6,15,27].

Reinforcement learning algorithms work in discrete time. It is natural when considering setting that are naturally discrete, such as video games or healthcare [18,31]. However, reinforcement learning is applied often to the problems that are continuous and it requires time discretization of the control process. Recent research shows that while finer discretization allows us to obtain better

B. Luo et al. (Eds.): ICONIP 2023, LNCS 14447, pp. 476–489, 2024.
https://doi.org/10.1007/978-981-99-8079-6_37

results, it is much more difficult, especially for algorithms that were not prepared specifically for fine time discretization setting [28,33].

In this work, we aim to utilize the benefits of both a simpler learning process with coarser time discretization and the possibility to obtain better policies using finer discretization. For this purpose, we propose an algorithm that can use experience collected by policy working in different discretization. The underlying policy uses a stochastic process to control current discretization by sustaining actions with a given probability. This approach separates the environment discretization, which is the finest discretization available, from the agent's discretization.

The contribution of this paper can be summarized as follows:

1. We introduce a framework for manipulating discretization during the reinforcement learning process via sustained actions.
2. We introduce an algorithm based on the Actor-Critic with Experience Replay that utilizes variable discretization. We call this algorithm Actor-Critic with Experience Replay and Sustained actions (SusACER).
3. We provide experimental results that compare the SusACER algorithm with state-of-the-art RL algorithms on simulated robotic control environments with continuous action spaces.

2 Problem Formulation

We consider typical reinforcement learning in a Markov Decision Process that is built on an underlying continous control process.

At the time step t the environment is in a state, s_t. An agent interacts with the environment by performing an action, a_t, at each time step. It causes the environment to change its state to s_{t+1} and the agent receives a reward, r_t. An episode is a single run of the agent-environment interaction, from an initial state to a terminal state.

The agent performs actions according to its policy $\pi(a|s)$ that determines the probability of each action a in the given state s.

The policy is optimized to maximize the expected total rewards' sum throughout an episode. A practical way to do that is to maximize the expected discounted rewards' sum $E(\sum_{i=0} \gamma^i r_{t+i} | x_t = x; \pi)$ for each state x in the state space, where $\gamma \in (0, 1)$ is a discount factor.

We note that the actor should select an optimal action in each time step to maximize the expected rewards sum. As such, we assume that the final policy obtained in a reinforcement learning process should make decisions in each time step to allow the best performance.

We assume that the Markov Decision Process is built on an underlying continuous control process. This control process is discretized in the time domain, making each time step of the environment last for a given short period of time. Our goal is to design an efficient learning algorithm for this setting.

3 Related Work

Actor-Critic Algorithms. The Actor-Critic approach to reinforcement learning was first introduced by Barto et al. [2]. The approach to use approximators to estimate discounted rewards sum was proposed by Kimura and Kobayashi [11]. The Actor-Critic with Experience Replay was introduced in [30] as an algorithm that combines the Actor-Critic structure with offline learning via replaying variable-length sequences of samples, called trajectories, stored in a buffer. This algorithm uses importance sampling to solve the problem of using trajectories obtained using different policies. [26] introduce constant length trajectories and soft truncation of importance sampling. Many state-of-the-art algorithms used for robotic control problems, such as Soft Actor-Critic (SAC) [9] and Proximal Policy Optimization (PPO) [22], use Actor-Critic structures.

Structured Exploration for Robotic Settings. Lillicrap et al. [14] introduced the Deep Deterministic Policy Gradient (DDPG) algorithm and shows the need for structured exploration in robotic environments. DDPG algorithm uses the Uhlenbeck-Ornstein process [29] to generate temporally correlated noise. Tallec et al. [28], Szulc et al. [26], and Łyskawa and Wawrzyński [33] show the importance of structured exploration for fine discretization controlling physical objects. Łyskawa and Wawrzyński [33] show that in fine discretization setting reinforcement learning algorithms should employ multiple-step trajectories for calculating approximators' updates.

Environment Discretization and Sustained Actions. Sustaining actions over constant-length number of frames was first introduced by Mnih et al. [18] for ATARI environments to reduce the number of times the policy has to be calculated in a setting where environment steps are relatively fast. This approach is used in later works as a standard preprocessing for Atari environments [13,19]. Kalyanakrishnan et al. [10] noted that for many video game environments, a higher frame skip parameter allows obtaining a higher score. [28] applies the Uhlenbeck-Ornstein process to the underlying values for calculating discrete action probabilities to provide a method for temporally-correlated actions in discrete action space settings. Dabney et al. [5] proposed ϵz-greedy exploration as a temporal extension of ϵ-greedy exploration, where the action duration is selected from a given distribution z to increase the probability of finding states outside policies similar to the greedy policy. However, this approach assumes action-value function estimation and single-step updates, which makes it not easily transferrable to algorithms that use value function estimation.

Learning Optimal Action Duration. Lakshminarayanan et al. [13] introduced Dynamic Action Repetition. This method works by including actions extended by a given number of steps in the available action space. Mann et al. [16] introduced Fitted Value Iteration algorithm. Similarly to the Dynamic Action Repetition, it extends the environment action space. It is however not limited to actions with increased duration. Instead it utilizes the framework of options, which are general sequences of actions and include both simple actions and actions with increased duration. Biedenkapp et al. [3] introduced a method based on the Q-Learning, called TempoRL, where action duration was introduced to the action space, resulting in an approach similar to some hierarchical reinforcement learning approaches [8]. Sharma et al. [23] introduced method called Figar, which uses an additional model to select one of the predefined action lengths. Yu et al. [32] introduced Temporally Abstract Actor-Critic, that includes additional model for determining if an action should be sustained. However, these works assume that the trained agent would make decisions only in selected time steps. Metelli et al. [17] points out that reducing control frequency results in performance loss. Thus, in this work we assume that outside training the agent selects the optimal action in each time step.

Action-Value Based RL in Fine-Time Discretization. Park et al. [21] utilise sustained actions to allow the usage of reinforcement learning algorithms that use action-value function estimators in fine-time discretization. Baird [1] notes that without increasing action duration the action-value function degrades to the value function, as the effect of a single very short action becomes negligible. This problem does not occur when using reinforcement learning algorithms that use the value function estimator [28], such as Actor-Critic with Experience Replay [30]

Summary. Most of the existing methods utilising action sustain use action-value function estimators and single step updates. As such, they are not as well-suited to fine-time discretization problems as algorithms that use value function estimators.

4 Variable Discretization

In this work, we consider two-level time discretization. The base discretization $T = \{1, 2, 3, ...\}$ is the finest available environment discretization, further referred to as environment discretization. The second discretization T_a is called the *agent discretization.* A single *agent time step* lasts for several *environment time steps.* The distribution of length of the *agent time step* at the *environment time step* t is determined by a geometric distribution[1] with a success (action finish) probability parameter p_t. In the geometric distribution, the probability of sustaining current action is the same regardless of how long the action already lasts, which is an useful property. The action selected at the beginning of an *agent time step* is sustained for the whole duration of the *agent time step.* We denote the expected duration of a sustained action as

$$E_t = 1 + \frac{1 - p_t}{p_t} = \frac{1}{p_t} \tag{1}$$

[1] Defined as the number of failures before the first success.

E_t is greater by 1 than the expected value of the geometric distribution as it also includes the environment step when the agent chooses the action.

Generally, p_t increases with t to 1, which means that the expected duration of actions decreases to 1. Initially, the actions are longer, and shorter combinations of them lead to high expected rewards. The space of these combinations is smaller, and the agent requires less experience to search it, thereby learning faster. Having learned to choose long-lasting actions, the agent is in a good position to learn dexterous behavior based on short-lasting actions.

We denote the underlying base agent policy $\pi_a(a|s;\theta)$, where θ is the vector of the policy parameters. It determines the probability of the action a being selected in the state s of the environment. The environment-time-step-level policy π is thus defined as

$$\pi(a|s_t, a_{t-1}; p_t, \theta_t) = \begin{cases} \pi_a(a|s_t;\theta_t) & \textit{if the agent must choose an action} \\ (1-p_t)U(a|a_{t-1}) + p_t\pi_a(a|s_t;\theta_t) & \textit{otherwise,} \end{cases} \tag{2}$$

where $U(a|a_{t-1})$ is a probability distribution of sustained action a_{t-1}, resulting in $a_t = a_{t-1}$. For discrete action space $U(a|a_t) = 1$ *if* $a = a_t$ *else* 0. For continuous action space $U(a|a_t) = \delta(a - a_t)$ where δ is a Dirac delta in the environment action space. The agent must choose an action if the previous action cannot be sustained, e.g. at the beginning of the episode.

The process described above in both the environment time discretization and the agent time discretization is Markovian. However, the policy in the environment discretization depends on both previous action and state.

4.1 Trajectory Importance Sampling

We consider a trajectory, $s_t, a_t, s_{t+1}, a_{t+1}, ...s_{t+n}$, where $t \in T_a$, i.e., a_t was selected from the actor's action probability distribution $\pi_a(\cdot|s_t)$. The following actions are selected or sustained independently in each environment step, thus the importance sampling of a trajectory is a product of density ratios for each environment step within this trajectory:

$$IS_t^n = \frac{\pi_a(a_t|s_t;\theta)}{\pi_a(a_t|s_t;\theta_t)} \prod_{\tau=t+1}^{t+n-1} \frac{\pi(a_\tau|s_\tau, a_{\tau-1}; p, \theta)}{\pi(a_\tau|s_\tau, a_{\tau-1}; p_\tau, \theta_\tau)}, \tag{3}$$

for registered data indexed with t and τ, and current policy parameter θ and success parameter p.

For discrete action distribution, all probabilities in Eq. 3 are finite. However, for continuous action spaces for $a_\tau = a_{\tau-1}$ and success parameter p the environment-time-step-level action probability density is equal to $(1-p)\delta(0) + p\pi_a(a|s_\tau;\theta_\tau)$. For $p < 1$ the expression $(1-p)\delta(0)$ is infinite. However, for $p_\tau < 1$ the infinite part is in both the nominator and denominator of this expression and the density ratio reduces to the following form.

$$\frac{\pi(a_\tau|s_\tau, a_{\tau-1}; p, \theta)}{\pi(a_\tau|s_\tau, a_{\tau-1}; p_\tau, \theta_\tau)} = \begin{cases} \frac{1-p}{1-p_\tau} & \text{iff } a_\tau = a_{\tau-1} \\ \frac{p}{p_\tau}\frac{\pi_a(a_\tau|s_\tau;\theta)}{\pi_a(a_\tau|s_\tau;\theta_\tau)} & \text{otherwise} \end{cases} \tag{4}$$

If the action is not sustained, the density ratio for the time step τ is equal to the densities ratio of the actor's probability distributions multiplied by the ratio of the probabilities that the actor will select the action. If the action is sustained, the importance sampling for the time step τ is equal to the ratio of probabilities that the action will be sustained. This value is greater than 0 for $p < 1$ and equal to 0 for $p = 1$. We can safely ignore the case when $a_\tau = a_{\tau-1}$ for $p_\tau = 1$ as the probability of drawing the same action from a continuous distribution twice is equal to 0.

As such, each sequence of sustained actions is non-negligible as long as $p < 1$. Furthermore, as we assume that the agent selected the first action of the trajectory from the underlying base agent policy π_a, a part of the trajectory is feasible even for $p = 1$. As such, the experience collected with any $p_t \in (0, 1]$ can be feasibly replayed for any other $p \geq p_t$.

4.2 Adaptation of Exploration to Sustained Actions

Sustaining actions over a number of steps increases the intensity of exploration [5]. In order to keep the exploration at the level defined by the underlying policy in control settings we propose the following solution. Let's assume that the underlying system is a Markovian continuous-time control process with continuous state and action spaces. Given continuous time τ and state s_τ, it can be described by a differential equation

$$\frac{ds_\tau}{d\tau} = F(s_\tau, a_\tau) \tag{5}$$

where a_τ is the action.

Let us assume that in short time $[\tau, \tau+\Delta]$ the F function can be approximated by an affine function,

$$F(s_\tau, a_\tau) \cong B + Ca_\tau. \tag{6}$$

If a single action with covariance matrix Σ is executed in time $[\tau, \tau + \Delta]$, the covariance matrix Σ_s of the state difference $s_{\tau+\Delta} - s_\tau$ equals

$$\Sigma_s = C\Sigma C^T \Delta^2. \tag{7}$$

However, if in time τ to $\tau + \Delta$ a sequence of n independent actions that have covariance matrices Σ' is performed, then the covariance Σ'_s of the state difference $s_{\tau+\Delta} - s_\tau$ equals

$$\Sigma'_s = nC\Sigma' C^T \left(\frac{\Delta}{n}\right)^2 = \frac{1}{n} C\Sigma' C^T \Delta^2 \tag{8}$$

Hence, for a constant action covariance, the amount of randomness in a state increases when the actions are sustained longer. However, we want to keep this amount of randomness in the state similar regardless of how long actions are sustained. In this order, we set the covariance of the action distribution inversely proportional to the expected time of sustaining actions.

5 SusACER: Sustained-Actions Actor-Critic with Experience Replay

We base our proposed SusACER algorithm on Actor-Critic with Experience Replay (ACER) [30]. We selected ACER as the base algorithm as it matches multiple requirements for efficient reinforcement learning in robotic control settings, namely uses state-dependant discounted rewards sum estimator, multiple-step updates, and experience replay. It was also demonstrated in [33] that it performs well in different discretization settings. As opposed to the original ACER algorithm, SusACER uses n-step returns for a constant n and soft truncation of density ratios, as proposed in [33].

SusACER uses two parameterized models, namely Actor and Critic. Actor specifies a policy, $\pi_a(\cdot|s;\theta)$. It takes as input the environment state s and it is parameterized by θ. Critic $V(s;\nu)$ estimates the discounted rewards sum for each state s and is parameterized by ν.

At each *environment time step* t the agent chooses an action according to the environment-level policy π. Then the experience samples $\langle s_t, a_t, r_t, s_{t+1}, \pi(a_t|s_t, a_{t-1}; p_t, \theta_t), \pi_a(a_t|s_t;\theta_t)\rangle$ are stored in the memory buffer of size M.

At each learning step the algorithm takes a trajectory of n samples starting at $\tau \in [t-M, t-n] \cap T_a$ and calculates updates $\Delta\theta$ and $\Delta\nu$ of parameters θ and ν.

The algorithm calculates m-step estimates of the temporal difference

$$A_\tau^m = \sum_{i=0}^{m-1} \gamma^i r_{\tau+i} + \gamma^m V(s_{\tau+m};\nu) - V(s_\tau;\nu) \tag{9}$$

for $m = 1, 2, \ldots, n$. To mitigate the non-stationarity bias, temporal difference estimates are weighted by importance sampling. Weights ρ_τ^m for each m-step estimate correspond to the change of the probability of the given experience trajectory according to Eq. 3. Following [26], we apply a soft-truncation function $\psi_b(x) = b\tanh(\frac{x}{b})$ to the calculated weights to improve the stability of the algorithm. Thus, the weight for an m-step estimate is given as

$$\rho_\tau^m = b\tanh(IS_\tau^m/b) \tag{10}$$

The algorithm calculates the unbiased temporal difference estimate d_τ^m for a sampled trajectory as an average of the m-step temporal difference estimates A_τ^m weighted by ρ_τ^m.

The algorithm calculates the update $\Delta\nu$ to train Critic to estimate the value function and the update $\Delta\theta$ to train Actor to maximize the expected discounted rewards' sum. The complete algorithm to calculate the updates is presented in Algorithm 1. We use ADAM [12] to apply the updates to the θ and ν parameters.

6 Empirical Study

In this section we present empirical results that show the performance of the SusACER algorithm. As benchmark problems we use a selection of simulated

Algorithm 1 Calculating parameters update from a single trajectory in Actor-Critic with Experience Replay and Sustained actions

Input: a trajectory of length n beginning at time step τ
Output: parameter updates $\Delta\theta$ and $\Delta\nu$

1: **for** $m \in \{1, 2, ..., n\}$ **do**
2: $\quad A_\tau^m \leftarrow \sum_{i=0}^{m-1} \gamma^i r_{\tau+i} + \gamma^m V(s_{\tau+m}; \nu) - V(s_\tau; \nu)$
3: $\quad$ Calculate IS_τ^m according to Eq. 3
4: $\quad \rho_\tau^m \leftarrow \psi_b(IS_\tau^m)$
5: **end for**
6: $d_\tau^n = \frac{1}{n} \sum_{m=1}^{n} A_\tau^m \rho_\tau^m$
7: $\Delta\nu \leftarrow \nabla_\nu V(s_\tau; \nu) d_\tau^n$
8: $\Delta\theta \leftarrow \nabla_\theta \ln \pi(a_t|s_\tau; \theta) d_\tau^n$

robotic environments, specifically Ant, HalfCheetah, Hopper and Walker2D. In our experiments we use the open source multiplatform PyBullet simulator [4].

On all benchmark problems we run experiments for $3 \cdot 10^6$ *environment time steps*. Each $3 \cdot 10^4$ steps we freeze the weights and evaluate the trained agents for 5 episodes. Learning curves in this section present the average results of evaluation runs over multiple runs and their standard deviations. For algorithm comparison we use the final obtained results and the area under the learning curve (AULC). AULC value is less influenced by noise and better reflects the learning speed.

We compare the results obtained using SusACER algorithm to the base ACER algorithm with constant trajectory length and two state-of-the-art algorithms, namely Soft Actor-Critic (SAC) [9] and Proximal Policy Optimization (PPO) [22]. We use the optimized hyperparameter values for SAC, PPO and ACER as provided in [33]. However, as ACER used in this study differs from ACER used in [33] by using constant trajectory length, we optimized the trajectory length with possible values set $\{2, 4, 8, 16, 32\}$ and the learning rates with possible values $\{1 \cdot 10^{-4}, 3 \cdot 10^{-4}, 1 \cdot 10^{-5}, 3 \cdot 10^{-5}, 1 \cdot 10^{-6}\}$.

For SusACER we used the same hyperparameters as for ACER where possible. We use the following environment discretization. The expected action sustain length E_t, as defined in Eq. 1, decreases linearly from E_0 to 1 over T_E steps. Specifically,

$$E_t = E_0 + (1 - E_0) \min\left\{\frac{t}{T_E}, 1\right\} \tag{11}$$

which directly translates into p_t (1). We use E_t instead of p_t as a parameter as we believe that it is more intuitive.

For SusACER we limit the maximum sustain length to the length of the trajectory used for calculating weight updates to avoid collecting and storing samples that would not be used for the training process.

Source code for experiments that we present in this section is available on github[2]. We list all hyperparameter settings in the Appendix A.

[2] https://github.com/lychanl/acer-release/releases/tag/SusACER.

6.1 Ablation Study

We present results that show the impact of different discretization settings for SusACER.

We ran experiments using 3 different initial expected action length values E_0, namely 2, 4, and 8. We also tested 3 different expected action length decrease times T_E, namely $3 \cdot 10^4$, $1 \cdot 10^5$, and $3 \cdot 10^5$.

Table 1 shows final results and AULC for these experiments. For Ant, the best results and AULC values are obtained for shorter sustain probability decay times and rather lower E_0 values. For HalfCheetah, the results vary, with the best results and AULCS obtained for medium E_0 value and long T_E value. For Hopper, the results are similar for all settings, with slightly better results for smaller initial expected action lengths. For Walker2D the best AULC values are obtained for smaller values of T_E, however the results have large standard deviation values.

For comparison with other algorithms we selected the discretization settings with the largest AULC values. Highest AULC values match the highest final results for all environments except of the Walker2D and for most discretization settings has lower standard deviation than the final result.

Table 1. Results and areas under the learning curves for Ant, HalfCheetah, Hopper, Walker2D for SusACER with different discretization settings. The bolded results have the highest AULC value and thus are used for comparison with other algorithms.

E_0	T_E	Ant		HalfCheetah		Hopper		Walker2D	
		Result	AULC	Result	AULC	Result	AULC	Result	AULC
2	$3 \cdot 10^4$	3311 ±218	2698 ±146	2837 ±444	2351 ±408	2218 ±315	2268 ±126	1477 ±699	1283 ±459
2	$1 \cdot 10^5$	3403 ±83	2730 ±86	2426 ±908	1935 ±705	2486 ±236	2278 ±133	**2059 ±520**	**1481 ±292**
2	$3 \cdot 10^5$	3274 ±128	2616 ±130	2558 ±668	1929 ±779	**2551 ±67**	**2357 ±113**	1061 ±721	1041 ±196
4	$3 \cdot 10^4$	**3427 ±244**	**2775 ±131**	2911 ±303	2261 ±419	2457 ±108	2287 ±65	1885 ±950	1367 ±483
4	$1 \cdot 10^5$	3351 ±167	2683 ±199	2699 ±470	1932 ±441	2551 ±133	2273 ±108	1856 ±923	1195 ±375
4	$3 \cdot 10^5$	3217 ±140	2532 ±107	**3059 ±151**	**2501 ±195**	2466 ±44	2131 ±85	1914 ±811	1275 ±171
8	$3 \cdot 10^4$	3281 ±216	2723 ±185	2887 ±385	2406 ±264	2310 ±339	2261 ±78	1768 ±778	1192 ±354
8	$1 \cdot 10^5$	3185 ±260	2374 ±218	2882 ±381	2289 ±444	2382 ±282	2259 ±154	2012 ±595	1177 ±236
8	$3 \cdot 10^5$	3301 ±384	2577 ±173	2682 ±413	1862 ±406	2418 ±187	2200 ±115	2228 ±399	1322 ±180

6.2 Experimental Results and Discussion

We compare the results obtained using SusACER algorithm to the results of the ACER, SAC, and PPO algorithms. Figure 1 shows learning curves for these algorithms. Table 2 shows the final results and AULC for these experiments.

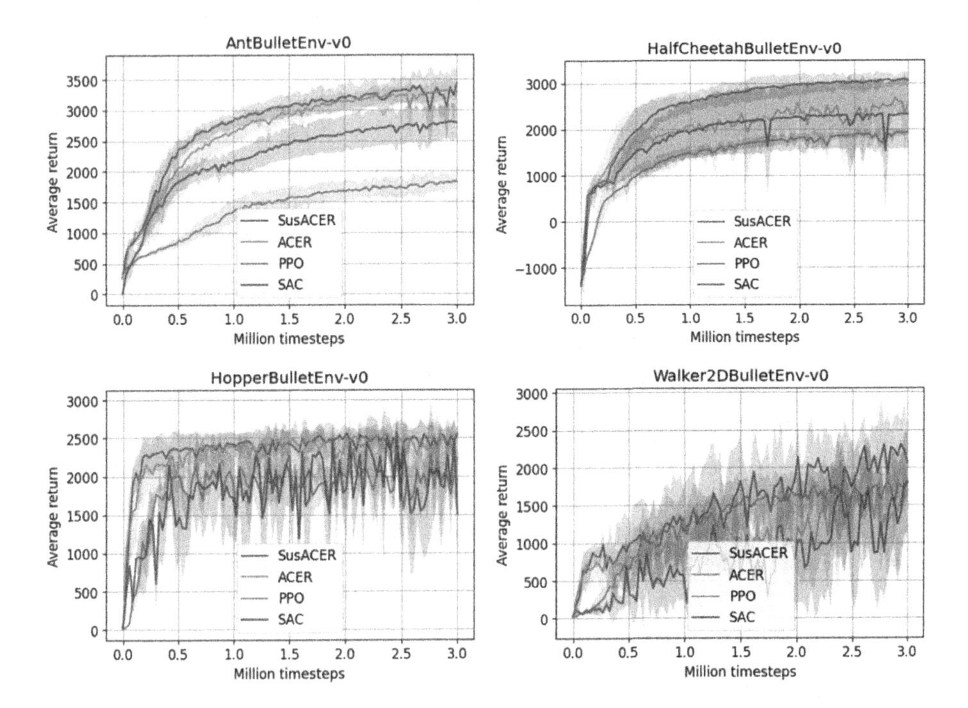

Fig. 1. Learning curves for SUSACER, SAC, PPO and FastACER for Ant (upper left), HalfCheetah (upper right), Hopper (lower left) and Walker2D (lower right) environments

SusACER obtains high results for all 4 environments. For HalfCheetah and Walker2D, it outperforms other algorithms by a large margin in terms of both training speed and final obtained results. For Ant and Hopper, it obtains similar final result as ACER. However, SusACER learns faster in the initial part of the training, which is reflected by higher AULC values.

When compared to the results obtained by ACER, the results obtained using SusACER with different inital discretizations and decay times are, for most combinations, similar or better than the results obtained by the ACER agloritham. It shows that the action sustain at the beginning of the training may easily improve the performance in simulated robotic problems.

The results presented in this section show that the impact of discretization setting may vary for each environment. For some environmnets, like Hopper,

Table 2. Results and areas under the learning curves for Ant, HalfCheetah, Hopper, Walker2D for SusACER, ACER, SAC and PPO. The bolded values are the highest final results and AULCs

	Ant		HalfCheetah		Hopper		Walker2D	
	Result	AULC	Result	AULC	Result	AULC	Result	AULC
SusACER	**3427 ±244**	**2775 ±131**	**3059 ±151**	**2501 ±195**	**2551 ±67**	**2357 ±113**	**2059 ±520**	**1481 ±292**
ACER	3289 ±124	2664 ±156	2562 ±362	2005 ±567	2230 ±412	2308 ±122	1700 ±578	1051 ±320
SAC	2788 ±263	2233 ±194	2329 ±753	1890 ±566	1496 ±664	1791 ±179	1801 ±674	785 ±160
PPO	1820 ±153	1385 ±140	1931 ±83	1417 ±95	1941 ±441	1816 ±84	1790 ±135	1271 ±215

this impact is negligible. For other, like HalfCheetah, correct discretization setting may greatly contribute to the algorithm performance. However, even if the impact is low, it may increase the speed of the learning process.

The optimal discretization setup varies between the environments. All tested environments require relatively fine discretization (with initial values of E_0 equal to 2 or 4), with more sensitive simulations, Hopper and Walker2D, requiring lower value than the two easier problems, Ant and HalfCheetah.

7 Conclusions and Future Work

In this paper, we have introduced SusACER, a reinforcement learning algorithm that manipulates time discretization to maximize learning speed in its early stages while simultaneously increasing the final results. In the early stages, the actions effectively last longer, which makes their sequences until the goal is reached shorter, and thus makes them easier to optimize. Eventually, the timespan of actions is reduced to their nominal length to allow finer control. Our experimental study with the robotic-like environments Ant, HalfCheetah, Hopper, and Walker2D confirms that this approach reaches its objectives: SusACER proves more efficient than state-of-the-art algorithms by a significant margin.

In this study, our approach to manipulating time discretization was combined with one of the most basic RL algorithms with experience replay, still giving high performance gain. The combination with other algorithms, such as SAC, could result in an even more efficient method.

We also show that optimal discretization varies between the environments. A possible next step in the research of the variable discretization setting would be to create a method to determine optimal discretization.

Ethical Statement. This work does not focus on processing personal data. The novel solutions presented in this paper cannot be directly used to collect, process, or infer personal information. We also believe that reinforcement learning methods, including SusACER, are currently not viable solutions for control processes used for policing or the military. This work does not have any ethical implications.

A Hyperparameters

In this section we provide hyperparameters used to obtain results in the Sect. 6. Table 3 contains common parameters for the offline algorithms, namely for SusACER, ACER and SAC. Table 4 contains shared parameters for SusACER and ACER algorithms. Tables 5 and 6 contain hyperparameters for SAC and PPO, respectively. Table 7 contains environment-specific reward scaling parameter values for the SAC algorithm.

Table 3. Common parameters for offline algorithms (SusACER, ACER, SAC).

Parameter	Value
Memory size	10^6
Minibatch size	256
Update interval	1
Gradient steps	1
Learning start	10^4

Table 4. SusACER and ACER hyperparameters.

Parameter	Value
Action std. dev.	0.4
Trajectory length n	4
b	3
Actor step-size	$3 \cdot 10^{-5}$
Critic step-size	10^{-4}

Table 5. SAC general hyperparameters. For environment-specific hyperparameters see Tab. 7

Parameter	Value
Target smoothing coef. τ	0.005
Learning start	10^4

Table 6. PPO hyperparameters.

Parameter	Value
GAE parameter (λ)	0.95
Minibatch size	64
Horizon	2048
Number of epochs	10
Value function clipping coef	10
Target KL	0.01
Step-size	$3 \cdot 10^{-4}$
Clip param	0.2

Table 7. SAC reward scaling.

Parameter	Value
Reward scaling for HalfCheetah env.	0.1
Reward scaling for Ant env.	1
Reward scaling for Hopper env.	0.03
Reward scaling for Walker2D env.	30

References

1. Baird, L.: Reinforcement learning in continuous time: advantage updating. In: Proceedings of 1994 IEEE International Conference on Neural Networks (ICNN 1994). vol. 4, pp. 2448–2453 (1994). https://doi.org/10.1109/ICNN.1994.374604
2. Barto, A.G., Sutton, R.S., Anderson, C.W.: Neuronlike adaptive elements that can learn difficult learning control problems. IEEE Trans. Syst. Man Cybern. B **13**, 834–846 (1983)
3. Biedenkapp, A., Rajan, R., Hutter, F., Lindauer, M.: Temporl: learning when to act. CoRR abs/2106.05262 (2021). https://arxiv.org/abs/2106.05262
4. Coumans, E., Bai, Y.: Pybullet, a python module for physics simulation for games, robotics and machine learning (2016–2021). https://pybullet.org
5. Dabney, W., Ostrovski, G., Barreto, A.: Temporally-extended (ϵ)-greedy exploration. CoRR abs/2006.01782 (2020). https://arxiv.org/abs/2006.01782
6. Dulac-Arnold, G., et al.: Challenges of real-world reinforcement learning: definitions, benchmarks and analysis. Mach. Learn. **110**(9), 2419–2468 (2021)
7. ElDahshan, K.A., Farouk, H., Mofreh, E.: Deep reinforcement learning based video games: a review. In: 2022 2nd International Mobile, Intelligent, and Ubiquitous Computing Conference (MIUCC), pp. 302–309 (2022). https://doi.org/10.1109/MIUCC55081.2022.9781752
8. Gürtler, N., Büchler, D., Martius, G.: Hierarchical reinforcement learning with timed subgoals (2021)
9. Haarnoja, T., Zhou, A., Abbeel, P., Levine, S.: Soft actor-critic: offpolicy maximum entropy deep reinforcement learning with a stochastic actor (2018). arXiv:1801.01290
10. Kalyanakrishnan, S., et al.: An analysis of frame-skipping in reinforcement learning (2021)
11. Kimura, H., Kobayashi, S.: An analysis of actor/critic algorithms using eligibility traces: reinforcement learning with imperfect value function. In: ICML (1998)
12. Kingma, D.P., Ba, J.: Adam: a method for stochastic optimization. In: Bengio, Y., LeCun, Y. (eds.) 3rd International Conference on Learning Representations, ICLR 2015, San Diego, CA, USA, 7–9 May 2015, Conference Track Proceedings (2015). https://arxiv.org/abs/1412.6980
13. Lakshminarayanan, A., Sharma, S., Ravindran, B.: Dynamic action repetition for deep reinforcement learning. In: Proceedings of the AAAI Conference on Artificial Intelligence, vol. 31, no. 1 (2017). https://doi.org/10.1609/aaai.v31i1.10918. https://ojs.aaai.org/index.php/AAAI/article/view/10918
14. Lillicrap, T.P., et al.: Continuous control with deep reinforcement learning (2016), arXiv:1509.02971
15. Liu, R., Nageotte, F., Zanne, P., de Mathelin, M., Dresp-Langley, B.: Deep reinforcement learning for the control of robotic manipulation: a focussed mini-review. Robotics **10**(1) (2021). https://doi.org/10.3390/robotics10010022. https://www.mdpi.com/2218-6581/10/1/22
16. Mann, T.A., Mannor, S., Precup, D.: Approximate value iteration with temporally extended actions. In: Proceedings of the 26th International Joint Conference on Artificial Intelligence, IJCAI 2017, pp. 5035–5039. AAAI Press (2017)
17. Metelli, A.M., Mazzolini, F., Bisi, L., Sabbioni, L., Restelli, M.: Control frequency adaptation via action persistence in batch reinforcement learning. CoRR abs/2002.06836 (2020). https://arxiv.org/abs/2002.06836

18. Mnih, V., et al.: Playing atari with deep reinforcement learning (2013). arXiv:1312.5602
19. Mnih, V., et al.: Human-level control through deep reinforcement learning. Nature **518**(7540), 522–533 (2015)
20. Akkaya, I., et al.: Solving Rubik's cube with a robot hand (2019)
21. Park, S., Kim, J., Kim, G.: Time discretization-invariant safe action repetition for policy gradient methods. CoRR abs/2111.03941 (2021). https://arxiv.org/abs/2111.03941
22. Schulman, J., Wolski, F., Dhariwal, P., Radford, A., Klimov, O.: Proximal policy optimization algorithms (2017). arXiv:1707.06347
23. Sharma, S., Srinivas, A., Ravindran, B.: Learning to repeat: fine grained action repetition for deep reinforcement learning (2020)
24. Singh, B., Kumar, R., Singh, V.P.: Reinforcement learning in robotic applications: a comprehensive survey. Artif. Intell. Rev. **55**(2), 945–990 (2022). https://doi.org/10.1007/s10462-021-09997-9
25. Sutton, R.S., Barto, A.G.: Reinforcement Learning: An Introduction, 2nd edn. The MIT Press, Cambridge (2018)
26. Szulc, M., Łyskawa, J., Wawrzyński, P.: A framework for reinforcement learning with autocorrelated actions. In: International Conference on Neural Information Processing, pp. 90–101 (2020)
27. Sünderhauf, N., et al.: The limits and potentials of deep learning for robotics. Int. J. Rob. Res. **37**(4–5), 405–420 (2018). https://doi.org/10.1177/0278364918770733
28. Tallec, C., Blier, L., Ollivier, Y.: Making deep q-learning methods robust to time discretization. In: International Conference on Machine Learning (ICML), pp. 6096–6104 (2019)
29. Uhlenbeck, G.E., Ornstein, L.S.: On the theory of the Brownian motion. Phys. Rev. **36**, 823–841 (1930). https://doi.org/10.1103/PhysRev.36.823
30. Wawrzyński, P.: Real-time reinforcement learning by sequential actor-critics and experience replay. Neural Netw. **22**(10), 1484–1497 (2009)
31. Yu, C., Liu, J., Nemati, S., Yin, G.: Reinforcement learning in healthcare: a survey. ACM Comput. Surv. **55**(1) (2021). https://doi.org/10.1145/3477600
32. Yu, H., Xu, W., Zhang, H.: TASAC: temporally abstract soft actor-critic for continuous control. CoRR abs/2104.06521 (2021). https://arxiv.org/abs/2104.06521
33. Łyskawa, J., Wawrzyński, P.: ACERAC: efficient reinforcement learning in fine time discretization. IEEE Trans. Neural Netw. Learn. Syst. (2022). https://doi.org/10.1109/TNNLS.2022.3190973

Scalable Bayesian Tensor Ring Factorization for Multiway Data Analysis

Zerui Tao[1,2], Toshihisa Tanaka[1,2], and Qibin Zhao[2,1](✉)

[1] Tokyo University of Agriculture and Technology, Tokyo, Japan
tanakat@cc.tuat.ac.jp
[2] RIKEN, AIP, Tokyo, Japan
{zerui.tao,qibin.zhao}@riken.jp

Abstract. Tensor decompositions play a crucial role in numerous applications related to multi-way data analysis. By employing a Bayesian framework with sparsity-inducing priors, Bayesian Tensor Ring (BTR) factorization offers probabilistic estimates and an effective approach for automatically adapting the tensor ring rank during the learning process. However, previous BTR [10] method employs an Automatic Relevance Determination (ARD) prior, which can lead to sub-optimal solutions. Besides, it solely focuses on continuous data, whereas many applications involve discrete data. More importantly, it relies on the Coordinate-Ascent Variational Inference (CAVI) algorithm, which is inadequate for handling large tensors with extensive observations. These limitations greatly limit its application scales and scopes, making it suitable only for small-scale problems, such as image/video completion. To address these issues, we propose a novel BTR model that incorporates a nonparametric Multiplicative Gamma Process (MGP) prior, known for its superior accuracy in identifying latent structures. To handle discrete data, we introduce the Pólya-Gamma augmentation for closed-form updates. Furthermore, we develop an efficient Gibbs sampler for consistent posterior simulation, which reduces the computational complexity of previous VI algorithm by two orders, and an online EM algorithm that is scalable to extremely large tensors. To showcase the advantages of our model, we conduct extensive experiments on both simulation data and real-world applications.

Keywords: Tensor decomposition · Tensor completion · Bayesian methods · Gibbs sampler

1 Introduction

Multi-way data are ubiquitous in many real-world applications, such as recommender systems, knowledge graphs, images/videos and so on. Since multi-way arrays are typically high-dimensional and sparse, tensor decomposition (TD) serves as a powerful tool for analyzing such data. Many elegant approaches for TD have been established, such as CANDECOM/PARAFAC (CP) [7], Tucker

B. Luo et al. (Eds.): ICONIP 2023, LNCS 14447, pp. 490–503, 2024.
https://doi.org/10.1007/978-981-99-8079-6_38

[20], Tensor Train (TT) [13], Tensor Ring (TR) [28] and so on. Among these TD models, TR decomposition and its descendants [10,19,21,24] have achieved great successes, due to its highly compact form and expressive power. In particular, Bayesian TR (BTR) [10,19] has many advantages. Firstly, by adopting fully probabilistic treatment, uncertainty estimates can be obtained, which is desirable for real-world applications. Secondly, with sparsity-inducing priors, it establishes a principled way of automatic rank adaption during the learning process and greatly avoid over-fitting. However, previous BTR are developed using Coordinate-Ascent Variational Inference (CAVI) [10] or Expectation-Maximization (EM) [19] algorithms, which can not handle large tensors with massive observations. Besides, both [10,19] did not consider discrete data, which are important in real-world applications.

To address the above issues, we proposed a novel BTR model which can deal with continuous or binary data and is scalable to large datasets. In particular, we firstly propose an weighted version of TR, which incorporates the nonparametric Multiplicative Gamma Process (MGP) prior [3] for automatic rank adaption. The Gaussian assumption is then extended with Pólya-Gamma augmentation [14] to deal with binary data. To simulate the posterior, we develop an efficient Gibbs sampler, which has smaller computational complexity than the CAVI algorithm in [10]. Moreover, an online version of EM algorithm is established to handle large tensors with massive observations. To showcase the advantages of our model, we conduct extensive experiments on both simulation data and real-world applications. The contributions are summarized as follows: (1) We propose a novel weighted TR decomposition with MGP prior. (2) The model is extended with Pólya-Gamma augmentation to deal with binary data. (3) An efficient Gibbs sampler and a scalable online EM algorithm is developed to deal with large datasets.

2 Related Work

To deal with multi-way data, many elegant tensor decompositions (TD) have been proposed, including CP [7], Tucker [20], tensor train [13], tensor ring [28] and their descendants [6,8]. Traditionally, TD factors are learnt through alternating least square (ALS) algorithms [8,21,28]. However, to handle large tensors with massive data, (stochastic) gradient-based optimization can also be applied [1,12,25]. Despite their success, Bayesian TDs have many advantages over traditional ones, such as uncertainty estimates, adaptive rank selection, a principled way of modeling diverse data types. For example, [26] proposed Bayesian CP to automatically select tensor ranks. [5,15,16] further derived online EM/VI algorithm for Bayesian CP to deal with both continuous and discrete data. Besides, it is also possible to derive Bayesian version of Tucker decomposition [17,27]. However, CP and Tucker have their limitations. CP format lacks flexibility to deal with complex real-world data, while Tucker suffers from the curse of dimensionality for high-order tensors. Tensor ring (TR) decomposition has shown predominant performances in many application, such image processing [21,24], model compression [22], generative models [9] and so on. To equip TR format with Bayesian framework, [10,18] developed a Bayesian TR with Automatic Relevance Determination (ARD) prior for sparsity and derived a Coordinate-Ascent Variational Inference (CAVI) algorithm to learn posteriors. [19] proposed a Bayesian TR model for factor analysis, which employs Multiplicative Gamma Process (MGP) [3] prior and EM algorithm. Other related works include Bayesian CP with generalized hyperbolic prior [4], Bayesian TT with Gaussian-product-Gamma prior [23]. Most of these works are not scalable and cannot handle discrete data.

3 Backgrounds

3.1 Notations

In machine learning community, the term "tensor" usually refers to multi-way arrays, which extends vectors and matrices. For consistency with previous literature, the style of notations follows the convention in [8]. We use lowercase letters, bold lowercase letters, bold capital letters and calligraphic bold capital letters to represent scalars, vectors, matrices and tensors, *e.g.*, $x, \boldsymbol{x}, \boldsymbol{X}$ and $\boldsymbol{\mathcal{X}}$. For an order-D tensor $\boldsymbol{\mathcal{X}} \in \mathbb{R}^{I_1 \times \cdots \times I_D}$, we denote its $(i_1, \ldots, i_D)$-th entry as $x_{\mathbf{i}}$. Moreover, $\mathcal{N}(\cdot, \cdot)$ denotes Normal distribution, $Ga(\cdot, \cdot)$ denotes Gamma distributions and $\mathbb{D}_{KL}(\cdot \| \cdot)$ denotes the Kullback-Leibler (KL) divergence.

3.2 Tensor Ring Decomposition

In this subsection, we introduce the tensor ring (TR) decomposition [28]. Given an order-D tensor $\boldsymbol{\mathcal{X}} \in \mathbb{R}^{I_1 \times \cdots \times I_D}$, TR format factorizes it into D core tensors,

$$x_{\mathbf{i}} = \operatorname{tr}\left(\boldsymbol{G}^{(1),i_1} \boldsymbol{G}^{(2),i_2} \cdots \boldsymbol{G}^{(D),i_D}\right), \tag{1}$$

where $\boldsymbol{G}^{(d),i_d} \in \mathbb{R}^{R_d \times R_d}, \forall i_d = 1, \ldots, I_d$ are i_d-th slices of the core tensor $\boldsymbol{\mathcal{G}}^{(d)} \in \mathbb{R}^{I_d \times R_d \times R_d}$. And the sequence $\{R_d\}_{d=1}^{D}$ is the rank for a TR format. Typically, we assume $R = R_1 = \cdots = R_d$ for convenience. We can denote a TR format as $\boldsymbol{\mathcal{X}} = TR(\boldsymbol{\mathcal{G}}^{(1)}, \ldots, \boldsymbol{\mathcal{G}}^{(D)})$. For ease of the derivation, we introduce the subchain of a TR format by contracting a subsequence of core tensors, namely, $\boldsymbol{\mathcal{G}}^{\neq d} \in \mathbb{R}^{I_{d+1} \times \cdots \times I_D \times I_1 \times I_{d-1} \times R \times R}$, where each slice becomes,

$$\boldsymbol{G}^{\neq d, i_{d+1} \cdots i_{d-1}} = \boldsymbol{G}^{(d+1),i_{d+1}} \cdots \boldsymbol{G}^{(d-1),i_{d-1}} \in \mathbb{R}^{R \times R}. \tag{2}$$

Then the TR format can be simplified as $x_{\mathbf{i}} = \operatorname{tr}\left(\boldsymbol{G}^{(d),i_d} \boldsymbol{G}^{\neq d, i_{d+1} \cdots i_{d-1}}\right)$.

4 Proposed Model

4.1 Weighted Tensor Ring Decomposition

To enable the rank inference, we employ a weight variable $\boldsymbol{\Lambda}$ that indicates the importance of each factor and rewrite Eq. 1 as,

$$x_{\mathbf{i}} = \operatorname{tr}\left(\boldsymbol{G}^{(1),i_1} \boldsymbol{\Lambda}^{(1)} \boldsymbol{G}^{(2),i_2} \boldsymbol{\Lambda}^{(2)} \cdots \boldsymbol{G}^{(D),i_D} \boldsymbol{\Lambda}^{(D)}\right), \tag{3}$$

where $\mathbf{\Lambda}^{(d)} = \text{diag}(\lambda_1^{(d)}, \ldots, \lambda_R^{(d)})$. Hence, if $\lambda_r^{(d)}$ becomes small, we can prune the corresponding redundant factor. Absorbing the weights into the core tensors, we can get an equivalent TR format $\boldsymbol{\mathcal{X}} = TR(\tilde{\boldsymbol{\mathcal{G}}}^{(1)}, \ldots, \tilde{\boldsymbol{\mathcal{G}}}^{(D)})$, where $\tilde{\boldsymbol{G}}^{(d),i_d} = \boldsymbol{G}^{(d),i_d}\mathbf{\Lambda}^{(d)}, \quad \forall i_d = 1, \ldots, I_d$. In real-world applications, the observations are typically corrupted by noises. In this paper, we consider both continuous and binary data, following Gaussian and Bernoulli distributions as follows. (1) Continuous data: Given a set of observations $\Omega = \{\mathbf{i}_i, \ldots, \mathbf{i}_N\}$, we assume the observations follow Gaussian distribution,

$$p(\boldsymbol{\mathcal{Y}}_\Omega \mid \boldsymbol{\mathcal{X}}) = \prod_{\mathbf{i}\in\Omega} \mathcal{N}(y_{\mathbf{i}} \mid x_{\mathbf{i}}, \tau^{-1}), \tag{4}$$

where τ is the noise precision and can be inferred under Bayesian framework. (2) Binary data: We employ the Bernoulli model,

$$p(\boldsymbol{\mathcal{Y}}_\Omega \mid \boldsymbol{\mathcal{X}}) = \prod_{\mathbf{i}\in\Omega} \left(\frac{1}{1+\exp(-x_{\mathbf{i}})}\right)^{y_{\mathbf{i}}} \left(\frac{\exp(-x_{\mathbf{i}})}{1+\exp(-x_{\mathbf{i}})}\right)^{1-y_{\mathbf{i}}}. \tag{5}$$

4.2 Probabilistic Model with Sparsity Inducing Prior

To induce low-rank sparsity, we add a Multiplicative Gamma Process (MGP) prior [3] on the weight parameters. For $d = 1, \ldots, D$ and $r = 1, \ldots, R$, let

$$\lambda_r^{(d)} \sim \mathcal{N}(0, (\phi_r^{(d)})^{-1}), \quad \delta_l^{(d)} \sim Ga(a_0, 1),$$

where $\phi_r^{(d)} = \prod_{l=1}^{r} \delta_l^{(d)}$ and $a_0 > 1$ to ensure sparsity. Then core tensors are assigned Gaussian distribution,

$$g_{r,r'}^{(d),i_d} \sim \mathcal{N}(0, (\psi_{r,r'}^{(d),i_d})^{-1}), \forall r, r' = 1, \ldots, R,$$

where $g_{r,r'}^{(d),i_d}$ are (r, r')-th elements of core tensor $\boldsymbol{G}^{(d),i_d}$. The precision can be either constant, *e.g.*, $\psi_{r,r'}^{(d),i_d} = 1$, or follow Gamma prior, $\psi_{r,r'}^{(d),i_d} \sim Ga(c_0, d_0)$. Finally, for continuous data, the noise precision follows a Gamma distribution, $\tau \sim Ga(\alpha_0, \beta_0)$. For binary data, we use an auxiliary variable $\boldsymbol{\mathcal{W}}$ for data augmentation, for which the details are presented in Sect. 5.2. The whole parameter set is denoted as $\mathbf{\Theta} = \{\tau, \boldsymbol{\mathcal{G}}, \mathbf{\Lambda}, \boldsymbol{\delta}, \boldsymbol{\mathcal{W}}\}$. Integrating all the prior distributions, we can get the joint prior distribution. For example, for continuous data,

$$\begin{aligned} p(\mathbf{\Theta}) = Ga(\tau \mid \alpha_0, \beta_0) \cdot \prod_{d=1}^{D} \prod_{i_d=1}^{I_d} \prod_{r,r'=1}^{R} \mathcal{N}(g_{r,r'}^{(d),i_d} \mid 0, (\psi_{r,r'}^{(d),i_d})^{-1}) \\ \cdot \prod_{d=1}^{D} \prod_{r}^{R} \mathcal{N}(\lambda_r^{(d)} \mid 0, (\phi_r^{(d)})^{-1}) \cdot Ga(\delta_r^{(d)} \mid a_0, 1). \end{aligned} \tag{6}$$

Then we can construct the whole generative model as,

$$p(\boldsymbol{\mathcal{Y}}, \mathbf{\Theta}) = p(\boldsymbol{\mathcal{Y}} \mid \boldsymbol{\mathcal{X}}) \cdot p(\tau, \boldsymbol{\mathcal{G}}, \mathbf{\Lambda}, \delta),$$

where $p(\boldsymbol{\mathcal{Y}} \mid \boldsymbol{\mathcal{X}})$ can be either Gaussian distribution Eq. 4 or Bernoulli distribution Eq. 5. Figure 1 shows the graphical illustration of the probabilistic model.

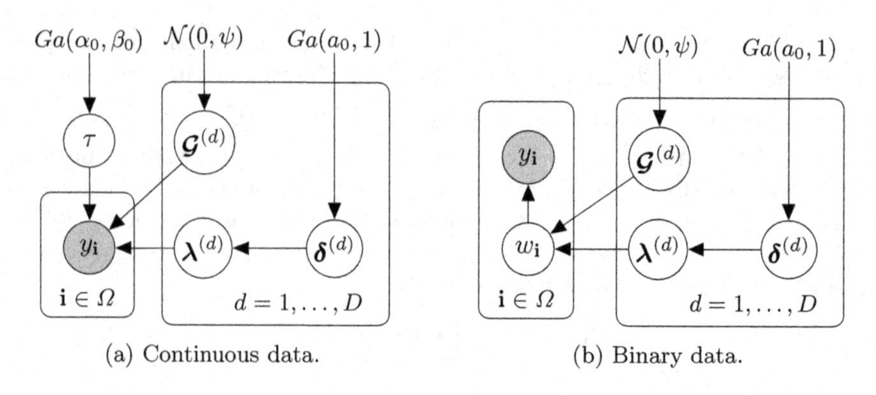

(a) Continuous data. (b) Binary data.

Fig. 1. Graphical models of Bayesian weighted tensor ring decomposition.

5 Gibbs Sampler

5.1 Sampling Rules

The basic idea of Gibbs sampler is to sequentially sample each parameter θ_j from conditional distribution $p(\theta_j \mid \theta_1, \ldots, \theta_{-j})$, where θ_{-j} means parameters exclude θ_j. It can be proved that the stationary distribution of this Markov chain is the true posterior. Due to the conjugate priors, all the conditional distributions can be derived analytically. More details can be find in tutorials such as [2].

Sample $\boldsymbol{\delta}$. To update the auxiliary variable $\boldsymbol{\delta}$, we sample from the conditional distribution

$$\delta_r^{(d)} \mid - \sim Ga\left(a_0 + \frac{1}{2}(R - r + 1), 1 + \frac{1}{2}\sum_{h=r}^{R} \lambda_h^{(d),r} \prod_{l=1, l\neq r}^{h} \delta_l^{(d)}\right). \tag{7}$$

Sample $\boldsymbol{\lambda}$. To update λ, we firstly rewrite the TR model Eq. 3 as

$$x_{\mathbf{i}} = \mathrm{tr}\left(\boldsymbol{\Lambda}^{(d)} \cdot \tilde{\boldsymbol{G}}^{\neq d, \mathbf{i}_{-d}} \cdot \boldsymbol{G}^{(d), i_d}\right) = a_{\mathbf{i}}^r \lambda_r^{(d)} + b_{\mathbf{i}}^r,$$

where $a_{\mathbf{i}}^r = \left(\tilde{\boldsymbol{G}}^{\neq d, \mathbf{i}_{-d}} \cdot \boldsymbol{G}^{(d), i_d}\right)_{rr}$ and $b_{\mathbf{i}}^r = \sum_{r' \neq r}\left(\tilde{\boldsymbol{G}}^{\neq d, \mathbf{i}_{-d}} \cdot \boldsymbol{G}^{(d), i_d}\right)_{r'r'} \lambda_{r'}^{(d)}$, with $\mathbf{i}_{-d}$ denoting $[i_{d+1}, \ldots, i_D, i_1, \ldots, i_{d=1}]$. Then we can sample from the conditional distribution,

$$\lambda_r^{(d)} \mid - \sim \mathcal{N}(\mu_r^{(d)}, (\sigma_r^{(d)})^{-1}), \tag{8}$$

where $\mu_r^{(d)} = (\sigma_r^{(d)})^{-1} \tau \sum_{\mathbf{i}} a_{\mathbf{i}}^r (y_{\mathbf{i}} - b_{\mathbf{i}}^r)$, and $\sigma_r^{(d)} = \phi_r^{(d)} + \tau \sum_{\mathbf{i}} (a_{\mathbf{i}}^r)^2$.

Sample core tensors $\mathcal{G}$. The derivation is similar with that in updating $\boldsymbol{\lambda}$. Particularly, the TR model Eq. 3 is rewritten as

$$x_{\mathbf{i}} = \mathrm{tr}\left(\boldsymbol{G}^{(d), i_d} \cdot \boldsymbol{\Lambda}^{(d)} \cdot \tilde{\boldsymbol{G}}^{\neq d, \mathbf{i}_{-d}}\right) = c_{r,r'}^{(k), i_d} g_{r,r'}^{(d), i_d} + d_{r,r'}^{(d), i_d},$$

where $c_{r,r'}^{(k),i_d} = \left(\mathbf{\Lambda}^{(d)} \cdot \tilde{\boldsymbol{G}}^{\neq d, \mathbf{i}_{-d}}\right)_{rr'}$ and $d_{r,r'}^{(d),i_d} = \sum_{(i,j)\neq(r,r')}^{R} \left(\mathbf{\Lambda}^{(d)} \tilde{\boldsymbol{G}}^{\neq d, \mathbf{i}_{-d}}\right)_{ij} g_{j,i}^{(d),i_d}$. Then the conditional distribution becomes,

$$\boldsymbol{G}_{r,r'}^{(d)} \mid - \sim \mathcal{N}(\boldsymbol{\mu}_{r,r'}^{(d)}, \mathbf{\Sigma}_{r,r'}^{(d)}). \tag{9}$$

The variance is computed by

$$\mathbf{\Sigma}_{r,r'}^{(d)} = (\text{diag}(\psi_{r,r'}^{(d)}) + \boldsymbol{T}_{r,r'}^{(d)})^{-1}, \tag{10}$$

where $\boldsymbol{T}_{r,r'}^{(d)} = \text{diag}(t_{r,r'}^{(d),i_1}, \ldots, t_{r,r'}^{(d),I_d})$ and $t_{r,r'}^{(d),i_d} = \tau \sum_{\mathbf{i}, i=i_d} (c_{r,r'}^{(d),i})^2$. Additionally, the mean parameter is computed as

$$\boldsymbol{\mu}_{r,r'}^{(d)} = \mathbf{\Sigma}_{r,r'}^{(d)} \boldsymbol{T}_{r,r'}^{(d)} \boldsymbol{\alpha}_{r,r'}^{(d)}, \tag{11}$$

where $\boldsymbol{\alpha}_{r,r'}^{(d)} = [\alpha_{r,r'}^{(d),i_d}, \ldots, \alpha_{r,r'}^{(d),I_d}]$, $\alpha_{r,r'}^{(d),i_d} = (t_{r,r'}^{(d),i_d})^{-1} \tau \sum_{\mathbf{i}, i=i_d} c_{r,r'}^{(k),i} (y_{\mathbf{i}} - d_{r,r'}^{(k),i})$.

Sample τ. To update the noise precision τ, we can simply sample from the Gamma distribution

$$\tau \mid - \sim Ga\left(\alpha_0 + \frac{|\Omega|}{2}, \beta_0 + \frac{1}{2} \sum_{\mathbf{i}\in\Omega} (y_{\mathbf{i}} - x_{\mathbf{i}})^2\right), \tag{12}$$

where $|\Omega|$ is the number of observed entries and α_0, β_0 are hyperparameters.

5.2 Binary Tensor

To deal with binary tensor, we employ the Pólya-Gamma (PG) augmentation [14] by introducing the auxiliary variable $\mathcal{W}$, which follows the PG distribution,

$$\omega_{\mathbf{i}} \sim PG(1, x_{\mathbf{i}}). \tag{13}$$

Due to properties of PG distribution, we can sample $\boldsymbol{\lambda}$ from

$$\lambda_r^{(d)} \mid - \sim \mathcal{N}(\mu_r^{(d)}, (\sigma_r^{(d)})^{-1}), \quad \forall r = 1, \ldots, R, \tag{14}$$

where $\mu_r^{(d)} = (\sigma_r^{(d)})^{-1} \sum_{\mathbf{i}} a_{\mathbf{i}}^r (y_{\mathbf{i}} - 0.5 - w_{\mathbf{i}} b_{\mathbf{i}}^r)$ and $\sigma_r^{(d)} = \phi_r^{(d)} + \sum_{\mathbf{i}} (a_{\mathbf{i}}^r)^2 w_{\mathbf{i}}$. The sampling rule of core tensors $\boldsymbol{\mathcal{G}}$ becomes,

$$\boldsymbol{G}_{r,r'}^{(d)} \mid - \sim \mathcal{N}(\boldsymbol{\mu}_{r,r'}^{(d)}, \mathbf{\Sigma}_{r,r'}^{(d)}), \tag{15}$$

where $\mathbf{\Sigma}_{r,r'}^{(d)}$ and $\boldsymbol{\mu}_{r,r'}^{(d)}$ are computed by Eq. 10 and Eq. 11 respectively, with $t_{r,r'}^{(d),i_d} = \sum_{\mathbf{i}, i=i_d} (c_{r,r'}^{(d),i})^2 w_{\mathbf{i}}$ and $\alpha_{r,r'}^{(d),i_d} = (t_{r,r'}^{(d),i_d})^{-1} \sum_{\mathbf{i}, i=i_d} c_{r,r'}^{(k),i} (y_{\mathbf{i}} - 0.5 - w_{\mathbf{i}} d_{r,r'}^{(k),i})$. For binary data, there is no need to sample τ. The overall procedure is presented in Algorithm 1.

Algorithm 1. Gibbs sampler

1: Input observation $\mathcal{Y}_\Omega$ a_0, α_0, β_0.
2: **for** $t \leq max\_step$ **do**
3: **for** $d = 1, \ldots, D$, $r = 1, \ldots, R$ **do**
4: Sample $\delta_r^{(d)}$ from Eq. 7. `// Sample` δ
5: **end for**
6: **for** $d = 1, \ldots, D$, $r = 1, \ldots, R$ **do**
7: Sample $\lambda_r^{(d)}$ from Eq. 8 or Eq. 14. `// Sample` λ
8: **end for**
9: **for** $d = 1, \ldots, D$, $r = 1, \ldots, R$, $r' = 1, \ldots, R$ **do**
10: Sample $\boldsymbol{G}_{r,r'}^{(d)}$ from Eq. 9 or Eq. 15. `// Sample` $\mathcal{G}$
11: **end for**
12: **if** $\mathcal{Y}$ is continuous **then**
13: Sample τ from Eq. 12. `// Sample` τ
14: **else if** $\mathcal{Y}$ is binary **then**
15: **for** i in Ω **do**
16: Sample w_{i} from Eq. 13. `// Sample` $\mathcal{W}$
17: **end for**
18: **end if**
19: Rank adaption as described in Section 5.3. `// Rank adaption`
20: **end for**

5.3 Tensor Ring Rank Adaption

In TD problems, one essential issue is choosing the appropriate rank. In Bayesian framework, the presence of sparsity-inducing priors allows us to adaptively search ranks during training. Previous BTR-VI model with ARD prior [10] proposed to start with a large TR rank R_0, and then prune redundant factors during training, which we refer to *truncation* strategy. Such strategy has two main limitations. Firstly, if we choose a large R_0, the computational complexity becomes extremely large at beginning. Secondly, if the predefined R_0 is smaller than the true rank, we can never learn it. Due to the nonparametric property of the MGP prior, the proposed model can potentially learn infinite number of latent factors. To achieve this, we adopt the *adaption* strategy, which is similar with [3,16]. In particular, the model starts with a relatively smaller R_0. Then, after each iteration, we prune factors with weights smaller than a threshold, e.g., $\lambda_r^{(d)} < \epsilon$. If all of the weights are larger than the threshold, a new factor is added with a probability $p(t) = \exp(\kappa_0 + \kappa_1 t)$, where t is the current iteration number. The *adaption* strategy has much lower computational burden at the beginning stage than the *truncation* strategy, and can learn higher ranks than the initial rank.

5.4 Complexity Analysis

We denote the tensor rank as R, the tensor order as D and the number of observations as M. Computing Eq. 7 has complexity $\mathcal{O}(R^2)$, hence the complexity of sampling δ is $\mathcal{O}(DR^3)$. Computing Eq. 8 has complexity $\mathcal{O}(MR^2)$, hence the complexity of sampling δ is $\mathcal{O}(DMR^3)$. Computing Eq. 9 has complexity $\mathcal{O}(MR^2)$, hence the complexity of sampling δ is $\mathcal{O}(DMR^4)$. The complexity of updating ϵ is $\mathcal{O}(M)$ and can be omitted. Hence, the overall complexity is $\mathcal{O}(DMR^4)$, which is smaller than the BTR-VI [10] ($\mathcal{O}(DMR^6)$) by two orders.

6 Online EM Algorithm

In real applications, tensors can be very large with massive observations. For such cases, the Gibbs sampler that uses the whole dataset is prohibited. To address the issue, we establish an online Variational Bayes EM (VBEM) algorithm that employs stochastic updates which is scalable to large tensors.

For illustration, we consider the binary case. The continuous case can be derived straightforwardly. Recall that the full probabilistic is

$$p(\mathcal{Y}, \Theta) = p(\mathcal{Y} \mid \mathcal{W})p(\mathcal{W} \mid \mathcal{G}, \mathbf{\Lambda})p(\mathbf{\Lambda} \mid \delta)p(\delta)p(\mathcal{G}).$$

To enable the VBEM algorithm, we assign a variational distribution $q(\mathcal{W})q(\delta) \approx p(\mathcal{W}, \delta \mid \mathcal{Y}, \mathbf{\Lambda}, \mathcal{G})$, with parametric forms as follows,

$$q(\mathcal{W}) = \prod_{\mathbf{i}} PG(1, x_{\mathbf{i}}), \quad q(\delta) = \prod_{d=1}^{D} \prod_{r=1}^{R} Ga(\alpha_{\delta}^{(d),r}, \beta_{\delta}^{(d),r}).$$

In the VBEM algorithm, we derive closed-form expectations of $\{\mathcal{W}, \delta\}$, and then update the rest parameters by gradient ascent of the free energy.

E-step. Using similar derivation with Sect. 5, we have the expectations,

$$\mathbb{E}_q[\delta_r^{(d)}] = \frac{a_\delta + \frac{1}{2}(R - r + 1)}{1 + \frac{1}{2}\sum_{h=r}^{R} \lambda_h^{(d),r} \prod_{l=1, l\neq r}^{h} \delta_l^{(d)}}, \quad \mathbb{E}_q[w_{\mathbf{i}}] = \frac{1}{2x_{\mathbf{i}}}\tanh(x_{\mathbf{i}}/2). \tag{16}$$

M-step. Parameters $\{\mathcal{G}, \mathbf{\Lambda}\}$ are then updated by maximizing the free energy,

$$\begin{aligned}
\mathcal{L} =& \mathbb{E}_q[\log p(\mathcal{Y}, \Theta)] \\
=& \sum_{\mathbf{i}\in\Omega} \mathbb{E}_q[\log p(y_{\mathbf{i}} \mid w_{\mathbf{i}}) + \log PG(w_{\mathbf{i}} \mid 1, x_{\mathbf{i}})] + \sum_{d=1}^{D} \sum_{r,r'=1}^{R} \log \mathcal{N}(\boldsymbol{G}_{r,r'}^{(d)} \mid \mathbf{0}, \boldsymbol{I}) \\
&+ \sum_{d=1}^{D} \sum_{r=1}^{R} \mathbb{E}_q[\log \mathcal{N}(\lambda_r^{(d)} \mid 0, (\phi_r^{(d)})^{-1})] + \sum_{d=1}^{D} \sum_{r=1}^{R} \mathbb{E}_q[\log Ga(\delta_r^{(d)} \mid \alpha_0, \beta_0)],
\end{aligned}$$

where the expectations can be computed via Eq. 16.

Since the free energy is factorized over observations $\mathbf{i}$, stochastic optimization with mini-batch samples can be adopted. The gradient can be computed through either analytical expressions or back propagation. The complexity of computing the free energy is $\mathcal{O}(BDR^4)$, where B is the mini-batch size. This complexity is the same order with the Gibbs sampler. However, when applying the online EM algorithm, we can choose small mini-batch size B to handle large-scale problems.

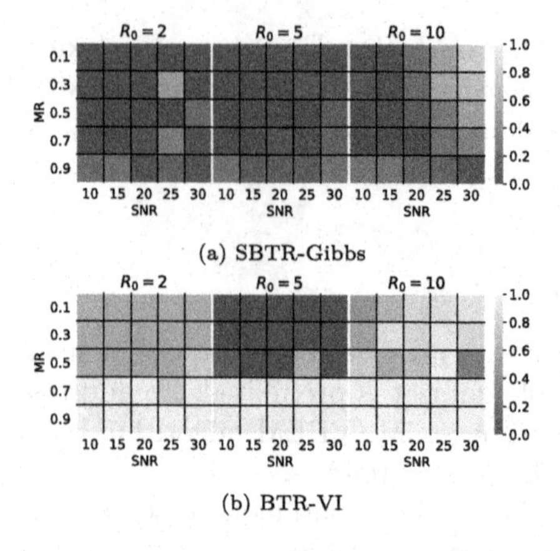

(a) SBTR-Gibbs

(b) BTR-VI

Fig. 2. Rank estimation results.

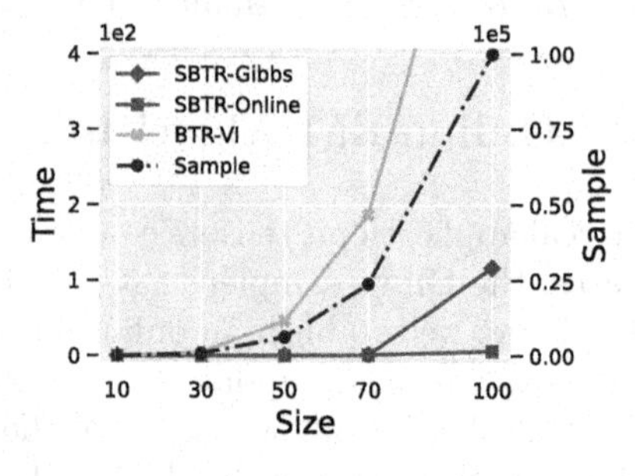

Fig. 3. Scalability results.

7 Experiments

In this section, we evaluate[1] the proposed Scalable Bayesian Tensor Ring (SBTR) model on synthetic data and several real-world datasets. The Gibbs sampler and online EM algorithm are denoted as SBTR-Gibbs and SBTR-Online, respectively. All the experiments are conducted on a Linux workstation with Intel Xeon Silver 4316 CPU@2.30GHz, 512GB RAM and an NVIDIA RTX A6000 GPU (48GB memory). Our model is implemented based on PyTorch. If not specified, we run SBTR on GPU and other baselines on CPU.

7.1 Simulation Study

Rank Estimation. Firstly, we conduct simulation study on Tensor Ring (TR) rank estimation. To show advantages of the MGP prior and Gibbs sampler, we compare with BTR-VI [10], which adopts ARD prior and variational inference. We consider the true low rank tensor $\boldsymbol{\mathcal{X}} \in \mathbb{R}^{10\times10\times10\times10}$ with TR rank 5. We firstly generate TR core tensors from standard Gaussian distribution, then standardize $\boldsymbol{\mathcal{X}}$ to have zero mean and unit variance. Finally, i.i.d. Gaussian noises and uniform masks are added on the underlying signal $\boldsymbol{\mathcal{X}}$. The signal-to-noise ratio (SNR) is chosen from $\{10, 15, 20, 25, 30\}$ and the missing rates (MR) vary from $\{0.1, 0.3, 0.5, 0.7, 0.9\}$. For BTR-VI, we use their default settings and run 500 iterations. For SBTR-Gibbs, we set hyperparameters $a_0 = 2.0, \alpha_0 = 1.0, \beta = 0.3$ and run 1500 burn-in steps to ensure convergence. The rank estimation performance is evaluated using the relative absolute error $\text{Err} = \sum_{d=1}^{4}|\hat{R}_d - R_d|/\sum_{d=1}^{4} R_d$, where $\hat{R}_d$ is the estimated rank on each mode.

[1] The code is available at https://github.com/taozerui/scalable_btr.

Figure 2 illustrates results under different initial rank R_0. SBTR-Gibbs is consistently better than BTR-VI, especially when the missing ratio becomes large.

Table 1. Continuous tensor completion results. Red and Blue entries indicate the best and second best, respectively.

Model	USHCN		Indian Pines		Alog	
	RMSE↓	MAE↓	PSNR↑	SSIM↑	RMSE↓	MAE↓
BayesCP	0.590±0.006	0.430±0.005	9.88±0.00	0.01±0.00	1.02±0.06	0.52±0.02
TTWOPT	0.761±0.001	0.570±0.001	27.13±0.07	0.81±0.00	2.84±1.38	0.80±0.06
BayesTR-VI	0.614±0.030	0.451±0.025	27.13±0.77	0.79±0.02	1.00±0.04	0.52±0.03
SBTR-Gibbs	0.587±0.007	0.430±0.003	29.71±0.12	0.84±0.00	0.98±0.05	0.51±0.02
SBTR-Online	0.560±0.004	0.405±0.004	28.07±0.04	0.83±0.00	0.92±0.03	0.47±0.01

Scalability. Then, we showcase the scalability of our model. In the experiment, an order-4 tensor of shape $I \times I \times \times I \times I$ is considered, where the tensor size I varies from $\{10, 30, 50, 70, 100\}$ and the missing rate is set to 99.9%. We compare the running time of BTR-VI, SBTR-Gibbs and SBTR-Online with TR rank 2. All the models are tested on the CPU in this experiment.

The results are presented in Fig. 3. The solid lines are average time costs of one iteration (or, epoch for SBTR-Online) and the dashed line plots the sample sizes. It shows that the BTR-VI algorithm does not scale well with the tensor size. On the contrary, both SBTR-Gibbs and SBTR-Online show superior scalability. The online EM algorithm is much faster than Gibbs sampler, since it can fully utilize parallel computation when choosing small batch sizes.

7.2 Continuous Data Completion

Datasets. We test our model on three continuous tensor for completion. (1) U.S. Historical Climatology Network (USHCN)[2], a climate dataset with shape $17 \times 125 \times 156$. We randomly select 10% points for training and predict the rest entries. (2) Indian Pines[3], a hyperspectral image with shape $145 \times 145 \times 200$. We randomly set 99% entries as missing and evaluate image completion performance. (3) Alog [29], a three-mode (*user* $\times$ *action* $\times$ *resource*) tensor of shape $200 \times 100 \times 200$, which is extracted from an access log of a file management system. This data is partially observed with about 0.33% nonzero entries. We use the same train/test split as in [29]. All the datasets are evaluated using 5-fold cross-validation with different random seeds.

[2] https://www.ncei.noaa.gov/products/land-based-station/us-historical-climatology-network.

[3] https://www.ehu.eus/ccwintco/index.php/Hyperspectral_Remote_Sensing_Scenes.

Competing Methods. We compare with the following baseline models. (1) Bayesian CP (BayesCP) [26], a CP decomposition adopting ARD prior and variational inference. (2) Tensor Train Weighted OPTimization (TTWOPT) [25], which uses (stochastic) gradient descent to learn TT decomposition. (3) Bayesian TR with variational inference (BTR-VI) [10] and ARD prior.

Table 2. Binary tensor completion results. Red and Blue entries indicate the best and second best, respectively.

	Kinship		Enron		DBLP	
Model	AUC↑	ACC↑	AUC↑	ACC↑	AUC↑	ACC↑
BayesCP	0.973±0.003	0.925±0.012	0.500±0.000	0.500±0.000	NA	NA
TTWOPT	0.943±0.008	0.774±0.156	0.819±0.049	0.787±0.017	0.912±0.004	0.809±0.051
BayesTR-VI	0.968±0.010	0.896±0.024	0.500±0.000	0.500±0.000	NA	NA
SBTR-Gibbs	0.962±0.002	0.921±0.002	0.834±0.018	0.758±0.019	0.925±0.003	0.869±0.002
SBTR-Online	0.973±0.003	0.940±0.005	0.911±0.039	0.811±0.042	0.957±0.001	0.889±0.003

Experimental Details and Results. The most important hyperparameter is the tensor rank. For BayesCP and BTR-VI, we set a large initial rank and prune redundant factors during training. For SBTR-Gibbs, we use the rank adaption strategy described in Sect. 5.3. The ranks of TTWOPT and SBTR-Online are chosen from $\{3, 5, 10\}$ via validation. Other settings of baselines are default in their codebases. The hyperparameters of SBTR-Gibbs are $a_0 = 2.0, \alpha_0 = 1.0, \beta = 0.3$. Moreover, we set batch size 512 and use Adam optimizer with learning rate 0.01 for SBTR-Online.

The completion results are shown in Table 1. For USHCN and Alog, we present the Root-Mean-Square Error (RMSE) and Mean Absolute Error (MAE), while for Indian, Peak Signal-to-Noise Ratio (PSNR) and Structural SIMilarity index (SSIM) are reported, since they are widely used to evaluate visual tasks. The proposed SBTR model achieve the best or second best results in all tasks.

7.3 Binary Data Completion

Datasets. We evaluate our model on three binary datasets. (1) Kinship [11], which is a multirelational data of shape $104 \times 104 \times 26$, representing 26 relations of 104 entities. (2) Enron [29], a tensor of shape $203 \times 203 \times 200$ extracted from the Enron email dataset. (3) DBLP [29], a tensor of shape $10000 \times 200 \times 10000$ depicting bibliography relationships in the DBLP website. Since these tensors are extremely sparse, we randomly sample the same number of zero and non-zeros entries to make the dataset balanced. Then the results are evaluated using 5-fold cross validation with different random seeds.

Experimental Details and Results. The settings are the same with those in Sect. 7.2. For baselines, we rescale their estimates to [0, 1] as final predictions. The prediction Area Under Curve (AUC) and ACCuracy (ACC) are presented in Table 2. In this experiment, SBTR-Online outperforms all other competitors. For Enron dataset, ALS-based baselines, including BayesCP and BayesTR-VI are unable to get faithful predictions, while our SBTR-Gibbs gets much better results. For DBLP dataset, BayesCP and BayesTR-VI exceed the maximum RAM of 512GB on our workstation. Meanwhile, SBTR-Gibbs, which is also an ALS-based algorithm using the whole dataset, costs only around 1.2GB memory with Float32 precision on GPU. All the results manifest the superior performance and scalability of the proposed SBTR.

8 Conclusion

This paper introduces a novel scalable Bayesian tensor ring factorization. To achieve this, a weighted tensor ring factorization is proposed. Then, a nonparametric multiplicative Gamma process is used to model the weights, enabling automatic rank adaption during training. A Gibbs sampler was proposed to obtained ALS-based updates, whose complexity is smaller than previous variational inference by two orders. Moreover, an online EM algorithm was established to scale to large datasets. We evaluate the proposed model on synthetic data, continuous tensor and binary tensor completion tasks. The simulation results show the superior performance of our model in rank estimation and scalability. Our model also outperforms all baselines in real-world tensor completion tasks.

Acknowledgements. This work was supported by the JSPS KAKENHI [Grant Number 20H04249, 23H03419]. Zerui Tao was supported by the RIKEN Junior Research Associate Program.

References

1. Acar, E., Dunlavy, D.M., Kolda, T.G., Mørup, M.: Scalable tensor factorizations for incomplete data. Chemom. Intell. Lab. Syst. **106**(1), 41–56 (2011)
2. Andrieu, C., De Freitas, N., Doucet, A., Jordan, M.I.: An introduction to MCMC for machine learning. Mach. Learn. **50**, 5–43 (2003)
3. Bhattacharya, A., Dunson, D.B.: Sparse Bayesian infinite factor models. Biometrika **98**(2), 291–306 (2011)
4. Cheng, L., Chen, Z., Shi, Q., Wu, Y.C., Theodoridis, S.: Towards flexible sparsity-aware modeling: automatic tensor rank learning using the generalized hyperbolic prior. IEEE Trans. Signal Process. **70**, 1834–1849 (2022)
5. Cheng, L., Wu, Y.C., Poor, H.V.: Scaling probabilistic tensor canonical polyadic decomposition to massive data. IEEE Trans. Signal Process. **66**(21), 5534–5548 (2018)
6. Cichocki, A., Lee, N., Oseledets, I., Phan, A.H., Zhao, Q., Mandic, D.P., et al.: Tensor networks for dimensionality reduction and large-scale optimization: Part 1 low-rank tensor decompositions. Found. Trends Mach. Learn. **9**(4–5), 249–429 (2016)

7. Hitchcock, F.L.: The expression of a tensor or a polyadic as a sum of products. J. Math. Phys. **6**(1–4), 164–189 (1927)
8. Kolda, T.G., Bader, B.W.: Tensor decompositions and applications. SIAM Rev. **51**(3), 455–500 (2009)
9. Kuznetsov, M., Polykovskiy, D., Vetrov, D.P., Zhebrak, A.: A prior of a googol gaussians: a tensor ring induced prior for generative models. In: Advances in Neural Information Processing Systems , vol. 32 (2019)
10. Long, Z., Zhu, C., Liu, J., Liu, Y.: Bayesian low rank tensor ring for image recovery. IEEE Trans. Image Process. **30**, 3568–3580 (2021)
11. Nickel, M., Tresp, V., Kriegel, H.P.: A three-way model for collective learning on multi-relational data. In: Proceedings of the 28th International Conference on International Conference on Machine Learning, pp. 809–816 (2011)
12. Oh, S., Park, N., Lee, S., Kang, U.: Scalable tucker factorization for sparse tensors-algorithms and discoveries. In: 2018 IEEE 34th International Conference on Data Engineering (ICDE), pp. 1120–1131. IEEE (2018)
13. Oseledets, I.V.: Tensor-train decomposition. SIAM J. Sci. Comput. **33**(5), 2295–2317 (2011)
14. Polson, N.G., Scott, J.G., Windle, J.: Bayesian inference for logistic models using pólya-gamma latent variables. J. Am. Stat. Assoc. **108**(504), 1339–1349 (2013)
15. Rai, P., Hu, C., Harding, M., Carin, L.: Scalable probabilistic tensor factorization for binary and count data. In: IJCAI, pp. 3770–3776 (2015)
16. Rai, P., Wang, Y., Guo, S., Chen, G., Dunson, D., Carin, L.: Scalable Bayesian low-rank decomposition of incomplete multiway tensors. In: International Conference on Machine Learning, pp. 1800–1808. PMLR (2014)
17. Schein, A., Zhou, M., Blei, D., Wallach, H.: Bayesian Poisson tucker decomposition for learning the structure of international relations. In: International Conference on Machine Learning, pp. 2810–2819. PMLR (2016)
18. Tao, Z., Zhao, Q.: Bayesian tensor ring decomposition for low rank tensor completion. In: International Workshop on Tensor Network Representations in Machine Learning, IJCAI (2020)
19. Tao, Z., Zhao, X., Tanaka, T., Zhao, Q.: Bayesian latent factor model for higher-order data. In: Asian Conference on Machine Learning, pp. 1285–1300. PMLR (2021)
20. Tucker, L.R.: Some mathematical notes on three-mode factor analysis. Psychometrika **31**(3), 279–311 (1966)
21. Wang, W., Aggarwal, V., Aeron, S.: Efficient low rank tensor ring completion. In: Proceedings of the IEEE International Conference on Computer Vision, pp. 5697–5705 (2017)
22. Wang, W., Sun, Y., Eriksson, B., Wang, W., Aggarwal, V.: Wide compression: tensor ring nets. In: Proceedings of the IEEE Conference on Computer Vision and Pattern Recognition, pp. 9329–9338 (2018)
23. Xu, L., Cheng, L., Wong, N., Wu, Y.C.: Probabilistic tensor train decomposition with automatic rank determination from noisy data. In: 2021 IEEE Statistical Signal Processing Workshop (SSP), pp. 461–465. IEEE (2021)
24. Yuan, L., Li, C., Mandic, D., Cao, J., Zhao, Q.: Tensor ring decomposition with rank minimization on latent space: an efficient approach for tensor completion. In: Proceedings of the AAAI conference on artificial intelligence, vol. 33, pp. 9151–9158 (2019)
25. Yuan, L., Zhao, Q., Cao, J.: Completion of high order tensor data with missing entries via tensor-train decomposition. In: Liu, D., Xie, S., Li, Y., Zhao, D., El-Alfy,

E.S. (eds.) Neural Information Processing. ICONIP 2017. LNCS, vol. 10634, pp. 222–229. Springer, Cham (2017). https://doi.org/10.1007/978-3-319-70087-8_24
26. Zhao, Q., Zhang, L., Cichocki, A.: Bayesian CP factorization of incomplete tensors with automatic rank determination. IEEE Trans. Pattern Anal. Mach. Intell. **37**(9), 1751–1763 (2015)
27. Zhao, Q., Zhang, L., Cichocki, A.: Bayesian sparse tucker models for dimension reduction and tensor completion. arXiv preprint arXiv:1505.02343 (2015)
28. Zhao, Q., Zhou, G., Xie, S., Zhang, L., Cichocki, A.: Tensor ring decomposition. arXiv preprint arXiv:1606.05535 (2016)
29. Zhe, S., Xu, Z., Chu, X., Qi, Y., Park, Y.: Scalable nonparametric multiway data analysis. In: Artificial Intelligence and Statistics, pp. 1125–1134. PMLR (2015)

Predefined-Time Event-Triggered Consensus for Nonlinear Multi-Agent Systems with Uncertain Parameter

Yafei Lu[1], Hui Zhao[1(✉)], Aidi Liu[2], Mingwen Zheng[3], Sijie Niu[1], Xizhan Gao[1], and Xiju Zong[1]

[1] School of Information Science and Engineering, Shandong Provincial Key Laboratory of Network Based Intelligent Computing, University of Jinan, Jinan 250022, China
hz_paper@163.com
[2] Collaborative Innovation Center of Memristive Computing Application, Qilu Institute of Technology, Jinan 250200, China
[3] School of Mathematics and Statistics, Shandong University of Technology, Zibo 255000, China

Abstract. In this paper, a novel predefined-time event-triggered control method is proposed, which achieved to the consistency of multi-agent systems with uncertain parameter. Firstly, a new predefined-time stability theorem is given, and the correctness and feasibility of this stability theorem are analyzed, the flexible preset time is more practical than the existed stability theorem. Compared with existing stability theorems, this theorem simplifies the conditions satisfied by Lyapunov function and is easier to implement in practical applications. Secondly, an event-triggered control strategy is designed to reduce control costs. Then, a new sufficient criterion is given to achieve the consistency of multi-agent systems with uncertain parameter based on the predefined-time stability theorem and event-triggered controller. In addition, the state consensus between nonlinear agents is completed in a predefined time, as well as the measurement error of the agent is converges to zero within the predefined time, respectively. Finally, the validity and feasibility of the given theoretical results are verified by a simulation example.

Keywords: Predefined-time Consensus · Event-triggered Control · Nonlinear Multi-agent Systems

1 Introduction

In the past decades, multi-agent systems have been widely applied in robot coordination and distributed optimization [1,2]. In the collective behavior of multi-agent systems, consensus problem is one of the basic problems in collective behavior, which has been widely studied. The speed and time of convergence are

B. Luo et al. (Eds.): ICONIP 2023, LNCS 14447, pp. 504–515, 2024.
https://doi.org/10.1007/978-981-99-8079-6_39

important subjects in the study of consensus problems. After a great deal of research, some results have been achieved [3,4].

In addition, most of the existing research focuses on the finite-time consensus [5,6] and fixed-time. For the finite-time consensus, it depends on initial conditions, which may limit practical application. In order to solve this practical problem and eliminate these limitations, the concept of fixed-time consensus was proposed [7,8]. The settling time of the fixed-time consensus is independent of the initial state. However, it relies on other system parameters, such as the eigenvalues of the Laplacian matrix, so there is a phenomenon of inflexible application. To solve these problems, the concept of predefined-time consensus [9–13] is proposed. The traditional fixed-time consensus systems can be improved by introducing predefined time parameters into the controller design process. Ref. [9] solved adaptive consensus in nonlinear multi-agent systems in switching topologies. Ref. [10] solved the predefined-time binary consistency control of output feedback for highorder nonlinear multi-agent systems. The predefined-time consensus problem of (T-S) fuzzy systems is studied in Ref. [11].

In the above research work on finite-time consensus, fixed-time consensus and predefined-time consensus, there is no way to avoid the constant communication and updating of the controller, which can lead to significant communication consumption. Event-triggered control [14–16] is a common control method, which can effectively avoid these disadvantages and has produced many important research results. In Ref. [14], dynamic event-triggered and self-triggered control of multi-agent systems were studied. A distributed dynamic event-triggered control method based on linear multi-agent system consistency of directed networks is studied in Ref. [15].

Inspired by these existing results, a predefined-time stability theorem was proposed. A controller with predefined time parameters is designed. This theorem ensuring that the settling time of the system does not depend on the initial value of the system and can be adjusted according to preset parameters. The work done in this paper has the following characteristics: Firstly, the settling time setting is more flexible, and this method is suitable for more systems and scenarios. Secondly, the design of the controller can reduce communication consumption and save resources. Finally, compared with finite-time consensus and fixed-time consensus, there is great scalability.

The rest of this article is organized as follows. The second section gives some basic graph theory, definitions and lemmas. In the third section, a new predefined-time stability theorem and a predefined-time event-triggered controller designed based on this theorem. The fourth section provides an example to demonstrate the correctness and feasibility of the above theorem and controllers. The fifth section provides conclusions and future work.

2 Preliminaries

First, graph theory is given. Then, some basic definitions and lemmas are introduced. Finally, a formulation of the problem is given.

Graph Theory. The communication topology between multiple agents is usually expressed in $\mathcal{G} = (\mathcal{V}, \mathcal{E})$, node set $\mathcal{V} = (1, \cdots, N)$, edge set $\mathcal{E} \subseteq \mathcal{V} \times \mathcal{V}$, edge (j, i) to indicate the presence of information flow from node j to node i, and the neighbor set of node i is denoted as $N_i = (j \in \mathcal{V} | (j, i) \in \mathcal{E}, j = i)$. Where $\mathcal{A} = [a_{ij}]$ in $R^{n \times n}$ is the adjacency matrix of $\mathcal{G}$ with the element representing the edge weight, where

$$a_{ij} = \begin{cases} 1, & if (i, j) \in \mathcal{E}, \\ 0, & otherwise. \end{cases}$$

The degree matrix is $\mathcal{D} = diag[d_1, d_2, \cdots, d_N], d_i = \sum_{j=1}^{N} a_{ij}$. The Laplacian matrix $\mathcal{L} = [l_{ij}]$ in $R^{N \times N}$ of $\mathcal{G}$ is defined as $\mathcal{L} = \mathcal{D} - \mathcal{A}$.

Suppose the origin is the equilibrium point of the following system:

$$\begin{cases} \dot{x}(t) = f(x(t), t), \\ x(0) = x_0, \end{cases} \tag{1}$$

where $f(x(t), t) : R^{+} \times R^{M} \to R^{M}$ is an unknown nonlinear function, and if $f(x(t), t)$ is not continuous, the solution of the above equation system (1) can be understood in the sense of Filippov.

Definition 1. [17]. If $x(t) = 0$ is asymptotically stable and $x(t)$ can reach 0 for a finite time for any $x(0) \in R^n$, then $x(t) = 0$ is finite-time stable. For any $x(0) \in R^n$, the settling time function is $T(x(0)) = inf \{T* : x(T) = 0, \forall T > T*\}$.

Definition 2. [18]. If the origin of the above system is asymptotically stable and there is a settling time $T(x_0) > 0$. If $\exists T_{max} > 0$ and the settling time $T(x_0) \leq T_{max}$ under any initial conditions, it is fixed-time stable.

Definition 3. [19]. $x(t) = 0$ is predefined-time (PDT) stable if two conditions are met:

(i) $x(t) = 0$ is finite-time stable;
(ii) For any constant $T_p > 0$, $sup_{x(0) \in R^n} T(x(0)) \leq T_p$. In this case, T_p is PDT.

Lemma 1. *[20]. For the strong connectivity graph $\mathcal{G}$, we have the following properties: $x^T L x = \frac{1}{2} \sum_{i=1}^{M} \sum_{j=1}^{M} a_{ij}(x_i - x_j)^2$, where $x = [x_1, x_2, \cdots, x_M]^T$. $\mathcal{L}$ is semipositive, assuming that the eigenvalue of $\mathcal{L}$ is labeled $0, \lambda_2, \cdots, \lambda_M$, and the second small eigenvalue $\lambda_2 > 0$. Also, if $1_n^T x = 0$, then $x^T L x \geq \lambda_2 x^T x$.*

Lemma 2. *[21]. $\Gamma(x) = \int_0^{+\infty} e^{-t} t^{x-1} dt$ represents the Gamma function. Suppose there exists a continuous function $V(\cdot) : R^n \to R_+ \cup \{0\}$, and three conditions are satisfied:*

(i) $V(x(t)) = 0 \Leftrightarrow x(t) = 0$;
(ii) $\|x(t)\| \to +\infty \Rightarrow V(x(t)) \to +\infty$;

(iii) For any non-zero $x(t) \in R^n$ and any constant $T_p > 0$,

$$\dot{V}(x(t)) \leq -\frac{\omega}{T_p}(\alpha V^q(x(t) + \beta V^r(x(t)))), \tag{2}$$

where $\alpha, \beta > 0, 0 < q < 1, r > 1$, and

$$\omega = \frac{\Gamma(\frac{1-q}{r-q})\Gamma(\frac{r-1}{r-q})}{\alpha(r-q)}(\frac{\alpha}{\beta})^{\frac{1-q}{r-q}}. \tag{3}$$

So $x(t) = 0$ is predefined-time stable, and T_p is predefined time.

Assumption 1. There is a positive known constant μ such that:

$$|f(x_i(t), t)| \leq \mu. \tag{4}$$

The nonlinear multi-agent system has M agents and its communication topology is connected undirected graph. The dynamics of the agent i has the following form:

$$\dot{x}_i(t) = u_i(t) + f(x_i(t), t), \tag{5}$$

where $i = 1, \cdots, M$. $x_i(t)$ represents the state of agent i, $u_i(t)$ as the control input, $f(x_i(t), t)$ is the uncertain nonlinear functions.

3 Main Results

3.1 A New Theorem for Predefined-Time Stability

Theorem 1. *For the above system (5), if there is a continuous positive definite function $V(x(t)) : R^n \to R, T_c$ is a user-defined parameter and meets the following two conditions:*

1. *$V(x(t)) = 0 \Leftrightarrow x(t) = 0$;*
2. *For any $V(x(t)) > 0$, $1 < r < 2, a, b > 0$, satisfied:*

$$\dot{V}(x(t)) \leq -\frac{G_c}{T_c}(aV(x(t)) + bV(x(t))^{r+sign(V(x(t)-1))}). \tag{6}$$

The system (5) can then achieve predefined-time stable, where:

$$G_c = \frac{1}{(2-r)a}ln\frac{a+b}{b} + \frac{1}{ra}ln\frac{a+b}{b}. \tag{7}$$

Proof. The settling time function can be expressed as:

$$\begin{aligned} T(x(0)) &= \int_0^{T(x(0))} dt, \\ &\leq \int_0^1 \frac{T_c}{Gc}\frac{dv}{aV + bV^{r-1}} + \int_1^\infty \frac{T_c}{G_c}\frac{dv}{aV + bV^{r-1}}. \end{aligned} \tag{8}$$

case 1:

Let $W = V^{2-r}, dV = \frac{V^{r-1}dW}{2-r}$,

$$\begin{aligned}\int_0^1 \frac{T_c}{G_c}\frac{dV}{aV+bV^{r-1}} &\leq \int_0^1 \frac{T_c}{G_c}\frac{1}{2-r}\frac{dW}{aW+b},\\ &= \frac{T_c}{G_c}\frac{1}{(2-r)a}ln\frac{a+b}{b}.\end{aligned} \tag{9}$$

case 2:

Let $W = V^{-r}, dV = \frac{V^{r+1}dW}{-r}$,

$$\begin{aligned}\int_1^\infty \frac{T_c}{G_c}\frac{dv}{aV+bV^{r-1}} &\leq \int_0^1 \frac{T_c}{G_c}\frac{1}{r}\frac{dW}{aW+b},\\ &= \frac{T_c}{G_c}\frac{1}{ra}ln\frac{a+b}{b}.\end{aligned} \tag{10}$$

Thus, we can get:

$$\begin{aligned}T(x_0) &\leq \frac{T_c}{G_c}(\frac{1}{(2-r)a}ln\frac{a+b}{b} + \frac{1}{ra}ln\frac{a+b}{b}),\\ &\leq T_c.\end{aligned} \tag{11}$$

3.2 Multi-agent Event-Triggered Consensus

Under an event-triggered strategy with continuous communication, exist a constant p, the control input of the agent i can be constructed as:

$$u_i(t) = \frac{G_c}{T_c}[-c_1 y_i(t_k^i) - c_2(y_i(t_k^i))^{2(r+sign(\sum_{i=1}^M y_i(t)^2)/p-1))-1}], \tag{12}$$

where $c_1, c_2 > 0$, t_k^i is latest triggering time for agent i, define $y_i(t)$:

$$y_i(t) = \sum_{j=1}^M a_{ij}(x_i(t) - x_j(t)), \tag{13}$$

thus,

$$\sum_{i=1}^M y_i(t)^2 = x^T(t)L^2x(t). \tag{14}$$

We can get the following result:

$$\lambda_M(L)x^T(t)Lx(t) \geq \sum_{i=1}^M y_i(t)^2 \geq \lambda_2(L)x^T(t)Lx(t), \tag{15}$$

where $\lambda_2(L)$ is the second smallest eigenvalue of matrix L, $\lambda_M(L)$ is the maximum eigenvalue of matrix L, there must be a constant $p \in [2\lambda_2(L), 2\lambda_M(L)]$ such that $\sum_{i=1}^M y_i(t)^2 = \frac{1}{2}px^T(t)Lx(t)$.

The measurement error of the agent i can be defined as:

$$\begin{aligned}E_i(t) &= c_1(y_i(t_k^i)) + c_2(y_i(t_k^i))^{2(r+sign(\sum_{i=1}^{M} y_i(t)^2/p-1))-1} \\ &- c_1 y_i(t) - c_2(y_i(t))^{2(r+sign(\sum_{i=1}^{M} y_i(t)^2/p-1))-1}.\end{aligned} \quad (16)$$

Combining the control input (12) and the measurement error (16), the control input can be sorted out:

$$u_i(t) = \frac{G_c}{T_c}[-E_i(t) - c_1 y_i(t) - c_2(y_i(t))^{2(r+sign(\sum_{i=1}^{M} y_i(t)^2/p-1))-1}]. \quad (17)$$

The event-triggered function of the agent i is constructed as:

$$g_i(t) = |E_i(t)| - \varepsilon c_1 |y_i(t)| - \varepsilon c_2 |y_i(t)|^{2(r+sign(\sum_{i=1}^{M} y_i(t)^2/p-1))-1}, \quad (18)$$

where $\varepsilon \in (0, 1)$ is the trigger parameter and can be selected later. Therefore, for the agent i, an event is fired when $g_i(t) \geq 0$. Its controller updates at its own event time $t_0^i, t_1^i, \cdots$.

Remark 1. In previous studies, commonly used trigger conditions have been designed based on error or communication time. When the error is too large or the communication time is too long, the trigger condition is reached, and the trigger occurs. The trigger condition used in this article is designed according to the error, and when $|E_i(t)| \geq c_1|y_i(t)| + c_2|y_i(t)|^{2(r+sign(\sum_{i=1}^{M} y_i(t)^2/p-1))-1}$, the system is triggered. Different from previous research, $c_2|y_i(t)|^{2(r+sign(\sum_{i=1}^{M} y_i(t)^2/p-1))-1}$ takes into account the superposition situation under small errors, and long-term small error superposition will also lead to event triggering, so it can be said that it is compatible with the limitations of large errors and long periods of untriggered.

Theorem 2. *Suppose Assumption 1 holds, and the following condition is satisfied:*

$$\mu \leq \frac{c_1 \lambda_2(L) G_c}{2T_c}. \quad (19)$$

Thus, the multi-agent system (5) is stable at a predefined time T_c.

Proof. Construct the following Lyapunov function:

$$\begin{aligned}V(t) &= \frac{1}{2}x^T(t)Lx(t), \\ &= \frac{1}{4}\sum_{i=1}^{M}\sum_{j=1}^{M} a_{ij}(x_i(t) - x_j(t))^2.\end{aligned} \quad (20)$$

For simplicity, let's write $V(t)$ as V.
According to the above equation:

$$\sum_{i=1}^{M}\sum_{j=1}^{M} a_{ij}(x_i(t) - x_j(t))^2 = 4V. \tag{21}$$

From Eqs. (14), (15) and (20), we get:

$$\sum_{i=1}^{M} y_i(t)^2 = pV. \tag{22}$$

Take the derivative of V:

$$\begin{aligned}
\dot{V} &= x^T(t)L\dot{x}(t),\\
&= \sum_{i=1}^{M} y_i(t)(u_i(t) + f(x_i(t), t)),\\
&= \frac{G_c}{T_c}\sum_{i=1}^{M} y_i(t)[-E_i(t) - c_1(y_i(t)) - c_2(y_i(t))^{2(r+sign(\sum_{i=1}^{M} y_i(t)^2/p-1))-1}]\\
&+ \sum_{i=1}^{M} y_i(t)f(x_i(t), t),\\
&\le -\frac{G_c}{T_c}[\sum_{i=1}^{M}|y_i(t)||E_i(t)| + c_1\sum_{i=1}^{M}(y_i(t))^2+\\
&c_2\sum_{i=1}^{M}(y_i(t))^{2(r+sign(\sum_{i=1}^{M} y_i(t)^2/p-1))}] + \sum_{i=1}^{M}\sum_{j=1}^{M} a_{ij}(x_i(t) - x_j(t))f(x_i(t), t),\\
&\le -\frac{G_c}{T_c}[\sum_{i=1}^{M}|y_i(t)||E_i(t)| + c_1\sum_{i=1}^{M}(y_i(t))^2+\\
&c_2\sum_{i=1}^{M}(y_i(t))^{2(r+sign(\sum_{i=1}^{M} y_i(t)^2/p-1))}] + \mu\sum_{i=1}^{M}\sum_{j=1}^{M} a_{ij}(x_i(t) - x_j(t))^2,\\
&\le -\frac{G_c}{T_c}[c_1\sum_{i=1}^{M}(y_i(t))^2 + c_2(pV)^{(r+sign(V-1))}] + 4\mu V,\\
&\le -\frac{G_c}{T_c}[c_1(2\lambda_2(L))V + c_2(pV)^{(r+sign(V-1))} - \frac{4\mu T_c}{G_c}V],\\
&= -\frac{G_c}{T_c}[(c_1(2\lambda_2(L) - \frac{4\mu T_c}{G_c})V + c_2(pV)^{(r+sign(V-1))}],\\
&= -\frac{G_c}{T_c}[a_1 V + c_2 p^{r+sign(V-1)}V^{r+sign(V-1)}],
\end{aligned} \tag{23}$$

where $a_1 = 2c_1\lambda_2(L)$, then, the following discussion needs to be done:

(1) when $V < 1$; $p^{r+sign(V-1)} = p^{r-1}$, $V^{r+sign(V-1)} = V^{r-1}$,

(2) when $V > 1$; $p^{r+sign(V-1)} = p^{r+1}, V^{r+sign(V-1)} = V^{r+1}$,
(3) when $V = 1$; $p^{r+sign(V-1)} = p^{r}, V^{r+sign(V-1)} = V^{r}$,

where $b_1 = c_2 min\left\{p^{r-1}, p^{r+1}, p^{r}\right\}$, then by using Theorem 1 and Theorem 2, it gives:

$$\dot{V} \leq \begin{cases} -\frac{G_c}{T_c}(a_1 V + b_1 V^{r-1}), & V < 1, \\ -\frac{G_c}{T_c}(a_1 V + b_1 V^{r+1}), & V > 1, \\ -\frac{G_c}{T_c}(a_1 V + b_1 V^{r}), & V = 1. \end{cases} \tag{24}$$

Thus,

$$\dot{V}(x(t)) \leq -\frac{G_c}{T_c}(a_1 V(x(t)) + b_1 V(x(t))^{r+sign(V(x(t)-1))}). \tag{25}$$

So, referring to Theorem 1, when $\dot{V} \leq -\dfrac{G_c}{T_c}(a_1 V + b_1 V^{r+sign(V-1)})$, the settling time $T \leq T_c$.

The proof is completed.

Remark 2. Compared with the finite-time consensus and fixed-time consensus, the predefined-time consensus has prominent advantages. First of all, the predefined-time is more flexible and can be set according to specific application scenarios and adjusted according to system requirements. Finite-time and fixed-time are usually fixed or can only be adjusted within a certain range. Secondly, the predefined-time is more adaptable to the dynamic system. However, the limited time and fixed-time may not adapt to the system changes in time. Therefore, this paper proposes a predefined-time event-triggered consensus.

4 Simulation Result

We will use an example to illustrate the usability of the proposed consensus algorithm. The undirected strong connectivity graph of the five agents is shown in Fig. 1.

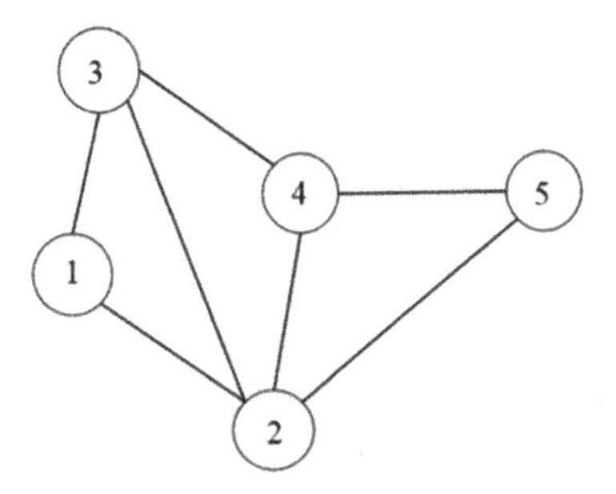

Fig. 1. An undirected graph of five agents.

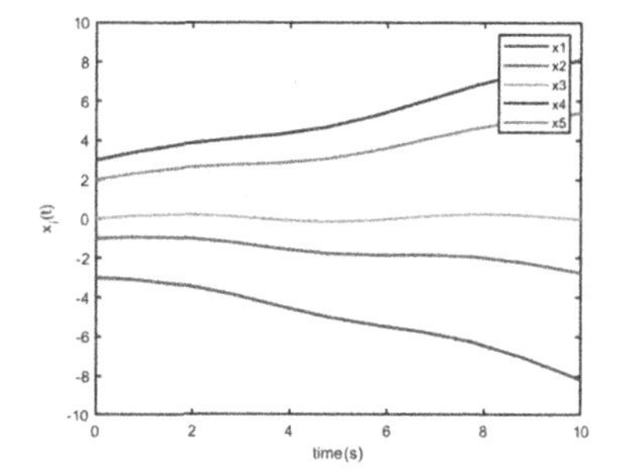

Fig. 2. The state trajectory of five agents without a controller when $T_c = 0.5$.

The dynamic model of agent i is written by:

$$\dot{x}_i(t) = u_i(t) + 0.5x_i(t) + 0.2sin(t). \tag{26}$$

Suppose the initial state of the five agents is $x(0) = [-3; -1; 0; 3; 2]^T$, and nonlinear function $f(x_i(t), t)$ satisfies Assumption 1 with $\mu = 0.2$. The state trajectories of the five agents without controllers are shown in Fig. 2. As shown in Fig. 2, when the five agents are not affected by the controller, the state cannot be consensus. It is possible that the state trajectory of an agent disappears. According to the undirected graph, L can be written as the following formula.

$$L = \begin{bmatrix} 2 & -1 & -1 & 0 & 0 \\ -1 & 4 & -1 & -1 & -1 \\ -1 & -1 & 3 & -1 & 0 \\ 0 & -1 & -1 & 3 & -1 \\ 0 & -1 & 0 & -1 & 2 \end{bmatrix}. \tag{27}$$

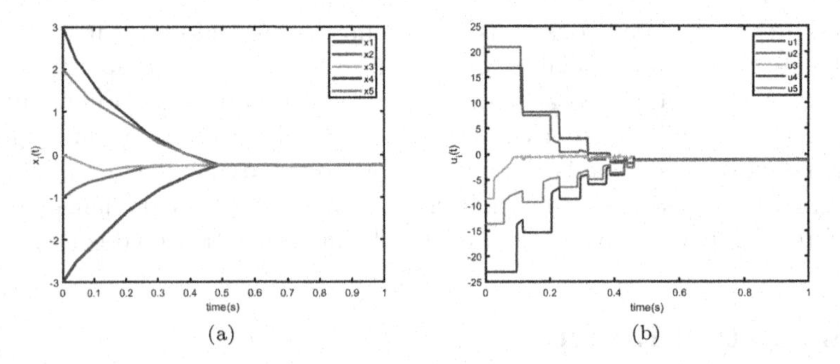

Fig. 3. (a) Represents the state trajectories of the five agents when $T_c = 1$, (b) represents the control input of the five agents when $T_c = 1$.

Under controller (12), select parameters $c_1 = 2; c_2 = 1; r = 1.6; \varepsilon = 0.6$. Fig. 3 and Fig. 4 respectively represent the state trajectories and control inputs of the five agents when T_c takes different values. It can be seen from Fig. 2 and Fig. 4 (a) that the system state of the agent can be consistent under the action of the controller. It can be seen from Fig. 3 (a) and Fig. 4 (a) that under the same conditions, the system state of the agent changes with different predefined-time parameter T_c, and both achieve consistency within T_c.

As can be seen from Fig. 3 (b) and Fig. 4 (b), the control inputs of the five agents are different when T_c takes different values. When the error is too large, the trigger condition (18) is reached, the trigger occurs, and the control input changes. As the controller takes effect, the error gradually decreases, and the control input will also reach stability and finally tend to 0. The figure below shows the fluctuation of the error measurement error of the five agents in $T_c = 1$, indicating that each of the five agents has its own event-triggered time.

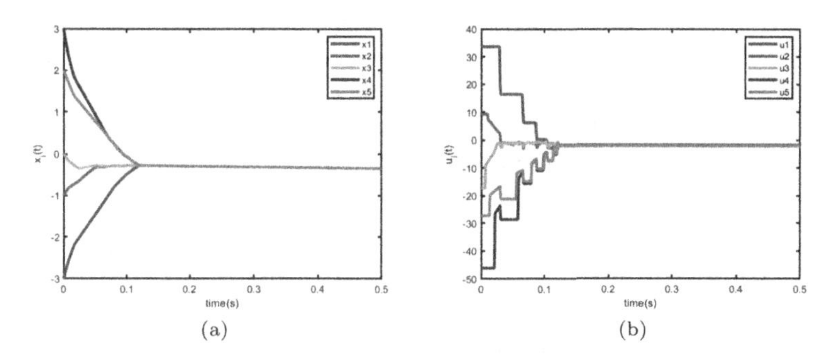

Fig. 4. (a) Represents the state trajectories of the five agents when $T_c = 0.5$, (b) represents the control input of the five agents when $T_c = 0.5$.

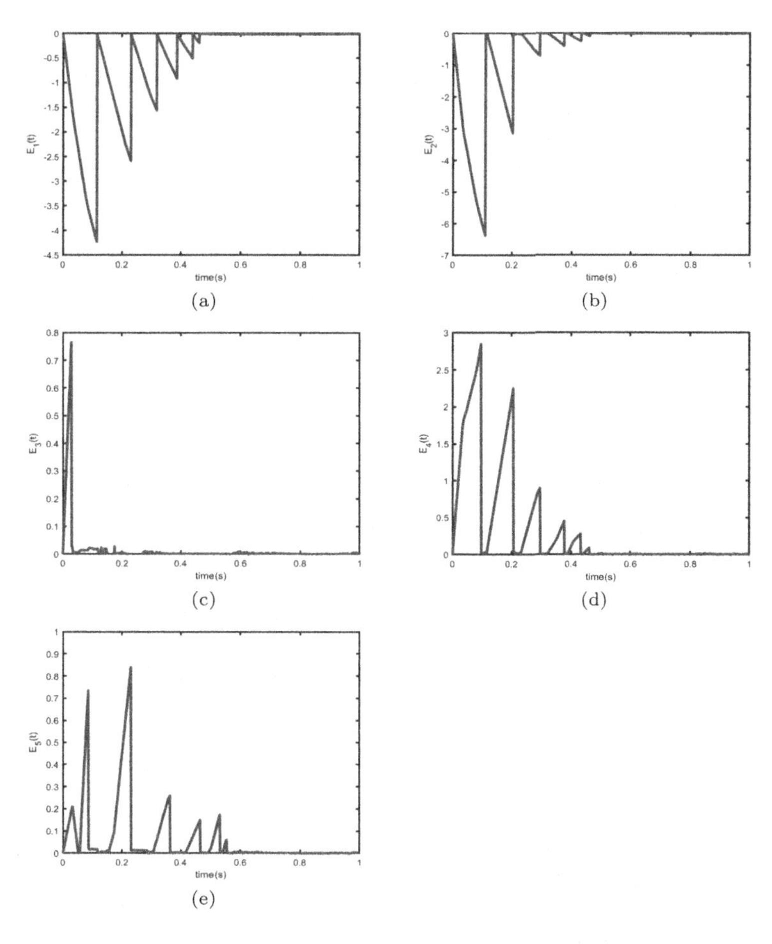

Fig. 5. The fluctuation of measurement error when the five agents communicate continuously under the controller.

As shown in the Fig. 5. First, under the action of the controller (12), the state trajectory of the multi-agent system can be consistent, and the control input of each agent is different due to the setting of trigger conditions. Second, when the predefined time parameters are different, the time to achieve consistency is also different. Finally, the measurement error of the five agents tend to 0 for a predefined time T_c.

5 Conclusion

This work investigates the predefined-time event-triggered consensus problem for nonlinear uncertain multi-agent systems. By designing the event-triggered controller, communication consumption is greatly reduced. The proposed predefined-time consensus differs from the existing finite-time consensus and fixed-time consensus in that it does not depend on the initial conditions and the parameters are adjustable. In addition, we demonstrate its feasibility for predefined-time stability conditions. In the future, we hope to investigate the consensus problem between predefined-time leaders-follower in linear multi-agent systems. First, the problem of leader-follower consistency is more practical than what this manuscript examines. Second, in the application, the leader-follower is easy to implement.

Acknowledgements. This work is supported by the National Natural Science Foundation of China (Grant Nos.62103165, 62101213), and the Natural Science Foundation of Shandong Province (Grant No.ZR2022ZD01).

References

1. Wang, Y., Cheng, Z., Xiao, M.: UAVs' formation keeping control based on multi-agent system consensus. IEEE Access **8**, 49000–49012 (2020)
2. Vazquez, T.: Robust formation control based on leader-following consensus in multi-agent systems with faults in the information exchange: application in a fleet of unmanned aerial vehicles. IEEE Access **9**, 104940–104949 (2021)
3. Amirkhani, A., Barshooi, A.H.: Consensus in multi-agent systems: a review. Artif. Intell. Rev. **55**(5), 3897–3935 (2022)
4. Li, Z., Duan, Z., Huang, L.: Leader-follower consensus of multi-agent systems. In: American Control Conference, pp. 3256–3261. IEEE (2009)
5. Dong, G., Li, H., Ma, H., et al.: Finite-time consensus tracking neural network FTC of multi-agent systems. IEEE Trans. Neural Netw. Learn. Syst. **32**(2), 653–662 (2021)
6. Bu, X., Zhu, P., Hou, Z., Liang, J.: Finite-Time consensus for linear multi-agent systems using data-driven terminal ILC. IEEE Trans. Circuits Syst. II Express Briefs **67**(10), 2029–2033 (2020)
7. Wang, H., Yu, W., Wen, G., et al.: Fixed-time consensus of nonlinear multi-agent systems with general directed topologies. IEEE Trans. Circuits Syst. II Express Briefs **66**(9), 1587–1591 (2018)

8. Wei, X., Yu, W., Wang, H., Yao, Y., Mei, F.: An observer-based fixed-time consensus control for second-order multi-agent systems with disturbances. IEEE Trans. Circuits Syst. II Express Briefs **66**(2), 247–251 (2019)
9. Zhu, Y., Wang, Z., Liang, H., et al.: Neural-network-based predefined-time adaptive consensus in nonlinear multi-agent systems with switching topologies. In: IEEE Transactions on Neural Networks and Learning Systems, pp. 1–11 (2023)
10. Li, K., Hua, C.: Output feedback predefined-time bipartite consensus control for high-order nonlinear multiagent systems. IEEE Trans. Circuits Syst. I Regul. Pap. **68**(7), 3069–3078 (2021)
11. Liang, C.D., Ge, M.F., Liu, Z.W., et al.: Predefined-time stabilization of t-s fuzzy systems: a novel integral sliding mode-based approach. IEEE Trans. Fuzzy Syst. **30**(10), 4423–4433 (2022)
12. Ni, J., Liu, L., Liu, C.: Predefined-time consensus tracking of second-order multiagent systems. IEEE Trans. Syst. Man Cybern. Syst. **51**(4), 2550–2560 (2021)
13. Wang, Z., Ma, J.: Distributed predefined-time attitude consensus control for multiple uncertain spacecraft formation flying. IEEE Access **10**, 108848–108858 (2022)
14. Yi, X., Liu, K., Dimarogonas, D.V., et al.: Dynamic event-triggered and self-triggered control for multi-agent systems. IEEE Trans. Autom. Control. **64**(8), 3300–3307 (2018)
15. Hu, W., Yang, C., Huang, T., et al.: A distributed dynamic event-triggered control approach to consensus of linear multi-agent systems with directed networks. IEEE Trans. Cybern. **50**(2), 869–874 (2018)
16. Hu, S., Yue, D., Chen, X., Dou, C.: Observer-based event-triggered control for networked linear systems subject to denial-of-service attacks. IEEE Trans. Cybern. **50**(5), 1952–1964 (2022)
17. Lin, X., Li, S., Zou, Y.: Finite-time stability of switched linear systems with subsystems which are not finite-time stable. IET Control Theory App. **8**(12), 1137–1146 (2014)
18. Polyakov, A.: Nonlinear feedback design for fixed-time stabilization of linear control systems. IEEE Trans. Autom. Control **57**(8), 2106–2110 (2012)
19. Chen, C., Mi, L., Zhao, D., et al.: A new judgment theorem for predefined-time stability and its application in the synchronization analysis of neural networks. Int. J. Robust Nonlinear Control **32**(18), 10072–10086 (2022)
20. Olfati-Saber, R., Murray, R.M.: Consensus problems in networks of agents with switching topology and time-delays. IEEE Trans. Autom. Control **49**(9), 1520–1533 (2004)
21. Aldana-López, R., Gómez-Gutiérrez, D., Jiménez-Rodríguez, E., et al.: Enhancing the settling time estimation of a class of fixed-time stable systems. Int. J. Robust Nonlinear Control **29**(12), 4135–4148 (2019)

Cascaded Fuzzy PID Control for Quadrotor UAVs Based on RBF Neural Networks

Zicheng Huang[1], Huiwei Wang[2(✉)], and Xin Wang[2]

[1] Westa College, Southwest University, Chongqing 400715, China
[2] College of Electronic and Information Engineering, Southwest University, Chongqing 400715, China
hwwang@swu.edu.cn

Abstract. Since quadrotor UAVs often need to fly in complex and changing environments, their systems suffer from slow smooth control response, weak self-turbulence capability, and poor self-adaptability. Thus, it is crucially important to carefully formulate a quadrotor UAV control system that can maintain high-precision control and high immunity to disturbance in complex environments. In this paper, an improved nonlinear cascaded fuzzy PID control approach for quadrotor UAVs based on RBF neural network is proposed. Based on the analysis and establishment of the UAV flight control model, this paper designs a control approach with an outer-loop fuzzy adaptive PID control and an inner-loop RBF neural network. The simulation results show that introducing RBF neural networks into the nonlinear fuzzy adaptive PID control can make it have better high-precision control and high anti-disturbance under the influence of different environmental variables.

Keywords: Quadrotor UAV · Adaptive fuzzy PID · RBF neural network · Flight control

1 Introduction

Today, quadrotor UAVs have a wide range of applications and specific functions in the military, commercial and civilian sectors. However, since quadrotor UAVs often need to fly in complex and changing environments and they are multi-input, multivariable, underdriven, semi-coupled and nonlinear systems, they are susceptible to the influence of external adverse variables. Therefore, it becomes critical to design quadrotor UAV control systems to maintain high accuracy control and high immunity to disturbances in complex environments.

Currently, the main research directions for quadrotor UAV control include nonlinear PID control, adaptive PID control, model prediction-based PID control, reinforcement learning-based PID control and deep learning-based PID control methods. At present, the research on PID control structure for quadrotor

*Supported by Natural Science Foundation of Chongqing cstc2020jcyj-msxmX0057.

B. Luo et al. (Eds.): ICONIP 2023, LNCS 14447, pp. 516–527, 2024.
https://doi.org/10.1007/978-981-99-8079-6_40

UAV mainly focuses on the optimization of unipolar PID control algorithm and string-level PID algorithm, which solves the problem of PID control's own simplicity and lack of adaptiveness to a certain extent, but fewer people introduce the fuzzy adaptive system based on RBF neural network into the string-level PID control, and on this basis, the correlation and update mode of internal and external loop parameters are modeled and theoretically explored. Most of the current studies are limited to enhancing the control system performance without introducing complex environmental variables (e.g., wind speed [1], terrain [2], altitude [3]) or considering the impact of human requirements (e.g., high immunity and high precision control) on the control algorithm performance.

In view of the shortcomings in the above research scheme, an improved RBF neural network-based nonlinear cascaded fuzzy PID control approach for quadrotor UAVs is proposed in this paper. In the meantime, by taking into account the impact of natural factors such as wind speed, terrain and altitude on the parameters associated with the control system of the UAV, certain artificial criteria such as strong immunity and precise control are quantified and incorporated into the system as environmental tendencies. The results from the experiments indicate a substantial enhancement in both control accuracy and immunity of the quadrotor UAV control system.

2 Literature Review

Currently, some of the more widely used iterative-based methods to achieve adaptive control of quadrotor UAVs at home and abroad include neural network-based model predictive control algorithms [4], non-monolithic fuzzy logic control algorithms with enhanced input uncertainty sensitivity [5], adaptive sliding mode control algorithms [6], and control algorithms based on swarm intelligence algorithms [7].

In the literature [4], feedforward neural networks were integrated into a model predictive controller (MPC), resulting in a 40 percent reduction in average tracking error compared to a conventional PID controller. However, a potential limitation of this study is the reliance on flight data collected from pre-designed trajectories, leaving its system lacking adaptive and iterative capabilities.

In the literature [5], real-time flight tests using vision-inertial simultaneous localization and mapping (SLAM) show that the Cen-NSFLC outperforms conventional PID controllers and singleton FLCs, exhibiting better control performance in uncertain environments. However, a potential limitation of this study is the lack of comparison with other advanced control methods or algorithms.

The literature [6] proposes an improved adaptive sliding mode control (IASMC) that achieves fast adaptation, robustness to uncertainty and disturbances. However, the method may involve complex control algorithms and parameter tuning processes, leading to increased difficulties in implementation and debugging.

In the literature [7], a system based on particle swarm optimization with series-level fuzzy PID control is proposed to determine the quantization factor, scaling factor. However, it fails to consider the impact of the environment on particle swarm optimization algorithms.

3 Quad Rotor UAV Dynamic Model

3.1 Movement Style

From the depicted Fig. 1, E_A and E_B are the body coordinate system, e is the world coordinate system, it is evident that the body of the quadrotor UAV adopts a solid crossover configuration where the upward force produced by four motors is proportional to the square of F_1, F_2, F_3, and F_4. By controlling the rotational speed of the four rotors on the two vertical frames, it tracks the desired speed value when the system is balanced to achieve six different states of its own motion.

The vectors ϕ, θ and ψ represent the three rotation variables of the UAV, respectively. Similarly, the vectors p, q and r represent the rotational angular velocities, respectively.

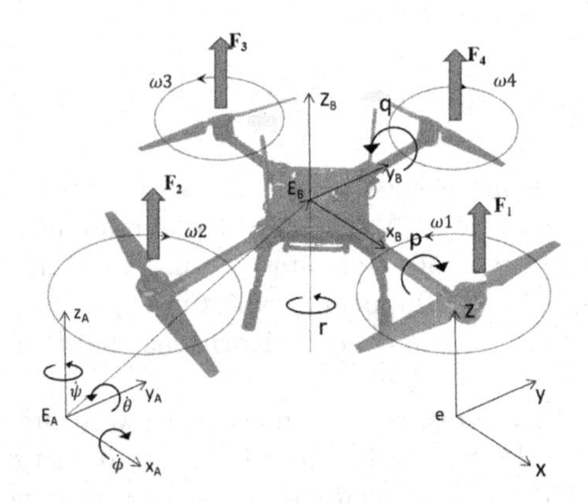

Fig. 1. Airframe - geographical coordinate system of quadrotor UAV

3.2 Model Building

From Fig. 1, it can be seen that once the UAV makes a traverse motion, pitch motion and yaw motion can be regarded as a rotation motion around the x, y and z axis, therefore, all attitude changes of the UAV can be represented by the rotation matrix:

$$R_E^e = \begin{bmatrix} \cos\phi\cos\theta & \sin\theta\cos\phi\sin\psi - \sin\phi\cos\psi & \sin\theta\cos\phi\cos\psi + \sin\phi\sin\psi \\ \sin\phi\cos\theta & \sin\theta\cos\phi\sin\psi + \sin\phi\cos\psi & \sin\theta\sin\phi\cos\psi - \cos\phi\sin\psi \\ -\sin\theta & \cos\theta\sin\psi & \cos\phi\cos\psi \end{bmatrix}. \quad (1)$$

Make following assumptions before modeling:

- The quadrotor UAV is a rigid body.
- There is no wind in the flight environment.
- The quadrotor UAV is not affected by other external disturbances.
- The quadrotor UAV has no energy loss during flight.

These assumptions help simplify the problem and establish a preliminary mathematical model for control system design and performance analysis. However, there may be factors in the actual situation that do not exactly match these assumptions, so that more refined modeling and control strategy design are required for specific situations in practical applications.

From the Newton-Euler modeling approach, the UAV linear motion model is as follows:

$$\begin{cases} \dot{v}_x = \dfrac{F_t}{m}(\cos\phi\cos\psi\sin\theta + \sin\phi\sin\psi) + \gamma_1, \\ \dot{v}_y = \dfrac{F_t}{m}(\cos\phi\sin\theta\sin\psi - \cos\psi\sin\phi) + \gamma_2, \\ \dot{v}_z = -g + \dfrac{F_t}{m}\cos\phi\cos\theta + \gamma_3, \end{cases} \tag{2}$$

where m is the UAV mass; g is the acceleration of gravity; F_t is the combined force of the lift provided by the UAV quadrotor; γ_1-γ_3 are the external disturbance, respectively. The angular motion model is given as follows:

$$\begin{cases} \dot{p} = \left(\dfrac{I_y - I_z}{I_x}\right) qr + \dfrac{\tau_p}{I_x} + \gamma_4, \\ \dot{q} = \left(\dfrac{I_z - I_x}{I_y}\right) pr + \dfrac{\tau_q}{I_y} + \gamma_5, \\ \dot{r} = \left(\dfrac{I_x - I_y}{I_z}\right) pq + \dfrac{\tau_r}{I_z} + \gamma_6, \\ \dot{\phi} = p + (r\cos\phi + q\sin\phi)\, tan\theta \\ \dot{\theta} = q\cos\phi - r\sin\phi \\ \dot{\psi} = \dfrac{1}{\cos\theta}(r\cos\phi + q\sin\phi) \end{cases} \tag{3}$$

where I_x, I_y, and I_z represent the moment of inertia for rotational motion around the corresponding axis, respectively; τ_p, τ_q, and τ_r are the moment term in each direction, respectively; γ_4-γ_6 are the external disturbance, and the moment terms are defined as follows

$$\begin{cases} \tau_p = -F_2 * l + F_4 * l \\ \tau_q = -F_1 * l + F_3 * l \\ \tau_r = -F_1 * d + F_2 * d - F_3 * d + F_4 * d \end{cases} \tag{4}$$

where l indicates the center-of-mass-rotor distance; d is the given torque coefficient, make the relationship between the combined force and the rotational speed as follows:

$$F_t = F_1 + F_2 + F_3 + F_4 = k\omega_1^2 + k\omega_2^2 + k\omega_3^2 + k\omega_4^2 \tag{5}$$

where k is the rotor lift coefficient and ω_1-ω_4 are the motor speed of the corresponding rotor.

4 Control System Design

In this paper, the self-anti-disturbance controller is improved. Based on the series-level PID controller, a fuzzy controller is used to iterate the outer-loop angle parameters, and an RBF neural network is introduced in the inner loop to optimize the global parameters in combination with environmental variables, the control system structure block diagram is shown in Fig. 2.

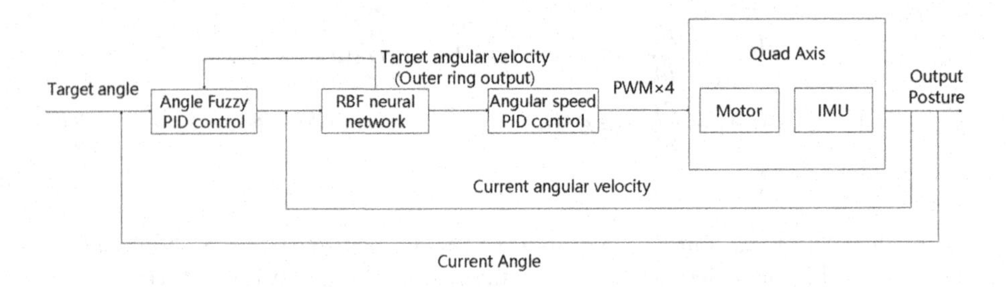

Fig. 2. Structure of RBF neural networks based series-level fuzzy PID control

4.1 Fuzzy PID Control

The tandem-level fuzzy PID control system has good adaptability and can automatically adjust the control parameters according to the changes of flight environment through the relationship between deviation E and deviation rate EC, the global parameters are optimized according to the fuzzy rules, the schematic diagram is as follows (Fig. 3):

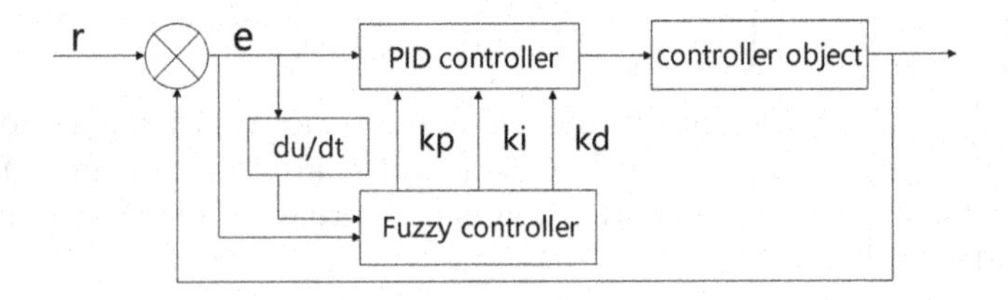

Fig. 3. Structure of series-level fuzzy PID controller

For the fuzzy PID controller, the following settings are made:

1) Determine the fuzzy domain. ΔKp, ΔKi, ΔKd are the proportional, integral, and differential variation quantities, respectively. The basic domain $(-5, 5)$ is defined, the actual domain of E, EC are $[-8, 8]$, the actual domain

Table 1. ΔKp, ΔKi, and ΔKd fuzzy rules.

E	EC																				
	ΔKp							ΔKi							ΔKd						
	NB	NM	NS	ZO	PS	PM	PB	NB	NM	NS	ZO	PS	PM	PB	NB	NM	NS	ZO	PS	PM	PB
NB	PB	PB	PM	PM	PS	ZO	ZO	NB	NB	NM	NM	N	ZO	ZO	PS	NS	NB	NB	NB	NM	PS
NM	PB	PB	PM	PS	PS	ZO	NS	NB	NB	NM	NM	NS	ZO	ZO	PS	NS	NB	NM	NM	NS	ZO
NS	PM	PM	PM	PS	ZO	NS	NS	NB	NM	NS	NS	ZO	PS	PS	ZO	NS	NM	NM	NS	NS	ZO
ZO	PM	PM	PS	ZO	NS	NM	NM	NM	NM	NS	ZO	PS	PM	PM	ZO	NS	NS	NS	NS	NS	ZO
PS	PS	PS	ZO	NS	NS	NM	NM	NM	NS	ZO	PS	PS	PM	PB	ZO	ZO	ZO	ZO	ZO	ZO	ZO
PM	PS	ZO	NS	NM	NM	NM	NB	ZO	ZO	PS	PS	PM	PB	PB	PB	NS	PS	PS	PS	PS	PB
PB	ZO	ZO	NM	NM	NM	NB	NB	ZO	ZO	PS	PM	PM	PB	PB	PB	PM	PM	PM	PS	PS	PB

of output ΔKp, ΔKi, and ΔKd are $[-15, 15]$, and the triangular subordination function is used.

2) Develop fuzzy control rules. The Mamdani method of reasoning is used, and 49 fuzzy rules are established, as shown in Table 1.

3) Frame ΔKp, ΔKi, ΔKd and adjust the law.

The values of ΔKp are used to adjust the gain of the output response. In the early stage of the system response, in order to maintain a fast response and decrease the absolute value of the error, a larger value of ΔKp should be selected appropriately, in the middle of the system response, the value of ΔKp can be relatively lower to prevent overshooting too much, and in the late stage of the system response, a larger value of ΔKp should be taken to promote the response speed [8].

The value of ΔKi is used to eliminate the gain of the integration error. In the early stage of the system response, the value of ΔKi should be selected relatively small to decrease the absolute value of the error, in the middle of the system response, decreasing the value of ΔKi to maintain a smooth system and prevent overshoot too large, the value of ΔKi is not easy to be too large, in the late stage of the system response, it should be reduced to avoid oscillations caused by the integral overshoot, or even the value of ΔKi can be canceled [9].

The value of ΔKd is used to nullify the amplification of the deviation's rate of change. In the early stage of system response, the value of ΔKd can be increased appropriately to avoid overshoot braking; in the middle stage of system response, when the error increases, a larger value of ΔKd should be taken; once the error reaches the maximum value and the system tends to the steady state, the value of ΔKd can be reduced or maintained appropriately to ensure that the system can maintain a fast response and prevent overshoot too much; in the late stage of system response, it should be increased to reduce overshoot [10].

4.2 RBF Neural Networks Optimized Fuzzy PID Controller Parameters

Although fuzzy PID control has a certain degree of adaptivity, fuzzy PID control still needs to set a reasonable set of initial parameters based on a priori knowledge, and the parameters still have to be continuously adjusted and optimized.

Therefore, RBF neural networks can be introduced to adjust the parameters of the fuzzy PID controller, so that the system can achieve more accurate, robust and efficient control and improve the safety and stability of the UAV.

In the topology of the set three-layer RBF neural network, the hidden layer and the output layer are expressed as follows:

$$h_j = \exp\left(-\frac{\|x - a_j\|^2}{2b_j^2}\right), j = 1, 2, \cdots, p \tag{6}$$

$$Z = \sum_{j=1}^{p} W_j h_j + \delta, j = 1, 2, \cdots, p \tag{7}$$

where a_j refers to the central vector of the jth neuron within the hidden layer, b_j denotes the width vector of the jth neuron within the hidden layer, h_j represents the output value of the jth neuron within the hidden layer, p refers to the count of neurons in the hidden layer, W_j denotes the approximated weight connecting the output layer to the jth neuron within the hidden layer, δ is the error of approximation Z of the ideal neural network, Z is a hyperparameter to balance the range of output values, is the network output.

Secondly, initialize the weights a_j, b_j, W_j using the exponential function of the second order parametrization of the input and hidden layers as the initial choice of W_j, thus reducing the range of variation of the gradient and facilitating better propagation of the gradient, as show:

$$W_j = \exp\left(\frac{p\|x - h_j\|^2}{h_{\max}^2}\right), j = 1, 2, \cdots, p \tag{8}$$

where h_{max} represents the highest value among the neurons in the hidden layer. The initial value of the width vector b_j is determined by (9). The center vector is selected directly from the given training sample set according to the least squares method based on the on-the-fly experience.

$$b_j = \frac{h_{\max}}{\sqrt{2p}}, 1, 2, \cdots, p \tag{9}$$

Subsequently, it becomes crucial to establish the input and loss function for the neural network, so that the expected value of the quantization factor E is E_d and the deviation of the quantization factor can be represented as $g_E = E - E_d$, and the derivative of its error expression can be obtained as $\dot{g_E} = \dot{E} - \dot{E_d}$, so let the input of the neural network be:

$$x = \left[g_E, \dot{g}_E\right] \tag{10}$$

Define the Lyapunov function [11]:

$$V\left(g_E\right) = \frac{1}{2} g_E^2 \tag{11}$$

Thus there are:

$$V(\dot{g}_E) = g_E \dot{g}_E = g_E \left(\dot{E} - \dot{E}_d\right) \tag{12}$$

According to Lyapunov's stability theorem, it is obtained that if going for the system to gradually stabilize, one needs to let $\dot{V}(g_E) \leq 0$. So adding the virtual control term $E^v = \dot{E}_d - kg_E$ in which $k > 0$, then the error variable g_E^v is depicted by the subsequent equation:

$$g_E^v = \dot{E} - E^v = \dot{E} - \dot{E}_d + kg_E \tag{13}$$

Bringing equation (13) into equation (12), we get:

$$V(\dot{g}_E) = g_E \dot{g}_E = g_E \left(\dot{E} - \dot{E}_d\right) = g_E \left(g_E^v - kg_E\right) = g_E g_E^v - kg_E^2 \tag{14}$$

From Eq. (14), it is known that when the positive direction $g_E g_E^v$ is close to 0, based on Lyapunov's stability theorem, the model of the network will tend to be more stable, based on this, the loss function of the neural network is set as the cross-entropy loss function to measure the disparity between the predicted probability and the true label, so as to improve the self-adaptive ability of the control system of the UAV in different environments, which can facilitate the updating of the weights of the network, which can provide a larger gradient signal and help to avoid the gradient disappearance problem, and $g_E g_E^v$ is introduced as a regularization parameter with the following expression:

$$\mathcal{L} = -\frac{1}{N} \sum_{i=1}^{N} \sum_{t=1}^{C} y_{i,t} \log(p_{i,t}) + g_E g_E^v \tag{15}$$

where N is the number of samples, C is the environmental category quantity, $y_{i,t}$ determines whether the ith sample belongs to the true category t, and $p_{i,t}$ denotes the likelihood of the category overlapping with the ith sample t. Ultimately, the relevant parameters will be updated according to (16):

$$\begin{aligned} W_j(t) &= w_j(t-1) - \eta \frac{\partial E}{\partial w_j(t-1)} + \alpha \left[w_j(t-1) - w_j(t-2)\right] \\ a_{ij}(t) &= a_{ij}(t-1) - \eta \frac{\partial E}{\partial a_{ij}(t-1)} + \alpha \left[a_{ij}(t-1) - a_{ij}(t-2)\right] \\ b_{ij}(t) &= b_{ij}(t-1) - \eta \frac{\partial E}{\partial b_{ij}(t-1)} + \alpha \left[b_{ij}(t-1) - b_{ij}(t-2)\right] \end{aligned} \tag{16}$$

where α and η represent the learning rate, and $\mathcal{L}$ represents the evaluation function utilized in the RBF neural network. The flow of RBF neural network optimized fuzzy PID controller is shown in Fig. 4.

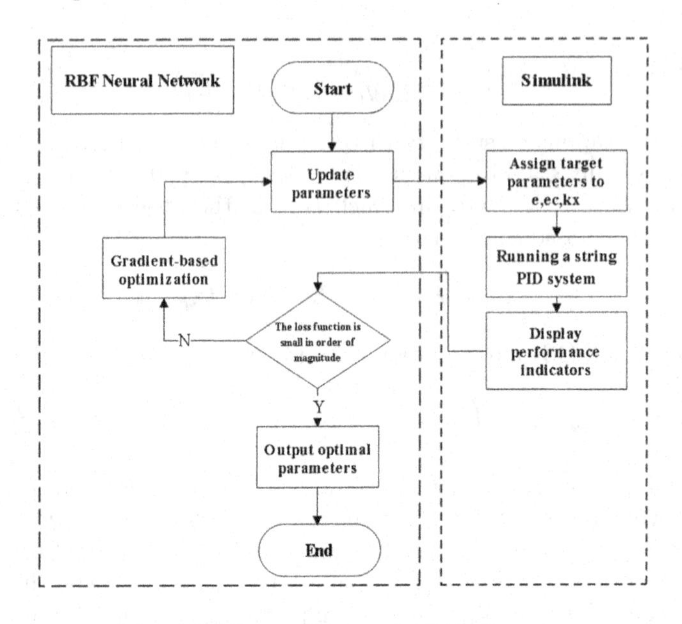

Fig. 4. RBF neural network block diagram

5 Control System Simulation and Analysis

First, the selected maximum number of iterations $n = 50$. E, EC, Kp, Ki, and Kd are taken in the range of $[0, 10]$, so that $E\,(t)$, $EC\,(t)$, $Kp\,(t)$, $Ki\,(t)$, and $Kd\,(t)$ represent the probability density function for each factor, respectively, the independent variables are also taken in the range of $[0, 10]$, its accuracy is two decimal places, the dependent variable is in the range of [0,1], so that the initial values of the dependent variable are $1/1000$, corresponding to the same probability of each value in the initial selection range of each factor, and the sum of all values in the range of each factor is equal to 1. The initial selection of the weight parameters of each layer is performed according to Eqs. (8), (9) and the least squares method. Let the learning factor α, η based on the flight experience, select the quantization parameters and proportional parameters when the UAV control system reaches stability under different wind speed, terrain and altitude environments, and iterate the weight parameters of each layer of the network through supervised learning.

When the UAV is flying, the neural network controls each parameter of the fuzzy controller based on gradient descent, and outputs the selected optimal solution when the number of iterations reaches a set number or the order of magnitude of the loss function is small enough. The Fig. 5 shows the optimization curve of the neural network on fuzzy controller.

From Fig. 5, the value of the parameters obtained by the RBF neural network optimization are: $E = 2.7$, $EC = 0.68$, $Kp = 0.9$, $Ki = 2.3$, $Kd = 0.8$ respectively. The optimal solution is obtained at 31 iterations of the neural network with the interference set to $D = sin\,(0.1t)$, the absolute value of the time error

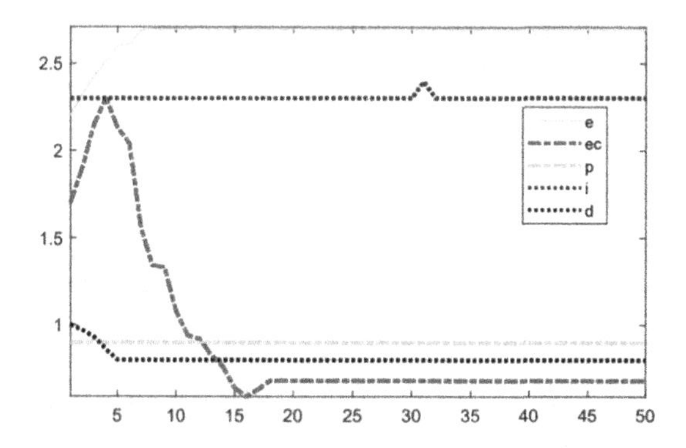

Fig. 5. Optimization curve of each parameter of fuzzy controller by neural network iteration

is then chosen as the fitness function ITAE criterion, and the adaptation value at this time is $J = 1.113$.

$$J = \int_0^{\infty} t\,|e|\,dt \tag{17}$$

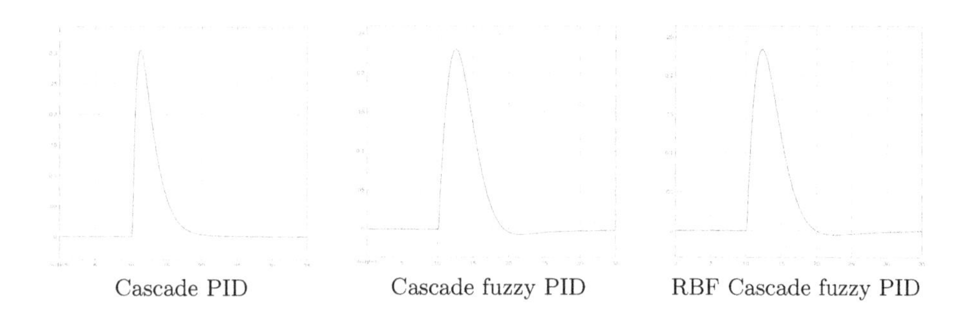

Fig. 6. Comparison of different control systems

Figure 6 shows the step response signal for velocity parameters corresponding to the cascade PID control, flight experience-based cascade fuzzy PID control and RBF neural network optimization-based cascade fuzzy PID control, and to reflect the response process more clearly, a delay of 1 s is added to each of the three systems and the time parameter is expanded by a factor of 10. The simulation curves show that when the system is affected by the step signal, all the three controlled systems return to the steady state within 3.5 s, which indicates that all the three systems have a certain degree of self-adaptability, the system subject to series PID control has a short settling time, but the lack of adaptivity leads to a significant overshoot amplitude; due to the fuzzy rules, the system subject to cascade fuzzy PID control avoids overshoot due to inertia and its response becomes smoother in recovery, but the settling time is longer;

Table 2. System Response Performance

Control strategy Response time	Settling time (sec)	Overshoot (%)	Peak time (sec)
String level PID control	1.78	30.7	1.12
String level fuzzy PID control	2.43	22.7	1.24
RBF neural network based fuzzy PID control	2.24	22.6	1.2

the series fuzzy PID control system, optimized by an RBF neural network, possesses enhanced nonlinear approximation capability, it effectively adapts to the nonlinear characteristics of the UAV while minimizing overshoot, furthermore, the memory function of the RBF neural network enables the system to reach a steady state more accurately and quickly, particularly in the presence of specific disturbances. The corresponding data are presented in Table 2.

If additional random number interference based on different environments are introduced into the inner loop of the system to evaluate the resilience of the inner loop, the simulation outcomes are depicted in Fig. 7.

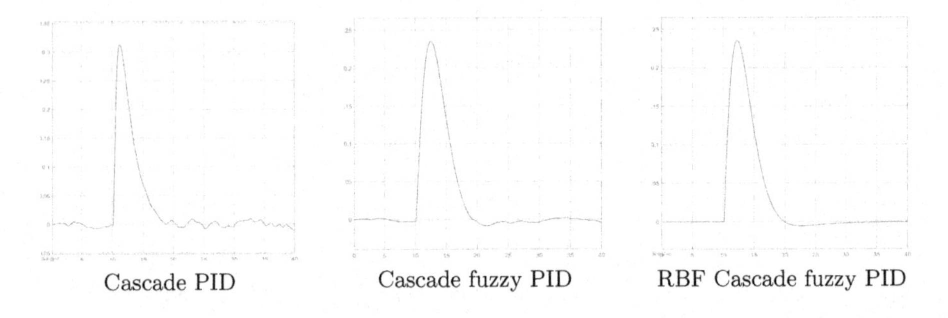

Fig. 7. Comparison of different control systems with interference

As can be seen in Fig. 7, all three control strategies can bring the system back to a stable attitude and maintain a small overshoot. Among them, fuzzy control plays a key role in mitigating the effects of multiple disturbances on the system's inner loop by adaptively adjusting its PID parameters according to the dynamics of the disturbances, thus ensuring that the system does not collapse in the event of a failure in the inner loop. Furthermore, by using RBF neural networks to optimize the string-level fuzzy PID control system, the system performance can be further optimized and the immunity to disturbances can be significantly improved, and the system can be observed to exhibit better stability and faster response at 2.5, 3 and 4 s. In summary, the optimization of the string-level fuzzy PID control system using RBF neural networks can effectively cope with the needs of UAVs facing external uncertainties, and make the control system of UAVs more resistant to interference and acquire precise control.

6 Conclusion

Synthesizing the above analysis, this paper proposed an improved nonlinear cascaded fuzzy PID control approach based on RBF neural networks for quadrotor UAVs, aiming to improve its performance in complex and variable environments. Through experimental simulation and comparative analysis, we conclude that the problems of slow control response smoothness, weak self anti-disturbance ability and poor self-adaptive ability can be solved to a certain extent, and the composite system significantly reduces the oscillation and jitter during UAV flight and effectively deals with the nonlinearity and time-variability during UAV flight. Meanwhile, the environmental trend factor is introduced to consider the influence of external factors, which further improves the adaptability and robustness of the control system.

References

1. Xiao, C., Mao, Y., Yuan, H., Wen, Y.: Design and simulation of intelligent control algorithm for quadcopter under windy conditions. Comput. Sci. **45**(05), 310–316 (2018)
2. Elmokadem, T., Savkin, A.V.: A method for autonomous collision-free navigation of a quadrotor UAV in unknown tunnel-like environments. Robotica **40**(4), 835–861 (2022)
3. Gao, H., Yan, S.: Research on flight attitude control technology of quadcopter in high-altitude environments. Electron. Des. Eng. **29**(15), 17–20,26 (2021)
4. Jiang, B., Li, B., Zhou, W., Lo, L.Y., Chen, C.K., Wen, C.Y.: Neural network based model predictive control for a quadrotor UAV. Aerospace **9**(8), 460 (2022)
5. Fu, C., Sarabakha, A., Kayacan, E., Wagner, C., John, R., Garibaldi, J.M.: Input uncertainty sensitivity enhanced nonsingleton fuzzy logic controllers for long-term navigation of quadrotor UAVs. IEEE/ASME Trans. Mechatron. **23**(2), 725–734 (2018)
6. Eltayeb, A., Rahmat, M.F.A., Basri, M.A.M., Eltoum, M.M., El-Ferik, S.: An improved design of an adaptive sliding mode controller for chattering attenuation and trajectory tracking of the quadcopter UAV. IEEE Access **8**, 205968–205979 (2020)
7. Shen, Y., Xing, H., Wang, S.: Particle swarm optimization-based cascaded fuzzy PID flight control system for unmanned aerial vehicles. Electron. Measure. Technol. **45**(01), 96–103 (2022)
8. Dong, J., He, B.: Novel fuzzy PID-type iterative learning control for quadrotor UAV. Sensors **19**(1), 24 (2018)
9. Abbasi, E., Mahjoob, M., Yazdanpanah, R.: Controlling of quadrotor UAV using a fuzzy system for tuning the pid gains in hovering mode. In: 10th International Conference on Advances in Computer Entertainment Technology, pp. 1–6. Tehran, Iran (2013)
10. Wang, Y., Chenxie, Y., Tan, J., Wang, C., Wang, Y., Zhang, Y.: Fuzzy radial basis function neural network PID control system for a quadrotor UAV based on particle swarm optimization. In: 2015 IEEE International Conference on Information and Automation, pp. 2580–2585. Lijiang, China (2015)
11. Yu, B., Gao, J., Wang, H., Fan, X., Li, X.: Research on RBF neural network attitude control of quadcopter. Mechanical Science and Technology, 1–8 (2023)

Generalizing Graph Network Models for the Traveling Salesman Problem with Lin-Kernighan-Helsgaun Heuristics

Mingfei Li, Shikui Tu(✉), and Lei Xu(✉)

Department of Computer Science and Engineering, Shanghai Jiao Tong University, Shanghai, China
{limingfei,tushikui,lxu}@sjtu.edu.cn

Abstract. Existing graph convolutional network (GCN) models for the traveling salesman problem (TSP) cannot generalize well to TSP instances with larger number of cities than training samples, and the NP-Hard nature of the TSP renders it impractical to use large-scale instances for training. This paper proposes a novel approach that generalizes well a pre-trained GCN model for a fixed small TSP size to large scale instances with the help of Lin-Kernighan-Helsgaun (LKH) heuristics. This is realized by first devising a Sierpinski partition scheme to partition a large TSP into sub-problems that can be efficiently solved by the pre-trained GCN, and then developing an attention-based merging mechanism to integrate the sub-solutions as a whole solution to the original TSP instance. Specifically, we train a GCN model by supervised learning to produce edge prediction heat maps of small-scale TSP instances, then apply it to the sub-problems of a large TSP instance generated by partition strategies. Controlled by an attention mechanism, all the heat maps of the sub-problems are merged into a complete one to construct the edge candidate set for LKH. Experiments show that this new approach significantly enhances the generalization ability of the pre-trained GCN model without using labeled large-scale TSP instances in the training process and also outperforms LKH in the same time limit.

Keywords: Graph convolutional networks · Subproblem partitioning · Traveling salesman problem · Combinatorial optimization

1 Introduction

As one of the most famous NP-hard combinatorial optimization problems, the Traveling Salesman Problem (TSP) has numerous real-world applications in a variety of domains, including route design, production process scheduling and transportation. The TSP is defined as follows: Given a set of cities as well as the distance d_{ij} between each pair of cities i and j, find the shortest tour that starts from an arbitrary city, visits each city exactly once, and finally returns to the beginning city.

B. Luo et al. (Eds.): ICONIP 2023, LNCS 14447, pp. 528–539, 2024.
https://doi.org/10.1007/978-981-99-8079-6_41

Owing to the theoretical and practical importance of the TSP, researchers developed numerous algorithms for it. With the advances in deep learning, learning-based methods train powerful deep neural networks [10,17,20] to learn complicated patterns from TSP examples created by certain distributions. However, the performance of most learning-based approaches is still substantially worse than strong classical heuristic algorithms [6–8], and these methods are also limited to small-scale problem instances.

Recently, it was suggested to develop learning-based methods for combinatorial optimization problems by combining with strong heuristic algorithms [3,23,24]. The neural network learning and the search algorithm are complementary and beneficial to each other. In this way, besides utilizing the strong representation ability of deep learning models to learn complex patterns from training data, researchers' highly optimized heuristics for well-studied famous problems like TSP can also be effectively used. The methods along this direction [16,21,22,25] improved the performance and generalization ability.

In this paper, we propose a novel learning-based method that generalizes small pre-trained Graph Convolutional Network (GCN) models via sub-problem partitioning and attention-based merging, then guides the efficient k-opt local search of LKH with the predictions of the GCN to search for valid TSP tours. Different from [21], we train the GCN on TSP instances of a small fixed size ω, with the obtained model denoted as GCN(ω), and also use it on the new TSP instances with the same number (ω) of cities. This setting leads to the best performance of the GCN(ω)'s prediction, i.e., accurately predicting the probability of each edge being selected by the TSP tour. To generalize it to arbitrarily large TSP instances, we partition the input into sub-problems of the same small fixed size ω, and construct the edge candidate sets for the k-opt local search of LKH by fusing the GCN(ω)'s predictions into one. To fulfill this goal, we devise a variant of Sierpinski partitioning to guarantee that the TSP sizes of the small sub-problems are all equal to ω. Moreover, as the sub-problems from different parts of the original input may have different impacts on the final tour construction, we present an attention mechanism to integrate the GCN(ω)'s predictions on the TSP sub-problems into one probability heatmap. The probability heatmap guides the high-quality construction of the edge candidate set for the LKH searching process, leading to substantially better edge candidate sets compared to LKH, as well as saving a considerable amount of time by replacing the subgradient optimization procedure in LKH. Following the evaluation process of existing learning-based approaches, we conduct comprehensive experiments to examine the performance of our approach.

The main contributions of this work are further summarized as follows:

- We propose a novel learning method to solve TSP. Our method divides the input into small sub-problems, predicts the edge selection probability of the sub-problems by a pre-trained GCN, and integrates the GCN predictions to construct high-quality edge candidate sets for the k-opt local search of LKH. We fix not only the number of cities of the TSP instances in the GCN's training set but also the ones in the partitioned sub-problems at the same

small number so that the GCN's predictions on the sub-problems are very accurate and efficient.
- We provide a variant of Sierpinski partitioning to ensure that the obtained sub-problems contain the same number of cities.
- It is noted that the sub-problems come from different parts of the original input TSP instance, and they have different impacts on constructing the final tour. We present an attention-based merging mechanism to precisely integrate the edge predictions for every sub-problem.
- Our method outperforms the original LKH and two state-of-the-art learning-based algorithms NeuroLKH, VSR-LKH for problem sizes ranging from 20–10000 under the same time limit and using the same training data set (if needed).

2 Related Works

The LKH algorithm [6–8] is considered as one of the state-of-the-art heuristics for the TSP. Based on the Lin-Kernighan (LK) algorithm [15], LKH has good generalization ability and is applicable for TSP instances with a wide range of scales and different distributions. LKH is especially powerful in solving immense problem instances and it has found the state-of-the-art solution of the famous World-TSP that consists of 1,904,711 cities.

Learning based methods for the TSP can be divided into two categories. The first category consists of supervised learning methods. [20] trained a sequence-to-sequence pointer network with supervised learning, and uses beam search to construct valid tours upon test. Following this framework, [17] trained a graph neural network (GNN) to produce the edge prediction matrix, [10] used residual gated graph convolutional networks to build edge predictions, and both find valid tours via beam search. [22] trained a GNN to learn the patterns from the input instances, then used Monte-Carlo Tree Search (MCTS) to construct TSP tours. [5] proposed an Att-GCN+MCTS method that generalizes small GCN models to large TSP instances by randomly sampling sub-graphs and merging GCN predictions of the sub-graphs by taking average values to guide the MCTS for valid tours. NeuroLKH [21] is one of the state-of-the-art supervised learning based methods for TSP, which trains a Sparse Graph Network with supervised learning to create the edge candidate set for the LKH searching process. Trained with different sizes of instances (ranging from 100–500), it can provide better results than LKH for randomly distributed instances with 100–5,000 nodes. The second category comprises the reinforcement learning based methods. [2,4] trained the networks with actor-critic reinforcement learning algorithm. [12] employed a reinforcement learning model to guide the selection of the next city for a partial tour until a valid tour is constructed. [14] implemented a graph attention network and trained an attention-based decoder with reinforcement learning to autoregressively build TSP tours. VSR-LKH [25] is the state-of-the-art reinforcement learning based method for the TSP, which uses reinforcement learning to guide the edge selection within the edge candidate set that is generated by the original LKH algorithm during the LKH searching process.

The supervised learning approach can be very efficient and precise once the models are trained with sufficient TSP instances with optimal solution. However, large-scale TSP instances with optimal solutions are very costly to generate, which renders supervised learning approaches hard to generalize to larger problem instance than the ones in the training set. Following previous heuristic algorithms [11,18,19] that break TSP instances into smaller sub-problems, we try to partition a large TSP instance into small fixed-size sub-problems, and propose an attention-based mechanism that uses different attention types to merge the edge predictions of the sub-problems made by the pre-trained GCN model to construct substantially better edge candidate set for the LK searching process in LKH. In this way, without losing much accuracy, only small scale instances are required in the training process. [5] also tried to generate sub-problems from a large TSP and merge the predictions of the sub-problems. However, the performance of their Att-GCN+MCTS approach is still much inferior to LKH; their random sampling - taking average strategy is rather straight forward and could be enhanced. Experimental results show that our method can produce substantially better solutions than LKH in the same time limit.

3 Methods

3.1 Overview of the Proposed Approach

We propose a novel approach that combines the heuristics from the LKH algorithm with a graph convolutional network model to solve TSP. An overview of our approach is given in Fig. 1. A GCN model (whose architecture is detailed in the full version of this paper[1]) is trained by supervised learning with a fixed input problem size ω, i.e., the number of cities, for predicting the probability of an edge being selected by the optimal tour.

To generalize the GCN model from ω to any size, a variant of Sierpinski partitioning scheme is presented to divide a large TSP instance into sub-problems, which are all constrained to have ω cities. The sub-problems are later fed to the pre-trained GCN which is supposed to be good at solving TSP instances of ω cities. Considering the fact that the sub-problems come from different parts of the original input, an attention-based merging mechanism is devised to merge the output edge probabilities of the sub-problems, in order to form a complete probability heat map of the original problem. The complete probability heat map is used to create the edge candidate set for the k-opt local searching process of LKH. Since the GCN model is trained and tested on the TSP instances of the same size equal to ω, the complete probability heat map would be very accurate, leading to better edge candidate set for improved performance. Besides, as a part of the GCN output, the encoded node features are also decoded and merged in the same manner as node penalties for the LKH searching process. Finally, a valid TSP tour is constructed via the LKH searching process. Notice that our

[1] https://github.com/CMACH508/Generalizing-Graph-Network-Models-for-the-TSP.

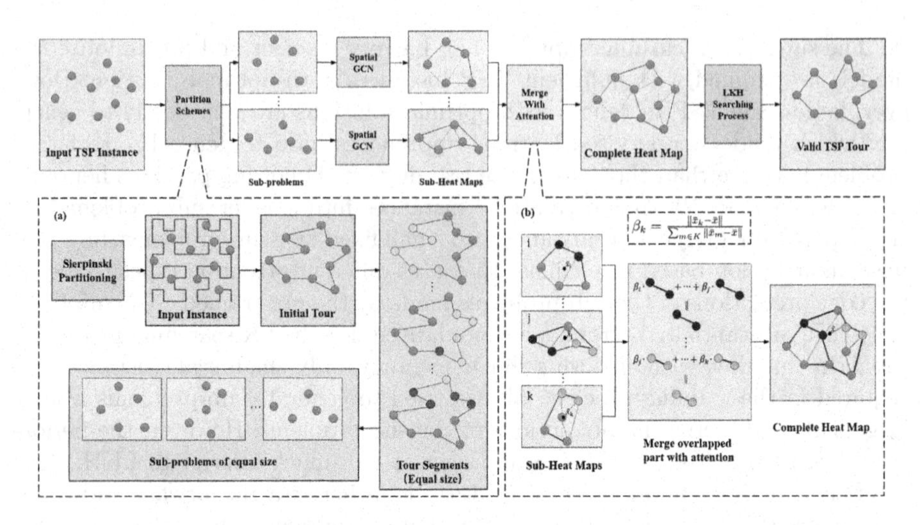

Fig. 1. Overview of the proposed approach and the illustration of (a) the proposed Sierpinski partitioning scheme, (b) the attention mechanism for heat maps merging. In sub-figure (a), the yellow nodes (nodes left for this tour segment) and red nodes (nodes "borrowed" from adjacent tour segments to maintain equal size) together compose a tour segment of a fixed size. Marked with blue (or orange), the overlapped edge predictions of the sub-problems in sub-figure (b) are added together after multiplying the attention scores β_i of the corresponding sub-problems. (Color figure online)

approach employs GCN to learn the complex patterns associated with the TSP instances, instead of the subgradient optimization process in LKH.

3.2 Sub-problem Partitioning Schemes

The pre-trained spatial GCN model does not generalize well to TSP instances of different sizes, so that it cannot be directly applied to large TSP instances for edge prediction. To overcome this drawback, as well as reduce the complexity of solving large-scale problem instances, we consider partitioning a problem into smaller sub-problems of a specific size ω.

We propose a variant of Sierpinski partitioning scheme, as well as two other schemes, and each of them guarantees to divide the original large TSP instance into small sub-problems of a fixed size ω. All the sub-problems are fed to the pre-trained graph convolutional network for edge prediction. Note that the pre-trained GCN model is scale sensitive and a sub-problem should be resized to $[0,1]^2$ upon input by simply multiplying its node coordinates by a coefficient $1/max(x_{max}-x_{min}, y_{max}-y_{min})$. The proposed partitioning schemes are detailed as follows:

A Variant of Sierpinski Partitioning. The original Sierpinski partitioning scheme [18] partitions a TSP tour induced by the Sierpinski space filling curve into tour segments of about equal size. The Sierpinski partitioning scheme is adopted by the LKH algorithm as one of its sub-problem partitioning schemes.

We propose a variant of the Sierpinski partitioning scheme to partition a problem instance (whose size is larger than ω) into sub-problems that each contains exactly ω nodes. The partitioning is started from one end of the Sierpinski tour, making sure that each segment has exactly ω nodes. The number of nodes left for the last segment n_{last} could be smaller than ω, if so, the closest $\omega - n_{last}$ nodes from other segments of this partition would also be assigned to the last segment. Each of these segments induces a sub-problem that consists of all the ω nodes in the segment. After all sub-problems of this partition have been treated, a revised partition of the Sierpinski tour is used where each new segment takes half its nodes from each of two adjacent old segments. The same strategy of sharing nodes with the neighbor segments is also used in this procedure such that the first and the last segment can both have exactly ω nodes.

Modified Karp Partitioning. The Karp partitioning scheme [11] subdivide the overall region containing all the nodes into rectangles that contains no more than a specific number of nodes. The rectangles are found in a manner similar to that used in the construction of the k-D tree data structure: Given a set S_0 of all the n points with a bounding rectangle R_0, the algorithm partitions S_0 into S_1 and S_2 by a boarder B_0 parallel to the shorter side of R_0 such that an equal number of points is on the both sides of B_0. The algorithm works recursively until each cluster S_i has no more than a specified number of points and each cluster induces a sub-problem.

The original Karp partitioning scheme also cannot guarantee that each sub-problem consists of exactly ω nodes. We propose a revised Karp partitioning scheme as follows: The first few steps of the recursive procedure stay the same as the above original version, but the last step is modified by using double boarders and allowing overlap. While partitioning the set S_k that satisfies $\omega < |S_k| \leq 2\omega$, a boarder B_k parallel to the shorter side of R_k is found such that one side of B_k forms a subset S_{2k+1} satisfying $|S_{2k+1}| = \omega$. Then, another boarder B'_k parallel to B_k is found in R_{2k+1} so that the number of nodes in set $S_k - S_{2k+1}$ plus the number of nodes between B_k and B'_k inside S_{2k+1} is ω. B'_k and B_k do not necessarily coincide, which means some nodes might be shared by S_{2k+1} and S_{2k+2}. After the last step, each rectangle induces a sub-problem consisting of all the ω nodes inside the rectangle.

Modified Rohe Partitioning. [19] proposed the following hybrid between the geometric and tour-based partitioning schemes: given the current tour, sub-problems consists of cities closely related in a geometric sense, for example all cities inside a given rectangle or the M nearest neighbors of a given city [9]. The Rohe partitioning scheme implemented in LKH employs random rectangles to partition the node set into disjoint subsets of about equal size.

To guarantee that each sub-problem has exactly ω nodes, we propose a modified version of the Rohe partitioning scheme: At first, all the nodes are marked as remaining nodes. Then, instead of generating a random rectangle that contains no more than ω nodes to form a sub-problem like LKH, a random remaining node is selected and its $\omega - 1$ nearest neighbors together with this node induces a sub-problem. All the nodes in this sub-problem are subtracted from the set of remaining nodes. This procedure repeats until the set of remaining nodes becomes empty.

3.3 Attention Mechanism for Merging Probability Heat Maps

As the original problem is partitioned into multiple sub-problems, and all the sub-problems are fed into the pre-trained graph network model to get the edge prediction, we will obtain multiple sub-heat maps and some of them may overlap. To merge all the sub-heat maps into a complete heat map, for each edge e_{ij} of the original problem, we estimate the probability of its presence p_{ij} in the optimal TSP tour E^{opt} as follows (node penalties π_i are merged in the same fashion):

$$\begin{aligned} p_{ij} &= \sum_{k \in K} \beta_k p_{ij}^{(k)}, \beta_k = \frac{\gamma(k)}{\sum_{m \in K} \gamma(m)}, \\ \pi_i &= \sum_{k' \in K'} \beta_{k'} \pi_i^{(k')}, \beta_{k'} = \frac{\gamma(k')}{\sum_{m \in K'} \gamma(m)}, \end{aligned} \tag{1}$$

where K is the set of the sub-problems that contains both node i and node j (K' the set of sub-problems containing node i), $p_{ij}^{(k)}$ denotes the probability of edge e_{ij} occurring in the optimal tour of sub-problem k predicted by the pre-trained spatial GCN model, $\pi_i^{(k')}$ the node penalty of node i in sub-problem k', and β_k for the attention score of sub-problem k, which is decided by its relative position in the original problem. We designed three different types of location-based weight function $\gamma(k)$ for attention score β_k:

$$\gamma_1(k) = \|\bar{x}_k - \bar{x}\|, \ \gamma_2(k) = \min(\bar{x}_k^{(0)}, 1 - \bar{x}_k^{(0)}, \bar{x}_k^{(1)}, 1 - \bar{x}_k^{(1)}), \ \gamma_3(k) = 1, \tag{2}$$

where $\bar{x}_k$ is the mean of all the nodes in sub-problem k , $\bar{x}$ is the mean of all the nodes in the original problem, and $\bar{x}_k^{(0)}, \bar{x}_k^{(1)}$ are the two dimensions of $\bar{x}_k$. Outer sub-problems get higher attention scores with $\gamma_1(k)$, while $\gamma_2(k)$ casts higher attention to inner sub-problems, and $\gamma_3(k)$ assigns equal attention to all the sub-problems.

By merging all the edge prediction heat maps of the sub-problems, we obtain a complete edge prediction heat map of the original problem, which is used to create the edge candidate set for the k-opt local search process of LKH algorithm. The edges affiliate to a node are sorted by their probability of being present in the TSP tour, and the top 5 edges with the highest probability forms the candidate set of a node.

4 Experiments

We conduct experiments on a large number of TSP instances whose scale range from 20–10000, and compare the performance of our method with exact solver

Concorde, heuristic algorithm LKH, as well as several learning based algorithms such as Att-GCN+MCTS [5], NeuroLKH [21], and VSR-LKH [25].

4.1 Datasets and Hyperparameters

The datasets we use consist of 2-dimensional TSP instances in the Euclidean distance space where both coordinates of each node are generated independently from a unit uniform distribution, following existing works like [10]. We use Concorde [1] to get the optimal edges E^{opt} for the supervised training of the GCN model. The training set of the spatial GCN model contains 800,000 instances of size ω. The testing set is composed of 1000 instances for problem size $n = 20, 50, 100, 200, 500, 1000$, as well as 128 instances for $n = 2000, 5000, 10000$.

We consider $k = 20$ nearest neighbors for each node as adjacent, as experiments show that for the TSP100 training dataset, this only causes 0.006% of the edges belonging to optimal tours missed in the edge candidate set. The hidden dimension is set to d = 300, and the GCN model contains L = 30 spatial graph convolutional Layers. For problem size $n \leq 100$, we set $\omega = 20$, and for problem size $n > 100$, we set $\omega = 100$. The GCN models are trained by Adam Optimizer [13] with an initial learning rate of 0.001. For each training epoch, 10000 instances are randomly selected from the training dataset, and the model is evaluated on a held-out validation set of 1000 instances after each training epoch. The learning rate is decayed by a factor of 0.99 if the validation loss has not decreased at least 1%. The deep learning models are trained and evaluated with one RTX-2080Ti GPU. The other parts of experiments without deep models for our method and other baselines are conducted on an Intel(R) Xeon(R) Gold 6130 CPU @ 2.10GHz. Hyperparameters for the LKH searching process are consistent with its default settings[2] (except that we need to specify the file names of edge candidate file and π file in the parameter file of LKH).

4.2 The Combination of Partitioning Schemes and Attention Types

Table 1. Average optimality gap of generalizing TSP20 model to TSP100 instances with different partitioning schemes and attention mechanisms. Attention types $\gamma_1(k)$, $\gamma_2(k)$ and $\gamma_3(k)$ are defined in Eq. (2). As a baseline, the optimality gap of directly applying the pre-trained TSP20 model to TSP100 instances is 69.63%.

Opt. Gap	$\gamma_1(k)$	$\gamma_2(k)$	$\gamma_3(k)$
Karp	3.42%	11.81%	3.82%
Sierbinski	2.81%	7.56%	3.77%
Rohe	3.37%	10.20%	4.28%

We conducted a series of experiments to generalize the pre-trained TSP20 model to TSP100 instances with all the possible combinations of partitioning schemes and attention types for heat maps merging. To emphasize the effect of different partitioning schemes and attention mechanisms for merging, a beam search decoder with beamsize $b = 1280$ is used instead of the k-opt local search process of LKH. The experimental results are shown in Table 1. The corner of the first column and the first line is the performance of

[2] http://akira.ruc.dk/~keld/research/LKH/.

directly applying the pre-trained TSP20 model to TSP100 testing set without the partitioning &merging process.

The optimality gap is with respect to the optimal tour generated by exact solver Concorde. As shown in Table 1, the optimality gap of directly applying the pre-trained TSP20 model to TSP100 instances is 69.63%, indicating very poor generalization capabilities. After including the partitioning & merging process in the whole procedure, the optimality gap can be lowered to 2.81% with the best combination, and 11.81% even with the worst combination, which is a solid roof that the partitioning schemes and attention mechanisms for heat maps merging are very effective for enhancing the generalization ability of the pre-trained spatial GCN model. Also, it's obvious that the best combination of the partitioning schemes and attention types for heat maps merging is Sierpinski partitioning and $\gamma(k) = \gamma_1(k)$, and this configuration is used in later experiments.

4.3 Comparative Study on TSP Instances of Different Sizes

For testing problem sizes range from $20 - 1000$, we set the max trials of the original LKH to 1000, and record the total elapsed time for LKH solving the 1000 instances as a time limit for all the methods except for the solver Concorde. Then, this time limit is imposed upon all the learning based methods, i.e. Att-GCN+MCTS, NeuroLKH, VSR-LKH and our method for solving the same 1000 instances for fair comparison. Note that for NeuroLKH, the SGN is trained with the same training set that our model used, i.e. only TSP20 instances are used in the training process for problem size $n \leq 100$ and only TSP100 instances for $n > 100$, therefore both methods gain the same amount of supervision. In the following tables, for each problem size, we report the average tour length and the average optimality gap of all the methods with respect to Concorde for the 1000 testing instances. The average optimality gap of a method is computed as

$$\frac{1}{m}\sum_{i}^{m}\frac{l_i - l_i^{opt}}{l_i^{opt}}, \tag{3}$$

where m is the size of the test set, l_i the tour length of instance i found by this method and l_i^{opt} is the length of the optimal tour for generated by concorde.

Table 2 summarizes the results of baseline algorithms and our approach on TSP20, TSP50 and TSP100 test set. We also use the exact solver Concorde on these instances to obtain the optimal solutions and compute the optimality gap for each method. Note that the SGN model in NeuroLKH is trained with only TSP20 instances as Att-GCN+MCTS and our method do for fair comparison, the results are inferior to its original version. As shown in the table, our method outperforms both LKH and other three learning based methods across different problem sizes in the same time limit. Our approach succeeds in matching the ground truth solutions for most of testing instances, corresponding to an average gap of $0.0000\%, 0.0000\%, 0.0000\%$ respectively for the testing instances of $n = 20, 50, 100$.

Table 2. Comparative results of problem size 20, 50, 100

Method	TSP20		TSP50		TSP100	
	Tour Len.	Opt. Gap	Tour Len.	Opt. Gap	Tour Len.	Opt. Gap
Concorde [1]	3.8365	0.0000%	5.6908	0.0000%	7.7612	0.0000%
LKH [8]	3.8365	0.0000%	5.6908	0.0001%	7.7612	0.0001%
Att-GCN+MCTS [5]	3.8365	0.0000%	5.6916	0.0145%	7.7640	0.0370%
VSR-LKH [25]	3.8365	0.0000%	5.6908	0.0001%	7.7613	0.0010%
NeuroLKH [21]	3.8365	0.0000%	5.6909	0.0010%	7.7614	0.0030%
TSP20Model+LKH(**Ours**)	**3.8365**	**0.0000%**	**5.6908**	**0.0000%**	**7.7612**	**0.0000%**

Table 3 shows the results of baseline algorithms and our approach on TSP200, TSP500 and TSP1000 test set. 800,000 TSP100 instances are used as the training set of the SGN model in NeuroLKH and the GCN model of our approach for problem size $n > 100$, such that both methods use the same training set and get the same amount of supervision. As the experimental result shows, under the same time limit, our approach outperforms the original LKH algorithm and the three learning based methods Att-GCN+MCTS, NeuroLKH and VSR-LKH for problem size $n = 200, 500, 1000$ with the pre-trained TSP100 model.

Table 3. Comparative results of problem size 200, 500, 1000

Method	TSP200		TSP500		TSP1000	
	Tour Len	Opt. Gap	Tour Len	Opt. Gap	Tour Len	Opt. Gap
Concorde [1]	10.7191	0.0000%	16.5458	0.0000%	23.1280	0.0000%
LKH [8]	10.7195	0.0005%	16.5461	0.0018%	23.1288	0.0035%
Att-GCN+MCTS [5]	10.8139	0.8844%	16.9655	2.5365%	23.8736	3.2238%
VSR-LKH [25]	10.7191	0.0003%	16.5459	0.0006%	23.1283	0.0011%
NeuroLKH [21]	10.7202	0.0013%	16.5474	0.0097%	23.1324	0.0192%
TSP100Model+LKH(**Ours**)	**10.7191**	**0.0001%**	**16.5458**	**0.0003%**	**23.1282**	**0.0007%**

4.4 Generalize to Problem Sizes Even Difficult for Solvers

We further generalized our method to larger TSP instances of larger size, i.e. TSP2000, TSP5000 and TSP10000. The test set of each problem size contains 128 instances. In these experiments, however, the exact solver Concorde failed to provide the optimal solutions for the problem instances in the test sets within a reasonable time, due to the NP-hard nature of the TSP. To compute the optimality gap of each algorithm, for each instance i in the test set, the length of the optimal solution l_i^{opt} is replaced by the length of the shortest tour generated by all methods l_i^s, and there is no guarantee that one of these methods can find the optimal tour. The time limit for a certain problem size is still the total running time of LKH solving all the instances in the corresponding test set with parameter MAX_TRIALS set as 1000.

Table 4. Comparative results of problem size 2000, 5000, 10000

Method	TSP2000		TSP5000		TSP10000	
	Tour Len	Gap	Tour Len	Gap	Tour Len	Gap
LKH [8]	32.4697	0.0046%	50.9725	0.0044%	71.8473	0.0041%
Att-GCN+MCTS [5]	33.6812	3.7360%	53.0230	4.0273%	75.1951	4.6639%
VSR-LKH [25]	32.4688	0.0018%	50.9725	0.0044%	71.8472	0.0040%
NeuroLKH [21]	32.5963	0.3944%	51.2820	0.6116%	72.5471	0.9782%
TSP100Model+LKH(**Ours**)	**32.4684**	**0.0005%**	**50.9707**	**0.0009%**	**71.8460**	**0.0023%**

The results summarized in Table 4 show that our approach generalizes well to larger problem sizes that is even difficult for solvers, and the improvement of our proposed approach over the baseline algorithms is still significant.

5 Conclusion

In this paper, we propose an algorithm that generalizes a small pre-trained graph convolutional network model to large TSP instances by partitioning strategies and attention-based merging techniques to solve TSP. This method significantly enhances the generalization ability of the pre-trained GCN model, and also outperforms the baseline algorithms in the same time limit without requiring additional large labeled TSP instances in the training process. In the future, we will try to combine this method with other powerful heuristic algorithms and extend this algorithm to other routing problems.

Acknowledgement.. This work was supported by the Shanghai Municipal Science and Technology Major Project (2021SHZDZX0102). Shikui Tu and Lei Xu are corresponding authors.

References

1. Applegate, D., Bixby, R., Chvatal, V., Cook, W.: Concorde TSP solver (2006). www.math.uwaterloo.ca/tsp/concorde
2. Bello, I., Pham, H., Le, Q.V., Norouzi, M., Bengio, S.: Neural combinatorial optimization with reinforcement learning. In: Workshop Track of the International Conference on Learning Representations (2017)
3. Bengio, Y., Lodi, A., Prouvost, A.: Machine learning for combinatorial optimization: a methodological tour d'horizon. Eur. J. Oper. Res. **290**(2), 405–421 (2021)
4. Emami, P., Ranka, S.: Learning permutations with sinkhorn policy gradient. arXiv preprint arXiv:1805.07010 (2018)
5. Fu, Z.H., Qiu, K.B., Zha, H.: Generalize a small pre-trained model to arbitrarily large tsp instances. In: Proceedings of the AAAI Conference on Artificial Intelligence, vol. 35, pp. 7474–7482 (2021)
6. Helsgaun, K.: An effective implementation of the lin-kernighan traveling salesman heuristic. Eur. J. Oper. Res. **126**(1), 106–130 (2000)

7. Helsgaun, K.: General k-opt submoves for the lin-kernighan tsp heuristic. Math. Program. Comput. **1**(2), 119–163 (2009)
8. Helsgaun, K.: An extension of the lin-kernighan-helsgaun tsp solver for constrained traveling salesman and vehicle routing problems. Roskilde: Roskilde University, pp. 24–50 (2017)
9. Johnson, D.S., McGeoch, L.A.: The traveling salesman problem: a case study in local optimization. Local Search Comb. Optim. **1**(1), 215–310 (1997)
10. Joshi, C.K., Laurent, T., Bresson, X.: An efficient graph convolutional network technique for the travelling salesman problem. arXiv preprint arXiv:1906.01227 (2019)
11. Karp, R.M.: Probabilistic analysis of partitioning algorithms for the traveling-salesman problem in the plane. Math. Oper. Res. **2**(3), 209–224 (1977)
12. Khalil, E., Dai, H., Zhang, Y., Dilkina, B., Song, L.: Learning combinatorial optimization algorithms over graphs. In: Advances in Neural Information Processing Systems 30 (2017)
13. Kingma, D.P., Ba, J.: Adam: a method for stochastic optimization. In: International Conference on Learning Representations (2015)
14. Kool, W., Hoof, H.V., Welling, M.: Attention, learn to solve routing problems! In: International Conference on Learning Representations (2019)
15. Lin, S., Kernighan, B.W.: An effective heuristic algorithm for the traveling-salesman problem. Oper. Res. **21**(2), 498–516 (1973)
16. Ma, H., Tu, S., Xu, L.: IA-CL: A deep bidirectional competitive learning method for traveling salesman problem. In: Tanveer, M., Agarwal, S., Ozawa, S., Ekbal, A., Jatowt, A. (eds.) Neural Information Processing. ICONIP 2022. Lecture Notes in Computer Science, vol. 13623, pp. 525–536. Springer International Publishing, Cham (2023). https://doi.org/10.1007/978-3-031-30105-6_44
17. Nowak, A., Villar, S., Bandeira, A.S., Bruna, J.: Revised note on learning quadratic assignment with graph neural networks. In: 2018 IEEE Data Science Workshop, DSW 2018, Lausanne, Switzerland, June 4–6, 2018, pp. 229–233. IEEE (2018)
18. Platzman, L.K., Bartholdi, J.J., III.: Spacefilling curves and the planar travelling salesman problem. J. ACM (JACM) **36**(4), 719–737 (1989)
19. Rohe, A.: Parallele heuristiken für sehr große travelling salesman probleme. diplom. de (1998)
20. Vinyals, O., Fortunato, M., Jaitly, N.: Pointer networks. In: Advances in Neural Information Processing Systems 28 (2015)
21. Xin, L., Song, W., Cao, Z., Zhang, J.: Neurolkh: combining deep learning model with lin-kernighan-helsgaun heuristic for solving the traveling salesman problem. Adv. Neural. Inf. Process. Syst. **34**, 7472–7483 (2021)
22. Xing, Z., Tu, S.: A graph neural network assisted monte Carlo tree search approach to traveling salesman problem. IEEE Access **8**, 108418–108428 (2020)
23. Xu, L.: Deep bidirectional intelligence: AlphaZero, deep IA-search, deep IA-infer, and TPC causal learning. Appl. Inform. **5**(1), 1–38 (2018). https://doi.org/10.1186/s40535-018-0052-y
24. Xu, L.: Deep IA-BI and five actions in circling. In: International Conference on Intelligent Science and Big Data Engineering, pp. 1–21. Springer, New York, NY (2019). https://doi.org/10.1007/0-387-23081-5_11
25. Zheng, J., He, K., Zhou, J., Jin, Y., Li, C.M.: Combining reinforcement learning with lin-kernighan-helsgaun algorithm for the traveling salesman problem. In: Proceedings of the AAAI Conference on Artificial Intelligence, vol. 35, pp. 12445–12452 (2021)

Communication-Efficient Distributed Minimax Optimization via Markov Compression

Linfeng Yang[1], Zhen Zhang[1], Keqin Che[1], Shaofu Yang[1(✉)], and Suyang Wang[2]

[1] School of Computer Science and Engineering, Southeast University, Nanjing, China
{linfengyang,zhang_zhen,kqche,sfyang}@seu.edu.cn

[2] Jiangsu Jinheng Information Technology Co., Ltd, Wujin, China
wangsy@njsteel.com.cn

Abstract. Recently, the minimax problem has attracted a lot of attention due to its wide applications in modern machine learning fields such as GANs. With the exponential growth of data volumes and increasing problem sizes, the design of distributed algorithms to train high-performance models has become imperative. However, distributed algorithms often suffer from communication bottlenecks. To address this challenge, in this paper, we propose a communication-efficient distributed compressed stochastic gradient descent ascent algorithm, abbreviated as DCSGDA, in a parameter-server setting. To reduce the communication cost, each client in DCSGDA transmits the compressed gradients of the primal and dual variables to the server at each iteration. In particular, we leverage a Markov compression mechanism that allows both unbiased and biased compressors to mitigate the negative effect of compression errors on convergence. Namely, we show theoretically that the DCSGDA algorithm can still achieve linear convergence in the presence of compression errors, provided that the local objective function is strongly-convex-strongly-concave. Finally, numerical experiments demonstrate the desirable communication efficiency and efficacy of the proposed DCSGDA.

Keywords: Minimax optimization · Parameter-server framework · Communication compression · Distributed optimization

1 Introduction

We consider the following distributed stochastic minimax problem

$$\min_{\mathbf{x}\in\mathbb{R}^{d_1}} \max_{\mathbf{y}\in\mathcal{Y}} f(\mathbf{x},\mathbf{y}) = \frac{1}{n}\sum_{i=1}^{n} f_i(\mathbf{x},\mathbf{y}), \tag{1}$$

This work was supported in part by the National Natural Science Foundation of China under Grant 62176056, and in part by the Young Elite Scientists Sponsorship Program by the China Association for Science and Technology (CAST) under Grant 2021QNRC001.

B. Luo et al. (Eds.): ICONIP 2023, LNCS 14447, pp. 540–551, 2024.
https://doi.org/10.1007/978-981-99-8079-6_42

where data is distributed among n clients so that each client will possess a local objective function $f_i(\mathbf{x}, \mathbf{y}) := \mathbb{E}_{\xi_i \sim \mathcal{D}_i}[\ell(\mathbf{x}, \mathbf{y}; \xi_i)]$. Here, ℓ represents the loss function and ξ_i refers to the data point sampled from the local data distribution $\mathcal{D}_i$. Such a problem can be seen in many modern machine learning areas including Generative Adversarial Networks (GANs) [5], robust learning [14] and multi-agent reinforcement learning [16], just to name a few. Many studies have addressed problem (1), including extragradient method (EG) [6], stochastic descent ascent (SGDA) [2,12,15], and quasi-Newton method [8,10]. Nevertheless, existing works mainly focus on solving the minimax problem at a single client, which is undesirable for settings with huge data volumes and large problem sizes, due to the limited computational capacity of one client. To overcome this challenge, distributed algorithms that leverage the collective computational power of multiple clients have become increasingly popular.

A simple yet popular distributed solution to solve problem (1) is the *parameter server* model. In this model, each client computes local gradients based on the model parameters received from the server, meanwhile, the server updates model parameters based on the local gradients received from the clients. Under this setting, several distributed minimax optimization algorithms have been reported [4,19,21]. In the works of [4,21], the authors propose local stochastic gradient descent ascent and its variant with momentum. Despite its popularity, the parameter server model suffers from communication bottlenecks due to the large amount of information exchanged between clients and the server, leading to overall performance degradation. Therefore, it is imperative to develop efficient communication algorithms that can effectively mitigate this challenge.

In the realm of communication-efficient distributed optimization algorithms, the existing works can be broadly classified into two main categories: (i) reducing the total communication rounds [4,9,13,21,23,26,28] and (ii) reducing the communication bits [7,18,24]. The first category relies on intermittent communication or improving convergence rate. In [4,21], the authors extend the local SGD to local SGDA for solving the minimax optimization problem, which achieves a desirable communication efficiency. The second category relies on the approaches that include quantization [27], sparsification [22], error-compensated compression [17], and their combinations [1], all of which have demonstrated a significant reduction in communication overhead. It is worth noting that compression will lead to "information distortion", which may sacrifice the convergence performance. Recently, Ref. [20] proposed an error-feedback method (EF21) that can achieve "distortion vanishes". Using EF21, SGD can achieve an analogous convergence rate as its uncompressed counterpart, indicating that compression methods can reduce communication costs without heavily degrading convergence performance.

Building on the above discussion, despite these advances, there is a lack of research focusing on communication compression in distributed minimax optimization. Though the works [4,21] develop communication-efficient algorithms for minimax optimization, they require accurate information transmission which may not be feasible for the scenarios with limited bandwidth resources. Hence, it is necessary to develop communication compression algorithms in distributed minimax optimization. To bridge this gap, in this paper, we propose DCSGDA, a distributed minimax algorithm that combines the renowned SGDA algorithm with communication compression techniques. The choice of SGDA is motivated by its favorable convergence properties, ease of execution, and applicability to a wide range of problems. To maintain a fast convergence rate, we utilize the Markov compressor proposed in EF21 [20] to compress the gradients before transmission. This compressor has demonstrated remarkable efficacy in mitigating information distortion. Our contributions are as follows:

- Building on SGDA, we propose a communication-efficient distributed stochastic communication compression minimax optimization algorithm termed DCSGDA, which aims to reduce the number of transmitted bits per communication round. In DCSGDA, we equip each client with a Markov compressor, which achieves the decrease of "information distortion" arising from compression. The Markov compressor used in DCSGDA encompasses a wide variety of prevalent compression techniques, enhancing its versatility.
- We theoretically show that DCSGDA can converge to the $\mathcal{O}(\sigma^2)$ neighborhood of the optimal solution at a linear rate for strongly-convex-strongly-concave objective functions. In contrast to EF21, since DCSGDA tackles a distinct class of minimax optimization problems, we adopt a different technique to analyze its convergence. Compared to SGDA, DCSGDA can reduce the number of transmitted bits per communication round while maintaining a linear convergence rate. Numerical experiments demonstrate the desirable communication efficiency of the proposed algorithm.

Notation. We use $\|\cdot\|$ to denote the Euclidean norm $\|\cdot\|_2$. Given a positive integer n, denote $[n]$ as the set $\{1, 2, \ldots, n\}$. The gradient vector is defined as $\nabla f_i(\mathbf{x}, \mathbf{y}) = [\nabla_{\mathbf{x}} f_i(\mathbf{x}, \mathbf{y})^\top, \nabla_{\mathbf{y}} f_i(\mathbf{x}, \mathbf{y})^\top]^\top$, and the stochastic gradient vector is defined as $\nabla f_i(\mathbf{x}, \mathbf{y}; \xi_i) = [\nabla_{\mathbf{x}} f_i(\mathbf{x}, \mathbf{y}; \xi_i)^\top, \nabla_{\mathbf{y}} f_i(\mathbf{x}, \mathbf{y}; \xi_i)^\top]^\top$, where ξ_i is a random variable.

2 Our Algorithm: DCSGDA

First, we introduce a common algorithm for minimax optimization, i.e., SGDA [2,12,15], which serves as the basis for our algorithm. The update of primal and dual variables $\mathbf{x}, \mathbf{y}$ in SGDA are as follows:

$$\mathbf{x}^{t+1} = \mathbf{x}^t - \gamma_{\mathbf{x}} \nabla_{\mathbf{x}} f(\mathbf{x}^t, \mathbf{y}^t; \xi^t), \qquad \mathbf{y}^{t+1} = \mathbf{y}^t + \gamma_{\mathbf{y}} \nabla_{\mathbf{y}} f(\mathbf{x}^t, \mathbf{y}^t; \xi^t),$$

where $\gamma_{\mathbf{x}}, \gamma_{\mathbf{y}}$ are learning rates. SGDA can be executed in a parameter-server framework, in which the stochastic gradients $\nabla_{\mathbf{x}} f$ and $\nabla_{\mathbf{y}} f$ are computed by multiple clients based on their own data, and the server updates the primal and dual variables based on received local gradients. Clearly, communication between the server and the clients is necessary.

In what follows, we will develop a communication-efficient distributed algorithm for minimax optimization problems based on SGDA. Before proceeding, we introduce the Markov compression mechanism used in our algorithm.

Algorithm 1. DCSGDA

1: **Input:** For server, step size $\gamma > 0$, model parameter $\mathbf{x}^0, \mathbf{y}^0 \in \mathbb{R}^d$, and $g_{\mathbf{x}}^{-1} = g_{\mathbf{y}}^{-1} = 0$. For client i, compute $g_{\mathbf{x},i}^0 = \mathcal{C}(\nabla_{\mathbf{x}} f_i(\mathbf{x}^0, \mathbf{y}^0; \xi_i^0))$, $g_{\mathbf{y},i}^0 = \mathcal{C}(\nabla_{\mathbf{y}} f_i(\mathbf{x}^0, \mathbf{y}^0; \xi_i^0)))$, and then send $\mathbf{u}_i^{-1} = g_{\mathbf{x},i}^0, \mathbf{v}_i^{-1} = g_{\mathbf{y},i}^0$ to the server.
2: **for** $t = 0, 1, 2 \ldots$ **do**
3: *Server:*
4: Receive $\mathbf{u}_i^{t-1}$, $\mathbf{v}_i^{t-1}$ from all clients and compute:
5: $g_{\mathbf{x}}^t = g_{\mathbf{x}}^{t-1} + \frac{1}{n} \sum_{i=1}^n \mathbf{u}_i^{t-1}$, $g_{\mathbf{y}}^t = g_{\mathbf{y}}^{t-1} + \frac{1}{n} \sum_{i=1}^n \mathbf{v}_i^{t-1}$;
6: Update model parameter: $\mathbf{x}^{t+1} = \mathbf{x}^t - \gamma_{\mathbf{x}} g_{\mathbf{x}}^t$ and $\mathbf{y}^{t+1} = \mathcal{P}_{\mathcal{Y}}(\mathbf{y}^t + \gamma_{\mathbf{y}} g_{\mathbf{y}}^t)$;
7: *Clients:*
8: **for** all clients $i = 1, \cdots, n$ in parallel **do**
9: $\mathbf{u}_i^t = \mathcal{C}(\nabla_{\mathbf{x}} f_i(\mathbf{x}^{t+1}, \mathbf{y}^{t+1}; \xi_i^{t+1}) - g_{\mathbf{x},i}^t)$ (compression)
10: $\mathbf{v}_i^t = \mathcal{C}(\nabla_{\mathbf{y}} f_i(\mathbf{x}^{t+1}, \mathbf{y}^{t+1}; \xi_i^{t+1}) - g_{\mathbf{y},i}^t)$ (compression)
11: Send $\mathbf{u}_i^t, \mathbf{v}_i^t$ to the server
12: Update $g_{\mathbf{x},i}^{t+1} = g_{\mathbf{x},i}^t + \mathbf{u}_i^t$ and $g_{\mathbf{y},i}^{t+1} = g_{\mathbf{y},i}^t + \mathbf{v}_i^t$
13: **end for**
14: **end for**

Definition 1 (Compression Operator). *Let $\mathcal{C} : \mathbb{R}^d \to \mathbb{R}^d$ be a compressed operator, if it satisfies $\mathbb{E} \|\mathbf{z} - \mathcal{C}(\mathbf{z})\|^2 \le (1 - \alpha) \|\mathbf{z}\|^2$ for any $\mathbf{z} \in \mathbb{R}^d$, where $\alpha \in (0, 1]$ is a constant.*

Definition 2 (Markov Compression [20]). *Let $\{\mathbf{z}^t\}_{t \ge 0}$ be a sequence of input vectors, and $\mathcal{C}(\cdot)$ be a compression operator, then the Markov compression mechanism can be written as*

$$\mathcal{M}(\mathbf{z}^0) = \mathcal{C}(\mathbf{z}^0), \qquad \mathcal{M}(\mathbf{z}^{t+1}) = \mathcal{M}(\mathbf{z}^t) + \mathcal{C}(\mathbf{z}^{t+1} - \mathcal{M}(\mathbf{z}^t)),$$

where $\mathcal{M}(\mathbf{z}^t)$ denotes the received information of $\mathbf{z}^t$.

Markov compression mechanism [20] is a refined error-feedback method with general contractive compression operators. It can be seen that the "information distortion" arising in compression can be measured by $\|\mathcal{M}(\mathbf{z}^t) - \mathbf{z}^t\|^2$. If $\mathbf{z}^t$ converges to the optimal $\mathbf{z}^*$, we can obtain that $\mathbb{E} \|\mathcal{M}(\mathbf{z}^t) - \mathbf{z}^t\|^2 \to 0$. In other words, the information distortion between $\mathcal{M}(\mathbf{z}^t)$ and $\mathbf{z}^t$ vanishes as $t \to \infty$.

By combining SGDA with Markov compression, we develop a communication-efficient distributed algorithm for minimax optimization, termed DCSGDA, whose pseudo-code is presented in Algorithm 1. The crucial processes of DCSGDA are the calculation of $g_{\mathbf{x}}^t$ and $g_{\mathbf{y}}^t$. The clients compute the Markov compressed stochastic gradients of primal and dual variables (see lines 9-10), and then send the compressed local gradients to the server. For each client i, let $g_{\mathbf{x},i}^t, g_{\mathbf{y},i}^t$ denote Markov compressed stochastic gradients, and $\mathbf{u}_i^t, \mathbf{v}_i^t$ denote the difference which needs to be transmitted of corresponding gradients between two successive updates. For the server, it aggregates the transmitted differences $\mathbf{u}_i^t, \mathbf{v}_i^t$ of all clients and obtains the global update of Markov compressed stochastic gradients $g_{\mathbf{x}}^t, g_{\mathbf{y}}^t$. It is worth noting that the server only has access to the transmitted difference and does not have access to the local gradient estimates of each client. By using this compression mechanism, DCSGDA reduces the communication cost and protects the privacy of local gradients. To be brief, compared with SGDA, the update rules of DCSGDA at the server can be described as: $\mathbf{x}^{t+1} = \mathbf{x}^t - \gamma_{\mathbf{x}} g_{\mathbf{x}}^t$, $\mathbf{y}^{t+1} = \mathcal{P}_{\mathcal{Y}}(\mathbf{y}^t + \gamma_{\mathbf{y}} g_{\mathbf{y}}^t)$.

3 Convergence Analysis

In this section, we present convergence result for DCSGDA. Specifically, DCSGDA can converge to a saddle point of problem (1), whose definition is given below.

Definition 3 (Saddle Point). *$(\mathbf{x}^*, \mathbf{y}^*)$ is a saddle point of a function f if $\forall \mathbf{x} \in \mathbb{R}^{d_1}, \mathbf{y} \in \mathcal{Y}$ satisfies $f(\mathbf{x}^*, \mathbf{y}) \leq f(\mathbf{x}^*, \mathbf{y}^*) \leq f(\mathbf{x}, \mathbf{y}^*)$.*

Before proceeding, some common and necessary assumptions are given below.

Assumption 1 (Smoothness). *Each f_i is L_i-smooth ($L_i > 0$), i.e., $\forall \mathbf{x}, \mathbf{x}' \in \mathbb{R}^{d_1}, \mathbf{y}, \mathbf{y}' \in \mathcal{Y}$, $\|\nabla f_i(\mathbf{x}, \mathbf{y}) - \nabla f_i(\mathbf{x}', \mathbf{y}')\| \leq L_i \|(\mathbf{x}, \mathbf{y}) - (\mathbf{x}', \mathbf{y}')\|$.*

Assumption 1 implies that f is L-smooth, where $L = \frac{1}{n} \sum_{i=1}^n L_i$.

Assumption 2 (Bounded Variance). *The stochastic gradient at each worker is unbiased with bounded variance, i.e., there exists a constant $\sigma > 0$ such that at each client $i \in [n]$, $\forall \mathbf{x} \in \mathbb{R}^{d_1}, \mathbf{y} \in \mathcal{Y}$,*

$$\mathbb{E}\nabla f_i(\mathbf{x}, \mathbf{y}; \xi_i) = \nabla f_i(\mathbf{x}, \mathbf{y}), \quad \mathbb{E} \|\nabla f_i(\mathbf{x}, \mathbf{y}; \xi_i) - \nabla f_i(\mathbf{x}, \mathbf{y})\|^2 \leq \sigma^2.$$

Assumption 3 (Strong Convexity). *$f(\mathbf{x}, \mathbf{y})$ is μ-strongly convex in $\mathbf{x}$, if there exists a $\mu > 0$ such that $\forall \mathbf{x}, \mathbf{x}' \in \mathbb{R}^{d_1}, \mathbf{y} \in \mathcal{Y}$ we have*

$$f(\mathbf{x}, \mathbf{y}) \geq f(\mathbf{x}', \mathbf{y}) + \langle \nabla_{\mathbf{x}} f(\mathbf{x}', \mathbf{y}), \mathbf{x} - \mathbf{x}' \rangle + \frac{\mu}{2} \|\mathbf{x} - \mathbf{x}'\|^2.$$

Assumption 4 (Strong Concavity). *$f(\mathbf{x}, \mathbf{y})$ is μ-strongly concave in $\mathbf{y}$, if there exists a $\mu > 0$ such that $\forall \mathbf{x} \in \mathbb{R}^{d_1}, \mathbf{y}, \mathbf{y}' \in \mathcal{Y}$ we have*

$$f(\mathbf{x}, \mathbf{y}) \leq f(\mathbf{x}, \mathbf{y}') + \langle \nabla_{\mathbf{y}} f(\mathbf{x}, \mathbf{y}'), \mathbf{y} - \mathbf{y}' \rangle - \frac{\mu}{2} \|\mathbf{y} - \mathbf{y}'\|^2.$$

3.1 Technical Lemmas

First, some technical lemmas of the compression mechanism are demonstrated.

Lemma 1 (Young's Inequality). *Given two vectors $\mathbf{a}, \mathbf{b} \in \mathbb{R}^d$, for any $s > 0$, we have $\|\mathbf{a} + \mathbf{b}\|^2 \leq (1 + s) \|\mathbf{a}\|^2 + (1 + s^{-1}) \|\mathbf{b}\|^2$.*

To analyze the convergence of DCSGDA, we then introduce a key lemma estimating the errors caused by compression mechanism.

Lemma 2 (Bounded Compression Distortion of Stochastic Gradients). *Let $G_{\mathbf{x},i}^t = \left\|g_{\mathbf{x},i}^t - \nabla_{\mathbf{x}} f_i(\mathbf{x}^t, \mathbf{y}^t)\right\|^2$, $G_{\mathbf{y},i}^t = \left\|g_{\mathbf{y},i}^t - \nabla_{\mathbf{y}} f_i(\mathbf{x}^t, \mathbf{y}^t)\right\|^2$ be the distortion of Markov compression on stochastic gradients. For any $a_1, a_2, a_3 > 0$, we have*

$$\mathbb{E}G_{\mathbf{x},i}^{t+1} \leq (1-\theta)G_{\mathbf{x},i}^t + \beta\left\|\nabla_{\mathbf{x}} f_i(\mathbf{x}^{t+1}, \mathbf{y}^{t+1}) - \nabla_{\mathbf{x}} f_i(\mathbf{x}^t, \mathbf{y}^t)\right\|^2 + c\sigma^2,$$
$$\mathbb{E}G_{\mathbf{y},i}^{t+1} \leq (1-\theta)G_{\mathbf{y},i}^t + \beta\left\|\nabla_{\mathbf{y}} f_i(\mathbf{x}^{t+1}, \mathbf{y}^{t+1}) - \nabla_{\mathbf{y}} f_i(\mathbf{x}^t, \mathbf{y}^t)\right\|^2 + c\sigma^2, \quad (2)$$

where $\theta = 1 - (1-\alpha)(1+a_1)(1+a_2)$, $\beta = (1-\alpha)(1+a_1)(1+a_2^{-1})(1+a_3)$, and $c = (1-\alpha)(1+a_1)(1+a_2^{-1})(1+a_3^{-1}) + (1+a_1^{-1})$.

Proof. According to Definitions 1 and 2, we can obtain

$$\begin{aligned}
&\mathbb{E}G_{\mathbf{x},i}^{t+1} \\
&= \mathbb{E}\left\|g_{\mathbf{x},i}^{t+1} - \nabla_{\mathbf{x}} f_i(\mathbf{x}^{t+1}, \mathbf{y}^{t+1})\right\|^2 \\
&\leq (1+a_1)\mathbb{E}\left\|g_{\mathbf{x},i}^t + \mathcal{C}(\nabla_{\mathbf{x}} f_i(\mathbf{x}^{t+1}, \mathbf{y}^{t+1}; \xi_i^{t+1}) - g_{\mathbf{x},i}^t) - \nabla_{\mathbf{x}} f_i(\mathbf{x}^{t+1}, \mathbf{y}^{t+1}; \xi_i^{t+1})\right\|^2 \\
&\quad + (1+a_1^{-1})\mathbb{E}\left\|\nabla_{\mathbf{x}} f_i(\mathbf{x}^{t+1}, \mathbf{y}^{t+1}; \xi_i^{t+1}) - \nabla_{\mathbf{x}} f_i(\mathbf{x}^{t+1}, \mathbf{y}^{t+1})\right\|^2 \\
&\leq (1-\alpha)(1+a_1)\left\|\nabla_{\mathbf{x}} f_i(\mathbf{x}^{t+1}, \mathbf{y}^{t+1}; \xi_i^{t+1}) - \nabla_{\mathbf{x}} f_i(\mathbf{x}^t, \mathbf{y}^t) + \nabla_{\mathbf{x}} f_i(\mathbf{x}^t, \mathbf{y}^t) - g_{\mathbf{x},i}^t\right\|^2 \\
&\quad + (1+a_1^{-1})\sigma^2 \\
&\leq (1-\alpha)(1+a_1)(1+a_2)G_{\mathbf{x},i}^t + (1+a_1^{-1})\sigma^2 + (1-\alpha)(1+a_1)(1+a_2^{-1})(1+a_3^{-1})\sigma^2 \\
&\quad + (1-\alpha)(1+a_1)(1+a_2^{-1})(1+a_3)\left\|\nabla_{\mathbf{x}} f_i(\mathbf{x}^{t+1}, \mathbf{y}^{t+1}) - \nabla_{\mathbf{x}} f_i(\mathbf{x}^t, \mathbf{y}^t)\right\|^2,
\end{aligned}$$

where the second and the last inequality use Assumption 2. By a similar analysis, we also have

$$\begin{aligned}
&\mathbb{E}G_{\mathbf{y},i}^{t+1} \\
&\leq (1-\alpha)(1+a_1)(1+a_2)G_{\mathbf{y},i}^t + (1+a_1^{-1})\sigma^2 + (1-\alpha)(1+a_1)(1+a_2^{-1})(1+a_3^{-1})\sigma^2 \\
&\quad + (1-\alpha)(1+a_1)(1+a_2^{-1})(1+a_3)\left\|\nabla_{\mathbf{y}} f_i(\mathbf{x}^{t+1}, \mathbf{y}^{t+1}) - \nabla_{\mathbf{y}} f_i(\mathbf{x}^t, \mathbf{y}^t)\right\|^2,
\end{aligned}$$

which completes the proof. ■

Lemma 2 shows that as the number of iterations increases, the second term in (2) will vanish, so the last term $c\sigma^2$ will dominate the compression error. According to Lemma 2, we can also obtain the global aggregation of compression distortions as follows:

$$\begin{aligned}
\mathbb{E}G_{\mathbf{x}}^{t+1} &= \frac{1}{n}\sum_{i=1}^{n}\mathbb{E}G_{\mathbf{x},i}^{t+1} \\
&\leq \frac{1}{n}\sum_{i=1}^{n}\left((1-\theta)G_{\mathbf{x},i}^t + \beta\left\|\nabla_{\mathbf{x}} f_i(\mathbf{x}^{t+1}, \mathbf{y}^{t+1}) - \nabla_{\mathbf{x}} f_i(\mathbf{x}^t, \mathbf{y}^t)\right\|^2 + c\sigma^2\right) \\
&\leq (1-\theta)G_{\mathbf{x}}^t + \beta\tilde{L}^2\left(\left\|\mathbf{x}^{t+1} - \mathbf{x}^t\right\|^2 + \left\|\mathbf{y}^{t+1} - \mathbf{y}^t\right\|^2\right) + c\sigma^2, \quad (3) \\
\mathbb{E}G_{\mathbf{y}}^{t+1} &\leq (1-\theta)G_{\mathbf{y}}^t + \beta\tilde{L}^2\left(\left\|\mathbf{x}^{t+1} - \mathbf{x}^t\right\|^2 + \left\|\mathbf{y}^{t+1} - \mathbf{y}^t\right\|^2\right) + c\sigma^2, \quad (4)
\end{aligned}$$

where $\mathbb{E}G_{\mathbf{y}}^{t+1}$ follows from the similar analysis as $\mathbb{E}G_{\mathbf{x}}^{t+1}$, and $\tilde{L}^2 = \frac{1}{n}\sum_{i=1}^{n} L_i^2$. The inequalities (3) and (4) play crucial roles in the convergence analysis of DCSGDA.

3.2 Main Result

Denote $\gamma_{\max} := \max\{\gamma_{\mathbf{x}}, \gamma_{\mathbf{y}}\}$ and $\gamma_{\min} = \min\{\gamma_{\mathbf{x}}, \gamma_{\mathbf{y}}\}$. Then, we arrive at our main result.

Theorem 1. *Under Assumptions 1, 2, 3, and 4, if there exists a constant* $a_4 > 0$ *such that* $\omega := 1 - (1 - \frac{\gamma_{\min}\mu}{2})(1 + a_4) \leq \frac{\theta^2}{2} - 4\delta\theta\beta\tilde{L}^2 + 16\delta\beta\tilde{L}^2$, *where* $\delta = \gamma_{\max}^2(1 + a_4^{-1})$, *and* $\gamma_{\max} \leq \frac{\sqrt{\gamma_{\min}\mu}}{2L}$, *then it holds that*

$$\Psi(T)+\frac{2\delta}{\theta - 8\delta\beta\tilde{L}^2}G(T) \leq \left(1 - \frac{\omega - 16\delta\beta\tilde{L}^2}{\theta - 8\delta\beta\tilde{L}^2}\right)^T \left(\Psi(0) + \frac{2\delta}{\theta - 8\delta\beta\tilde{L}^2}G(0)\right)+\rho\sigma^2,$$

where $\Psi(t) = \|\mathbf{x}^t - \mathbf{x}^*\|^2 + \|\mathbf{y}^t - \mathbf{y}^*\|^2$, $G(t) = G_{\mathbf{x}}^t + G_{\mathbf{y}}^t$, $\rho = \frac{4\delta c}{\omega - 16\delta\beta\tilde{L}^2}$ *and* θ, β, c *are defined in Lemma 2.*

Proof. According to the update rules, we have

$$\begin{aligned}
\mathbb{E}\left\|\mathbf{x}^{t+1} - \mathbf{x}^*\right\|^2 &= \mathbb{E}\left\|\mathbf{x}^t - \gamma_{\mathbf{x}} g_{\mathbf{x}}^t + \gamma_{\mathbf{x}}\nabla_{\mathbf{x}} f(\mathbf{x}^t, \mathbf{y}^t) - \gamma_{\mathbf{x}}\nabla_{\mathbf{x}} f(\mathbf{x}^t, \mathbf{y}^t) - \mathbf{x}^*\right\|^2 \\
&\leq (1 + a_4)\left\|\mathbf{x}^t - \gamma_{\mathbf{x}}\nabla_{\mathbf{x}} f(\mathbf{x}^t, \mathbf{y}^t) - \mathbf{x}^*\right\|^2 \\
&\quad + (1 + a_4^{-1})\left\|\gamma_{\mathbf{x}} g_{\mathbf{x}}^t - \gamma_{\mathbf{x}}\nabla_{\mathbf{x}} f(\mathbf{x}^t, \mathbf{y}^t)\right\|^2 \\
&\leq (1 + a_4)\Big((1 - \gamma_{\mathbf{x}}\mu)\left\|\mathbf{x}^t - \mathbf{x}^*\right\|^2 - 2\gamma_{\mathbf{x}}\left(f(\mathbf{x}^t, \mathbf{y}^t) - f(\mathbf{x}^*, \mathbf{y}^t)\right) \\
&\quad + \gamma_{\mathbf{x}}^2 L^2\left(\left\|\mathbf{x}^t - \mathbf{x}^*\right\|^2 + \left\|\mathbf{y}^t - \mathbf{y}^*\right\|^2\right)\Big) + \gamma_{\mathbf{x}}^2(1 + a_4^{-1})G_{\mathbf{x}}^t,
\end{aligned} \tag{5}$$

where the last inequality uses the μ-convexity and L-smoothness of the $f(\cdot, \mathbf{y})$, and the fact that $\nabla_{\mathbf{x}} f(\mathbf{x}^*, \mathbf{y}^*) = 0$.

With a similar analysis, we have

$$\begin{aligned}
\mathbb{E}\left\|\mathbf{y}^{t+1} - \mathbf{y}^*\right\|^2 &= \mathbb{E}\left\|\mathcal{P}_{\mathcal{Y}}(\mathbf{y}^t + \gamma_{\mathbf{y}} g_{\mathbf{y}}^t) - \mathcal{P}_{\mathcal{Y}}(\mathbf{y}^*)\right\|^2 \\
&\leq \mathbb{E}\left\|\mathbf{y}^t + \gamma_{\mathbf{y}} g_{\mathbf{y}}^t + \gamma_{\mathbf{y}}\nabla_{\mathbf{y}} f(\mathbf{x}^t, \mathbf{y}^t) - \gamma_{\mathbf{y}}\nabla_{\mathbf{y}} f(\mathbf{x}^t, \mathbf{y}^t) - \mathbf{y}^*\right\|^2 \\
&\leq (1 + a_4)\Big((1 - \gamma_{\mathbf{y}}\mu)\left\|\mathbf{y}^t - \mathbf{y}^*\right\|^2 - 2\gamma_{\mathbf{y}}\left(f(\mathbf{x}^t, \mathbf{y}^*) - f(\mathbf{x}^t, \mathbf{y}^t)\right) \\
&\quad + \gamma_{\mathbf{y}}^2 L^2\left(\left\|\mathbf{x}^t - \mathbf{x}^*\right\|^2 + \left\|\mathbf{y}^t - \mathbf{y}^*\right\|^2\right)\Big) + \gamma_{\mathbf{y}}^2(1 + a_4^{-1})G_{\mathbf{y}}^t.
\end{aligned} \tag{6}$$

Then by adding (5) and (6), we can obtain

$$\begin{aligned}
&\mathbb{E}\left\|\mathbf{x}^{t+1}-\mathbf{x}^*\right\|^2+\mathbb{E}\left\|\mathbf{y}^{t+1}-\mathbf{y}^*\right\|^2 \\
&\leq (1+a_4)\left((1-\gamma_{\min}\mu+2\gamma_{\max}^2L^2)\left(\left\|\mathbf{x}^{t+1}-\mathbf{x}^*\right\|^2+\left\|\mathbf{y}^{t+1}-\mathbf{y}^*\right\|^2\right)\right. \\
&\quad - 2\gamma_{\min}\left(f(\mathbf{x}^t,\mathbf{y}^t)-f(\mathbf{x}^*,\mathbf{y}^t)+f(\mathbf{x}^t,\mathbf{y}^*)-f(\mathbf{x}^t,\mathbf{y}^t)\right)) \\
&\quad + \gamma_{\max}^2(1+a_4^{-1})(G_{\mathbf{x}}^t+G_{\mathbf{y}}^t).
\end{aligned}$$

Here, $\gamma_{\max}^2L^2 \leq \frac{\gamma_{\min}\mu}{4}$ holds since $\gamma_{\max} \leq \frac{\sqrt{\gamma_{\min}\mu}}{2L}$. Then we have

$$\begin{aligned}
&\mathbb{E}\left\|\mathbf{x}^{t+1}-\mathbf{x}^*\right\|^2+\mathbb{E}\left\|\mathbf{y}^{t+1}-\mathbf{y}^*\right\|^2 \\
&\leq (1-\frac{\gamma_{\min}\mu}{2})(1+a_4)\left(\left\|\mathbf{x}^{t}-\mathbf{x}^*\right\|^2+\left\|\mathbf{y}^{t}-\mathbf{y}^*\right\|^2\right) - 2\gamma_{\min}(1+a_4)\left(f(\mathbf{x}^t,\mathbf{y}^t)\right. \\
&\quad \left. - f(\mathbf{x}^*,\mathbf{y}^t)+f(\mathbf{x}^t,\mathbf{y}^*)-f(\mathbf{x}^t,\mathbf{y}^t)\right) + \gamma_{\max}^2(1+a_4^{-1})(G_{\mathbf{x}}^t+G_{\mathbf{y}}^t). \qquad (7)
\end{aligned}$$

According to Definition 3, we can get

$$\begin{aligned}
&f(\mathbf{x}^t,\mathbf{y}^t)-f(\mathbf{x}^*,\mathbf{y}^t)+f(\mathbf{x}^t,\mathbf{y}^*)-f(\mathbf{x}^t,\mathbf{y}^t) \\
&= f(\mathbf{x}^t,\mathbf{y}^*)-f(\mathbf{x}^*,\mathbf{y}^*)+f(\mathbf{x}^*,\mathbf{y}^*)-f(\mathbf{x}^*,\mathbf{y}^t) \geq 0.
\end{aligned}$$

So (7) can be rearranged as

$$\begin{aligned}
&\mathbb{E}\left\|\mathbf{x}^{t+1}-\mathbf{x}^*\right\|^2+\mathbb{E}\left\|\mathbf{y}^{t+1}-\mathbf{y}^*\right\|^2 \\
&\leq (1-\omega)\left(\left\|\mathbf{x}^{t}-\mathbf{x}^*\right\|^2+\left\|\mathbf{y}^{t}-\mathbf{y}^*\right\|^2\right)+\delta(G_{\mathbf{x}}^t+G_{\mathbf{y}}^t), \qquad (8)
\end{aligned}$$

where $\omega = 1-(1-\frac{\gamma_{\min}\mu}{2})(1+a_4)$, and $\delta = \gamma_{\max}^2(1+a_4^{-1})$.

Let $\Psi(t) = \|\mathbf{x}^t-\mathbf{x}^*\|^2+\|\mathbf{y}^t-\mathbf{y}^*\|^2$, and $G(t) = G_{\mathbf{x}}^t+G_{\mathbf{y}}^t$. Then by adding $\frac{2\delta}{\theta}G(t+1)$ on both sides of (8), we can obtain

$$\begin{aligned}
&\Psi(t+1)+\frac{2\delta}{\theta}G(t+1) \\
&\leq (1-\omega)\Psi(t)+\frac{2\delta}{\theta}(1-\frac{\theta}{2})G(t)+\frac{4\delta\beta\tilde{L}^2}{\theta}\left(\left\|\mathbf{x}^{t+1}-\mathbf{x}^t\right\|^2+\left\|\mathbf{y}^{t+1}-\mathbf{y}^t\right\|^2\right)+\frac{4\delta c}{\theta}\sigma^2 \\
&\leq (1-\omega)\Psi(t)+\frac{2\delta}{\theta}(1-\frac{\theta}{2})G(t)+\frac{4\delta\beta\tilde{L}^2}{\theta}\left(2\left\|\mathbf{x}^{t+1}-\mathbf{x}^*\right\|^2+2\left\|\mathbf{x}^{t}-\mathbf{x}^*\right\|^2\right. \\
&\quad \left.+2\left\|\mathbf{y}^{t+1}-\mathbf{y}^*\right\|^2+2\left\|\mathbf{y}^{t}-\mathbf{y}^*\right\|^2\right)+\frac{4\delta c}{\theta}\sigma^2.
\end{aligned}$$

Then we have

$$\begin{aligned}
&\Psi(t+1)+\frac{2\delta}{\theta-8\delta\beta\tilde{L}^2}G(t+1) \\
&\leq \left(1-\frac{\omega-16\delta\beta\tilde{L}^2}{\theta-8\delta\beta\tilde{L}^2}\right)\Psi(t)+\frac{2\delta}{\theta-8\delta\beta\tilde{L}^2}(1-\frac{\theta}{2})G(t)+\frac{4\delta c}{\theta-8\delta\beta\tilde{L}^2}\sigma^2 \\
&\leq \left(1-\frac{\omega-16\delta\beta\tilde{L}^2}{\theta-8\delta\beta\tilde{L}^2}\right)\left(\Psi(t)+\frac{2\delta}{\theta-8\delta\beta\tilde{L}^2}G(t)\right)+\frac{4\delta c}{\theta-8\delta\beta\tilde{L}^2}\sigma^2, \qquad (9)
\end{aligned}$$

where the last inequality holds since $\omega \le \frac{\theta^2}{2} - 4\delta\theta\beta\tilde{L}^2 + 16\delta\beta\tilde{L}^2$.

By summing (9) from 0 to $T-1$, we have

$$\begin{aligned}
&\Psi(T) + \frac{2\delta}{\theta - 8\delta\beta\tilde{L}^2} G(T) \\
&\le \left(1 - \frac{\omega - 16\delta\beta\tilde{L}^2}{\theta - 8\delta\beta\tilde{L}^2}\right)^T \left(\Psi(0) + \frac{2\delta}{\theta - 8\delta\beta\tilde{L}^2} G(0)\right) + \frac{4\delta c\sigma^2}{\theta - 8\delta\beta\tilde{L}^2} \sum_{k=0}^{T-1} \left(1 - \frac{\omega - 16\delta\beta\tilde{L}^2}{\theta - 8\delta\beta\tilde{L}^2}\right)^k \\
&\le \left(1 - \frac{\omega - 16\delta\beta\tilde{L}^2}{\theta - 8\delta\beta\tilde{L}^2}\right)^T \left(\Psi(0) + \frac{2\delta}{\theta - 8\delta\beta\tilde{L}^2} G(0)\right) + \rho\sigma^2,
\end{aligned}$$

where $\rho = \frac{4\delta c}{\omega - 16\delta\beta\tilde{L}^2}$, which completes the proof. ∎

It can be observed that DCSGDA achieves a linear convergence rate in $\mathcal{O}(\sigma^2)$ neighborhood of the optimal solution. Thus, we can recover the convergence rate of SGDA in the strongly-convex-strongly-concave setting. If $\sigma^2 = 0$, i.e., with full gradient information for calculation, then our algorithm would have an exact linear convergence. We can also conclude that the convergence rate is influenced by the objective function, the compressor, and the batch size. If we properly choose ρ, the error term $\rho\sigma^2$ would have no great impact on the convergence of the algorithm.

4 Experiments

In this section, we present numerical experiments to demonstrate the performance of DCSGDA. Consider a distributed minimax logistic regression problem on "a9a" dataset [3] as below:

$$\min_{\mathbf{x}_i \in \mathbb{R}^{d_1}} \max_{\mathbf{y} \in \mathcal{Y}} \frac{1}{n} \sum_{i=1}^{n} \underbrace{\frac{1}{m} \sum_{j=1}^{m} (y_i l_{ij}(\mathbf{x}_i, \mathbf{y}_i) + \frac{1}{2}\lambda_1 \|\mathbf{x}_i\|^2 - \frac{1}{2}\lambda_2 \|n\mathbf{y}_i - \mathbf{1}\|^2)}_{:=F_i(\mathbf{x}_i, \mathbf{y}_i)}, \tag{10}$$

where $F_i(\mathbf{x}_i, \mathbf{y}_i)$ refers to the objective function of agent i and depends on its own datasets $\{(\mathbf{a}_{ij}, b_{ij})\}_{j=1}^m$ with $\mathbf{a}_{ij} \in \mathbb{R}^{d_1}$ and $b_{ij} \in \{1, -1\}$ being the feature and its associated label of the jth sample, respectively. $l_{ij}(\mathbf{x}_i, \mathbf{y}_i) := \log(1 + \exp(-b_{ij}\mathbf{a}_{ij}^\top \mathbf{x}_i))$ is the loss function, $\frac{1}{2}\lambda_1 \|\mathbf{x}_i\|^2$ and $\frac{1}{2}\lambda_2 \|n\mathbf{y}_i - \mathbf{1}\|^2$ are strongly-convex and strongly-concave regularizer, respectively. Clearly, each F_i is strongly-convex-strongly-concave.

The Setting of Simulation. The number of agents is set as $n = 50$. Let $\lambda_1 = 10^{-4}$, $\lambda_2 = 1/n^2$, and $\mathcal{Y} = [-1/n, 1/n]^n$. The following two compressors are selected: **(i)** ***p*-bits biased quantizer** [25]:

$$\mathcal{C}_1(\mathbf{z}) = \left\lfloor \mathbf{z} + \frac{2^p \|\mathbf{z}\|_\infty \mathbf{1}_d}{2^p - 1} \right\rfloor - \|\mathbf{z}\|_\infty \mathbf{1}_d,$$

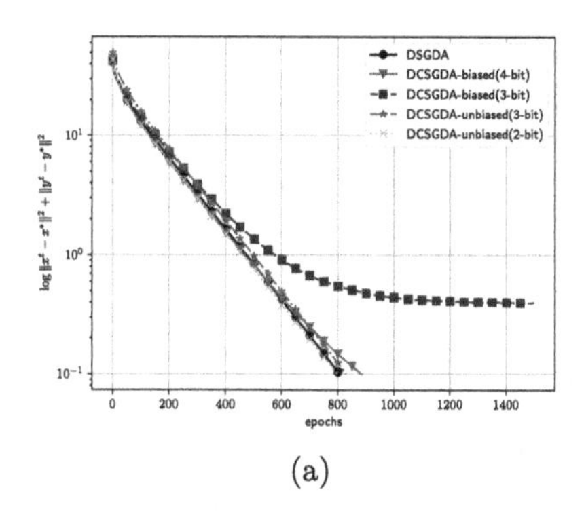

(a)

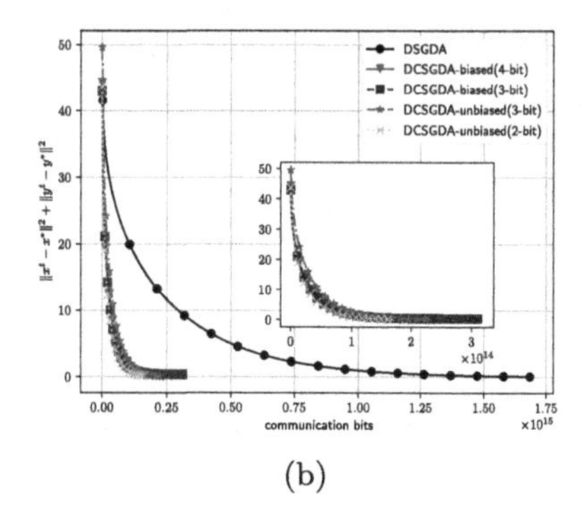

(b)

Fig. 1. Comparison between DSGDA and DCSGDA with different compressiopg methods.

where $\lfloor \cdot \rfloor$ is floor function; **(ii)** l**-bits unbiased quantizer** [11]:

$$\mathcal{C}_2(\mathbf{z}) = \frac{\|\mathbf{z}\|_\infty}{2^{l-1}} \operatorname{sign}(\mathbf{z}) \circ \left\lfloor \frac{2^{l-1}|\mathbf{z}|}{\|\mathbf{z}\|_\infty} + \varpi \right\rfloor,$$

where $\circ$ denotes the Hadamard product and ϖ is a random perturbation vector uniformly sampled from $[0, 1]^d$. Clearly, if we transmit a scalar with b bits, then the transmission of $\mathcal{C}_1(\mathbf{z})$ and $\mathcal{C}_2(\mathbf{z})$ need $dp+b$ and $(l+1)d+b$ bits, respectively, where d is the dimension of $\mathbf{z}$.

Simulation Results. We run DCSGDA with different compression ratios. In addition, we also run DSGDA (no compression) for comparison. The simulation results are presented in Fig. 1. Figure 1a depicts the relationship between convergence accuracy and epochs. It can be seen that DCSGDA with unbiased compressors can achieve comparable linear convergence rates as DSGDA. For DCSGDA with biased compressors, a linear convergence rate can also be achieved when the compression bits are chosen appropriately, but more epochs are needed to achieve the same accuracy. We also find that the biased compressors would have a more negative effect on the convergence of DCSGDA compared to the unbiased compressors with the same compression ratios. Figure 1b shows the relationship between convergence accuracy and communication bits. We can see that DCSGDA can significantly reduce the communication cost compared to DSGDA. With the same convergence accuracy, DSGDA uses around 7 times more communication bits than DCSGDA algorithm.

5 Conclusion

In this paper, we proposed a communication-efficient first-order distributed minimax algorithm based on SGDA. Markov compression mechanism is implemented to reduce the communication bits per iteration. Theoretical analysis has shown that our algorithm can obtain linear convergence in the neighborhood of the optimal solution under strongly-convex-strongly-concave and smooth conditions. And numerical experiments have demonstrated the communication efficiency and convergence of the algorithm. Future work will focus on extending the algorithm to nonconvex settings and combining other communication-efficient methods.

References

1. Basu, D., Data, D., Karakus, C., Diggavi, S.: Qsparse-local-sgd: distributed sgd with quantization, sparsification and local computations. In: Advances in Neural Information Processing Systems, vol. 32 (2019)
2. Beznosikov, A., Gorbunov, E., Berard, H., Loizou, N.: Stochastic gradient descent-ascent: unified theory and new efficient methods. In: Proceedings of The 26th International Conference on Artificial Intelligence and Statistics, vol. 206, pp. 172–235. PMLR (2023)
3. Chang, C.C., Lin, C.J.: Libsvm: a library for support vector machines. ACM Trans. Intell. Syst. Technol. **2**(3), 1–27 (2011)
4. Deng, Y., Mahdavi, M.: Local stochastic gradient descent ascent: convergence analysis and communication efficiency. In: Proceedings of the 24th International Conference on Artificial Intelligence and Statistics, vol. 130, pp. 1387–1395. PMLR (2021)
5. Goodfellow, I., et al.: Generative adversarial nets. In: Advances in Neural Information Processing Systems, vol. 27 (2014)
6. Korpelevich, G.: An extragradient method for finding saddle points and for other problems. Ekonomika i Matematicheskie Metody **12**, 747–756 (1976)
7. Lin, Y., Han, S., Mao, H., Wang, Y., Dally, W.J.: Deep gradient compression: reducing the communication bandwidth for distributed training. In: International Conference on Learning Representations (2018)
8. Liu, C., Bi, S., Luo, L., Lui, J.C.: Partial-quasi-Newton methods: efficient algorithms for minimax optimization problems with unbalanced dimensionality. In: Proceedings of the 28th ACM SIGKDD Conference on Knowledge Discovery and Data Mining, pp. 1031–1041 (2022)
9. Liu, C., Chen, L., Luo, L., Lui, J.: Communication efficient distributed Newton method with fast convergence rates. arXiv preprint arXiv:2305.17945 (2023)
10. Liu, C., Luo, L.: Quasi-Newton methods for saddle point problems. In: Advances in Neural Information Processing Systems, vol. 35, pp. 3975–3987 (2022)
11. Liu, X., Li, Y., Wang, R., Tang, J., Yan, M.: Linear convergent decentralized optimization with compression. arXiv preprint arXiv:2007.00232 (2020)
12. Loizou, N., Richtárik, P.: Momentum and stochastic momentum for stochastic gradient, newton, proximal point and subspace descent methods. Comput. Optim. Appl. **77**(3), 653–710 (2020)
13. McMahan, B., Moore, E., Ramage, D., Hampson, S., Arcas, B.A.y.: Communication-efficient learning of deep networks from decentralized data. In: Proceedings of the 20th International Conference on Artificial Intelligence and Statistics, vol. 54, pp. 1273–1282. PMLR (2017)
14. Namkoong, H., Duchi, J.C.: Stochastic gradient methods for distributionally robust optimization with f-divergences. In: Advances in Neural Information Processing Systems, vol. 29 (2016)
15. Nesterov, Y.: Primal-dual subgradient methods for convex problems. Math. Program. **120**(1), 221–259 (2009)

16. Omidshafiei, S., Pazis, J., Amato, C., How, J.P., Vian, J.: Deep decentralized multi-task multi-agent reinforcement learning under partial observability. In: Proceedings of the 34th International Conference on Machine Learning, vol. 70, pp. 2681–2690. PMLR (2017)
17. Qian, X., Richtárik, P., Zhang, T.: Error compensated distributed sgd can be accelerated. In: Advances in Neural Information Processing Systems, vol. 34, pp. 30401–30413 (2021)
18. Rabbat, M., Nowak, R.: Quantized incremental algorithms for distributed optimization. IEEE J. Sel. Areas Commun. **23**(4), 798–808 (2005)
19. Rasouli, M., Sun, T., Rajagopal, R.: Fedgan: federated generative adversarial networks for distributed data. arXiv preprint arXiv:2006.07228 (2020)
20. Richtárik, P., Sokolov, I., Fatkhullin, I.: Ef21: a new, simpler, theoretically better, and practically faster error feedback. In: Advances in Neural Information Processing Systems, vol. 34, pp. 4384–4396 (2021)
21. Sharma, P., Panda, R., Joshi, G., Varshney, P.: Federated minimax optimization: Improved convergence analyses and algorithms. In: Proceedings of the 39th International Conference on Machine Learning, vol. 162, pp. 19683–19730. PMLR (2022)
22. Shi, S., et al.: A distributed synchronous sgd algorithm with global top-k sparsification for low bandwidth networks. In: 2019 IEEE 39th International Conference on Distributed Computing Systems, pp. 2238–2247 (2019)
23. Stich, S.U.: Local SGD converges fast and communicates little. In: International Conference on Learning Representations (2019)
24. Ström, N.: Scalable distributed dnn training using commodity gpu cloud computing. In: Proceedings of the Annual Conference of the International Speech Communication Association. pp. 1488–1492 (2015)
25. Sun, J., Chen, T., Giannakis, G., Yang, Z.: Communication-efficient distributed learning via lazily aggregated quantized gradients. In: Advances in Neural Information Processing Systems, vol. 32 (2019)
26. Wang, S., et al.: Adaptive federated learning in resource constrained edge computing systems. IEEE J. Sel. Areas Commun. **37**(6), 1205–1221 (2019)
27. Yu, Y., Wu, J., Huang, L.: Double quantization for communication-efficient distributed optimization. In: Advances in Neural Information Processing Systems, vol. 32, pp. 4438–4449 (2019)
28. Zhang, Z., Yang, S., Xu, W., Di, K.: Privacy-preserving distributed admm with event-triggered communication. IEEE Trans. Neural Networks Learn. Syst., 1–13 (2022)

Multi-level Augmentation Boosts Hybrid CNN-Transformer Model for Semi-supervised Cardiac MRI Segmentation

Ruohan Lin[1(✉)], Wangjing Qi[1], and Tao Wang[2]

[1] Guangxi Key Laboratory of Image and Graphic Intelligent Processing, Guilin University of Electronic Technology, Guilin Guangxi 541004, China
rhlinnn@gmail.com

[2] College of Computer Science and Software Engineering, Shenzhen University, Shenzhen, China

Abstract. Over the past few years, many supervised deep learning algorithms based on Convolutional Neural Networks (CNN) and Vision Transformers (ViT) have achieved remarkable progress in the field of clinical-assisted diagnosis. However, the specific application of these algorithms e.g. ViT which requires a large amount of data in the training process is greatly limited due to the high cost of medical image annotation. To address this issue, this paper proposes an effective semi-supervised medical image segmentation framework, which combines two models with different structures, i.e. CNN and Transformer, and integrates their abilities to extract local and global information through a mutual supervision strategy. Based on this heterogeneous dual-network model, we employ multi-level image augmentation to expand the dataset, alleviating the model's demand for data. Additionally, we introduce an uncertainty minimization constraint to further improve the model's robustness, and incorporate an equivariance regularization module to encourage the model to capture semantic information of different categories in the images. In public benchmark tests, we demonstrate that the proposed method outperforms the recently developed semi-supervised medical image segmentation methods in terms of specific metrics such as Dice coefficient and 95% Hausdorff Distance for segmentation performance. The code will be released at https://github.com/swaggypg/MLABHCTM.

Keywords: Semi-supervised Learning · Cardiac MRI · Image Segmentation

1 Introduction

In clinical diagnosis, Magnetic Resonance Imaging (MRI) is widely used for non-invasive qualitative and quantitative assessment of cardiac anatomy and function. Cardiac MRI segmentation, which divides the image into multiple semantic

B. Luo et al. (Eds.): ICONIP 2023, LNCS 14447, pp. 552–563, 2024.
https://doi.org/10.1007/978-981-99-8079-6_43

regions, facilitates subsequent tasks such as cardiac disease diagnosis and treatment planning, cardiac pathology research, cardiac surgery planning and navigation, drug development, and treatment monitoring [5]. Ideally, deep learning-based methods can achieve precise cardiac semantic segmentation by leveraging a large amount of annotated data [2,17,19], even surpassing the accuracy of clinical experts. However, the high cost of annotating medical images remains a challenge. In contrast, a large amount of unlabeled data is readily available and easily accessible. In this context, researchers have proposed semi-supervised learning methods [7], which significantly improve the utilization of unlabeled data in real clinical scenarios, and reduce the workload of clinical experts in data annotation.

The key to semi-supervised learning methods lies in effectively mining the latent information within unlabeled data. Over the past years, semi-supervised learning has rapidly evolved and has given rise to many outstanding strategies, such as self-training [6], consistency regularization [3], entropy minimization [9] and self-ensembling [10]. However, most of these methods suffer from two main issues: (1) insufficient training data volume and (2) limited perspectives of the backbone network, resulting in a lack of comprehensive feature information. These issues often lead to higher epistemic uncertainty.

To address the aforementioned issues, we propose a heterogeneous models consistency learning method. Our method incorporates three modules. The first module is heterogeneous model cross-supervision. In each iteration, a combination of strong and weak data perturbations is applied to the images, which are then inputted into both the CNN and Transformer models for segmentation. After that, the predictions of the weakly augmented images from the CNN model are utilized as pseudo-labels to supervise the predictions of the Transformer model for the strongly augmented data, and vice versa. The second module is uncertainty minimization constraint. We introduce an uncertainty minimization term as a constraint to reduce prediction entropy and enhance the model's robustness. Both the CNN and Transformer models are employed to segment the original images. The image-level uncertainty values are then calculated by averaging the segmentation results obtained from these two models. This calculation captures the overall prediction entropy generated by the framework. The third module is equivariance regularization which enhance the model's learning ability of image features by randomly rotating the input images and letting the model predict the rotation count.

In summary, the main contributions of this work are as follows:

- Proposing an effective semi-supervised learning method that leverages cross-pseudo supervision applied to heterogeneous models and employs multi-level data augmentation strategies to fully satisfy the training's data requirements.
- Combining multi-model structures, uncertainty minimization strategy, and equivariance regularization to enhance model's robustness and the effectiveness of feature information extracted by the encoder from images.

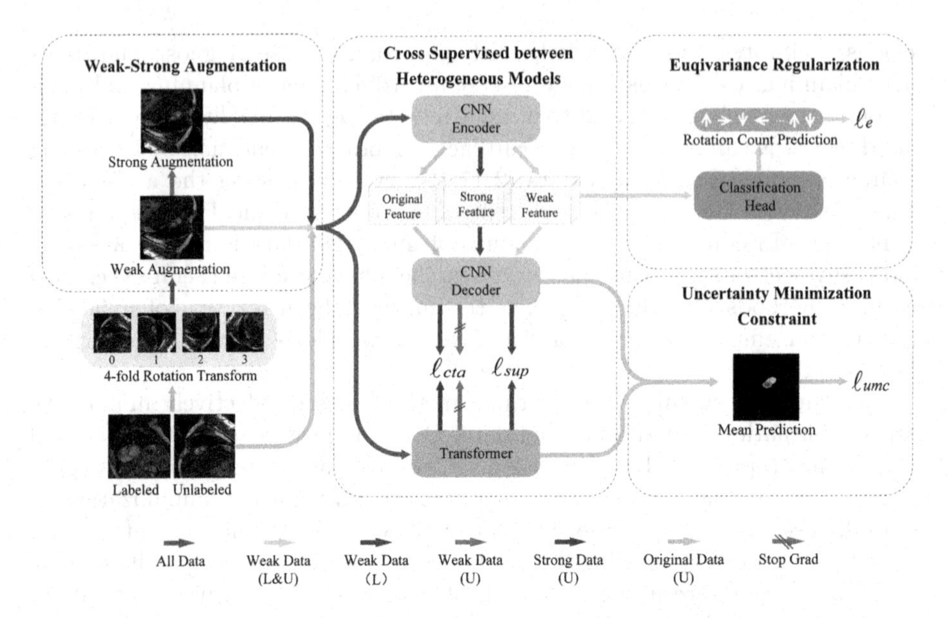

Fig. 1. Overview of our Method. In the data flow of the equivariance regularization module, the labeled data and unlabeled data, after undergoing 4-fold rotation and weak augmentation, are inputted into the encoder of the CNN to extract feature information. The extracted features are then passed to the 4-way classification head to determine the number of rotations applied.

2 Methodology

In this study, we employed a semi-supervised learning approach by dividing the training dataset into a labeled dataset $D_L = \{(x_i, y_i), i = 1, \ldots, N\}$ and an unlabeled dataset $D_U = \{x_j, j = 1, \ldots, M\}$, where N and M represent the quantities of labeled and unlabeled data, and M is much larger than N. For the labeled dataset, weak augmentation is applied to each x_i and input into the network, utilizing the ℓ_{sup} loss function for supervised learning. For the unlabeled dataset, two different intensity levels of image augmentation are applied to generate weakly and strongly augmented data. These data, along with the original image data, are utilized in three modules: Cross Supervised between Heterogeneous Models, Uncertainty Minimization Constraint, and Equivariance Regularization. Figure 1 illustrates the proposed semi-supervised learning framework.

2.1 Cross Supervised Between Heterogeneous Models

CNN models are known for their inductive bias, which allows them to effectively extract local information but may lack a holistic understanding of the input data. In contrast, Transformer models leverage self-attention mechanisms to capture global positional information and long-range dependencies in feature extraction.

Inspired by [11], we propose a heterogeneous dual-network structure. However, we need to address a challenge: Transformer models typically have a larger number of parameters compared to CNN models, requiring more data to avoid overfitting. Moreover, when processing sequential data, Transformer models need to capture the relationships between different positions in the sequence, which further necessitates a larger dataset to learn these dependencies. In the field of medical image processing, labeled data is often limited, leading to suboptimal performance of the trained Transformer network. Inspired by [14], we employ a strategy of strong and weak augmentation to expand the dataset. By utilizing a cross-supervision strategy, we encourage the two heterogeneous models to complement each other in capturing different features and improve the quality of pseudo-labels, thereby optimizing the overall segmentation performance of the models.

Weak-Strong Augmentation. In our approach, we utilize weak and strong augmentation strategies. The weak strategy involves operations such as rotation and cropping applied to the images. The strong strategy includes adjustments to contrast, brightness, and modifications to the pixel value range, etc. Similar to FixMatch [14], we adopt CTAugment [1] to select the augmentation strategy for each iteration. CTAugment randomly selects two weak augmentation strategies and two strong augmentation strategies from predefined pools and sets the magnitude for each transformation. It is worth noting that while the numerical limits for the magnitude of each image transformation in CTAugment are manually set, the method dynamically learns the magnitude for each transformation based on the average loss value of each epoch. For an input image x_i, the weak and strong image augmentation modules produce two augmented images: $a(x_i)$ and $A(x_i)$, where $a(.)$ represents a random combination of weak augmentation functions, and $A(.)$ represents a random combination of strong augmentation functions.

Cross Teaching Learning. In our approach, we train two neural networks with different structures: a CNN network denoted as $f_\theta^c(.)$ and a Transformer network denoted as $f_\theta^t(.)$. In each iteration, we input the augmented versions of two unlabeled data samples, $a(x_i)$ and $A(x_i)$, into both $f_\theta^c(.)$ and $f_\theta^t(.)$, resulting in four sets of predictions:

$$p_i^{cw} = f_\theta^c\left(a\left(x_i\right)\right); \quad p_i^{cs} = f_\theta^c\left(A\left(x_i\right)\right); \tag{1}$$

$$p_i^{tw} = f_\theta^t\left(a\left(x_i\right)\right); \quad p_i^{ts} = f_\theta^t\left(A\left(x_i\right)\right); \tag{2}$$

where p_i^{cw} and p_i^{cs} represent the predictions of the CNN for the weakly augmented and strongly augmented versions of the image, respectively, and p_i^{tw} and p_i^{ts} represent the predictions of the Transformer for the weakly augmented and strongly augmented versions of the image, respectively. Because the weakly augmented data usually have smaller aleatoric uncertainty, the predictions for these images are more reliable. Therefore, we use the predictions from the weakly augmented images to compute pseudo-labels:

$$y_i^{cp} = Argmax\left(p_i^{tw}\right); \quad y_i^{tp} = Argmax\left(p_i^{cw}\right) \tag{3}$$

Here, y_i^{cp} is the pseudo-label for training the CNN network, computed from the Transformer's predictions, and y_i^{tp} is the pseudo-label for training the Transformer network, computed from the CNN's predictions. In summary, for the augmented unlabeled data, the Cross Teaching with Augmentation (cta) loss function is defined as:

$$\ell_{cta} = \mathcal{L}_{Dice}\left(p_i^{cs}, y_i^{cp}\right) + \mathcal{L}_{Dice}\left(p_i^{ts}, y_i^{tp}\right) \tag{4}$$

where $\mathcal{L}_{Dice}$ represents the Dice loss function.

2.2 Uncertainty Minimization Constraint

Our method employs multiple image augmentation techniques to expand the size of the training dataset, but it also introduces a certain amount of noise, which can lead to performance degradation of the CNN model. To mitigate this issue, we aim to reduce uncertainty and enhance the model's robustness by minimizing uncertainty.

A previous studie [16] has demonstrated that reducing prediction entropy can increase the information gain in each training iteration and reduce the probability of misclassifications. Inspired by [12], we propose the uncertainty minimization constraint as a loss function to reduce the prediction entropy of both the CNN and Transformer models. It is worth noting that this module only utilizes the original unlabeled data for computing the prediction entropy. Compared to the augmented data, the original images exhibit lower aleatoric uncertainty, resulting in a lower probability of the model being overconfident in incorrect predictions and accumulating noise during the training process. Specifically, for an input image x_i, we use the CNN and Transformer to make predictions, and then calculate the average probability distribution:

$$p_c = f_\theta^c\left(x_i\right); \quad p_t = f_\theta^t\left(x_i\right); \quad \hat{p} = \frac{1}{2}\left(p_c + p_t\right) \tag{5}$$

Then, we compute the uncertainty of the probability distribution for each pixel along the channel dimension, defined as follows:

$$E_i = -\sum_{j=0}^{C} \hat{p}_{i,j} \log\left(\hat{p}_{i,j} + \epsilon\right) \tag{6}$$

Here, C represents the number of classes in the segmentation task, $\hat{p}_{i,j}$ denotes the probability value of pixel i in class j, and ϵ is a very small value used to prevent singularity. In this work, we set ϵ to 1e-6. Finally, we define the image-level entropy of the average probability distribution as the loss function:

$$\ell_{umc} = \frac{1}{P}\sum_{i=0}^{P} E_i \tag{7}$$

Here, P represents the total number of pixels in the uncertainty map.

2.3 Equivariance Regularization

One common strategy in self-supervised learning methods is to set up auxiliary tasks that are related to the main task, encouraging the network to extract meaningful features and thereby improving learning efficiency and segmentation accuracy.

Inspired by [18], we introduce the equivariance regularization module to enhance the effectiveness of the feature information extracted from the data. It is worth noting that a U-shape Encoder-Decoder style CNN network is used in our method, which is represented as $f(.) = f_\gamma(f_\xi(.))$, where $f_\xi(.)$ represents the encoder and $f_\gamma(.)$ represents the decoder. A previous study [8] has demonstrated that training the encoder $f_\xi(.)$ to be sensitive to certain geometric transformations $G(.)$ is beneficial for improving the image representation quality and has a significant impact on downstream tasks, particularly with four-fold rotation.

Specifically, we separate the four-fold rotation transformation from the weak-strong augmentation module and apply it individually to the original images. We also record the count of rotations applied to each image as a label for training. For each input image x_i, we apply r clockwise rotation transformations and additionally introduce a random rotation within the range of [-20, 20], resulting in $G_r(x_i)$. We employ a classification predictor $P_\varphi(.)$ to determine the rotation count r for $G_r(x_i)$ by inputting the feature map extracted from $G_r(x_i)$ through $f_\xi(.)$. To complete such a task, the model needs to understand the semantic information of each class. The overall equivariance regularization loss function is defined as follows:

$$\ell_e = \mathcal{L}_{CE}\left(P_\varphi\left(f_\xi\left(G_r\left(x_i\right)\right)\right), r\right) \tag{8}$$

where $\mathcal{L}_{CE}$ denote the Cross Entropy loss.

2.4 The Overall Objective Function

Our method aims to jointly optimize the supervised loss on labeled data and the unsupervised loss on unlabeled data. For the supervised loss on labeled data x_i, we combine two popular loss functions:

$$\ell_{sup}^c = \frac{1}{2}\mathcal{L}_{CE}\left(p_i^{cw}, a\left(y_i\right)\right) + \frac{1}{2}\mathcal{L}_{Dice}\left(p_i^{cw}, a\left(y_i\right)\right) \tag{9}$$

$$\ell_{sup}^t = \frac{1}{2}\mathcal{L}_{CE}\left(p_i^{tw}, a\left(y_i\right)\right) + \frac{1}{2}\mathcal{L}_{Dice}\left(p_i^{tw}, a\left(y_i\right)\right) \tag{10}$$

$$\ell_{sup} = \ell_{sup}^c + \ell_{sup}^t \tag{11}$$

where ℓ_{sup}^c and ℓ_{sup}^t represent the supervised loss function for the CNN and Transformer networks, respectively. $\mathcal{L}_{Dice}$ denote the Dice loss. p_i^{cw} and p_i^{tw} correspond to the predictions of the CNN and Transformer models on the weakly augmented version of the labeled data $a(x_i)$. $a(y_i)$ represents the weakly augmented label of x_i.

The overall objective function of our method is defined as:

$$\ell = \ell_{sup} + \lambda_t (\ell_{cta} + 2\ell_{umc}) + \lambda \ell_e \tag{12}$$

where λ_t is a temperature parameter that gradually increases from 0 to 0.1. λ is a parameter that controls the weight of the equivariance regularization, and we set it to 0.1. Furthermore, the uncertainty minimization constrained loss is doubled to enhance its effectiveness.

3 Experiments

3.1 Experimental Setup

Datasets and Evaluation Metrics. All experiments and comparisons in this work are based on the publicly available Automatic Cardiac Diagnosis Challenge (ACDC) dataset. This dataset consists of 200 annotated short-axis cardiac MR-Cine images from 100 patients. We randomly select 140 scans for training, 20 scans for validation, and 40 scans for testing, following the setup described in CTCT [11]. Prior to input the images into the networks, we resize them to 224 × 224 pixels to match the input requirements of the networks. Additionally, due to the large inter-slice spacing of this dataset, we perform slice-wise segmentation by dividing each case into individual slices instead of directly performing 3D segmentation. During the inference stage, we use two commonly used metrics for evaluating the performance of medical image semantic segmentation methods: Dice Coefficient (denoted as *Dice*); 95% Hausdorff Distance (denoted as HD_{95}).

Implementation Details. This method is implemented on an Ubuntu 22.04 system using an NVIDIA GTX 4090 GPU and the PyTorch framework. Due to ACDC's anisotropic resolution, we use UNet [13] as the CNN network and Swin-UNet [4] as the Transformer network. Swin-UNet utilized the same pretraining weights as the official implementation, while CNN was not pretrained. Both networks are trained using the SGD optimizer with a weight decay of 0.0001 and momentum of 0.9. The initial learning rate is set to 0.01 and decreased gradually using the poly learning rate strategy until reaching 0.001. The batch size is set to 8, where 4 images are labeled. The total iterations is 30k.

In each training iteration, two augmentation strategies are randomly selected from the weak augmentation strategy pool and the strong augmentation strategy pool to form the weak and strong augmentation strategy for that iteration. Specifically, the weak augmentation strategy pool includes: Identity, Cutout, Rescale, Shear_x, Shear_y, Translate_x, and Translate_y. The strong augmentation strategy pool includes: Identity, Autocontrast, Brightness, Contrast, Posterize, and Solarize. The definition and implementation of these strategies are consistent with the approach used in Fixmatch [14].

3.2 Results

In the experiments of "Quantitative Comparison" section, for reliable results, we conducted an 8-fold cross-validation on the training and validation sets. We selected the model weights from the last epoch of each fold experiment and evaluated the performance on a fixed testing set. The average and standard deviation of the segmentation results for the performance measures from the eight test results are presented. In the experiments of other sections, we use fixed training, validation, and testing sets, and each experiment is conducted only once. The average and standard deviation of the segmentation results for each case in the testing set are presented.

Compared Methods. We compared our method with several classic and recently proposed semi-supervised segmentation methods, including Mean-Teacher (MT) [15], Uncertainty-Aware Mean Teacher (UA-MT) [20], Cross Pseudo Supervision (CPS) [6], Cross Teaching between CNN and Transformer (CTCT) [11], and FixMatch [14]. Except for FixMatch and our method, which use specific image augmentation strategies, all methods are trained with the same hyperparameter settings to ensure fairness in the evaluation process. Additionally, they are implemented under the same environment and backbone.

Table 1. Quantitative comparison of different methods on ACDC dataset. All results are based on the same backbone (UNet) with a fixed seed. |XL| represents the ratio of labeled images. Mean and standard variance (in parentheses) are presented in this table.

Method	$ACDC_{\|XL\|=5\%}$		$ACDC_{\|XL\|=10\%}$	
	Dice [%]	HD_{95}	Dice [%]	HD_{95}
-	61.14 (0.038)	29.87 (4.909)	83.06 (0.013)	12.74 (1.142)
MT	61.61 (0.067)	39.97 (10.479)	83.90 (0.006)	14.88 (1.485)
UA-MT	66.38 (0.058)	26.09 (5.221)	83.27 (0.020)	14.02 (0.954)
CPS	71.99 (0.041)	15.04 (2.237)	85.74 (0.003)	12.36 (1.459)
CTCT	79.64 (0.031)	9.56 (1.289)	87.03 (0.006)	9.41(1.911)
FIXMATCH	78.58 (0.022)	12.34 (2.738)	87.55 (0.003)	6.72 (1.426)
Ours	**86.08 (0.003)**	**6.76 (1.072)**	**88.78 (0.005)**	**5.19 (1.327)**

Quantitative Comparison. In the experiments conducted on the ACDC dataset, we performed quantitative comparisons. Table 1 presents the performance comparison of our proposed method with other methods in scenarios where the number of labeled patients is 3 (5% labeled) and 7 (10% labeled). The first row of the table represents the performance of the baseline model trained only with the labeled data. All methods' results are evaluated using the weights of the UNet model from the last epoch.

Our method achieves outstanding performance according to the experimental results. In the scenario with a data labeling rate of 10%, compared to the second-best method FixMatch, our method shows a 1.23% improvement in Dice and a 1.53 reduction in HD_{95}. In the scenario with a data labeling rate of 5%, our method outperforms the second-best method CTCT with a remarkable 6.44% increase in Dice and a 2.8 reduction in HD_{95}.

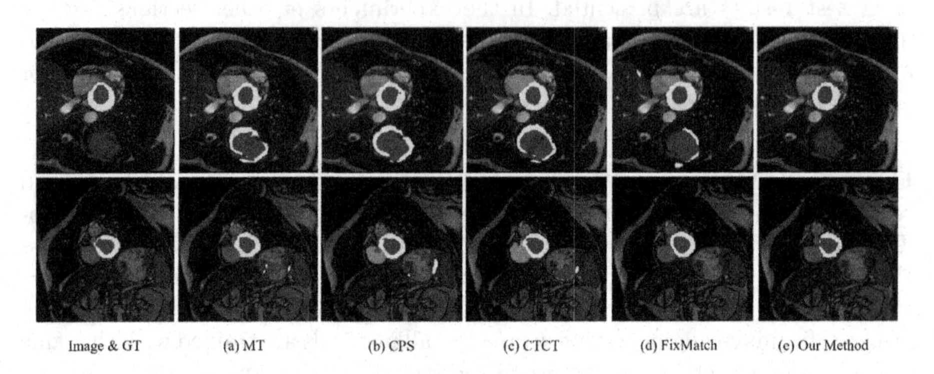

Fig. 2. Qualitative comparison of the segmentation results produced by different methods on ACDC dataset with 10% Labeled Data.

Qualitative Comparisons. Figure 2 displays the segmentation effect of our proposed method, the ground truth and two other methods, in a scenario where only 10% of the data is labeled. For comparison, we select MT [15], a classic method in the field of semi-supervised medical image segmentation, as well as three newer methods, CPS [6], FixMatch [14], and CTCT [11], which have demonstrated promising performance in quantitative experiments. The experimental results indicate that our method achieves better prediction results with reduced false predictions compared to other methods.

Ablation Study. Table 2 presents the results of the ablation experiments on the ACDC dataset, with a labeled data ratio set to 10%. We use the segmentation performance of the UNet model and the Swin-UNet model, trained only on labeled data, as baselines. We gradually add components to demonstrate their effectiveness. It should be noted that ℓ_{ct}, as introduced in the CTCT [11], represents cross pseudo-labeling between CNN and Transformer with simple image augmentation (random rotation and flipping). We independently add the three main components of our method to the baseline to test their effectiveness. Then, we combine ℓ_{umc} and ℓ_e with ℓ_{cta} to examine the possibility of mutual exclusivity between components.

Table 2. Quantitative results of ablation study on ACDC dataset when using 10% cases as labeled. The first line represents the baseline results of UNet and Swin-UNet models that only use labeled data for supervised learning.

ℓ_{ct}	ℓ_{cta}	ℓ_{umc}	ℓ_e	CNN		Transformer	
				Dice [%]	HD_{95}	Dice [%]	HD_{95}
		-		83.06 (0.088)	12.48 (9.683)	78.85 (0.110)	11.45 (7.551)
✓				86.38 (0.062)	8.24 (10.566)	85.44 (0.080)	7.62 (11.574)
	✓			87.16 (0.057)	9.16 (16.326)	87.04 (0.045)	6.12 (7.484)
		✓		85.87 (0.069)	10.24 (9.122)	81.20 (0.076)	10.99 (8.476)
			✓	86.40 (0.063)	7.74 (8.202)	81.51 (0.102)	10.89 (8.463)
	✓	✓		88.24 (0.044)	6.19 (7.321)	88.58 (0.036)	4.32 (4.470)
	✓		✓	88.61 (0.038)	**5.01 (5.663)**	88.43 (0.034)	4.61 (4.098)
	✓	✓	✓	**89.17 (0.039)**	6.57 (8.644)	**88.98 (0.035)**	**3.98 (3.884)**

Table 3. Quantitative comparison of equivariance regularization on different geometric transformations. These results are based on our proposed method.

Augmentation Strategy	CNN		Transformer	
	Dice [%]	HD_{95}	Dice [%]	HD_{95}
Horizontal Flips	87.98 (0.051)	8.44 (11.777)	88.65 (0.034)	4.29 (4.110)
Vertical Flips	88.26 (0.049)	8.37 (12.174)	88.78 (0.035)	4.78 (5.538)
8-fold Rotations	89.13 (0.037)	**5.96 (7.114)**	88.79 (0.036)	4.23 (3.865)
4-fold Rotations	**89.17 (0.039)**	6.57 (8.644)	**88.98 (0.035)**	**3.98 (3.884)**

From the results, it can be observed that compared to ℓ_{ct}, ℓ_{cta} exhibits a certain degree of improvement. Furthermore, the uncertainty minimization constraint module and equivariance regularization module significantly enhance the model's performance. The experimental results also demonstrate that the combination of these modules complements each other, further enhancing the overall performance of the model.

Effects of Geometric Transformations in Equivariance Regularization. By conducting experiments using four common geometric transformations, namely 4-fold rotation, 8-fold rotation, vertical flipping, and horizontal flipping, we investigate the effectiveness of employing different geometric transformation strategies in the equivariance regularization module to improve model performance. Table 3 presents the performance of the CNN and Transformer in the framework under four different methods.

It can be observed that encouraging equivariance to 4-fold rotations leads to a greater improvement in the quality of the image features outputted by the encoder, thus enhancing the segmentation performance of the final model.

4 Conclusion

In conclusion, this paper presents a semi-supervised learning method that incorporates heterogeneous models consistency learning, multi-level image augmentation, uncertainty constraints, equivariance regularization module. The proposed method extends the training data size through various image augmentation strategies to meet the data requirements of the Transformer, thereby improving the performance of the dual-network structure. Additionally, we introduce uncertainty minimization constraints to enhance model robustness and utilizes equivariance regularization to improve the quality of image features outputted by the encoder. Experimental results demonstrate that this method outperforms most popular semi-supervised medical image segmentation methods on the 2D MRI cardiac dataset. In the future, we plan to further explore the adaptation relationship between image augmentation and different datasets to improve the generalizability of the method.

References

1. Berthelot, D., et al.: Remixmatch: semi-supervised learning with distribution alignment and augmentation anchoring. arXiv preprint arXiv:1911.09785 (2019)
2. Bian, Y., et al.: Artificial intelligence to predict lymph node metastasis at ct in pancreatic ductal adenocarcinoma. Radiology **306**(1), 160–169 (2023)
3. Bortsova, G., Dubost, F., Hogeweg, L., Katramados, I., de Bruijne, M.: Semi-supervised medical image segmentation via learning consistency under transformations. In: Shen, D., Liu, T., Peters, T.M., Staib, L.H., Essert, C., Zhou, S., Yap, P.-T., Khan, A. (eds.) MICCAI 2019. LNCS, vol. 11769, pp. 810–818. Springer, Cham (2019). https://doi.org/10.1007/978-3-030-32226-7_90
4. Cao, H., Wang, Y., Chen, J., Jiang, D., Zhang, X., Tian, Q., Wang, M.: Swin-unet: Unet-like pure transformer for medical image segmentation. In: Computer Vision-ECCV 2022 Workshops: Tel Aviv, Israel, October 23–27, 2022, Proceedings, Part III. pp. 205–218. Springer (2023)
5. Chen, C., et al.: Deep learning for cardiac image segmentation: a review. Front. Cardiovascular Med. **7**, 25 (2020)
6. Chen, X., Yuan, Y., Zeng, G., Wang, J.: Semi-supervised semantic segmentation with cross pseudo supervision. In: Proceedings of the IEEE/CVF Conference on Computer Vision and Pattern Recognition, pp. 2613–2622 (2021)
7. Cheplygina, V., de Bruijne, M., Pluim, J.P.: Not-so-supervised: a survey of semi-supervised, multi-instance, and transfer learning in medical image analysis. Med. Image Anal. **54**, 280–296 (2019)
8. Dangovski, R., et al.: Equivariant contrastive learning. arXiv preprint arXiv:2111.00899 (2021)
9. Hang, W., et al.: Local and global structure-aware entropy regularized mean teacher model for 3D left atrium segmentation. In: Martel, A.L., et al. (eds.) MICCAI 2020. LNCS, vol. 12261, pp. 562–571. Springer, Cham (2020). https://doi.org/10.1007/978-3-030-59710-8_55
10. Li, X., Yu, L., Chen, H., Fu, C.W., Xing, L., Heng, P.A.: Transformation-consistent self-ensembling model for semisupervised medical image segmentation. IEEE Trans. Neural Networks Learn. Syst. **32**(2), 523–534 (2020)

11. Luo, X., Hu, M., Song, T., Wang, G., Zhang, S.: Semi-supervised medical image segmentation via cross teaching between cnn and transformer. In: International Conference on Medical Imaging with Deep Learning, pp. 820–833. PMLR (2022)
12. Luo, X., et al.: Semi-supervised medical image segmentation via uncertainty rectified pyramid consistency. Med. Image Anal. **80**, 102517 (2022)
13. Ronneberger, O., Fischer, P., Brox, T.: U-Net: convolutional networks for biomedical image segmentation. In: Navab, N., Hornegger, J., Wells, W.M., Frangi, A.F. (eds.) MICCAI 2015. LNCS, vol. 9351, pp. 234–241. Springer, Cham (2015). https://doi.org/10.1007/978-3-319-24574-4_28
14. Sohn, K., et al.: Fixmatch: simplifying semi-supervised learning with consistency and confidence. Adv. Neural. Inf. Process. Syst. **33**, 596–608 (2020)
15. Tarvainen, A., Valpola, H.: Mean teachers are better role models: Weight-averaged consistency targets improve semi-supervised deep learning results. Advances in neural information processing systems 30 (2017)
16. Vu, T.H., Jain, H., Bucher, M., Cord, M., Pérez, P.: Advent: adversarial entropy minimization for domain adaptation in semantic segmentation. In: Proceedings of the IEEE/CVF Conference on Computer Vision and Pattern Recognition, pp. 2517–2526 (2019)
17. Wang, R., Lei, T., Cui, R., Zhang, B., Meng, H., Nandi, A.K.: Medical image segmentation using deep learning: a survey. IET Image Proc. **16**(5), 1243–1267 (2022)
18. Wang, T., Lu, J., Lai, Z., Wen, J., Kong, H.: Uncertainty-guided pixel contrastive learning for semi-supervised medical image segmentation. In: Proceedings of the Thirty-First International Joint Conference on Artificial Intelligence, IJCAI, pp. 1444–1450 (2022)
19. Yao, J., et al.: Deep learning for fully automated prediction of overall survival in patients undergoing resection for pancreatic cancer: a retrospective multicenter study. Annals of Surgery, pp. 10–1097 (2022)
20. Yu, L., Wang, S., Li, X., Fu, C.-W., Heng, P.-A.: Uncertainty-aware self-ensembling model for semi-supervised 3D left atrium segmentation. In: Shen, D., et al. (eds.) MICCAI 2019. LNCS, vol. 11765, pp. 605–613. Springer, Cham (2019). https://doi.org/10.1007/978-3-030-32245-8_67

Wasserstein Diversity-Enriched Regularizer for Hierarchical Reinforcement Learning

Haorui Li[1,2], Jiaqi Liang[1(✉)], Linjing Li[1,2], and Daniel Zeng[1,2]

[1] State Key Laboratory of Multimodal Artificial Intelligence Systems, Institute of Automation, Chinese Academy of Sciences, Beijing, China

[2] School of Artificial Intelligence, University of Chinese Academy of Sciences, Beijing, China

{lihaorui2021,liangjiaqi2014,linjing.li,dajun.zeng}@ia.ac.cn

Abstract. Hierarchical reinforcement learning composites subpolicies in different hierarchies to accomplish complex tasks. Automated subpolicies discovery, which does not depend on domain knowledge, is a promising approach to generating subpolicies. However, the degradation problem is a challenge that existing methods can hardly deal with due to the lack of consideration of diversity or the employment of weak regularizers. In this paper, we propose a novel task-agnostic regularizer called the Wasserstein Diversity-Enriched Regularizer (WDER), which enlarges the diversity of subpolicies by maximizing the Wasserstein distances among action distributions. The proposed WDER can be easily incorporated into the loss function of existing methods to boost their performance further. Experimental results demonstrate that our WDER improves performance and sample efficiency in comparison with prior work without modifying hyperparameters, which indicates the applicability and robustness of the WDER.

Keywords: Hierarchical Reinforcement Learning · Subpolicy · Diversity · Wasserstein Regularizer

1 Introduction

Hierarchical reinforcement learning (HRL) decomposes the tasks to be addressed into distinct subtasks and organizes them in a hierarchical structure, where the high-level policies solve complex tasks by recombining the low-level subpolicies. Through this way, the skills, knowledge, or experience learned by HRL can be shared and reused among different tasks [25]. The transferable ability makes HRL an effective approach to dealing with complex and sparse tasks, such as multi-level decision-making and fine-grained control over long-horizon manipulation [10,25], which have made notable progress in recent years.

The generation of subpolicies is the most crucial part of HRL since the quality and diversity of the subpolicies directly affect the performance of the combined

B. Luo et al. (Eds.): ICONIP 2023, LNCS 14447, pp. 564–577, 2024.
https://doi.org/10.1007/978-981-99-8079-6_44

policy. Subpolicies can be established by domain experts or be learned automatically. Human-designed subpolicies are highly dependent on domain-specific knowledge and meticulously crafted auxiliary pseudo-rewards, as a result, it is difficult to generalize the obtained subpolicies to new tasks. By contrast, automated subpolicies discovery aims to learn subpolicies based on simulations with limited input data. The automated approach is more demanded as a learning framework that could be applied to various tasks after a little work of adaption. However, it suffers greatly from the degradation problem that all subpolicies degenerate to a common subpolicy in the later stage of the training phase. The cause of degradation can be imputed to the lack of explicit constraints on the diversity of subpolicies. Thus, regularization and rewards reshaping have been employed to mitigate the degradation problem by incorporating information-theoretic objectives, such as maximizing mutual information (MI) and Jensen-Shannon (JS) divergence. Nevertheless, the maximum diversity is restricted since both MI and JS are bounded from above and fail to provide an effective gradient when the distributions are supported on non-overlapping domains. As a result, the degradation problem is still a challenge for training HRL agents.

This paper proposes a Wasserstein Diversity-Enriched Regularizer (WDER) to increase the diversity of subpolicies learned in HRL, which differs from those methods based on the information-theoretic objectives mentioned above. Wasserstein distance (WD) can accurately measure the distribution distance [7], and it provides a geometry-aware topology than traditional f-divergences (such as those based on KL divergence). By incorporating a WD-based regularization term in the loss function, the "distance" between sequenced subpolicies can be enlarged as far as possible, which not only promotes the diversity of the learned subpolicies but also enhances the exploration ability of the composite policy. The main contributions of this paper are four folds:

- We propose a task-agnostic regularizer utilizing WD to enhance the diversity of the learned subpolicies.
- The proposed regularizer can be easily integrated into various existing HRL methods with a fixed number of subpolicies.
- We propose a method that applies WDER to two different frameworks: meta-reinforcement learning (Meta-RL) [5] and the option framework [1].
- We evaluate the effectiveness of WDER through two HRL tasks in both discrete and continuous action spaces. The experimental results indicate that our approach outperforms information-theoretic-based methods.

2 Related Works

This paper is closely related to automated subpolicies generation and WD; thus, we review related work on both in this section. The formal definition of and how to estimate WD are also introduced to facilitate the expression of our work.

2.1 Automated Subpolicies Generation

In RL, the methods for automated subpolicy discovery can be categorized into two families: Unified Learning of Feudal Hierarchy (ULFH) and Unified Learning of Policy Tree (ULPT) [15]. In ULFH, a higher-level network called the "Manager" samples a subgoal in a learned latent subspace, then a lower-level network called the "Worker" must learn a subpolicy to achieve this subgoal [22]. Within ULPT, the option framework is a widely applied method that discovers a fixed number of subpolicies in accordance with the learning of a hierarchical policy [1,9]. Meanwhile, Meta Learning Shared Hierarchies (MLSH) [5], an algorithm similar to the ULPT, has been proposed in meta-learning. MLSH contains a master policy and multiple subpolicies. The master policy employs the same subpolicies in related tasks to accelerate the learning process on unseen tasks. Our work is more closely related to enhancing the diversity of subpolicies in ULPT and MLSH, as the main challenge of ULFH is how to design subgoals, which is not along the line of ULPT, MLSH, and our work.

However, the option framework and MLSH training could suffer from the lack of diversity in subpolicies, i.e., different subpolicies converge to nearly the same one. Moreover, in the option framework, the high-level policy may predominantly use only one subpolicy in the entire episode. Some studies have investigated diversity-driven regularizers or reward reshaping through information-theoretic objectives to mitigate this degradation phenomenon. Florensa et al. [4] introduced a regularizer based on MI between the latent variable and the current state, where the latent variable follows a categorical distribution with uniform weights, in order to increase the diversity of the stochastic neural network policy. Haarnoja et al. [6] obtained diverse policies by maximizing the expected entropy of the trajectory distribution in the reinforcement learning objective. In addition, "DIAYN" [3] forces policies to be diverse and distinguishable by encouraging skills to explore a part of the state space far away from other skills by maximizing entropy in unsupervised RL tasks. Huo et al. [8] proposed a method using direct JS divergence regularization on the action distributions and emphasized the connection between the visited environment states of subpolicies.

2.2 Wasserstein Distance

The measurement of discrepancy or distance between two probability distributions can be treated as a transport problem [23]. Let p be a probability distribution defined on domain $\mathcal{X} \subseteq \mathbb{R}^n$ and q be a distribution defined on $\mathcal{Y} \subseteq \mathbb{R}^m$. Let $\Gamma[p,q]$ be the set of all distributions on the product space $\mathcal{X} \times \mathcal{Y}$, with their marginal distributions on $\mathcal{X}$ and $\mathcal{Y}$ being p and q, respectively. Thus, given an appropriate cost function $c(x,y) : \mathcal{X} \times \mathcal{Y} \rightarrow \mathbb{R}$ which represents the cost of moving a unit "mass" from x to y, the WD is defined as

$$W_c(p,q) = \inf_{\gamma \in \Gamma[p,q]} \int_{\mathcal{X} \times \mathcal{Y}} c(x,y) d\gamma. \tag{1}$$

The optimal transport is the one that minimizes the above transport cost. The smoothed WD is introduced to address the challenge of super-cubic complexity:

$$\widetilde{W}_c(p,q) = \inf_{\gamma\in\Gamma[p,q]} \left[\int_{\mathcal{X}\times\mathcal{Y}} c(x,y)d\gamma + \beta KL(\gamma \mid p,q) \right]. \tag{2}$$

Equation (2) can be estimated either by the primal or dual formulation. In this paper, we calculate the WD by the dual formulation. The dual formulation is based on the Fenchel-Rockafellar duality [23], which provides a convenient neural way to estimate WD. Let set $\mathcal{A} = \{(u,v) \mid \forall(x,y) \in \mathcal{X}\times\mathcal{Y} : u(x)-v(y) \leq c(x,y)\}$, where $\mu : \mathcal{X} \to \mathbb{R}$ and $\nu : \mathcal{Y} \to \mathbb{R}$ are continuous functions and the cost function $c(x,y)$ may not be smooth, then the dual formulation estimation of WD is

$$W_c(p,q) = \sup_{(\mu,\nu)\in\mathcal{A}} \mathop{\mathbb{E}}_{x\sim p(x), y\sim q(y)} [\mu(x) - \nu(y)]. \tag{3}$$

Theoretically, the maximum value obtained from the dual formulation aligns with the minimum value of the original formulation. Accordingly, the dual formulation of the smoothed WD is

$$\widetilde{W}_c(p,q) = \sup_{\mu,\nu} \mathop{\mathbb{E}}_{x\sim p(x), y\sim q(y)} \left[\mu(x) - \nu(y) - \beta \exp\left(\frac{\mu(x) - \nu(y) - c(x,y)}{\beta} \right) \right]. \tag{4}$$

The dual formulation is more convenient, as it does not impose any constraints on the functions μ and ν.

WD has been widely used to quantify distribution differences in representation learning and generative modeling. In the context of RL, WD has been used to quantify the difference between policies [26]. Compared to traditional KL and other f-divergences, WD has shown to be a versatile measure. Pacchiano et al. [14] adopted WD to enhance the performance of trust region policy optimization and evolutionary strategies, Dadashi et al. [2] showed its effectiveness in imitation learning by minimizing it between the state-action distributions of the expert and the agent. Moskovitz et al. [13] used it as a divergence penalty with the local geometry to speed optimization. Furthermore, WD was employed as a metric for unsupervised RL to encourage the agent to explore the state space extensively to generate diverse subpolicies [7]. Different from these studies, our WDER utilizes an action distributions-based regularization term, WD is employed to measure the differences in the learning phase of different subpolicies.

3 Methodology

3.1 Standard RL and HRL

This paper adopts the standard RL setting, which is built upon the theory of Markov decision processes (MDPs) [20]. A MDP can be formalized as a tuple $\langle \mathcal{S}, \mathcal{A}, \mathcal{R}, \mathcal{P}, \gamma \rangle$, where $\mathcal{S}$ is a finite set of states, $\mathcal{A}$ is a finite set of actions, $\mathcal{R} : \mathcal{S} \times \mathcal{A} \to [R_{min}, R_{max}]$ is the reward function, $\mathcal{P} : \mathcal{S} \times \mathcal{A} \to \Delta(\mathcal{S})$ is the

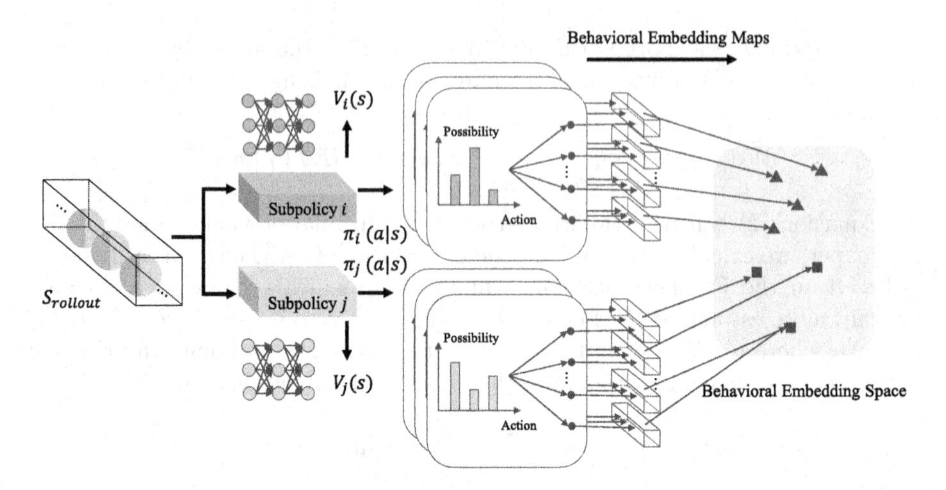

Fig. 1. Estimation of the WD between two subpolicies π_i and π_j. The actions generated by subpolicy π_i and π_j at the same states in the state set $S_{rollout}$ are sampled and mapped to feature vectors in the behavior embedding space according to the behavioral embedding maps. Two sampled action vectors can be mapped to the same point in the behavior embedding space. After the mapping is completed, we can estimate the WD by utilizing Algorithm 1 and Eq. (8).

transition function, where $\Delta(\mathcal{S})$ denotes the set of probability distributions over $\mathcal{S}$, and $\gamma \in [0, 1)$ is the horizon discount factor. This paper focuses on a standard RL agent with a two-level architecture, and there are K low-level subpolicies denoted as $\pi_1, \pi_2, \cdots, \pi_K$, and a high-level master policy π_m that decides which subpolicy to be used in the current state. At each time step t, the currently selected subpolicy π_k samples an action a_t based on the observed state s_t with respect to the distribution $\pi_k(s_t)$. The execution of the action a_t results in the generation of an environmental reward r_t and the transition of the system to a new state s_{t+1} according to the transition probability $p(s_{t+1} \mid s_t, a_t)$. This process continues iteratively until the master policy π_m selects another subpolicy.

3.2 Wasserstein Distance Between Subpolicies

To estimate the WD between two subpolicies, we need to obtain the policy embeddings [14] of these two subpolicies. Figure 1 depicts the process involved. In the context of HRL, each subpolicy π_k can be fully defined by its action probability distribution. In settings with discrete action spaces, the action probability distribution of a subpolicy typically follows a categorical distribution. While in the case of continuous action spaces, the action probability distribution of a subpolicy commonly follows a Gaussian distribution, which means that the agent samples an action $a_t \sim \mathcal{N}(\mu, \sigma)$ at each time step. The mean μ and the standard deviation σ are obtained by fitting a normal distribution according to the outputs of the last dense layer of π_k's policy network.

Algorithm 1: Random Features Wasserstein SGD

Input: kernels κ, ℓ over $\mathcal{X}$, $\mathcal{Y}$ respectively with corresponding random feature maps ϕ_κ, ϕ_ℓ, smoothing parameter γ, gradient step size η, number of optimization rounds M, initial dual vectors $\mathbf{p}_0^\mu$, $\mathbf{p}_0^\nu$.

for $t = 0, \cdots, M$ **do**

  Sample $(x_t, y_t) \sim \mu \otimes \nu$.

  Update $\mathbf{p}_t^\mu$, $\mathbf{p}_t^\nu$ using Eq. (7).

end

Output: $\mathbf{p}_M^\mu$, $\mathbf{p}_M^\nu$

Next, we elucidate the sampling procedure, outlined in Algorithm 1 [14], when estimating the WD between subpolicy π_k and π_l. First, we extract a set of T states, denoted as $S_{rollout}$, from the trajectories generated by these two subpolicies. The states within $S_{rollout}$ are represented as $s_1, s_2, \cdots, s_T$. For each state s_t, we then sample B actions from the action probability distributions generated by the policy networks of π_k and π_l. These sampled actions are denoted as $\{a_{kti}\}_{i=1}^B$ and $\{a_{lti}\}_{i=1}^B$, respectively. To mitigate the variance of the cost function $c(x, y)$ in Eq. (4), we employ the same sequence of random numbers, i.e., the common random numbers [18], during the generation of both $\{a_{kti}\}_{i=1}^B$ and $\{a_{lti}\}_{i=1}^B$.

Second, we map the sampled actions to the embedding space by a radial basis function (RBF) kernel using random Fourier feature maps [16]. This process is referred to as the behavioral embedding map (BEM):

$$\Phi : \Gamma[p_k, q_l] \to \mathcal{E}, \tag{5}$$

where $\Gamma[p_k, q_l]$ bears the same meaning as $\Gamma[p, q]$ defined in Sect. 2.2. In this context, the notations p_k and q_l denote the action probability distributions corresponding to subpolicies π_k and π_l. While $\mathcal{E}$ corresponds to the embedding space, wherein each action (of dimension m) is meticulously mapped into a vector of features (of dimension $d > m$). The BEM Φ induces a corresponding pushforward distribution, i.e., the resulting distribution, on $\mathcal{E}$. For subpolicies π_k and π_l, we denote their pushforward distributions as $\mathbb{P}_{\pi_k}^\Phi$ and $\mathbb{P}_{\pi_l}^\Phi$, respectively.

For subpolicies π_k and π_l, we define $\mu(x)$ and $\nu(y)$ in Eq. (4) as

$$\mu(x) = (\mathbf{p}^\mu)^\top \phi(x), \quad \nu(y) = (\mathbf{p}^\nu)^\top \phi(y), \tag{6}$$

where $\mathbf{p}^\mu$, $\mathbf{p}^\nu \in \mathbb{R}^m$ are vectors with m random features, $\phi(x)$ is defined as $\phi(x) = \frac{1}{\sqrt{m}} \cos(x\mathbf{G} + \mathbf{b})$, x and y belong to $\mathbb{R}^{h \times d}$, $h = B \times T_{\text{minibatch}}$ represents the number of actions in a minibatch, d denotes the dimensionality of the action space. $\mathbf{G} \in \mathbb{R}^{d \times m}$ is a Gaussian matrix with iid entries sampled from $\mathcal{N}(0, 1)$. The vector $\mathbf{b} \in \mathbb{R}^m$ is composed of independently sampled elements from the uniform distribution $\mathrm{U}[0, 2\pi]$, and the $\cos(\cdot)$ function is applied elementwise.

Then we can find the optimal dual estimation of WD by Algorithm 1 employing Random Features Wasserstein Stochastic Gradient Descent (SGD). Given the input kernels κ, ℓ, and a fresh sample $(x_t, y_t) \sim \mu \otimes \nu$, where $\otimes$ represents the tensor product, the parameters w.r.t. the current iteration should satisfy:

$$F\left(\mathbf{p}_t^{\mu}, \mathbf{p}_t^{\nu}, x_t, y_t\right) = \exp\left(\frac{\left(\mathbf{p}_t^{\mu}\right)^{\top} \phi_{\kappa}(x_t) - \left(\mathbf{p}_t^{\nu}\right)^{\top} \phi_{\ell}(y_t) - C(x_t, y_t)}{\gamma}\right),$$
$$\begin{pmatrix} \mathbf{p}_{t+1}^{\mu} \\ \mathbf{p}_{t+1}^{\nu} \end{pmatrix} = \begin{pmatrix} \mathbf{p}_t^{\mu} \\ \mathbf{p}_t^{\nu} \end{pmatrix} + \left(1 - F\left(\mathbf{p}_t^{\mu}, \mathbf{p}_t^{\nu}, x_t, y_t\right)\right) v_t, \tag{7}$$

where $v_t = \eta \left(\phi_{\kappa}\left(x_t\right) - \phi_{\ell}\left(y_t\right)\right)^{\top}$. Let M be the maximum number of iteration, $\mathbf{p}_M^{\mu}$ and $\mathbf{p}_M^{\nu}$ be the output of Algorithm 1. We can estimate the WD between two subpolicies π_k and π_l using $\mathbf{p}_M^{\mu}$ and $\mathbf{p}_M^{\nu}$ as

$$\mathrm{WD}_{\gamma}\left(\mathbb{P}_{\pi_k}^{\Phi}, \mathbb{P}_{\pi_l}^{\Phi}\right) = \hat{\mathbb{E}}\left[\left(\mathbf{p}_M^{\mu}\right)^{\top} \phi_{\kappa}(x_i) - \left(\mathbf{p}_M^{\nu}\right)^{\top} \phi_{\ell}(y_i) - \frac{F\left(\mathbf{p}_M^{\mu}, \mathbf{p}_M^{\nu}, x_i, y_i\right)}{\gamma}\right], \tag{8}$$

where $\hat{\mathbb{E}}$ denotes the empirical expectation over C iid action samples $\{(x_i, y_i)\}_{i=1}^{C}$, x_i and y_i correspond to the sampled actions of subpolicies π_k and π_l, respectively.

3.3 HRL with Wasserstein Diversity-Enriched Regularizer

Regularization is an effective and convenient framework to help generate diverse subpolicies, the maximum value of the regularizer and the distributions taken into consideration by the regularization are the most important determinants. Information-theoretic measures, such as MI and JS divergence, are the commonly encountered regularization form [3,4,6]. However, these measures are bounded from above, which intrinsically limits the diversity that can be achieved. Taking the classical MI as an illustration, the MI between two random variables S and Z is $I(S;Z) = H(Z) - H(Z|S)$, which is bounded by $H(Z)$, where $H(\cdot)$ is the Shannon entropy. The MI reaches its maximum value when $H(Z|S) = 0$, i.e., the support of S and Z do not overlap. The JS divergence has the same limitation when adopted to encourage diverse subpolicies. With this upper bound, the distance between two distributions cannot be enlarged; even their essential difference can still be amplified. Different from information-theoretic measures, the upper bound of WD can be set to the predefined value

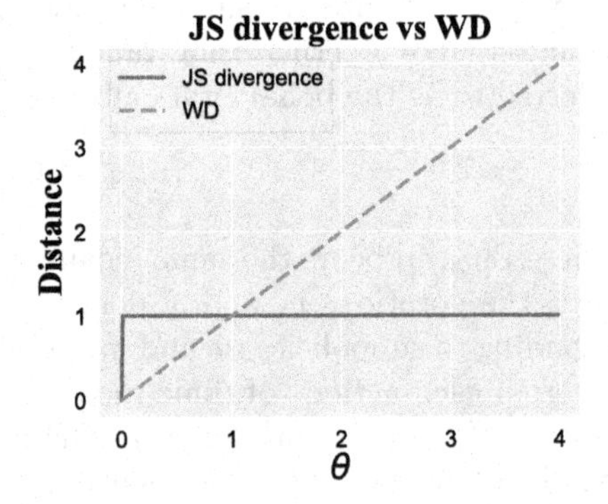

Fig. 2. Examples of JS divergence and WD between distributions P and Q, where $\forall(x,y) \in P$, $x = 0$ and $y \sim U(0,1)$ and $\forall(x,y) \in Q$, $x = \theta, \theta \geq 0$ and $y \sim U(0,1)$. The WD provides useful distance information between P and Q, while the JS divergence fails.

by choosing the appropriate cost function $c(x, y)$ according to Eq. (1). This is a highly demanded property to encourage diverse subpolicies in HRL [7]. In addition, WD can provide smooth and informative gradients for updating parameters, regardless of whether the distributions of the two subpolicies overlap or not, while MI and JS cannot. Figure 2 illustrates the changes of the JS divergence and WD between distributions P and Q with respect to θ. When the distributions do not overlap, the WD still provides useful information about the distance between the distributions, while the JS divergence does not.

As to the distributions, few works directly took the action distributions as the inputs [8,9]. Differentiating strategies based on action probability distribution has advantages that are not exhibited by state probability distribution-based strategies. It helps the agent explore more action choices, leading to the discovery of better subpolicies. It also allows the agent to adjust its action choices according to different situations, which improves its ability to adapt to various environments and states. Moreover, this approach enhances the agent's capability to handle complex policy spaces by providing it with a rich range of policy expressions. To sum up, in this paper, we devise the WDER and take the action distributions as the input for automatically generating highly diverse subpolicies in HRL.

Given $N \geq 2$ subpolicies, for an arbitrarily chosen subpolicy π_k, to make it distinct from other subpolicies, we want to maintain a distance (the larger, the better) of π_k with respect to others. Based on WD, We adopt the regularizer as

$$\mathrm{WD}_{min}(\pi_k) = \min_{j \neq k} \mathrm{WD}_\gamma \left(\mathbb{P}^{\Phi}_{\pi_k}, \mathbb{P}^{\Phi}_{\pi_j} \right), \tag{9}$$

to make π_k away from its nearest subpolicy. By WDER, the actor networks are encouraged to converge in different local maxima [14]. For subpolicy π_k, let $\theta_{\pi k}$ and θ_{vk} be the parameters of the policy network and the value network in the actor-critic framework, respectively. The modified actor network incorporates $\mathrm{WD}_{min}(\pi_k)$ as the regularization term in its loss function

$$L_{new}\left(\theta_{\pi k}\right) = L_{old}\left(\theta_{\pi k}\right) - \alpha \mathrm{WD}_{min}(\pi_k), \tag{10}$$

where $L_{old}\left(\theta_{\pi k}\right)$ is depended on the baseline and α is a hyperparameter.

The choice of backend training algorithm for RL can be different according to the nature and settings of the specific problems. As the PPO (Proximal Policy Optimization) [19] is adaptable to both discrete and continuous action space, we employ it as the backend RL algorithm. Specifically, Algorithm 2 outlines the proposed method, where $\mathrm{HRL}_{\mathrm{base}}$ denotes the input baseline. The algorithm outputs the trained model with the parameters of the subpolicies described by $\theta_{\pi k}$ and θ_{vk}, $k = 1, 2, \cdots, K$, together with the parameters of the master policy π_m updated according to the loss function of the baseline master policy.

As to the complexity of the proposed method, the distance computing time grows in the order of $O(N^2)$ as the number N of subpolicies increase [7]. When N is large, the WD can be approximated by the sliced or projected WD [17,24], and some experiments indict that HRL algorithm with two subpolicies achieves

Algorithm 2: Training Algorithm for the WDER Approach

Input: The baseline method HRL_{base}, and regularization hyperparameters α
while *not convergence* **do**
  while *episode not terminates* **do**
    Sample a subpolicy π_k from the master policy π_m according to HRL_{base};
    while π_k *not terminates* **do**
      Sample a_t according to the subpolicy $\pi_k(a_t \mid s_t)$, and collect the critic network output $v_k(s_t)$ from π_k;
      Perform action a_t and receive reward r_t and get the next state from the transition probability $p(s_{t+1} \mid s_t, a_t)$;
      Collect experience s_t, a_t, r_t, s_{t+1}, $v_k(s_t)$;
    end
  end
  update the actor network and the critic network of the master policy π_m according to the loss function of the baseline master policy
  for *each subpolicy* π_k *with parameters* $\{\theta_{\pi k}, \theta_{vk}\}$ **do**
    update parameter $\theta_{\pi k}$ following the Eq. (10)
    update parameter θ_{vk};
  end
end
Output: The trained model by the WDER-augmented HRL_{base}

the best performance in most practical applications [9]. Based on this result, we use two subpolicies in our experiments.

4 Experiment

We examine our method on two typical RL domains and select the corresponding state-of-the-art approaches for comparison: 1) MLSH for Meta-RL [5]; 2) OC (option-critic) for the option framework [1].

4.1 Variant Algorithms and Experimental Setup

WDER-MLSH. The MLSH architecture contains a master policy and several subpolicies. Our proposed WDER method can be easily integrated into the loss functions of the MLSH subpolicies. We compare the generalization performance of our WDER-MLSH with the original MLSH approach on a discrete 2-D navigation task. We also evaluate the ability to adapt to new tasks of our approach with α setting to 0.5. Other hyperparameters are the same as used in [5].

WDER-OC. As a significant component of HRL, the option framework has its own learning and optimization system. To test our WDER approach on high-dimensional input tasks, we evaluate its performance with the original OC on complex robot tasks. For these experiments, we set the value of the corresponding parameter α as 0.2 and use two options. We adopt the hyperparameters and convolution structure settings used in [1].

4.2 Performance Evaluation

We first investigate the performance of subpolicy discovery in Meta-RL by evaluating our WDER-MLSH algorithm on MovementBandits, viz a 2-D navigation task [5]. In this task, an agent is placed in a planar world, and it has already known its current location and the candidate target locations. The agent can take discrete actions to move in four directions or remain stationary. The environment sends 1 to the agent as a reward if the agent is at a certain distance from the correct target point and 0 otherwise. We use two subpolicies to train the MLSH baseline, MLSH-JS with JS divergence regularizer implemented following [8] and our WDER-MLSH method, with the duration of each subpolicy being ten timesteps. We vary the coefficient of the WDER term α from 0.2 to 0.6 with step 0.1.

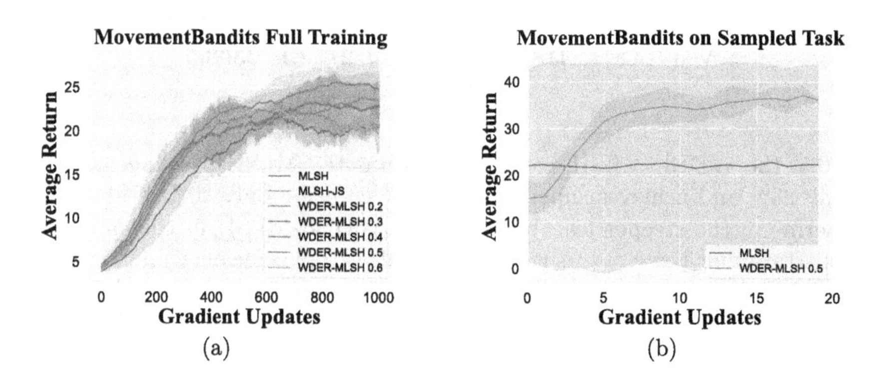

Fig. 3. (a) Average return curves of WDER-MLSH with different α during the training phase in the MovementBandits environment. (b) Average return curves for the newly sampled MovementBandits task. Performance corresponds to the ability to adapt to new tasks.

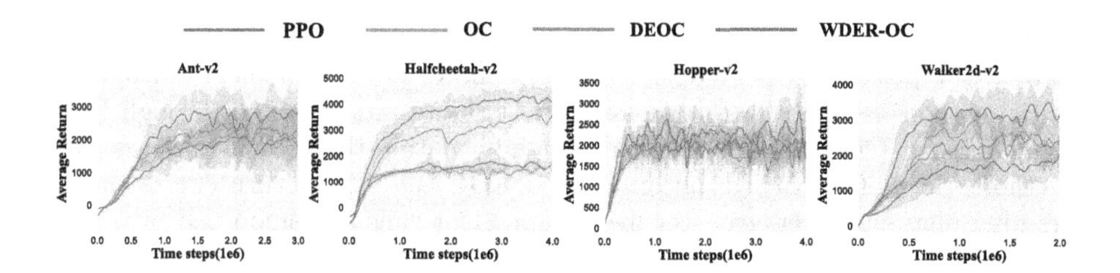

Fig. 4. Average return curves of WDER-OC on standard Mujoco tasks.

The average return curves of different agents during the training phase are shown in Fig. 3a, where each line is averaged over three runs, and the shaded areas represent one standard deviation. Our WDER-MLSH agents outperform the MLSH agent with respect to the average return. Especially when setting

Table 1. The max average returns on four Mujoco tasks over 4 million timesteps

	Ant	Halfcheetah	Hopper	Walker2d
PPO	2103.2	1799.0	2172.7	2083.1
OC	2310.6	1682.3	2498.6	3012.6
DEOC	2471.5	3640.4	2267.2	2778.0
WDER-OC	**3023.9**	**4322.8**	**2823.7**	**3510.1**

Table 2. The average returns on four classical Mujoco tasks over 1 million timesteps

	Ant	Halfcheetah	Hopper	Walker2d
DAC+PPO	985.8	1830.1	1702.2	1968.0
AHP+PPO	1359.3	1701.7	1993.6	1520.6
MOPG	907.4	**3446.7**	1955.3	1856.9
WDER-OC	**2418.9**	3119.5	**2128.1**	**2368.5**

α as 0.5, the WDER-MLSH agent outperforms the MLSH baseline by approximately 25% and achieves superior performance using only 30% of all samples. Furthermore, the steeper learning curves of WDER-MLSH in the early stages demonstrate that these agents have remarkable sample efficiency and can quickly improve their overall performance by mastering some fundamental skills.

Further, we assess the transferability of each agent by examining the diversity and effectiveness of the subpolicies learned in new tasks. Specifically, we train the agents in a MovementBandits environment until the cumulative reward stabilizes, then freeze the learned subpolicies, and then fine-tune the high-level strategy. We conduct six independent runs to compare the results with the original MLSH. Figure 3b indicates that our approach surpasses the baseline by converging faster and achieving superior performance in nearly 20 steps. These results suggest that our method improves Meta-RL performance for new tasks.

Next, we evaluate the generalization ability of our WDER algorithm in the option framework on four classic Mujoco tasks [21]. We compare it with a standard OC approach with two options. Furthermore, as we implement our option-critic method using PPO [19], we also report the results obtained through PPO as a reference. Our comparisons include an information-theoretic intrinsic reward method (diversity-enriched option-critic, DEOC) [9]. The return curves in the training phase are averaged over five independent runs and smoothed by a sliding window of size 20 (Fig. 4). Table 1 shows the max average returns over 4 million timesteps. The results indicate that WDER-OC surpasses all three baselines, particularly on the Halfcheetah task, where the performance is 157% higher than the original OC baseline and with less variance. Hence, our WDER algorithm is shown to be more effective in improving the performance of the option framework.

In order to validate whether WDER-OC can outperform other option variants and non-option baselines, we compared WDER-OC with DAC+PPO [27], AHP+PPO [11], and MOPG [12]. Since MOPG uses the least timesteps among these algorithms, we compare the performance of these algorithms with WDER-OC after running for 1 million time steps as shown in Table 2. The results for all algorithms except WDER-OC are the same as reported in the MOPG paper. It can be observed from Table 2 that our WDER-OC achieves the highest average returns in three out of four Mujoco tasks and is competitive in the remaining task, which demonstrates the improvement of WDEC-OC over the original OC regarding the performance and the sample efficiency.

5 Conclusion

This paper proposed a novel solution to the automated subpolicies discovery problem in HRL by introducing a task-agnostic regularizer, WDER, based on Wasserstein distance. Theoretically, the upper bound of the diversity of subpolicies generated by our approach is far larger than that of other algorithms utilizing information-theoretic objectives, and the gradients are more stable and effective throughout the updating process. We also demonstrated the effectiveness of our approach through extensive evaluations in two popular HRL task domains. The experimental results demonstrated that our method's robustness and generalization ability is higher than existing algorithms. Our future work will focus on an efficient Wasserstein distance estimation method to deal with situations involving more subpolicies.

Acknowledgements. This work was supported in part by the National Key Research and Development Program of China under Grant 2020AAA0103405, the National Natural Science Foundation of China under Grants 72293573 and 72293575, as well as the Strategic Priority Research Program of Chinese Academy of Sciences under Grant XDA27030100.

References

1. Bacon, P., Harb, J., Precup, D.: The option-critic architecture. In: Proceedings of the AAAI Conference on Artificial Intelligence, pp. 1726–1734 (2017)
2. Dadashi, R., Hussenot, L., Geist, M., Pietquin, O.: Primal Wasserstein imitation learning. In: Proceeding of the International Conference on Learning Representations (2021)
3. Eysenbach, B., Gupta, A., Ibarz, J., Levine, S.: Diversity is all you need: learning skills without a reward function. In: Proceeding of the International Conference on Learning Representations (2019)
4. Florensa, C., Duan, Y., Abbeel, P.: Stochastic neural networks for hierarchical reinforcement learning. In: Proceeding of the International Conference on Learning Representations (2017)
5. Frans, K., Ho, J., Chen, X., Abbeel, P., Schulman, J.: Meta learning shared hierarchies. In: Proceeding of the International Conference on Learning Representations (2018)

6. Haarnoja, T., Hartikainen, K., Abbeel, P., Levine, S.: Latent space policies for hierarchical reinforcement learning. In: Proceedings of the International Conference on Machine Learning, vol. 80, pp. 1846–1855 (2018)
7. He, S., Jiang, Y., Zhang, H., Shao, J., Ji, X.: Wasserstein unsupervised reinforcement learning. In: Proceeding of the AAAI Conference on Artificial Intelligence, pp. 6884–6892 (2022)
8. Huo, L., Wang, Z., Xu, M., Song, Y.: A task-agnostic regularizer for diverse subpolicy discovery in hierarchical reinforcement learning. IEEE Trans. Syst. Man Cybern. Syst. **53**(3), 1932–1944 (2023)
9. Kamat, A., Precup, D.: Diversity-enriched option-critic. arXiv preprint arXiv:2011.02565 (2020)
10. Levy, A., Konidaris, G.D., Jr., R.P., Saenko, K.: Learning multi-level hierarchies with hindsight. In: Proceeding of the International Conference on Learning Representations (2019)
11. Levy, K.Y., Shimkin, N.: Unified inter and intra options learning using policy gradient methods. In: Sanner, S., Hutter, M. (eds.) EWRL 2011. LNCS (LNAI), vol. 7188, pp. 153–164. Springer, Heidelberg (2012). https://doi.org/10.1007/978-3-642-29946-9_17
12. Li, C., Song, D., Tao, D.: Hit-MDP: learning the SMDP option framework on MDPs with hidden temporal embeddings. In: Proceeding of the International Conference on Learning Representations (2023)
13. Moskovitz, T., Arbel, M., Huszar, F., Gretton, A.: Efficient Wasserstein natural gradients for reinforcement learning. In: Proceeding of the International Conference on Learning Representations (2021)
14. Pacchiano, A., Parker-Holder, J., Tang, Y., Choromanski, K., Choromanska, A., Jordan, M.I.: Learning to score behaviors for guided policy optimization. In: Proceedings of the International Conference on Machine Learning, vol. 119, pp. 7445–7454 (2020)
15. Pateria, S., Subagdja, B., Tan, A., Quek, C.: Hierarchical reinforcement learning: a comprehensive survey. ACM Comput. Surv. **54**(5), 109:1–109:35 (2022)
16. Rahimi, A., Recht, B.: Random features for large-scale kernel machines. In: Proceedings of the Annual Conference on Neural Information Processing Systems, pp. 1177–1184 (2007)
17. Rowland, M., Hron, J., Tang, Y., Choromanski, K., Sarlos, T., Weller, A.: Orthogonal estimation of Wasserstein distances. In: The 22nd International Conference on Artificial Intelligence and Statistics, pp. 186–195. PMLR (2019)
18. Schulman, J., Levine, S., Abbeel, P., Jordan, M.I., Moritz, P.: Trust region policy optimization. In: Proceedings of the International Conference on Machine Learning, vol. 37, pp. 1889–1897 (2015)
19. Schulman, J., Wolski, F., Dhariwal, P., Radford, A., Klimov, O.: Proximal policy optimization algorithms. arXiv preprint arXiv:1707.06347 (2017)
20. Sutton, R.S., Barto, A.G.: Reinforcement Learning: An Introduction. MIT Press, Cambridge (2018)
21. Todorov, E., Erez, T., Tassa, Y.: MuJoCo: a physics engine for model-based control. In: Proceeding of the International Conference on Intelligent Robots and Systems, pp. 5026–5033 (2012)
22. Vezhnevets, A.S., et al.: Feudal networks for hierarchical reinforcement learning. In: Proceedings of the International Conference on Machine Learning, vol. 70, pp. 3540–3549 (2017)
23. Villani, C., et al.: Optimal Transport: Old and New, vol. 338. Springer, Heidelberg (2009). https://doi.org/10.1007/978-3-540-71050-9

24. Wu, J., et al.: Sliced Wasserstein generative models. In: Proceeding of the IEEE Conference on Computer Vision and Pattern Recognition, pp. 3713–3722 (2019)
25. Yang, X., et al.: Hierarchical reinforcement learning with universal policies for multistep robotic manipulation. IEEE Trans. Neural Networks Learn. Syst. **33**(9), 4727–4741 (2022)
26. Zhang, R., Chen, C., Li, C., Carin, L.: Policy optimization as Wasserstein gradient flows. In: Proceedings of the International Conference on Machine Learning, vol. 80, pp. 5741–5750 (2018)
27. Zhang, S., Whiteson, S.: DAC: the double actor-critic architecture for learning options, vol. 32 (2019)

An Adaptive Detector for Few Shot Object Detection

Jiming Yan, Hongbo Wang(✉), and Xinchen Liu

State Key Laboratory of Networking and Switching Technology, Beijing University of Posts and Telecommunications, Beijing 100876, China
{buptyjm,hbwang,xinchen_liu}@bupt.edu.cn

Abstract. Few-shot object detection has made progress in recent years. However, most research assumes that base and new classes come from the same domain. In real-world applications, they often come from different domains, resulting in poor adaptability of existing methods. To address this problem, we designed an adaptive few-shot object detection framework. Based on the Meta R-CNN framework, we added an image domain classifier after the backbone's last layer to reduce domain discrepancy. To avoid class feature confusion caused by image feature distribution alignment, we also added a feature filter module (CAFFM) to filter out features irrelevant to specific classes. We tested our method on three base/new splits and found significant performance improvements compared to the base model Meta R-CNN. In base/new split2, mAP50 increased by $\pm 8\%$, and in the remaining two splits, mAP50 improved by $\pm 3\%$. Our method outperforms state-of-the-art methods in most cases for the three different base/new splits, validating the efficacy and generality of our approach.

Keywords: Few shot · Object detection · CNN

1 Introduction

Object detection involves identifying the location and class of objects in images or videos. With the development of deep learning, significant progress has been made in this field. However, most existing object detection methods rely on public datasets with extensive annotations. These datasets have limited classes and may not meet the diverse detection requirements of real-world applications. Annotating new data requires significant manpower and resources, limiting the application and promotion of object detection.

To address the limitations of existing object detection methods, researchers have focused on few-shot object detection [1,6–9,11,12,15–17]. It is similar to the few-shot classification problem but more complicated because it also involves object localization. This problem is known as the C-way K-shot problem, where C represents the number of new classes and K refers to the number of shots in each new class.

B. Luo et al. (Eds.): ICONIP 2023, LNCS 14447, pp. 578–591, 2024.
https://doi.org/10.1007/978-981-99-8079-6_45

It is challenging to learn knowledge directly from a small number of samples. As a result, most few-shot object detection methods first train on base class samples with sufficient annotations and then transfer the learned detection knowledge to new classes.

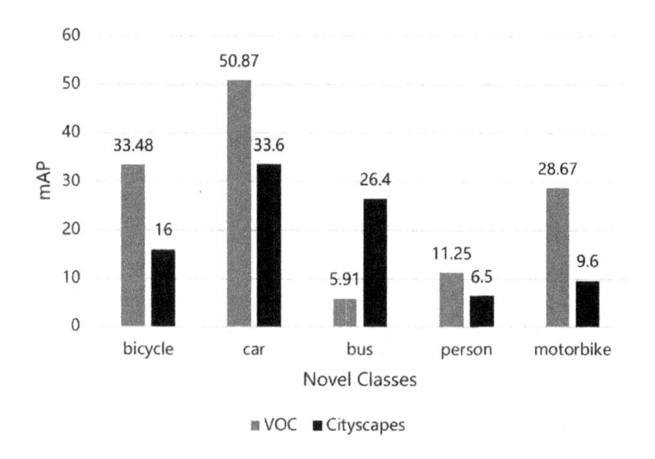

Fig. 1. When the base classes samples are same and the new samples come from different domains, the test results differ greatly

There are three mainstream few-shot object detection methods: metric learning, fine-tuning, and meta-learning-based.

Metric learning-based methods try to learn a metric function during training to build a model of the metric between samples. During testing, the class of the test sample is predicted by the metric between the test set and the support set. Fine-tuning-based methods first train on a large number of source-domain datasets and then fine-tune on a small number of target-domain samples. Meta-learning-based methods generally consist of a basic model and a meta-learner. The meta-learner learns the basic parameters during training and then adjusts the network's weight and optimizes its structure during testing.

Although few-shot object detection has made progress, most methods assume that base and new classes come from the same domain. However, in real-world scenarios, base and new classes often come from different domains, resulting in the unsatisfactory performance of existing methods in practical applications.

This paper verifies the poor generalization performance of existing few-shot object detection methods through an experiment. To facilitate comparison, we selected 15 classes as base classes from the public Pascal VOC object detection dataset and used the remaining 5 classes as new classes. All base class annotation samples in the Pascal VOC dataset were retained as the base class dataset. If an image contained both base and new class objects, the new class object annotation was ignored. Only 10 shots were selected for each new class, consistent with existing few-shot object detection methods' experimental settings. The same number and classes of new class shots were also extracted from another object detection dataset (Cityscapes) as new class samples in different domains. For this experiment, we selected Meta R-CNN [20]. We first meta-trained on base class samples and then fine-tuned on PASCAL VOC and Cityscapes new class samples. Finally, we tested on the PASCAL VOC and Cityscapes test sets. The mAP of new classes is shown in Fig. 1. It can be seen that when Meta R-CNN is directly applied to the target domain, the mAP of all new classes except for the BUS class dropped significantly. This is because the Cityscapes dataset is an urban street view dataset that includes a large number of vehicles.

The problem addressed in this paper is defined as few-shot object detection in cross-domain scenarios. Given a training set D_{train} that contains D_b and D_n, where D_b has class set C_b with enough shots for each class and D_n has class set C_n with only a small number of shots for each class (usually less than 10). C_b and C_n do not overlap $C_b \cap C_n = \emptyset$. Unlike the usual few-shot object detection problem, D_b and D_n come from different domains in this paper. For the test set, in the usual few-shot object detection problem, $C_{test} \in C_b \cup C_n$. However, this paper focuses on the adaptability of few-shot object detection methods, so in our experiment, $C_{test} = C_n$. Our goal is to improve the model's adaptability and robustness by improving the accuracy of new class samples.

In summary, our work and contributions are as follows:

Firstly, we identified and addressed the poor performance of few-shot object detection in cross-domain scenarios. To solve this problem, we designed an adaptive few-shot object detection network that greatly improves the adaptability and robustness of few-shot object detection.

Secondly, we introduced an image-level unsupervised domain adaptation module to align global image feature distribution and reduce domain displacement. To avoid confusion about class features, we also introduced Channel Attention Feature Filtering Module (CAFFM) to filter out features unrelated to specific classes.

Finally, we verified our method on three base/new splits from different public datasets. Compared to the base model, our model performed better and achieved state-of-the-art (SOTA) results compared to the latest few-shot object detection methods.

2 Related Work

2.1 General Object Detection

Deep learning-based general object detection methods are divided into two branches: one-stage detectors and two-stage detectors. Two-stage detectors generally generate a series of bounding boxes using the Region Proposal Network (RPN) algorithm and then classify samples using a convolutional neural network. Common two-stage detectors include the R-CNN [1] series. One-stage detectors do not generate bounding boxes and instead transform the object localization problem into a regression problem. Common one-stage detectors include the Yolo [2] series, SSD [3], RetinaNet [4], etc. These detection models all require a large number of annotated datasets. While public datasets can meet the requirements in terms of sample quantity, their classes are limited. As a result, these approaches are most suitable for detecting classes contained in public datasets.

2.2 Few Shot Object Detection

Few-shot object detection aims to detect new classes using a small number of new class samples based on the original model, reducing the cost of labeling data. Recent studies in vision are mainly classified into three streams based on metric learning, fine-tuning, and meta-learning.

Metric learning-based methods classify samples by calculating their similarity. However, this approach depends on a good metric method. [5] proposed a representational distance metric learning method that represents each class with a mixture model with multiple modalities and uses the centers of these modalities as the class's representation vector. [6] proposed a metric learning method for few-shot object detection using a contrast network. After training, the model can detect new class targets without adjusting its parameters.

The fine-tuning method first trains on datasets composed of sufficient base samples and then fine-tunes on a subset composed of a small number of base and new samples to achieve object detection under few-shot conditions. TFA [16] adopted the Faster R-CNN detection model with Feature Pyramid Network (FPN) as the basic detection framework. They first pre-trained the detection network on a large amount of base class data, then fine-tuned the model using a small number of support samples to achieve good generalization performance in both base and new classes. FSCE [17] found that fine-tuning only the classification and regression network could achieve a higher recall rate but was prone to class confusion. Therefore, they fine-tuned the FPN network, RPN regression, and classification networks to maintain differences between classes and reduce class confusion.

Few-shot object detection based on meta-learning changes the traditional approach of learning for a specific task to learning for learning. [7] introduced a meta-learner and a lightweight feature re-weighting module to the YOLOv2 detector to enable it to quickly adapt to new classes. [8] realized parameter conversion by introducing a parameterized weight prediction meta-model. Meta R-CNN [9] was proposed based on Mask R-CNN [10] and used a support branch to obtain a class attention vector that was fused with region-of-interest features as a new prediction feature for detection or segmentation. [11] further improved the fusion network based on Meta R-CNN to achieve better detection performance.

[12] argued that existing meta-learning methods were mainly limited to region-level prediction and depended on well-positioned initial region proposals. To address this problem, they combined Transformer [14], which has recently become popular, with meta-learning based on Deformable DETR [13] and proposed an image-level meta-learning few-shot object detection model. They replaced heuristic components such as Non-Maximum Suppression (NMS) and

anchor frames with encoders and decoders to achieve image-level object localization and classification.

[15] used a parameter-sharing feature extraction network to extract features from query and support images and performed intensive feature matching through a density relation extraction module to activate coexisting features in the input image. These features were then sent into the RPN network to extract ROIs. For the pooled part of the region of interest, pooling was first performed at different scales before fusion and then sent to the detector head for classification and positioning.

2.3 Unsupervised Domain Adaptation

Unsupervised domain adaptation involves transferring empirical knowledge learned from the source domain to the target domain when the target domain data has no labels. Unsupervised domain adaptation for object detection is mainly classified into two streams based on adversarial learning and sample generation.

Adversarial learning-based domain adaptation methods introduce the idea of Generative Adversarial Networks (GAN) [18] into the domain adaptation problem. [19] applied the idea of unsupervised domain adaptation to object detection networks for the first time. Based on the Faster R-CNN [23] network, they added an image-level and an instance-level domain discriminator to train the network so that the discriminator could not distinguish which domain features came from. [20] considered that different domains have different data distributions and object combinations in object detection tasks. This method strongly aligns local features and weakly aligns high-level feature space.

Sample generation-based domain adaptation methods involve combining labeled target domain samples with source domain samples and training the target domain network with composite samples. [21] used Cycle-GAN [22] to generate multiple intermediate domain images between the source and target domains and shared labels with the source domain. Generated, source domain and target domain data were sent to Faster R-CNN [23], and then all images were sent to the domain classifier together to learn invariant features between multiple domains.

[24] found through experiments that few-shot classification performance degraded in cross-domain scenarios and that this degradation became more pronounced as the difference between the source and target domains increased. [25] introduced the idea of unsupervised domain adaptation to improve few-shot classification performance in cross-domain scenarios for the first time.

Unsupervised domain adaptation aims to transfer a model trained on a large number of labeled source domain samples to the target domain with a large number of unlabeled target domain samples and achieve good results on the target domain. The classes of the source and target domains are the same. In the few-shot object detection problem, the target domain has only a few shots, and the classes of the source and target domains are inconsistent.

3 Methods

In this section, we will elaborate on our method, the network structure is shown in Fig. 2:

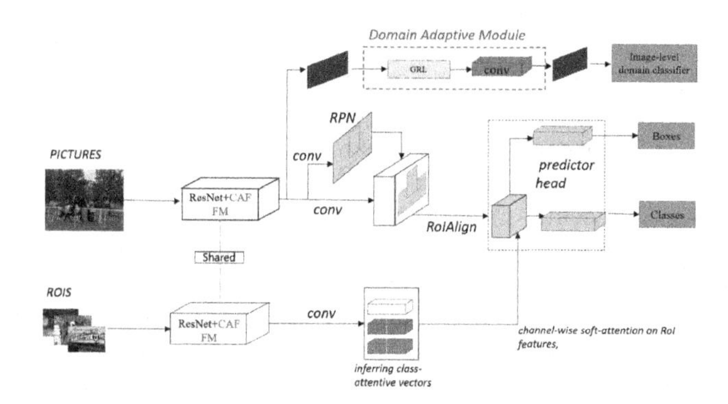

Fig. 2. The network of our methods Adaptive Meta R-CNN. It is extended from Meta R-CNN, a domain adaptive module is introduced after backbone to align distributions of features across source and target domains. Meanwhile, an channel attention feature filtering module is added to avoid class feature confusion

3.1 Meta R-CNN

In our model, we chose Meta R-CNN as the base model. Meta R-CNN is a classic few-shot object detection framework based on meta-learning. It introduces a Predictor-head Remodeling Network (PRN) based on Faster R-CNN that shares trunk parameters with Faster R-CNN. PRN receives objects from base or new samples and obtains class attention vectors for each class. Each vector performs a channel-wise attention operation on all RoI features and outputs detection results through prediction headers.

Meta R-CNN also introduces a meta-loss function L_{Meta} eta to make the class attention vector fall on each object's class, implemented using cross-entropy loss. The meta-loss function greatly improves Meta R-CNN's performance. In our method, we followed Meta R-CNN's meta-loss function L_{Meta}.

3.2 Image Level Domain Adaptive Module

In general unsupervised domain adaptive object detection methods, both image-level and instance-level feature distributions are aligned to reduce domain displacement. The output of the backbone is usually defined as the image-level feature distribution and normalized RoI features are defined as the instance-level feature distribution.

However, the problem addressed in this paper is fundamentally different from unsupervised domain adaptive object detection. In few-shot object detection, the classes of the target and source domains are not the same. Aligning instance-level features could cause class confusion. Therefore, we only consider image-level feature distribution alignment in this work.

In few-shot object detection applications, the base and new classes often come from different domains. To eliminate the difference in feature distribution between the source and target domains, we added an image-level domain adaptive module. Specifically, we trained a domain classifier from each activation of the feature map. Since each receptive field corresponds to an input image patch, the domain classifier actually predicts the domain label of each image patch. This approach has two benefits: 1) Aligning image-level feature representation helps reduce differences caused by global image features such as image style, proportion, and lighting conditions, etc. 2) Due to the use of high-resolution input, the training network's batch size is usually set to be small. This patch-based design helps increase the number of training samples for training domain classification.

As the network trains, the domain classification loss value gradually decreases and the domain classifier's performance improves. It can accurately classify source and target domain image features, separating their features. However, our goal is to confuse source and target domain feature distributions, extract domain-invariant features, and make the classifier unable to distinguish which domain the current features come from. Therefore, using only a classifier to achieve feature distribution alignment is not feasible.

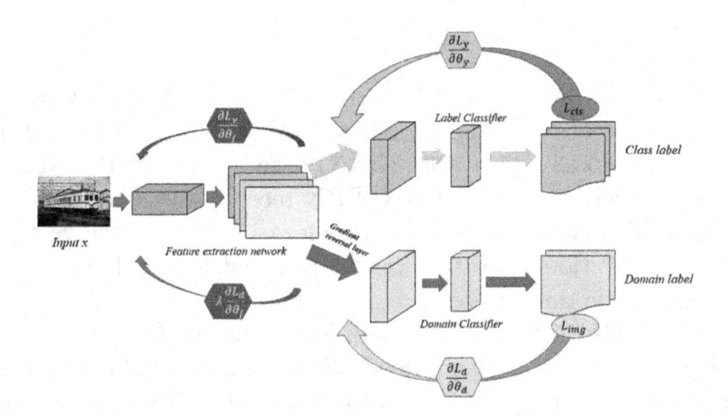

Fig. 3. Gradient Reversal Layer. The GRL layer of the domain adaptation module calculates the gradient of current layer by multiplying the loss passed by the previous layer network to this layer by a negative number $-\lambda$, the gradient of this layer is also a negative number, so that the loss value increases with the network training. Gradually increase, the source and target domain features are confused. At the same time, the loss value of the label classifier becomes smaller and smaller with the network training, which achieves the adversarial training.

We implemented our approach using Gradient Reversal Layer (GRL), as illustrated in Fig. 3. In deep learning, forward propagation performs the weighted compilation and other operations on input features before outputting them. Backpropagation calculates the partial derivative of the objective function with respect to neuron values using gradient descent to obtain the difference between predicted and real values and transfers this difference backward layer by layer. Each network layer calculates its gradient based on the returned loss and updates its parameters accordingly. GRL multiplies the error transmitted from the previous layer network to this layer by a negative number $-\lambda$ to calculate the current layer's gradient, resulting in a negative gradient that achieves the opposite effect of adversarial training. The domain classification loss function is shown in Formula 1:

$$L_{img} = -\sum_{i,u,v}[D_i log P_i^{(u,v)} + (1 - D_i) log(1 - P_i^{(u,v)})] \tag{1}$$

where D_i is defined as the domain label of the training picture i, $P_i^{(u,v)}$ represents the prediction result of the picture i at position (u, v). The whole module maximizes the loss of domain classification and minimizes the loss of basic network, realizes the confrontation effect, makes the domain classifier difficult to distinguish the feature of source domain and target domain, and then aligns the feature distribution of source domain and target domain.

3.3 Feature Filtering Module CAFFM

As mentioned earlier, existing unsupervised domain adaptive object detection methods uses image and instance level feature alignment usually. However, in the problem to be solved in this article, the object classes between source domain and target domain are completely different. If instance features are also aligned, classes would be confused. Therefore, instance features cannot be forcibly aligned. In addition, this article only aligns image level features, which can indirectly lead to instance classes confusion, as instance level features also come from feature maps extracted on the entire image.

To avoid this problem, we introduced CAFFM, a channel attention feature filtering module, as shown in Fig. 4. CAFFM makes the network pay more attention to features unrelated to specific classes and reduces the risk of class confusion.

Attention methods has excellent effectiveness and flexible adaptability. When seeing a picture, people quickly focus their attention on areas where objects are prominent, while ignoring other irrelevant background areas. This feature enables people to accurately process information in a short period of time. Researchers have been inspired by the human visual system and proposed attention mechanisms. In object detection tasks, attention methods can be divided into spatial domain attention, channel domain attention, and mixed domain attention methods.

The spatial attention method focuses on the regions with the most prominent features in the image in object detection tasks. The channel attention method models the importance of each channel feature, which can be weighted based on input for different tasks, focusing on channel features with higher weights. The mixed domain attention mechanism combines the above two attention models.

In this paper, the feature extraction network needs to pay attention to domain invariant features, which are obviously more focused on channel character. Essentially, the invariant characteristics between two domains are not directly related to their spatial positions, therefore, the spatial attention mechanism is not applicable in this article.

In object detection networks, the features extracted by feature extraction networks, where each channel feature is related to a specific target class and different channels are interrelated. It is possible to discover relatively important channel features by utilizing the correlation between channel features, and then screen out these features.

For the convolutional neural network, the core computing process is to learn from the input feature map to a new one through the convolutional kernel Feature map. Essentially, convolution fuses the features of a local area, which includes feature fusion in space (H and W dimensions) and between channels (C dimensions)

One of the functions of convolution operation is to improve the receptive field and integrate deeper and more features. In the process of channel dimension feature fusion, the convolution operation often defaults that each channel is equally important. However, in actual tasks, the importance of each feature in a specific category is often different. In the research task of this paper, because the categories of objects in the source domain and the target domain are completely different, the channel features extracted in the source domain contain the responses of the features of the base class, and these features will cause some interference to the class features of the target domain. Therefore, this paper introduces a feature filtering module CAFFM based on channel attention. The module can learn the importance of different channel features, so as to pay more attention to features unrelated to specific categories.

Since convolution only operates in local space, it is difficult for feature maps to extract the relationship between channels. This is more pronounced for earlier network layers because their receptive fields are relatively small. Therefore, CAFFM first encodes the entire spatial feature on a channel as a global feature using global average pooling:

$$z_c = \frac{1}{H \times W} \sum_{i=1}^{H} \sum_{j=1}^{W} u_c(i,j) \tag{2}$$

After global average pooling, CAFFM gets the global description feature. Next, we need to get the relationship between channels. For this reason, the gating mechanism in the form of sigmoid is used:

$$\mathbf{s} = \sigma\left(\mathbf{W}_2 \delta\left(\mathbf{W}_1 \mathbf{z}\right)\right) \tag{3}$$

where W_1 is the weight set of conversion layer for down sampling, W_2 is the weight set of conversion layer for up sampling, σ refers to Sigmoid and δ refers to the ReLu function. In order to reduce the complexity of the model and improve the generalization ability, a bottleneck structure containing two fully connected layers is used here. The first FC layer plays the role of dimensionality reduction, and the connection reduction number r is a hyperparameter, and ReLU is used to activate it. The final FC layer restores the original dimension.

Finally, the learned activation value (0∼1) of each channel is multiplied by the original feature.

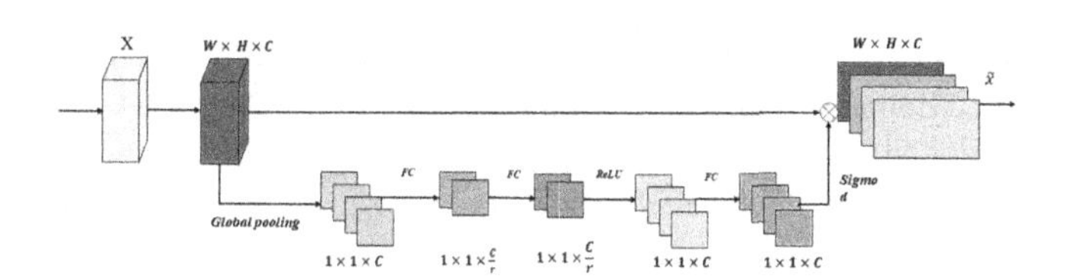

Fig. 4. CAFFM Module. CAFFM adopts Squeeze and Excitation similar to SENet [26]. The output of the network is the weighted channel features, so that the model is dedicated to features that are not related to a specific class and avoids class feature confusion.

This series of operations can be regarded as learning the weight coefficients of each channel, thus making the model more capable of distinguishing the characteristics of each channel.

Obviously, our method is a multi-task, in which the joint loss function is defined as:

$$Loss = L_{rpn} + L_{cls} + L_{loc} + L_{meta} + \lambda L_{img} \tag{4}$$

where L_{meta} is derived from Meta R-CNN, L_{img} refers to image-level domain classification loss, where $\lambda \in [0, 1]$, is the weight of L_{img}.

4 Experiments

4.1 Datasets Setting

To verify the effectiveness of our method, we designed three base/new splits using four public object detection datasets. We conducted experiments on these splits with 1, 2, 3, 5, and 10 shots respectively. The details of these splits are shown in Table 1. These three splits represent different styles of cross-domain scenarios commonly used in reality. In our experiments, we chose classes with a large number of samples as the base class to reflect real-world conditions. Base/new split 1 represents the cross-domain from natural style to urban landscape(The source domain dataset is from PASCAL VOC, with 15 classes as the base classes. The target domain dataset is composed of Cityscapes, with 5 classes as new classes), while base/new split 2 represents the cross-domain from the real world to the cartoon world(The source domain dataset is also from PASCAL VOC, with 15 classes as the base classes. The target domain dataset is composed of Clipart, with 5 classes as new classes). Base/new split 3 represents cross-domains of different camera angles(tThe source domain data is composed of KITTI, with 5 classes as the base classes. The target domain dataset is composed of Cityscapes, with 3 classes as the new classes). Our evaluation focused on the mAP50 of new samples in the test set. The split method of the dataset refers to the method of FSCE.

Table 1. The split of the experimental dataset. In split1, the source-domain samples are from PASCAL VOC [27], the target-domain samples are from Cityscapes [28]. In split2, the source-domain samples are from PASCAL VOC, the target-domain samples are from Clipart [29]. In split3, the source-domain samples are from KITTI [30], the target-domain samples are from Cityscapes.

split	base classes															new classes				
1	aeroplane	bird	bottle	chair	cow	diningtable	dog	horse	pottedplant	train	tvmonitor	boat	cat	sheep	sofa	bicycle	car	bus	person	motorbike
2	aeroplane	bird	bottle	chair	diningtable	horse	pottedplant	tvmonitor	boat	cat	sheep	sofa	car	person	motorbike	train	dog	cow	bicycle	bus
3	car	person	tram	truck	van											bicycle	bus	motorbike		

Table 2. Few-shot detection evaluation results(mAP50 of new classes) on Clipart with ResNet-50.

Methods/shot	split1					split2					split3				
	1	2	3	5	10	1	2	3	5	10	1	2	3	5	10
Meta R-CNN	6.4	7.3	7.5	**10.3**	14.8	9.6	10.7	15.4	21.1	29.8	5.4	**10.0**	**11.3**	11.9	15.4
TFA	2.6	3.7	1.1	1.1	2.3	8.1	9.0	16.5	15.9	17.7	8.2	5.4	7.7	7.8	10.9
FSCE	5.0	3.5	4.9	4.9	11.6	**19.4**	18.6	24.4	23.4	29.6	8.5	4.9	9.5	**15.5**	**17.4**
adaptive Meta R-CNN(Ours)	**11.2**	**8.6**	**8.8**	9.1	**20.9**	17.7	**20.2**	**25.3**	**31.2**	**31.7**	**9.3**	8.5	10.6	12.6	16.8

Table 3. Ablation results

Methods / shots	1	2	3	5	10
Meta R-CNN	9.6	10.7	15.4	21.1	29.8
Meta R-CNN+DA	13.1	12.6	15.5	27.0	29.0
Adaptive Meta R-CNN(OURS)	**17.7**	**20.2**	**25.3**	**31.2**	**31.7**

4.2 Experiments Details

Our experiment was conducted on a single 1080Ti GPU with a batch size of 4. The λ is set to 0.5 in three experiments. We used SGD as our optimizer and decreased the learning rate by a factor of 0.1 after every 3 turns. After 6 epochs of meta-training on a dataset composed of base class samples, we performed meta-fine-tuning for 6 epochs on a balanced subset composed of K new shots and 3*K base shots. The resulting model was then tested on the test set.

4.3 Experiments Results

We took the average value of five experimental results as the final result, which is shown in Table 2.

Our method generally showed improvement compared to the baseline, with an increase of about 8% in base/new split 2 and about 3% in the other two splits. The improvement was more significant when the difference between the source and target domains was larger, as in base/new split 2.

Our method also achieved better results compared to TFA. FSCE performed well in some cases, particularly in base/new split 3 where there were fewer base and new classes and the contrastive branch of FSCE was effective.

We also conducted ablation experiments to verify the effectiveness of each module we designed. On base/new split 2, we verified the effectiveness of each module separately. The results are shown in Table 3.

Compared to the Base model, the mAP50 improved after adding the DA module and further improved after adding the CAFFM. Both modules had a positive effect on the model. We also noted that the DA module was more effective when there was a larger style difference between the source and target datasets, such as in base/new split 2.

5 Conclusion

In this work, we found that existing few-shot object detection methods have poor generalization performance and verified this problem through experiments. To our knowledge, we are the first to identify and solve this problem. To address it, we designed the adaptive Meta R-CNN. Based on Meta R-CNN, we added an image-level unsupervised domain adaptation module based on adversity learning to align the global image feature distribution. Additionally, we introduced a feature filtering module based on channel attention to avoid class confusion. We verified the effectiveness of our method using three base/new splits from different public datasets.

References

1. Girshick, R., Donahue, J., Darrell, T., et al.: Rich feature hierarchies for accurate object detection and semantic segmentation. In: Proceedings of the IEEE Conference on Computer Vision and Pattern Recognition, pp. 580–587 (2014)
2. Redmon, J., et al.: You only look once: unified, real-time object detection. In: Proceedings of the IEEE Conference on Computer Vision and Pattern Recognition (2016)
3. Liu, W., et al.: SSD: single shot multibox detector. In: Leibe, B., Matas, J., Sebe, N., Welling, M. (eds.) ECCV 2016. LNCS, vol. 9905, pp. 21–37. Springer, Cham (2016). https://doi.org/10.1007/978-3-319-46448-0_2
4. Lin, T.-Y., et al.: Focal loss for dense object detection. In: Proceedings of the IEEE International Conference on Computer Vision (2017)

5. Karlinsky, L., Shtok, J., Harary, S., et al.: RepMet: representative-based metric learning for classification and few-shot object detection. In: 2019 IEEE/CVF Conference on Computer Vision and Pattern Recognition (CVPR). IEEE (2019)
6. Zhang, T., Zhang, Y., Sun, X., et al.: Comparison network for one-shot conditional object detection. arXiv preprint arXiv:1904.02317 (2019)
7. Kang, B., Liu, Z., Wang, X., et al.: Few-shot object detection via feature reweighting. In: 2019 IEEE/CVF International Conference on Computer Vision (ICCV). IEEE (2020)
8. Wang, Y.X., Ramanan, D., Hebert, M.: Meta-learning to detect rare objects. In: Proceedings of the IEEE/CVF International Conference on Computer Vision, pp. 9925–9934 (2019)
9. Yan, X., Chen, Z., Xu, A., et al.: Meta R-CNN: towards general solver for instance-level low-shot learning. In: Proceedings of the IEEE/CVF International Conference on Computer Vision, pp. 9577–9586 (2019)
10. He, K., Gkioxari, G., Dollár, P., et al.: Mask R-CNN. In: Proceedings of the IEEE International Conference on Computer Vision, pp. 2961–2969 (2017)
11. Zhou, X., Wang, D., Krähenbühl, P.: Objects as points. arXiv preprint arXiv:1904.07850 (2019)
12. Zhang, G., Luo, Z., Cui, K., et al.: Meta-DETR: few-shot object detection via unified image-level meta-learning. arXiv preprint arXiv:2103.11731 (2021)
13. Zhu, X., Su, W., Lu, L., et al.: Deformable DETR: deformable transformers for end-to-end object detection. arXiv preprint arXiv:2010.04159 (2020)
14. Vaswani, A., Shazeer, N., Parmar. N., et al.: Attention is all you need. In: Advances in Neural Information Processing Systems, pp. 5998–6008 (2017)
15. Hu, H., Bai, S., Li, A., et al.: Dense relation distillation with context-aware aggregation for few-shot object detection. In: Proceedings of the IEEE/CVF Conference on Computer Vision and Pattern Recognition, pp. 10185–10194 (2021)
16. Wang, X., et al.: Frustratingly simple few-shot object detection. arXiv preprint arXiv:2003.06957 (2020)
17. Sun, B., Li, B., Cai, S., et al.: FSCE: few-shot object detection via contrastive proposal encoding. In: Proceedings of the IEEE/CVF Conference on Computer Vision and Pattern Recognition, pp. 7352–7362 (2021)
18. Goodfellow, I., Pouget-Abadie, J., Mirza, M., et al.: Generative adversarial nets. In: Advances in Neural Information Processing Systems 27 (2014)
19. Chen, Y., et al.: Domain adaptive faster R-CNN for object detection in the wild. In: Proceedings of the IEEE Conference on Computer Vision and Pattern Recognition (2018)
20. Saito, K., Ushiku, Y., Harada, T., et al.: Strong-weak distribution alignment for adaptive object detection. In: Proceedings of the IEEE/CVF Conference on Computer Vision and Pattern Recognition, pp. 6956–6965 (2019)
21. Kim, T., Jeong, M., Kim, S., et al.: Diversify and match: a domain adaptive representation learning paradigm for object detection. In: Proceedings of the IEEE/CVF Conference on Computer Vision and Pattern Recognition, pp. 12456–12465 (2019)
22. Zhu, J.Y., Park, T., Isola, P., et al.: Unpaired image-to-image translation using cycle-consistent adversarial networks. In: Proceedings of the IEEE International Conference on Computer Vision, pp. 2223–2232 (2017)
23. Ren, S., He, K., Girshick, R., et al.: Faster R-CNN: towards real-time object detection with region proposal networks. Adv. Neural. Inf. Process. Syst. **28**, 91–99 (2015)
24. Chen, W.-Y., et al.: A closer look at few-shot classification. arXiv preprint arXiv:1904.04232 (2019)

25. Zhao, A., et al.: Domain-adaptive few-shot learning. In: Proceedings of the IEEE/CVF Winter Conference on Applications of Computer Vision (2021)
26. Hu, J., Shen, L., Sun, G.: Squeeze-and-excitation networks. In: Proceedings of the IEEE Conference on Computer Vision and Pattern Recognition, pp. 7132–7141 (2018)
27. Everingham, M., Van Gool, L., Williams, C.K.I., et al.: The pascal visual object classes (voc) challenge. Int. J. Comput. Vision **88**(2), 303–338 (2010)
28. Cordts, M., Omran, M., Ramos, S., et al.: The cityscapes dataset for semantic urban scene understanding. In: Proceedings of the IEEE Conference on Computer Vision and Pattern Recognition, pp. 3213–3223 (2016)
29. Venkateswara, H., Eusebio, J., Chakraborty, S., et al.: Deep hashing network for unsupervised domain adaptation. In: Proceedings of the IEEE Conference on Computer Vision and Pattern Recognition, pp. 5018–5027 (2017)
30. Geiger, A., Lenz, P., Urtasun, R.: Are we ready for autonomous driving? The Kitti vision benchmark suite. In: 2012 IEEE conference on Computer Vision and Pattern Recognition, pp. 3354–3361. IEEE (2012)

Author Index

B. Luo et al. (Eds.): ICONIP 2023, LNCS 14447, pp. 593–595, 2024.
https://doi.org/10.1007/978-981-99-8079-6

Printed in the United States
by Baker & Taylor Publisher Services